THE NIDOVIRUSES
(CORONAVIRUSES AND ARTERIVIRUSES)

ADVANCES IN EXPERIMENTAL MEDICINE AND BIOLOGY

Recent Volumes in this Series

Volume 484
PHYLOGENETIC PERSPECTIVES ON THE VERTEBRATE IMMUNE SYSTEM
Edited by Gregory Beck, Manickam Sugumaran, and Edwin L. Cooper

Volume 485
GENES AND PROTEINS UNDERLYING MICROBIAL URINARY TRACT
VIRULENCE: Basic Aspects and Applications
Edited by Levente Emődy, Tibor Pál, Jörg Hacker, and Gabriele Blum-Oehler

Volume 486
PURINE AND PYRIMIDINE METABOLISM IN MAN X
Edited by Esther Zoref-Shani and Oded Sperling

Volume 487
NEUROPATHOLOGY AND GENETICS OF DEMENTIA
Edited by Markus Tolnay and Alphonse Probst

Volume 488
HEADSPACE ANALYSIS OF FOODS AND FLAVORS: Theory and Practice
Edited by Russell L. Rouseff and Keith R. Cadwallader

Volume 489
HEMOPHILIA CARE IN THE NEW MILLENNIUM
Edited by Dougald M. Monroe, Ulla Hedner, Maureane R. Hoffman, Claude Negrier,
Geoffrey F. Savidge, and Gilbert C. White II

Volume 490
MECHANISMS OF LYMPHOCYTE ACTIVATION AND IMMUNE REGULATION VIII
Edited by Sudhir Gupta

Volume 491
THE MOLECULAR IMMUNOLOGY OF COMPLEX CARBOHYDRATES—2
Edited by Albert M. Wu

Volume 492
NEUROIMMUNE CIRCUITS, DRUGS OF ABUSE, AND INFECTIOUS DISEASES
Edited by Herman Friedman, Thomas W. Klein, and John J. Madden

Volume 493
NUTRITION AND CANCER PREVENTION: New Insights into the Role of
Phytochemicals
Edited under the auspices of the American Institute for Cancer Research

Volume 494
THE NIDOVIRUSES (CORONAVIRUSES AND ARTERIVIRUSES)
Edited by Ehud Lavi, Susan R. Weiss, and Susan T. Hingley

A Continuation Order Plan is available for this series. A continuation order will bring delivery of each new volume immediately upon publication. Volumes are billed only upon actual shipment. For further information please contact the publisher.

THE NIDOVIRUSES
(CORONAVIRUSES AND ARTERIVIRUSES)

Edited by

Ehud Lavi
Susan R. Weiss
University of Pennsylvania School of Medicine
Philadelphia, Pennsylvania

and

Susan T. Hingley
Philadelphia College of Osteopathic Medicine
Philadelphia, Pennsylvania

Springer Science+Business Media, LLC

Library of Congress Cataloging-in-Publication Data

The nidoviruses (coronaviruses and arteriviruses)/edited by Ehud Lavi, Susan R. Weiss,
and Susan T. Hingley.
      p. ; cm. — (Advances in experimental medicine and biology; 494)
     "Proceedings of the VIII International Symposium on Nidoviruses (Coronaviruses and
  Arteriviruses), held May 20–25, 2000, in Lake Harmony, Pennsylvania"—T.p. verso.
  Includes bibliographical references and index.
  ISBN 978-1-4613-5498-7   ISBN 978-1-4615-1325-4 (eBook)
  DOI 10.1007/978-1-4615-1325-4
     1. Coronaviruses—Congresses. I. Lavi, Ehud. II. Weiss, Susan R. III. Hingley, Susan
T. IV. International Symposium on Nidoviruses (Coronaviruses and Arteriviruses) (8th:
2000: Philadelphia, Pa.) V. Series.
     [DNLM: 1. Arterivirus—Congresses. 2. Coronavirus—Congresses. QW 168.5.C8 N664 2000]
  QR399 .N53 2001
  616′.0194—dc21

                                                  2001042723

Proceedings of the VIII International Symposium on Nidoviruses (Coronaviruses and Arteriviruses), held May
20–25, 2000, in Lake Harmony, Pennsylvania.

ISBN 978-1-4613-5498-7

©2001 Springer Science+Business Media New York
Originally published by Kluwer Academic/Plenum Publishers, New York in 2001
Softcover reprint of the hardcover 1st edition 2001

http://www.wkap.nl/

10 9 8 7 6 5 4 3 2 1

A C.I.P. record for this book is available from the Library of Congress

PREFACE

In 1996 the International Committee for Taxonomy of Viruses (ICTV) recognized the name *Nidovirales,* as the formal name for *Coronaviridae* and *Arteriviridae.* In recognition of this change, and in response to the wishes of our colleagues we named this meeting for the first time "The International Symposium of Nidoviruses". The meeting in the wooded environment of Lake Harmony, Pennsylvania, provided a stimulating opportunity for assessing the progress made in the field since the last meeting in Segovia Spain in 1997. Over 150 scientists from academia and industry attended the meeting. The meeting hosted senior members of the Nidovirus community, some of whom have been studying the subject for over 20 years, as well as younger scientists, the next generation of Nidovirologists. The traditional informal format, the shared meals, the social activities and the relatively inexpensive venue made the meeting a popular adventure.

In her opening remarks Susan Weiss showed pictures from previous meetings, reminding us how young we used to look. Neal Nathanson was our keynote speaker at the opening night, giving an overview on how viral pathogenesis studies helped in shaping the evolution of viral research and vaccine development. The scientific program of the meeting was divided into 9 sessions including 10 keynote presentations.

The meeting opened with a session on epidemiology, evolution and genome structure. Sasha Gorbalenya shared with us insights gained from comparative sequence analysis, emphasizing the unifying traits among nidovriuses, but also pointed out the remaining "black holes". Jeff Cowley introduced a new member to the nidovirus family, a virus that infects prawns. This is the first intervertebrate nidovirus to be identified.

Before the session on pathogenesis Stuart Siddell said a few words in memory of Helmut Wege who passed away prematurely, a few months before the meeting. One of the founders of coronavirus research, Helmut made important observations about MHV pathogenesis. Michael Buchmeier described new developments in the field of coronavirus-induced demyelination, highlighting the role of cytokines and chemokines in MHV pathogenesis. The field of coronavirus pathogenesis received a tremendous boost by the recent adaptation of the techniques of targeted RNA recombination. The role of apoptosis in pathogenesis emerged in several presentations.

In the session on virus entry, receptors, fusion and glycoprotein functions Kay Holmes described the dynamics of spike-receptors interactions. In addition to MHV, the entry of 229E, EAV and TGEV were also discussed in this session. Tom Gallagher discussed the importance of the coronavirus spike glycoprotein in virus entry and pathogenesis. In the session on the replicase gene and protein, Eric Snijder dissected the RNA synthesis of arteriviruses.

The exciting news at this meeting was three reports describing the generation of infectious coronavirus clones. One report by Thiel, Herold and Siddell described the autonomous replication and transcription of a recombinant human coronavirus RNA 229E. Another report by the Enjuanes lab described the engineering of the largest RNA virus genome (TGEV) as an infectious bacterial artificial chromosome. The third report came from Baric, Yount and Curtis describing the assembling of infectious full-length clones of TGEV. We are eagerly waiting for similar achievements in the field of MHV.

In the session on immunology and vaccine development Conni Bergmann described the role of CD8(+) T cell mediated immunity to neurotropic MHV infection. The importance of cytokines, matrix metaloproteinase, interferons, CD8(+) T cells, B cells, N-glycans and cellular and humoral immunity to viral pathogenesis were further discussed in individual presentations. In the session on RNA synthesis David Brian discussed transcriptional pathways and cis-acting signals of coronaviruses. Three papers discussed the role of nuclear ribonucleoprotein A1 in RNA replication, two suggesting an important role while another suggesting that its role in non-essential. We are still not sure what the final answer is.

In the session on viral assembly, Carolyn Machamer discussed trafficking through the cellular organelles, and the assembly of infectious bronchitis virus membrane proteins. Various presentations further discussed the roles of the M and E proteins in viral assembly. Finally, in the session on virus-cell interaction Mark Denison presented a fantastic videotape of live cell imaging and three-dimensional reconstructions of MHV infection.

The organizers of the meeting wish to thank all of those who helped to make the meeting a success and the completion of the book possible. We thank Amanda D'Zurilla and Jan Harkin for their tremendous help in attending to all the fine details of the registration and organization of the meeting. We thank Li Fu and Joanna Phillips for their help during the meeting, Dorothy Saunders for help with the mailing list, and Judy Lavi and Donna Bauhof for assistance with the final assembly of the book. We thank our sponsors, the departments of Microbiology and Pathology of the University of Pennsylvania, the National Multiple Sclerosis Society, Centocor, Elanco Animal Health, a division of Eli Lilly and Company, and Sorvall. We wish you a productive period and we will hopefully see you again in Leiden, the Netherlands in 2003.

Ehud Lavi, Susan R. Weiss, Susan T. Hingley
The Editors

CONTENTS

1. Epidemiology, evolution and genome structure

2.   Pathogenesis

3. Viral entry, receptors, fusion, and glycoprotein functions

4. Replicase

5. Immunology and vaccine development

6. Replication: RNA synthesis

7. Replication: viral assembly

8. Virus-cell interaction

Big Nidovirus Genome
When count and order of domains matter

ALEXANDER E. GORBALENYA
Advanced Biomedical Computing Center, 430 Miller Dr. Rm. 235, SAIC/NCI-Frederick, Frederick, MD 21702-1201, USA

1.   INTRODUCTION

For a long time, the major hallmark of nidoviruses (coronaviruses), which clearly distinguished them from other positive (+) RNA viruses, had been that the viral subgenomic (sg) mRNAs are derived from noncontinuous segments of the viral genome. Nidovirus sg mRNAs are, thus, mosaic and, in this respect, resemble eukaryotic mRNAs generated by splicing. Each nidovirus sg mRNA matches the genomic RNA in the 5' and 3' terminal sequences and carries a specific, 3'-nested subset of the open reading frames (ORFs) found in the genomic RNA (Spaan *et al.*, 1983; Lai *et al.*, 1984). Commonly, only the most 5'-proximal ORF is expressed from a sg mRNA. Even though both the spliced eukaryotic mRNAs and the nidovirus sg mRNAs relate to the transcribed genomes in similar ways, the underlying mechanisms that generate these RNAs differ dramatically. The nidovirus sgRNAs are generated during a unique discontinuous transcription, which does not require breaking and ligation of phosphodiester bonds, as is the case with RNAs generated by splicing. The discontinuous transcription is rather a variant of similarity-assisted RNA recombination and has been established to operate in coronaviruses and arteriviruses, the two families of nidoviruses (Sawicki and Sawicki, 1995; van Marle *et al.*, 1999). It, however, remains to be determined if it is universally conserved in other nidoviruses, i.e., the poorly characterized mammalian toroviruses (Snijder and Horzinek, 1993) and the invertebrate okaviruses (Cowley *et al.*, 2000).

The Nidoviruses (Coronaviruses and Arteriviruses).
Edited by Ehud Lavi *et al.*, Kluwer Academic/Plenum Publishers, 2001.

1

The latter two groups, the former with the rank of a genus in the coronavirus family and the other with a yet-to-be established rank, also strikingly differ from the other nidoviruses in virion morphology and composition. Regardless of whether discontinuous transcription is employed by *all* or only a fraction of the nidoviruses, there is another firm ground to recognize nidoviruses as a distinct group - the conservation of a unique genetic plan that is expressed in a conserved fashion (Gorbalenya and Koonin, 1993; Snijder *et al.*, 1993). This paper will give a brief and historically inclined overview of the progress made in establishing a collinearity among the nidovirus genomes by using theoretical and experimental approaches.

To give a proper perspective, nidoviruses are to be compared to other +RNA viruses. The majority of these viruses belong to one of the two large supergroups - Picornavirus-like and Alphavirus-like - that have been recognized in the middle of the 1980s (Goldbach, 1987; Strauss and Strauss, 1988). Each of these supergroups is characterized by a specific and ordered set of three conserved domains forming the backbone of the replicative polyproteins. For viruses of the Picornavirus-like supergroup these domains include a (putative) RNA-helicase of the so-called superfamily 2 or 3 (HEL2 and HEL3, respectively), an unusual cysteine proteinase that adopts a chymotrypsin-like fold and cleaves at conserved junctions commonly composed of Gln (Glu) at the P1 position and a small amino acid at the P1' position (3C-like proteinase - 3CLpro; after 3C of picornaviruses), and an RNA-dependent RNA polymerase (RdRp). For viruses of the Alphavirus-like supergroup this set consists of a methyltransferase of a unique superfamily, an RNA-helicase of superfamily I (HEL1) and an RdRp (reviewed in Gorbalenya and Koonin, 1993).

2. NIDOVIRUS GENOMES - OVERVIEW

Complete or nearly complete nidovirus genomes are currently available for four species of coronaviruses, porcine transmissible gastroenteritis virus (TGEV) (Eleouet *et al.*, 1995) and human coronavirus strain 229E (HCoV) (Herold *et al.*, 1993) that both belong to group 1 coronaviruses, mouse hepatitis virus (MHV, three strains; group 2 coronavirus) (Lee *et al.*, 1991; Bonilla *et al.*, 1994), avian infectious bronchitis virus (IBV; group 3 coronavirus) (Boursnell *et al.*, 1987), and three species of arteriviruses, equine arteritis virus (EAV) (den Boon *et al.*, 1991), lactate dehydrogenase-elevating virus (LDV, 2 strains) (Godeny *et al.*, 1993; Palmer *et al.*, 1995) and porcine reproductive and respiratory sindrome virus (PRRSV, 3 strains) (Meulenberg *et al.*, 1993; Allende *et al.*, 1999; Nelsen *et al.*, 1999). Additionally, a considerable number of genomes were only partly sequenced. Among those, large fragments of the torovirus Berne virus

(BEV) (Bredenbeek *et al.*, 1990b; Snijder *et al.*, 1990; Snijder *et al.*, 1991) and the okavirus gill-associated virus (GAV) (Cowley *et al.*, 2000) genomes provide convincing evidence (see below) to cluster these viruses with other nidoviruses.

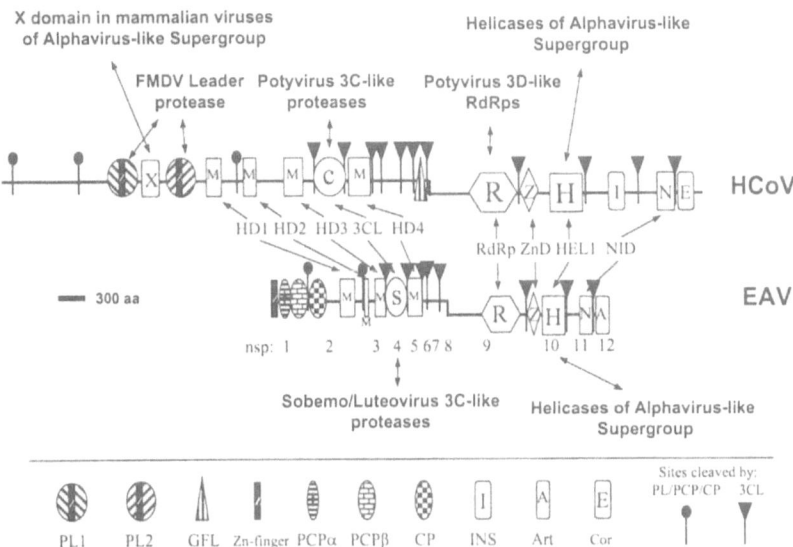

Figure 1. The domain layout and distant relationships of the nidovirus replicative pp1ab polyproteins. Shown are the polyproteins of HCoV and EAV whose domain organization is typical for corona- and arteriviruses, respectively. Domains and cleavage sites discussed in the text are depicted. The position of the pp1a/pp1b junction, which results from the ribosomal frameshifting, is indicated by a vertical bar upstream of the RdRp. Some nidoviruses may have domain organizations which deviate from that shown. Domains forming the backbone are specified between two polyproteins. The relationships with other +RNA viruses are indicated by two-headed arrows above and below the polyproteins. INS, a unique domain conserved in corona/toro/okaviruses between the HEL1 and NID domains; Art and Cor, the most C-terminal domain in pp1ab of arteriviruses and corona/toro/okaviruses, respectively. For the designation of the other domains see text.

The basic nidovirus genetic plan includes non-translated regions at 5' and 3' ends flanking from 6 to 11 ORFs, some of which are partly overlapping (Lai and Cavanagh, 1997; Snijder and Meulenberg, 1998). The two most 5'-proximal ORFs, ORF1a and ORF1b, are the largest and overlap. The ORF1a encodes pp1a polyprotein, and ORF1a+ORF1b direct, through ribosomal frame-shifting, the synthesis of pp1ab polyprotein. Both polyproteins are organized around a conserved framework of a dozen replicative domains and autocatalytically processed to numerous products at the conserved interdomain junctions (Fig. 1) (Ziebuhr *et al.*, 2000). The other ORFs, whose number varies in different viruses, are located further downstream and direct the synthesis of the capsid proteins and, optionally, accessory proteins. To express these ORFs, a nested set of sg mRNAs is generated. This genetic plan is executed in the described fashion by all nidoviruses - from the smallest arterivirus EAV (12.7 kb) up to the largest coronavirus MHV (31.5 kb).

3.   DOMAIN ORGANIZATION OF THE REPLICATIVE POLYPROTEINS OF CORONAVIRUSES IS COMPLEX AND UNIQUE: INITIAL INSIGHTS LEARNED FROM IBV AND MHV

The first nidovirus genome to be fully sequenced was that of IBV (Boursnell *et al.*, 1987). The British group, which has accomplished this breakthrough, correctly predicted the two different modes used by IBV to express, respectively, ORF1a/ORF1b and the other, downstream located ORFs in the genome. It was also predicted that the activities that are crucial to drive the IBV replicative cycle are associated with two giant and totally uncharacterized polyproteins - pp1a (called F1) and frameshifted pp1ab (F1F2) of 3951 and 6629 amino acids, respectively. A comparison of pp1a/pp1ab with a sequence databank in a pairwise mode returned only one reliable hit - that between regions in the middle of IBV pp1b and the alphavirus replicative nsP2 protein.

This initial observation was significantly extended through thorough pairwise and profile comparisons of the IBV pp1a/pp1ab with replicative proteins of +RNA viruses and selected cellular proteins (Gorbalenya *et al.*, 1988; Hodgman, 1988; Gorbalenya *et al.*, 1989). In total, eight domains and thirteen sites of proteolytic autoprocessing were tentatively identified in pp1a/pp1ab (Fig. 1). This analysis proved to be quite accurate - the identification of six domains and nine cleavage sites was confirmed by subsequent studies. Upstream of the pp1a/pp1b junction, 3CLpro were correctly recognized[1]. Furthermore, the 3CLpro was predicted to be flanked by two highly hydrophobic, presumably trans-membrane domains HD3 and HD4. Downstream of the pp1a/1b junction, two domains, RdRp and HEL1,

[1] The domains designations used in this paper may deviate from those found elsewhere.

the latter linked to an upstream-positioned Cys/His-rich domain predicted to form triple mono-nuclear Zn-fingers (ZnD), were correctly identified. Unlike the above predictions, the assignment of two domains listed below was partly revised more recently. Not every Cys residue, which was initially thought to be a counterpart of one of the Cyss forming essential Cys-Cys bridges in a family of growth factors, proved later to be conserved in the growth factor-like (GFL) domain. Accordingly, this Cys-rich domain mapped immediately upstream of the pp1a/pp1b junction may not have a growth factor-like activity. Another sequence in the N-terminal region of pp1a, which was marginally similar to a papain-like protease from Streptococcus pneumoniae (called SPL), later proved to be not conserved. In a surprising twist, this sequence of pp1a does overlap with a real papain-like protease (PLpro) located slightly upstream and identified later (see below).

The identification of 3CLpro was corroborated by a tentative mapping of the cognate cleavage sites at the interdomain junctions including those that separate most of the domains described above (Gorbalenya *et al.*, 1989). The identification of 3CLpro cleavage sites became possible partly due to the extended conservation of these sites, which include Leu or another bulky hydrophobic amino acid at the P2 position, Gln at the P1 and a small residue at the P1' (Ziebuhr *et al.*, 2000). The vast majority of these sites were mapped to the C-terminal half of the pp1ab polyprotein covering the region downstream of the N-terminus of HD3 (Fig. 1). Hence, 3CLpro was specifically implicated in the control of the expression of the C-terminal part of pp1a/1ab, while another proteinase might be mainly responsible for the processing of the N-terminal half.

Three domains identified in IBV, 3CLpro, RdRp and HEL1, were (distant) variants of domains that are ubiquitous in the replicative polyproteins of +RNA viruses (Fig. 1). Like their counterparts in other viruses, the IBV RdRp and HEL1 were postulated to play a central role in RNA-synthesis. These domains, along with 3CLpro, were found to possess unique structural properties (Gorbalenya *et al.*, 1989). RdRp has a replacement of Gly by Ser in the GDD signature box that is otherwise absolutely conserved in +RNA viruses. HEL1 is uniquely associated with ZnD in the same protein. 3CLpro has a replacement of Gly by Tyr in the GxH signature of the substrate pocket which is otherwise unprecedented at this position. Also, the flanking of 3CLpro by hydrophobic domains from both sides has not been observed elsewhere. Furthermore, these and other (unique) conserved domains line in a very specific order, which was never observed in the polyproteins of other +RNA viruses. This observation confirmed that coronaviruses must prototype a separate supergroup.

Since most unique features of the proposed IBV pp1a/pp1ab structural organization were derived from the analysis of (very) weak sequence similarities, an independent validation of these findings remained urgent. Thus, comparative analyses of IBV and two strains of another coronavirus, MHV, reported soon afterward (Bredenbeek *et al.*, 1990a; Gorbalenya *et al.*, 1991; Lee *et al.*, 1991; Bonilla *et al.*, 1994) were essential. These analyses confirmed the conservation of seven out of eight domains and nine out of thirteen 3CLpro cleavage sites provisionally described for IBV. A tenth, new, 3CLpro site was identified at the C-terminal border of HD4 slightly downstream of a fortuitous site originally assigned to IBV. It was also shown that the most N-terminal ~2000 aa of pp1a/pp1ab are poorly conserved between these two coronaviruses (Lee *et al.*, 1991; Bonilla *et al.*, 1994). This region contained SPL in IBV, whose predicted essential residues were not conserved in MHV. Instead, a new X domain flanked by two paralogous PLpro domains in MHV (PL1pro and PL2pro, respectively) and one PLpro, from the C-terminus, in IBV were delineated in this region (Fig. 1) (Gorbalenya *et al.*, 1991). The X domain was shown to be a homologue of a functionally uncharacterized domain previously identified in the vicinity of the HEL1 domain in alpha- and rubivirus replicative polyproteins (Dominguez *et al.*, 1990). The PL1pro domain overlapped with a region thought to encode a protease responsible for the release of the N-terminal p28 in MHV (Baker *et al.*, 1989). It was predicted and later confirmed, that PL1pro mediates the p28 production from the polyproteins (Lee *et al.*, 1991; Baker *et al.*, 1993).

4.   TOROVIRUS BEV REPLICATIVE POLYPROTEIN HAS A CORONAVIRUS-LIKE DOMAIN ORGANIZATION THAT IS DECORATED WITH NEW ELEMENTS

Further insight into the organization and evolution of coronavirus replicative polyproteins was gained by Dutch researchers in a study of the torovirus BEV (Bredenbeek *et al.*, 1990b; Snijder *et al.*, 1990; Snijder *et al.*, 1991). They demonstrated that BEV has a coronavirus-like genome organization, the characteristic RdRp-ZnD-HEL1 domain layout in pp1b, and a frameshifting signal at the pp1a/pp1b junction, although the sequence for the most part of pp1a remains non-determined. These results unequivocally established the relationship between BEV and coronaviruses, even though these viruses have different virion morphologies. The comparison of BEV with coronaviruses also led to new assignments. First, in the C-terminal part of pp1b, a unique conserved domain with an unknown function, was delineated. Second, a surprising similarity was found between the C-terminal region of pp1a immediately upstream of the pp1a/pp1b junction of BEV and two accessory proteins produced from sg mRNAs 2 and 4 in group 2 coronaviruses. This observation indicated that some replicative proteins: i) are optional (since they are not conserved in all corona- and toroviruses), ii) form a pool that can be subject to reshuffling through recombination, and iii) can be expressed by different means (e.g., through proteolytic processing of the replicative polyproteins or by translation of distinct sg mRNAs).

5. ARTERIVIRUS EAV IS A 'MINI' CORONAVIRUS: THE BIRTH OF THE *NIDOVIRALES*

Shortly after finishing the analysis of BEV, the group led by Eric Snijder described the genome organization of the arterivirus EAV (den Boon *et al.*, 1991). At that time, this virus, together with alphaviruses and rubiviruses, was considered to be part of the togavirus family. Although EAV was shown to have a distinct virion morphology and the genome size approximately twice as small as that of an average coronavirus genome, it proved to share numerous features with coronaviruses. Particularly, the frameshifted pp1b polyprotein contained distant homologs of four conserved domains that had previously been identified in both corona- and toroviruses. Likewise, the pp1a polyprotein contained a characteristic domain set consisting of PLpro (called papain-like cysteine protease, PCP) as well as 3CLpro embedded between two HDs (Fig. 1).

Arteriviruses also possessed specific features. All conserved arterivirus domains were found to be smaller than their coronavirus counterparts, and the sequence similarities between pp1a-locating domains of corona- and arteriviruses were so weak that they could not be proved rigorously (Ziebuhr *et al.*, 2000). Unlike their coronavirus counterparts and also picorna-like enzymes, the arterivirus 3CLpro was demonstrated to employ Ser as the catalytic nucleophile (Snijder *et al.*, 1996).

Importantly, the replicase domains of arteriviruses, like those of the corona- and toroviruses before them, were found to have no strong sequence similarities to other virus groups. All these observations were most compatible with a pronounced divergent evolution of arteri- and coronaviruses from the common root. Accordingly, these families were united in a Coronavirus-like supergroup (Gorbalenya and Koonin, 1993; Snijder *et al.*, 1993), which was subsequently recognized as the *order* Nidovirales (Cavanagh, 1997).

6.   REFINING THE NIDOVIRUS GENETIC PLAN FOR REPLICATIVE POLYPROTEINS - CORONAVIRUSES

Before Nidovirales won formal approval, new theoretical and experimental data had substantiated the original findings described above. The genome analysis of two coronaviruses, HCoV (Herold *et al.*, 1993) and TGEV (Eleouet *et al.*, 1995), revealed the conservation of all domains and 3CLpro cleavage sites recognized before in the replicative polyproteins of MHV (Fig. 1). Subsequently, the functionality of all proteases, including the long-time elusive PL2pro, as well as the identity of the 3CLpro sites have been confirmed for IBV, MHV and HCoV mainly through efforts led by Susan Baker, Mark Denison, Jens Herold, Ding Liu, Susan Weiss and John Ziebuhr (reviewed in (Ziebuhr *et al.*, 2000); see also (Kanjanahaluethai and Baker, 2000; Ziebuhr *et al.*, 2001)). In the course of the cleavage site mapping, a new, previously overlooked 3CLpro site with an otherwise uncharacteristic Asn at the P1' position was identified in the region between HD4 and GFL domains (Liu *et al.*, 1997; Lu *et al.*, 1998; Ziebuhr and Siddell, 1999). The bioinformatics analysis of all PLpros of coronaviruses, which was supported by the results of subsequent experiments, implied that these proteases adopt a papain-like fold, in which two domains are uniquely connected by a Zn-finger (Herold *et al.*, 1999). This unprecedented protease architecture was provisionally implicated in the control of viral RNA synthesis.

In contrast to the 3CLpro sites, the initial mapping of the interdomain junctions recognized by PLpros was accomplished without theoretical guidance in experiments using MHV PL1pro (Dong and Baker, 1994; Hughes *et al.*, 1995; Bonilla *et al.*, 1997) and IBV PLpro (Lim and Liu, 1998; Lim *et al.*, 2000). Each of these proteases was shown to cleave two sites that included at least one small amino acid. In MHV, both sites were mapped upstream of the cognate PL1pro, and for IBV, one site was found to be upstream and another downstream of the PLpro. A recent comparative sequence analysis of the N-terminal half of pp1a of four coronaviruses has revealed the conservation of 3 sites (predicted to be) cleaved by various PLpros (Fig. 1) (Ziebuhr *et al.*, 2001). The most N-terminal site is conserved in all coronaviruses except IBV, while two other sites are conserved in all coronaviruses. In HCoV, the two N-terminal sites were shown to be cleaved by PL1pro and/or PL2pro (Herold *et al.*, 1998; Ziebuhr *et al.*, 2001). Interestingly, the third, most C-terminal site is located in a region flanked by two hydrophobic domains (HD1 and HD2) (Fig. 1) (Gorbalenya, unpublished data).

7. REFINING THE NIDOVIRUS GENETIC PLAN FOR THE REPLICATIVE POLYPROTEINS - ARTERIVIRUSES

While the work on the processing map of the replicative polyproteins for three coronaviruses is currently in its final stage, it has already been completed for the arterivirus EAV (van Dinten *et al.*, 1999). With slight deviations in the N-terminal part, it is conserved in all arteriviruses. Thus, it was feasible to develop a uniform nomenclature for the replicative proteins of arteriviruses. Each protein in pp1ab was assigned a name consisting of the prefix nsp (non-structural protein) and a digit between 1 and 12, reflecting the order of encoding (Fig. 1) (Snijder and Meulenberg, 1998).

An initial draft of the processing map of the arterivirus pp1ab was developed by a group which analyzed LDV (Godeny *et al.*, 1993). This map included three sites in pp1a that were subsequently confirmed, and three sites in pp1b that had to be revised later. The complete 3CLpro-driven proteolytic processing map, consisting of five sites in pp1a and three sites in pp1b, has been unequivocally established for EAV through the combination of theoretical and experimental approaches (Fig. 1) (Snijder *et al.*, 1996; van Dinten *et al.*, 1996; Wassenaar *et al.*, 1997; van Dinten *et al.*, 1999). With one exception, the arterivirus 3CLpro cleavage sites have Glu at the P1 position and a small amino acid at the P1' position. Upstream of the P1 position, these sites are less conserved than their coronavirus counterparts (Ziebuhr *et al.*, 2000), which explains the difficulties experienced with their prediction.

The analysis of the arterivirus LDV and PRRSV genomes also confirmed the conservation of the replicative domains identified in EAV. In the N-terminal part of pp1a, two divergent copies of PCPs were identified in these arteriviruses. The most N-terminal one was initially claimed to be an ortholog to the EAV PCP and the C-terminal one was proposed to be unique (Godeny *et al.*, 1993). This assignment was revised by our group when a defective PCP (PCPα) was identified upstream of the active PCP (renamed to PCPβ) in EAV (den Boon *et al.*, 1995). Thus, the collinearity of a pair of PCP domains, PCPα and PCPβ, was established for three arteriviruses (Fig. 1). The EAV PCPβ and PRRSV PCPα and PCPβ were proved to cleave their own C-terminus and the PCPβ cleavage site was shown to include at least one Gly (Snijder *et al.*, 1992; den Boon *et al.*, 1995; Ziebuhr *et al.*, 2000). Downstream of PCPβ, another cysteine proteinase domain associated with the nsp2 protein (CP2) was predicted and confirmed for EAV (Fig. 1) (Snijder *et al.*, 1995). It is distantly related to PCPs and, like EAV PCPβ, cleaves between two Gly residues at the nsp2|nsp3 junction. It contains three additional conserved Cys residues that are essential for the proteolytic activity and might be involved in metal-binding. Furthermore, a small domain with a Zn-finger signature covalently linked to the downstream-located PCPα was predicted for all arteriviruses (Fig. 1). A crucial role of this putative Zn-finger and its association with the protease domains were recently demonstrated for sg mRNA synthesis in EAV (Tijms *et al.*, 2001). Also, the predicted membrane association of proteins containing HDs has been proven for EAV (van der Meer *et al.*, 1998; Pedersen *et al.*, 1999).

The structural organization of the region located at the opposite side of the pp1ab polyprotein was also revised. Thus, it became clear that the very C-terminal region of pp1ab is occupied by two conserved domains (rather than one domain as initially proposed in den Boon *et al.*, 1991), which are separated by a 3CLpro cleavage site. The upstream nsp11-associated domain was shown to be conserved in all nidoviruses (NID, nidovirus-specific domain), while the C-terminally located, nsp12-associated domain was found to be arterivirus-specific (Fig. 1) (Godeny *et al.*, 1993; de Vries *et al.*, 1997)(Gorbalenya, unpublished data). On the basis of a detailed mutagenesis analysis of the EAV HEL1-associated ZnD (nsp10), the original model, which predicted three mononuclear Zn-fingers, has been revised. The new model implies that this domain may adopt a unique multinuclear organization and bind four Zn^{++}s (van Dinten *et al.*, 2000).

8. THE COLLINEARITY BETWEEN REPLICATIVE POLYPROTEINS OF ARTERI- AND CORONAVIRUSES - THE CURRENT STATUS

Comparisons of pp1a/1ab of nidoviruses, particularly coronaviruses and arteriviruses, showed a remarkable collinearity in their design, contrasting with an overall low sequence similarity. As the coronavirus and arterivirus genetics plans have been refined in the last years, the nidovirus-specific backbone became elaborated with new elements (Fig. 1).

In the pp1b portion of the pp1ab polyprotein, downstream of the pp1a/pp1ab frameshifting signal, four domains, RdRp, ZnD, HEL1 and NID, separated by three 3CLpro cleavage sites, are conserved in all nidoviruses. Importantly, the functionality of the HEL1 domain was recently proved for both corona- and arteriviruses (Seybert *et al.*, 2000a; Seybert *et al.*, 2000b). Coronaviruses have an additional domain, INS, which is inserted between HEL1 and NID and proteolytically released as a separate protein (de Vries *et al.*, 1997). At the C-terminus, the pp1b is capped by a family-specific domain that is separated from NID by a 3CLpro cleavage site. Two distinct nidoviruses, the mammalian BEV and invertebrate GAV, have the coronavirus-like rather than arterivirus-like domain organization of pp1b, although their 3CLpro cleavage site maps remain to be established (Snijder and Horzinek, 1993; de Vries *et al.*, 1997; Cowley *et al.*, 2000) (Gorbalenya, unpublished data).

The backbone of the pp1a polyprotein of nidoviruses is formed by the following structural elements (from N- to C-terminus): the N-terminally positioned PLpros including at least one Zn-finger; four HDs, the last two of which flank the 3CLpro from both sides; and a variable number of 3CLpro cleavage sites separating family-specific small proteins. Arteriviruses and coronaviruses employ very specific variants of the nidovirus genetic plan upstream of the HD1 and downstream of the HD4; so a one-to-one sequence alignment may not be possible to produce for these regions. For GAV, only a 3CLpro, which is flanked by two HDs, was recognized so far (Cowley *et al.*, 2000); this virus may deviate further from the plan outlined above. Consistent with these observations, the most fast changing regions in pp1ab of both coronaviruses and arteriviruses were identified in the N-terminal half of pp1a upstream of the HD1 (Lee *et al.*, 1991; Bonilla *et al.*, 1994; Nelsen *et al.*, 1999) (Gorbalenya, unpublished data).

9. HOW DID THE NIDOVIRUS GENETIC PLAN COME TO LIFE?

How have nidoviruses evolved replicative polyproteins of such outstanding complexity? There are two basic ways to increase complexity - either by acquiring domains from an outside source or by a duplication of a pre-existing domain of the parental genome using intergenomic or intragenomic recombination, respectively.

Most nidoviruses encode two or more papain-like domains that form family-specific branches well separated from their relatives of other origin (Ziebuhr *et al.*, 2000). This implies that an amplification of PL domains has contributed to the enlargement of the nidovirus genome. However, it remains an open question how many amplifications occurred and where they took place - in the ancestral nidovirus lineage and/or later, in separate nidovirus branches.

The nidovirus replicative domains also showed a remote sequence affinity to replicative proteins encoded by different +RNA viruses, mostly belonging to the Alphavirus-like or Picornavirus-like supergroups (Fig. 1). In particular, the coronavirus X domain and nidovirus HEL1 are clustered with homologs encoded by viruses of the Alphavirus-like supergroup (Gorbalenya *et al.*, 1988; Gorbalenya *et al.*, 1991). In contrast, coronavirus

RdRp and 3CLpro showed a specific affinity to the homologous enzymes of plant potyviruses (Gorbalenya *et al.*, 1989) (Gorbalenya, unpublished data), and the coronavirus PLpros hit the leader protease (Lpro) of animal aphthoviruses (Gorbalenya *et al.*, 1991). Both poty- and aphthoviruses belong to the Picornavirus-like supergroup. The sequence affinity to the potyvirus 3CLpro was also documented for the homologous GAV enzyme (Cowley *et al.*, 2000). Furthermore and in apparent contradiction to the relationships cited above, the 3CLpros of arteriviruses are clustered with homologs encoded by plant sobemo- and luteoviruses (Snijder *et al.*, 1996; Ziebuhr *et al.*, 2000). All these proteins also have cellular homologs (prototypes). These mosaic relationships might indicate that the replicative domains form several common pools from which different +RNA viruses build their genomes by reshuffling. This hypothesis implies that the nidovirus replicative polyprotein may be of chimeric origin. However, it must be stressed, that no evidence has so far been provided that any of the nidovirus domains is evolutionarily interleaved with related domains of other origins. Thus, it remains possible that all or some of these across-borders' similarities have been preserved in the course of a profound divergent and *continuous* evolution of +RNA viruses, including nidoviruses, from the common root (Gorbalenya and Koonin, 1993). More analyses and new sequences are needed to reconstruct the evolution of the nidovirus replicative polyproteins in greater detail.

10. CONCLUDING REMARKS

The most distinguishing feature of the nidovirus genome is the conservation of the specific domain arrangement in the replicative polyproteins, which are expressed by the multi-protease-mediated and ribosomal frame-shifting mechanisms. There are other supergroups of +RNA viruses, e.g., Picornavirus-like and Flavivirus-like supergroups, that heavily rely on proteases. What is, however, unique for nidoviruses is an unusual complexity of the conserved backbone of the replicative polyproteins, which is tolerant to twofold-plus variations of the genome size and pronounced sequence divergency. This complexity is likely to be translated into the elaborated nidovirus life cycle through an ordered release of the replicative subunits upon genome expression. We may expect the key features of this cycle to be conserved for different nidoviruses, regardless of their genome size and host range. The identification of force(s) that drive the selection of the unique nidovirus genetic plan and the understanding of *how* this plan shapes the life cycle of nidoviruses are among the most exciting goals for future studies.

ACKNOWLEDGEMENTS

This paper is based on a talk given during a meeting of the nidovirus community at Split Rock, PA, May 2000, at the kind invitation from the Organizers. I am most grateful to all colleagues with whom I have been privileged to collaborate on studying nidoviruses. My special thanks are to Eric Snijder and John Ziebuhr for the permission to cite unpublished results, and to John Ziebuhr for the valuable comments on the review. The help of Maritta Grau, editor, Publications Department, NCI-Frederick, was instrumental in preparing this text. My current studies of RNA viruses are supported with Federal funds from the National Cancer Institute, National Institutes of Health, under Contract No. NO1-CO-56000. The content of this publication does not necessarily reflect the views or policies of the Department of Health and Human Services, nor does mention of trade names, commercial products, or organization imply endorsement by the U.S. Government.

REFERENCES

Allende, R., Lewis, T.L., Lu, Z., Rock, D.L., Kutish, G.F., Ali, A., Doster, A.R. and Osorio, F.A. (1999) North American and European porcine reproductive and respiratory syndrome viruses differ in non-structural protein coding regions. *J Gen Virol*, **80**, 307-15.

Baker, S.C., Shieh, C.K., Soe, L.H., Chang, M.F., Vannier, D.M. and Lai, M.M. (1989) Identification of a domain required for autoproteolytic cleavage of murine coronavirus gene A polyprotein. *J Virol*, **63**, 3693-9.

Baker, S.C., Yokomori, K., Dong, S., Carlisle, R., Gorbalenya, A.E., Koonin, E.V. and Lai, M.M. (1993) Identification of the catalytic sites of a papain-like cysteine proteinase of murine coronavirus. *J Virol*, **67**, 6056-63.

Bonilla, P.J., Gorbalenya, A.E. and Weiss, S.R. (1994) Mouse hepatitis virus strain A59 RNA polymerase gene ORF1a: heterogeneity among MHV strains. *Virology*, **198**, 736-740.

Bonilla, P.J., Hughes, S.A. and Weiss, S.R. (1997) Characterization of a second cleavage site and demonstration of activity in trans by the papain-like proteinase of the murine coronavirus mouse hepatitis virus strain A59. *J Virol*, **71**, 900-9.

Boursnell, M.E.G., Brown, T.D.K., Foulds, I.J., Green, P.F., Tomley, F.M. and Binns, M.M. (1987) Completion of the sequence of the genome of the coronavirus avian infectious bronchitis virus. *J. Gen. Virol.*, **68**, 57-77.

Bredenbeek, P.J., Pachuk, C.J., Noten, A.F., Charite, J., Luytjes, W., Weiss, S.R. and Spaan, W.J. (1990a) The primary structure and expression of the second open reading frame of the polymerase gene of the coronavirus MHV-A59; a highly conserved polymerase is expressed by an efficient ribosomal frameshifting mechanism. *Nucleic Acids Res*, **18**, 1825-32.

Bredenbeek, P.J., Snijder, E.J., Noten, F.H., den Boon, J.A., Schaaper, W.M., Horzinek, M.C. and Spaan, W.J. (1990b) The polymerase gene of corona- and toroviruses: evidence for an evolutionary relationship. *Adv Exp Med Biol*, **276**, 307-16.

Cavanagh, D. (1997) Nidovirales: a new order comprising Coronaviridae and Arteriviridae. *Arch Virol*, **142**, 629-33.

Cowley, J.A., Dimmock, C.M., Spann, K.M. and Walker, P.J. (2000) Gill-associated virus of Penaeus monodon prawns: an invertebrate virus with ORF1a and ORF1b genes related to arteri- and coronaviruses. *J Gen Virol*, **81 Pt 6**, 1473-84.

de Vries, A.A.F., Horzinek, M.C., Rottier, P.J.M. and de Groot, R.J. (1997) The genome organization of the Nidovirales: similarities and differences between Arteri-, Toro-, and Coronaviruses. *Semin. Virology*, **8**, 33-47.

den Boon, J.A., Faaberg, K.S., Meulenberg, J.J., Wassenaar, A.L., Plagemann, P.G., Gorbalenya, A.E. and Snijder, E.J. (1995) Processing and evolution of the N-terminal region of the arterivirus replicase ORF1a protein: identification of two papainlike cysteine proteases. *J Virol*, **69**, 4500-5.

den Boon, J.A., Snijder, E.J., Chirnside, E.D., de Vries, A.A., Horzinek, M.C. and Spaan, W.J. (1991) Equine arteritis virus is not a togavirus but belongs to the coronaviruslike superfamily. *J Virol*, **65**, 2910-20.

Dominguez, G., Wang, C.Y. and Frey, T.K. (1990) Sequence of the genome RNA of rubella virus: evidence for genetic rearrangement during togavirus evolution. *Virology*, **177**, 225-38.

Dong, S. and Baker, S.C. (1994) Determinants of the p28 cleavage site recognized by the first papain-like cysteine proteinase of murine coronavirus. *Virology*, **204**, 541-549.

Eleouet, J.F., Rasschaert, D., Lambert, P., Levy, L., Vende, P. and Laude, H. (1995) Complete sequence (20 kilobases) of the polyprotein-encoding gene 1 of transmissible gastroenteritis virus. *Virology*, **206**, 817-22.

Godeny, E.K., Chen, L., Kumar, S.N., Methven, S.L., Koonin, E.V. and Brinton, M.A. (1993) Complete genomic sequence and phylogenetic analysis of the Lactate Dehydrogenase-Elevating virus (LDV). *Virology*, **194**, 585-596.

Goldbach, R. (1987) Genome similarities between plant and animal RNA viruses. *Microbiol. Sci.*, **4**, 197-201.

Gorbalenya, A.E. and Koonin, E.V. (1993) Comparative analysis of the amino acid sequences of the key enzymes of the replication and expression of positive-strand RNA viruses. Validity of the approach and functional and evolutionary implications. *Sov. Sci. Rev. D. Physicochem. Biol.*, **11**, 1-84.

Gorbalenya, A.E., Koonin, E.V., Donchenko, A.P. and Blinov, V.M. (1988) A novel superfamily of nucleoside triphosphate-binding motif containing proteins which are probably involved in duplex unwinding in DNA and RNA replication and recombination. *FEBS Lett.*, **235**, 16-24.

Gorbalenya, A.E., Koonin, E.V., Donchenko, A.P. and Blinov, V.M. (1989) Coronavirus genome: prediction of putative functional domains in the non-structural polyprotein by comparative amino acid sequence analysis. *Nucleic Acids Res.*, **17**, 4847-4861.

Gorbalenya, A.E., Koonin, E.V. and Lai, M.M.C. (1991) Putative papain-related thiol proteases of positive-strand RNA viruses. *FEBS Letters*, **288**, 201-205.

Herold, J., Gorbalenya, A.E., Thiel, V., Schelle, B. and Siddell, S.G. (1998) Proteolytic processing at the amino terminus of human coronavirus 229E gene 1-encoded polyproteins: identification of a papain-like proteinase and its substrate. *J Virol*, **72**, 910-8.

Herold, J., Raabe, T., Schelle-Prinz, B. and Siddell, S.G. (1993) Nucleotide sequence of the human coronavirus 229E RNA polymerase locus. *Virology*, **195**, 680-91.

Herold, J., Siddell, S.G. and Gorbalenya, A.E. (1999) A human RNA viral cysteine proteinase that depends upon a unique Zn2+-binding finger connecting the two domains of a papain-like fold [published erratum appears in J Biol Chem 1999 Jul 23;274(30):21490]. *J Biol Chem*, **274**, 14918-25.

Hodgman, T.C. (1988) A new superfamily of replicative proteins. *Nature*, **333**, 22-23.

Hughes, S.A., Bonilla, P.J. and Weiss, S.R. (1995) Identification of the murine coronavirus p28 cleavage site. *J Virol*, **69**, 809-13.

Kanjanahaluethai, A. and Baker, S.C. (2000) Identification of mouse hepatitis virus papain-like proteinase 2 activity. *J Virol*, **74**, 7911-21.

Lai, M.M., Baric, R.S., Brayton, P.R. and Stohlman, S.A. (1984) Characterization of leader RNA sequences on the virion and mRNAs of mouse hepatitis virus, a cytoplasmic RNA virus. *Proc Natl Acad Sci U S A*, **81**, 3626-30.

Lai, M.M.C. and Cavanagh, D. (1997) The molecular biology of coronaviruses. *Adv. Virus Res.*, **48**, 1-100.

Lee, H.J., Shieh, C.K., Gorbalenya, A.E., Koonin, E.V., La Monica, N., Tuler, J., Bagdzhadzhyan, A. and Lai, M.M. (1991) The complete sequence (22 kilobases) of murine coronavirus gene 1 encoding the putative proteases and RNA polymerase. *Virology*, **180**, 567-82.

Lim, K.P. and Liu, D.X. (1998) Characterization of the two overlapping papain-like proteinase domains encoded in gene 1 of the coronavirus infectious bronchitis virus and determination of the C-terminal cleavage site of an 87-kDa protein. *Virology*, **245**, 303-312.

Lim, K.P., Ng, L.F. and Liu, D.X. (2000) Identification of a novel cleavage activity of the first papain-like proteinase domain encoded by open reading frame 1a of the coronavirus Avian infectious bronchitis virus and characterization of the cleavage products. *J Virol*, **74**, 1674-85.

Lu, X.T., Sims, A.C. and Denison, M.R. (1998) Mouse hepatitis virus 3C-like protease cleaves a 22-kilodalton protein from the open reading frame 1a polyprotein in virus-infected cells and in vitro. *J Virol*, **72**, 2265-71.

Meulenberg, J.J., Hulst, M.M., de Meijer, E.J., Moonen, P.L., den Besten, A., de Kluyver, E.P., Wensvoort, G. and Moormann, R.J. (1993) Lelystad virus, the causative agent of porcine epidemic abortion and respiratory syndrome (PEARS), is related to LDV and EAV. *Virology*, **192**, 62-72.

Nelsen, C.J., Murtaugh, M.P. and Faaberg, K.S. (1999) Porcine reproductive and respiratory syndrome virus comparison: divergent evolution on two continents. *J Virol*, **73**, 270-80.

Palmer, G.A., Kuo, L., Chen, Z., Faaberg, K.S. and Plagemann, P.G. (1995) Sequence of the genome of lactate dehydrogenase-elevating virus: heterogenicity between strains P and C. *Virology*, **209**, 637-42.

Pedersen, K.W., van der Meer, Y., Roos, N. and Snijder, E.J. (1999) Open reading frame 1a-encoded subunits of the arterivirus replicase induce endoplasmic reticulum-derived double-membrane vesicles which carry the viral replication complex. *J Virol*, **73**, 2016-26.

Sawicki, S.G. and Sawicki, D.L. (1995) Coronaviruses use discontinuous extension for synthesis of subgenome-length negative strands. *Adv Exp Med Biol*, **380**, 499-506.

Seybert, A., Hegyi, A., Siddell, S.G. and Ziebuhr, J. (2000a) The human coronavirus 229E superfamily 1 helicase has RNA and DNA duplex-unwinding activities with 5'-to-3' polarity. *Rna*, **6**, 1056-68.

Seybert, A., van Dinten, L.C., Snijder, E.J. and Ziebuhr, J. (2000b) Biochemical characterization of the equine arteritis virus helicase suggests a close functional relationship between arterivirus and coronavirus helicases. *J Virol*, **74**, 9586-93.

Snijder, E.J., den Boon, J.A., Bredenbeek, P.J., Horzinek, M.C., Rijnbrand, R. and Spaan, W.J. (1990) The carboxyl-terminal part of the putative Berne virus polymerase is expressed by ribosomal frameshifting and contains sequence motifs which indicate that toro- and coronaviruses are evolutionarily related. *Nucleic Acids Res*, **18**, 4535-42.

Snijder, E.J., den Boon, J.A., Horzinek, M.C. and Spaan, W.J. (1991) Comparison of the genome organization of toro- and coronaviruses: evidence for two nonhomologous RNA recombination events during Berne virus evolution. *Virology*, **180**, 448-52.

Snijder, E.J. and Horzinek, M.C. (1993) Toroviruses: replication, evolution and comparison with other members of the coronavirus-like superfamily. *J Gen Virol*, **74**, 2305-16.

Snijder, E.J., Horzinek, M.C. and Spaan, W.J. (1993) The coronaviruslike superfamily. *Adv Exp Med Biol*, **342**, 235-44.

Snijder, E.J. and Meulenberg, J.J. (1998) The molecular biology of arteriviruses. *J Gen Virol*, **79**, 961-79.

Snijder, E.J., Wassenaar, A.L., Spaan, W.J. and Gorbalenya, A.E. (1995) The arterivirus Nsp2 protease. An unusual cysteine protease with primary structure similarities to both papain-like and chymotrypsin- like proteases. *J Biol Chem*, **270**, 16671-6.

Snijder, E.J., Wassenaar, A.L., van Dinten, L.C., Spaan, W.J. and Gorbalenya, A.E. (1996) The arterivirus nsp4 protease is the prototype of a novel group of chymotrypsin-like enzymes, the 3C-like serine proteases. *J Biol Chem*, **271**, 4864-71.

Snijder, E.J., Wassenaar, A.L.M. and Spaan, W.J.M. (1992) The 5' end of the equine arteritis virus replicase gene encodes a papain-like protease. *J. Virol.*, **66**, 7040-7048.

Spaan, W., Delius, H., Skinner, M., Armstrong, J., Rottier, P., Smeekens, S., van der Zeijst, B.A. and Siddell, S.G. (1983) Coronavirus mRNA synthesis involves fusion of non-contiguous sequences. *Embo J*, **2**, 1839-44.

Strauss, J.H. and Strauss, E.G. (1988) Evolution of RNA viruses. *Ann. Rev. Microbiol.*, **42**, 657-683.

Tijms, M.A., van Dinten, L.C., Gorbalenya, A.E. and Snijder, E.J. (2001) A zinc finger-containing papain-like protease couples subgenomic mRNA synthesis to genome translation in a positive-stranded RNA virus. *Proc. Natl. Acad. Sci. USA, in press*.

van der Meer, Y., van Tol, H., Locker, J.K. and Snijder, E.J. (1998) ORF1a-encoded replicase subunits are involved in the membrane association of the arterivirus replication complex. *J Virol*, **72**, 6689-98.

van Dinten, L.C., Rensen, S., Gorbalenya, A.E. and Snijder, E.J. (1999) Proteolytic processing of the open reading frame 1b-encoded part of arterivirus replicase is mediated by nsp4 serine protease and is essential for virus replication. *J Virol*, **73**, 2027-37.

van Dinten, L.C., van Tol, H., Gorbalenya, A.E. and Snijder, E.J. (2000) The predicted metal-binding region of the arterivirus helicase protein is involved in subgenomic mRNA synthesis, genome replication, and virion biogenesis. *J Virol*, **74**, 5213-23.

van Dinten, L.C., Wassenaar, A.L., Gorbalenya, A.E., Spaan, W.J. and Snijder, E.J. (1996) Processing of the equine arteritis virus replicase ORF1b protein: identification of cleavage products containing the putative viral polymerase and helicase domains. *J Virol*, **70**, 6625-33.

van Marle, G., Dobbe, J.C., Gultyaev, A.P., Luytjes, W., Spaan, W.J. and Snijder, E.J. (1999) Arterivirus discontinuous mRNA transcription is guided by base pairing between sense and antisense transcription-regulating sequences. *Proc Natl Acad Sci U S A*, **96**, 12056-61.

Wassenaar, A.L., Spaan, W.J., Gorbalenya, A.E. and Snijder, E.J. (1997) Alternative proteolytic processing of the arterivirus replicase ORF1a polyprotein: evidence that NSP2 acts as a cofactor for the NSP4 serine protease. *J Virol*, **71**, 9313-22.

Ziebuhr, J. and Siddell, S.G. (1999) Processing of the human coronavirus 229E replicase polyproteins by the virus-encoded 3C-like proteinase: identification of proteolytic products and cleavage sites common to pp1a and pp1ab. *J Virol*, **73**, 177-85.

Ziebuhr, J., Snijder, E.J. and Gorbalenya, A.E. (2000) Virus-encoded proteinases and proteolytic processing in the Nidovirales. *J Gen Virol*, **81 Pt 4**, 853-79.

Ziebuhr, J., Thiel, V. and Gorbalenya, A.E. (2001) Two paralogous RNA viral papain-like proteases recognize the same cleavage site in the virus-encoded polyprotein. *to be published.*

Molecular Epidemiology and Evolution of Equine Arteritis Virus

UDENI B. R. BALASURIYA, JODI F. HEDGES, AND N. JAMES MACLACHLAN
Bernard and Gloria Salick Equine Viral Disease Laboratory, Department of Pathology, Microbiology and Immunology, School of Veterinary Medicine, One Shields Avenue, University of California, Davis, CA 95616

1. INTRODUCTION

EVA is an infectious disease of equids that occurs in many parts of the world and is caused by EAV (Timoney and McCollum, 1993; Glaser *et al.*, 1997). Most EAV infections are subclinical but occasional outbreaks of disease occur that are characterised by any combination of influenza-like illness in adult horses, abortion in pregnant mares, and interstitial pneumonia in very young foals. Up to 60% of stallions infected with EAV become persistently infected carriers and shed virus continuously in semen. Persistence of EAV in the male reproductive tract is testosterone-dependent, and persistently infected stallions function as a natural reservoir that can disseminate virus to susceptible mares at breeding (Timoney and McCollum, 1993).

The two principal modes of EAV transmission are horizontal transmission by aerosolization of infectious respiratory tract secretions from acutely infected horses and venereal transmission during natural or artificial insemination with infective semen from persistently infected stallions (Timoney and McCollum, 1993; Glaser *et al.*, 1997). EAV also can be transmitted through indirect contact with formites or personnel. Congenital infection results from transplacental transmission (vertical transmission) of virus when pregnant mares are infected late in gestation.

There is only one serotype of EAV and all field isolates evaluated

The Nidoviruses (Coronaviruses and Arteriviruses).
Edited by Ehud Lavi *et al.*. Kluwer Academic/Plenum Publishers. 2001.

thus far are neutralised by polyclonal antiserum raised against the VBS53 strain, however, geographically and temporally distinct strains of EAV differ in the severity of the clinical disease they induce and in their abortigenic potential (Timoney and McCollum, 1993). There is considerable genetic variation amongst field strains of EAV (Balasuriya, 1995, 1998 and 1999; Hedges *et al.*, 1996 and 1999; Patton *et al.*, 1998; Stadejek *et al.*, 1999; Chirnside *et al.*, 1994; Archambault *et al.*, 1997).

We have focused our recent studies on the molecular epidemiology and evolution of EAV to better characterise the role of the carrier stallion in the emergence of novel genotypic and phenotypic variants of the virus.

2.　　MATERIALS AND METHODS

Viral RNA directly isolated from nasal swabs, mononuclear cells and from the semen of two stallions that were experimentally inoculated with the virus derived from the infectious cDNA clone of EAV was used to characterise the genetic heterogeneity of EAV during transient infection of horses (Balasuriya *et al.*, 1999). To study viral heterogeneity during acute infections, viral RNA was isolated directly from homogenates of placenta from one aborted foetus and six congenitally infected foals, and from tissues of two foals that died during an outbreak of EVA (Balasuriya *et al.*, 1999). Viral RNA was also isolated from the semen of carrier stallions naturally infected with EAV (Balasuriya *et al.*, 1999; Hedges *et al.*, 1999). ORFs 2-7 and flanking portions of the EAV genome (2923 bp) were RT-PCR amplified by RT-PCR using Superscript II and *Pfu* Turbo DNA polymerase, as previously described (Balasuriya *et al.*, 1999; Hedges *et al.*, 1999). Individual cDNA amplicons of ORF5 within each RT-PCR sample were cloned into a vector and 14-20 clones were compared to determine the quasispecies heterogeneity of each sample.

3.　　RESULTS

The master sequence of the virus present in the nasal swabs, buffy coat cells and semen collected from one stallion after experimental infection with the recombinant virus was identical to that of the cloned EAV030H virus used to infect the horses. There was little heterogeneity (2/20) in the clones derived from the virus used to inoculate the horses or in the mononuclear cells collected at 8 DPI (2/20 clones), whereas there were substitutions in 15/20 clones of ORF5 of EAV in the mononuclear cells collected from 14 days post inoculation (DPI) and in all 20 clones from the mononuclear cells collected from the second stallion at 10 DPI. The data indicate that although the master sequence of the recombinant virus was conserved during infection of the two stallions, the genetic heterogeneity of the virus population increased in the course of acute transient infection.

We also investigated an extensive outbreak of EVA on a Warmblood breeding farm to characterize the molecular epidemiology of EAV infection and to determine the genetic heterogeneity that was generated during this outbreak. Sequence and phylogenetic analysis of ORFs 2-7 of the master sequences of viruses that were present in foal tissues and sequential semen samples from four persistently infected stallions on the farm demonstrated that a virus in the semen of one carrier stallion initiated the outbreak. The virus in the lung of the first foal that died during the outbreak was very similar (99.5% identity) to virus in a semen sample that was collected from one carrier stallion 4 days prior to the outbreak. There were no obvious sequence differences between viruses present in the tissues of the aborted foal, foals that developed pneumonia or systemic illness, and in the placentas of foals that exhibited no clinical evidence of EAV infection (>99.9% identity) during the outbreak. In contrast, viral sequences derived from the semen of the two stallions that became persistently infected during the outbreak differed by 44 nucleotides (98.4% identity) after 12 months of persistent infection, indicating that each virus evolved independently during the course of persistent infection. The ORF5 sequences of clones derived from the RT-PCR products directly amplified from the semen of three carrier stallions (the stallion that initiated the outbreak and two stallions that became persistently infected following the outbreak) and tissues from three foals, were compared to further characterize the genetic heterogeneity of the EAV quasispecies during horizontal and vertical transmission. A total of 78 clones that included the entire ORF5 sequence were analyzed, including 14 to 20 clones from each semen sample and 28 clones from the various foal tissues. The clones from the semen of two stallions that became persistently infected following the outbreak showed increased genetic heterogeneity during the course of persistent infection. In contrast, the clones from the outbreak were homogeneous.

To characterize the genetic heterogeneity of EAV during persistent infection, detailed sequence analysis of the structural protein genes (ORFs 2-7) was performed with viral RNA purified directly from semen collected sequentially over a 10-year period from two Thoroughbred carrier stallions that were infected during an EVA outbreak in Kentucky in 1984 (Hedges *et al.*, 1999). Sequence analysis of ORFs 2-7 clearly indicated that the RNA in the semen of the two stallions was distinct from year to year. The master sequence of the virus population shed in the semen of individual stallions varied by approximately 1% per year. ORFs 2-7 of major EAV variants (master sequences) present in the semen of the two carrier stallions evolved independently of each other. Overall, ORFs 3 and 5 evolved most rapidly,

ORFs 2 and 4 moderately, and ORFs 6 and 7 were substantially more conserved during persistent EAV infection.

Variation occurred in the G_L protein during persistent infection of stallions primarily within a specific section of the V1 region that expresses three of the major virus neutralization sites. The neutralization phenotype of selected sequential EAV isolates from the semen of the two carrier stallions was evaluated by neutralization assay with a large panel of monoclonal antibodies and polyclonal equine antisera. There were significant differences in the neutralization phenotype of some of the sequential isolates. The data clearly indicate that selective pressures exerted during the course of persistent EAV infection of stallions significantly influence the evolution of specific regions of the G_L protein as variants with altered neutralization phenotypes emerge.

To further characterize the EAV quasispecies present during persistent infection of stallions, ORF 5 was RT-PCR amplified and cloned from viral RNA isolated directly from the semen collected in 1984 and 1994 from one stallion (17 and 16 clones respectively). Most ORF5 clones were distinct, indicating that EAV exists as a mixture of related genomes. The clones from semen collected in 1994 were more variable than the clones derived from semen collected from the same stallion 10 years earlier. Sequence and phylogenetic analysis of the various clones clearly showed that the heterogeneity of the quasispecies increased during the course of long-term persistent EAV infection of the stallion.

4.   DISCUSSION

In our studies of the molecular evolution of EAV, we have shown that virus derived from an infectious cDNA clone is genetically stable during acute, transient infection. The genetic stability of viruses that circulated during an outbreak of EVA contrasts markedly with the heterogeneous virus populations present in the semen of persistently infected stallions. Selective pressures exerted during the course of persistent infection of the reproductive tract of the carrier stallion clearly can be responsible for genotypic divergence and emergence of phenotypically novel EAV variants, and likely compensate for the relatively limited diversity of EAV that is generated during EVA outbreaks. Novel viral strains can emerge that are capable of initiating outbreaks of EVA. Thus, the persistently infected carrier stallions acts as a reservoir that harbors the EAV between breeding seasons, during which viral genetic diversity increases.

ACKNOWLEDGMENTS

These studies were supported by the Grayson-Jockey Club Research Foundation Inc., the Center for Equine Health at the University of California-Davis, the Bernard and Gloria Salick Endowment, and USDA Competitive Grant 97-35204-4736. We would like to thank Dr. Eric J. Snijder for providing the infectious cDNA clone of EAV and Drs. William H. McCollum and Peter J. Timoney for providing various field stains of the virus.

REFERENCES

Archambault, D., Laganiere, G., Carman, S., and St-Laurent, G.(1997). Comparison of nucleic and amino acid sequences and phylogenetic analysis of open reading frames 3 and 4 of various equine arteritis virus isolates. *Vet.Res.* **28**, 505-516.

Balasuriya, U.B.R., Evermann, J.F., Hedges, J.F., McKeirnan, A.J., Mitten, J.Q., Beyer, J.C., McCollum, W.H., Timoney, P.J., and MacLachlan, N.J.(1998). Serologic and molecular characterization of an abortigenic strain of equine arteritis virus derived from infective frozen semen and an aborted equine fetus. *J.Am.Vet.Med.Assoc.* **213**, 1586-1589.

Balasuriya, U.B.R., Hedges, J.F., Timoney, P.J., McCollum, W.H., and MacLachlan, N.J.(1999). Genetic stability of equine arteritis virus during horizontal and vertical transmission in an outbreak of equine viral arteritis. *J.Gen.Virol.* **80**, 1949-1958.

Balasuriya, U.B.R., Snijder, E.J., van Dinten, L.C., Heidner, H.W., Wilson, W.D., Hedges, J.F., Hullinger, P.J., and MacLachlan, N.J.(1999). Equine arteritis virus derived from an infectious cDNA clone is attenuated and genetically stable in infected stallions. *Virology* **260**, 201-208.

Balasuriya, U.B.R., Timoney, P.J., McCollum, W.H., and MacLachlan, N.J.(1995). Phylogenetic analysis of open reading frame 5 of field isolates of equine arteritis virus and identification of conserved and nonconserved regions in the G_L envelope glycoprotein. *Virology* **214**, 690-697.

Chirnside, E.D., Wearing, C.M., Binns, M.M., and Mumford, J.A.(1994). Comparison of M and N gene sequences distinguishes variation amongst equine arteritis virus isolates. *J.Gen.Virol.* **75**, 1491-1497.

Hedges, J.F., Balasuriya, U.B.R., Timoney, P.J., McCollum, W.H., and MacLachlan, N.J.(1996). Genetic variation in open reading frame 2 of field isolates and laboratory strains of equine arteritis virus. *Virus Res.* **42**, 41-52.

Hedges, J.F., Balasuriya, U.B.R., Timoney, P.J., McCollum, W.H., and MacLachlan, N.J.(1999). Genetic divergence with emergence of phenotypic variants of equine arteritis virus during persistent infection of stallions. *J.Virol.* **73**, 3672-3681.

McCollum, W.H., Prickett, M.E., and Bryans, J.T.(1971). Temporal distribution of equine arteritis virus in respiratory mucosa, tissues and body fluids of horses infected by inhalation. *Res.Vet.Sci.* **12**, 459-464.

Patton, J.F., Balasuriya, U.B.R., Hedges, J.F., Schweidler, T.M., Hullinger, P.J., and MacLachlan, N.J.(1998). Phylogenetic characterization of a highly attenuated strain of equine arteritis virus from the semen of a persistently infected standardbred stallion. *Arch.Virol.* **144**, 817-827.

Stadejek, T., Bjorklund, H., Ros Bascunana, C., Ciabatti, I.M., Scicluna, M.T., Amaddeo, D., McCollum, W.H., Autorino, G.L., Timoney, P.J., Paton, D.J., Klingeborn, B., and Belak, S.(1999). Genetic diversity of equine arteritis virus. *J.Gen.Virol.* **80**, 691-699.

Timoney, P.J. and McCollum, W.H.(1993). Equine viral arteritis. *Vet.Clin.N.Am., Eq.Pract.* **9**, 295-309.

Genetic and Antigenic Stability of PRRS Virus in Pigs
Field and experimental prospectives

KYOUNG-JIN YOON, CHIH-CHENG CHANG, JEFF ZIMMERMAN, AND KAREN HARMON
Department of Veterinary Diagnostic and Production Animal Medicine, College of Veterinary Medicine, Iowa State University, Ames, IA, U.S.A.

1. INTRODUCTION

Unrecognised prior to 1991, porcine reproductive and respiratory syndrome (PRRS) is one of the most economically significant diseases of swine in the world today. A newly emerged arterivirus, PRRS virus causes reproductive losses in adult animals and respiratory disease in pigs of all ages (Collins *et al* 1992, Wensvoort *et al* 1991). In acute outbreaks, economic losses from PRRS virus have been estimated to range from $236 to $502 per sow in farrow-to-finish and breeding stock operations (Polson *et al* 1992). In response to the economic effects of PRRS, various management strategies and vaccination protocols have been tested for controlling PRRS. At present, the definitive solution to the prevention and control of PRRS has not been found.

Several characteristics of PRRS virus have been identified. The virus is highly infectious (Yoon *et al* 2000) and preferentially replicates in host macrophages (Wensvoort *et al* 1991). Infection results in humoral and cellular immune responses, but infectious virus can be recovered from pigs for several months following the initial exposure (Wills *et al* 1997, Zimmerman et al 1992). Subclinically infected carrier animals are considered to be the key to the perpetuation of PRRS virus in endemically infected herds. At present, the exact mechanism by which PRRS virus evades the immune system is unknown. However, in other RNA viruses, persistent infections appear to be based on continuous mutations that select

The Nidoviruses (Coronaviruses and Arteriviruses).
Edited by Ehud Lavi *et al.*, Kluwer Academic/Plenum Publishers, 2001.

for viral "quasi-species" best adapted to continuous replication in certain cells (Domingo and Holland 1997). That is, diversity occurs over the course of infection within an individual. The observation that PRRS virus field isolates vary genetically and antigenically suggest that a similar mechanism might be involved (Andreyev *et al* 1997, Murtaugh *et al* 1998, Wensvoort *et al* 1992, Yang *et al* 1999; Yang *et al* 2000, Yoon *et al* 1997). To evaluate the dynamics of PRRS virus in individual animals and in populations, field-based and experimental studies were conducted.

2.   GENOTYPIC AND PHENOTYPIC DIVERSITY AMONG FIELD ISOLATES OF PRRS VIRUS

Field isolates of PRRS virus were obtained from clinical submissions to the Iowa State University Veterinary Diagnostic Laboratory between 1996 and 2000. Isolates were selected for genetic and/or antigenic analysis based on two criteria: the origin of isolates (i.e., viruses from the same herd or farm) or the restriction fragment length polymorphism (RFLP) pattern.

Six isolates were chosen based on the first criterion. These isolates were recovered in a two-month period from acutely affected pigs at different sites on the same farm. To avoid the introduction of PRRS virus strains into the herd, the management only accepted replacement animals from a single PRRS virus-negative source.

The 6 isolates were first assayed for genetic variability using a fingerprinting technique described by Wesley *et al* (1998) and then for antigenic differences using a panel of PRRS virus-specific monoclonal antibodies. Individual monoclonal antibodies represented distinct epitopes on the major structural (N, M, E) and 43kD proteins of PRRS virus (Yang *et al* 1999, Yang *et al* 2000). The fingerprinting analysis demonstrated that all

Table 1. Diversity of PRRS virus isolated from one herd as determined by monoclonal antibody analysis

PRRSV isolate	Month	Monoclonal antibody panels										
		15kD					19kD		25kD			43kD	
		A	B	C	D	E	A	B	A	B	C	A	B
VDL 98-16736	5	+	+	+	+	+	+	-	+	+	+	-	-
VDL 98-17455	5	+	+	+	+	+	+	+	+	+	+	+	-
VDL 98-19622	6	+	+	+	+	-	+	+	+	+	+	+	+
VDL 98-22148	7	+	+	+	+	+	+	+	+	+	+	+	-
VDL 98-24002	7	+	+	+	+	+	+	+	+	+	+	+	+
VDL 98-24104	7	+	+	+	+	+	+	+	+	+	+	+	+

isolates had the RFLP pattern designated 1-4-2. In contrast, monoclonal antibody analysis (Table 1) categorised the 6 isolates into 4 antigenic groups, indicating extensive phenotypic variability among these isolates.

In a second study, a total of 21 field isolates with RFLP cutting pattern 1-4-2 were selected based on the second criterion. All isolates were recovered from swine herds in the state of Iowa. Open reading frame (ORF) 5 of these isolates was sequenced and compared. ORF5 encodes for the major envelope glycoprotein (25kD) and is known to be the most variable among isolates (Andreyev *et al* 1997; Murtaugh *et al* 1998). The percent sequence homology among the isolates ranged from 84% to 98%. Amino acid substitutions occur more frequently in N terminal ends. A computer-aided phylogenetic analysis revealed two genotypic clusters (Fig 1), suggesting that the isolates were from two distinct origins. However, genotypic variability among the isolates and between 2 clusters did not correlate with geographical proximity or chronological order of isolation.

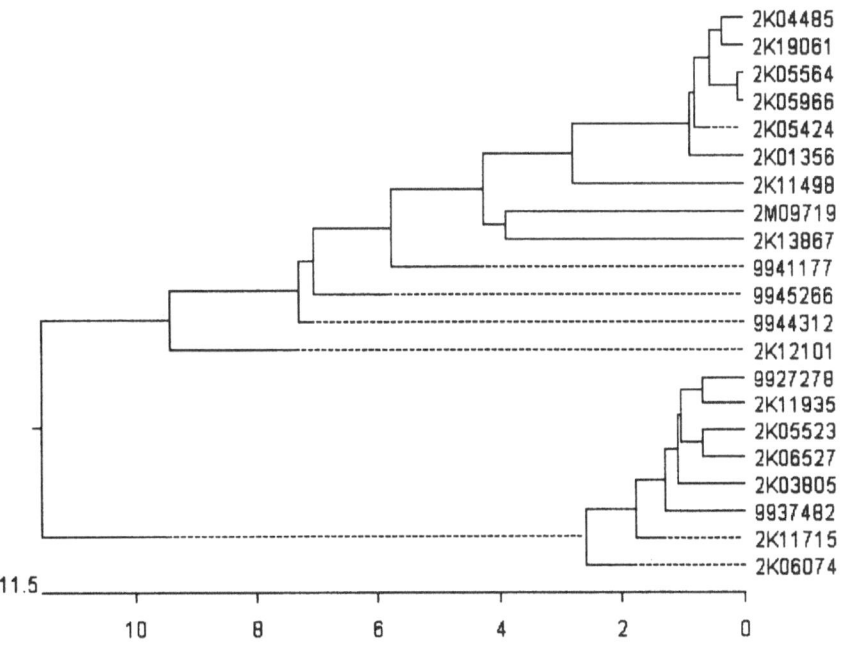

Figure 1. Variability of ORF5 nucleotide sequence among PRRS virus field isolates which had the same RFLP cutting pattern

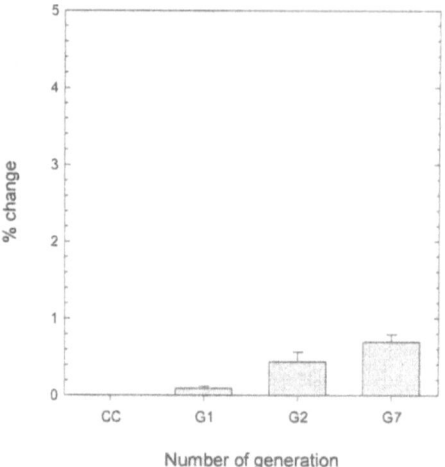

Figure 2. Rate of nucleotide change in ORF5 (603 nucleotide long) of PRRS virus in pig over time. Each vertical bar is the mean of 45 clones at each generation. Error bars are SEM.

3.   GENETIC AND ANTIGENIC CHANGES OF PRRS VIRUS IN PIGS OVER TIME

Field isolates of PRRS virus show a remarkable degree of genetic and antigenic variability, but the degree and rate of virus mutation in pigs over time has not been characterized. To address this question, a study consisting of a series of pig passages (n = 7) of PRRS virus was conducted. Each passage consisted of 4 pigs, with each animal individually housed in a HEPA-filtered isolation unit. In passage 1, 3 pigs were inoculated with a plaque cloned PRRS virus derived from ATCC VR-2332, the prototypic North American isolate, and one pig served as a mock-infected control. The pigs were kept for 60 days post inoculation. During this time, serum samples were collected periodically from all pigs for virological and serological monitoring. After 60 days, each pig in passage 2 was inoculated with tissue homogenate filtrates from the corresponding pig in passage 1. This process was repeated for each subsequent passage. All inoculated pigs harboured infectious virus at 60 days post inoculation, i.e., transmission to the subsequent passage was successful. Monitoring of viremia and antibody response at each passage did not reveal significant differences in virus replication during passages. Plaque clones (n = 15) from the original inoculum and clones (n = 135) from pigs at passages 1, 2, and 7 were collected and compared with respect to ORF5 nucleotide sequence and their susceptibility to neutralising activity of antiserum collected at the end of the first passage. Genetic changes in ORF5 were detected over time (Fig 2), although the degree of changes were relatively small as compared to that found among field isolates. In addition, preliminary analysis of one-way cross neutralisation data suggested that escape mutants appeared during animal passage.

4. CONCLUSION

A descriptive study of field isolates clearly demonstrated that PRRS viruses vary genetically and antigenically. Subsequent experiments demonstrated that PRRS virus "quasi-species" appear over time as the virus replicates in animals. Although the degree of genotypic changes in ORF5 was less than expected based on field observations; some mutants appear to have been changed sufficiently to escape serum neutralizing antibodies conferred by the initial infection. Collectively, our observations indicate that viral mutation may be a mechanism of PRRS virus persistence. However, the rate, type, and "hot spots" of mutation remain to be addressed. Other questions remain to be addressed, as well. In particular, what is the actual role of viral mutation in PRRS virus persistence? And could the presence of various phenotypic endogenous strains within the same farm or herd account for the apparent ineffectiveness of PRRS control by monovalent vaccine?

REFERENCES

Andreyev, V.G., Wesley, R.D., Mengeling, W.L., Vorwald, A.C., and Lager, K.M., 1997, Genetic variation and phylogenetic relationships of 22 porcine reproductive and respiratory syndrome virus (PRRSV) field strains based on sequence analysis of open reading frame 5. Arch. Virol. 142: 993-1001.

Collins, J.E., Benfield, D.A., Christianson, W.T., Harris, L., Hennings, J.C., Shaw, D.P., Goyal, S.M., McCullough, S., Morrison, R.B., Joo, H.S., Gorcyca, D., and Chladek, D., 1992. Isolation of swine infertility and respiratory syndrome virus (isolate TCC VR-2332) in North America and experimental reproduction of the disease in gnotobiotic pigs. J. Vet. Diagn. Invest. 4: 117-126.

Domingo, E. and Holland, J.J., 1997, RNA virus mutations and fitness for survival. Annu. Rev. Microbiol. 51: 151-178.

Murtaugh, M.P., Faaberg, K.S., Laber, J., Elam, M., and Kapur, V., 1998. Genetic variation in the PRRS virus. Adv. Exp. Med. Biol. 440: 787-794.

Polson, D.D., Marsh, W.E., Dial, G.D., and Christianson, W.T., 1992, Financial impact of porcine epidemic abortion and respiratory syndrome (PEARS). Proceedings of 12th Int. Pig Vet. Soc. Congr. 1:132.

Wesley, R.D., Mengeling, W.L., Lager, K.M., Clouser, D.F., Landgraf, J.G., and Frey, M.L., 1998, Differentiation of a porcine reproductive and respiratory syndrome virus vaccine strain from North American field strains by restriction fragment length polymorphism analysis of ORF 5. J. Vet. Diagn. Invest. 10: 140-144.

Wensvoort, G., de Kluyver, E.P., Luijtze, E.A., den Besten, A., Harris, L., Collins, J.E., Christianson, W.T., and Chladek, D., 1992, Antigenic comparison of Lelystad virus and swine infertility and respiratory syndrome (SIRS) virus. J. Vet. Diagn. Invest. 4: 134-138.

Wensvoort, G., Terpstra, C., Pol, J.M.A., ter Laak, E.A., Bloemraad, M., de Kluyver, E.P., Kragten, C., van Buiten, L., den Besten, A., Wagenaar, F., Broekhuijsen, J.M., Moonen, P.L.J.M., Zetstra, T., de Boer, E.A., Tibben, H.J., de Jong, M.F., van't Veld, P., Groenland, G.J.R., van Gennep, J.A., Voets, M.Th., Verheijden, J.H.M., Braamskamp, J., 1991, Mystery swine disease in The Netherlands: the isolation of Lelystad virus. Vet. Q. 13: 121-130.

Wills, R.W., Zimmerman, J.J., Yoon, K.-J., Swenson, S.L., McGinley, M.J., Hill, H.T., Platt, K.B., Christopher-Hennings, J., and Nelson, E.A., 1997, Porcine reproductive and respiratory syndrome virus: a persistent infection. Vet. Microbiol. 55: 231-240.

Yang, L., Frey, M.L., Yoon, K.-J., Zimmerman, J.J., and Platt, K.B., 2000. Categorization of North American porcine reproductive and respiratory syndrome viruses: Epitopic profiles of the 15, 19, 25 and 45 kD proteins and susceptibility to neutralization. Arch. Virol. (*in press*).

Yang, L., Yoon, K.-J., Li, Y., Lee, J.-H., Zimmerman, J.J., Frey, M.L., Harmon, K.M., and Platt, K.B., 1999, Antigenic and genetic variations of the 15 kD nucleocapsid protein of porcine reproductive and respiratory syndrome virus isolates. Arch. Virol. 144: 525-546.

Yoon, K.-J., Wu, L.-L., Zimmerman, J.J., and Platt, K.B., 1997. Field isolates of porcine reproductive and respiratory syndrome virus (PRRSV) vary in their susceptibility to antibody dependent enhancement (ADE) of infection. Vet. Microbiol. 55: 277-287.

Yoon, K.J., Zimmerman, J.J., Chang, C.C., Cancel-Tirado, S., Harmon, K.M., and McGinley, M.J., 2000, Effect of challenge dose and route on porcine reproductive and respiratory syndrome virus (PRRSV) infection in young swine. Vet. Res. 30: 629-638.

Zimmerman J; Sanderson T; Eernisse KA; Hill HT; Frey ML., 1992, Transmission of SIRS virus in convalescent animals to commingled penmates under experimental conditions. Am. Assoc. Swine Pract. Newsl. 4: 25.

Appearance of Novel PRRSV Isolates by Recombination in the Natural Environment

MICHAEL P. MURTAUGH, SHISHAN YUAN, AND KAY S. FAABERG
Department of Veterinary PathoBiology, University of Minnesota, 1971 Commonwealth Avenue, St. Paul, MN

1. INTRODUCTION

Porcine reproductive and respiratory syndrome virus (PRRSV) emerged in North America and Europe in the 1980's as a new viral disease of swine. PRRSV now is endemic in swine-rearing regions essentially worldwide. A better understanding of the mechanisms of genetic change may help to elucidate its evolutionary history and emergence as a swine pathogen.

1.1 Initial Comments

PRRSV appeared at approximately the same time in Europe and North America as the causative agent of reproductive disease in sows and interstitial pneumonia in young pigs (Collins et al. 1992, Wensvoort et al. 1991). Surprisingly, the genomic sequences of European and North American viruses were substantially divergent (Nelsen et al. 1999), suggesting that PRRSV was present for an unknown time in an undetected form, and also that it might be capable of rapid genetic change. Nucleotide sequence analysis of PRRSV isolates in Europe and North America revealed substantial genetic and antigenic variation within the viral populations in Europe and North America (Drew et al. 1997, Kapur et al. 1996). The genetic differences between European and North American PRRSV is nearly as great as the differences between either genotype and lactate dehydrogenase-elevating virus (Murtaugh et al. 1994). This striking

The Nidoviruses (Coronaviruses and Arteriviruses).
Edited by Ehud Lavi *et al.*, Kluwer Academic/Plenum Publishers, 2001.

diversity suggests that PRRSV has evolved independently on two continents for an extended period or is evolving very quickly.

2.   GENETIC DRIFT AND THE APPEARANCE OF NEW PHENOTYPES

The population structure of PRRSV is changing rapidly. Genotypes of isolates present in the early 1990's are largely extinct or are present as minor groups. Figure 1 shows that the large majority of recent PRRSV isolates have evolved from a limited set of ancestral genotypes consistent with genetic drift. This pattern of genetic drift is observed regardless of the genetic region used for analysis, and in European forms of the virus.

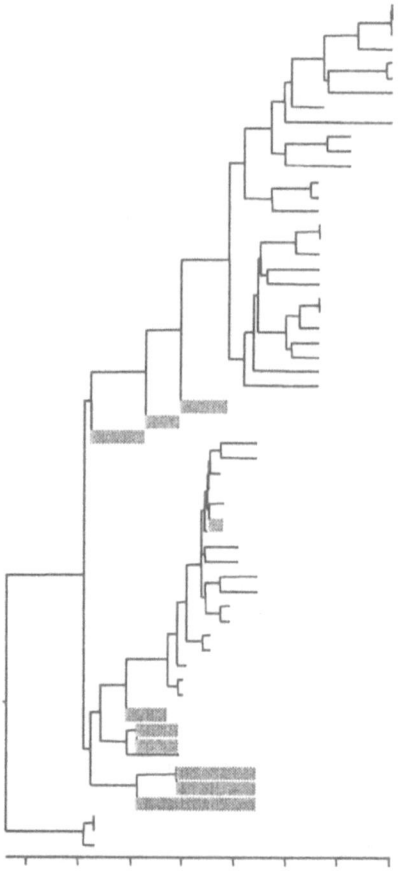

Figure 1. Genetic drift in PRRSV. Clustal analysis of the ORF 5 of nonredundant North American isolates submitted to the Minnesota Veterinary Diagnostic Laboratory in 1997-98. Filled rectangles are 1989-92 isolates described in Kapur et al. 1996.

New phenotypes also are emerging. In the U.S., neurovirulent PRRSV was isolated in 1996 from farms with a history of PRRSV infection and previous or ongoing use of modified-live vaccine (Rossow et al. 1999). In 1999, a European-like PRRSV with approximately 95% similarity to Lelystad virus was discovered in the midwestern U.S. on a farm previously positive for PRRS (K.D. Rossow and K.S. Faaberg, unpublished observations). The emergence of phenotypes with new pathogenic characteristics and the geographic redistribution of existing strains illustrates the existence of broad genetic variation from which novel genotypes might arise.

2.1 Evolutionary Considerations

Three mechanisms of genetic change contribute to the evolution of viruses: small local changes in nucleotide sequence due to natural mutation and polymerase infidelity, introduction of new genetic information by horizontal gene transfer from other organisms, and intra- or intergenomic reshuffling of subgenomic nucleic acids by recombination (Arber, 2000). In PRRSV, recombination has the potential to rapidly create new genotypes and phenotypes in the natural environment since the virus is distributed worldwide, because numerous genetically distinct variants coexist in local environments, and since greatly divergent genotypes characteristic of North American and European forms of the virus are now redistributed.

3. HOMOLOGOUS RECOMBINATION GENERATES NEW PRRSV ISOLATES

To determine if recombination was contributing to genetic diversity in PRRSV, nucleotide sequencing was performed on a panel of 50 U.S. isolates and phylogenetic dendrograms from the ORF 5 and ORF 6 regions of the genomes were compared. A set of three natural isolates from the same geographic region appeared to be directly related according to the analysis of ORF 5, whereas a different relationship was inferred from analysis of ORF 6. Direct analysis of the nucleotide sequences of the three strains showed that the isolate NC-93-14 was a recombination between the NC-93-15 and NC-93-20 within ORF 5 (Figure 2). In this example, the predicted envelope glycoprotein of the recombinant virus contained ectodomains of isolate NC-93-14 and a cytoplasmic tail composed largely of isolate NC-93-20. In ORF 6 the sequences of isolates NC-93-14 and NC-93-20 are identical and different from isolate NC-93-15. All three isolates were obtained within a six week time period in adjacent counties in North Carolina, USA, making recombination between these strains possible.

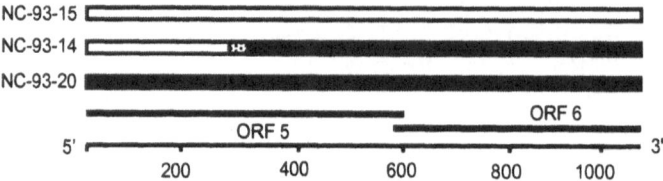

Figure 2. Diagrammatic representation of putative parental and recombinant PRRSV isolates obtained from the field and containing a recombination site in ORF 5. The stippled region shows the site at which recombination occurred.

Several European-type isolates obtained in recent years from Denmark demonstrate marked deletions or frank nonhomologous recombination as compared to earlier Danish isolates (Figure 3).

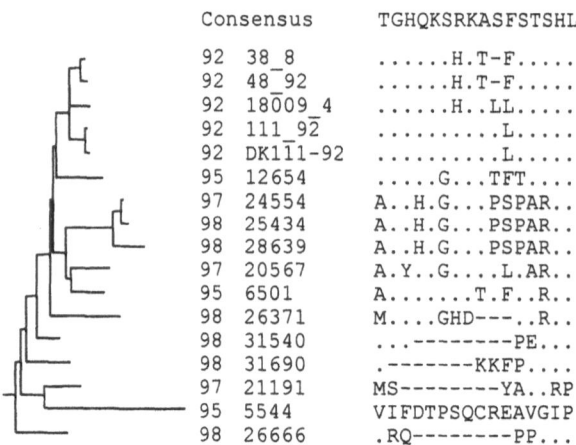

```
              Consensus        TGHQKSRKASFSTSHL

          92  38_8             ......H.T-F.....
          92  48_92            ......H.T-F.....
          92  18009_4          ......H..LL.....
          92  111_92           ..........L.....
          92  DK111-92         ..........L.....
          95  12654            .....G...TFT....
          97  24554            A..H.G...PSPAR..
          98  25434            A..H.G...PSPAR..
          98  28639            A..H.G...PSPAR..
          97  20567            A.Y..G....L.AR..
          95  6501             A.......T.F..R..
          98  26371            M....GHD---..R..
          98  31540            ...-------PE...
          98  31690            .-------KKFP....
          97  21191            MS-------YA..RP
          95  5544             VIFDTPSQCREAVGIP
          98  26666            .RQ-------PP...
```

Figure 3. Amino acid sequence at residues 237-252 in ORF 3 of Danish PRRSV isolates. Phylogenetic relationships as determined by Clustal analysis, year of isolation, and isolate identification in Genbank are indicated. Dots represent the consensus sequence. Dashes represent amino acid sequence that is missing in the indicated isolates.

Five PRRSV samples isolated since 1997 contained deletions of 3-8 amino acids near the carboxyl terminus of the standard 265 residue protein. The deletion pattern is different for every isolate, indicating that each sample arose independently. In addition, isolate 5544, from 1995, had a unique sequence, due to an upstream fusion of ORF 3 and ORF 4, so that the indicated residues actually are derived from ORF 4. These observations extend the findings of Drew et al. (1997) regarding the high degree of variation in this region. Deletions of this magnitude in PRRSV ORFs have not been reported previously, suggesting that nonhomologous recombination within or among viral strains is involved in the generation of genetic diversity.

4. CONCLUSION

Reshuffling of genomic information due to recombination provides a high probability for rapid genotypic and phenotypic change in viruses due to the large scale introduction of new genetic information into a genome. Rapid evolution is a feature of PRRSV in all regions where pigs are intensively reared. Moreover, nucleotide and amino acid sequence differences between European and North American forms of the virus are found throughout the genome. We have demonstrated that recombination, occurring in the natural environment, has contributed to the genetic diversity of PRRSV. Local nucleotide changes due to spontaneous mutation and low fidelity of the arteriviral polymerase also occur at a relatively high rate. The combination of small local changes, genetic recombination, and geographic redistribution of European and North American genotypes are likely to result in continued rapid evolution of PRRSV and the further emergence of new viral phenotypes in the environment.

ACKNOWLEDGMENTS

The work was supported in part by grants and contracts from Boehringer Ingelheim Vetmedica. Beverly Schmitt of the National Veterinary Service Laboratory, Ames, IA generously provided 50 field isolates for these studies.

REFERENCES

Arber, W. 2000. Genetic variation: molecular mechanisms and impact on microbial evolution. *FEMS Microbiol. Rev.* **26**:1-7.

Collins, J.E., Benfield, D.A., Christianson, W.T., Harris, L., Hennings, J.C., Shaw, D.P., Goyal, S.M., McCullough, S., Morrison, R.B., Joo, H.S., Gorcyca, D.E., and Chladek, D.W. 1992. Isolation of swine infertility and respiratory syndrome virus (isolate ATCC VR-2332) in North America and experimental reproduction of the disease in gnotobiotic pigs. *J. Vet. Diagn. Invest.* **4**:117-126.

Drew, T.W., Lowings, J.P., and Yapp. F. 1997. Variation in open reading frames 3, 4, and 7 among porcine reproductive and respiratory syndrome virus isolates in the UK. *Vet. Microbiol.* **55**:209-221.

Kapur, V., Elam, M.R., Pawlovich, T.M., and Murtaugh, M.P. 1996. Genetic variation in porcine reproductive and respiratory syndrome virus isolates in the midwestern United States. *J. Gen. Virol.* **77**:1271-1276.

Murtaugh, M.P. Elam, M.R., and Kakach, L.T. 1995. Comparison of the structural protein coding sequences of the VR-2332 and Lelystad virus strains of the PRRS virus. *Arch. Virol.* **140**:1451-1460.

Nelsen, C.J., Murtaugh, M.P. and Faaberg, K.S. 1999. Porcine reproductive and respiratory syndrome virus comparison: divergent evolution on two continents. *J. Virol.* **73**:270-280.

Pirzadeh, B., Gagnon, C.A., and Dea, S. 1998. Genomic and antigenic variations of porcine reproductive and respiratory syndrome virus major envelope GP5 glycoprotein. *Can. J. Vet. Res.* **62**:170-177.

Rossow, K.D., Shivers, J.L., Yeske, P.E., Polson, D.D., Rowland, R.R.R., Lawson, S.R., Murtaugh, M.P., Nelson, E.A., and Collins, J.E. 1999. Porcine reproductive and respiratory syndrome virus infection in neonatal pigs characterised by marked neurovirulence. *Vet. Rec.* **144**:444-448.

Wensvoort, G., Terpstra, C., Pol, J.M.A., ter Laak, E.A., Bloemraad, M., de Kluyver, E.P., Kragten, C., van Buiten, L., den Besten, A., Wagenaar, F., Broekhuijsen, J.M., Moonen, P.L.J.M., Zetstra, T., de Boer, E.A., Tibben, H.J., de Jong, M.F., van't Veld, P., Groenland, G.J.R., van Gennep, J.A., Voets, M.T., Verheijden, J.H.M., and Braamskamp, J. 1991. Mystery swine disease in the Netherlands: the isolation of Lelystad virus. *Vet. Q.* **13**:121-130.

Predicted RNA Folding Suggests PRRSV Major and Heteroclite Subgenomic Transcripts Result from Polymerase Switching at Unpaired Nucleotides

KAY S. FAABERG, MICHAEL P. MURTAUGH, SHISHAN YUAN

Department of Veterinary PathoBiology, University of Minnesota, Veterinary Science Building, 1971 Commonwealth Avenue, Saint Paul, MN 55108

1. INTRODUCTION

Porcine reproductive and respiratory syndrome virus (PRRSV) has emerged as a major disease of swine worldwide, causing stillbirths and abortions in pregnant sows and reproductive failure in young swine. Even though modified-live vaccines have been used for many years, PRRSV continues to thrive and spread. Previous work has shown that one of the mechanisms which PRRSV uses to evade host defenses is to undergo viral recombination at a high frequency (Yuan *et al* 1999). The high frequency nidovirus recombination mechanism occurs by discontinuous transcription (Baric *et al* 1987) that appears to be guided by basepairing between sense and antisense strands at sites of secondary structure. Support for this model has been recently been obtained for equine arteritis virus subgenomic (sg) mRNAs, which utilize a common leader body junction site sequence which undergoes sequential basepairing at two distinct sites on the genome of the opposite strand (Marle *et al* 1999). In this report, we document that besides standard PRRSV sg mRNAs, other subgenomic RNAs, which utilize alternative junction sites, are transcribed during PRRSV infection . We also provide evidence that these novel subgenomic RNAs are produced by a similar basepairing mechanism, yet the newly identified RNAs appear to be generated in a less-stringent manner. Thus, the PRRSV polymerase complex appears to be more promiscuous than other nidoviruses complexes in deciding where and when to switch templates during transcription. The identification of these novel subgenomic RNAs during normal PRRSV infection will allow further characterization of the specific mechanism for nidovirus discontinuous transcription.

The Nidoviruses (Coronaviruses and Arteriviruses).
Edited by Ehud Lavi *et al.*, Kluwer Academic/Plenum Publishers, 2001.

2.   NOVEL SUBGENOMIC RNA SPECIES

Defective RNA species have been identified for many coronaviruses and for one arterivirus (Makino *et al* 1984; Snijder *et al* 1991; Penzes *et al* 1994; Chang *et al* 1994; Mendez *et al* 1996; Marle *et al* 2000). In an effort to identify potential defective RNA species for PRRSV strain VR-2332, we passaged virus at both high and low multiplicity of infection (MOI) and surveyed each passage by Northern blot analysis with an ORF1a specific probe, assuming that probe would identify defective RNAs. To our surprise, regardless of the passage number or the MOI utilized, Northern analysis of infected cell RNA using the ORF1a oligonucleotide probe showed that many ORF1a-containing subgenomic RNA species were present during viral infection. They were found to consist of only two portions of the genome and contained the full 5'-leader and 3'-untranslated region (UTR). We have termed these species as heteroclite, meaning "deviating from common forms or rules".

3.   TRANSCRIPTION OF SUBGENOMIC RNA

PRRSV strain VR-2332 is 15.4 kb in length, possesses a 190 base 5'-leader sequence ending in the sequence UUAACC and has a 154 base 3'-UTR. Strain VR-2332 produces two transcripts for mRNA7 during infection, termed mRNA7-1 and 7-2 (Nelsen *et al* 1999). The major species, mRNA7-1, utilizes a leader-body junction sequence (AUAACC) located 123 bases upstream from the start site for ORF7 translation, and the minor species, mRNA 7-2, utilizes a junction sequence (UAAACC) located 9 bases upstream of the ORF7 initiation codon (Figure 1).

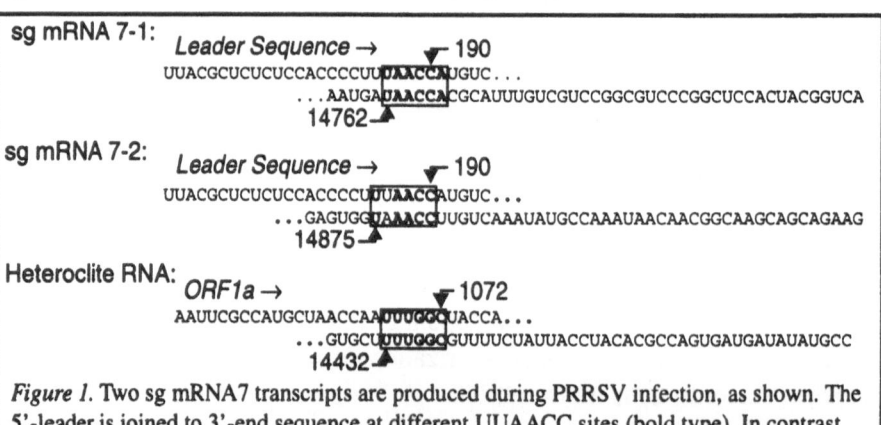

Figure 1. Two sg mRNA7 transcripts are produced during PRRSV infection, as shown. The 5'-leader is joined to 3'-end sequence at different UUAACC sites (bold type). In contrast, heteroclite RNA utilizes UUUGGC to join bases 1067-1072 to bases 14432-14437.

These two transcripts, along with one heteroclite RNA which utilizes UUUGGC as a junction sequence to join bases 1067-1072 to base 14432-14437 during transcription (Figure 1), were utilized to examine potential mechanisms of discontinuous transcription. Toward this aim, we generated a model (Figure 2) to scrutinize predicted positive strand RNA folding structures and their function in putative negative strand RNA discontinuous transcription, as recent work has suggested subgenomic mRNAs are formed only during negative strand synthesis (Marle *et al* 1999).

We examined whether negative strand discontinous transcription might be explained by the model presented in Figure 2. Sgro and Palmenberg (1998) have provided evidence that the genome of RNA viruses are folded three dimensionally such that the 5'-end is juxtaposed near the 3'-end. Thus, the junction sites for all subgenomic RNAs would be in close proximity to one another. In this model, transcribing negative strand subgenomic RNAs would stall at hairpins of stem loop structures, fall off of the 3'-end of the positive strand template and realign with similar sequences located in hairpins of stem loop structures located near the 5'-end of the positive strand template. The formation of both major subgenomic transcripts (mRNA 7-1 and 7-2) and heteroclite subgenomic RNA would be explained by this model.

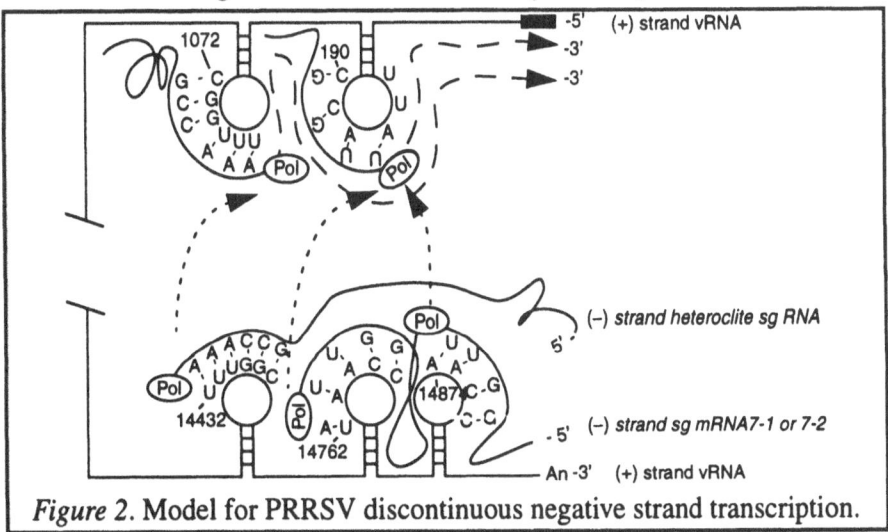

Figure 2. Model for PRRSV discontinuous negative strand transcription.

4.   PREDICTED FOLDING PATTERNS SUGGEST BOTH SUBGENOMIC RNA SPECIES ARE FORMED BY A SIMILAR MECHANISM

Portions of the positive-strand genome, representing 5'- and 3'-terminal nucleotides of PRRSV, were analyzed for predicted RNA folding structure (GCG Mfold; Zuker 1989). From the 5'- terminal predictions, it was apparent that the nucleotide sequence surrounding base 190 (consensus leader-body junction site UUAACC) were predicted to form an extended stem-loop structure. Likewise, nucleotide sequence surrounding the heteroclite junction site (UUUGGC) were predicted to form a similar structure. This suggested that both junction sequences were located at or near loop termini (hairpins) and might be readily available as a landing pad for a partially synthesized negative strand transcript bound to the PRRSV polymerase complex. Similarly, the relevant 3'-end leader-body junction sequences formed analogous hairpins, except for the heteroclite RNA which was predicted to be part of a stem structure (Figure 3).

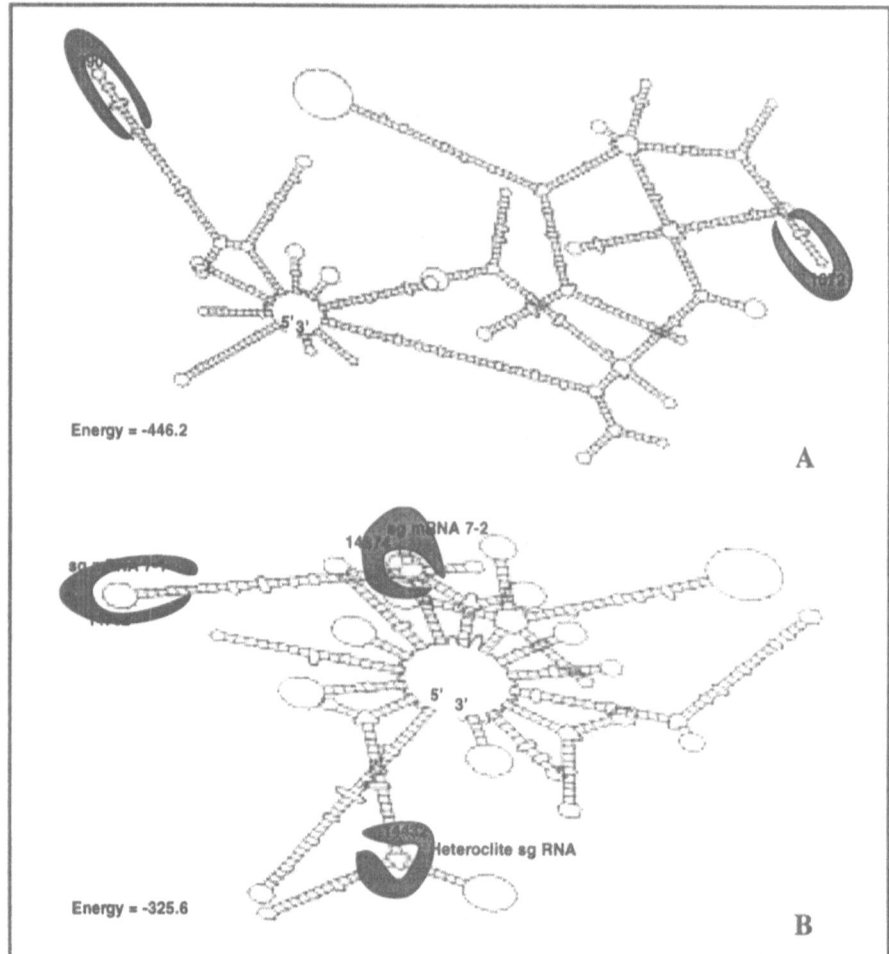

Figure 3. Predicted positive strand viral RNA folding. A. PRRSV nucleotides 1-1400. B. PRRSV nucleotides 14301-15412. Structures were generated using the Mfold program of GCG (default parameters).

According to the model, negative strand sgmRNA7-1 would stall in transcription at nucleotide 650 (UGGUU), where the positive strand nucleotides ACCAA were predicted to be located at loop termini by Mfold analysis. At this juncture, the transcribing negative strand RNA was predicted to form a loose stem structure, held together by just three nucleotide interactions. Thus, two short sequences of unpaired nucleotides would be predicted to interact at the point where the nascent transcript falls off of the template. The transcript would then realign itself with unpaired nucleotides located in another hairpin stem loop on the positive strand template (186-190, ACCAA) (Figure 4).

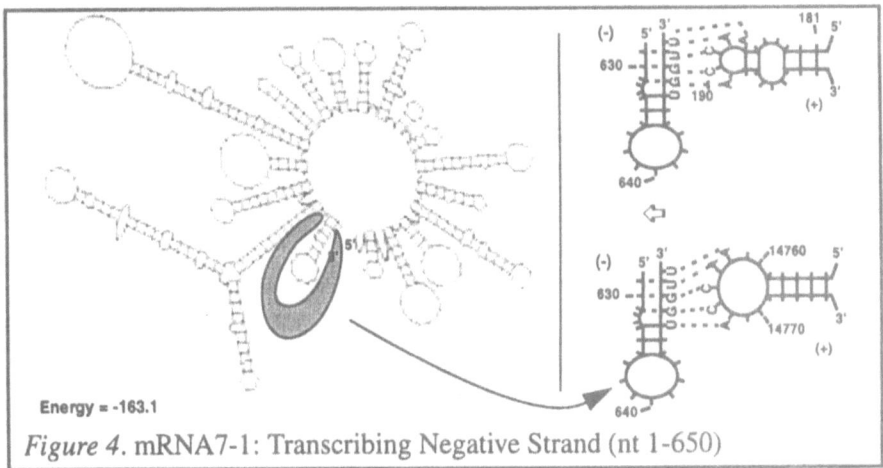

Figure 4. mRNA7-1: Transcribing Negative Strand (nt 1-650)

However, during the transcription of negative strand heteroclite RNA, the polymerase complex may stall at nucleotide 979, where the nucleotides (GCCAAA) are predicted to be unpaired with other nascent negative strand bases. The predicted positive strand RNA structure suggested that nucleotides 14432-14437 (UUUGGC) were not located in hairpin loops, but rather were involved in stem structure interactions with other nucleotides (Figure 5). Thus, heteroclite RNA formation could not be explained by the model presented in Figure 2.

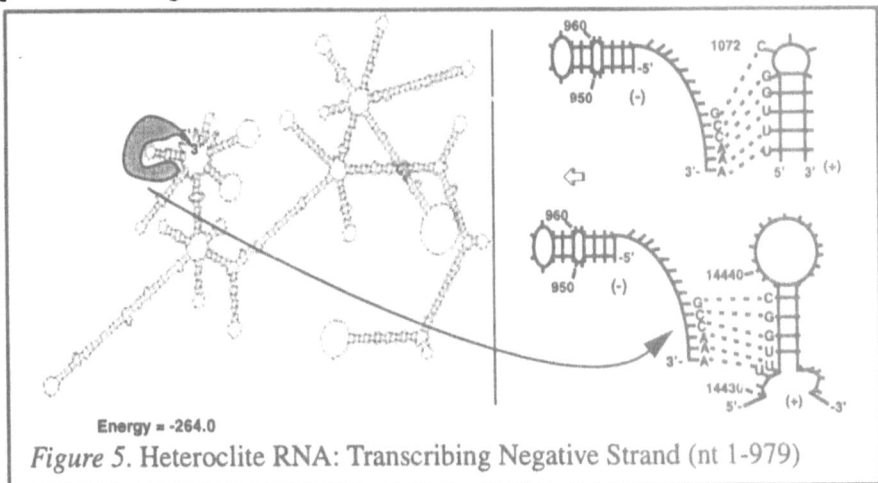

Figure 5. Heteroclite RNA: Transcribing Negative Strand (nt 1-979)

5.   CONCLUSION

Predicted RNA folding patterns suggested that the major subgenomic mRNAs are formed by interaction between unpaired nucleotides on the transcribing negative strand RNA with unpaired nucleotides located at hairpin termini of extended stem-loop structures of positive strand RNA. The interaction between these short sequences might be unstable due to secondary structure. Thus, transcription terminates and then reinitiates at a similar structure of positive strand RNA located at position 186-190. The model of transcription could not explain heteroclite RNA formation, however. This suggests that other factors may be involved in junction site formation, such as RNA-protein interactions and alternative RNA secondary and tertiary structures which are not yet elucidated.

REFERENCES

Baric, R.S., Shieh, C.K., Stohlman, S.A., and Lai, M.M., 1987, Analysis of intracellular small RNAs of mouse hepatitis virus: evidence for discontinuous transcription.*Virology.* **156**, 342-54.

Chang, R. Y., Hofmann, M. A., Sethna, P. B., and Brian, D. A., 1994, A cis-acting function for the coronavirus leader in defective-interfering RNA replication. *J. Virol.* **68**, 8223–8231.

Makino, S., Taguchi, F., and Fujiwara, K., 1984, Defective interfering particles of mouse hepatitis virus.*Virology* **133**, 9-17.

Mendez, A., Smerdou, C., Izeta, A., Gebauer, F. and Enjuanes, L., 1996, Molecular characterization of transmissible gastroenteritis coronavirus defective interfering genomes: packaging and heterogeneity. *Virology* **217**, 495–507.

Nelsen, C.J., Murtaugh, M.P., and Faaberg, K.S., 1999, Porcine reproductive and respiratory syndrome virus comparison: divergent evolution on two continents. *J. Virol.* **73**, 270-280.

Penzes, Z., Tibbles, K., Shaw, K., Britton, P., Brown, T. D. K. and Cavanagh, D., 1994, Characterization of a replicating and packaged defective RNA of avian coronavirus infectious bronchi-tis virus. *Virology* **203**, 286–293.

Sgro, J.-Y. and Palmenberg, A.C., 1998, 2D folding of large RNA genomes: confidence level, information content and evolutionary selection. 16th Annual Meeting of the American Society for Virology, W26-7.

Snijder, E. J., den Boon, J. A., Horzinek, M. C. and Spaan, W. J. M., 1991, Characterization of defective interfering RNAs of Berne virus. *J. Gen. Virol.* **72**, 1635–1643.

van Marle, G., Dobbe, J.C., Gultyaev, A.P., Luytjes, W., Spaan, W.J., and Snijder, E.J., 1999, Arterivirus discontinuous mRNA transcription is guided by base pairing between sense and antisense transcription-regulating sequences *Proc. Natl. Acad. Sci. U S A* **96**, 12056-12061.

Yuan, S., Nelsen, C.J., Murtaugh, M.P., Schmitt, B. J., and Faaberg, K. S., 1999, Recombination between North American strains of porcine reproductive and respiratory syndrome virus. *Virus Res.* **61**, 87-98.

Zuker M, 1989, On finding all suboptimal foldings of an RNA molecule. *Science* **244**, pp. 48-52.

Gill-Associated Virus Of *Penaeus Monodon* Prawns
Molecular evidence for the first invertebrate nidovirus

J.A. COWLEY, C.M. DIMMOCK, K.M. SPANN, AND P.J. WALKER
CRC for Aquaculture, CSIRO Tropical Agriculture, Long Pocket Laboratories, Indooroopilly, QLD 4068, Australia

1.   INTRODUCTION

Gill-associated virus (GAV) is a rod-shaped, enveloped virus that infects *Penaeus monodon* (black tiger prawns) in Australia (Spann *et al* 1997). The morphology and pathology of GAV closely resemble that of yellow head virus (YHV) which has caused significant production losses to prawn aquaculture industry in Thailand (Chantanachookin *et al* 1993). We have recently shown that theses viruses share a high level of sequence identity and can be considered as geographic topotypes (Cowley *et al* 1999).

YHV has been shown to possess a long >22 kb ssRNA genome and four major structural proteins (~170, 135, 67 and 22 kDa) of which the 135 kDa protein is glycosylated (Wongteerasupaya *et al* 1995, Nadala *et al* 1997). There is conflicting evidence for the polarity of the ssRNA genome of YHV (Nadala *et al* 1997, Tang and Lightner *et al* 1999) and speculation that it may be related to rhabdoviruses or coronaviruses. In order to establish an appropriate taxonomic classification of these viruses, we are investigating the genome sequence, organisation and replication strategy of GAV.

Sequence analysis of the GAV genome has identified a ~20 kb gene that encodes two long overlapping open reading frames (ORFs) equivalent to the ORF1a and ORF1b coding regions of corona-, toro- and arteriviruses (Cowley *et al* 2000). Although ORF1b translation is reliant on a −1 ribosomal frameshift site quite distinct from those employed in mammalian nidoviruses,

The Nidoviruses (Coronaviruses and Arteriviruses).
Edited by Ehud Lavi *et al.*, Kluwer Academic/Plenum Publishers, 2001.

43

sequence relationships of ORF1 motifs and the overall genome organisation indicate that GAV, and the closely related YHV, represent the first invertebrate viruses of the Order *Nidovirales*. However, the number, order and sequence of the structural genes, and the nature of putative transcription initiation sites, do not conform with either of the established families, *Coronaviridae* or *Arteriviridae*.

2.   MORPHOLOGY AND MORPHOGENESIS

GAV replication occurs in the cell cytoplasm, primarily in the prawn lymphoid organ, gills, haemocytes and connective tissues (Spann *et al* 1997, unpubl). The tubular helical nucleocapsids (~16 nm dia.) of GAV can vary considerably in length (166-435 nm) and its rod-shaped enveloped virions (~34 nm x 192 nm) appear to mature primarily by budding at the endoplasmic reticulum, often forming arrays in cytoplasmic vesicles (Fig 1). GAV virions also possess regularly-spaced surface projections ~8 nm in length. YHV virions possess similar morphology and purified particles appear to be highly flexible (Nadala *et al* 1997)

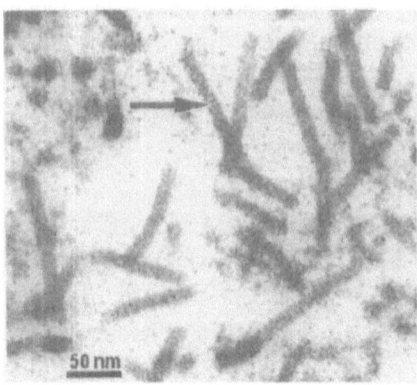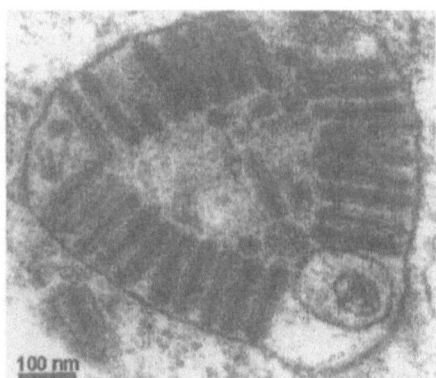

Figure 1. Electron micrograph of tubular helical nucleocapsids (arrow, left) and rod-shaped enveloped virions within a cytoplasmic vesicle in a lymphoid organ cell (right) (Spann *et al* 1997, Reproduced with the permission of Inter-Research Science Publisher).

3.   GENOME ORGANISATION

The majority of the sequence of the GAV genome was obtained from cDNA clones generated from a >22 kbp dsRNA purified from lymphoid organ tissue of infected *P. monodon*. It is likely that this dsRNA represents a replicative intermediate. Gaps in the sequence were filled using clones generated by RT-PCR with specific primers and putative 5'- and 3'-terminal sequences were determined using clones generated by 5'-RACE and RT-PCR with an anchored oligo-dT primer, respectively (Cowley *et al* 2000). A schematic diagram of the 26.2 kb ssRNA GAV genome is shown in Fig 2.

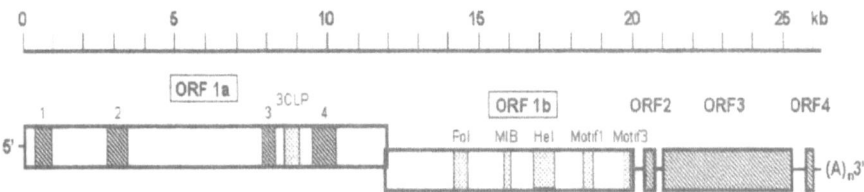

Figure 2. Putative gene organisation in the 26.2 kb (+)ssRNA genome of GAV.

The 5'-terminal 20 kb of the genome comprises a 68 nucleotide (nt) non-coding region proceeded by two long ORFs of 4060 amino acids (aa) and 2646 aa. These contain functional motifs indicating they are equivalent to the ORF1a and ORF1b coding regions of nidoviruses (de Vries *et al* 1997, Lai and Cavanagh 1997, Snijder and Spaan 1995). The C-terminus of GAV ORF1a overlaps the N-terminus of ORF1b by 33 codons (99 nt). ORF1a contains four significant hydrophobic domains comprising multiple putative transmembrane sequences. The region encompassed by the two C-terminal hydrophobic domains encodes a 3C-like cysteine protease (3CLP). Interestingly, database searches indicated a closer relationship to plant potyvirus rather than to coronavirus 3CLPs. Moreover, although the GAV 3CLP contains a cysteine as the catalytic nucleophile as in coronaviruses, alignment of upstream aspartic acid and threonine and downstream histidine and glycine residues believed to be important to formation of the substrate-binding site suggests some structural similarity to the 3C-like serine proteases of arteriviruses (Snijder *et al* 1996). Characterisation of the cleavage site specificity of the GAV 3CLP should help define the evolutionary links among these viruses.

GAV ORF1b contains homologues of the polymerase, metal ion binding (MIB), helicase and Motif 1 and 3 domains conserved among toro- and coronaviruses (Cowley *et al* 2000). The motifs are ordered similarly but their spatial distribution is distinct. Significantly, all absolutely conserved amino acids in the 8 domains of Supergroup 1 (+)ssRNA virus polymerases, including the 'SDD' core motif, are preserved in GAV (Koonin, 1991). The three MIB domains are defined by conservation of potentially active cysteine and histidine residues, and by positioning of aromatic residues in common with numerous RNA/DNA binding proteins (Gorbalenya *et al* 1989). The helicase contains the conserved purine NTP-binding motifs A (<u>GPPGTGKT</u>) and B (<u>DE</u>) (Gorbalenya and Koonin 1989). Phylogenetic relationships determined for the helicase placed GAV on a branch marginally more closely related to coronaviruses than the Berne equine torovirus. Similar analyses with the polymerase, however, placed GAV on a separate branch from corona-, toro- or arteriviruses.

Only three long ORFs, ORF2 (144 aa), ORF3 (1640 aa) and ORF4 (83 aa) reside between GAV ORF1a-1b and a 3'-poly(A) sequence. ORF2 possesses a high proportion of proline residues (13%), is highly hydrophilic and, in size and structure, shows some resemblance to the nucleocapsid (N) protein of toroviruses (Snijder *et al* 1989). ORF3 possesses 15 potential N-linked glycosylation sites and six probable transmembrane domains. A stretch of basic amino acids (<u>KVHARHHK</u>) representing a potential trypsin-like cleavage site resides between the two central transmembrane domains. ORF3 appears functionally equivalent to the S glycoprotein of corona- and toroviruses. Its predicted molecular weight (182 kDa) suggests processing into S1 and S2 components is required to generate a product equivalent to the 135 kDa YHV glycoprotein (Nadala *et al* 1997). The N-terminal 230 aa of the putative S1 component contains three potential transmembrane domains. As there appears to be no discrete membrane (M) gene in GAV, this region may function similarly to the triple membrane-spanning M proteins of toroviruses (Den Boon *et al* 1991) and coronaviruses (Rottier 1995). The 0.6 kb region between ORF3 and the putative 3'-poly(A) tail contains only one ORF of 83 aa for which no potential function has yet been assigned.

Intergenic sequences upstream of ORF2 (93 nt) and ORF3 (57 nt) contain regions in which 29/32 nt are conserved. Sequencing of 5'-RACE clones has identified identical mRNA transcription start sites within these conserved regions.

4.   ORF1A-1B FRAMESHIFT ELEMENT

The GAV ORF1a-1b overlap contains a slippery heptanucleotide AAAUUUU associated with -1 ribosomal frameshifting followed 3 nt downstream by a highly stable 131 nt complex stem-loop structure. Base-pairing of a 7 nt sequence 40 nt downstream of the stem-loop with a predicted single-stranded bulge in the stem would allow formation of a complex RNA pseudoknot (Fig 3).

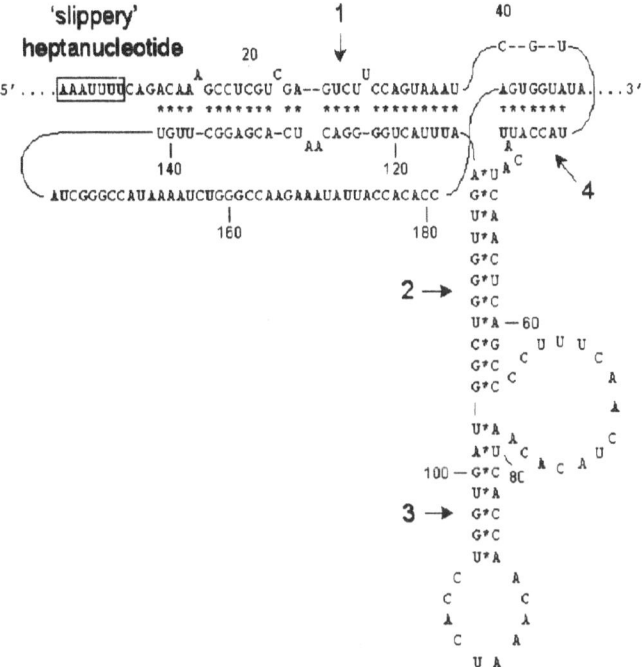

Figure 3. Putative RNA pseudoknot affording ribosomal frameshifting and readthrough of the GAV ORF1a-ORF1b overlap. Predicted helices 1 to 4 are indicated (Cowley *et al* 2000, Reproduced with the permission of the Society for General Microbiology).

Although unlike the UUUAAAC slippage sites and H-type pseudoknots of corona-, toro- and arteriviruses (Brierley, 1995), translational analysis of this GAV element indicates that -1 frameshifting occurs at a similar efficiency (~23%) (Cowley *et al* 2000).

5. CONCLUSION

The replication site and genome organisation of the 26235 nt ssRNA genome of the gill-associated virus (GAV) of penaeid shrimp indicates a distant relationship to coronaviruses, toroviruses and arteriviruses. Thus GAV represents the first nidovirus of invertebrate origin. However, unique features of the virion morphology, the number, order and structural similarity of the gene coding regions and the nature of the ORF1a-1b ribosomal frameshift element and the internal transcription initiation sites do not conform with either of the established families (*Coronaviridae* and *Arteriviridae*) of the *Nidovirales*. We propose GAV as the type species of a new taxon 'Okavirus' based on virus replication in the shrimp lymphoid or 'Oka' organ. Although this may initially be classified as a floating genus, the establishment of a new nidovirus family may also be warranted.

REFERENCES

Brierley, I., 1995, Ribosomal frameshifting on viral RNAs. *J. Gen. Virol.* **76**: 1885-1892.

Chantanachookin, C., Boonyaratpalin, S., Kasornchandra, J., Sataporn, D., Ekpanithanpong, U., Supamataya, K., Sriurairatana, S., and Flegel, T.W., 1993, Histology and ultrastructure reveal a new granulosis-like virus in *Penaeus monodon* affected by yellow-head disease. *Dis. Aquat. Org.* **17**: 145-157.

Cowley, J.A., Dimmock, C.M., Wongteerasupaya, C., Boonsaeng, V., Panyim, S., and Walker, P.J., 1999, Yellow head virus from Thailand and gill-associated virus from Australia are closely related but distinct viruses. *Dis. Aquat. Org.* **36**: 153-157.

Cowley, J.A., Dimmock, C.M., Spann, K.M., and Walker, P.J., 2000, Gill-associated virus of *Penaeus monodon* prawns: an invertebrate virus with ORF1a and ORF1b genes related to arteri- and coronaviruses. *J. Gen. Virol.* **81**: 1473-1484.

De Vries, A.A.F., Horzinek, M.C., Rottier, P.J.M., and De Groot, R.J., 1997, The genome organization of the *Nidovirales*: similarities and differences between arteri-, toro-, and coronaviruses. *Seminars in Virology* **8**: 33-47.

Den Boon, J.A., Snijder, E.J., Krijnse-Locker, J., Horzinek, M.C., and Rottier, P.J.M., 1991, Another triple-spanning envelope protein among intracellularly budding viruses: The torovirus E protein. Virology **182**: 655-663.

Gorbalenya, A.E., and Koonin, E.V., 1989 Viral proteins containing the purine NTP-binding sequence pattern. *Nucl. Acids Res.* **17**: 8413-8440.

Gorbalenya, A.E., Koonin, E.V., Donchenko, A.P., and Blinov, V.M., 1989, Coronavirus genome: predictive functional domains in the non-structural polyprotein by comparative amino acid sequence analysis. *Nucl. Acids Res.* **17**: 4847-4861.

Koonin, E.V., 1991, The phylogeny of RNA-dependent RNA polymerases of positive-strand RNA viruses. *J. Gen. Virol.* **72**: 2197-2206.

Lai, M.M.C., and Cavanagh, D., 1997, The molecular biology of coronaviruses. *Adv. Virus Res.* **48**: 1-100.

Nadala, E.C.B. Jr., Tapay, L.M., and Loh, P.C., 1997, Yellow-head virus: a rhabdovirus-like pathogen of penaeid shrimp. *Dis. Aquat. Org.* **31**: 141-146.

Rottier, P.J.M., 1995, The Coronavirus Membrane Glycoprotein. In *The Coronaviridae* (S.G. Siddell, ed.), Plenum Press, London, pp.115-139.

Snijder, E.J., and Spaan, W.J.M., 1995, The Coronaviruslike Superfamily. In *The Coronaviridae* (S.G. Siddell, ed.), Plenum Press, London, pp.239-255.

Snijder, E.J., Den Boon, J.A., Spaan, W.J.M., Verjans, G.M.G.M., and Horzinek, M., 1989, Identification and primary structure of the gene encoding the Berne virus nucleocapsid protein. *J. Gen. Virol.* **70**: 3363-3370.

Snijder, E.J., Wassenaar, A.L.M., van Dinten, L.C., Spaan, W.J.M., and Gorbalenya, A.E., 1996, The arterivirus Nsp4 protease is the prototype of a novel group of chymotrypsin-like enzymes, the 3C-like serine proteases. *J. Biol. Chem.* **271**: 4864-4871.

Spann, K.M., Cowley, J.A., Walker, P.J., and Lester, R.J.G., 1997, A yellow-head-like virus from *Penaeus monodon* cultured in Australia. *Dis. Aquat. Org.* **31**: 169-179.

Tang, K.F.-J., and Lightner D.V., 1999, A yellow head virus gene probe: nucleotide sequence and application for *in situ* hybridization. *Dis. Aquat. Org.* **35**: 165-173.

Wongteerasupaya, C., Sriurairatana, S., Vickers, J.E., Akrajamorn, A., Boonsaeng, V., Panyim, S., Tassanakajon, A., Withyachumnarnjul, B., and Flegel, T.W., 1995, Yellow-head virus of *Penaeus monodon* is an RNA virus. *Dis. Aquat. Org.* **22**: 45-50.

Elucidation of the Genomic Nucleotide Sequence of Bovine Coronavirus and Analysis of Cryptic Leader mRNA Fusion Sites

VLADIMIR N. CHOULJENKO, TIMOTHY P. FOSTER, XIAOQING LIN, JOHANNES STORZ, AND KONSTANTIN G. KOUSOULAS

Department of Veterinary Microbiology and Parasitology, School of Veterinary Medicine,Louisiana State University, Baton Rouge, LA 70803

1. INTRODUCTION

Over the last few years numerous respiratory bovine coronaviruses (RBCV) were isolated from respiratory infections of cattle including two shipping fever epizootics (Storz *et al.*, 1996; Storz *et al.*, 2000). The RBCV strain 98TXSF-110-LU (LU) and the enteric strain 98TXSF-110-ENT (ENT) were isolated from lung and intestinal samples, respectively, from an animal experiencing fatal respiratory disease during a 1998 shipping fever epizootic. We obtained and compared the entire genomic sequences of the LU and ENT isolates to delineate nucleotide (nt) and amino acid (aa) differences between them. In addition, we analyzed the leader-mRNA junction sequences of the S and 12.7 kDa mRNAs specified by different BCV strains.

2. MATERIALS AND METHODS

The LU and ENT BCV strains were plaque purified, gradient purified, total RNA was extracted from infected HRT-18G cells, and cDNA was produced as described previously (Chouljenko *et al.*, 1998). Primers for sequencing of unknown BCV RNA polymerase were designed based on the

The Nidoviruses (Coronaviruses and Arteriviruses).
Edited by Ehud Lavi *et al.*, Kluwer Academic/Plenum Publishers, 2001.

sequence of MHV-A59 c12 mutant (accession number AF 029248). To characterize the leader-mRNA junction sequences of the S and 12.7 kDa mRNAs, individual junctions were amplified by RT-PCR, cloned into plasmid pCR 2.1 (Invitrogen, Inc, Carlsbad, CA) and sequenced. Primer LBCV annealing to the leader sequence in combination with primers 3B21, 3B11, and 3B26 annealing to coding sequences of S, were used to produce PCR products containing the putative leader-body mRNA junctions (Chouljenko and Kousoulas, 2000).

3.   RESULTS AND DISCUSSION

3.1   Genomic sequences of ENT and LU strains

Sequence comparison between the intestinal and lung isolates derived from the same animal revealed that these viruses contained numerous single nucleotide substitutions. In general, the structure of the BCV RNA polymerase gene was found to be similar to the other coronaviruses. ORF 1a overlapped by 31 nt with ORF 1b located in the −1 reading frame with respect to the upstream ORF 1a. A total of 32 nt changes, 5 of which caused aa changes were detected within the predicted ORF 1a cDNA sequence between the LU and ENT isolates. Fifteen nt changes were detected within the ORF 1b and none caused aa changes. Fifty-three nucleotide changes were detected between the LU and ENT strains, 17 of which caused amino acid changes located within a 9.5 kb sequence spanning all the non-polymerase ORFs. Thus, genomic sequencing revealed that two genetically different BCV strains were isolated from lung and intestinal samples of the same animal.

3.2   Intergenic leader-mRNA junctions of S mRNA specified by different BCV strains

Thirty cloned DNA fragments containing leader-body junction sequences of the S mRNA were obtained after RT-PCR using primer pair LBCV/3B21. The majority of these mRNA junctions utilized the canonical consensus sequence located within the 5′ UTR of S. Specifically, 29 of 30 clones contained the consensus intergenic sequence TCTAAAC as part of the S leader-mRNA junction (type 1, Fig. 1). One clone contained the noncanonical leader-body sequence TA located downstream from the initiation codon of S at nucleotide positions 63-64 (type 2; Fig. 1).

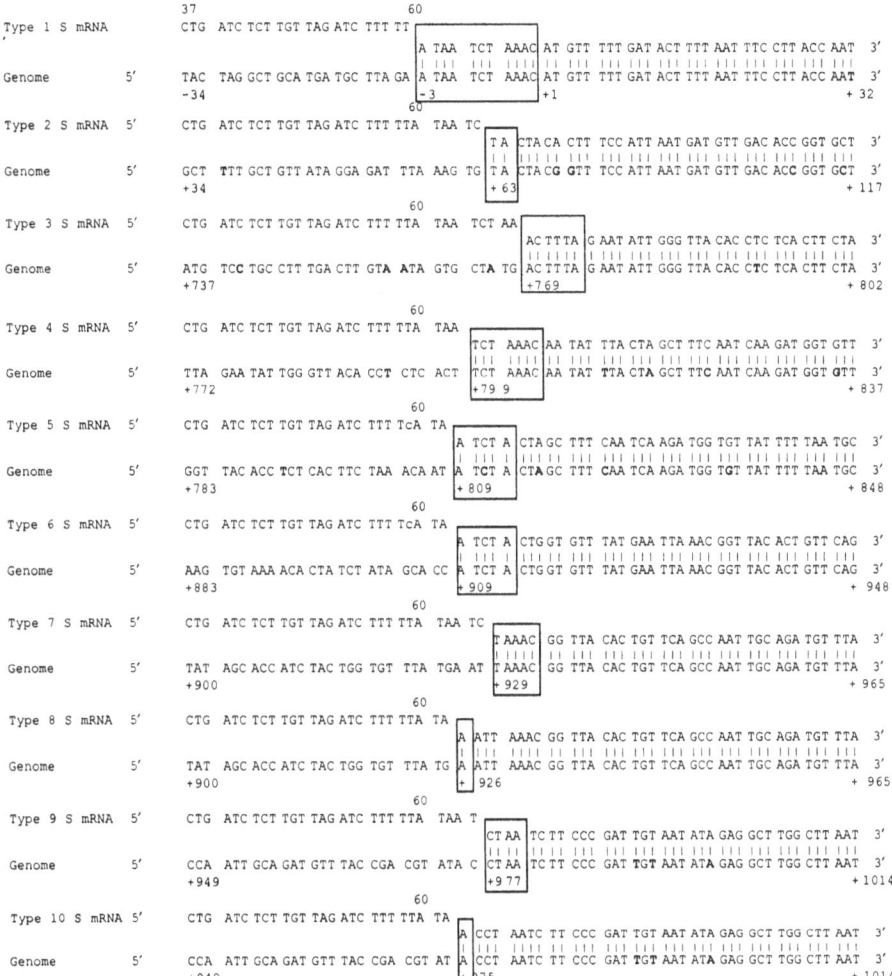

Figure 1. **Cryptic leader-mRNA junction and flanking sequences of different mRNAs detected within the 5′ end of S.** Leader and genome sequences are shown for the EBCV strain L9 except types 5 and 6 detected only for the RBCV OK strain (Chouljenko *et al.*, 1998). Nucleotide sequences of subgenomic RNAs (sgRNAs) are shown in alignment with the genome sequence. Numbers above each sgRNA correspond to the leader sequence. Numbering on the genome sequence begins from the +1 (base A of the start codon of S). The putative leader-body sequences are boxed. Numbers inside the boxes denote the first nt of the putative junction sequences counting from the start codon of S. Nucleotide variations among different BCV strains noted previously (Chouljenko *et al.*, 1998; Zhang *et al.*, 1991) are shown as bold letters. The RBCV-specific change within the leader sequence is shown with a small letter.

3.3   Cryptic leader-mRNA junctions sequences within the 5′ end of the S ORF

The S ORF contains three heptameric nucleotide sequences that resemble the consensus intergenic (IG) sequence of S mRNA located within the 5′ UTR (Zhang *et al.,* 1991). The oligonucleotide primer 3B11 was used to specifically assess whether the intragenic IG homolog sequence located at nucleotide positions 799-805 of the S gene was utilized for the generation of mRNAs. A "shotgun" cloning of the heterogeneous LBCV/3B11 RT PCR products yielded a total of 33 DNA fragments encompassing S sequences ranging in size from 0.3 to 1.1 kb. Five DNA fragments contained the predicted size of RT-PCR products, while the other 28 contained shorter junctions of different sizes. DNA sequence analysis of these junctions revealed 9 different types, all of which contained the leader fused to S coding sequences (Fig. 1). The majority of the cloned DNA fragments represented type 4 junctions (14 clones) containing the targeted consensus sequence. The remaining junctions utilized nonconsensus sequences located both upstream and downstream from this consensus sequence analog. Similarly, different types of junction sequences were detected at the 3′ end of the S mRNA proximal to the other IG-like sequence, as well as within the 12.7 kDa ORF (Chouljenko and Kousoulas, 2000).

Overall, RT-PCR analysis using primer pairs that targeted two of the three intragenic consensus analog sequences located at the 5′ and 3′ ends of S, respectively, confirmed that they were utilized for the production of mRNAs in infected cells. In addition, leader mRNA junctions containing nonconsensus sequences were also detected. Therefore, leader-mRNA junction formation is not strictly dependent on sequence homology between the 3′ end of the leader and the sequences of the mRNA fusion sites or on the location of these fusion sites within the mRNA. However, sequence complementarity seems to increase the efficiency by which homologous sites are used because canonical mRNAs are the predominant species detected by northern analysis (not shown).

Cryptic or alternative junction sequences were reported for a recombinant MHV mutant in which the green fluorescence protein (GFP) gene was inserted in the viral genome (Fischer *et al.,* 1997). However, these junctions may have formed due to the GFP gene insertion within the MHV genome and in this regard, differ from our findings that cryptic junctions can be detected in infections with wild-type BCV strains.

Additional control experiments were performed to assure that the detected cryptic leader-mRNA fusion sequences were not produced by copy-choice template switching of the reverse transcriptase or Taq polymerase during the RT-PCR procedures. In these experiments, we selected the cell-adapted L9 and respiratory OK strains, because they contain a number of nucleotide differences within the genomic region bracketing the leader-mRNA junction sequences of the 12.7 kDa ORF as well as a single nucleotide substitution between OK and L9 leader sequences. RNA was extracted from a control sample, which was prepared by mixing 1:1 virus stocks of strains L9 and OK containing approximate the same number of plaque forming units (PFU). Fourteen different DNA fragments were cloned and sequenced. Seven DNA fragments contained L9-specific leader and part of the 12.7 kDa body-mRNA sequences. Seven additional cloned DNA fragments contained OK-specific leader and body-mRNA, while two of these DNA fragments represented cryptic junction sequences. Overall, none of these cloned DNA fragments contained hybrid L9/OK sequences (not shown) proving that RT-PCR did not generate the observed cryptic junctions.

3.4   Location of cryptic fusion sites on the predicted secondary structure of the S mRNA

We generated the predicted global minimum energy secondary structure of the entire S mRNA using RNAdraw (Matzura and Wennborg, 1996). The predicted secondary structure of the BCV S mRNA revealed an unusually high level of internal homology causing extensive formation of double stranded RNA as well as multiple single stranded loops of RNA (Fig. 2). Interestingly, all cryptic leader-mRNA junctions as well as the junction within the 5′ UTR were predominantly located within single stranded portions of the S mRNA. Therefore, single stranded regions of mRNAs may be required for leader-mRNA junction formation.

3.5   Single nucleotide substitutions proximal to junction sequences within the S and 12.7 kDa ORFs specified by different virus isolates

We found that different BCV strains contained a single nucleotide substitutions at the 5′ end of S, which coincided with the location of 6 cryptic in a total of 9 detected junctions (Fig. 1). In addition, a single nucleotide substitution was detected immediately proximal to the consensus heptameric sequence within the 3′ of the S ORF specified by RBCV strains associated with the 1998 Texas "shipping fever" epizootic, while the intestinal isolates remained unchanged (Chouljenko and Kousoulas, unpublished). Similarly, the intergenic region between the 4.9 and 12.7 kDa ORFs, where the 12.7 kDa ORF cryptic junctions were detected, is the most mutated area when RBCV strains are compared among them as well as with EBCV strains (Chouljenko *et al.*, 1998). It is conceivable that the cryptic junction sequences also represent "hotspots" for recombination that may lead to sequence variation at these sites.

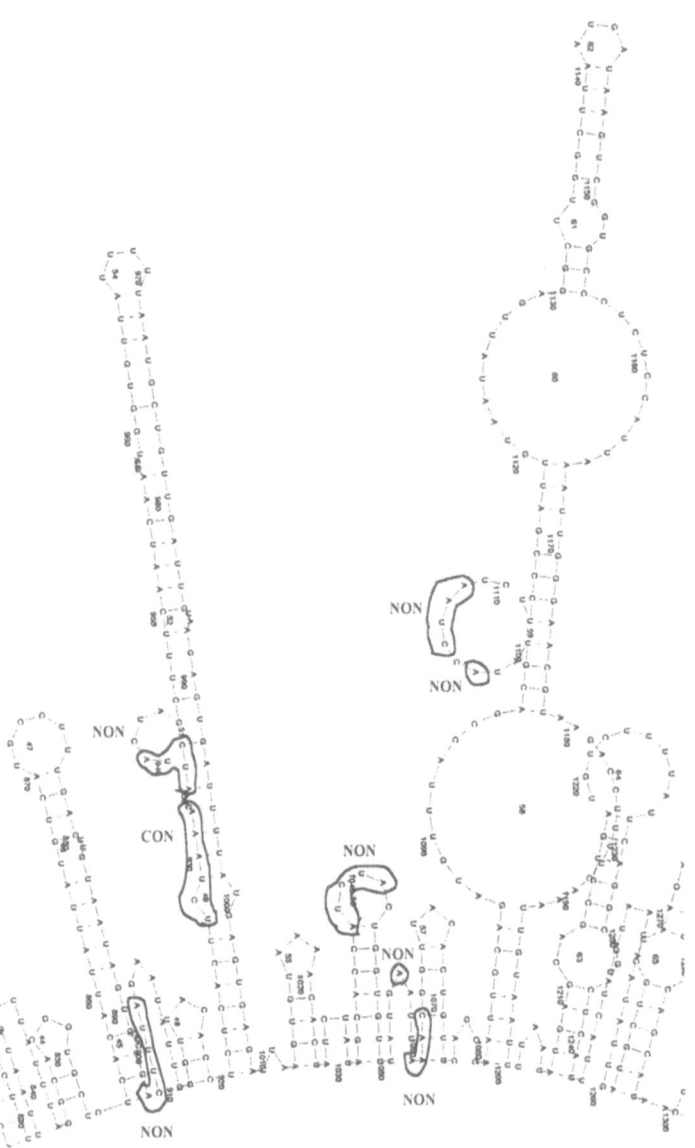

Figure 2. **Predicted secondary structures of the S mRNA.** A portion of the entire predicted secondary structure of the RBCV OK S mRNA is shown. Leader-mRNA body junctions utilizing the consensus and nonconsensus sequences are boxed and marked CON and NON, respectively.

REFERENCES

Chouljenko, V. N., and K. G. Kousoulas. 2000. Cryptic leader-mRNA fusion sites within the bovine coronavirus S and 12.7 kDa coding sequences (Submitted).

Chouljenko, V. N., K. G. Kousoulas, X. Lin, and J. Storz. 1998. Nucleotide and predicted amino acid sequences of all genes encoded by the 3' genomic portion (9.5 kb) of respiratory bovine coronaviruses and comparisons among respiratory and enteric coronaviruses. Virus Genes. **17**:33-42.

Fischer, F., C. F. Stegen, C. A. Koetzner, and P. S. Masters. 1977. Analysis of a recombinant mouse hepatitis virus expressing a foreign gene reveals a novel aspect of coronavirus transcription. J. Virol. **71**:5148-60.

Storz, J., C. W. Purdy, X. Lin, M. Burrell, R. E. Truax, R. E. Briggs, G. H. Frank, R. W. Loan. Isolation of respiratory bovine coronavirus, other cytocidal viruses, and Pasteurella spp from cattle involved in two natural outbreaks of shipping fever. J. Am. Vet. Med. Assoc. **216**:1601-1606.

Storz, J., L. Stine, A. Liem, and G. A. Anderson. 1996. Coronavirus isolation from nasal swap samples in cattle with signs of respiratory tract disease after shipping. J. Am. Vet. Med. Assoc. **208**:1452-5.

Zhang, X. M., K. G. Kousoulas, and J. Storz. 1991. Comparison of the Nucleotide and Deduced Amino Acid Sequences of the S-Genes Specified by Virulent and Avirulent Strains of Bovine Coronaviruses. Virology. **183**:397-404.

Matzura, O., and A. Wennborg. 1996. RNAdraw:an integrated program for RNA secondary structure calculation and analysis under 32-bit Microsoft Windows. Comp. Applic. Bioscie. (CABIOS). **12**:247-249.

Biological and Molecular Characteristics of an HEV Isolate Associated with Recent Acute Outbreaks of Encephalomyelitis in Quebec Pig Farms

A. MARIE-JOSÉE SASSEVILLE, ANNE-MARIE GÉLINAS, NICOLE SAWYER, MARTINE BOUTIN, AND SERGE DEA
Centre de Recherche en Microbiologie et Biotechnologie, INRS-Institut Armand-Frappier, Université du Québec, Laval, Québec, H7V 1B7, Canada

1.   INTRODUCTION

Porcine hemagglutinating encephalomyelitis virus (HEV) is a member of the antigenic subgroup of hemagglutinating coronaviruses, including human respiratory coronavirus HCV-OC43, bovine and turkey coronaviruses, and mouse hepatitis virus (Spaan et al., 1988). These viruses share antigenic determinants located on their homologous structural proteins N, M, HE and S. The HEV specifically infects swine, having a strong tropism for epithelial cells of the upper respiratory tract and for the central nervous system (CNS) (Andries and Pensaert, 1980). The virus causes two clinical syndromes: the vomiting and wasting disease (VWD) and non-suppurative encephalomyelitis (Pensaert and Andries, 1993). The mode of transmission is through nasal secretions and the virus spread via the peripheral nerves to the CNS. The HEV infection is believed to be widespread in many pig producing countries as a subclinical condition.

Since fall 1998, several clinical manifestations suggestive of HEV infections have been reported in Quebec and Ontario. A cytopathogenic coronavirus, IAF-404, was cultivated in HRT-18G cells from the brain of a paretic weaned piglet. This paper describes the clinical and pathological findings associated with these recent outbreaks of encephalitis in newborn

The Nidoviruses (Coronaviruses and Arteriviruses).
Edited by Ehud Lavi *et al.*, Kluwer Academic/Plenum Publishers, 2001.

and weaned piglets, as well as the biological and molecular characteristics of the HEV isolate involved.

2.   METHODOLOGY AND RESULTS

2.1   Clinical and pathological findings

Two clinical manifestations of HEV infection were observed in a farrowing unit of 650 sows and a postweaning unit of 2000 piglets on the same farm in Southern Quebec. Some of the newborn piglets showed the vomiting and wasting form of the disease, but in all cases early weaned pigs (2 to 5%) manifested clinical signs of CNS involvement, mainly a posterior paralysis, with mortality that averaged 100 %. Two paretic weaned piglets were submitted for necropsy. Histologically, perivascular mononuclear cell cuffing, and focal gliosis were observed in the brain and spinal cord neuropil with minimal occasional malacia and spongiosis. Serological titres against HEV for the two necropsied piglets varied between 1/80 and 1/160 by HAI tests. By indirect immunofluorescence (IIF), anti-HEV titers of 1/320 were obtained by testing serums on HEV-infected HRT-18G cells.

2.2   Virological findings

The isolation of the cytopathogenic coronavirus, IAF-404, was done as followed: brain homogenate of a paretic weaned piglet was prepared in RPMI 1640 medium supplemented with antibiotics, clarified by low speed centrifugation, and used to inoculate HRT-18G cells (kindly provided to us by J. Storz, Lousiana State Univ.). The Quebec HEV isolate, as well as the reference HEV-67N (ATCC VR740) were propagated in the presence of bovine pancreatic trypsin (10 U/ml). The IAF-404 isolate induced a distinct cytopathic effect (CPE) characterized by the presence of multinucleated giant cells (syncytia), as observed also with the reference HEV-67N strain. The syncytia were usually apparent after 24 h post-infection (pi) with maximal CPE being reached by 96 h pi. Aliquots of infected supernatants of HRT-18G cells were spotted on formvar-carbon-coated grids and stained negatively with sodium phosphotungstate (pH 7.0). The extracellular viral particles observed displayed a double fringe of surface projections charac-teristic of hemagglutinating coronaviruses. Primary cultures of porcine lung (PPL) cells were also shown to be permissive to both the Quebec IAF-404 and the reference 67N strains of HEV. Viral antigens in the cytoplasm of infected cells could be demonstrated by IIF using a pool of MAbs raised against the prototype HEV-67N strain. Most of these MAbs appeared to be directed against the N protein of HEV-67N and cross-reacted with BCV, TCV and HCV-OC43. The MAb 2D5-1, which was found to be specific to HEV, also gave a positive fluorescence with HRT-18G cells infected with the Quebec IAF-isolate, thus confirming the serological identification of the virus.

The IAF-404 strain displayed HA and AE activities, and a RDE activity that decreased with the number of passages in cell cultures. With the HEV-67N strain, no significant difference was noted in the HA titers estimated at both incubation temperatures.

Table 1. Hemagglutination titers of HEV-infected HRT-18G supernatant fluids

Viral isolate	# passages on HRT18G cells	HA titers at 4°C	HA titers at 37°C	RDE activity	Infectivity titers ($\log_{10}TCID_{50}$/.1ml)
HEV 67N	2	< 20	< 20	1	NA
	3	40	80	1	NA
	4	320	640	1	7.125
IAF-404	2	80	< 20	≥ 4	NA
	3	320	< 20	≥ 16	5.750
	4	1280	1280	1	9.250
	5	2560	5120	1	8.500

The HA titers were determined as the reciprocal of highest dilution of clarified supernatant fluids showing complete agglutination of rat erythrocytes (0.4% suspension). The receptor-destroying enzyme (RDE) was determined by placing plates at 37^{0}C for 2h following their incubation at 4^{0}C, which resulted in the activation of the enzyme leading to elution of the virus from the surface of erythrocytes. The RDE activity was evaluated as the ratio of the HA titer at 4^{0}C to that obtained at 37^{0}C.

Table 2. Acetyl Esterase Activity (AE)

Viral isolate	# of passages on HRT18G cells	Infectivity titers ($\log_{10}TCID_{50}$/.1ml)	AE (15 :1 purified virus)
HEV 67N	4	9.25	2.46
IAF-404	3	9.00	2.09

The AE activity was determined after a 5 min incubation at room temperature, in the presence of 1 mM p-nitrophenyl acetate (PNPA) The optical density (OD) was read at 405 nm

2.3   Genomic comparison

Genomic RNA was extracted from concentrated virus by tripure (Boehringer Manheim). The approach of reverse transcription (RT) followed by PCR was used to amplify the HE, ORF4/5 and MN regions of the genome. Primers were choosen according to the cDNA sequence of the Mebus strain of BCV(Abraham et al., 1990; Lapps et al., 1987). cDNA fragments were subcloned in T vector pCR2I (Invitrogen Co) and sequencing was performed in an Automated Laser Fluorescent DNA sequencer

(Pharmacia LKB). The sequences were computer analysed with the Mac Vector 6.5.3 program (Oxford Molecular Group). Only minor variations were identified between both HEV isolate. Indeed, preliminary studies of the aa variations between both HEV strains revealed only 1 aa change in the 12.7 kDa protein (ORF5), 3 aa changes in the M protein and 3 aa changes in the N protein, 100% homology being demonstrated for the 4.9 kDa NS protein encoded by ORF4. Two different clones of HEV 67N or IAF-404 were sequenced, except for the MN fragment of HEV-67N. The sequences used for comparison were: BCV-Mebus (M30612, M16620); HEV-67N (X89861); HCV-OC43 (M76373, M99576, M93390) and Kamahora et al.(1989). For the N protein, insertion of an E residue (glutamate) was demonstrated at aa position 384 of both HEV strains. The later may be part of the specific HEV epitope recognized by MAb 2D5-1. As there are still ambiguities in the HE sequence of HEV, further analysis are needed to confirm the data obtained (Table 3). The sequence obtained for the 4.9 kDa protein encoded by the ORF4 confirmed that IAF-404 derived from HEV 67N (Fig. 1).

Table 3. Positions of aa changes between HEV 67N et IAF-404, BCV-Mebus and HCVOC43

Protein	IAF-404 ≠ HEV 67N	HEV≠BCV HEV=HCV BCV≠HCV	HEV=BCV HEV≠HCV BCV≠HCV	HEV≠BCV HEV≠HCV BCV≠HCV	HEV≠BCV HEV≠HCV BCV=HCV	HEV sequence uncleared
9.8 kDa		8, 53	29			
12.7 kDa	68	8, 32, 57,	14, 52, 74, 85		5, 25, 29, 43 44, 72, 80 82, 83, 102, 104	
M	54, 61, 100,	156, 180, 183, 192	85, 114, 132, 167, 168, 174, 194	4, 12,	8, 11, 33, 57, 98, 147, 216	67N : 21, 49, IAF : 129
N	94, 136, 410	15, 53, 162, 262, 341,	22, 33, 81, 107, 146, 204, 257, 313, 400, 414		43, 45, 58, 155, 205, 206, 209, 224, 226, 375, **384**, 390, 405,	
HE			8, 20, 30, 47, 50, 53, 63, 114, 115, 177, 178, 181, 247, 250, 344, 379, 382			67N : 2, 5, 49, 66, 67, 103, 249, 297, 367, 376, 392, 400 IAF: 2,51, 124, 365

3. CONCLUSION

These recent outbreaks of encephalomyelitis in Quebec pig farms could be attributed to an acute HEV infection, the strain involved being genomically and antigenically closely related to the reference HEV-67N strain. Serological studies, as well as sequencing analyses, showed that this porcine coronavirus remained antigenically and genetically stable since its first isolation in Canada in 1962. The propagation of this wild isolate in PPL cells may allow us to study the pathogenesis of this strain in swine with a better efficiency than with the HEV67N strain adapted (attenuated) on human cell cultures.

```
Mebus(ATCC)  MTTKFVFDLL APDDILHPFN HVKLIIRPIE VEHIIIATTM PAV  43
HEV-NT9      ......I.........V...... TL...H...VTLSI CQSF               24
HEV-VW572    ...................... TL.......VTLSI CQSF                24
HEV-IAF404   ......IN.........--- -GFH...VTLSI CQSF                    20
HEV-67N      ......IN.........--- -GFH...VTLSI CQSF                    20
HCV-OC43     ......................S... H                              11
```

Figure 1. Comparison of a.a. sequences of the ORF4 region encoding the 4.9 kDa protein. (Corresponding GenBank sequence accession numbers : Mebus (M30612), HEV-NT9 (X89863), HEV-VW572 (X89862), HEV-67N (X89861) et HCV-OC43 (M99576)

ACKNOWLEDGMENTS

This work was supported by the National Science and Engineering Research Council of Canada and the Ministère de l'Agriculture, des Pêches et de Alimentation du Québec. A. M.-J. S. is a recipient of a J.-L. Lévesque fellowship from the Fondation Armand-Frappier.

REFERENCES

Abraham, S., Kienzle, T. E., Lapps, W. and Brian, D. A. (1990). Deduced sequence of the bovine coronavirus spike protein and identification of the internal proteolytic cleavage site *Virology* **176**, 296-301.

Andries, K., Pensaert, M.B. (1980) Virus isolated and immunofluorescence in different organs of pigs infected with hemagglutinating encephalomyelitis virus. *Am J Vet Res* **41**, 215-218.

Kamahora, T., Soe, L. H. and Lai, M. M. (1989). Sequence analysis of nucleocapsid gene and leader RNA of human coronavirus OC43. *Virus Res* **12**, 1-9.

Lapps, W., Hogue, B. G. and Brian, D. A. (1987) Sequence analysis of the bovine coronavirus nucleocapsid and matrix protein genes. *Virology* **157**, 47-57.

Mounir, S. and Talbot, P. J. (1993). Human coronavirus OC43 RNA 4 lacks two open reading frames located downstream of the S gene of bovine coronavirus. *Virology* **192**, 355-360.

Pensaert, M. and Andries, K. (1993). *In* "Diseases of Swine, Viral Diseases" (Iowa state Univ. Press, 7th Edition, pp. 268-273, Ames, Iowa.

Spaan, W., Cavanagh, D. and Horzinek, M. C. (1988). Coronaviruses: structure and genome expression. *J Gen Virol* **69**, 2939-2952.

Identification of Specific Variations within the HE, S1, and ORF4 Genes of Bovine Coronaviruses Associated with Enteric and Respiratory Diseases in Dairy Cattle

ANNE-MARIE GÉLINAS, AM-J SASSEVILLE, AND SERGE DEA
Centre de Recherche en Microbiologie et Biotechnologie, INRS-Institut Armand-Frappier, Université du Québec, Laval, Québec, H7V 1B7.

1. INTRODUCTION

The bovine coronavirus (BCV) is the causative agent of enteric diseases including neonatal calf diarrhea (NCD) (Dea et al., 1980; Mebus et al., 1973), winter dysentery (WD) (Benfield and Saif, 1990; Dea et al., 1995), and chronic shedding in adult cattle (AD) (Tsunemitsu et Saif, 1995). BCV can also infect the upper respiratory tract of growing calves causing pneumonia (Chouljenko et al., 1998; Reynolds et al., 1985). The respiratory BCV (RBCV) is now recognized as an important agent associated to shipping fever (Storz et al., 1996). The virion possesses a single stranded, non segmented RNA genome of positive polarity, and is made of 5 structural proteins: the nucleocapsid phosphoprotein (N: 52 kDa), the matrix glycoprotein (M: 25 kDa), the peplomer glycoprotein (S: 200 kDa), the small membrane protein (E: 9.5 kDa) and the hemagglutinin esterase glycoprotein (HE: 140 kDa) (Spaan et al., 1988). Both S and HE are able to agglutinate red blood cells and elicit production of neutralizing antibodies (Dea et Tijssen, 1989; Deregt and Babiuk, 1987; Vautherot et al., 1992). The S protein has an important role to play in tropism being responsible for binding to the host cell (Spaan et al., 1988). The regions situated between the S and M genes (ORF4 and ORF5) may also be implicated in the tropism of coronaviruses (Mounir et al., 1992).

The Nidoviruses (Coronaviruses and Arteriviruses).
Edited by Ehud Lavi *et al.*, Kluwer Academic/Plenum Publishers, 2001.

Considering that genomic variations may lead to antigenic diversity, changes in tropism and virulence of field BCV strains, the S (S1 subunit) and HE genes, and the ORF4 and ORF5 regions of enteric BCV and RBCV strains, have been cloned and sequenced. Data were compared to those of the reference Mebus strain associated with NCD.

2.   METHODS

The Mebus strain of BCV (VR 874), as well as the porcine hemagglutinating encephalomyelitis virus HEV67N strain (VR 740) were obtained from the ATCC. The reference RBCV strain OK-0514 was provided by J. Storz, Louisiana State University. The OC43 strain of human coronavirus was obtained from P.J. Talbot, INRS-Institut Armand-Frappier, Laval, Canada. Field strains of BCV associated either to NCD (BCQ.571, BCQ.1523) or WD (BCQ.2590 and BCQ.7373) were obtained from Quebec dairy herds (Dea et al., 1995). RBCV BCQ.3994 strain was isolated from the lung of a dysphneic calf from a dairy farm in Southern Quebec, whereas BCO.43277 and BCO.44175 isolates, associated with pneumonia in Ontario dairy farms, were obtained from S. Carman, OVC, University of Guelph, Ontario. BCV isolates were grown in the presence of bovine pancreatic trypsin in HRT18G cells, a cell line derived from human rectal adenocarcinoma.

For serological studies, the MAbs 1D6-2, 1D6-3 and 9F2-1R directed to the HE of Quebec WD reference strain BCQ.2590 (Milane *et al.,* 1997), as well as rabbit hyperimmune sera to the Mebus (5801) and the BCQ.2590 strains, were tested for their capacity to inhibit the HA activity of the various BCV isolates. The HA, acetyl esterase (AE) and RDE activities of the BCV isolates tested were determined as previously described (Dea et al., 1995).

For genomic studies, genomic RNA was extracted from concentrated virus by the method described by Chomczynski and Sacchi (1987). The approach of reverse transcription (RT) followed by PCR was used to amplify the S1 subunit and HE gene, as well as the ORF4 et ORF5 regions, of the coronaviruses (Dea et al., 1995). Primers were choosen according to the cDNA sequence of the Mebus strain of BCV (Parker et al., 1989). cDNA fragments were cloned in T vector pCR2I (Invitrogen Co) and sequenced using an Automated Laser Fluorescent DNA sequencer (Pharmacia LKB).

3.   RESULTS AND DISCUSSION

The capacity of BCV isolates to bind to 9-O-acetylated sialic residues present at the surface of erythrocytes at different incubation temperatures is a biological property which permits to distinguish among BCV field strains

(Dea et al, 1995). In our study, HA titers of BCV strains associated to WD, which varied between 1280 and 2560, decreased significantly when the incubation T^o switched from 4^0C to 37^0C, due to the activation of the higher RDE activity of these strains. For RBCV and NCD strains (HA titers of 640 to 5120), the capacity to destroy the host cell receptors was not influenced by the variations of incubation T, except for the BCQ.1523 strain which appeared to possess minimal RDE activity. The AE activities of the WD and RBCV strains were comparable, and appeared superior to that of NCD strains. Anti-HE MAbs 1D6-2, 1D6-3 and 9F2-1R have shown little activity against BCV Mebus and BCQ.571 strains, IHA titers being lower than 1/40. However, IHA titers of at least 2560 were obtained with MAbs 1D6-2 and 1D6-3 for all the other strains tested. MAbs 9F2-1R reacted only weakly against Ontario RBCV strains BCO.43277 and BCO.44175.

Table 1: Primers used for the amplication of the S (S1 subunit) and HE genes and ORF4 and ORF5 regions of BCV strains

Primer	Sequence	Amplified region
S1.Ad5 sens	5' **GGATCC GGATCC** <u>GCC GCC GCC</u> **ATG** TTT TTG ATA CTT TTA ATT TCC 3'	S1 subunit of S
S1. Ad5 AS	5' **GGATCC GGATCC** TCA TCT ACG ACT TCG TCT TTT TG 3'	S1 subunit of S
HE.Ad5 sens	5' **GGATCC GGATCC** <u>GCC GCC GCC</u> ATG TTT TTG CTT CTT AGA TTT GTT C 3'	HE glycoprotein
HE.Ad5 AS	5' **GGATCC GGATCC** TCA CTA AGC ATC ATG CAG CCT AGT ACC 3'	HE glycoprotein
BCV.5327 sens	5' **GGATCC GGATCC** ATG TGG TGG TTG TTG TGA TGA 3'	ORF4 and ORF5
BCV.ORF4 AS	5' **GAATTC GAATTC** AAC GTC ATC CAC ATC AAG AAC 3'	ORF4 and ORF5

For the HE gene, the majority of the nucleotidic variations were silent mutations. The substitution of a leucine (L) by a proline(P) residue at aa position 5, and the substitution of a serine (S) by a proline (P) at aa position 367 for all of the BCV and RBCV strains studied, when compared to the reference Mebus strain, had already been reported (Dea et al., 1995; Zhang et al., 1991). The substitution of an aspartic acid (D) by a glycine (G) at aa position 66 seems to be specific to the RBCV strains.

For the S1 subunit of the S gene, a total of 7 aa changes were observed among the enteropathogenic strains. Only one of these variations (aa 509) was located within the hypervariable region (aa 452 to 593). The inclusion of an Asp residue (aa 196), observed only for both WD strains, did not result in the appearance of an additional glycosylation site. A Met residue at aa position 256 appeared to be specific to enteropathogenic strains. Five substitutions were identified in the S1 subunit of S gene of BCV strains having a respiratory tropism. Two of these aa substitutions (aa 510 and 578) had already been reported (Chouljenko et al., 1998), and the 3 other (aa 256, 362 and 691) were observed only in case of Ontario RBCV strains. Chouljenko et al (7) had identified 5 other variations specific to RBCV strains situated in the N-terminal of the S1 gene (aa 11, 115, 118, 173 and 179). The latter were also present in the S1 sequences of some enteropathogenic strains as well as the substitutions reported at aa positions 531 and 543.

For all BCV isolates studied, the region located between S and M genes (ORF4) apparently encodes for two non-structural proteins of 4,9 and 4,8 kDa and comprises a total of 302 nt. In comparison to BCV isolates, the HEV and HCV-OC43 showed a major deletion in the region corresponding to ORF4. However, a specific anti-sense substitution was identified at aa position 88 of the putative 4,9 kDa protein of RBCV isolates resulting in 29 rather than 43 aa residues. The ORF5, encoding for a 12,7 NS protein and the 9,5 kDa E protein, was highly conserved among BCV isolates studied.

ACKNOWLEDGMENTS

This work was supported by the National Science and Engineering Research Council of Canada

REFERENCES

Benfield DA, Saif LJ (1990) Cell culture propagation of a coronavirus isolated from cows with winter dysentery. *J Clin Microbiol* 28: 1454-1457

Chomczynski P, Sacchi N (1987) Single-step method of RNA isolation by acid guanidium thiocyanate-phenol-chloroform extraction. *Ann Biochem* 162: 156-159

Chouljenko VN, Kousoulas KG, Lin X, Storz, J (1998) Nucleotide and predicted amino acid sequences of all genes encoded by the 3' genomic portion (9.5 kb) of respiratory bovine coronaviruses and comparisons among respiratory and enteric coronaviruses. *Virus genes* 17 :33-42

Dea S, Roy RS, Begin, ME (1980) Bovine coronavirus isolation in cell cultures. *Am J Vet* Res **41,** 30-38.

Dea S, Michaud L, Milane G (1995) Comparison of bovine coronavirus isolates associated with neonatal calf diarrhoea and winter dysentery in adult dairy cattle in Québec. *J Gen Virol* **76,** 1263-1270

Dea S, Tijssen P (1989) Antigenic and polypeptide structure of turkey enteric coronaviruses as defined by monoclonal antibodies. *J Gen Virol* **70**:,1725-1741

Deregt D, Babiuk LA (1987) Monoclonal antibodies to bovine coronavirus: characteristics and topographical mapping of neutralizing epitopes on the E2 and E3 glycoproteins. *Virology* **161,** 410-420

Mebus CA, Stair EL, Rhodes MB, Twiehaus MJ (1973) Neonatal calf diarrhea: propagation, attenuation. and characteristics of a coronavirus-like agent. *Am J Vet Res* **34,** 145-150

Milane G, Kourtesis AB, Dea S (1997) Characterization of monoclonal antibodies to the hemagglutinin-esterase glycoprotein of a bovine coronavirus associated with winter dysentery and cross-reactivity to field isolates. *J Clin Microbiol* **35**: 33-40.

Mounir S., and Talbot, P.J. (1993) Human coronavirus OC43 RNA lacks two open reading framed located downstream of the S gene of bovine coronavirus. *Virology* **192**, 355-360.

Parker MD, Yoo D, Babiuk LA (1990) Expression and secretion of the bovine coronavirus hemagglutinin-esterase glycoprotein by insect cells infected with recombinant baculoviruses. *J Virol* **64,** 1625-1629,

Reynolds DJ, Debney T J, Hall GA, Thomas LH, Parsons K R (1985) Studies on the relationship between coronaviruses from the intestinal and respiratory tracts of calves. *Arch Virol* **85,** 71-83,

Spaan WD, Cavanagh D, Horzinek MC (1988) Coronaviruses: structure and genome expression. *J Gen Virol* **69**, 2939-2952

Storz J, Stine L, Liem A, et al. (1996) Coronavirus isolation from nasal swabs samples in cattle with signs of respiratory tract disease after shipping. JAVMA **208,** 1452-1454

Tsunemitsu H, Saif LJ (1995) Antigenic and biological comparisons of bovine coronaviruses derived from neonatal calf diarrhea and winter dysentery of adult cattle. *Arch Virol* **140,** 1303-1311

Vautherot JF, Madelaine MF, Boireau P, Laporte J (1992). Bovine coronavirus peplomer glycoproteins: detailed antigenic analysis of S1, S2 and HE. *J Gen Virol* **73**, 1725-1737.

Genetic Variation of ORFs 3 and 4 of Equine Arteritis Virus

JODI F. HEDGES[1], UDENI B.R. BALASURIYA[1], JB TOPOL[1], DUSTIN W. LEE[2], AND N. JAMES MACLACHLAN[1]

[1]*The Bernard and Gloria Salick Equine Viral Diseases Laboratory, Department of Pathology, Microbiology and Immunology, School of Veterinary Medicine, and* [2]*Department of Computer Science, University of California, Davis, CA 95616, USA*

1. INTRODUCTION

The open reading frames (ORFs) 3 and 4 of equine arteritis virus (EAV) encode the GP3 and GP4 proteins, respectively. The GP3 protein of EAV is an extensively glycosylated membrane protein that is likely anchored by the uncleaved signal sequence (Hedges et al., 1999a). ORF 4 is predicted to encode a membrane glycoprotein. The goal of this study was to determine the variation in ORFs 3 and 4 and their encoded GP3 and GP4 proteins amongst a large number of EAV strains, including those amplified directly from the semen of carrier stallions.

2. GENETIC AND PROTEOMIC VARIATION

ORFs 3 and 4 of 70 field isolates and laboratory strains of EAV were compared. The description and passage history of 18 field isolates and one laboratory strain (EAVATCC) of EAV used in these studies, as well as methods have been previously described (Hedges et al., 1996). Also included in the analyses were 29 strains of EAV amplified directly from the semen of 10 different carrier stallions and 14 viruses present in the tissues of foals affected in recent outbreaks of EVA (Hedges et al., 1999b; Patton et

The Nidoviruses (Coronaviruses and Arteriviruses).
Edited by Ehud Lavi *et al.*, Kluwer Academic/Plenum Publishers, 2001.

al., 1998; Balasuriya et al., 1999). The published sequences of ORFs 3 and 4 from eight strains of EAV were also included (Archambault et al., 1997). The number of variable sites and type of substitution in the ORFs 3 and 4 of the 70 strains of EAV were determined by comparison to the prototype EAV ATCC sequence using a program developed in our laboratory (Table 1). The variability of ORFs encoding the four known structural proteins of most of the same strains of EAV was also determined. The ORF 5 was more variable than the other ORFs and had a high proportion of non-synonymous changes. In every ORF, specific synonymous changes occurred more frequently in individual rather than multiple strains, indicating that these changes occur randomly. In contrast, in ORFs 3, 4, 5 and 6 most (77.9%, 63%, 52.2% and 55% respectively) non-synonymous nucleotide changes occurred in more than one virus strain suggesting that many coding nucleotide changes that occur during the evolution of EAV are not random.

Table 1. Percentages of variable sites, non-synonymous and synonymous changes in EAV ORFs as compared to the prototype sequence (EAVATCC)

ORF	Variable Nucleotide Sites (%)	Non-synonymous Changes (%)	Synonymous Changes (%)
2b	26.6	34.6	65.4
3	40	60.3	39.7
4	37.4	37.9	62.1
5	67.8	59.7	40.3
6	24.9	30.5	69.9
7	14.7	31.5	68.5

Hypervariable regions of the GP3 protein were amino acids 1-41 and 88-131. Only two of the six prototypical asparagine (N)-linked glycosylation sites are conserved amongst all 70 strains of EAV; at position 96, where position 98 wobbles between serine (S) and threonine (T), thus preserving the glycosylation site requirements, and at position 106. There is a new glycosylation site at position 39 of the GP3 protein in many strains, and at position 115 where the N substitution occurs only where there is a T or an S in position 117 in most cases, suggesting that the N substitution is advantageous when it is also a glycosylation site. We have determined that the lack of glycosylation of N^{28} and N^{29} had no obvious effect on either the signal sequence cleavage or membrane association (Hedges et al., 1999a). The glycosylation site at asparagine 96, the cysteine residues at positions 13, 33 and 109 and, the tryptophan residue at position 146 are conserved in the GP3 proteins of all 70 strains of EAV and of the other three prototype *Arteriviruses*. Some EAV strains had altered hydrophobicity profiles in the signal sequence region.

a.

```
MGRAYSGPVA LLCFFLYFCF ICGSVGSNNT TICMHTTSDT SVHLFYAANV TFPSHFQRHF
AAAQDFVVHT GYEYAGVTML VHLFANLVLT FPSLVNCSRP VNVFANASCV QVVCSHTNST
TGLGQLSFSF VDEDLRLHIR PTLICWFALL LVHFLPMPRC RGSQFYLH
```

b.

```
MKIYGCILGL LLFVGLPCCW CTFYPCHAAE ARNFTYISHG LGHVHGHEGC RNFINVTHSA
FLYLNPTTLT APAITHCLLL VLAAKMEHPN ATIWLQLQPF GYHVAGDVIV NLEENKRHPY
FKLLRAPALP LGFVAIVYVL LRLVRWAQQC YL
```

Figure 1. Amino acid sequences of the GP3 protein (a) and the GP4 protein (b) of the prototype EAVATCC. Bold letters indicate amino acids that varied amongst 70 strains of EAV, italicises indicate conservation amongst all *Arteriviruses*, glycosylation sites are underlined, strikethrough letters indicate amino acids found on the GP3 protein of the modified live virus vaccine only.

The GP4 protein was more conserved than was the GP3 protein. The GP4 protein of the EAV ATCC strain has three potential glycosylation sites, but only the site at amino acid 90 was conserved amongst all isolates. The majority of the cysteines in the GP4 protein were conserved and the cysteine at position 19 is conserved in those of the prototype strains of the other *Arteriviruses* as well.

3. CONCLUSION

We have determined the genetic and proteomic diversity of the uncharacterised ORFs 3 and 4 and encoded GP3 and GP4 proteins of EAV. Variation in the predicted signal sequence region amongst the GP3 proteins of different EAV strains may influence membrane interaction, as is perhaps the case with divergent strains of the closely related porcine *Arterivirus* (van Nieuwstadt et al., 1996; Mardassi et al., 1998). There were a large number of non-synonymous substitutions in both ORF 3 and ORF 5 of EAV strains. Amino acid changes in the ORF5 encoded G_L protein alter the neutralisation phenotype of EAV strains (Balasuriya et al., 1997; Hedges et al., 1999b). The GP3 protein is under strong selective pressure during long term persistent infection of stallions, and carrier stallions and horses vaccinated with the live attenuated vaccine virus develop antibodies specific for the GP3 protein (Hedges et al., 1999a,b). The large number of non-synonymous substitutions in ORF 3 amongst strains of EAV is consistent with a strong selective pressure exerted on the GP3 protein. Clearly, individual EAV proteins are under unique selective pressure and structural/functional constraint and the corresponding ORFs evolve independently. Regions of the GP3 protein that vary may be important for the maintenance of persistent infection of carrier stallions, whereas conserved regions are likely important for protein function in virus replication. The significance of nucleotide and amino acid substitutions identified in this study can now be determined using the infectious cDNA clone of EAV.

ACKNOWLEDGMENTS

These studies were supported by the Center for Equine Health at the University of California-Davis, the Bernard and Gloria Salick Endowment and USDA National Research Initiative Competitive Grant 97-35204-4736.

REFERENCES

Archambault, D., Laganiere, G., Carman, S., and St-Laurent, G. (1997). Comparison of nucleic and amino acid sequences and phylogenetic analysis of open reading frames 3 and 4 of various equine arteritis virus isolates. Vet. Res. 28, 505-516.

Balasuriya, U.B.R., Hedges, J.F., Timoney, P.J., McCollum, W.H., and MacLachlan, N.J. (1999). Genetic stability of equine arteritis virus during horizontal and vertical transmission in an outbreak of equine viral arteritis. J. Gen. Virol. 80, 1949-1958.

Balasuriya, U.B.R., Patton, J.F., Rossitto, P.V., Timoney, P.J., McCollum, W.H., and MacLachlan, N.J. (1997). Neutralization determinants of laboratory strains and field isolates of equine arteritis virus: Identification of four neutralization sites in the amino-terminal ectodomain of the G_L envelope glycoprotein. Virology 232, 114-128.

Hedges, J.F., Balasuriya, U.B.R., Timoney, P.J., McCollum, W.H., and MacLachlan, N.J. (1996). Genetic variation in open reading frame 2 of field isolates and laboratory strains of equine arteritis virus. Virus Res. 42, 41-52.

Hedges, J.F., Balasuriya, U.B.R., and MacLachlan, N.J. (1999a). The open reading frame 3 of equine arteritis virus encodes an immunogenic glycosylated, integral membrane protein. Virology 264, 92-98.

Hedges, J.F., Balasuriya, U.B.R., Timoney, P.J., McCollum, W.H., and MacLachlan, N.J. (1999b). Genetic divergence with emergence of phenotypic variants of equine arteritis virus during persistent infection of stallions. J. Virol. 73, 3672-3681.

Mardassi, H., Gonin, P., Gagnon, C.A., Massie, B., and Dea, S. (1998). A subset of porcine reproductive and respiratory syndrome virus GP3 glycoprotein is released into the culture medium of cells as a non-virion-associated and membrane-free (soluble) form. J. Virol. 72, 6298-6306.

Patton, J.F., Balasuriya, U.B.R., Hedges, J.F., Schweidler, T.M., Hullinger, P.J., and MacLachlan, N.J. (1998). Phylogenetic characterization of a highly attenuated strain of equine arteritis virus from the semen of a persistently infected standardbred stallion. Arch. Virol. 144, 817-827.

van Nieuwstadt, A.P., Meulenberg, J.J.M., van Essen-Zandbergen, A., Peterson-den Besten, A., Bende, R.J., Moormann, R.J., and Wensvoort, G. (1996). Proteins encoded by open reading frames 3 and 4 of the genome of lelystad virus (Arteriviridae) are structural proteins of the virion. J. Virol. 70, 4746-4772.

Full-Length Genomic Sequence of Bovine Coronavirus (31kb)

Completion of the open reading frame 1a/1b sequences

DONGWAN YOO* and YANLONG PEI

*Department of Pathobiology, Ontario Veterinary College, University of Guelph, Ontario N1G 2W1 Canada. *Corresponding author: Email:dyoo@uoguelph.ca*

1. INTRODUCTION

Bovine coronavirus (BCV) is an important veterinary pathogen which causes neonatal diarrhea in newborn calves and winter dysentery in adult cattle. Recent studies indicate that BCV also infects the respiratory tract of cattle producing severe respiratory problems, especially in feedlot cattle. BCV belongs to the antigenic group II of coronaviruses and shares the antigenic and genetic similarities with mouse hepatitis virus (MHV), human coronavirus strain OC43, turkey coronavirus, and hemagglutinating encephalomyelitis virus of pigs. Approximately 10 kb of the 3' most BCV genome has been sequenced, and this region is known to encode all the structural proteins plus the 32k nonstructural protein which resides immediately upstream of the hemagglutinin-esterase (HE) glycoprotein gene. The comparative studies indicate that the genome organization of BCV is similar to but distinct from MHV. In BCV, the HE gene is functional coding for a major envelop protein while it is an optional gene in MHV. Similarly, the 32k protein gene is absent in some of the MHV variants. The region between the spike (S) protein gene and the small membrane (E) protein gene also shows a significant sequence divergence among coronaviruses. Two thirds of the BCV genome remain largely undetermined.

To date, coronaviruses of which the full length genomic sequences are available include mouse hepatitis virus (MHV) (Bredenbeek et al, 1990; Lee etal, 1991; Bonilla et al, 1994), human coronavirus 229E (Herold et al,

The Nidoviruses (Coronaviruses and Arteriviruses).
Edited by Ehud Lavi *et al.*, Kluwer Academic/Plenum Publishers. 2001.

73

1994), porcine transmissible gastroenteritis virus (TGEV) (Almazan et al, 2000), and avian infectious bronchitis virus (IBV) (Boursnell et al 1987). Of these viruses, human coronavirus 229E and TGEV fall within the same antigenic group (group I), and IBV is a group III virus. In the antigenic group II which BCV belongs to, MHV is the only virus that the full-length sequence has been determined. My laboratory has recently completed sequencing of the entire genome of BCV, and in this report we present the structural organization of the BCV genome.

2.   MATERIALS AND METHODS

2.1   cDNA Cloning

The Quebec strain of BCV was propagated in Mardin-Darby bovine kidney (MDBK) cells. Genomic RNA was prepared from the purified virions, and double stranded cDNA was synthesized by standard methods. After addition of Bam HI linkers, cDNAs were ligated into the Bam HI site of pTZ19R and used transform E. coli strain DH1. Specific clones were identified by colony hybridization using a ^{32}P-labeled cDNA probe prepared by randomly primed reverse transcription of viral genomic RNA or by PCR. For RT-PCR, total cellular RNA was extracted from virus-infected cells using Trizol and used as a template for cDNA synthesis.

2.2   Reverse Transcription Polymerase Chain Reaction

For first-stranded cDNA synthesis, Superscript II RNase H Negative Reverse Transcriptase (GIBCO BRL) was used. The first strand cDNA was PCR-amplified using Vent DNA polymerase (New England Biolab) in GeneAmp Thermocycler PE2400 (Perkin Elmer).

2.3   DNA Sequencing

The nucleotide sequences were determined either by manual sequencing or BY using an automated DNA sequencer. For manual sequencing, a series of overlapping clones were generated and the overlapping clones were sequenced using universal primers. For automated sequencing, new primers were generated based on the determined sequence to walk along the unsequenced region. Sequences were assembled and analyzed using the GCG Wisconsin sequence analysis package. The sequence has been deposited to the GenBank database under accession number AF220295.

3. RESULTS AND DISCUSSION

The overlapping sequences were assembled as a single contiguous sequence to represent the full-length genomic sequence of BCV. The BCV genome appeared to comprise of 31043 nucleotides, excluding the 3' terminal polyadenylation tail. The BCV sequence was 291 nucleotides shorter than the genome of MHV A59, and the two viruses shared an overall 71% sequence identity at the nucleotides level of the full-length genome. Open reading frame 1a (ORF1a) was able to code for a polypeptide of 4383 amino acids. ORF1a and 1b overlapped by 28 nucleotides, and this overlap was significantly shorter than that of MHV A59. A slippery sequence was identified within the overlapping region suggesting that a 1a/1b fusion protein of 7061 amino acids would be produced (Figure 1).

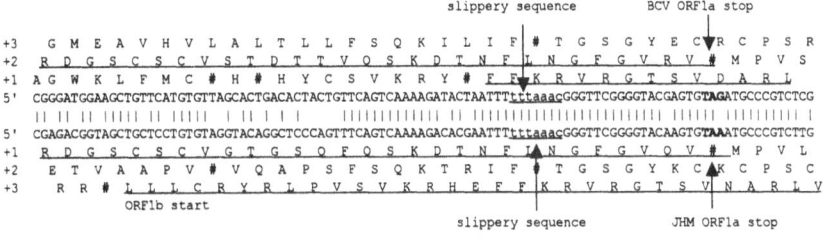

Figure 1. Sequence comparison of the ORF1a/ORF1b ribosomal frame shifting region of BCV and MHV JHM. Slippery sequences are illustrated in lower case and indicated by arrows. Stop codons for ORF1a (TAG for BCV and TAA for JHM) are depicted in boldface and indicated by arrow. Frames are indicated in numbers, and amino acid sequences encoded in the open reading frames 1a and 1b are underlined.

The BCV full-length genomic sequence was compared with that of other coronaviruses (Table 1). The BCV genome exhibited the highest sequence homology with the genome of MHV A59 (70.8%), but the homologies were only 48.3%, 52.5%, and 53%, when compared with IBV, HCV 229E, and TGEV, respectively. The 1a/1b polyprotein of BCV retained a similar level of amino acid identity to that of A59 (72.9%). The 1a sequence diversed more with only 68% homology while the 1b homology was 86%. The sequence conservation of the 1a/1b polyprotein was low in IBV, 229E, and TGEV, with an overall homology of only 36% (Table 1). Nevertheless, all the putative functional domains that were identified in all other coronaviruses were found to have been conserved in BCV, including the motifs for two separate papain-like cysteine proteinases, a 3C-like chymotrypsin proteinase, a cysteine rich murine growth factor-like, a RNA dependent RNA polymerase, a zinc finger-like, and a helicase.

Putative proteolytic cleavage sites specific for the papain-like proteinase were predicted at RG246/V and A851/G. In addition to these sites, 11 additional cleavage sites (9 sites in 1a and 4 sites in 1b) were identified as potential catalytic sites for 3C-like proteinase: Q3247/S, Q3549S, Q3836/S, Q3928/S. Q4122/N, Q4232/A, Q4369/S, Q5299/S, Q5902/C, Q6423/S, and Q6792/A.

Table 1. Sequence identities of the BCV genome with other coronaviruses

Virus	Full genomic sequence (nts)	1a/1b fusion protein sequence (aa)
MHV A59	70.8%	72.9%
TGEV	53.0%	36.4%
HCV 229E	52.5%	35.7%
IBV	48.3%	36.5%

ACKNOWLEDGMENTS

Authors are grateful to Drs. M. D. Parker and G. J. Cox for their contribution in completing this sequence. We thank Dr. E. J. Snijder for his help with predictions of the proteolytic cleavage sites of the ORF1a and 1b proteins. This research was supported by Medical Research Council of Canada.

REFERENCES

Almazan, F., Gonzalez, J. M., Penzes, Z., Izeta, A., Calvo, E., Plana_Duran, J., and Enjuanes, L. 2000. *Proc. Natl. Acad. Sci. U.S.A.* **97**, 5516_5521.

Bonilla, P. J., Gorbalenya, A. E. and Weiss, S. R. 1994. *Virology* **198**, 736-740.

Boursnell, M. E., Brown, T. D., Foulds, I. J., Green, P. F., Tomley, F. M. and Binns, M. M. 1987. *J. Gen. Virol.* **68**, 57-77.

Bredenbeek, P. J., Pachuk, C. J., Noten, A. F., Charite, J., Luytjes, W., Weiss, S. R. and Spaan, W. J. 1990. *Nucl. Acids Res.* **18**, 1825-1832.

Herold, J., Raabe, T., Schelle_Prinz, B. and Siddell, S. G. 1993. *Virology* **195**, 680-691.

Lee, H. J., Shieh, C. K., Gorbalenya, A. E., Koonin, E. V., La Monica, N., Tuler, J., Bagdzhardzhyan, A. and Lai, M. M. C. 1991. *Virology* **180**, 567-582.

Storz, J., Stine, L., Liem, A., Anderson, G. A. 1996. *J. Am. Vet. Med. Assoc.* **208**:1452-1455.

Analysis of CNS Inflammatory Responses to MHV

Role of spike determinants in initiating chemokine and cytokine responses

JULIA D. REMPEL AND MICHAEL J. BUCHMEIER
The Scripps Research Institute, La Jolla, CA.

1.   INTRODUCTION

Development of demyelination in the MHV system is associated with inflammatory cytokine responses in the brain. It is likely that these responses reflect the influx of T cells into the brain following infection. Previous work has described a central role for CD4+ T cells in viral clearance and subsequent demyelination (Lane, Liu et al. 1999; Lane, Liu et al. 2000; Wu, Dandekar et al. 2000). While CD8+ T cells are necessary for viral clearance, their involvement in demyelination is less defined (Castro, Evans et al. 1994; Stevenson, Belz et al. 1999). Glial cells have also been implicated in the disease process, contributing to cytokine production in the central nervous system (Sun, Grzybicki et al. 1995; Lane, Asensio et al. 1998).

We were interested in examining how the host immune response against an extensively demyelinating virus (MHV-A59) and a highly encephalitic virus (JHM/MHV-4) might contribute to disease outcome. Furthermore, previous studies have emphasized the importance of the spike glycoprotein as a determinant of viral pathogenesis (Phillips, Chua et al. 1999) and as a target for immune responses (Bergmann, Yao et al. 1996). To determine how the JHM spike protein participates in regulating host responses seen upon JHM infection, immune responses against a recombinant virus containing the JHM spike protein in the context of a MHV-A59-like genome (Phillips, Chua et al. 1999) were examined.

The Nidoviruses (Coronaviruses and Arteriviruses).
Edited by Ehud Lavi *et al.*, Kluwer Academic/Plenum Publishers, 2001.

2.   JHM INFECTION INDUCES IL-1 AND IL-6 MESSENGER RNA EXPRESSION

MHV-A59 and JHM (MHV-4) are neurotropic MHV strains that differ in pathology. Infections with high doses of the less neurovirulent virus MHV-A59 result in acute encephalitis and extensive demyelination (Lavi, Gilden et al. 1986). In contrast, mice infected with a low dose of the highly neurovirulent JHM die within a week from acute encephalitis (Dalziel, Lampert et al. 1986). To investigate how host immune responses induced by these viruses might participate in disease outcome, mice were infected with doses of JHM (10 PFU) and MHV-A59 (1000 PFU) that produced similar degrees of acute encephalitis. Isolation of cells from the brains of mice infected with JHM consistently resulted in recovery of 1.7 fold more cells during acute encephalitis (day 7 post infection) compared to MHV-A59 infected mice. However, CD8 T cells were decreased approximately 3 fold upon JHM infection as compared to MHV-A59 as determined by flow cytometry. CD4 T cell numbers were also reduced but to a lesser degree. This translated into approximately 1.4 percent of the total cells isolated upon JHM infection being T cells; by comparison with MHV-A59 infection about 5.1% of the total cells were T cells. Thus, the increase of total cells isolated from the brains of JHM infected mice did not reflect an overall increase in recruitment of T cells from the periphery.

The disparity in T cell recruitment was not reflected in a difference in RANTES mRNA regulation as determined by RNA protection assay (RPA) analysis (Figure 1). A two-fold increase in MIP-1α, MIP-3 and IP-10 mRNA expression relative to MHV-A59 infection was evident suggesting that JHM infection may result in a greater influx of other peripheral mononuclear cells, such as macrophages. This possibility will be explored by cell surface staining with macrophage markers.

Differential responses seen by RPA analysis of cytokine transcription also implicated non-lymphocyte populations, possibly resident glial cells in participating in disease outcome. JHM and MHV-A59 infection produced similar TNFα, TGFβ1, TGFβ3 transcription (Table 1). This indicated that while these cytokines played a critical role in the immune response against MHV infection, they probably did not contribute to the differences in neurovirulence between JHM and MHV-A59. A striking difference was, however, observed in IFNα/β, IL-6 and IL-1β mRNA message (Table1). JHM infection resulted in strong expression of IFNα/β, IL-6 and IL-1β mRNA's; whereas, MHV-A59 did not. IL-1 and IL-6 are pro-inflammatory cytokines that can act synergistically in the periphery and in the central nervous system. High levels of IL-6 in the brain were shown to contribute to astrogliosis (Campbell, Abraham et al. 1993; Campbell, Stalder et al. 1998). More importantly, TNFα, IL-6 and IL-1β production was previously attributed to glial cells in JHM models of acute encephalitis and demyelination (Sun, Grzybicki et al. 1995).

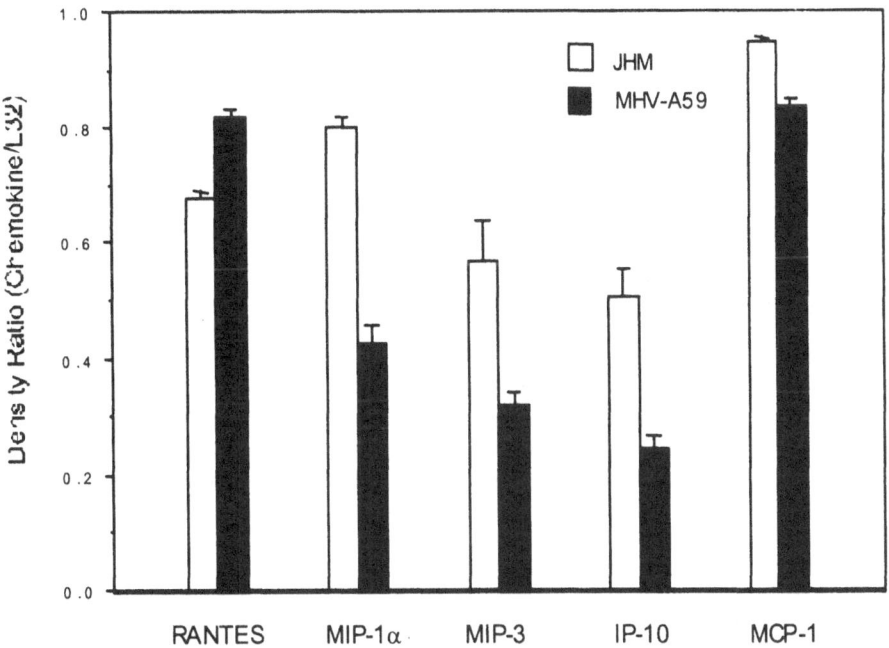

Figure 1. Chemokine mRNA expression following JHM and MHV-A59 infection. Mice were injected intracranially with either 10 PFU of JHM or 1000 PFU of MHV-A59. Brains were harvested on day 7. Total RNA was isolated and analyzed by RNase protection assay as previously described (Stalder and Campbell 1994). Means of 4 mice ± SE are shown.

Table 1. JHM infection induces greater expression of IL-1 and IL-6 mRNA[a].

Virus[b]	LTB	TNFα	IL-1	IL-6	IFNγ	IFNα/β	TGFβ1	TGFβ3
JHM	0.11	0.39	0.40	0.36	0.16	0.49	0.32	0.47
MHV-A59	0.13	0.31	0.15	0.10	0.25	0.11	0.25	0.38

a. Means of 4 mice shown representing band density of cytokine normalized against the corresponding band density of the housekeeping gene. Standard error < ± 0.07.

b. Mice were injected intracranially. Brains were harvested on day 7 and total RNA was isolated.

Astrocytes were also shown to synthesize chemokines after in vitro infection with V5A13.1 (a JHM variant) suggesting that astrocytes play an important role in participating in immunity against MHV infection (Lane, Asensio et al. 1998). As such, they could influence the direction of innate and adaptive immune activity within the brain. Clearly, differences in how viruses affect chemokine and cytokine responses by these cells may be a crucial early event in determining the pathogenesis of infection.

3.   CYTOKINE PRODUCTION MAY BE REGULATED BY SPIKE DETERMINANTS

It is known that specific viral proteins can directly affect immune responses and pathology (Yoshimoto, Rosenfeld et al. 1991; Hoshino, Jones et al. 1998). The MHV spike glycoprotein has been established as an important target for adaptive immune responses. It contains neutralizing antibody and T cell epitopes, some of which are shared between JHM and MHV-A59 (Castro, Evans et al. 1994; Bergmann, Yao et al. 1996; Xue and Perlman 1997). However, differing spike determinants appear critical to the variation in disease outcome seen between JHM and MHV-A59. Using MHV-A59 variant chimeric viruses that differed only in their spike protein, the variation in the severity of acute encephalitis between JHM and MHV-A59 was attributed to differences in this glycoprotein (Phillips, Chua et al. 1999). Specifically, the S4R22 chimera with a MHV-A59-like background and the JHM spike protein displayed a neurovirulence more reflective of JHM than MHV-A59.

Infection with S4R22 produced an increased number of total cells and T cells as compared to JHM or MHV-A59 infections, most likely reflecting coordinated MHV-A59 background and JHM spike activity. While the significance has not been established definitively, it was interesting that S4R22 infection resulted in a similar percentage of infiltrating T cells as MHV-A59. Preliminary evaluation of the chemokine responses elicited by S4R22 infection did not reveal any marked differences (Table 2). Examination of cytokine responses also indicated similar TNFα, TGFβ1, TGFβ3 transcription. However, in this study S4R22 infection resulted in IL-6 and IFNα/β mRNA responses similar to those seen in MHV-A59 infection suggesting that the background, non-spike, genes of these viruses may participate in the induction of these cytokines. This is not surprising in light of the presence of known CD4 T cell epitopes on the M protein and potential for others within the large remaining uncharacterized genome. In contrast, IL-1 expression from these animals reflected that of JHM, perhaps indicating that spike protein interactions can trigger this cytokine. The presence of IL-1 may be significant in that it has been suggested to alter the antigen presenting capacity of astrocytes (Williams, Dooley et al. 1995). The possibility that the neurovirulence of JHM may be in part due to astrocyte responses against its spike and non-spike genes is currently being investigated.

Table 2. Cytokine and chemokine mRNA expression following S4R22 infection[a, b].

Chemokine	RANTES	MIP-1α	MIP-3	IP-10	MCP-1
	0.83	0.54	0.43	0.28	0.98
Cytokine	TNFα	IL-1	IL-6	IFNα/β	TGFβ1
	0.38	0.36	0.13	0.14	0.37

a. Means of 2 mice shown representing band density of virus normalized against the corresponding band density of the housekeeping gene. Standard error $< \pm 0.07$.
b. Mice were injected intracranially. Brains were harvested on day 7 and total RNA was isolated.

4. CONCLUSION

Our studies with the virulent JHM, the mildly neurovirulent MHV-A59 and the recombinant S4R22 (containing the JHM spike gene on a MHV-A59-like background) suggested that differential chemokine and cytokine responses to infection may participate in the disease outcome. Most prominently, IL-1 and IL-6 up regulation was associated with high neurovirulence. We are currently investigating how early sentinel responses of glial cells to these viruses might impact IL-6 and IL-1 production and alter subsequent disease.

ACKNOWLEDGMENTS

J.D.R. is supported by fellowships from the MS Society of Canada and the National Multiple Sclerosis Society. This work was also supported by N.I.H. Grants AI 43103 and AI 25913 to M.J.B. The authors thank Shannon Murray for technical support.

REFERENCES

Bergmann, C. C., Q. Yao, et al. ,1996, The JHM strain of mouse hepatitis virus induces a spike protein-specific Db-restricted cytotoxic T cell response. *J Gen Virol* 77: 315-25.

Campbell, I. L., C. R. Abraham, et al. ,1993, Neurologic disease induced in transgenic mice by cerebral overexpression of interleukin 6. *Proc Natl Acad Sci U S A* 90: 10061-5.

Campbell, I. L., A. K. Stalder, et al. ,1998, Transgenic models to study the actions of cytokines in the central nervous system. *Neuroimmunomodulation* **5**: 126-35.

Castro, R. F., G. D. Evans, et al. ,1994, Coronavirus-induced demyelination occurs in the presence of virus- specific cytotoxic T cells. *Virology* **200**: 733-43.

Dalziel, R. G., P. W. Lampert, et al. ,1986, Site-specific alteration of murine hepatitis virus type 4 peplomer glycoprotein E2 results in reduced neurovirulence. *J Virol* **59**: 463-71.

Hoshino, Y., R. W. Jones, et al. ,1998, Serotypic characterization of outer capsid spike protein VP4 of vervet monkey rotavirus SA11 strain. *Arch Virol* **143**: 1233-44.

Lane, T., M. T. Liu, et al. ,1999, A central role for CD4+ T cells and RANTES in virus-induced central nervous system inflammation and demyelination. *J. Virol.* **74**: 1415-1424.

Lane, T. E., V. C. Asensio, et al. ,1998, Dynamic regulation of alpha- and beta-chemokine expression in the central nervous system during mouse hepatitis virus-induced demyelinating disease. *J Immunol* **160**: 970-8.

Lane, T. E., M. T. Liu, et al. ,2000, A central role for CD4(+) T cells and RANTES in virus-induced central nervous system inflammation and demyelination. *J Virol* **74**: 1415-24.

Lavi, E., D. H. Gilden, et al. ,1986, The organ tropism of mouse hepatitis virus A59 in mice is dependent on dose and route of inoculation. *Lab Anim Sci* **36**: 130-5.

Phillips, J. J., M. M. Chua, et al. ,1999, Pathogenesis of chimeric MHV4/MHV-A59 recombinant viruses: the murine coronavirus spike protein is a major determinant of neurovirulence. *J Virol* **73**: 7752-60.

Stalder, A. K. and I. L. Campbell ,1994, Simultaneous analysis of multiple cytokine receptor mRNAs by RNase protection assay in LPS-induced endotoxemia. *Lymphokine Cytokine Res* **13**: 107-12.

Stevenson, P. G., G. T. Belz, et al. ,1999, Changing patterns of dominance in the CD8+ T cell response during acute and persistent murine gamma-herpesvirus infection. *Eur J Immunol* **29**: 1059-67.

Sun, N., D. Grzybicki, et al. ,1995, Activation of astrocytes in the spinal cord of mice chronically infected with a neurotropic coronavirus. *Virology* **213**: 482-93.

Williams, K. C., N. P. Dooley, et al. ,1995, Antigen presentation by human fetal astrocytes with the cooperative effect of microglia or the microglial-derived cytokine IL-1. *J Neurosci* **15**: 1869-78.

Wu, G. F., A. A. Dandekar, et al. ,2000, CD4 and CD8 T cells have redundant but not identical roles in virus-induced demyelination. *J Immunol* **165**: 2278-86.

Xue, S. and S. Perlman ,1997, Antigen specificity of CD4 T cell response in the central nervous system of mice infected with mouse hepatitis virus. *Virology* **238**: 68-78.

Yoshimoto, K., S. Rosenfeld, et al. ,1991, A second neutralizing epitope of B19 parvovirus implicates the spike region in the immune response. *J Virol* **65**: 7056-60.

Analysis of Nonessential Gene Function in Recombinant MHV-JHM

Gene 4 knockout recombinant virus

EVELENA ONTIVEROS[1], LILI KUO,[2] PAUL MASTERS,[2] AND STANLEY PERLMAN[1,3]

[1]*Interdisciplinary Program in Immunology and* [3]*Departments of Pediatrics and Microbiology, University of Iowa, Iowa City, Iowa 52242.* [2]*David Axelrod Institute, Wadsworth Center for Laboratories and Research, New York State Department of Health, Albany, New York 12201*

1. INTRODUCTION

The large size of the coronavirus genome has made reverse genetics difficult. Targeted recombination, a technique developed by Masters and colleagues, has facilitated the introduction of mutations into the coronavirus genome (Kuo et al., 2000). Previous work by Skinner and Siddell showed that MHV-JHM ORF 4 encodes a 15 kDa protein composed of 139 amino acids. This protein is relatively rich in threonines and includes a hydrophobic region. The N-terminus contains a potential membrane-anchoring region and the C-terminus has a possible RNA binding region (Skinner and Siddell, 1985). MHV-S, a natural variant, does not encode a functional ORF 4 suggesting that the ORF 4 product was not necessary for growth in tissue culture cells or animals. Additionally this strain contained a deletion within ORF 5a (Yokomori and Lai, 1991). Lack of mRNA 4 synthesis most likely resulted from a point mutation in the intergenic sequence (UCUAAAC to UUUAAAC). In this study, targeted recombination was used to genetically disrupt ORF 4. This recombinant virus was then analyzed in a murine model of encephalitis.

The Nidoviruses (Coronaviruses and Arteriviruses).
Edited by Ehud Lavi *et al.*, Kluwer Academic/Plenum Publishers, 2001.

2.   MATERIALS AND METHODS

2.1   Recombinant Virus

Initially, targeted recombination (Kuo et al., 2000) was used to construct a chimeric virus containing MHV-JHM genes 1 and 2, the surface (S) glycoprotein of feline coronavirus (FCoV) and MHV-A59 genes 4-7 (designated FCV-MHV-JHM clone B3b) (Figure 1).   Subsequently, mutations were introduced into this virus using targeted recombination.

2.2   Construction of plasmids

A plasmid containing gene 3 from MHV-JHM and genes 4-7 from MHV-A59 (pMH54-4) was kindly provided by Dr. Susan Weiss, University of Pennsylvania.   PCR was used to replace genes 4-7 in this plasmid with homologous sequences from the JHM strain (Figure 1). JHM-specific DNA fragments were prepared either from plasmids encoding JHM proteins (a gift from Dr. S. Siddell, University of Wurzburg) or from JHM-specific RNA harvested from infected tissue culture cells (Pewe et al., 1996). The fidelity of all PCR products was confirmed by sequencing.

2.3   Measurement of LD$_{50}$

Pathogen-free C57Bl/6 mice were purchased from the National Cancer Institute (NCI).   Dilutions of each virus were prepared in phosphate buffered saline and each was delivered in 30 μl under anesthesia to 3-5 mice.   Virus was inoculated into the left cerebrum.   LD$_{50}$ values were calculated by the Reed-Muench method (Reed and Muench, 1938).

2.4   Viral RNA analysis

Virus from infected brains was amplified by passage through 17Cl-1 cells as described previously.   RNA was harvested from 17Cl-1 cells infected with this amplified virus (m.o.i. 0.5-1.0) and analyzed by Northern blot analysis as described previously (Perlman et al, 1990, Pewe et al, 1996). Processed blots were exposed to film or analyzed using a phosphoimager.

3. RESULTS AND DISCUSSION

3.1 Construction of recombinant MHV-JHM

In all previous cases, targeted recombination was used to introduce mutations into the MHV-A59 background. MHV-JHM is useful for studies of neuropathogenesis. Consequently we adapted targeted recombination to MHV-JHM (Figure 1). Initially, a chimeric MHV-JHM containing the FCoV surface glycoprotein was developed using standard techniques (Kuo et al., 2000). Next a donor plasmid was constructed using an MHV-A59 based plasmid (pMH54-4). This plasmid contained the JHM S gene sequences with MHV-A59 genes 4-7. MHV-JHM genes 4-7 were cloned using PCR products amplified from plasmids or cDNA derived from MHV-JHM and inserted into the MHV-A59-based plasmid.

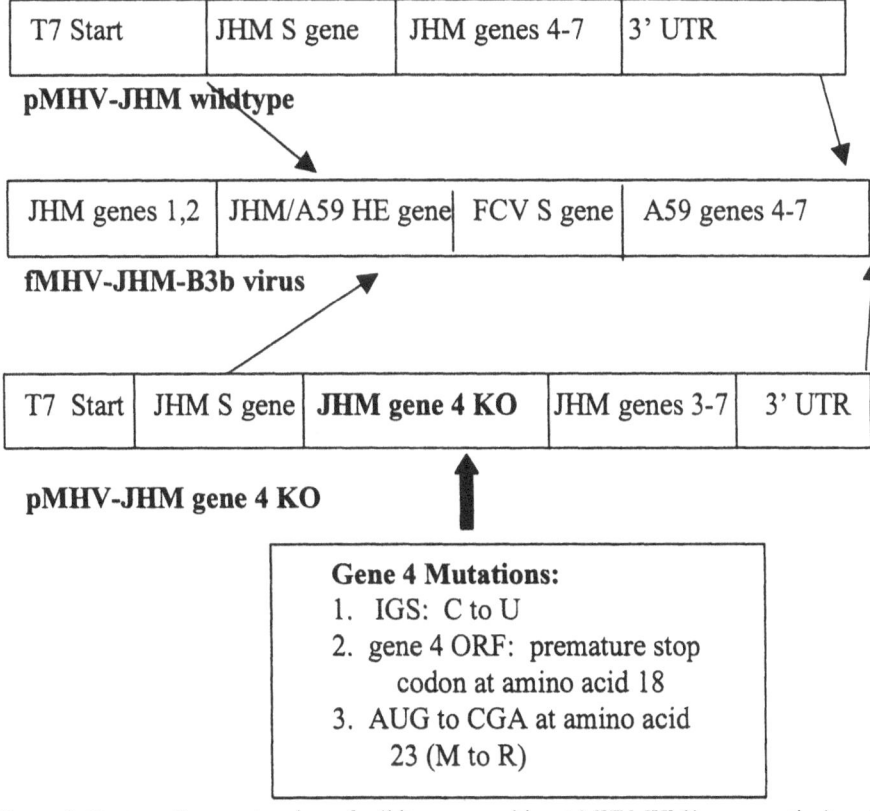

Figure 1. Strategy for construction of wildtype recombinant MHV-JHM(upper portion) and MHV-JHM gene 4 knockout recombinant(lower portion). Recipient virus and donor plasmid were constructed as described in Materials and Methods. Gene 4 was genetically disrupted by insertion of three mutations as shown above.

The strategy for knocking out expression of gene 4 involved introducing three separate mutations (Figure 1). The same point mutation which had been identified in the intergenic sequences of MHV-S, the natural variant lacking gene 4 mRNA, was introduced into the IGS of our MHV-JHM gene 4 (Yu and Leibowitz, 1994). This change should eliminate transcription initiating at gene 4. In addition, a premature stop codon at amino acid 18 was introduced. This mutation should result in premature termination of any translation occurring from residual gene 4 RNA. Finally, an AUG downstream from this termination codon was also mutated. Though not used by MHV-JHM, this AUG may be functional in MHV-A59 infected cells.

The MHV-JHM wildtype and MHV-JHM gene 4 knockout plasmids were transcribed into an 8 KB RNA product. This RNA was electroporated into feline FCWF cells (American Type Culture Collection) previously infected with fMHV-JHM-B3b. Recombinants were selected and the three mutations in gene 4 were confirmed.

3.2   Mouse model of neuropathogenesis

C57Bl/6 infected intracranially or intranasally with wild type MHV-JHM develop an invariably fatal acute encephalitis. In initial studies, the MHV-JHM wildtype and gene 4 knockout recombinants were used to infect 6 week old C57Bl/6 mice. Intranasal inoculation of mice with undiluted stocks of virus demonstrated that both the recombinant wildtype MHV-JHM and the MHV-JHM-gene 4 KO recombinant were capable of causing acute encephalitis with similar kinetics.

These results show that there was not a high level of attenuation after genetic disruption of gene 4. In order to define more precisely any attenuation of the gene 4 KO recombinant, LD_{50} after intracranial inoculation was determined. Serial dilutions of virus were prepared and inoculated into 3-4 mice per dilution. The recombinant wildtype MHV-JHM had an LD_{50} of 9 pfu, similar to what we observed for nonrecombinant wildtype MHV-JHM (3 pfu). The LD_{50} of MHV-JHM gene 4 KO recombinant was 73, higher than the wild type recombinant.

Table 1. LD_{50} after intracranial inoculation.

	LD_{50}
JHM wildtype	3
JHM wildtype Recombinant	9
JHM-Gene 4 KO Recombinant	73

3.3 Analysis of recombinant viral RNA

Since the measurements of LD_{50} indicated that the gene 4 KO recombinant was only minimally attenuated, the next step was to confirm that the mutations into gene 4 were maintained after passage through mice. RNA, harvested from infected brains, was used to prepare cDNA (Pewe et al., 1996). Sequencing of this cDNA confirmed that the point mutations introduced into gene 4 were still present. In other experiments, virus from infected brains was harvested and passaged once through 17Cl-1 cells. RNA was harvested from cells infected with these amplified viruses and assessed by Northern blot analysis (Figure 2).

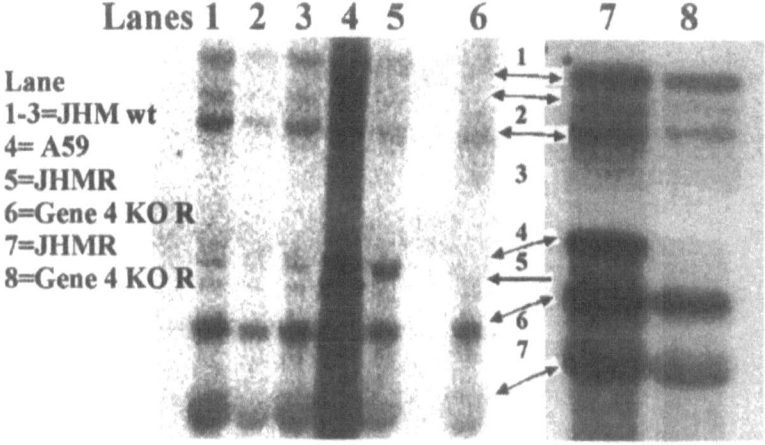

Figure 2. Northern blot analysis of RNA harvested from infected cells. RNA was prepared from cells infected with brain-derived virus as described in text. Bands corresponding to mRNA 4 and 5 were present in cells infected with MHV-JHM and MHV-A59. mRNA 4 was detected in cells infected with recombinant wildtype MHV-JHM but not in those infected with MHV-JHM gene 4 KO recombinant. mRNA 5 was not detected in either of these viruses but could be amplified by PCR. Levels of other mRNAs were not consistently changed by mutations introduced into gene 4.

As expected, two bands corresponding to mRNA 4 and 5 were easily visualized from MHV-A59 infected cells. Two bands, albeit fainter, are also visible in cells infected with wildtype MHV- JHM. Surprisingly only a band corresponding to mRNA 4 was detected in cells infected with recombinant wildtype MHV-JHM. The protein encoded by gene 5b is essential (Yu and

Leibowitz, 1994) making it unlikely the gene 5 transcript would be completely absent. Supporting this conclusion, a product corresponding to gene 5 could be amplified by PCR from viral RNA. In addition, we consistently observed that the mRNA 4 band is more intense than what we observe for that band in our wildtype virus. Introduction of the Sbf I site into the 3' end of the S gene alters the upstream context of the gene 4 intergenic sequence so that gene 4 transcription is enhanced (unpublished data). As shown in Figure 2, no gene 4 RNA is detected in cells infected with the MHV-JHM gene 4 KO recombinant. The pattern seen with the recombinant viruses is shown most clearly in Lanes 7-8 of Figure 2.

4.   CONCLUSIONS

In summary, we have adapted targeted recombination to introduce mutations into the MHV-JHM strain. Our initial results show that genetic disruption of gene 4 results in mild attenuation in a murine model of acute encephalitis. We are in the process of confirming the lack of expression of gene 4 protein in cells infected with MHV-JHM gene 4 KO recombinants.

ACKNOWLEDGMENTS

We thank Joanna Phillips and Susan Weiss for advice and reagents. This research was supported in part by grants from the NIH (NS 36592) and National Multiple Sclerosis Society. E.O. was supported by a N.I.H. predoctoral National Research Service Award (F31 AI10348).

REFERENCES

Kuo, L., Godeke, G.J., Raamsman, M.J., Masters, P.S. and P.J. Rottier. 2000. Retargeting of coronavirus by substitution of the spike glycoprotein ectodomain: crossing the host cell species barrier. *J. Virol.* 74(3):1393-406.

Perlman, S. Jacobsen, G., Olson, A.L., and A. Afifi. 1990. Identification of the spinal cord as a major site of persistence during chronic infection with a murine coronavirus. *Virology.* 175: 418-426.

Pewe, L., Wu, G.F., Barnett, E.M., Castro, R.F. and S. Perlman. 1996. Cytotoxic T cell-resistant variants are selected in a virus-induced demyelinating disease. *Immunity.* 5:253-262.

Reed, L.J. and H. Muench. 1938. A simple method of estimating fifty percent points. *Am. J. Hygiene.* 27:493-497.

Skinner, M.A. and S.G. Siddell. 1985. coding sequence of coronavirus MHV-JHM mRNA 4. *J. Gen. Virol.* 66:593-596.

Weiss, S.R., Zoltick, P.W., and J.L. Leibowitz. 1993. The ns 4 gene of mouse hepatitis virus (MHV), strain A 59 contains two ORFs and thus differs from ns 4 of the JHM and S strains. *Arch.Virol.* 129:301-309.

Yokomori, K., and M.M.C. Lai. 1991. Mouse hepatitis virus S RNA sequence reveals that nonstructural proteins ns4 and ns5a are not essential for murine coronavirus replication. *J. Virol.* 65(10):5605-5608.

Yu, X., Bi, W., Weiss, S.R. and J.L. Leibowitz. 1994. Mouse hepatitis virus gene 5b protein is a new virion envelope protein. *Virology.* 202:1018-23.

Persistence of Porcine Reproductive and Respiratory Syndrome in Pigs

DENNIS HORTER, CHIH-CHEN CHANG, ROMAN POGRANICHNYY, JEFF ZIMMERMAN, AND KYOUNG-JIN YOON
Department of Veterinary Diagnostics and Production Animal Medicine, Iowa State University College of Veterinary Medicine, Ames IA

1. INTRODUCTION

First reported in the United States in 1987 and in Europe in 1990, porcine reproductive and respiratory syndrome has rapidly become the most important infectious disease problem in the North American swine industry. The etiological agent, porcine reproductive and respiratory syndrome virus (PRRSV), is a member of the family *Arterivirdae* in the order *Nidovirales*. PRRSV is a single-stranded, enveloped, positive-sense RNA virus with a genome of approximately 15 kb in length containing 8 open reading frames (ORFs) which are expressed as subgenomic mRNA. At present, PRRSV is present in at least 60 percent of North American swine herds (USDA:APHIS:VS, 1997)

Clinical signs exhibited by pigs infected with PRRSV are age dependent. Newborn and nursing pigs (one to 21 days old) are at risk of increased pre-weaning mortality with accompanying clinical signs of fever and lethargy. Infected growing pigs (4 to 12 weeks old) exhibit respiratory signs typically resulting from PRRSV in combination with secondary bacterial or viral infections. "Falling-back," i.e., failure to thrive, is another problem associated with PRRSV infection in growers. In breeding stock, PRRSV infection causes late-term abortions, an increase in the rate of stillborn piglets, and poor conception rates. Unless the degree of secondary involvement is severe, animals eventually recover clinically, but they still

The Nidoviruses (Coronaviruses and Arteriviruses).
Edited by Ehud Lavi *et al.*, Kluwer Academic/Plenum Publishers, 2001.

harbor infectious virus. These clinically normal, but persistently infected animals are "chronic carriers."

2.   PERSISTENT INFECTION OF PRRSV

Persistent PRRSV infection is well documented. Wills et al. (1997) isolated PRRSV from oropharyngeal scrapings collected 157 days post inoculation. Albina et al. (1994) reported that, at 22 weeks of age, animals originally infected with PRRSV *in utero* transmitted PRRSV to susceptible animals through direct contact. Similarly, Benfield et al. (1997) reported detection of PRRSV RNA by PCR in serum 210 days after farrowing following transplacental exposure

Although persistence of PRRSV in individual animals has been clearly established, the dynamics of PRRSV persistence in populations has not been investigated. With that in mind, we conducted an animal trial to characterize the carrier state of PRRSV infected animals over time. The experiment consisted of 180 3-to 4-week-old pigs in 2 treatment groups: infected (N=90) and uninfected (N=90). Animals in the infected group were inoculated with PRRSV VR-2332, the North American prototype, via intranasal exposure. After inoculation, animals from each group were periodically euthanized and examined for the presence of infectious PRRSV in serum and tissues using virus isolation and swine bioassays. We found that 100% of animals experimentally inoculated with PRRSV harbored infectious virus at day 63 post inoculation (PI). The proportion of persistently infected animals decreased over time, but 90% of the animals were still carrying infectious virus 105 days PI.

3.   DISCUSSION OF CARRIER STATE

PRRSV carriers pose a serious problem for the prevention and control of the disease. Carrier animals play a major role in the epidemiology of PRRSV. Transmission from chronic carriers to susceptible animals perpetuates PRRSV in endemically infected herds. Transport of infected animals is widely recognized as an important means of moving virus between herds and even continents (Dewey et al., 2000).

In our study, identifying carrier animals economically and reliably appeared to be difficult using the diagnostic technology currently available. Probably as a direct result of the low level of PRRSV in persistently infected individuals, an RT-nPCR assay was not sufficiently sensitive to detect all animals known to be carrying infectious virus on the basis of virus isolation or bioassay results. Nor was it possible to discriminate between carriers and inoculated animals that cleared PRRSV on the basis of quantitative results produced by a commercial PRRS ELISA.

The existence of chronic carriers raises serious questions about the virology and immunology of PRRSV. PRRSV persists even in the face of an active immune response, both humoral and cell-mediated. Antibody production against PRRSV starts by 7 days PI (Yoon, et. al. 1996). Neutralizing antibodies are detectable as early as day 10 PI at low levels and peak between day 35-50 PI (Yoon, et. al. 1996). PRRSV-specific lymphocyte blastogenesis is detected by 28 days PI (Bautista, et. al. 1994). PRRSV-specific interferon gamma producing T-cells can be detected beginning about day 65 PI and rise slowly until day 200 (Zuckerman, et. al. 1999). Despite this immune response, PRRSV still persists in carrier animals, although the mechanisms by which the virus is able to persist are not clear.

4. CONCLUSION

PRRSV infection results in persistent infections. Carriers appear to be the norm, rather than the exception. Approximately 98% of animals harbor infectious virus at day 60 PI, and approximately 90% are still carriers at day 105 PI. These carrier animals are difficult to detect using current diagnostic technology. Further work needs to be done to extend these estimates beyond day 105 PI and to determine the mechanisms of persistence.

REFERENCES

Albina E, Madec F, Cariolet R, Torrison J. Immune response and persistence of the porcine reproductive and respiratory syndrome virus in infected pigs and farm units. *Vet Rec* 1994:134:567-573.

Bautista EM, Molitor TW. Cell-mediated immunity to porcine reproductive and respiratory syndrome virus in swine. *Viral Immunol.* 1997;10:83-94.

Benfield DA, Christopher-Hennings J, Nelson EA, Rowland RRR, Nelson JK, Chase CCL, Rossow KD, Collins JE. Persistent fetal infection of porcine reproductive and respiratory syndrome (PRRS) virus. *Proc Am Assoc Swine Pract* 1997:455-458.

Dewey D, Charbonneau G, Carman S, Nayar H, Nayar G, Friendship R, Eernisse K, Swenson S. Lelystad-like strain of porcine reproductive and respiratory syndrome virus (PRRSV) identified in Canadian swine. *Can Vet J.* 2000 41:

USDA:APHIS:VS. Prevalence of PRRS virus in the United States. National Animal Health Monitoring System, 1997, Info Sheet N225.197.

Wills RW, Zimmerman JJ, Yoon KJ, Swenson SL, McGinley MJ, Hill HT, Platt KB, Christopher-Hennings J, Nelson EA. Porcine reproductive and respiratory syndrome virus: a persistent infection. *Vet Microbiol.* 1997 55:231-40.

Yoon KJ, Zimmerman JJ, Swenson SL, McGinley MJ, Eernisse KA, Brevik A, Rhinehart LL, Frey ML, Hill HT, Platt KBCharacterization of the humoral immune response to porcine reproductive and respiratory syndrome (PRRS) virus infection. J Vet Diagn Invest. 1995 Jul;7(3):305-12.

Zuckerman FA, Zuckerman FA, Osorio F, Husmann RJ, et al. 1998. Analysis of the cell-mediated immune response of pigs to PRRSV. Proc Am Assoc Swine Pract. 1998:399-400.

The Severity of Hepatic Lesion after Intraperitoneal JHMV Infection in IFN-gamma Deficient Mice is Parallel to Viral Replication in Hepatocytes *in Vitro*

[1]SHIGERU KYUWA, [2]SEIJI KAWAMURA, [3]SHINWA SHIBATA, [4]KENJI MACHII, [5]YOH-ICHI TAGAWA, [3]YOICHIROH IWAKURA, AND [1]TORU URANO

[1]*Division of Microbiology and Genetics, Center for Animal Resources and Development, Kumamoto University, Kumamoto:* [2]*Department of Biomedical Science, Faculty of Agriculture, University of Tokyo, Tokyo:* [3]*Center for Experimental Medicine, Institute of Medical Science, Univerisity of Tokyo, Tokyo:* [4]*Department of Veterinary Public Health, Institute of Public Health, Tokyo:* [5]*Institute of Experimental Animals, Shinshu University Schoool of Medicine, Matsumoto, Japan*

1. INTRODUCTION

Several factors affect MHV infection in mice (Table 1). They can be classified into two groups, viral factors and host factors. Although JHMV induces a fatal encephalitis in mice after intracerebral infection, it does a mild hepatitis when it is inoculated intraperitoneally. On the other hand, host genetic factor is a critical determinant. Among MHV researchers, it is well known that SJL mice are relatively resistant to intracerebral infection with JHMV (Stohlman & Frelinger 1978). In addition, the immune system is one of the key players that determine MHV infection in mice.

Table 1. Factors that affect MHV infection

Viral factor	Host factor
Viral strain	Strain of mice (Genetical background)
Viral dose	Age
Route of infection	Immune status

The Nidoviruses (Coronaviruses and Arteriviruses).
Edited by Ehud Lavi *et al.*. Kluwer Academic/Plenum Publishers. 2001.

1.1   Subacute Peritonitis in IFN-gamma Deficient B6 Mice after Intraperitoneal JHMV Infection

Comparing wild type B6 mice with genetically engineered IFN-gamma (IFN-g) deficient B6 mice (B6-GKO), we demonstrated that IFN-gamma was involved in viral clearance after intraperitoneal JHMV infection (Kyuwa *et al.* 1998). Although the virus was cleared within 10 days from the liver in wild-type B6 mice, it persisted for 50 days in the liver in B6-GKO mice. B6-GKO mice survived in the acute phase, but some began to die from 2 weeks after infection.

We therefore examined the pathological changes in JHMV-infected B6-GKO mice in the subacute phase. In all the mice, a pseudomembrane was observed on the surface of the livers and organs in the abdominal cavity had adhered to each other and to the peritoneum. Approximately a half of the mice had an accumulation of a viscous fluid in the abdominal and thoracic cavities. Microscopically, the disease was characterized by disseminated granulomatous inflammation and exudative fibrinous serositis with plasma cells and eosinophils. The histopathological examination revealed that the liver lesion in B6-GKO mice was not progressive. This form of disease was not expected but reminded us a feline infectious peritonitis, another coronavirus-induced disease in cats (Olsen 1993).

1.2   Acute Hepatic Failure in IFN-g Deficient BALB/c Mice after Intraperitoneal JHMV Infection

To see further the role of IFN-g in intraperitoneal JHMV infection in mice, IFN-g deficient mice with BALB/c background (BALB-GKO) as well as B6-GKO mice were inoculated intraperitoneally with 10^6 PFU of JHMV (Kyuwa *et al.* in preparation). B6-GKO mice survived in the acute phase but suffered subacute peritonitis as described above. In contrast to B6-GKO mice, all the BALB-GKO mice died before 7 days postinfection.

The viral titer in the liver, serum alanine aminotranferase (ALT) activity and histopathological changes in the liver from B6-GKO and BALB-GKO mice at 5 days postinfection were examined. The viral titer in BALB-GKO mice was 100-fold higher than that in B6-GKO mice. Similarly, the ALT activity in BALB-GKO mice was 27-fold higher than that in B6-GKO mice. Histopathologically, the lesion in B6-GKO mice was larger than that in wild-type B6 mice, but restricted. In contrast, all the hepatocytes in BALB-GKO mice appeared to be necrotic. These data strongly suggest that BALB-GKO mice died as a result of acute hepatic failure caused by viral infection.

2. JHMV REPLICATION *IN VITRO* SYSTEM

Although the viral replication rate in hepatocytes is the most vital element that determines the viral growth in the liver, other extrinsic factor such as the immune response also influences the viral titer. In fact, we previously reported that T cells played a critical role in viral clearance in the liver (Kyuwa *et al.* 1996). To examine viral replication in the absence of the immune system, we set up JHMV replication systems with the primary hepatocytes and peritoneal macrophages *in vitro*.

2.1 JHMV Replication in the Primary Hepatocytes

The primary hepatocyte culture was prepared from B6-GKO mice, BALB-GKO mice and their wild type counterparts (Arnheiter 1980), with a minor modification. They were cultured in the presence or absence of recombinant IFN-g in 24-well plates overnight. After washing three times, they were inoculated with JHMV at an m.o.i. 0.1. The culture supernatants were collected 12 and 24 hrs postinfection and the viral titer was determined by plaque assay on DBT cells.

In the absence of IFN-g, the viral titer of the supernatant from BALB-GKO mice was 100-fold higher than that from B6-GKO mice. That from wild type BALB/c mice was almost equivalent to that from BALB-GKO mice and was also significantly higher than that from wild type B6 mice. These results suggest that the viral replication rate in BALB/c hepatocytes is significantly higher than in B6 hepatocytes and that the phenomenon is not dependent on IFN-g.

The pre-treatment of IFN-g clearly inhibited viral replication in dose dependent fashion and 100 U/ml of IFN-g completely inhibited the viral replication in the primary hepatocytes from both BALB-GKO and B6-GKO mice.

2.2 JHMV Replication in Macrophages

Since a significant difference was observed in JHMV replication in the primary hepatocytes from BALB-GKO and B6-GKO mice, the replication of JHMV in macrophages, another MHV-susceptible cell type in the mouse, was examined *in vitro*. Peritoneal exudates cells were collected from BALB-GKO and B6-GKO mice, and cultured in 24-well plates overnight. Non-adherent cells were removed and the adherent cells were inoculated with JHMV at an m.o.i. 0.1. Culture supernatants were harvested at 12 and 24 hrs postinfection and their viral titers were determined by plaque assay on DBT cells.

In contrast to the case of the primary hepatocytes, JHMV growth in macrophages from BALB-GKO was nearly equal to that from B6-GKO mice.

3.   CONCLUSION

In this study, we demonstrated that unlike B6-GKO mice, BALB-GKO mice died within a week after intraperitoneal JHMV infection, due to acute hepatic failure. It is very interesting that JHMV induces a different type of disease in IFN-g deficient mice of different genetic backgrounds (Figure 1). Secondly, JHMV replication in the primary hepatocytes from BALB/c mice was significantly higher than that from B6 mice. Thirdly, JHMV replication in the primary hepatocytes was inhibited by IFN-g in dose-dependent manner. However, there was not a significant difference in JHMV replication in macrophages from both strains of mice. These results suggest that some intrinsic factors in hepatocytes are involved in the regulation of JHMV replication and a virus-induced disease in mice. The elucidation of the mechanism may offer the key to an understanding of the susceptibility to JHMV.

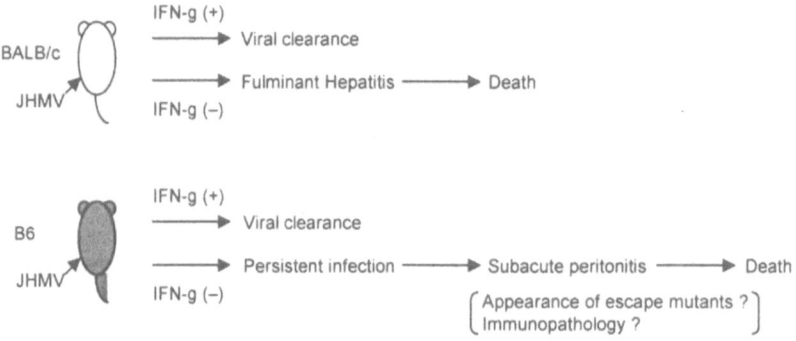

Figure 1. The clinical consequences of intraperitoneal JHMV infection in BALB-GKO mice, B6-GKO mice and their wild type counterparts. Whereas wild type BALB/c and B6 mice recover from a mild acute hepatitis, BALB-GKO and B6-GKO mice die after distinct consequences.

ACKNOWLEDGMENTS

This work was supported in part by a grant-in-aid for scientific research from Japan Society for the Promotion of Science.

REFERENCES

Arnheiter, H., 1980, Primary monolayer culture of adult mouse hepatocytes—a model for the study of hepatotropic viruses. *Arch. Virol.* **63**: 11-22.

Kyuwa, S., Machii, K., and Shibata, S., 1996, Role of CD4$^+$ and CD8$^+$ T cells in mouse hepatitis virus infection in mice. *Exp. Anim.* **45**: 81-83.

Kyuwa, S., Tagawa, Y., Shibata, S., Doi, K., Machii, K., and Iwakura, Y., 1998, Murine coronavirus-induced subacute fatal peritonitis in C57BL/6 mice deficient in gamma interferon. *J. Virol.* **72**: 9286-9290.

Olsen, C. W., 1993, A review of feline infectious peritonitis virus: molecular biology, immunopathogenesis, clinical aspects, and vaccination. *Vet Microbiol* **36**: 1-37.

Stohlman, S. A., and Frelinger, J. A., 1978, Resistance to fatal central nervous system disease by mouse hepatitis virus, strain JHM. I. Genetic analysis. *Immunogenetics* **6**: 277-281.

Susceptibility of Murine CNS to OC43 Infection

HÉLÈNE JACOMY AND PIERRE J. TALBOT

Laboratory of Neuroimmunovirology, Human Health Research Center, INRS-Institut Armand-Frappier, Université du Québec, 531 boulevard des Prairies, Laval, Québec, CANADA H7V 1B7

1.   INTRODUCTION

Multiple sclerosis (MS), one of the most common neurological diseases of young adults, is accompanied by inflammatory demyelination and loss of oligodendrocytes in the central nervous system (CNS). Although its etiology remains unclear, a generally accepted hypothesis is that virus infections could initiate a CNS-directed immune process in a genetically predisposed host (Oldstone 1997). Amongst various animal models of virus-induced demyelination, studies on murine hepatitis virus (MHV) have revealed that this virus is capable of causing direct oligodendrocyte cytopathology, but may also elicit a variety of immunopathological responses (Lane and Buchmeier 1997). Given that MHV causes MS-like CNS demyelination, the related human coronaviruses (HCoV) represent a logical target of investigation as a potential microbial agent involved in MS pathogenesis.

HCoV are respiratory pathogens responsible for 10 to 35% of common colds (McIntosh 1990). They have occasionally been associated with other pathologies such as pneumonia or meningitis (Riski and Hovi 1980). We have reported their ability to replicate and persist in human brain cells (Bonavia *et al*. 1997; Arbour *et al*. 1999a, b) and even human brains (Arbour *et al*. 2000). Such neurotropic and neuroinvasive properties and analogies with MHV-induced MS-like disease in mice and rats have stimulated research on the possible implication of coronaviruses in MS.

The goal of the present study was to develop and characterize an experimental mouse model of HCoV-associated neuropathology. We first

The Nidoviruses (Coronaviruses and Arteriviruses).
Edited by Ehud Lavi *et al*., Kluwer Academic/Plenum Publishers, 2001.

investigated *in vitro* the ability of primary cultures of murine cells to be infected by the two known viral serotypes (229E and OC43). We then studied the infection of mouse CNS by HCoV-OC43.

2.   MATERIALS AND METHODS

2.1   Virus and cell culture

The 229E and OC43 strains of HCoV were originally obtained from ATCC, plaque-purified and grown on the human embryonic lung cell line L132 or the human rectal carcinoma cell line HRT-18, respectively. Virus stocks with a titer of about 5×10^6 TCID$_{50}$/mL were kept at -80°C. Primary neural cell cultures were obtained by dissecting out the cortical brain of 1-day-old mice. A cell suspension was produced by passing the tissue several times through a syringe fitted with a 2mm gauge needle, then filtered on a 82 μm mesh. Cells were grown as monolayers at 37°C, in a humidified atmosphere containing 5% (v/v) CO_2, in DMEM supplemented with 10% (v/v) heat-inactivated fetal calf serum.

After 10 days of growth, three series of four Petri dishes, containing approximately 1.5×10^6 cells each, were infected by a 2-hour contact with a dilution of HCoV-229E or HCoV-OC43 virus stock to yield a MOI of 1. Infectious viral particle production was monitored for 8 weeks after infection. During this period, the entire volume of medium covering the cells was collected at weekly intervals and placed at -80°C until infectious virus titers could be determined. After each sampling, cell cultures were extensively washed in sterile PBS before adding new DMEM.

Infectious virus titers were determined by an indirect immunoperoxidase assay on susceptible cells (Arbour *et al.* 1999a). Simple and double immunofluorescence labeling identified cell types and which of them were infected by HCoV. A rabbit-anti-glial-fibrillary acidic protein antibody (GFAP, DAKO) was used to identify astrocytes and an ascites fluid from rat anti-MACII (ATCC) used to stain microglia/macrophages. For detection of viral antigens, ascites fluids containing MAbs directed to the N protein of HCoV-OC43 or -229E were used (Bonavia *et al.* 1997).

2.2   Animal model

Based on the results obtained in cell cultures, we only inoculated mice with HCoV-OC43. BALB/c mice were inoculated intracerebrally with 10 μl of virus at 10^3 TCID$_{50}$/mL at the age of 8-days post-natal (P8). Infected sucklings were replaced with their mother until experimentation. Littermates of the same age were injected with 10 µl of PBS to serve as control animals.

2.2.1 Immunohistochemistry and electron microscopy

Mice were perfused by intraventricular injection of 4% paraformaldehyde under deep ketamine-xylazine anesthesia. Serial 50 µm-thick sections of brain or spinal cord tissue blocks were incubated overnight in anti-OC43 MAb (1/1000), then rinsed and processed in Vectastain ABC Kit (Vector Laboratories, Burlingame, CA, USA). Labeling was revealed with 0.03% 3,3'-diaminobenzidine tetrahydrochloride (Sigma) and 0.01% H_2O_2. Samples for electron microscopy were postfixed for 2 hours with 2% osmium tetraoxide in 0.1M phosphate buffer, dehydrated in graded ethanol series, and embedded in Epon.

2.2.2 Infectious Tests

At 3, 4, 6, 8 and 10 days post-infection, four animals were sacrificed. Brain and spinal cords were dissected out homogenized in sterile PBS and tissue passed through a 0.22 µm filter-fitted syringe. The filtered extracts were processed for the presence of infectious virus.

2.2.3 Western Blot Analyses

Tissues were homogenized in SUB buffer, as previously described (Jacomy *et al.* 1999). Samples (5 µg total proteins) were fractionated on a 7.5% polyacrylamide gel and either visualized by Coomassie Blue staining or transferred to nitrocellulose for Western blot analysis. Membranes were incubated for 2 hours with the OC43-specific antibody. After several rinses with TS buffer containing 0.05% (v/v) Tween 20, membranes were incubated for one hour with peroxidase-conjugated anti-mouse IgG (1/1000, DAKO). Bands were visualized using a Western blot chemoluminescent kit (Super Signal, PIERCE, Rockford, MD, USA).

2.2.4 Preparation of RNA and RT-PCR

CNS tissues were dissected out at different time points post-infection. Total RNA was extracted by homogenization in Trizol (GibcoBRL, Burlington, CA, USA). For RT-PCR, one pair of HCoV-OC43 primers was designed to amplify a region containing 920 nucleotides (primers O1 and O9) of the gene coding for the N protein (Arbour *et al.* 1999b).

Approximately 5 µg of RNA were reverse transcribed and the product was added to a PCR mix followed by 30 amplification cycles of 2 min at 72°C, 1 min at 95°C and 2 min at 60°C. Ten µl of reaction product was loaded onto a 1.2% (w/v) agarose gel containing 5 µl (v/v) ethidium bromide.

3.   RESULTS AND DISCUSSION

3.1   Cell Cultures

Using double immunofluorescence detection, we showed numerous GFAP-positive astrocytes also staining for HCoV-OC43 antigens. Some of the MAC-II-stained microglia/macrophages also contained viral antigens. These results demonstrate that both murine microglia/macrophages and astrocytes were susceptible to an infection by HCoV-OC43. Moreover, measurements of infectious virus titers in cell culture supernatants demonstrated that glial cell cultures produced infectious viral particles and were therefore productively infected. At 4 days post-infection, titers reached about 10^7 $TCID_{50}$/ml (Table 1). Relatively stable amounts of infectious virions were detected at all time points. Titers of about 10^5 $TCID_{50}$/mL observed after 8 weeks are consistent with a persistent infection of primary murine CNS glial cell cultures by HCoV-OC43.

Table 1: Yield of infectious virions from persistent HCoV-OC43 infections of primary glial cell cultures. The highest level of infectious virus titer was obtained at 4 days post-infection. Glial cells produced infectious virions at least until 8 weeks.

Weeks post-infection	Yield of infectious virions ($\log_{10}$ $TCID_{50}$/mL)
0.5 (~ 4 days)	7.02 (± 0.90)
2	5.49 (±0.86)
4	5.02 (±0.03)
6	4.25 (±0.00)
8	5.13 (±0.63)

Pearson and Mims (1985) reported that only neuronal cell types were able to produce infectious HCoV-OC43 virions, and that astrocytic cells could produce viral antigens but not infectious virus. Our culture conditions avoided the possibility of neurons playing a role in the production of infectious viral particles. Indeed, no neurons were able to live in the culture conditioned medium used and no neurons could survive after 2 months of culture. Thus, this is the first report of a productive and persistent HCoV-OC43 infection of murine astrocytes and microglial/macrophages.

No infection could be detected with HCoV-229E, probably due to the absence on murine cell membranes of the aminopeptidase-N (CD13) molecule, which is necessary for infection (Lachance *et al.*, 1998).

3.2   Mouse model

Empiric experimentation was used to obtain a sub-lethal infection by HCoV-OC43 in the hope of establishment of life-long persistence. Given a previous report on the age-dependent resistance of mice to HCoV-OC43 infection (Pearson and Mims, 1983), intracerebral inoculations were performed on mice at various post-natal ages, using 10 μl of various dilutions of the virus stock, to determine the 50% lethal dose (LD_{50}). BALB/c mice were selected in view of their relative susceptibility to both respiratory and enteric strains of MHV. Intracerebral inoculations were used to favor a CNS infection. Finally, we determined the optimal experimental conditions to be as follows: 10 μl of a diluted virus solution at about 100 $TCID_{50}$ inoculated intracerebrally at 8 days post-natal. In these conditions, about 50% of the pups died within the first 8 days post-infection. All inoculated mice developed signs of acute disease characterized by apathy, hunched posture, ruffled fur and tremor, comparable to pathological signs described after MHV infection (Kristensson *et al.* 1986). Mice which survived after 8 days showed clinical remission and have remained well for 10 months.

Viral infections initiated by intracerebral inoculations were quickly disseminated. Indeed, cells positive for viral antigens were detected throughout the CNS. At the electron microscopic level, virions were observed in the spinal cord 8 days after injection in the brain. These particles were mostly localized in the cell cytoplasm, closely associated with the Golgi apparatus, or in extracellular spaces, between dendrites and the axonal endings. This reflects a replication of the virus in the cytoplasm of the cells and probably cell-to-cell transport of the virions, as was described after MHV infection (Lavi *et al.*, 1988; Sun and Perlman, 1995).

Infectious virus could be isolated from the CNS between 2 and 8 days post-infection. The highest level of infectious virions was obtained at 6 days post-infection; this titer could reach about 10^7 $TCID_{50}$/g for the brain and about 10^5 $TCID_{50}$/g for the spinal cord extracts. These high CNS virus titers were indicative of a generalized infection during the first week post-inoculation. No infectious virions could be detected starting at 10 days post-infection, as was also reported with MHV (Woyciechowska *et al.*, 1984).

Viral proteins were detected in the brain and the spinal cord between 4 to 8 days post-infection. Coomassie blue-stained gels of protein extracts revealed the presence of new protein bands in infected compared to control animals, indicative of viral protein synthesis (Fig. 1A). Indeed, proteins transferred onto nitrocellulose membranes and stained with antibodies against the N protein of HCoV-OC43 recognized these new bands in all the isolates of virus-infected animals, without staining any proteins in control animals (Fig. 1B). Moreover, Western blot analysis revealed a second N protein band in infected mice, compared to the single N protein band of the viral stock solution. This second band probably represents a phosphorylated form of the N protein which may account for high viral synthesis in infected mice. Viral proteins were found in 90% of mice at 4 days after infection, but only in 22% of the animals at 8 days. Viral proteins were not detectable after 10 days.

Having demonstrated persistent infection in murine CNS cell cultures and a generalized infection of CNS during the first 8 days post-infection, we wanted to detect a possible persistent infection in the CNS of infected mice. By RT-PCR analyses, viral RNA could be found in some animals at least until 5 months post-infection (Fig. 1C).

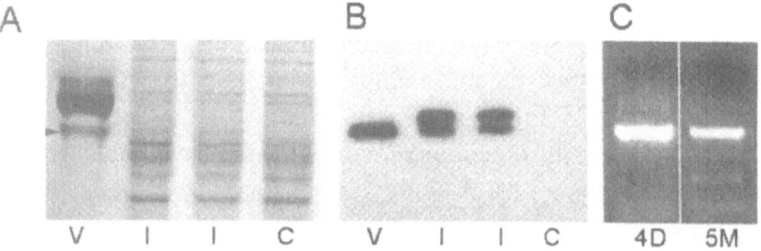

Figure 1: HCoV-OC43 nucleocapsid protein and RNA in mouse brain. (A) Coomassie blue stained gel; I: infected mice and C: control animal. (B) Western blot confirmation of the presence of N protein in infected mouse brain. (C) RT-PCR: viral RNA was detected at 4 days (4D) but also 5 months (5M) post-infection. V: virion controls.

4. CONCLUSIONS

These results demonstrate that the CNS of BALB/c mice is susceptible to an acute and persistent HCoV-OC43 infection, with viral spread and replication, including a productive infection of astrocytes and macrophages/microglial. Viral persistence observed in the CNS may play a role in the development of chronic pathologies, as it was suggested for MHV infection. It will be interesting to study demyelination and immunopathology in this new animal model.

ACKNOWLEDGMENTS

This work was supported by a grant from the Medical Research Council of Canada to PJT and a fellowship from the *Fondation Armand-Frappier* to HJ. We thank Tina Miletti for critically reading the manuscript.

REFERENCES

Arbour, N., Ekandé, S., Côté, G., Lachance, C., Chagnon, F., Tardieu, M. Cashman, N.R., and Talbot, P.J., 1999a, Persistent infection of human oligodendrocytic and neuroglial cell lines by human coronavirus 229E, *J. Virol.* **73** :3326-3337.

Arbour, N., Côté, G., Lachance, C. Tardieu, M. Cashman, N.R. and Talbot, P.J., 1999b, Acute and persistent infection of human neural cell lines by human coronavirus OC43, *J. Virol.* **73**:3338-3350.

Arbour, N., Day, R., Newcombe, J., and Talbot, P.J., 2000, Neuroinvasion by human respiratory coronaviruses, *J. Virol.,* in press.

Bonavia, A., Arbour, N., Yong, V.W., and Talbot, P.J., 1997, Infection of primary cultures of human neural cells by human coronavirus 229E and OC43, *J. Virol.* **71**:800-806.

Kristensson, K., Holmes, K.V., Duchala, C.S., Zeller, N.K., Lazzarini, R., and Dubois-Dalcq, M., 1986, Increased levels of myelin basic protein transcripts in virus-induced demyelination, *Nature* **322**:544-547.

Jacomy, H., Zhu, Q., Couillard-Desprès, S., Beaulieu, M., and Julien, J.-P., 1999, Disruption of type IV intermediate filament network in mice lacking medium and heavy subunits, *J. Neurochem.* **73**:972-984.

Lachance, C., Arbour, N., Cashman, N.R., and Talbot, P.J., 1998, Involvement of aminopeptidase N (CD13) in infection of human Neural cells by human coronavirus 229E, *J. Virol.* **72**:6511-6519.

Lane, T.E., and Buchmeier, M.J., 1997, Murine coronavirus infection : a paradigm for virus-induced demyelinating disease, *Trends Microbiol.* **5**:9-14.

Lavi, E., Fishman P.S, Highkin M.K., and Weiss S.R. (1988) : Limbic encephalitis after inhalation of murine coronavirus, *Lab. Invest.* **58**:31-36.

McIntosh K., 1990, In "Virology, 2nd edn.", B.N. Fields, D.M. Knipe *et al.,* eds., p. 857, Raven Press, New York.

Oldstone, M.B.A., 1997, Viruses and autoimmune diseases, *Scand. J. Immunol.* **46**:320-325

Pearson J., and Mims C.A, 1983, Selective vulnerability of neural cells and age-related susceptibility to OC4 virus in mice, *Arch. Virol.* **77**:109-118.

Pearson, J., and Mims, C.A., 1985, Differential susceptibility of cultured neural cells to human coronavirus OC43, *J. Virol.* **53**:1016-1019.

Riski H., and Hovi, T., 1980, Coronavirus infections of man associated with diseases other than the common cold, *J. Med. Virol.* **6**:259-265.

Sun, N., and Perlman, S., 1995, Spread of neurotropic coronavirus to spinal cord white matter via neurons and astrocytes, *J. Virol.* **69**:633-641.

Woyciechowska, J., Trapp, B.; Patrick, D., Shekarchi, I.; Leinikki, P., and Server, J., 1984, Acute and subacute demyelination induced by mouse hepatitis virus stain A59 in C3H mice, *J. Exp. Neurol.* **4**:295-306.

Caspase Inhibitors Block MHV-3 Induced Apoptosis and Enhance Viral Replication and Pathogenicity

JULIAN L. LEIBOWITZ AND ELENA BELYAVSKAYA
Department of Pathology and Laboratory Medicine, Texas A&M University System Health Science Center, College Station, Texas 77843-1114

1. INTRODUCTION

Infection with mouse hepatitis virus strain 3 (MHV-3) results in lethal hepatitis in BALB/c mice, compared to the minimal disease observed in A/J mice (Levy, Leibowitz, and Edgington, 1981). The response of the macrophage to MHV infection is a key determinant of the outcome of MHV-induced hepatitis (Levy, Leibowitz, and Edgington, 1981; Shif and Bang, 1969). We have previously shown that infection of A/J macrophages with MHV-3 triggers apoptosis in most of the infected cells (Belyavskyi *et al.*, 1998). In contrast, infection of BALB/c macrophages results in only a small percentage of the cells undergoing apoptosis. In this report we have utilized caspase inhibitors to investigate the hypothesis that the rapid induction of apoptosis in macrophages derived from A/J mice limits MHV replication and hepatic injury during MHV infection.

The Nidoviruses (Coronaviruses and Arteriviruses).
Edited by Ehud Lavi *et al.*, Kluwer Academic/Plenum Publishers, 2001.

2.   MATERIALS AND METHODS

2.1   Virus and Cells

Stocks of MHV-3 were grown as described (Levy, Leibowitz, and Edgington, 1981). Peritoneal macrophages were harvested from 8-10 week old mice and put into culture and infected with MHV-3 (MOI=1) as described (Belyavskyi *et al.*, 1998). At various times post infection (p.i.) cells were fixed for TUNEL staining, frozen for viral titration, or harvested for DNA fragmentation assays.

2.2   TUNEL assay

Macrophages from A/J mice were plated on Lab-Tek Chamber Slides (3.3×10^5 cells/well), mock or MHV-3 -infected, and fixed at various times after infection. TUNEL assays were performed as described (Belyavskyi *et al.*, 1998).

2.3   DNA laddering

At 12 h p.i. DNA was extracted from 3.2×10^6 mock-infected and MHV-3 infected macrophages as described (Hsu *et al.*, 1999). Samples were analyzed agarose gel electrophoresis and Syber Gold (Molecular Probes) staining.

2.4   Caspase Inhibitors

Caspase inhibitors were purchased from Calbiochem and Enzyme Research. Stock solutions of inhibitors were prepared in 20 mM DMSO in PBS.

2.5   Mice

Strain A/J mice were obtained from Jackson Labs and allowed to rest for at least 3 days following shipping. For *in vivo* experiments, 6 week old mice were injected intraperitoneally (IP) with either 333 μg of the pan-caspase inhibitor Z-VAD-FMK or with vehicle alone. Fours hours later both groups of mice were infected with 10^6 PFU of MHV-3 by IP injection. Mice in the

Z-VAD-FMK treated group were given also given 333 µg Z-VAD-FMK at 24 and 48 h p.i. Mice in the control group received 20 mM DMSO in PBS at these times. Groups of 3 mice were sacrificed at 1,2 and 3 days p.i. and portions of their livers were prepared for histological examination and virus titration.

3. RESULTS

To determine if the induction of apoptosis during MHV-3 infection of A/J macrophages limits virus replication and spread, we employed the caspase inhibitor Ac-YVAD-FAOM. This inhibitor is relatively specific for caspase-1, although it also inhibits caspase-4 with lesser effectiveness (Margolin *et al.*, 1997). Ac-YVAD-FAOM reduced the number of TUNEL positive macrophages; 80%-90% inhibition of apoptosis was achieved at 200 µg/ml (Figure 1). Cultures treated with the caspase inhibitor underwent extensive cell fusion, an event not observed in MHV-infected A/J macrophages in the absence of inhibitor. A different inhibitor of caspase-1, Z-YVAD-FMK, also inhibited MHV induced apoptosis as did the caspase-3 inhibitor Z-DEVD-FMK (not shown). Treatment of MHV-3 infected cultures with Ac-YVAD-FAOM, resulted in a 35-fold increase in viral yield (Table 1) and approximated levels reached during MHV-3 infection of BALB/c macrophages. Thus blocking the induction of apoptosis with a caspase inhibitor appeared to allow MHV-3 to replicate to similar levels in A/J derived macrophages as that observed during infection of BALB/c macrophages.

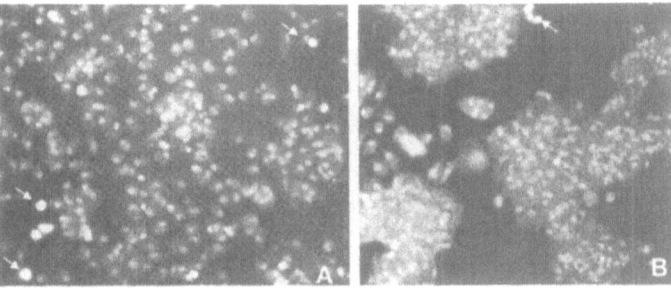

Figure 1. Effect of AC-YVAD-FAOM on MHV-3 infection. A/J macrophages were infected with MHV-3 and incubated with 200 µg/ml of drug (B) or with vehicle (A) until 10 h p.i. when they were stained by the TUNEL method. Arrows indicate brightly fluorescent TUNEL positive cells.

Table 1. Enhancement of MHV-3 replication by Ac-YVAD-FAOM†

Time (p.i.)	Control	100µg/ml Inhibitor	200µg/ml Inhibitor
8h	5.0×10^3	1.0×10^4	2.0×10^6
10h	2.0×10^4	1.0×10^5	2.2×10^5
12h	1.0×10^5	2.0×10^6	3.5×10^6

†Results are shown as pfu/ml. Caspase inhibitor was added to culture following infection with MHV-3 and was present in the cultures for the entire infection.

To explore the role of apoptosis in restricting MHV-3 infection *in vivo*, we administered the pan-caspase inhibitor Z-VAD-FMK to A/J mice, and infected the mice with MHV-3. Mice were sacrificed at 1, 2, and 3 days post infection. Treatment with Z-FAD-FMK lead to a 3-fold increase in virus present in the liver relative to control mice as early as day1, a 4.5-fold increase by day 2, and a 9-fold increase by day 3 (Table 2). Mice treated with caspase inhibitor also had more extensive pathologic changes. At day 3 (Figure 2) livers from Z-VAD-FMK treated mice contained many easily identified foci of necrotic hepatocytes. Livers from mice which did not receive Z-VAD-FMK contained rare very small foci of necrotic hepatocytes, consistent with the hypothesis that rapid apoptosis may contribute to the resistance of A/J mice to fulminant MHV-3 induced hepatitis.

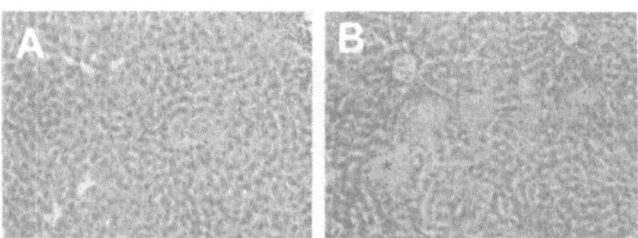

Figure 2. The effect of Z-VAD-FMK on MHV-3 infection. A/J mice were treated with either Z-VAD-FMK (B) or vehicle (A) for 3 days, sacrificed and histologic sections prepared. The asterisk indicates a focus of hepatocellular necrosis.

Table 2. In vivo enhancement of MHV-3 replication by Z-VAD-FMK

Treatment	Virus Recovered†		
	1 Day P.I.	2 Days P.I.	3 Days P.I.
Vehicle Alone	1×10^4	1.3×10^5	1.2×10^6
Z-VAD-FMK	3×10^4	6.0×10^5	1.0×10^7

†Results are shown as pfu/gram of liver.

4.   DISCUSSION

Destruction of virus-infected cells by programmed cell death is an important mechanism for limiting viral replication and spread (Teodoro and Branton, 1997). However, this is not uniformly the case. Inhibition of apoptosis during infection with Sindbis decreased viral yields (Levine *et al.*, 1993). We have shown that inhibition of apoptosis during MHV-3 infection of A/J macrophages increased yields and enhanced spread of the infection to adjacent cells by cell fusion. This contrasts with previously reported results in fibroblasts (An *et al.*, 1999). In these cells MHV triggered apoptosis occurred late in infection, after peak viral titers had been achieved. Thus inhibiting apoptosis had little effect on virus yield. During MHV infection of A/J macrophages apoptosis can be detected by 8 h p.i. (Belyavskyi *et al.*, 1998), long before peak viral yields are obtained. The difference in timing between the onset of apoptosis and peak viral yields in these cell types likely accounts for the different results in these studies.

The pathways by which MHV infection triggers apoptosis are currently unknown. Our data implicates capases-1 and -3. Caspase-3 is activated during the execution phase of apoptosis, thus it is not surprising that inhibitors of caspase-3 (An *et al.*, 1999) block MHV triggered apoptosis. Caspase-1 takes part in apoptosis triggered via the TNFR family of receptors (Thome *et al.*, 1998), suggesting that these pathways play a role in MHV triggered apoptosis.

The induction of apoptosis in macrophages derived from strain A/J mice contributes to resistance of this strain of mice to MHV-3 induced hepatitis. Although the extent of viral replication in A/J mice was considerable, the titer achieved was less and the kinetics slower than in susceptible BALB/c mice (Dindzans *et al.*, 1985). Administration of Z-VAD-FMK to MHV-3 infected A/J mice increased hepatic viral loads to levels comparable to those observed in BALB/c mice and increased the number and size of lesions. Virus spread from infected macrophages and Kupffer cells to hepatocytes may be limited in A/J mice by the failure of infected macrophages to form syncytia, as well as by the relatively modest amount of virus produced in these cells. The induction of apoptosis in macrophages may have an additional protective effect. BALB/c macrophages elaborate fgl2 prothrombinase and TNF-α in response to MHV-3 infection (Pope *et al.*, 1995) which contribute to hepatic injury in this model. This suggests the possibility that in A/J macrophages, the rapid development of apoptosis during MHV-3 infection prevents the expression of the fgl2 prothrombinase protein which results in fibrin deposition and hepatic necrosis in BALB/c mice (MacPhee *et al.*, 1985). Our data is consistent with this model, and thus apoptosis could represent a mechanism of eliminating cells expressing a host gene which is potentially harmful to the survival of the organism.

ACKNOWLEDGMENTS

This work was supported in part by NIH grant AI 43368.

REFERENCES

An, S., Chen, C.-J., Yu, X., Leibowitz, J. L., and Makino, S. (1999). Induction of apoptosis in murine coronavirus-infected cultured cells and demonstration of E protein as an apoptosis inducer. *J. Virol.* **73**, 7853-7859.

Belyavskyi, M., Belyavskaya, E., Levy, G. A., and Leibowitz, J. L. (1998). Coronavirus MHV-3 induced apoptosis in macrophages. *Virology* **250**, 41-49.

Dindzans, V. J., MacPhee, P. J., Fung, L. S., Leibowitz, J. L., and Levy, G. A. (1985). The immune response to mouse hepatitis virus: Expression of monocyte procoagulant activity and plasminogen activator during infection in vivo. *J. Immunol.* **135**, 4189-4197.

Hsu, S. L., Yin, S. C., Liu, M. C., Reichert, U., and Ho, W. L. (1999). Involvement of cyclin-dependent kinase activities in CD437-induced apoptosis. *Exp Cell Res* **252**(2), 332-41.

Levine, B., Huang, Q., Issacs, J., Reed, J., Griffin, D., and Hardwick, J. (1993). Conversion of lytic to persistent alphavirus infection by the bcl-2 cellular oncogene. *Nature* **361**, 739-742.

Levy, G. A., Leibowitz, J. L., and Edgington, T. S. (1981). Induction of monocyte procoagulant activity by murine hepatitis virus type 3 parallels disease susceptibility in mice. *J. Exp. Med.* **154**, 1150-1163.

MacPhee, P. J., Dindzans, V. J., Fung, L. S., and Levy, G. A. (1985). Acute and chronic changes in the microcirculation of the liver in inbred strains of mice following infection with mouse hepatitis virus type 3. *Hepatology* **5**, 649-660.

Margolin, N., Raybuck, S. A., Wilson, K. P., Chen, W., Fox, T., Gu, Y., and Livingston, D. J. (1997). Substrate and inhibitor specificity of interleukin-1 beta-converting enzyme and related caspases. *J Biol Chem* **272**(11), 7223-8.

Pope, M., Rotstein, O., Sinclair, S., Parr, R., Cruz, B., Fingerote, R., Chung, S., Gorczynski, R., Fung, L., Leibowitz, J., Rao, Y. S., and Levy, G. (1995). Pattern of disease after murine hepatitis virus strain 3 infection correlates with macrophage activation and not viral replication. *J. Virol.* **69**, 5252-5260.

Shif, I., and Bang, F. (1969). In vitro interaction of mouse hepatitus virus and macrophages from genetically resistant mice. *J. Exp. Med.* **131**, 851-862.

Teodoro, J. G., and Branton, P. E. (1997). Regulation of apoptosis by viral gene products. *J Virol* **71**(3), 1739-46.

Thome, M., Hofmann, K., Burns, K., Martinono, F., Bodmer, J.-L., Mattmann, C., and Tschopp, J. (1998). Identification of CARDIAK, a RIP-like kinase that associates with caspase-1. *Current Biology* **8**, 885-888.

MHV Neuropathogenesis: The Study of Chimeric S Genes and Mutations in the Hypervariable Region

JOANNA J. PHILLIPS AND SUSAN R. WEISS
Department of Microbiology, University of Pennsylvania School of Medicine, Philadelphia, PA 19104-6076.

1. INTRODUCTION

Mouse hepatitis virus (MHV) is studied as a model system for both acute and chronic virus-induced neurologic disease. Infection of susceptible mice with MHV results in a potentially fatal acute encephalomyelitis, and survivors of this acute disease often go on to develop a chronic demyelinating disease. Using targeted recombination we have demonstrated that the MHV spike (S) glycoprotein is a major determinant of the severity of the acute disease (Phillips et al., 1999). Isogenic recombinant viruses differing exclusively in the spike gene were found to differ dramatically in their neurovirulence. Recombinants containing the spike of the highly neurovirulent MHV-4 virus were highly neurovirulent while recombinants containing the spike of the mildly neurovirulent MHV-A59 virus exhibited a similarly mild degree of neurovirulence.

The MHV spike, found on the virion envelope and on the plasma membrane of infected cells, plays a major role in viral entry and the immune response to infection. The S protein can be divided both structurally and functionally into two subunits, S1 and S2. S1, thought to form the globular head of the spike, is responsible for binding to the viral receptor, and S2, thought to form the stalk of the spike, mediates membrane fusion. Within the S1 subunit there is a region termed the

The Nidoviruses (Coronaviruses and Arteriviruses).
Edited by Ehud Lavi *et al.*, Kluwer Academic/Plenum Publishers, 2001.

hypervariable region (HVR) that exhibits a great deal of diversity among viral strains. A number of studies have associated mutations or deletion in the HVR with alterations in pathogenesis, however, the specific role of this region in pathogenesis is not well understood (Dalziel et al., 1986, Fazakerley et al., 1992, Fleming et al., 1986, Parker et al., 1989, Wege et al., 1988).

Despite the dramatic difference in virulence conferred by the MHV-4 and MHV-A59 spike proteins they are highly homologous. Overall they share more than 92% amino acid identity. The most striking difference between the two spikes is that the MHV-A59 spike contains a deletion of 52 amino acids in the HVR relative to MHV-4. To examine the role of various regions of the MHV-4 spike gene in neurovirulence, we generated a series of recombinnt viruses containing exchanges between the MHV-4 and MHV-A59 spike genes. Using these recombinant viruses containing chimeric spike genes we were able to address the importance of specific regions of the spike in neuroviulence and viral replication in the brain.

2.   RECOMBINANTS WITH CHIMERIC MHV-4/MHV-A59 SPIKE GENES

To determine if either the S1 or the S2 subunit of the MHV-4 S protein could confer an increase in neurovirulence we generated recombinant viruses that contained exchanges between the MHV-4 and MHV-A59 subunits. Chimeric spikes were constructed by introducing a silent mutation creating an EcoRV restriction site at codon 775 (MHV-4 sequence). Using targeted recombination we then generated recombinant viruses containing either the S1 of MHV-4 and the S2 of MHV-A59, named $S1_4R70$ or R71, or the S1 of MHV-A59 and the S2 of MHV-4, named $S2_4R81$ and R82 (Kuo et al., 2000). The members of each pair of recombinants exhibited similar phenotypes in vitro and in vivo, and the data from only one member of each pair is shown. As controls we also generated recombinants containing the MHV-4 spike gene, S_4R29, and the MHV-A59 spike gene, $S_{A59}R16$. To verify that the selected recombinants could replicate efficiently in cell culture we performed growth curves on L2 cell monolayers. The virus titers (16 hours post infection) shown in Table 1, demonstrate that all of the recombinants exhibited efficient replication. To examine the virulence of these recombinants we determined intracranial $LD_{50}s$ in C57Bl/6 mice. The results are shown in Table 1. As expected, S_4R29 was extremely neurovirulent and $S_{A59}R16$ was only mildly neurovirulent. The recombinants with chimeric spike genes, however, were highly attenuated for virulence. To determine if the decrease in virulence was attributable to decreased efficiency of replication in the brain, we examined the virus titers in the brain following intracranial inoculation with 10 PFU of virus. The virus titers at the peak of virus replication, day 5 post infection, are shown in Table 1. Both the $S1_4$ and the $S2_4$ recombinants exhibited ineffficient replication in the brain. Thus despite their ability to function well in cell culture, it appeared that recombinants with chimeric spike genes between MHV-4 and MHV-A59 replicated inefficiently in vivo. Moreover, this suggested that in vivo homotypic S1/S2 interactions are required for efficient replication and high neuroviulence.

Table 1. In vivo and in vitro properties of recombinant viruses containing exchanges between the S1 and S2 subunits of MHV-4 and MHV-A59.

Recombinant virus	Log (LD_{50})	Peak Brain Virus Titers (Log(PFU/g))[a]	Virus Replication in Cell Culture (Log(PFU/mL))
S_4R29	0.6	4.8	4.0
$S1_4R70$	>4.7	3.2	5.3
$S2_4R81$	>3.9	2.5	5.0
$S_{A59}R16$	3.4	4.9	6.4

[a] Virus replication on day 5 post intracranial inoculation, n= 6 animals.

[b] Virus replication on L2 cell monolayers at 16 hours post infection, MOI 1.8 PFU/cell .

3.   RECOMBINANTS WITH ALTERATIONS IN THE HYPERVARIABLE REGION

Various mutations and deletions in the MHV hypervariable region have been associated with neuroattenuation. To specifically examine the role of the HVR in pathogenesis and neurovirulences we generated a series of recombinant viruses. To determine if the MHV-4 HVR was necessary to confer an increase in neurovirulence, we deleted the MHV-4 HVR. Based on a previously identified neutralizing monoclonal antibody escape mutant (Parker et al., 1989), we generated $S_4\Delta HVR160$ and R161, which contained a 142 amino acid deletion in the HVR ($\Delta434-575$). In addition, we replaced a segment of the MHV-4 HVR (488-600) with the corresponding region from the MHV-A59 HVR (488-548), S_4HV-A59R131 and R133. To determine if the MHV-4 HVR was sufficient to alter the neurovirulence of a recombinant virus, we generated $S_{A59}HV$-4R51 and R52 in which a region of the MHV-A59 HVR (412-562) was replaced with the corresponding region from the MHV-4 HVR (412-614). Each pair of viruses exhibited similar properties in vitro and in vivo and the data is presented for only one member of each pair. As shown in

Table 2, all of the recombinants exhibited efficient replication in cell culture. When we examined the intracranial virulence of these recombinants, however, we found that alterations in the HVR had a profound affect on pathogenesis. Deletion or replacement of the MHV-4 HVR resulted in a dramatic decrease in virulence as demonstrated by $S_4\Delta HVR160$ and $S_4HV-A59R131$. The MHV-4 HVR was not sufficient, however, to confer an increase in neurovirulence as demonstrated by the similar virulence of $S_{A59}HV-4R51$ and $S_{A59}R16$. Furthermore, these data suggest that the MHV-A59 spike can tolerate manipulations within the HVR with little affect on neurovirulence.

Table 2. In vivo and in vitro properties of recombinant viruses with alterations in the spike hypervariable region.

Recombinant Virus	$Log(LD_{50})$	Peak Brain Virus Titers $(Log(PFU/g))$ [a]	Virus Replication in Cell Culture $(Log(PFU/mL))$ [b]
S_4R29	0.6	4.8	4.8
$S_4\Delta HVR160$	>4.1	2.3	4.8
$S_4HV-A59R131$	>4.0	2.4	5.8
$S_{A59}HV-4R51$	3.7	3.4	7.0
$S_{A59}R16$	3.4	4.9	6.7

[a] Virus replication on day 5 post intracranial inoculation, n=4-6 animals.

[b] Virus replication on L2 cell monolayers at 16 hours post infection. MOI 0.4 PFU/cell.

4.   CONCLUSION

The MHV spike is a major determinant of neurovirulence. Using targeted RNA recombination, we have now begun to examine the role of specific regions of the spike in neuropathogenesis. From our studies with recombinants containing exchanges between the MHV-4 and MHV-A59 spike it appears that multiple regions of the MHV-4 spike are required to confer increased neurovirulence. Despite efficient replication in vitro the recombinants containing chimeric spike genes exhibited attenuated replication and virulence in vivo suggesting that homotypic S1/S2 interactions are required for an efficient in vivo infection. Furthermore, alterations in the HVR revealed that the MHV-4 hypervariable region was necessary but not sufficient to confer an increase in neurovirulence.

ACKNOWLEDGMENTS

This work was supported by Public Health Service grants NS-30606 and NS-21954 as well as grant RG-2585 from the National Multiple Sclerosis Society. J.J.P. was supported in part by training grant GM-07229. We would also like to thank Paul S. Masters, Lili Kuo, and Peter J.M. Rottier for providing fMHV, pMH54, nd FCWF cells, Jean Tsai for kindly providing $S_{A59}R16$, and Su-hun Seo for technical assistance.

REFERENCES

Dalziel, R. G., P. W. Lampert, P. J. Talbot, and M. J. Buchmeier. 1986. Site-specific alteration of murine hepatitis virus type 4 peplomer glycoprotein E2 results in reduced neurovirulence. J. Virol. 59:463-471.

Fazakerley, J. K., S. E. Parker, F. Bloom, and M. J. Buchmeier. 1992. The V5A13.1 envelope glycoprotein deletion mutant of mouse hepatitis virus type-4 is neuroattenuated by its reduced rate of spread in the central nervous system. Virol. 187:178-188.

Fleming, J. O., M. D. Trousdale, F. A. K. El-Zaatari, S. A. Stohlman, and L. P. Weiner. 1986. Pathogenicity of antigenic variants of murine coronavirus JHM seleced with monoclonal antibodies. J. Virol. 58:869-875.

Kuo, L., G. -J. Godeke, M. J. B. Raamsman, P. S. Masters, and P. J. M. Rottier. 2000. Retargeting of coronavirus by substitution of the spike glycoprotein ectodomain: Crossing the host cell species barrier. J. Virol. 74:1393-1406.

Parker, S. E., T. M. Gallagher, and M. J. Buchmeier. 1989. Sequence analysis reveals extensive polymorphism and evidence of deletions within the E2 glycoprotein gene of several strains of murine hepatitis virus. Virology 173:664-673.

Phillips, J. J., M. M. Chua, E. Lavi, and S. R. Weiss. 1999. Pathogenesis of chimeric MHV4/MHV-A59 recombinant viruses: the murine coronavirus spike protein is a major determinant of neurovirulence. J. Virol. 73:7752-7760.

Wege, H., J. Winter, and R. Meyermann. 1988. The peplomer protein E2 of coronavirus JHM as a determinant of neurovirulence: Definition of critical epitopes by variant analysis. J. Gen. Virol. 69:87-98.

Mutation of the Immunodominant CD8+ Epitope in the MHV-4 Spike Protein

[1]MING MING CHUA, [1]JOANNA J. PHILLIPS, [1]SU-HUN SEO, [2]EHUD LAVI, AND [1]SUSAN R. WEISS

[1]*Department of Microbiology and* [2]*Department of Pathology and Laboratory Medicine, University of Pennsylvania School of Medicine, Philadelphia, PA 19104-6076.*

1.   INTRODUCTION

Our lab focuses on understanding the mechanism of murine coronavirus mouse hepatitis virus (MHV) pathogenesis in the central nervous system (CNS). MHV strains MHV-4 (JHM) and A59 induce acute encephalitis, followed by chronic demyelination in weanling C57Bl/6 (B6) mice; virus is cleared from the CNS by about 2 weeks post infection and demyelination occurs in the absence of infectious virus. We have previously used targeted recombination to select isogenic viruses differing only in the spike gene, expressing either the A59 spike ($S_{A59}R16$) or the MHV-4 spike (S_4R29) in the background of the A59 genome (Phillips et al., 1999). While the MHV-4 spike expressing virus (S4R29) is, like MHV-4, highly neurovirulent, the $S_{A59}R16$, like MHV-A59, is significantly less neurovirulent

Recently we have been interested in the role of the T cell response in acute pathogenesis and demyelination; we have begun to investigate the role of the immunodominant CD8+ T cell spike protein (H-2D^b) epitope S510-518 in pathogenesis of MHV in B6 mice. It is likely that response to this epitope may play a role in MHV pathogenesis for the following reasons. Pewe et al., (1996) demonstrated that in a suckling mouse model of MHV-JHM (MHV-4) infection, the development of demyelination requires the selection of CTL escape mutants to this epitope and resulting persistent viral infection; and 2) The highly neurovirulent MHV-4 spike gene expressing

The Nidoviruses (Coronaviruses and Arteriviruses).

Edited by Ehud Lavi *et al.*, Kluwer Academic/Plenum Publishers, 2001.

viruses and the less neurovirulent A59 spike gene expressing viruses differ
in their CD8+ T cell epitomes. The immunodominant S510-518 epitope lies
within a 52 amino acid hypervariable (HVR) domain that is present in the
MHV-4 spike and absent in the A59 spike protein (Figure 1). Both spike
proteins have the subdominant S598-605 epitope. (Perlman, 1998). Thus, in
order to begin to understand the role of S510-518 in pathogenesis, we have
selected viruses in which the S510-518 epitope has been inactivated within
the MHV-4 spike protein.

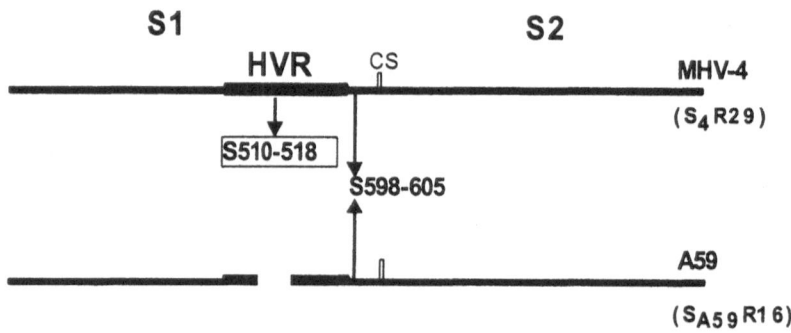

Figure 1. *MHV-4 and A59 spike genes.* S 1 and S2 subunits and the cleavage site (CS) are indicated. The 52 amino acid
deletion within the hypervariable region (HVR) of the A59 spike is shown relative to the MHV-4 spike. The CD8+
epitopes (arrows) and the CD4+ epitopes (open bars) within the spike protein are shown.

2.   RESULTS AND DISCUSSION

Recombinant viruses with mutations in the immunodominant S510-518
CD8+ epitope, within the MHV-4 spike gene, were selected using targeted
recombination (Kuo et al., 2000; Phillips et al., 1999). We selected two
independent recombinants for each of two mutations, one a single amino
acid substitution of the MHC class I anchor asparagine residue (N514S,
recombinants $S_4R141,143$) and the other a two amino acid deletion replaced
by one amino acid (NG to R; recombinants $S_4R145,146$) (see Table 1). The
spike genes S4R141 and S4R143 were sequenced and demonstrated to have
no additional amino acid substitutions. (The substitution in S4R141,145 was
found in naturally selected CTL escape mutants isolated by Pewe et al.,
(1996) who showed that peptides with such mutations were not recognized
by CD8+ T lymphocytes isolated from wild type virus infected mice.)

We compared the immune response to both CD8+ T cell epitopes in
these mutants with control virus, S_4R29, which contains the wild type
MHV-4 spike protein. We isolated spleen cells from animals infected with
S_4R29 as well as S_4R141 and S_4R143; we then determined the percentage of
CD8+ T cells that were positive also for intracellular IFNγ staining in

response to peptides containing the each of the CD8+T cell epitopes (Table 1). These data showed that the response to S510-518 is indeed not present in mice infected by either of S_4R141 or S_4R143. A similar level of cells specific for S598-605 was present in animals infected with S_4R29, each of the epitope mutants and MHV-A59 (Table 1 below). Thus A59 does indeed induce a CD8+ T cell response against S598-605, contrary to the report of Heemskirk et al., (1995) who did not detect this response in A59 infected animals, using a chromium release assay to detect cytolytic activity. Interestingly, there is a similar level of cells responding to S598-605 in the absence and presence of the S510-518 response, suggesting that this does not become a more dominant epitope in the absence of S510-518.

Table 1. Intracellular IFNγ assay of splenocytes to quantitate epitope specific CD8+ splenocytes. These intracellular IFNγ+ assays were carried out using splenocytes from infected animals 8 days after intraperitoneal inoculation with 10^4 PFU of virus. Splenocytes were incubated with epitope specific peptides and assayed as described by (Mural-Krishna et al., 1998). The values shown here are averages of two animals. This assay has been repeated twice with similar results.

VIRUS	% IFNγ+ cells (S510-518)	% IFNγ+ cells (S598-605)
S_4R29	20	4
S_4R141	0	8
S_4R143	0	6
$S_{A59}R16$	0	8

We have begun to examine the pathogenesis of these recombinant viruses with mutant S510-518 epitope. $S_4R141,143,145,146$ were slightly attenuated, although not dramatically so, as compared to S4R29. (The data are shown for S_4R141 and S_4R143 (Table 1); the data were similar for R145 and R146.) these small differences in LD_{50} were reproducible with all four mutants. Interestingly, at one month post infection, a significant percentage of the surviving mice that had been infected with mutant viruses were moderately to severely paralysed (more severe in the case of S_4R143), while all the surviving S_4R29 infected mice appeared normal. The extent of demyelination (the percentage of spinal cord quadrants with demyelination,) in these animals at one month post infection correlated with the severity of paralysis, with the average extent of demyelination being 62.5% for paralyzed mice and 10% for asymptomatic mice. Thus, these inactivating mutations in the CD8+ T cell epitope appear to shift the infection to be less lethal in the acute stage and result in more demyelination in the late stage. There was no detectable infectious virus in the brains or spinal cords of any of the surviving mice, whether paralyzed or asymptomatic when examined at one month post infection. Thus, the mutant viruses, like the S_4R29, are effectively cleared from the CNS.

Table 2. Pathogenesis of S510-518 mutant viruses. LD_{50} assays were carried out as previously described (Phillips et al., 1999). The LD_{50} assay was repeated twice; these are values from one representative experiment. Paralyzed mice were found among survivors at all doses of virus used (1-100 PFU). Demyelination was quantitated by counting spinal cord quadrants with demyelination and expressed here as the average value for the groups of paralyzed or normal mice for each virus, at one month post infection (Leparc-Goffart et al., 1997).

VIRUS	S510-518	LD_{50} (PFU)	Paralyzed/survivors
S_4R29	CSLWNGPHL	2.1	0/5
S_4R141	CSLWSGPHL	36	3/13 moderate
S_4R143	CSLWRPHL	12	5/10 severe
$S_{A59}R16$	Not present	3000	Not Done

We cannot distinguish as yet whether the higher levels of demyelination observed in the S510-518 mutant infected mice are due to the fact that more animals survive acute disease and are available to develop demyelination or that, in the absence of the S510-518 response, there is a change in the acute disease that results in more demyelination and paralysis. To address this question, we will investigate the cell types infected and the types of immune cells infiltrating the brain in S_4R29 and mutant infected animals.

3.   CONCLUSION

Recombinant viruses containing inactivating mutations in the immuodominant S510-518 CD8+ T cells epitope were compared with parental S_4R29 virus containing the wild type MHV-4 spike in the background of the MHV-A59 genome. The S510-518 epitope mutants did not induce a response against S510-518 while maintaining a similar level of CD8+ T cell response to S598-605 as compared with parental S_4R29. The epitope mutants were slightly attenuated during acute infection but displayed higher levels of paralysis and demyelination compared with S4R29.

ACKNOWLEDGMENTS

This work was supported by and National Multiple Sclerosis Society grant RG 2585B5/1 and NIH grant NS-30606.

REFERENCES

Heemskerk, M. . H. M., Schoemaker, H. M., Spaan, W. J. M., and Boog, C. J. P.(1995). Predominance of MHC class II restricted CD4+ T cells against mouse hepatitis virus A59. *Immunol.* **84**, 521-527.

Kuo, L, Godeke, G.J., Raamsman, M.J,. Masters, P.S., and Rottier, P.J. (2000). Retargeting of coronavirus by substitution of the spike glycoprotein ectodomain: crossing the host cell species barrier. *J Virol* **74**, 1393-406.

Leparc-Goffart, I., Hingley, S. T., Chua, M. M., Jiang, X., Lavi, E., and Weiss, S. R.(1997). Altered pathogenesis of a mutant of the murine coronavirus MHV-A59 is associated with a Q159L amino acid substitution in the spike protein. *Virology* **239**, 1-10.

Murali-Krishna, K., Altman, J.D., Suresh, M., Sourdive, D.J.D., Zajac, A.J., Miller, J.D., Slansky, J. and Ahmed, R.1998. Counting antigen-specific CD8 T cells: a reevaluation of bystander activation during viral infection.

Perlman, S.(1998). Pathogenesis of coronavirus-induced infections. Review of pathological and immunological aspects. *Adv Exp Med Biol* **440**, 503-513.

Pewe, L., Wu, G. f., Barnett, E. M., Castro, R. F., and Perlman, S.(1996). Cytotoxic T cell-resistant variants are selected in a virus-induced demyelinating disease. *Immunity* **5**, 253-262.

Phillips, J.J.,Chua, M.M., Lavi, E., Weiss, S. R. 1999. Pathogenesis of chimeric MHV-4/MHV-A59 recombinant viruses: the murine coronavirus spike protein is a major determinant of neurovirulence. J. Virol. **73**:7752-7760.

Spread of Hemagglutinating Encephalomyelitis Virus (HEV) in the CNS of Rats Inoculated by Intranasal Route

[1]NORIO HIRANO, [3,4]KOUJIRO TOHYAMA, [2]HIDEHARU TAIRA, AND [4]TSUTOMU HASHIKAWA

[1]Department of Veterinary Microbiology, and [2]Department of Bioscience and Technology, Iwate University, Morioka 020-8550; [3]The Center for Electron Microscopy and Bio-imaging Research and Department of Neuroanatomy, Iwate Medical University, Morioka 020-8505; [4]Laboratory of Neural Architecture, Brain Science Institute, RIKEN, Wako 351-0198, Japan

1. INTRODUCTION

HEV is known to cause vomiting and wasting disease or encephalomyelitis in piglets (Andries et al., 1980,1981; Mengeling et al., 1972; Roe et al., 1958). In experimental oronasal infection of piglets, the virus spread to the CNS predominantly via the nerve pathways (Andries et al., 1980, 1981). In our previous study, 4-week-old rats died of encephalitis after inoculation by different routes including intranasal route (Hirano et al., 1993). In the infected animals, the virus reaches the CNS through the nerve pathways from peripheral nerve. Our previous studies have demonstrated that HEV propagated through nervous route and its infection was restricted to neurons after inoculation into sciatic nerve or footpad of rats (Hirano et al., 1995, 1998). Our findings suggest that HEV is useful as a trans-synaptic tracer for analyzing neuronal connections in the CNS.

In the present study, we attempt to examine and evaluate the usefulness of HEV as a trans-synaptic tracer in the olfactory pathways. In addition, we examine the possibility whether this system provides a model for analysing the natural infection in pigs.

The Nidoviruses (Coronaviruses and Arteriviruses).
Edited by Ehud Lavi et al., Kluwer Academic/Plenum Publishers, 2001.

2.   MATERIALS AND METHODS

Plaque-purified HEV 67N strain (Mengeling et al., 1972) was propagated and assayed for infectivity in SK-K cells as described previously (Hirano et al., 1990). SPF of 4- to 6-week-old Wistar male rats were used. Twenty µl of the virus (1×10^5 PFU) was inoculated into the right nasal cavity of rats. On days 1 to 7 after inoculation, rats were killed under deep anesthesia with pentobarbital and the brain and spinal cord were dissected out and assayed. For immunostaining, rats were perfused with 4% paraformaldehyde in 0.1M phosphate buffer under deep anesthesia on day 3 to 13. The brain was dissected out and sections were obtained on a freezing microtome, treated with anti-HEV 67N mouse antibody (1:1000) at 4C overnight, and reacted with FITC-conjugated goat serum (anti-mouse IgG) at room temperature for 2 hours. The stained sections were examined under confocal laser scanning microscope.

3.   RESULTS

Inoculated rats became hypersensitive on day 7 to 10 and began to die showing CNS signs on day 10. On day 2, HEV was first isolated from the ipsilateral brain but not from the contralateral. On day 3, the virus was isolated from the contralateral brain and its titer was about one tenth of the ipsilateral brain. On day 4, the virus was detected from the spinal cord. Until day 5 to 7, the infectivity titers of the contralateral side reached nearly the same level as in the ipsilateral brain, 10^6 to 10^7 PFU/0.2g. HEV antigen was found in the olfactory bulb ipsilateral to the inoculated nasal cavity on day 3. On day 4 in main olfactory bulb (MOB), many positive neurons located in mitral cell and external flexiform layers (Fig.1a). Many immuno-positive neurons were found in anterior olfactory nucleus (AON) (Fig.1b) and in the olfactory cortex (OC) (Fig.1c). Day 6, additional area including septal (SNT) and paraventricular nuclei (PVT) of thalamus contained many antigen positive neurons (Fig. 2a). Neurons in the superior colliculus (SC) and a few Purkinje and granular cells in the cerebellar cortex (CC) exhibited antigen-positive (data not shown). Day 13, pyramidal cells in hippocampus (CA) and granular cells in dentate gyrus (DG) became infected (Fig.2b), which differ from the findings on day 6 (Fig.2a). In OC, the number of antigen positive neurons did not increase. Neurons appeared to be antigen positive in the inferior colliculus (IC) where no immunostained neurons were found on day 6. The antigen was detected predominantly in neurons but not in glial cells. No cytopathological evidence was found in HEV infected neurons.

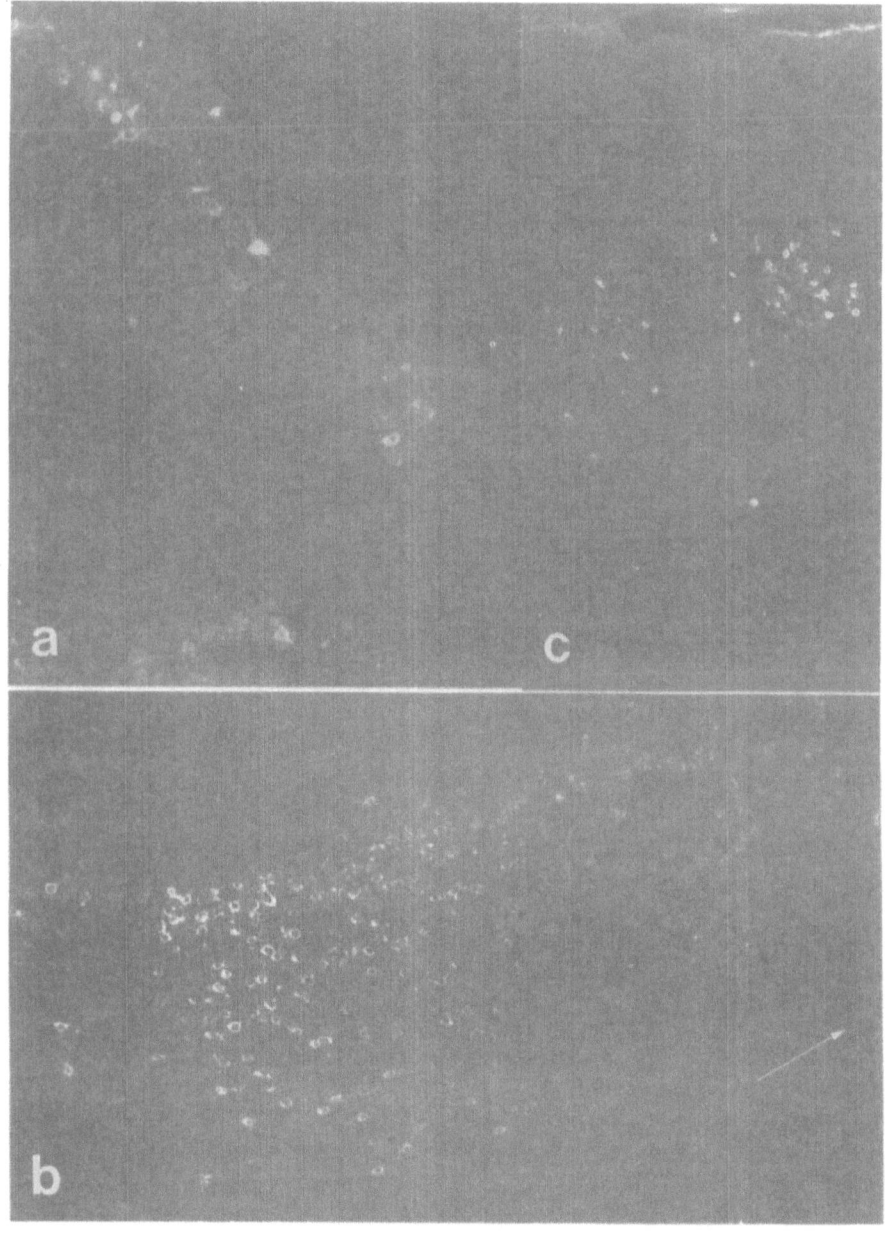

Figure 1. On day 4, a) HEV positive neurons including mitral cells in ipsilateral olfactory bulb; b) anterior olfactory nucleus containing numerous positive neurons; c) primary olfactory cortex.

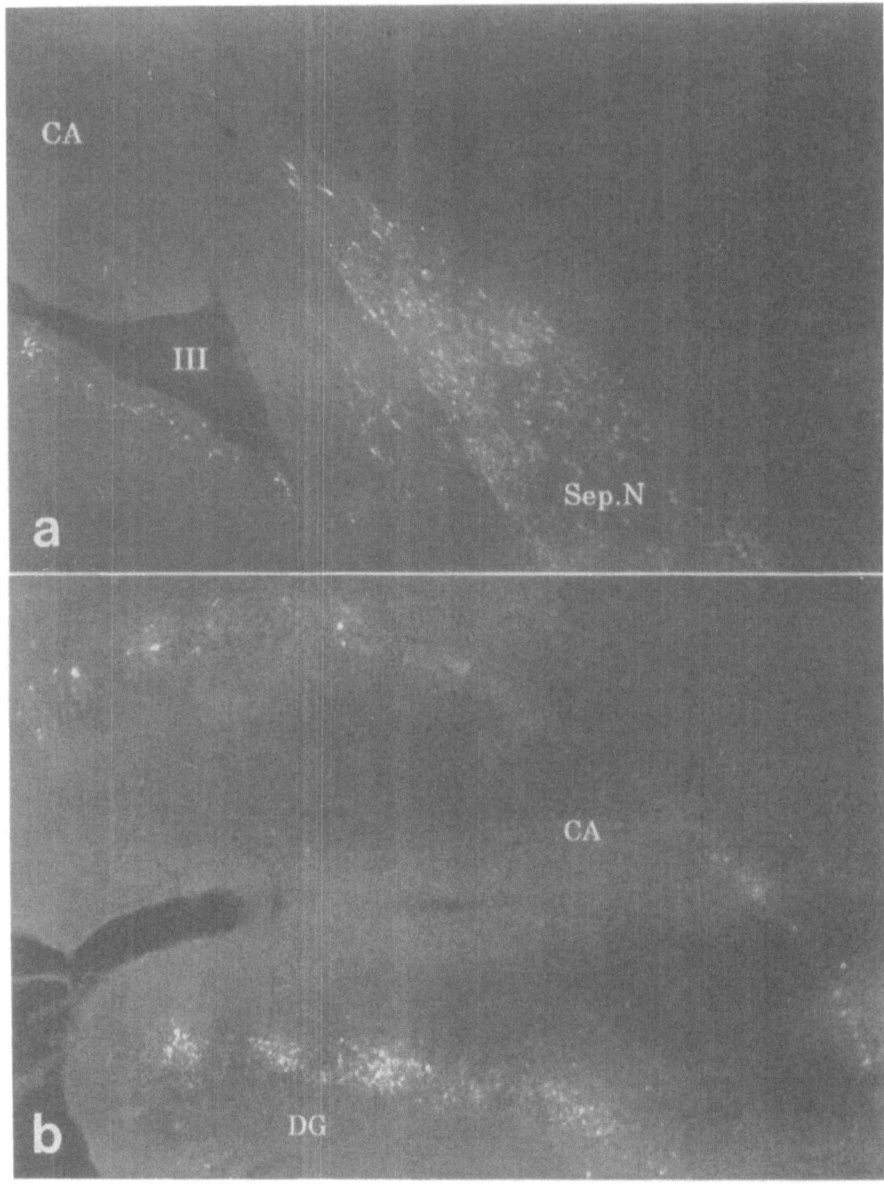

Figure 2. On day 6 (a) and 13(b), a) antigen positive neurons in the septal nucleus (Sep.N) and in paraventricular thalamic nucleus but not in the hippocampus (CA), III: 3rd ventricle; b) antigen positive pyramidal cells in the hippocampus (CA) and granular cells in dentate gyrus (DG).

4.   DISCUSSION

In rats, two components of the olfactory system are recognised: the main and accessory olfactory systems. In the present study, HEV positive neurons distributed in MOB, AON, OC, SNT, PVT, SC, CC, CA, DG, and IC. MOB and OC are well known structures of the main olfactory pathway. MOB, AON, and some limbic nuclei including SNT, PVT, CA and DG belong to accessory olfactory system. Positive neurons in other components (SC, CC and IC) may infect through trigeminal nerve route.

Andrires and Pensaert (1980) reported the trigeminal nerve route as well as olfactory pathway in piglets after oronasal inoculation of HEV. Lavi et al. (1988) described the virus spread through the limbic system in mice after intranasal inoculation of mouse hepatitis virus (MHV). Our observations in the present study are coincident with the results of these reports. HEV antigen was detected predominantly in neurons, but not in glial cells, of olfactory pathways as well as trigeminal route in the present study. This result indicates that HEV is confirmed to be extremely neurotropic virus different from others including MHV and herpes simplex virus, which infects not only neurons but also surrounding glial cells (Lavi et al., 1988; Cook et al., 1973). In other word, HEV is a causative agent, which has strict neurotropism. In addition, no distinct degenerating changes were found in HEV infected neurons, suggesting that the fiber connections along olfactory pathways and other related fiber connections kept intact as far as the period examined. In general, such properties of HEV have the advantage of analyzing neuroanatomical connections as a trans-synaptic tracer. In the olfactory pathways, however, some other related fiber connections make it difficult to investigate anatomical connections using HEV. Still, it retains the possibility that more precise study may resolve this complication and reveal the systems combined with other tracers.

ACKNOWLEDGMENTS

This study was supported by the Grands-in-aid (No. 11660309) from the Ministry of Education and Culture of Japan.

REFERENCES

Andries, K., and Pensaert, M.B. 1980. Immunofluorescence studies on the pathogenesis of hemagglutinating encephalomyelitis virus infection in pigs after oronasal inoculation. *Am. J. Vet. Res.* **41**: 1372-1385.

Andries, K., and Pensaert, M.B. 1981. Vomiting and wasting disease; a coronavirus infection of pigs. *Adv. Exp. Med. Biol.* **142**: 399-408.

Hirano, N., Haga, S., and Fujiwara, K. 1993. The route of transmission of hemagglutinating encephalomyelitis virus (HEV) 67N strain in 4-week-old rats. *Adv. Exp. Med. Biol.* **342**: 333-338.

Hirano, N., Nomura, R., Tawara, K., Ono, K., and Iwasaki, Y. 1995. Neuronal spread of hemagglutinating encephalomyelitis virus (HEV) 67N strain in 4- week-old rats. *Adv. Exp. Med. Biol.* **390**: 117-119.

Hirano, N., Ono, K., Takasawa, H., and Haga, S. 1990. Replicaiton and plaque formation of swine hemagglutinating encephalomyelitis virus (67N) in swine cell line, SK-K culture. *J. Virol. Methods* **27**:91-100.

Hirano, N., Tohyama, K., and Taira, H. 1998. Spread of hemagglutinating encephalomyelitis virus from peripheral nerve to the CNS. *Adv. Exp. Med. Biol.* **440**: 601-607.

Lavi, E., Fishman, P., Highkin, M., and Weiss, S. 1988. Limbic encephalitis after inhalation of a murine coronavirus. *Lab. Invest.* **98**: 31-36.

Mengeling, W.L., Boothe, A.D., and Richite, A.E. 1972.Characterization of a coronavirus (Strain 67N) of pigs. *Am. J. Vet. Res.* **33**: 297-308.

Mengeling, W.L., and Cutlip, R.C. 1976. Pathogenicity of field isolates of hemagglutinating encephalomyelitis virus for neonatal pigs. *J. Am. Vet. Assoc.* **68**: 236-239.

Roe, C.K., and Alexander, T.J.L. 1958. A disease of nursing pigs previously unreported in Ontario. *Canad. Comp. Med.* **22**: 305-307.

Demyelination Determinants in the S Gene of MHV

[1]JAYASRI DAS SARMA, [1]LI FU, [2]SUSAN R. WEISS, AND [1]EHUD LAVI
[1]*Department of Pathology and Laboratory Medicine, and* [2]*Department of Microbiology, University of Pennsylvania School of Medicine, Philadelphia, PA 19104-6076.*

1.   INTRODUCTION

Demyelination is the hallmark pathologic process of several viral-induced, and inflammatory, immune-mediated diseases of the central nervous system (Allen and Bradkin 1993). Several experimental animal model systems are used to study demyelination, including MHV-induced demyelination in mice (Stohlman *et al* 1981; Wege *et al* 1982; Lavi *et al* 1984; Lavi *et al* 1984b; Lavi and Weiss 1989; Houtman *et al* 1996; Lavi *et al* 1999). Previous studies with this model suggested that the S protein and gene may be important for pathogenesis since the spike (S) protein participates in many functions related to virus-host interactions (Dalziel *et al* 1986). Thus we wanted to explore the possibility that the S gene contains molecular determinants of demyelination. We used targeted RNA recombination as previously described (Peng *et al* 1995; Leparc-Goffart *et al* 1998; Phillips *et al* 1999) to insert the S gene of a non-demyelinating virus (MHV-2) into the background of a demyelinating virus (MHV-A59). We then studied the pathogenesis of the new recombinant viruses.

2.   GENERATION OF RECOMBINANT VIRUSES

We first cloned the S gene of MHV-2 into a pGEM-T[(a)] Vector. We then removed the MHV-2 S gene from the pGEM-T[(a)] vector by digestion with AvrII and SbfI. The S gene of MHV-2 was then gel purified and subcloned

The Nidoviruses (Coronaviruses and Arteriviruses).
Edited by Ehud Lavi *et al.*, Kluwer Academic/Plenum Publishers, 2001.

into the corresponding site in pMH54 (Kuo *et al* 2000), which contains the entire 3' end of MHV-A59 downstream to the HE gene. The new plasmid was labeled pMHV2. We sequenced the S gene portion of pMHV2 in order to verify that it contained the exact sequence of the MHV-2 S gene. We then carried out targeted RNA recombination between Alb4 (a temperature sensitive MHV-A59, obtained from Dr. Paul Masters, containing an 87 nucleotide deletion in the N gene) and synthetic capped RNAs transcribed from pMHV2. MHV-2 S gene recombinant viruses were selected by antibody neutralization treatment. The released viruses were treated with A2.1 and A2.3, anti-S monoclonal antibodies (obtained from Dr. John Fleming) specific for the S protein of MHV-A59. The antibody treatment neutralized the parent viruses Alb4 and MHV-A59, but not MHV-2 or recombinant viruses containing the S gene derived from MHV-2. Viruses were then identified by small plaque morphology. The presence of AvrII and SbfI restriction sites was then confirmed in putative recombinant viruses and selected recombinants were plaque purified two additional times. The identity of the S gene in the selected recombinant viruses and the absence of additional mutations were confirmed by sequencing. The new recombinant viruses were labelled Penn98-1 and Penn98-2.

3.   PATHOGENESIS OF THE RECOMBINANT VIRUSES

3.1   Virulence

The LD50 experiments revealed that the virulence of both Penn98-1 and Penn98-2 was higher than that of both parental viruses, but was closer to that of MHV-2 (5 PFU = 1 LD50). Penn 98-1 and Penn 98-2 replicated efficiently in both brain and liver. The kinetics of replication was closer to that of MHV-2, suggesting that the S gene contains determinants of virulence and hepatotropism.

3.2.   Pathology

Histopathological studies revealed that Penn98-1 and Penn98-2 produced acute meningoencephalitis similar to MHV-A59. Brain pathology included focal acute encephalitis, which was characterized by inflammatory mononuclear cell infiltrates, predominantly lymphocytes. Microglial proliferation, microglial nodules, and neuronophagia were also identified. Areas of involvement included the regions of the brain typically susceptible to MHV-A59 infection. Liver pathology of moderate to severe hepatitis following Penn98-1 and Penn98-2 infection was characterized by multiple foci of necrosis throughout the liver. Each area of necrosis consisted of degenerating hepatocytes, polymorphonuclear and lymphocytic inflammatory infiltrates, and cellular debris. The extent and distribution of the hepatitis caused by 5 PFU of Penn98-1 and Penn98-2 was similar to the hepatic changes produced by 1000 PFU of MHV-2 and was more severe than the changes produced by 5000 PFU of MHV-A59.

3.2   Immunohistochemistry

Viral antigen was analyzed by immunohistochemistry on tissue sections obtained from mice during acute infection with the recombinant viruses (Penn98-1, Penn98-2) and was compared to sections obtained from mice infected with MHV-A59 and MHV-2. In MHV-A59 infected mice, viral antigen was distributed in focal areas of the brain parenchyma concomitant with the distribution of inflammatory infiltrates. In MHV-2 infected mice, viral antigen was detected in the meninges, choroid plexus, and ependymal cells. There was no viral antigen detected in neurons. The involvement of glial cells was minimal and restricted to the subependymal location. In Penn98-1 and Penn98-2 infected mice, the distribution of viral antigen was similar to that seen in MHV-A59 infection.

3.3   Demyelination

The ability of recombinant viruses Penn 98-1 and Penn 98-2 to induce demyelination, was examined and compared to wild type recombinant wtR13 and wild type viruses MHV-A59 and MHV-2. Penn 98-1 and Penn 98-2 did not produce demyelination in any of the 7 mice injected with each virus. Wild type recombinant virus wtR13, containing an S gene derived from A59, produced demyelination in 100% of the mice (5/5), similar to wild type MHV-A59. All three viruses (wtR13 and Penn98-1 and Penn 98-2) were given at the same dose (5 PFUs). The recombinant virus wtR13, when given at 2500 PFUs, produced larger demyelinating lesions than at 5PFU, but with both doses of virus 100% of the mice produced demyelination.

3.4   Viral persistence

In order to investigate whether differences in the ability of viruses to cause demyelination were associated with differences in viral persistence, we amplified viral RNA from livers, brains and spinal cords of mice infected with demyelinating and non-demyelinating viruses. During the acute infection, PCR products with a band size consistent with MHV RNA

(601bp), were detected at 5 days post infection, in all mice infected with each one of the viruses. Viral RNA was detected in all 3 anatomic locations examined (liver, brain and spinal cord). However, at 30 days post infection, PCR products corresponding to viral RNA were detected only in the MHV-A59 infected spinal cords, but not in the liver or brain of the same mice. There were no detectable PCR products in the livers, brains and spinal cords of mice infected with MHV-2 and another non-demyelinating virus, Penn 97-1, or in organs of control uninfected mice. Using a second pair of primers, RT-PCR amplified a fragment of the predicted size of 147 bp only in the sample of MHV-A59 infected spinal cord, but not in the liver and brain of the same mouse. The spinal cord had the most abundant viral transcript during the chronic phase. This is consistent with previous reports suggesting that the spinal cord is the major site of viral persistence during chronic infection with JHM and MHV-A59. Mice infected with non-demyelinating viruses MHV-2 and Penn97-1, and uninfected controls, were negative.

4. CONCLUSIONS

Using targeted RNA recombination, the present study provides direct evidence that a molecular determinant of demyelination maps to the S gene of MHV. However, we cannot rule out that demyelination may depend on the integrity of additional, non-S determinants, within the viral genome. Viral persistence appears to be an important factor and may even be a prerequisite for MHV–induced demyelination. However, based on the studies presented here, viral persistence per se is insufficient to induce demyelination. The findings presented here pave the way for further studies to investigate in more detail the potential role of the viral envelope S glycoproteins in autoimmunity and demyelination. These studies are potentially relevant to other forms of demyelination including MS.

ACKNOWLEDGMENTS

The work from our laboratory was supported by grants from the National Multiple Sclerosis Society (RG 2615-B-2) and the NIH (NS30606). We thank Dr. Paul Masters for the gifts of pMH54 and Alb4 virus, Dr. John Fleming for the gift of anti-S monoclonal antibodies, Donna Bauhof for critical review of the manuscript, and Elsa Aglow for histology expertise.

REFERENCES

Allen, I., and B. Brankin. 1993. Pathogenesis of multiple sclerosis - the immune diathesis and the role of viruses. J. Neuropath. Exp. Neurol. **52**:95-105.

Dalziel, R. G., P. W. Lampert, P. J. Talbot, and M. J. Buchmeier. 1986. Site specific alteration of murine hepatitis virus type 4 peplomer glycoprotein S results in reduced neurovirulence. J. Virol. **59**:463-471.

Houtman, J. J., and J. O. Fleming. 1996. Pathogenesis of mouse hepatitis virus-induced demyelination. J. Neurovirol. **2**:361-376.

Kuo, L., G.-J. Godeke, M. J. B. Raamsman, P. S. Masters, and P. J. M. Rottier. 2000. Retargeting of coronavirus by substitution of the spike glycoprotein ectodomain: crossing the host cell species barrier. J. Virol. **74**:1393-1406.

Lavi, E., D. H. Gilden, M. K. Highkin, and S. R. Weiss. 1984. Persistence of MHV-A59 RNA in a slow virus demyelinating infection in mice as detected by in situ hybridization. J. Virol. **51**:563-566.

Lavi, E., D. H. Gilden, Z. Wroblewska, L. B. Rorke, and S. R. Weiss. 1984b. Experimental demyelination produced by the A59 strain of mouse hepatitis virus. Neurology. **34**:597-603.

Lavi, E., T. Schwartz, Y. P. Jin, and L. Fu. 1999. Nidovirus infections: experimental model systems of human neurologic diseases. J. Neuropathol. Exp. Neurol. **58**:1197-1206.

Lavi, E., and S. R. Weiss. 1989. Coronaviruses., p. 101-139. *In* D. H. Gilden and H. L. Lipton (ed.), Clinical and molecular aspects of neurotropic viral infections. Kluwer, Academic Publishers, Boston.

Leparc-Goffart, I., S. T. Hingley, M. M. Chua, J. Phillips, E. Lavi, and S. R. Weiss. 1998. Targeted recombination within the spike gene of murine coronavirus mouse hepatitis virus-A59: Q159 is a determinant of hepatotropism. J. Virol. **72**:9628-9636

Peng, D., C. A. Koetzner, T. McMahon, Y. Zhu, and P. S. Masters. 1995. Construction of murine coronavirus mutants containing interspecies chimeric nucleocapsid proteins. J. Virol. **69**:5475-5484.

Phillips, J. J., M. M. Chua, E. Lavi, and S. R. Weiss. 1999. Pathogenesis of chimeric MHV-4/MHV-A59 recombinant viruses: the murine coronavirus spike protein is a major determinant of neurovirulence. J. Virol. **73**:7752-7760.

Stohlman, S. A., and L. P. Weiner. 1981. Chronic central nervous system demyelination in mice after JHM virus infection. Neurology. **31**:38-44.

Wege, H., S. Siddell, and V. ter Meulen. 1982. The biology and pathogenesis of coronaviruses. Adv. Virol. Immunol. **99**:165-200.

Role of the Spike Protein in Murine Coronavirus Induced Hepatitis: An *in vivo* Study Using Targeted RNA Recombination

[1]SONIA NAVAS , [1]SU-HUN SEO, [1]MING MING CHUA, [2]JAYASRI DAS SARMA, [3]SUSAN T. HINGLEY, [2]EHUD LAVI, AND [1]SUSAN R. WEISS
[1]*Department of Microbiology, and* [2]*Department of Pathology and Laboratory Medicine, University of Pennsylvania School of Medicine, Philadelphia, Pennsylvania;* [3]*Department of Microbiology, Philadelphia College of Osteopathic Medicine, Philadelphia, Pennsylvania.*

1.   INTRODUCTION

Various strains of the murine coronavirus, mouse hepatitis virus (MHV), have been shown to display different organ tropism and pathogenesis, including enteritis, encephalitis, demyelination and hepatitis in C57BL/6 mice (Perlman et al., 1998). Infection of mice by MHV is an experimental model of chronic demyelinating diseases, such as multiple sclerosis (Buchmeier et al., 1999). Furthermore, some MHV strains can be considered as a model for studying acute and chronic hepatitis of viral etiology (Ding et al., 1997). It is well established that the severity of MHV-induced hepatitis is dependent on virus strain. MHV-A59 induces moderate to severe hepatitis, whereas MHV-4 (an isolate of MHV-JHM strain) produces none to minimal hepatitis. MHV-2 is a highly hepatotropic strain that causes severe hepatitis. However, the viral determinants which explain the differences in pathogenicity are poorly understood.

Until very recently, the study of viral determinants of pathogenesis was limited to the comparison of different strains and mutant viruses. This

The Nidoviruses (Coronaviruses and Arteriviruses).
Edited by Ehud Lavi *et al.*, Kluwer Academic/Plenum Publishers. 2001.

139

type of approach has demonstrated that mutations in the spike (S) gene were associated with altered pathogenicity (Fleming et al., 1989; Rowe et al., 1997; Leparc-Goffart et al., 1997). We have previously characterized fusion-defective mutants of MHV-A59, which were derived from persistently infected glial cell cultures (Gombold et al., 1993). C12, one of these fusion-defective A59 mutants, was attenuated and displayed an altered hepatotropism (Hingley et al., 1994). The spike gene of C12 contains only two mutations: i) one in the cleavage site of the S protein, H716D was previously shown to be responsible of the fusion defect phenotype; ii) the second mutation (Q159L) was found in the receptor binding domain of the S protein and its presence was associated with the loss of the hepatotropic phenotype (Leparc-Goffart et al., 1997). We further derived hepatitis-producing revertants through 8 in vivo passages of C12 through the livers of mice (Hingley et al., 1995). Interestingly, both Q159L as well as H716D were maintained but, two additional mutations in the spike gene appeared: R654H and E1035D.

Genetic manipulation of MHV genome has been recently available with the development of targeted RNA recombination (Koetzner et al.; Masters, 1999). Using this technique, we have recently demonstrated that the spike protein is a major determinant of neuropathogenic properties of MHV (Phillips et al., 1999). In a previous study, we found that a Q159L amino acid substitution introduced by targeted RNA recombination into the spike gene of MHV-A59 alter the ability of the virus to replicate efficiently in the liver and induce hepatitis (Leparc-Goffart et al., 1998). By means of targeted RNA recombination we are currently beginning to map viral determinants of hepatotropism within the spike protein.

2.   METHODS

2.1   Targeted RNA recombination

We have compared three isogenic recombinant viruses that differ only in the spike gene, expressing the spike protein of A59 ($S_{A59}R13$), MHV-4 (S_4R21) and MHV-2 (Penn98-1), all in the A59 background. The construction of this recombinant viruses has been described elsewhere (Phillips et al., 1999; Das Sarma et al., 2000).

The hepatitis revertants (HR2) recombinants (R173, R174) were constructed following a recent modification (Kuo et al., 2000).

2.2. Virulence assays

Fifty-percent lethal dose (LD_{50}) assays were carried out as described previously (Leparc-Goffart et al., 1998). Mice were inoculated intracranially with fivefold serial dilutions of wild-type and recombinant viruses. Mice were examined for signs of disease or death on a daily basis for up to 21 days postinfection. LD50 values were calculated by the Reed-Muench method (Smith et al., 1997).

2.3. Viral replication in infected tissue

C57BL/6 mice were infected intrahepatically (ih) with 500 plaque forming units (PFU) of virus as described previously (Hingley et al., 1994). Virus in homogenized liver tissue was titered, in duplicate, by plaque assay on L2 cell monolayers in 6-well micotiter plates.

2.4 Liver histology and immunohistochemistry

Mice were sacrificed at selected times postinfection and perfused with 10 ml of PBS and livers were removed. Formalin-fixed tissue was embedded in paraffin, sectioned and stained with hematoxylin and eosin (H&E) for histological diagnosis. Hepatitis was scored as mild (1), moderate (2) and severe (3). Immunohistochemical analysis was performed by the avidin-biotin-immunoperoxidase technique (Vector Laboratories, Burlingame, Calif.). As a primary antibody we used a monoclonal antibody (Mab) against the nucleocapsid protein (N) of MHV-JHM strain (Mab clone 1-16-1, kindly provided by J.L. Leibowitz). All slides were read blindly and mock-infected controls were included.

3. RESULTS AND DISCUSSION

Preliminary data obtained in this study are shown in *Table 1*. After i.h. inoculation with 500 PFU per virus, the peak of replication was at day 5 postinfection (p.i.). The replication pattern of each recombinant was similar to the corresponding parental virus. Penn98-1 showed the higher viral titers (range $10^7 - 10^8$ PFU/g tissue) than $S_{A59}R13$ ($10^6 - 10^7$ PFU/g tissue, whereas S_4R21 presented a very low level of hepatic replication ($10^3 - 10^4$ PFU/g). Regarding the hepatic damage, Penn 98-1 produced severe hepatitis, whereas S_4R21 induced only none to mild hepatitis. The liver damage produced by $S_{A59}R13$ was moderate to severe hepatitis. Viral antigen staining colocalized with necrotic areas in the Penn 98-1 and $S_{A59}R13$. In S_4R21, specific viral staining was found in hepatocytes with conserved cellular morphology.

Table 1. Intrahepatic replication, hepatic injury and viral staining observed in recombinant viruses.

Recombinant virus	Spike gene	ih replication[a]	Liver histology[b]	Viral staining[c]
$S_{A59}R13$	A59	++	2-3	++
Penn98-1	MHV-2	+++	3	+++
S_4R21	MHV-4	+/-	0-1	+/-
R173	A59*	+++++	ND[¶]	ND[¶]

[a] viral titers in liver at day 5 p.i.: ++++++ (range $10^9 - 10^{10}$); +++ ($10^7 - 10^8$); ++ ($10^6 - 10^7$); + ($10^4 - 10^6$); +/- ($10^3 - 10^4$); ranges are expressed as PFU/g tissue.
[b] Hepatitis was scored as: 1 (mild), 2 (moderate) and 3 (severe).
[c] Immunohistochemistry showed viral antigen staining in the hepatic parechyma. The extension and the intensity of the staining are scored as +/- (low), + (medium) and high (+++).
*A59 hepatitis revertant spike gene has only four mutations in comparison with A59 spike: Q159L, R654H, H716D and E1035D.
[¶] ND, not done.

Recombinants viruses with the spike of the hepatitis revertants (HR2) were also constructed.   The HR2 recombinants showed high virulence and replication ability in the liver (Table 1 and 2).

Table 2. Virulence and type of spike protein of various parental and recombinant viruses.

Virus	Spike protein	Log (LD50)
MHV-A59	A59	3.8
$S_{A59}R16$	A59	3.8
C12	Q159L , H716D	6
R36	Q159L , H716D	5.1
HR2	Q159L, H716D R654H, E1035D	1.2
R173	Q159L, H716D R654H, E1035D	1.0

Taking together these data, underline a direct role of the MHV spike glycoprotein in the development of hepatitis in infected mice. The role of the S protein in the pathogenesis is consistent with its biological function. The S protein, found on the virion envelope and on the plasma membrane of infected cells is responsible for the attachment to cellular receptor and virus-cell fusion during viral entry and for cell-to-cell fusion later during infection. One interesting hypothesis is that mutations in the spike may account for an altered interaction with the cellular receptor. In particular, the Q159L mutation is in the putative receptor-binding domain of the spike. We previously demonstrated by targeted recombination that the Q159 is a determinant of hepatotropism (Leparc-Goffart et al., 1998). The hepatitis revertant virus (HR2) were originated by in vivo passages of the C12 through the liver of C57BL/6 mice. Sequencing of the spike of this HR2 virus showed that two new mutations (R654H, E1035D) in the spike appeared whereas those observed in the C12 (Q159L, H716D) were maintained. We have constructed a recombinant virus with the HR2 spike mutants in the A59 background. Interestingly, this HR2-spike recombinant virus replicates in the liver to 2 logs over the A59 and even 3 logs over the highly hepatotropic MHV-2 virus in the liver. As recombinants virus A59R16, R36 and HR2R173 are isogenic virus that differs only in the above mentioned mutations in the spike gene, it can be argued that the R654H, E1035D are playing an important role in the dramatic differences observed in liver replication. Thus, in hypothesis, one or both of the two mutations may account for the highly hepatotropic phenotype of the hepatitis revertant virus.

In summary, by means of targeted RNA recombination we found direct correlation between mutations in the murine coronavirus spike protein and hepatotropism.

REFERENCES

Buchmeier, M.J., and Lane, T.E. Viral-induced neurodegenerative disease. Curr Opinion Microbiol 1999; **2**:398-402.

Das Sarma, J., Fu L., Tsai, J., Weiss, S.R., and Lavi, E. Demyelination determinants map to the spike gene of coronavirus mouse hepatitis virus. J Virol 74:9206-9213.

Ding, J.W., Ning, Q., Liu, M.F., Lai, A., Leibowitz, J., Peltekian, K.M., Cole, E.H., Fung, L.S., Holloway, C., Marsden, P.A., Yeger, H., Phillips, M.J., Levy, G.A. Fulminant hepatic failure in murine hepatitis virus strain 3 infection: tissue-specific expression of a novel *fgl2* prothrombinase. J Virol 1997; **71**:9223-92-30.

Fleming. J.O., Trousdale, M.D., El-Zaatari, F.A.K., Stohlmam, S.A., Weiner, L.P. Pathogenicity of antigenic variants of murine coronavirus JHM selected with monoclonal antibodies. J Virol 1989; **58**:869-875.

Gombold, J.L., Hingley, S.T., Weiss, S.R. Fusion-defective mutants of mouse hepatitis virus A59 contain a mutation in the spike protein cleavage signal. J Virol 1993; **67**:4504-4512.

Hingley, S.T., Gombold, J.L., Lavi, E., Weiss S.R. MHV-A59 fusion mutants are attenuated and display altered hepatotropism. Virology 1994; **200**:1-10.

Hingley, S.T., Gombold, J.L., Lavi, E., Weiss, S.R. Hepatitis mutants of mouse hepatitis virus strain A59. Adv Exp Med Biol 1995; **380**:577-582.

Koetzner, CA., Parker, M.M., Richard, C..S., Sturman, S., Masters, P.S. Repair and mutagenesis of the genome of a deletion mutant of the murine coronavirus mouse hepatitis virus by targeted RNA recombination. J Virol 1992; **66**:1841-1848.

Kuo l., Godeke, G.J., Raamsman, J.B., Masters, P.S., Rottier, P. Retargeting of coronavirus by substitution of the spike glycoprotein ectodomain: crossing the host cell species barrier. J Virol 2000; **74**:1393-1406.

Leparc-Goffart, I., Hingley, S.T., Chua, Jiang, X., Lavi, E., Weiss, S.R. Altered pathogenesis of a mutant of the murine coronavirus MHV-A59 is associatted with a Q159L amino acid substitution in the spike protein. Virology 1997; **239**:1-10.

Leparc-Goffart, I., Hingley, S.T., Chua, M.M., Phillips J., Lavi, E., Weiss, S.R. Targeted recombination within the spike gene of murine coronavirus mouse hepatitis virus-A59: Q159 is a determinant of hepatotropism. J Virol 1998; **72**:9628-9636.

Masters, P.S. Reverse genetics of the largest RNA viruses. Advances Virus Research 1999; **53**:245-264.

Perlman, S. Pathogenesis of coronavirus-induced infections. Review of pathological and immunological aspects. In: Coronaviruses and Arteriviruses. L. Enjuanes, S.G. Sidell and W. Spaan, ed. Plenum Press, New York, 1998.

Phillips, J.J., Chua M.M., Lavi, E., Weiss, S.R. Pathogenesis of MHV4/MHV-A59 recombinant viruses: the murine coronavirus spike protein is a major determinant of neurovirulence. J Virol 1999; **73**:7752-7760.

Rowe, C.L., Baker, S.C., Nathan, M.J., Fleming, J.O. Evolution of mouse hepatitis virus: detection and characterization of spike deletion variants during persistent infection. J Virol 1997; **71**:2959-2969.

Smith, A.L., Barthold, S.W. Methods in viral pathogenesis. In: N. Nathanson (ed.), Viral pathogenesis. Lippincott-Raven, Philadelphia, 1997.

Neurovirulence for Mice of Soluble Receptor-Resistant Mutants of Murine Coronavirus JHMV

[1]SHUTOKU MATSUYAMA, [2]RIHITO WATANABE, AND [1]FUMIHIRO TAGUCHI
[1]*National Institute of Neuroscience, NCNP,4-1-1 Ogawahigashi, Kodaira Tokyo 187-8502;*
[2]*Institute of Life Science, Soka University, 1-236 Tangi Hachiohji, Tokyo192-8577, Japan*

1. INTRODUCTION

Soluble receptor-resistant (srr) mutants isolated from a highly neurotropic mouse hepatitis virus (MHV), JHMV cl-2 strain, have a single mutation in either S1 or S2 subunit of spike (S) protein (Saeki et al., 1997). These mutants were shown to grow and induce syncytia in cultured cells as efficiently as wild type (wt) virus (Saeki et al., 1997). Since it has been reported that the S protein is a key determinant in the viral virulence for animals (Dalziel et al., 1986; Fleming et al., 1986; Philips et al., 1999), we have attempted to see whether the mutations observed in srr mutants affect the neurovirulence of JHMV.

2. MATERIALS AND METHODS

2.1 Virus and animals

Neurotropic MHV JHMV cl-2 strain (wt) (Taguchi et al, 1985) and the soluble receptor-resistant (srr) mutants, srr7, srr11 and srr18, derived from cl-2 (Saeki et al., 1997) were used. Srr7, srr11 and srr18 have a mutation at amino acid position 1114 (Leu to Phe), 65 (Leu to His) and 1163 (Cys to

The Nidoviruses (Coronaviruses and Arteriviruses).
Edited by Ehud Lavi *et al.*, Kluwer Academic/Plenum Publishers, 2001.

Phe), respectively. These viruses were propagated and plaque-assayed as previously reported (Taguchi et al., 1981). Four-week-old male ICR mice, 21 to 25g body weights, proven free from the infection of MHV and other murine pathogens, were purchased from Charles River Japan. They were inoculated intracerebrally (i.c.) with various doses of wt or srr mutants in 50 μl phosphate buffered saline, pH 7.2 (PBS) and fed with commercial pellets and water *ad libitum* before and after infection.

3.   RESULTS

3.1   Neurovirulence of wt and srr mutants for ICR mice as examined by survival time and LD_{50}

Mice were inoculated i.c. with various doses of wt and srr mutants and the mortality was daily checked for 9 days after infection. All mice inoculated with $3x10^3$ PFU of wt virus and srr mutants died within 9 days postinfection (p.i.), though the mice inoculated with srr7 and srr11 survived longer than those infected with wt or srr18. Upon $3x10^2$ and $3x10$ PFU inoculation, all mice infected with wt and srr18 died, while some mice infected with srr11 and srr7 survived over the observation period. The LD50s examined on day 9 p.i. were 3 PFU, 95 PFU, 5 PFU and 3 PFU for wt virus, srr7, srr11 and srr18, respectively. These results indicate that srr7 and srr11 have reduced neurovirulence for ICR mice and srr18 is similar to wt virus.

Clinically, mice infected with wt virus developed central nervous system (CNS) symptoms, such as hypersensitivity to physical stimulants, trembling and finally paralysis, from second to third day p. i. Mice infected with srr7 or srr11 mutants started displaying such CNS symptoms from day 5 or later and showed mild and prolonged CNS symptoms. Some of those mice suffering CNS symptoms such as hind leg paralysis survived over the observation period.

3.2   Viral growth in the brain and spinal cord

Viral growth in the brain and spinal cord was examined after inoculation with 10^3 PFU of wt virus, srr7 and srr11 mutants. On day 2 p.i. the wt virus titer in the brain was about 1 and 2.5 log10 PFU/g higher than those of srr11 and srr7, respectively. Wt virus reached a plateau, 6.2 log10 PFU/g on day 2 p.i., while srr mutants failed to attain such a high titer even when they attained the plateau on day 4 p.i. Especially, srr7 grew inefficiently in both brain and spinal cord relative to the wt virus. Also, similar difference was detected in the virus titer in the liver between wt-infected mice and those infected with srr mutants.

3.3 Histopathological examination

We have compared the histopathological changes of the mice inoculated with10^3 PFU of the least virulent srr7 and those with wt virus. On day 2 p.i., there was minimal neuropathological lesions in the mice. Mild infiltration in the meningeal space and small necrotic foci scattered in parenchyma were found in the most of the infected animals. In spite of the variations among individuals, the necrosis and inflammatory changes were slightly more intensive in the wt-infected mice than in those infected with srr7. In srr7 infected mice, the clusters of small vacuolation were formed in the preventircular region and thalamus (Fig. 1A). The vacuolating lesion became obvious as a huge spongeotic lesion on day 5 or later of infection (Fig. 1B). Immunohistochemical analysis also revealed a slight difference in the localization and the intensity of viral antigens. In the brain of wt-infected mice, neurons containing viral antigen were detected in the wide area of the cortex (Fig. 1C), while the distribution of the antigen was restricted in the white matter of periventricular area in the brain of some of the srr7-infected mice (Fig. 1D). A significant difference in pathology between wt and srr7 infection was the occurrence of apoptosis in the parenchyme, which was detected by a tunnel method. Large number of apoptotic cells were observed in the periventricular and the deep cortex of the srr7-infected mice, while only a small number of apoptotic cells were encountered in the brain parenchyme of the wt-infected mice.

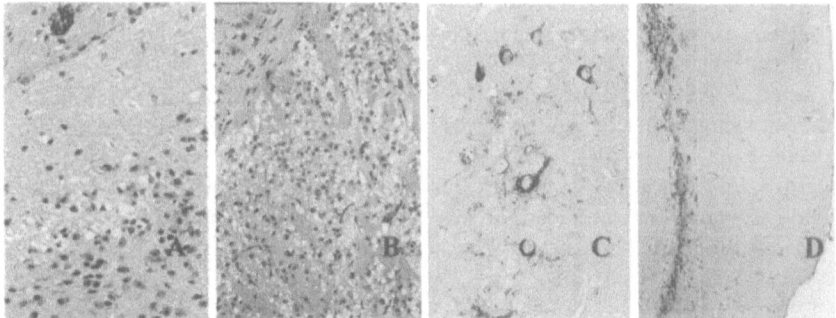

Figure 1. Vacuolating degeneration observed on day 2 p.i. in srr7 infected mouse (A). A large spongiotic change in basal nucleus 8 days after srr infection (B). Viral antigens detected in the brain of wt infected (C) and srr infected (D) mice on day 2 p.i.

4.   CONCLUSION

Three srr mutants derived from a highly neurotropic JHMV have a single amino acid mutation in the S protein. To see whether these amino acids are involved in the neurovirulence, we have examined their neuropathogenicity for ICR mice. Srr7 and srr11 caused alleviated neurological diseases relative to wt virus in terms of the LD50 and survival time, while srr18 showed virulence not significantly different from wt virus. The growth of srr7 and srr11 with reduced virulence was 1 to 2 log10 PFU/g lower than that of wt virus in the brain and spinal cords. Histopathological changes in the brain of srr7-infected mice were slightly less remarkable relative to those caused by wt virus. Vacuolation was prominent in the brain of srr7-infected mice on day 5 or later of infection. Many apoptic cells were found in srr7-infected brain while it was less evident in wt-infected mouse brain. These results suggest that the amino acids at positions 1114 and 65 mutated in srr7 and srr11 S proteins, respectively, influence the neurovirulence of JHMV.

REFERENCES

Dalziel,R.G., P.W.Lampert, P.J.Talbot, and M.J.Buchmeier. 1986. Site-specific alteration of murine hepatitis virus type 4 peplomer glycoprotein E2 results in reduced neurovirulence. J.Virol. **59**: 463- 471.

Fleming,J.O., M.D.Trousdale, F.A.K.El-Zaatari, S.A.Stohlman, and L.P.Weiner. 1986. Pathogenicity of antigenic variants of murine coronavirus JHM selected with monoclonal antibodies. J.Virol. **58**: 869-875.

Phillips, J. J., M. M. Chua, E. Lavi and S. R. Weiss. 1999. Pathogenesis of chimeric MHV4/MHV-A59 recombinant viruses: the murine coronavirus spike protein is a major determinant of neurovirulence. J. Virol. **73**: 7752-7760

Saeki, K., N. Ohtsuka, and F. Taguchi. 1997. Identification of spike protein residues of murine coronavirus responsible for receptor-binding activity by use of soluble receptor-resistant mutants. J. Virol. **71**: 9024-9031

Taguchi,F., S.G.Siddell, H.Wege, and V.ter Meulen. 1985. Characterization of a variant virus selected in rat brain after infection by coronavirus mouse hepatitis virus JHM. J. Virol. **54**: 429-435.

Taguchi,F., A.Yamada, and K.Fujiwara. 1980. Resistance to highly virulent mouse hepatitis virus acquired by mice after low-virulence infection: enhanced antiviral activity of macrophages. Infect. Immun. **29**: 42-49.

Infection of Hemagglutinating Encephalomyelitis Virus (HEV) at the Visual Pathways of Rats

[1]NORIO HIRANO, [2,3]KOUJIRO TOHYAMA, [2]NOBUYO OOTANI, AND [3]TSUTOMU HASHIKAWA

[1]Department of Veterinary Microbiology, Iwate University, Morioka 020-8550; [2]Center for Electron Microscopy and Bio-Imaging Research, and Department of Neuroanatomy, Iwate Medical University, Morioka 020-8505; [3]Laboratory of Neural Architecture, Brain Science Institute, RIKEN, Wako 351-0198, Japan

1.   INTRODUCTION

Hemagglutinating encephalomyelitis virus (HEV) is neurotropic coronavirus causing vomiting and wasting disease or encephalitis in piglets (Andries and Pensaert. 1981). In the infected animals, the virus reaches the CNS through the nerve pathways from peripheral nerve. Our previous studies have demonstrated that HEV propagated through nervous route and its infection was restricted to neurons after inoculation into sciatic nerve or footpad of rats (Hirano et al., 1995, 1998). Our findings suggest that HEV is useful as a trans-synaptic tracer for analyzing neuronal connections in the CNS. The present study was performed to examine and evaluate the usefulness of HEV as a trans-synaptic tracer in the visual pathway.

2.   MATERIALS AND METHODS

Plaque-purified HEV 67N strain was propagated and assayed for infectivity in SK-K cells as described previously (Hirano et al., 1990). Specific pathogen free of 6- to 8-week-old Wistar male rats were used. Using microsyringe, 5 µl (5 x 10^4 PFU) was inoculated into vitreous body of

The Nidoviruses (Coronaviruses and Arteriviruses).

Edited by Ehud Lavi et al., Kluwer Academic/Plenum Publishers, 2001.

149

the right eye of rats under deep anesthesia with halothane. Animals were perfused with a fixative containing 4% paraformaldehyde in 0.1M phosphate buffer under deep anesthesia with pentobarbital. The brain sections were obtained on a freezing microtome, reacted with anti-HEV 67N mouse antibody (1:1000) at 4C overnight, and then stained with FITC-conjugated goat serum (anti-mouse IgG) at room temperature for 2 hours. The stained sections were examined under a confocal laser scanning microscope.

3.   RESULTS

On day 3 after inoculation, the virus antigen was found in the retina (RT), dorsal lateral geniculate nucleus (DLG), superior colliculus (SC), and primary visual cortex (VC). In the inoculated eye, immuno-reaction was detected in vitreous body. Retinal ganglion cells as well as neurons in the inner nuclear layer were antigen positive (Fig. 1a). No antigen was found in other layers including pigmental epithelial cells. A small number of neurons in DLG were shown to be positive (Fig.1b). A few neurons in SC were immuno-positive (Fig. 2a). The morphology of antigen positive neurons in SC and DLG exhibited typical shape demonstrated with Golgi-impregnation as described in a textbook (Sefton and Dreher, 1994). Only few neurons in VC were positive (Fig. 2b). On day 7, antigen-positive neurons in SC and VC increased in number (Figs. 3a and 3b). In SC, the antigen-positive neurons and their dendrites located in the optic layer and superficial grey (Fig. 3a). In VC, positive neurons distributed in the cortical column with about 300μm in width in a patchy like fashion (Fig. 3b). The virus antigen was detected predominantly in neurons but not in glial cells including ependymal cells. During this experiment, HEV infected neurons appeared to be without cytopathological changes.

4.   DISCUSSION

After direct inoculation of HEV into eyeball, the antigen positive neurons were found in the RT, DLG, SC and VC on day 3. Latter, the number of HEV-positive neurons in DLG, SC, and VC increased with time dependent pattern. These nuclei are known to be retino-geniculo-cortical visual pathway. Such findings described above demonstrate that HEV propagation and progression is based on anterograde/trans-synaptic transport.

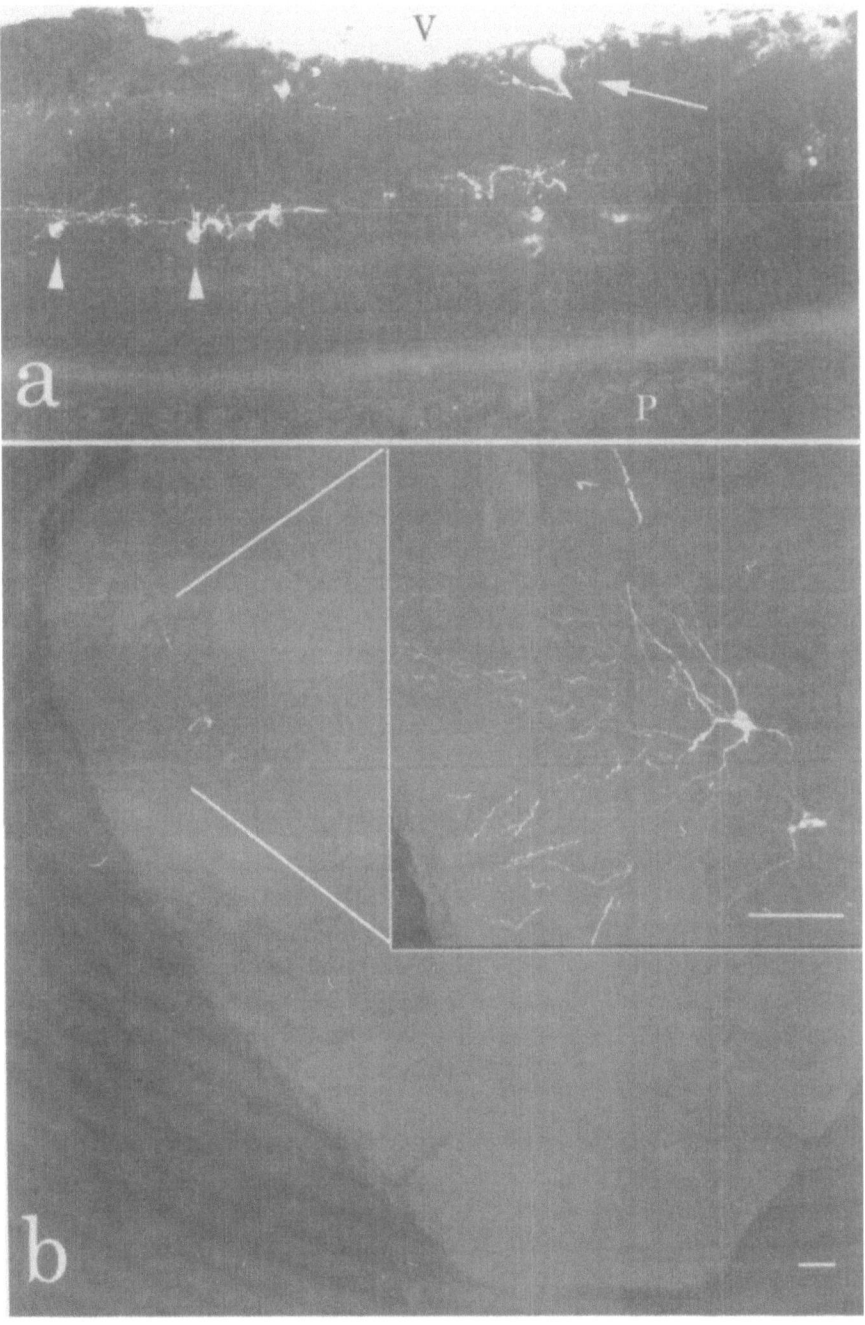

Figure 1: Day 3, a) Immuno-positive ganglion cells (arrow) and neurons in the inner nuclear layer (arrow heads) of retina. P: pigmented epithelial cells; V: vitreous body; b) A typical Immuno-positive neuron in dorsal lateral geniculate nucleus, Bar: 100 μm

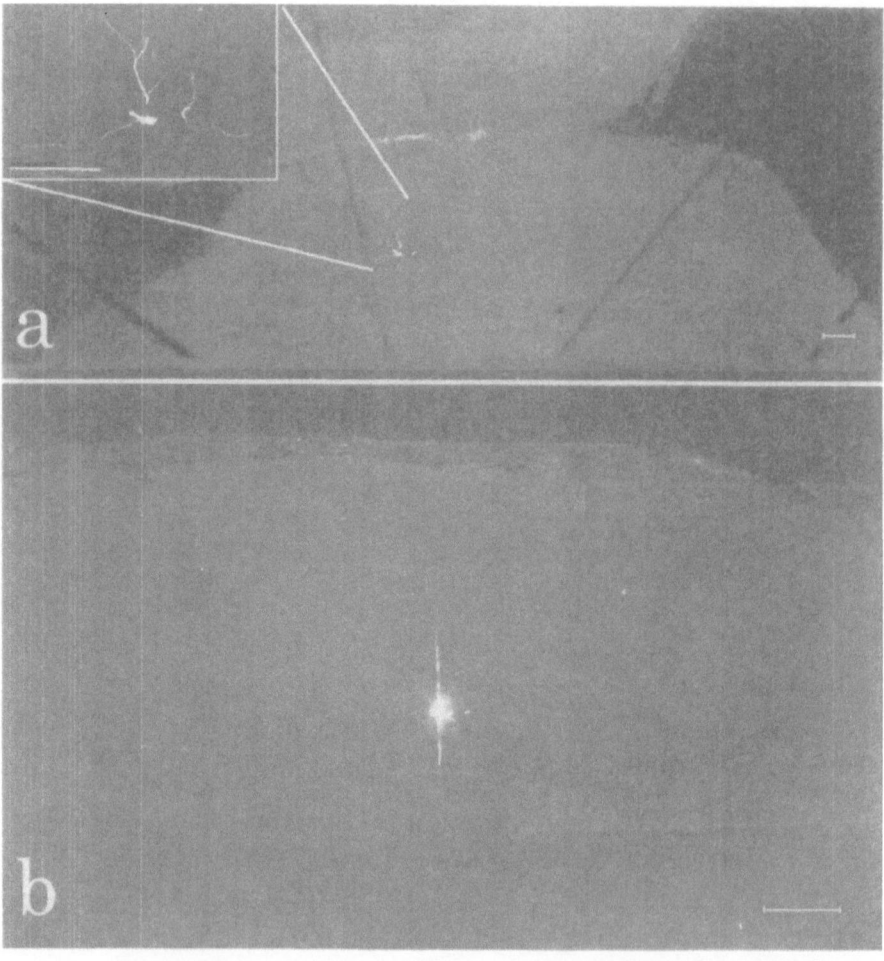

Figure 2: Day 3, a) An antigen positive neuron in the superior colliculus. b) Primary visual cortex, Bar: 100 μm

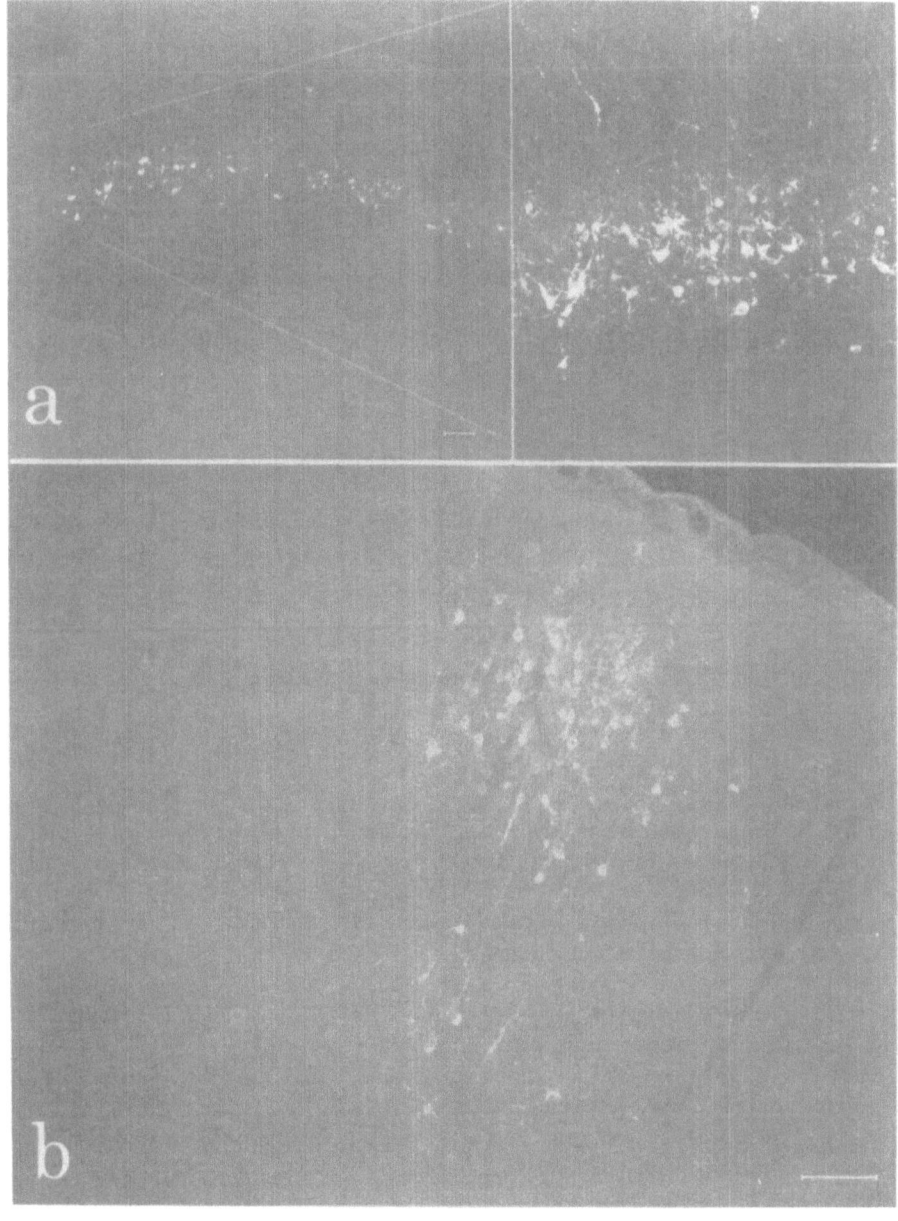

Figure 3: Day 7, a) Superior colliculus; b) Primary visual cortex. Immuno-positive neurons increased in number (see Fig. 2). Bar: 100 μm

Our observations in the present study is coincident with the results of the report on the visual system of mice infected by intraocular route with herpes simplex virus (Sun et al., 1996). HEV antigen was detected predominantly in neurons, but not in glial cells, of visual pathway nuclei in the present study. This result indicates that HEV is confirmed to be extremely neurotropic different from other viruses including mouse coronavirus (MHV), which infects not only neurons but also surrounding glial cells (Lavi et al., 1988). In other word, HEV is a causative agent which has strict neurotropism. In addition, no distinct degenerating changes were found in HEV infected neurons, suggesting that the fiber connections along visual pathway kept intact as far as the period examined. Such properties of HEV have the advantage of analyzing neuroanatomical connections in the visual pathways as a trans-synaptic tracer.

ACKNOWLEDGMENTS

This study was supported by Grands-In-Aids (No.11660309) from the Ministry of Education and Cultures of Japan and by Japan Society for the Promotion of Science.

REFERENCES

Andries, K., and Pensaert, M.B. 1981. Vomiting and wasting disease; a coronavirus infection in pigs. *AdV. Exp. Med. Biol.* **142**: 399-408.

Hirano, N,,Nomura, R., Tawara, K., Ono, K., and Iwasaki, Y. 1995. Neuronal spread of hemagglutinating encephalomyelitis virus (HEV) 67N strain in 4-week-old rats. *Adv. Exp. Med. Biol.* **390**: 117-119.

Hirano, N., Ono,K., Takasawa, H., and Haga, S. Replication and plaque formation of swine hemagglutinating encephalomyelitis virus (67N) in swine cell line, SK-K culture. *J. Virol. Methods* **27**: 91-100.

Hirano, N., Tohyama,K., Taira, H., and Hashikawa, T. 1998. Spread of   hemagglutinating encephalomyelitis virus from peripheral nerve to the   CNS. *Adv. Exp. Med. Biol.* **440**; 601-607.

Lavi, E., Fishman, P.S., Highkin, M.K., and Weiss, S.R. 1988. Limbic encephalitis after inhalation of a murine coronavirus. *Lab. Invest.* **58**: 31-36.

Sefton, A.J. and Dreher,B. 1995. Visual System. In *The rat nervous system.* 2nd ed. By Paxinos,G. pp.833-898. Academic Press, Inc. New York.

Sun, N., Cassell, M.D., and Perlman, S. 1996. Anterograde, transneuronal transport of herpes simplex virus type 1 strain H129 in the murine visual system. *J. Virol.* **70** : 5405-5413.

The Effect of the T1087N S Gene Mutation on MHV-A59 Pathogenesis

JAYASRI DAS SARMA, LI FU, AND EHUD LAVI
Department of Pathology and Laboratory Medicine, University of Pennsylvania, School of Medicine, Philadelphia, Pennsylvania, USA

1.   INTRODUCTION

Infection with mouse hepatitis virus (MHV) strain A59 produces acute hepatitis, encephalitis and chronic demyelination in mice (Lavi *et al* 1984). A closely related strain, MHV-2, produces acute meningitis and hepatitis without encephalitis and chronic demyelination (Das Sarma *et al*, submitted). ML-10 is a recombinant virus between MHV-2 and LA-7, a temperature-sensitive mutant of MHV-A59 (Keck *et al* 1985). It produces acute encephalitis, without hepatitis and demyelination. Sequencing of the entire genome of ML-10 revealed that ML-10 is a product of recombination between the leader of MHV-2 and the genome of MHV-A59 except for several mutations. Of all the differences between ML-10 and MHV-A59 only 11 nucleotide differences in the leader sequence and 4 amino acids substitutions occured in all other demyelination negative viruses. Of these, the only amino acid substitution in the S gene that was common to all demyelination negative viruses was a T1087N mutation. The objective of this study was to produce the T1087N mutation by targeted recombination and study the effect of this mutation on pathogenesis.

The Nidoviruses (Coronaviruses and Arteriviruses).
Edited by Ehud Lavi *et al.*, Kluwer Academic/Plenum Publishers. 2001.

2.   GENERATION OF THE T1087N MUTATION IN A PLASMID VECTOR

The transcription vector pMH54 (obtained from Dr. Paul Masters) was used in this study. It contains the 3' of the MHV-A59 genome downstream to the HE gene. The S gene ORF sequences of pMH54 are identical to our wild type strain MHV-A59. In addition, silent nucleotide substitutions were made at codons 12 and 13 of the S gene; this generated an AvrII site. An SbfI site was created 11 bp downstream to the termination codon of the S gene ORF (Kuo *et al* 2000). The AvrII/SbfI sites were useful for introducing different S genes into the background of pMH54. We mutated the S gene of MHV-A59 in position 1087 (T to N) by using the Quickchange™ Site-directed mutagenesis kit from Stratagene.

3.   RECOMBINATION BETWEEN THE MUTATED VECTOR AND MHV-A59

The mutated pMH54 plasmid was transfected into L2 cells along with infection of the culture with Alb4, a temperature sensitive mutant of MHV-A59, which has an N gene 87-nucleotide deletion. Upon recombination between the virus and the plasmid, the culture was switched to 39^0C in order to select against parental temperature-sensitive viruses. Only recombinant viruses that acquired the repaired N gene and temperature stability could grow in this condition. Recombinant viruses were selected, plaque purified, and further characterized. Characterization included determination of the repair of gene 7, the presence of the two restriction sites AvrII and Sbf1, and complete sequencing of the S gene to confirm the presence of the mutated amino acid in position 1087 and the absence of additional mutations. Thus we generated two recombinant viruses labelled Penn99-1 and Penn99-2, which contained all of these conditions.

4.   CHARACTERIZATION OF THE PATHOGENESIS OF THE RECOMBINANT VIRUSES

The newly generated viruses were injected into 4-week-old B6 mice and the virulence and growth kinetics in mouse organs along with organ tropism was characterized as previously described (Das Sarma *et al* 2000). The LD50 dose of both viruses was determined as 20,000 PFU, as compared to 4,000 PFU of the wt MHV-A59. Thus, the virulence of these viruses was attenuated in 4-week-old mice. Mice developed acute hepatitis and encephalitis, but required much higher doses. Viral titers were recovered from both of these organs during the acute phase of the disease. During the chronic phase, demyelination was seen in all of the mice examined.

5. CONCLUSION

In conclusion, the T1087N point mutation in MHV-A59 has an attenuating effect on virulence upon infection of 4-week-old B6 mice. However, the T1087N mutation alone is not sufficient to abolish demyelination or hepatitis.

ACKNOWLEDGMENTS

The work from our laboratory was supported by grants from the National Multiple Sclerosis Society (RG 2615-B-2) and the NIH (NS30606). We thank Dr. Paul Masters for the gifts of pMH54 and Alb4 virus, Donna Bauhof for critical review of the manuscript, and Elsa Aglow for histology expertise.

REFERENCES

Das Sarma J, L. Fu, J.C. Tsai, S.R. Weiss, and E. Lavi. 2000. Demyelination determinants map to the spike glycoprotein gene of coronavirus mouse hepatitis virus. *J. Virol.* **74**: 9206-9213.

Keck, J. G., L. H. Soe, S. Makino, S. A. Stohlman, and M. M. C. Lai. 1988. RNA recombination of murine coronavirus: recombination between fusion-positive mouse hepatitis virus A59 and fusion-negative mouse hepatitis virus 2. *J. Virol.* **62**: 1989-1998.

Kuo, L., G.-J. Godeke, M. J. B. Raamsman, P. S. Masters, and P. J. M. Rottier. 2000. Retargeting of coronavirus by substitution of the spike glycoprotein ectodomain: crossing the host cell species barrier. *J. Virol.* **74**: 1393-1406.

Lavi, E., D. H. Gilden, Z. Wroblewska, L. B. Rorke, and S. R. Weiss. 1984. Experimental demyelination produced by the A59 strain of mouse hepatitis virus. *Neurology.* **34**: 597-603.

Pathogenesis of Fusion Deficient Recombinant Mouse Hepatitis Viruses

L. DE GROOT[1], J. D. PIÑÓN[1], J. PHILLIPS[1], E. LAVI[2], AND S. R. WEISS[1]
Departments of Microbiology[1] and Pathology[2], University of Pennsylvania School of Medicine, Philadelphia, PA 19104-6076

1. INTRODUCTION

It has been shown that MHV-4 (JHM) can utilize both endosomal and nonendosomal pathways for entry into cells, depending on the strain of the virus and the nature of the cell being infected (Nash and Buchmeier, 1997). Wildtype MHV-4 fuses at the cell surface. However, persistent MHV-4 infection of a neuronal cell line (OBL21a) gave rise to the acid-dependent fusion variant OBLV60. Previous studies demonstrated that there were 8 amino acid differences between the S genes of MHV-4 and OBLV60. Elimination of neutral pH fusion was dependent only on amino acid alterations at positions 1067 (Q to H), 1094 (Q to H), and 1114 (L to R) within the first heptad repeat of the S2 subunit (Gallagher *et al*, 1991). Additionally, it was shown that intranasal inoculation with MHV-4 resulted in fatal encephalitis as a consequence of widespread distribution of infection in the brain. In contrast, the OBLV60 variant grew preferentially in the glomerular and mitral layers of the olfactory bulb and no fatal encephalitis was observed (Pearce *et al.*, 1994).

To test the hypothesis that altered entry into cells, caused by the inability to induce fusion at neutral pH, affects virulence and pathogenesis, we used targeted RNA recombination to select recombinant viruses containing the MHV-4 spike protein (S_4), carrying the three point mutations. We describe here our initial characterization of these recombinant viruses.

The Nidoviruses (Coronaviruses and Arteriviruses).
Edited by Ehud Lavi *et al.*, Kluwer Academic/Plenum Publishers, 2001.

2.   RESULTS AND DISCUSSION

We used the technique of targeted RNA recombination (Masters *et al.*, 1994), to create recombinant MHV-A59 viruses that carried either the wildtype S_4 gene or the S_4 gene encoding the Q1067H, Q1094H and L1114R mutations (Figure 1).

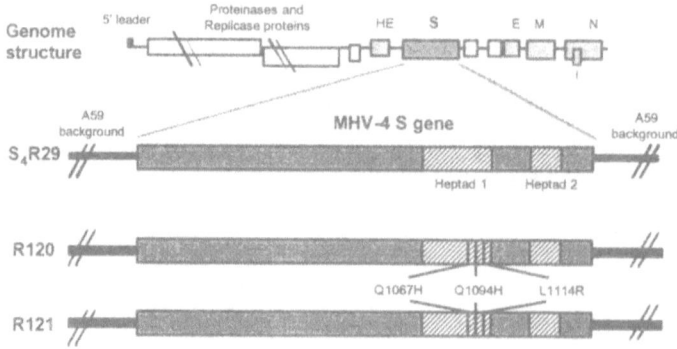

2.1   *In vitro* replication of recombinant viruses

In cell culture, the R120 and R121 recombinants showed a more efficient replication than the wildtype MHV-4 S recombinant (S_4), S_4R29. At 24 hours post-infection, R120 and R121 reached viral titers of 3.9×10^7 and 1.3×10^7 PFU/ml respectively, while S_4R29 titers were at 2.4×10^3 PFU/ml. Because the R120/R121 recombinants carry the 3 amino acid alterations (Q1067H, Q1094H, and L1114R) that are associated with a fusion negative phenotype at neutral pH, it is possible that the cells infected with these recombinants remain viable longer, due to decreased cytopathicity.

2.2   *In vitro* fusion assays

To demonstrate that these three amino acid mutations indeed confer a fusion negative phenotype to R120/R121, the ability of our recombinants to induce cell-to-cell fusion was assayed *in vitro* on L2 and DBT cells. For comparison, wildtype MHV-4 virus and the OBLV60 variant were also assayed for their ability to induce cell-to-cell fusion *in vitro*. Syncytia were visible 6 hours postinfection in cells infected with either wildtype MHV-4 virus or the S_4R29 recombinant. In contrast, in cells infected with viruses carrying the Q1067H, Q1094H, and L1114R mutations (R120/R121 and OBLV60), no fusion was observed even 20 hours postinfection (data not shown). These results confirm previous data that these 3 amino acid alterations are associated with a fusion negative phenotype at neutral pH (Gallagher *et al.*, 1991).

2.3 Effect of ammonium chloride on viral replication

We tested the effect of ammonium chloride (NH_4Cl), a lysomotropic agent, on the replication of our recombinants *in vitro*. Vesicular stomatitis virus (VSV), which is known to enter the cells via the endosomal pathway, was used as a control in our assays. The presence of NH_4Cl inhibited the replication of OBLV60 and R120 to a similar extent as VSV, and to a much greater degree than wildtype MHV-4 or the S_4R29 recombinant. At an ammonium chloride concentration of 60 mM (the highest concentration used in this study), VSV, OBLV60 and R120 viral replication was completely inhibited. At this concentration, the replication of MHV-4 and S_4R29 was only 50% inhibited. These results suggest that R120, like OBLV60 and VSV, enters cells via the endosomal pathway.

2.4 Virulence of recombinant viruses

Mice inoculated either intracranially (IC) or intranasally (IN), with serial dilutions of S_4R29 or R120/R121, were observed for disease and death and an LD_{50} value was calculated for each virus. The results are shown in Table 1. Recombinant viruses R120/R121, carrying the 3 amino acid alterations, displayed a higher LD50, or a less virulent phenotype, than S_4R29, which carries the wildtype S_4 gene.

Table 1. Virulence and replication in brain of recombinant viruses

	Virus	LD_{50} (Log)	Viral titers in brain (Log)	
			3 d.p.i.	5 d.p.i.
intracranial	S_4R29	0.45	4.39	5.69
	R120/R121	> 3.70	3.01	5.37
intranasal	$S_4R29/R22$[a]	2.74	4.49	4.68
	R120	> 5.00	3.91	6.36

[a] S4R22 (Phillips *et al.*, 1999) is an MHV-4 spike recombinant in the background of A59 independently isolated from S4R29. It has the same sequence and phenotype as S4R29 and was used to determine the LD50 by intranasal inoculation.

2.5 Replication in brain

To determine whether the observed attenuation with R120/R121 corresponded with a decreased viral replication *in vivo*, we inoculated mice either IC (10 PFU) or IN (5000 PFU). The amount of virus in the brain was determined at various times postinfection by plaque assay on L2 cell monolayers. S_4R29 and R120/R121 replicated to equivalent levels in the brain following IC inoculation (Table 1). Following IN inoculation, our initial results suggested that R120/R121 replicated to a higher level than S_4R29. These results suggest that the attenuation observed with the R120/R121 viruses is not a result of lower levels of replication *in vivo*.

3.   SUMMARY

In this study we have demonstrated that recombinant viruses carrying the amino acid mutations Q1067H, Q1094H, and L1114R were unable to induce fusion at neutral pH, replicated more efficiently in L2 cells, and that infection was delayed by ammonium chloride. These results suggest that the R120/R121 recombinants most likely use the endosomal pathway to enter cells. In this sense they are similar to the pH-dependent MHV-4 variant OBLV60.

We were able to observe an attenuated virulence *in vivo*, despite the fact that our R120/R121 recombinants replicated to comparable (IC) or higher (IN) titers than the S_4R29 recombinant in the brain. Preliminary results showed that the level of inflammation observed in infected mice is consistent with the attenuated virulence, but they cannot be explained by the high titers of replication.

REFERENCES

Gallagher, T.M., C. Escarmis, M.J. Buchmeier. 1991. Alterations of pH dependence of coronavirus-induced cell fusion: effect of mutations in the spike glycoprotein. *J. Virol.* **65**:1916-1928.

Masters, P.S., C.A. Koetzner, C.A. Kerr, and Y. Heo. 1994. Optimization of targeted RNA recombination and mapping of a novel nucleocapsid gene mutation in the coronavirus mouse hepatitis virus. *J.Virol.* **68**:328-337.

Nash, T.C., and M.J. Buchmeier. 1997. Entry of mouse hepatitis virus into cells by endosomal and nonendosomal pathways. *Virology* **233**: 1-8

Pearce, B.D., M.V. Hobbs, T.S. McGraw and M.J. Buchmeier. 1994. Cytokine induction during T-cell-mediated clearance of mouse hepatitis virus from neurons in vivo. *J.Virol.* **68**:5483-5495

Phillips, J.J., M.M. Chua, E. Lavi, and S.R. Weiss. 1999. Pathogenesis of chimeric MHV-4/MHV-A59 recombinant viruses: the murine coronavirus spike protein is a major determinant of neurovirulence. *J. Virol.* **73**: 7752-7760

Programmed Cell Death in MHV-Induced Demyelination

TALYA SCHWARTZ, LI FU, AND EHUD LAVI

Department of Pathology and Laboratory Medicine, University of Pennsylvania, School of Medicine, Philadelphia, PA, USA

1. INTRODUCTION

Coronavirus mouse hepatitis virus (MHV) strain A59 causes severe acute hepatitis, focal meningoencephalitis, and chronic demyelinating disease of the spinal cord. It serves as an experimental model for multiple sclerosis. MHV-2 causes acute hepatitis and meningitis. The pathogenesis of MHV-induced demyelination in mice is not clear and a potential mechanism of apoptosis was suggested. Previous studies showed apoptotic T cells, astrocytes, and oligodendrocytes in demyelinating areas following MHV-JHM infection (Barac-Latas *et al* 1997). However, the distribution of demyelination has been suggested to correlate better with macrophage infiltration than with the apoptotic cells (Wu and Perlman 1999). In the chronic demyelinating disease induced by Theiler's virus, apoptotic astrocytes were found in demyelinating lesions (Palma *et al* 1999). In experimental allergic encephalitis, the disease was much milder in mice lacking Fas-Fas ligand molecules, suggesting that apoptosis plays a role in the disease (Waldner *et al* 1997).

The Nidoviruses (Coronaviruses and Arteriviruses).
Edited by Ehud Lavi *et al.*, Kluwer Academic/Plenum Publishers, 2001.

2.   METHODS AND MATERIALS

2.1   Viruses

MHV-A59 and MHV-2 were grown and titered on L2 cells. Viral titers were determined by standard plaque assays.

2.2   Mice

4-week-old male C57BL/6 and B6MRL-Fas-lpr (lpr) mice were obtained from The Jackson Laboratory (Bar Harbor, Maine). C57BL/6 mice were injected intracerebrally (I.C.) with 2.5×10^3, 25 and 5 PFU of MHV-A59, 25 and 5 PFU of MHV-2. B6MRL-Fas-lpr mice were injected I.C. with 25 and 5 PFU of MHV-A59. Mock-infected mice were injected I.C. with L2 cell lysate.

2.3   Histology

Mice were sacrificed on days 1,3,5,7,9,11,and 30 (2-3 mice per time point per virus, during the acute stage and 52 mice on day 30). Brain, spinal cord and liver were removed and placed in 10% normal buffered formalin and embedded in paraffin. Five μm thick sections were stained with LFB and H&E.

2.4   TUNEL

Detection of in situ DNA fragmentation was done with fluorescein *in situ* cell death detection kit (Boehringer Mannheim, Indianapolis, Indiana) as specified by the manufacturer.

2.5   Double labeling

Double labeling for TUNEL with viral antigen, and TUNEL with specific markers for astrocytes, oligodendrocytes, and macrophages was performed by immunohistochemistry. The analysis was performed by the avidin-biotin-phosphatase based technique (Biomeda, Foster City, CA) using fast-red or vector blue (Vector laboratories, Burlingame, CA) as a staining substrate and a 1:200 dilution of rabbit anti-MHV-A59 polyclonal antibody, 1:100 dilution of monoclonal glial fibrillary acidic protein (Lee *et al.* 1984),

1:500 carbonic anhydrase II (Cammer *et al* 1985) or 1:50 rat anti mouse F4/80 antigen (Serotec Inc., Raleigh, NC), respectively.

3.   RESULTS

3.1   Acute Stage

TUNEL staining was detected in the brains and livers of mice infected with MHV-A59 and MHV-2. Extensive liver apoptosis was observed in both MHV-2 and MHV-A59 infections. Co-localization of A59 viral antigen and TUNEL staining was detected in hepatocytes. Apoptosis was found in the brain parenchyma of MHV-A59 infected mice and meningial apoptosis was found in both infections. The kinetics, intensity and pattern of apoptosis correlated with the inflammatory events. Mock-infected mice were apoptosis-negative in all tissues. No apoptosis was identified in the spinal cord during the acute stage.

3.2   Chronic stage

TUNEL staining was observed exclusively in the spinal cords with demyelinating lesions of 15 mice infected with MHV-A59. No apoptosis was detected in the spinal cords of 3 MHV-A59 infected mice without demyelination, 28 MHV-2 infected mice and 10 control mice. TUNEL staining was negative in the brains and the livers of all infected mice.

3.3   Fas knock-out mice

To assess the role of the apoptotic cascade in the demyelination process, lpr mice were infected with MHV-A59 virus. Two of 5 mice in both groups developed demyelination, however the extent of demyelination in the spinal cord was significantly less in the lpr mice compared to wt mice (10% and 34% demyelinating quadrants respectively).

3.4   Apoptotic cell types

Double labeling for TUNEL-positive nuclei and specific markers for astrocytes, oligodendrocytes and macrophages was observed. Quantification studies demonstrated 3-5% oligodendrocytes, 1-2% astrocytes and 70% macrophages, double stained.

4.   DISCUSSION

Apoptosis was observed in mice following infection with MHV-A59 and MHV-2. The kinetics, intensity and pattern of the apoptotic staining correlated well with the distribution of inflammation. However, apoptosis was found in both inflammatory cells (macrophages and possibly lymphocytes), and parenchymal cells such as hepatocytes, oligodendrocytes and astrocytes.

Several findings suggest that apoptosis may play an active role in the process of tissue damage and demyelination in MHV infection. 1. The finding of apoptosis in hepatocytes, oligodendrocytes and astrocytes; 2. The high prevalence of apoptotic cells in the demyelinating lesions; 3. The absence of CNS apoptosis in mice infected with a non-demyelinating virus MHV-2 and control mice; 4. The significantly lower extent of demyelination in lpr mice. A recent study did not find any difference in the extent of demyelination between wt and lpr mice 13 days following JHM infection (Parra *et al* 2000). However, the different results may be due to the differences in strains of the virus or time points examined.

In conclusion, apoptosis may play an important role in both acute and chronic MHV disease. The relationship between apoptosis, inflammation and tissue damage is yet to be defined, but can possibly be defined through further dissection of the apoptotic cascade.

ACKNOWLEDGMENTS

The work from our laboratory was supported by grants from the National Multiple Sclerosis Society (RG 2615-B-2) and the NIH (NS30606). We thank Donna Bauhof for critical review of the manuscript and Elsa Aglow for histology expertise.

REFERENCES

Barac-Latas V., G. Suchanken, H. Breitschop, A. Stuehler, H. Wege, H. Lassmann. Patterns of oligodendrocytes pathology in coronavirus induced sub-acute demyelinating encephalomyelitis in the Lewis rat.1997. *Glia* **19**:1-12.

Cammer W., Sacchi R., Sapirstein V. Immunohistochemical locaiozation of carbonic anhydrase in the spinal cords of normal nd mutant (shiverer) adult mice with comparison among fixation methods.1985 *J Histochem Cytochem.* **33**(1):45-54.

Lee VM, Page CD., Wu HL., Schlaepfer WW. Monoclonal antibodies to gel-excised glial filament protein and their reactivities with other intermediate filament proteins.1984. *J Neurochem.* **42**(1):25-32.

Palma JP, Yauch RL., Lang S., Kim BS. Potential role of CD4+ T cell mediated apoptosis of activated astrocytes in Theiler's virus induced demyelination. 1999. *J. Immunol* **162**:6543-6551.

Parra B., Lin MT, Stohlman SA, Bergmann C., Atkinson R., Hinton D. Contribution of Fas-Fas ligand interaction to the pathogenesis of mouse hepatitis virus in the central nervous system. 2000. *J Virol.* **74**(5):2447-2450.

Sungwhan AN., Chen CJ, Yu X., Leibowitz JL., Makino S. Induction of apoptosis in murine coronavirus infected cultured cells and demonstration of E protein as an apoptosis inducer. 1999. *J Virol.* **73**(9):7853-7859.

Waldner H., R.A. Sobel, E. Howard, V. K. Kuchroo. Fas- and FasL deficient mice are resistant to induction of autoimmune encephalomyelitis. 1997 *J. Immunol.* **159**:3100-3103.

Wu GF, Perlman S. Macrophages infiltration, but not apoptosis, is correlated with immune mediated demyelination following murine infection with a neurotropic coronavirus.1999. *J Virol* **73**(10):8771-8780.

In vitro Properties and Pathogenesis of A59/MHV4 Chimeric Mouse Hepatitis Viruses

JEAN C. TSAI AND SUSAN R. WEISS

Department of Microbiology, University of Pennsylvania School of Medicine, Philadelphia, Pennsylvania,19104-6076, U.S.A

1.   INTRODUCTION

Early in infection, the mouse hepatitis virus (MHV) envelope glycoprotein spike (S) binds a cell surface receptor glycoprotein and is postulated to undergo a conformational change to reveal a hydrophobic domain, promoting the fusion of the viral envelope and the cell membrane. The actual steps of MHV entry and their effects on pathogenesis are largely unknown. By binding truncated S1 peptides to receptor glycoproteins blotted on membranes, the putative receptor binding domain (RBD) was mapped to the N-terminal 330 amino acids of S (Kubo et al., 1994). In this study, we have generated two groups of chimeric recombinant viruses in an isogenic background in order to study the role of the RBD in pathogenesis.

MHV strains A59 and MHV-4 (JHM) provide a frame-work to study virus entry and pathogenesis. Although both strains replicate to similar levels in the brain, MHV-4 induces a more severe encephalitis in extent and intensity (Phillips et al., 1999). Interestingly, both seem to infect similar cell types. On the other hand, MHV-4 barely replicates in the liver and causes minimal hepatitis, whereas A59 replicates well and causes moderate to severe hepatitis. The outcome of infection is vastly different; the LD_{50} of MHV-4 is three logs lower than that of A59.

The Nidoviruses (Coronaviruses and Arteriviruses).
Edited by Ehud Lavi *et al.*, Kluwer Academic/Plenum Publishers, 2001.

2.   RESULTS

2.1   Selection of Recombinant Viruses

All of the recombinant viruses, shown in Figure 1, contain either the A59, MHV-4, or an A59/MHV-4 chimeric spike, in the A59 background, and were selected independently from each other. All viruses grow in L2 cell cultures with similar kinetics and to similar extents.

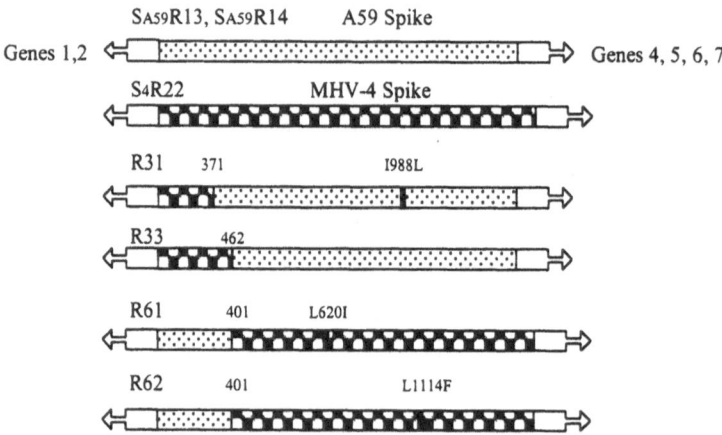

Figure 1. Schematic of the fully sequenced S genes of the Recombinant Viruses. Shown are the 5' end amino acid substitutions in the chimeric spikes and the secondary mutations. The methods for selection were described previously (Masters et al., 1994).

2.2   *In vitro* Characteristics

The recombinant viruses with chimeric spike glycoproteins are likely to exhibit *in vitro* characteristics which are different from the recombinant viruses with parental spike glycoproteins, resulting in altered functionality in virus entry and/or subsequent steps in pathogenesis. We have begun to identify these changes through a number of assays, including tissue culture assays, in order to correlate the *in vitro* properties to some functional role during infection.

Thermolability measures the loss of infectious virus over time when incubated at a given temperature. In our study, a possible mechanism is the spike glycoprotein's tendency to undergo a conformational change which impairs its function in virus entry and/or a subsequent step in the infectious cycle. Interestingly, all the chimeric recombinant viruses, except R33, are highly stable compared to the recombinant viruses which contain the parental spike glycoproteins (Table 1).

Soluble receptor (sR) neutralization further characterizes the spike glycoprotein's ability to alter conformations, this time, in the presence of receptor glycoproteins. Replacing the S_4R22 RBD with the A59 RBD cause these chimeric recombinant viruses, R61 and R62 to resist neutralization. We do not see the same stabilizing shift when the $S_{A59}R14$ RBD is replaced with the MHV-4 RBD (Table 1).

Receptor-independent fusion is the phenomenon that certain infected cells, depending on the virus strain, can form syncytia with neighboring cells which do not express the cell surface receptor glycoprotein. These results parallel those of sR neutralization (Table 1).

Table 1. The stability of the viruses and virus-spike interactions

	Thermolability[a]	sR Neutralization[b]	R-indep. fusion[c]
$S_{A59}R14$	2.7	0.4	-
R31	0.8	0.5	-
R33	2.3	1.3	-
S_4R22	3.7	3.9	+
R61	1.2	0.4	-
R62	1.9	0.5	-

[a] Viral titres are taken after 30 minutes of incubation at pH 8.7 and 42°C. Both thermolability and sR Neutralization are presented here as the difference between the starting and the residual log titres.
[b] MHVR-Ig, constructed with domain 1 of CEACAM1[a] ligated to the immunoglobulin Fc domain (Gallagher 1997), is the soluble receptor (sR) used to neutralize viruses. The log titre after one hour of incubation at 37°C is subtracted from that incubated without receptors at 4°C.
[c] Infected L2 cells are overlaid onto BHK cell and syncytium observed after 24 hours.

2.3   Pathogenesis

We studied viral replication and virulence in C57Bl/6 mouse to determine whether or not the *in vitro* phenotypes attributed to S can correlate to or provide a mechanism for their pathogenesis (Table 2).

Table 2. Viral virulence and replication

	LD$_{50}$ [log(pfu)]	Replication [log(pfu/ml)]	
		Brain	Liver
$S_{A59}R13$	3.4	6.0	4.8
R31	>5.4	3.1	2.4
R33	5.1	4.8	2.9
S_4R22	0.1	nd	nd
R61	>4.4	3.3	2.7
R62	>4.0	3.3	2.3

Methods are as described (Phillips et al., 1999), except that 5000 pfu were injected for replication.

S$_4$R22 replicates similarly to S$_{A59}$R13 in the brain and minimally in the liver (data not shown). The chimeric recombinant viruses are attenuated in both replication and virulence to varying degrees.

3.   DISCUSSION

The RBD alone is not sufficient to confer the ability of the spike to undergo certain receptor-dependent or independent conformational changes. Thermolability, receptor-independent fusion, and soluble receptor neutralization all depend on mulitple regions of the spike.

Although the mechanism is unclear, it is interesting that the level of increased thermal stability of the chimeras, except R33, compared to the parental viruses, correlates with the level of *in vivo* attenuation.

The loss of S$_4$R22's ability for receptor-independent fusion and soluble receptor neutralization after the substitution of its MHV-4 spike RBD, in chimera R61 and R62, may result from a nonspecific disruption of the spike structure. Nevertheless, the fact that the loss of these *in vitro* properties correlate with *in vivo* attenuation leads us to hypothesize that these properties of S may be important for the pathogenesis of MHV-4.

REFERENCES

Gallagher, T. M., 1997, A role for naturally occurring variation of the murine coronavirus spike protein in stabilizing association with the cellular receptor. *J. Virol.* **71**:3129-3137.

Kubo, H., Yamada, Y. K., and Taguchi, F., 1994, Localization of neutralizing epitopes and the receptor-binding site within the amino-terminal 330 amino acids of the murine coronavirus spike protein. *J. Virol.* **68**:5403-5410.

Masters, P. S., Koetzner, C. A., Kerr, A. C., and Heo, Y., 1994, Optimization of targeted RNA recombination and mapping of a novel nucleocapsid gene mutation in the coronavirus mouse hepatitis virus. *J. Virol.* **68**:328-337.

Phillips, J. J., Chua, M. M., Lavi, E., and Weiss, S. R., 1999, Pathogenesis of chimeric MHV4/MHV-A59 recombinant viruses: the murine coronavirus spike protein is a major determinant of neurovirulence. *J. Virol.* **73**:7752-7760.

Receptor Specificity and Receptor-Induced Conformational Changes in Mouse Hepatitis Virus Spike Glycoprotein

[1]KATHRYN V. HOLMES, [1]BRUCE D. ZELUS, [1]JEANNE H. SCHICKLI, AND [2]SUSAN R. WEISS
[1]*University of Colorado Health Sciences Center, Denver, Colorado;* [2]*University of Pennsylvania School of Medicine, Philadelphia, Pennsylvania*

1.   INTRODUCTION

Coronavirus spike (S) glycoproteins bind to specific glycoproteins on host cell membranes that serve as virus receptors. Receptors for S glycoproteins of several coronaviruses have been identified (Dveksler, *et al.*, 1991; Dveksler, *et al.*, 1993; Chen, *et al.*, 1995; Yeager, *et al.*, 1992; Delmas, *et al.*, 1992; Tresnan, *et al.*, 1996; Benbacer, *et al.*, 1997). The specificity of virus/receptor interactions is an important determinant of the species-specificity of coronavirus infection, and may play a role in the tissue tropism and virulence of coronavirus diseases (Kolb, *et al.*, 1997; Ballesteros, *et al.*, 1997; Sanchez, *et al.*, 1999). The mechanism of entry has been studied in great detail for several enveloped viruses including influenza A virus, avian leukosis viruses and HIV-1 (Kemble, *et al.*, 1994; Stegmann, *et al.*, 1990; Chen, *et al.*, 1999; Gilbert, *et al.*, 1995; Hernandez and White, 1998; Zhang, *et al.*, 1999; Turner and Summers, 1999). Binding of the spike glycoprotein on the viral envelope to the receptor on the cell membrane may induce specific, pre-programmed conformational changes in the spike protein and/or the receptor that bring together the lipid bilayers of the viral envelope and the cell membrane. A fusion pore is created that expands to permit the entry of the viral nucleocapsid into the cytoplasm, leading to virus

The Nidoviruses (Coronaviruses and Arteriviruses).
Edited by Ehud Lavi *et al.*, Kluwer Academic/Plenum Publishers, 2001.

infection. This chapter will summarize the specificity of the interactions of murine coronavirus MHV and mutants of MHV with cellular receptors, and discuss evidence that receptor binding induces a temperature-dependent conformational change in the MHV S glycoprotein that may play a role in virus entry and MHV-induced cell fusion.

2.   MHV RECEPTORS

As receptors for entry into susceptible murine cells, MHV strains use isoforms of several cellular membrane proteins in the immunoglobulin superfamily, formerly called MHVR or Bgp, that are now called CEACAMs (Beauchemin, *et al.*, 1999). The CEACAM1a gene of mice has 4 isoforms that consist of either two or four Ig-like exodomains linked by a transmembrane domain to either a short or a long cytoplasmic tail (Nedellec, *et al.*, 1995). The long tail isoform can be phosphorylated on a tyrosine residue, and both long and short tail isoforms can be phosphorylated on serine residues. It is not known whether phosphorylation of CEACAM1 affects its MHV receptor activity. MHV-resistant adult SJL/J mice express isoforms of a different allele of the receptor, called CEACAM1b (Yokomori and Lai, 1992; Dveksler, *et al.*, 1993). The CEACAM1 glycoproteins are expressed on apical membranes of epithelial cells of the respiratory and enteric tracts, proximal renal tubules, vascular endothelium, and on B cells, macrophages and activated T cells (Godfraind, *et al.*, 1995; Coutelier, *et al.*, 1994). CEACAMs are homophilic cell adhesion proteins that can play a role in signalling (Turbide, *et al.*, 1991; Huber, *et al.*, 1999). A CEA-related murine pregnancy specific glycoprotein called bCEA also has receptor activity for MHV-JHM (Chen, *et al.*, 1995). Several human CEACAM glycoproteins serve as receptors for virulent strains of *Neisseria meningitidis, Neisseria gonorrhoeae,* and *Hemophilus influenzae* (Virji, *et al.*, 1999; Virji, *et al.*, 2000).

3.   MECHANISM OF RECEPTOR-DEPENDENT, TEMPERATURE-DEPENDENT VIRUS NEUTRALIZATION

Purified, soluble, anchorless murine CEACAM glycoproteins with a 6 histidine tag can neutralize the infectivity of MHV at 37°C in a concentration-dependent manner (Zelus, *et al.*, 1998; Kubo, *et al.*, 1994). We found that the receptor isoforms had different virus neutralizing activities and that MHV strains differed in the specificity of receptor interactions. We have used two complementary approaches to study the mechanism of temperature-dependent, receptor-dependent virus neutralization: a liposome flotation assay to detect changes in the hydrophobicity of virions, and a protease susceptibility assay to identify the domain of the viral spike protein that undergoes conformational change.

3.1 Liposome Flotation Assay for Conformational Change in S

We investigated whether binding at $37^{O}C$ of soluble receptor glycoproteins made MHV-A59 virions more hydrophobic so that they would bind to liposomes and float up in a sucrose density gradient. The location of virions in the gradients was determined by immunolabeling of gradient fractions or by assaying for ^{3}H-uridine label from RNA in virions. Table 1 shows that binding of the soluble CEACAM1a receptor with 4 Ig-like domains to MHV-A59 virions at $37^{O}C$ and pH 6.5 markedly increased the hydrophobicity of virions. No increase in hydrophobicity was observed following incubation without receptor, or at $4^{O}C$, pH 6.5, with receptor.

We previously used reactivity with anti-S monoclonal antibodies to show that incubation of MHV-A59 virions at $37^{O}C$ and pH 8.0, but not pH 6.5, induced a conformational change in the availability of epitopes on the viral S protein (Weismiller, *et al.*, 1990; Sturman, *et al.*, 1990). Therefore, we performed the liposome flotation assay on MHV-A59 virions that had been incubated at $37^{O}C$ for 30 minutes at pH 8.0 without added soluble receptor. As summarized in Table 1, we found that the MHV-A59 virions increased in hydrophobicity when treated at pH 8.0, $37^{O}C$ for 30 minutes.

Table 1. Receptor binding or pH 8.0 at $37^{O}C$ increases hydrophobicity of MHV-A59 Virions*

Incubation Mix	pH	Temperature (^{O}C)	Location of virions in gradient	Hydrophobicity
MHV-A59 + LS	6.5	4	Bottom	Unchanged
		37	Bottom	Unchanged
MHV-A59 + CEACAM1a + LS	6.5	4	Bottom	Unchanged
		37	**Top**	**Increased**
MHV-A59 + LS	8.0	4	Bottom	Unchanged
		37	**Top**	**Increased**

* LS indicates liposomes

3.2 Protease Susceptibility Assay for Conformational Change in S Glycoprotein

We previously showed that incubation of MHV-A59 virions with high concentrations of trypsin cleaved the viral S glycoprotein only at one of many potential trypsin cleavage sites, and that this cleavage was associated with increased cell fusion activity of S (Sturman and Holmes, 1977; Frana, *et al.*, 1985). After incubating MHV-A59 virions for 30 minutes at 37°C, pH 6.5, with soluble MHV receptor CEACAM1a, the mixture was chilled and incubated at 4°C for 30 minutes with 10 ug/ml trypsin. Analysis of the viral S protein and its fragments on SDS-PAGE gels by immunoblotting with monoclonal anti-S antibodies showed that the S2 domain of the spike glycoprotein was proteolytically degraded, while the S1 domain that binds to the receptor was more resistant to protease degradation. Similarly, immunoblotting with monoclonal antibodies showed that S2, but not S1, was degraded after incubation of virions for 30 minutes at 37°C, pH 8.0, without soluble receptor, followed by incubation with trypsin for 30 minutes at 4°C (Table 2). These observations from the protease susceptibility assay indicate that a specific conformational change in the S2 domain of the viral spike glycoprotein can be triggered either by incubation with soluble receptor at pH 6.5, 37°C, or by incubation without receptor at pH 8.0, 37°C.

Table 2. Receptor binding or pH 8.0 at 37°c makes S2 protein of MHV-A59 virions susceptible to degradation by trypsin at 4°c.

Incubation Mix	pH	Temperature (°C)	S1 Protein Degraded?	S2 Protein Degraded?
MHV-A59	6.5	4	No	No
		37	No	No
MHV-A59 + CEACAM1a	6.5	4	No	No
		37	No	**Yes**
MHV-A59	8.0	4	No	No
		37	No	**Yes**

These experiments show that the S glycoprotein of MHV-A59 has a dynamic structure that undergoes pre-programmed conformational change(s) following binding to receptor at 37°C or exposure to alkaline pH at 37°C. These changes are likely to be important for the functions of the S protein including fusion of the viral envelope with cell membranes and/or virus-induced cell fusion.

3.3   Alternative Trypsin Cleavage Site in H716D S Glycoprotein

Viruses isolated from persistently infected glial cells had a H716D mutation adjacent to the cleavage site in the MHV-A59 spike glycoprotein and had a small plaque, delayed fusion phenotype on the 17 Cl 1 line of murine fibroblasts (Gombold, *et al.*, 1993). Targeted RNA recombination was used to introduce this mutation into the genome of wild type MHV-A59 (Kuo, *et al.*, 2000; Weiss and Hingley, in preparation). We studied the effects of soluble receptor at 37°C on the hydrophobicity of mutant virions and on trypsin cleavage of the viral S protein. We found that this virus, which differs from wild type MHV-A59 by only one amino acid, behaved quite differently from the wild type virus in both the liposome flotation assay and the protease assay. The H716D mutant virions floated part way to the top of the gradient at 4°C, pH 6.5, even without any soluble receptor. The protease cleavage assay detected a conformational change in S of H716D that was dependent upon soluble receptor and 37°C, or pH 8.0 and 37°C. As expected, trypsin treatment of H716D virions did not generate S1 and S2 peptides of approximately 90 kDa. In the presence of the receptor or pH 8.0 at 37°C, H716D S protein was cleaved at a new site to form a peptide of approximately 125 kDa. Immunoblotting with mapped monoclonal antibodies to S showed that this peptide contains epitopes in S1 and the N-terminal region of S2. This alternative or secondary cleavage site in S that is exposed by treatment with soluble receptor or pH 8.0 at 37°C may facilitate fusion of the viral envelope with host cell membranes, and/or for virus-induced cell fusion. The small plaque, delayed fusion phenotype of this virus may be due to the effect of the mutation in S on the dynamics of the mutant S glycoprotein.

4.   ROLE OF S GLYCOPROTEIN IN ALTERED SPECIES SPECIFICITY OF MHV FROM PERSISTENTLY INFECTED CELLS

Coronaviruses can cause persistent infection *in vitro* and *in vivo*. Persistent infection of murine cell lines with MHV results in a carrier culture that is characterized by downregulation of the expression of the MHV receptor CEACAM1 and selection of small plaque, sometimes temperature sensitive viruses, many of which have mutations in the S glycoprotein (Holmes and Behnke, 1981; Sawicki, *et al.*, 1995; Schickli, *et al.*, 1997; Lavi, *et al.*, 1987; Gombold, *et al.*, 1993; Gallagher, 1991; Chen and Baric, 1996). A small plaque virus (MHV/BHK) isolated from murine cells persistently infected with MHV-A59 utilizes the CEACAM1a receptor less efficiently than wild type virus (Schickli, *et al.*, 1997). Several other MHV isolates from persistently infected murine cells also have a broadly extended host range and can infect cell lines from hamsters, rats, monkeys, humans, and/or cats in addition to mice (Schickli, *et al.*, 1997; Baric, *et al.*, 1997; Baric, *et al.*, 1999). Multiple mutations were identified in the S proteins of these viruses. MHV/BHK virus differed markedly from the parental MHV-A59 virus in the liposome flotation and protease susceptibility assays. The marked conformational changes induced in MHV-A59 S protein by soluble receptor at 37°C were not observed for MHV/BHK, and the S protein of MHV/BHK showed a conformational change at pH 5.0 and 37 °C, rather than pH 8.0 like MHV-A59.

We co-infected murine cells with wild type MHV-A59 and with the MHV/BHK virus, and selected recombinants that grew in hamster cells at 39.5°C. These recombinant viruses were characterized for plaque morphology, cell fusing activity, reactivity with anti-S monoclonal antibodies, restriction fragment pattern in the S gene, susceptibility to protease cleavage of S, and host range. The S genes were sequenced to identify the site of recombination in S between the two parental genomes. The recombinant that had the least amount of the MHV/BHK S gene but still could be propagated in hamster cells had the N-terminal 1493 nucleotides of MHV/BHK with the remainder of the gene identical to MHV-A59. This region contained a 21 base insertion that encoded amino acids TRTKKVP as well as 24 point mutations upstream of the insert. Thus, the N-terminal approximately 410 amino acids of the S protein are responsible for the extended host range of MHV/BHK. This is similar to the observation that MHV-JHM binds to the CEACAM1 receptor on murine cells by the N-terminal 330 amino acids of the S glycoprotein (Kubo, *et al.*, 1994). Interspecies transmission of murine coronavirus depends upon mutations in the N-terminal domain of the S protein. It will be interesting to determine which of the mutations in the recombinant virus containing this domain of MHV/BHK is/are responsible for the extended host range of this virus. The evolution of coronaviruses to generate viruses that can infect and become adapted to new hosts probably depends on selection of mutations in this region of the S protein. How likely a receptor-jumping event is to occur in nature may depend upon how many mutations are required to recognize a receptor in the new host and what selective pressures in growth of the virus favor the replication of the variant virus over the parental virus.

ACKNOWLEDGEMENTS

The authors are grateful for the excellent technical assistance of Fenna Phibbs. Isabel Laparc Goffart selected the H716D mutant. This work was funded by NIH grants AI 25231NS30606 and NS 21954.

REFERENCES

Ballesteros, M. L., C. M. Sanchez, and L. Enjuanes. 1997. Two amino acid changes at the N-terminus of transmissible gastroenteritis coronavirus spike protein result in the loss of enteric tropism. Virology. 227:378-388.

Baric, R. S., B. Yount, L. Hensley, S. A. Peel, and W. Chen. 1997. Episodic evolution mediates interspecies transfer of a murine coronavirus. J. Virol. 71:1946-1955.

Baric, R. S., E. Sullivan, L. Hensley, B. Yount, and W. Chen. 1999. Persistent infection promotes cross-species transmissibility of mouse hepatitis virus. J. Virol. 73:638-649.

Beauchemin, N., T. Chen, P. Draber, G. Dveksler, P. Gold, S. Gray-Owen, F. Grunert, S. Hammarstrom, K. V. Holmes, A. Karlson, M. Kuroki, S. H. Lin, L. Lucka, S. M. Najjar, M. Neumaier, B. Obrink, J. E. Shively, K. M. Skubitz, C. P. Stanners, P. Thomas, J. A. Thompson, M. Virji, S. von Kleist, C. Wagener, S. Watt, and W. Zimmermann. 1999. Redefined nomenclature for members of the carcinoembryonic antigen family. Exp. Cell Res. 252:243-249.

Benbacer, L., E. Kut, L. Besnardeau, H. Laude, and B. Delmas. 1997. Interspecies aminopeptidase-N chimeras reveal species-specific receptor recognition by canine coronavirus, feline infectious peritonitis virus, and transmissible gastroenteritis virus. J. Virol. 71:734-737.

Chen, D. S., M. Asanaka, K. Yokomori, F. Wang, S. B. Hwang, H. P. Li, and M. M. Lai. 1995. A pregnancy-specific glycoprotein is expressed in the brain and serves as a receptor for mouse hepatitis virus. Proc. Natl. Acad. Sci U. S. A. 92:12095-12099.

Chen, J., J. J. Skehel, and D. C. Wiley. 1999. N- and C-terminal residues combine in the fusion-pH influenza hemagglutinin HA(2) subunit to form an N cap that terminates the triple- stranded coiled coil. Proc. Natl. Acad. Sci U. S. A. 96:8967-8972.

Chen, W. and R. S. Baric. 1996. Molecular anatomy of mouse hepatitis virus persistence: coevolution of increased host cell resistance and virus virulence. J. Virol. 70:3947-3960.

Coutelier, J. P., C. Godfraind, G. S. Dveksler, M. Wysocka, C. B. Cardellichio, H. Noel, and K. V. Holmes. 1994. B lymphocyte and macrophage expression of carcinoembryonic antigen-related adhesion molecules that serve as receptors for murine coronavirus. Eur. J. Immunol. 24:1383-1390.

Delmas, B., J. Gelfi, R. L'Haridon, L. K. Vogel, H. Sjostrom, O. Noren, and H. Laude. 1992. Aminopeptidase N is a major receptor for the entero-pathogenic coronavirus TGEV. Nature 357:417-420.

Dveksler, G. S., M. N. Pensiero, C. B. Cardellichio, R. K. Williams, G. S. Jiang, K. V. Holmes, and C. W. Dieffenbach. 1991. Cloning of the mouse hepatitis virus (MHV) receptor: expression in human and hamster cell lines confers susceptibility to MHV. J. Virol. 65:6881-6891.

Dveksler, G. S., C. W. Dieffenbach, C. B. Cardellichio, K. McCuaig, M. N. Pensiero, G. S. Jiang, N. Beauchemin, and K. V. Holmes. 1993. Several members of the mouse carcinoembryonic antigen-related glycoprotein family are functional receptors for the coronavirus mouse hepatitis virus-A59. J. Virol. 67:1-8.

Frana, M. F., J. N. Behnke, L. S. Sturman, and K. V. Holmes. 1985. Proteolytic cleavage of the E2 glycoprotein of murine coronavirus: host-dependent differences in proteolytic cleavage and cell fusion. J. Virol. 56:912-920.

Gallagher, T. M. 1991. Alteration of the pH dependence of coronavirus-induced cell fusion: effect of mutations in the spike glycoprotein. J. Virol. 65:1916-128..

Gilbert, J. M., L. D. Hernandez, J. W. Balliet, P. Bates, and J. M. White. 1995. Receptor-induced conformational changes in the subgroup A avian leukosis and sarcoma virus envelope glycoprotein. J. Virol. 69:7410-7415.

Godfraind, C., S. G. Langreth, C. B. Cardellichio, R. Knobler, J. P. Coutelier, M. Dubois-Dalcq, and K. V. Holmes. 1995. Tissue and cellular distribution of an adhesion molecule in the carcinoembryonic antigen family that serves as a receptor for mouse hepatitis virus. Lab. Invest. 73:615-627.

Gombold, J. L., S. T. Hingley, and S. R. Weiss. 1993. Fusion-defective mutants of mouse hepatitis virus A59 contain a mutation in the spike protein cleavage signal. J. Virol. 67:4504-4512.

Hernandez, L. D. and J. M. White. 1998. Mutational analysis of the candidate internal fusion peptide of the avian leukosis and sarcoma virus subgroup A envelope glycoprotein. J. Virol. 72:3259-3267.

Holmes, K. V. and J. N. Behnke. 1981. Evolution of a coronavirus during persistent infection in vitro. Adv Exp Med Biol 142:287-299.

Huber, M., L. Izzi, P. Grondin, C. Houde, T. Kunath, A. Veillette, and N. Beauchemin. 1999. The carboxyl-terminal region of biliary glycoprotein controls its tyrosine phosphorylation and association with protein-tyrosine phosphatases SHP-1 and SHP-2 in epithelial cells. J. Biol Chem. 274:335-344.

Kemble, G. W., T. Danieli, and J. M. White. 1994. Lipid-anchored influenza hemagglutinin promotes hemifusion, not complete fusion. Cell 76:383-391.

Kolb, A. F., A. Hegyi, and S. G. Siddell. 1997. Identification of residues critical for the human coronavirus 229E receptor function of human aminopeptidase N. J. Gen. Virol. 78:2795-2802.

Kubo, H., Y. K. Yamada, and F. Taguchi. 1994. Localization of neutralizing epitopes and the receptor-binding site within the amino-terminal 330 amino acids of the murine coronavirus spike protein. J. Virol. 68:5403-5410.

Kuo, L., G. J. Godeke, M. J. Raamsman, P. S. Masters, and P. J. Rottier. 2000. Retargeting of coronavirus by substitution of the spike glycoprotein ectodomain: crossing the host cell species barrier. J. Virol. 2000. Feb;74(3):1393-406. 74:1393-1406.

Lavi, E., A. Suzumura, M. Hirayama, M. K. Highkin, D. M. Dambach, D. H. Silberberg, and S. R. Weiss. 1987. Coronavirus mouse hepatitis virus (MHV)-A59 causes a persistent, productive infection in primary glial cell cultures. Microb. Pathog. 3:79-86.

Nedellec, P., C. Turbide, and N. Beauchemin. 1995. Characterization and transcriptional activity of the mouse biliary glycoprotein 1 gene, a carcinoembryonic antigen-related gene. Eur. J. Biochem. 231:104-114.

Sanchez, C. M., A. Izeta, J. M. Sanchez-Morgado, S. Alonso, I. Sola, M. Balasch, J. Plana-Duran, and L. Enjuanes. 1999. Targeted recombination demonstrates that the spike gene of transmissible gastroenteritis coronavirus is a determinant of its enteric tropism and virulence. J. Virol. 73:7607-7618.

Sawicki, S. G., J. H. Lu, and K. V. Holmes. 1995. Persistent infection of cultured cells with mouse hepatitis virus (MHV) results from the epigenetic expression of the MHV receptor. Journal of Virology 69:5535-5543.

Schickli, J. H., B. D. Zelus, D. E. Wentworth, S. G. Sawicki, and K. V. Holmes. 1997. The murine coronavirus mouse hepatitis virus strain A59 from persistently infected murine cells exhibits an extended host range. J. Virol. 71:9499-9507.

Stegmann, T., J. M. White, and A. Helenius. 1990. Intermediates in influenza induced membrane fusion. EMBO J. 9:4231-4241.

Sturman, L. S. and K. V. Holmes. 1977. Characterization of coronavirus II. Glycoproteins of the viral envelope: tryptic peptide analysis. Virology. 77:650-660.

Sturman, L. S., C. S. Ricard, and K. V. Holmes. 1990. Conformational change of the coronavirus peplomer glycoprotein at pH 8.0 and 37 degrees C correlates with virus aggregation and virus-induced cell fusion. J. Virol. 64:3042-3050.

Tresnan, D. B., R. Levis, and K. V. Holmes. 1996. Feline aminopeptidase N serves as a receptor for feline, canine, porcine, and human coronaviruses in serogroup I. J. Virol. 70:8669-8674.

Turbide, C., M. Rojas, C. P. Stanners, and N. Beauchemin. 1991. A mouse carcinoembryonic antigen gene family member is a calcium-dependent cell adhesion molecule. J. Biol Chem. 266:309-315.

Turner, B. G. and M. F. Summers. 1999. Structural biology of HIV. J. Mol. Biol 285:1-32.

Virji, M., D. Evans, A. Hadfield, F. Grunert, A. M. Teixeira, and S. M. Watt. 1999. Critical determinants of host receptor targeting by Neisseria meningitidis and Neisseria gonorrhoeae: identification of Opa adhesiotopes on the N-domain of CD66 molecules. Mol. Microbiol. 34:538-551.

Virji, M., D. Evans, J. Griffith, D. Hill, L. Serino, A. Hadfield, and S. M. Watt. 2000. Carcinoembryonic antigens are targeted by diverse strains of typable and non-typable Haemophilus influenzae. Mol. Microbiol. 2000. May;36(4):784-95. 36:784-795.

Weismiller, D. G., L. S. Sturman, M. J. Buchmeier, J. O. Fleming, and K. V. Holmes. 1990. Monoclonal antibodies to the peplomer glycoprotein of coronavirus mouse hepatitis virus identify two subunits and detect a conformational change in the subunit released under mild alkaline conditions. J. Virol. 64:3051-3055.

Yeager, C. L., R. A. Ashmun, R. K. Williams, C. B. Cardellichio, L. H. Shapiro, A. T. Look, and K. V. Holmes. 1992. Human aminopeptidase N is a receptor for human coronavirus 229E. Nature. 357:420-422.

Yokomori, K. and M. M. Lai. 1992. The receptor for mouse hepatitis virus in the resistant mouse strain SJL is functional: implications for the requirement of a second factor for viral infection. J. Virol. 66:6931-6938.

Zelus, B. D., D. R. Wessner, R. K. Williams, M. N. Pensiero, F. T. Phibbs, M. deSouza, G. S. Dveksler, and K. V. Holmes. 1998. Purified, soluble recombinant mouse hepatitis virus receptor, Bgp1(b), and Bgp2 murine coronavirus receptors differ in mouse hepatitis virus binding and neutralizing activities. J. Virol. 72:7237-7244.

Zhang, W., G. Canziani, C. Plugariu, R. Wyatt, J. Sodroski, R. Sweet, P. Kwong, W. Hendrickson, and I. Chaiken. 1999. Conformational changes of gp120 in epitopes near the CCR5 binding site are induced by CD4 and a CD4 miniprotein mimetic. Biochemistry 38:9405-9416.

Murine Coronavirus Spike Glycoprotein
Receptor binding and membrane fusion activities

THOMAS M. GALLAGHER

Department of Microbiology and Immunology, Loyola University Medical Center, Maywood, IL 60153

1.   INTRODUCTION

The mature coronavirus spike is a large, oligomeric, type I integral membrane glycoprotein that projects about 20 nm from the surface of infected cells and virions. In this extracellular position, the spikes function to bind the cellular receptors extending from opposing membranes. Following receptor binding, largely hypothetical structural changes take place to generate spike conformations that are capable of mediating fusion of the juxtaposed membranes. Membrane fusion creates pores for entrance of genomes into uninfected cells.

In their role as agents of viral genome entry and dissemination, the spikes encounter varied physical, chemical and biological conditions. These different environmental conditions can impact spike protein structure, often impairing their participation in the infection process. For example, spikes may complex with antibodies (Dalziel *et al.*, 1986), or spikes may become exposed to changes in pH and temperature (Sturman *et al.*, 1990), in ways that can render them incompetent during virus entry. To persist in these changing environments, coronavirus populations adapt by natural selection and this process fixes mutations into the spike genes. With extended selection in different niches, spikes collectively establish a rich genetic diversity that can be exploited by coronavirologists. For example, researchers interested in the tropism and pathogenesis of the coronaviruses have discovered that naturally-occurring spike mutations can lead to

The Nidoviruses (Coronaviruses and Arteriviruses).
Edited by Ehud Lavi *et al.*, Kluwer Academic/Plenum Publishers, 2001.

dramatic changes in the targeting of *in vivo* infection (Leparc-Goffart *et al.*, 1998, Sanchez *et al.*, 1999).

This chapter describes results extending from investigations on the entry of a unique strain of murine coronavirus, the JHM strain of murine hepatitis virus (MHV) (Cheever *et al.*, 1949). MHV strain JHM is extremely pathogenic to mice, causing acute, fatal encephalitis. In tissue culture, infectious JHM viruses are difficult to obtain in abundance because the spikes are unstable at 37°C temperature and slightly basic pH. Amplification of JHM in tissue culture naturally selects for variants of greater stability. These variants contain mutations in spike genes that correlate with reduced *in vivo* pathogenicity (Fazakerley *et al.*, 1992) and reduced spike-mediated membrane fusion activity (Gallagher *et al.*, 1990). JHM and its tissue-culture variants provide the source material to study the molecular basis of coronavirus entry, dissemination and pathogenesis.

2.   METHODS AND RESULTS

Spike protein synthesis begins with the co-translational insertion of nascent chains into the lumen of the endoplasmic reticulum (ER). Newly synthesized proteins assemble into oligomers, probably dimers (Vennema *et al.*, 1990) and then proceed through the exocytic pathway of the infected cell. Spikes are heavily glycosylated and are palmitoylated near their carboxy-terminal transmembrane anchors. The most notable additional spike modification takes place within the trans-Golgi network, where cellular endoprotease(s) recognize a central stretch of multibasic residues found within spikes of a selected group of strains (Cavanagh, 1995). This proteolysis generates an amino-terminal, peripheral S1 of 769 amino acids, and carboxy-terminal, integral-membrane S2 of 607 amino acids (Fig. 1). These two fragments remain noncovalently associated for variable time periods, depending on the strain under investigation.

During a natural coronavirus infection, most spikes that leave the ER associate with other viral components to become incorporated into secreted virions (Opstelten *et al.*, 1995). To avoid this loss of spike proteins from cells via their incorporation into secreted virions, most of our studies involve spike biosynthesis in the absence of other coronavirus proteins, a process that is accomplished by expressing spike cDNA from vaccinia vectors (Gallagher, 1997). We synthesize many different spike mutants ("strains") in this way, and the spikes are routinely investigated using a series of straightforward assays. First, spikes are monitored for synthesis and post-translational transport through the exocytic pathway by pulse-chase radiolabelling with [^{35}S] methionine, immunoprecipitation and autoradiographic detection after electrophoresis. Second, spikes inside cells or on plasma membranes are assayed for their ability to bind receptors by measuring their adsorption to a soluble form of the murine hepatitis virus receptor (sMHVR). Third, spikes on plasma membranes are further assayed for their ability to mediate membrane fusion by measuring syncytium formation following incubation with suitable target cells. We have used these three approaches to identify structure-function relationships in the coronavirus spike.

2.1 Biosynthesis of spike proteins: Formation of the MHVR-binding site

One question about coronavirus entry concerns the binding sites for MHVR on the spike proteins. Binding of MHVR to spikes is thought to promote the formation of spike conformations capable of membrane fusion. However, the way that this hypothetical induction might occur, or even how the MHVR interacts with spikes, is largely unknown. Our studies of binding between soluble MHVR proteins and spikes are aimed at addressing these questions.

The sMHVR-Fc (Fig. 1) is a bifunctional binding agent. Its amino-terminal region is identical to the "N" domain of mouse CEACAM1a (Beauchemin *et al.*, 1999) and will therefore bind MHV spike proteins (Dveksler *et al.*, 1993), while its carboxy-terminal region is a human IgG1 Fc and will therefore bind to protein A or G in immunoprecipitations. Using sMHVR-Fc in immunoprecipitations of metabolically labeled spikes, we discovered that spikes synthesized within a 10 min pulse period were not efficiently captured. Chase periods of 1 h or more were required for spikes to develop structures capable of recognition by sMHVR-Fc (see, for example, Fig. 4). Newly-synthesized spikes were sedimented down sucrose gradients, and the spikes in gradient fractions were captured by either sMHVR-Fc or by polyclonal anti-spike antiserum. The results revealed that the antiserum captured a spectrum of spike conformations ranging from ~ 5S to ~16S, while the sMHVR-Fc only captured the small amount of ~15S material that is consistent with the sedimentation rate of spike dimers (Fig. 2). Thus the formation of the sMHVR-Ig binding site was correlated with formation of spike protein oligomers.

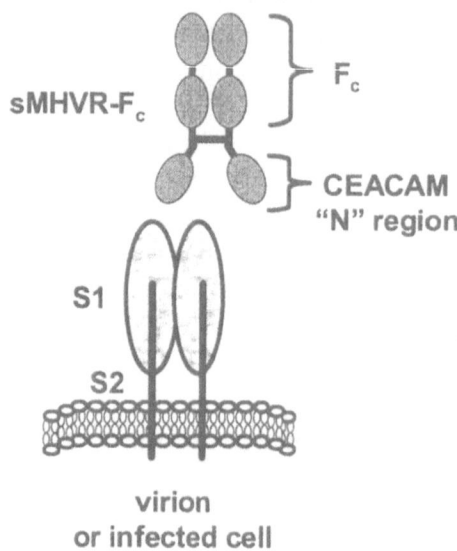

Figure 1. Depiction of spike and sMHVR-Fc proteins. The S1 subunit is attached to S2 through noncovalent bonding. The "N" region of CEACAM binds directly to S1. sMHVR-Fc is a disulfide-linked homodimer.

The results of Kubo *et al.* (1994) indicate that cDNAs encoding only the amino-terminal 330 residues of spike could be expressed from vaccinia vectors, and the secreted peptide fragments could bind to immobilized MHVR proteins. If the MHVR-binding site was dependent on spike oligomerization, then the S_{330} fragment should harbor oligomerization determinant(s). We tested this possiblity by expressing two cDNAs within a single cell culture; one encoded S_{330} and the other encoded a complete S1 fragment, S_{769}. The existence of $S_{769}:S_{330}$ hetero-oligomers would provide support for the hypothesis. Indeed, a monoclonal antibody (mAb J.2.6; Fleming *et al.*, 1983) with specificity for residues between 510-540 (data not shown), did indeed immunoprecipitate both S_{769} and S_{330} from the culture lysate (Fig. 3; lane 4). The same mAb captured only S_{769} when produced alone (lane 2), and failed to recognize S_{330} when produced alone (lane 3). This data supported the contention that an oligomerization region lies within the amino-terminal 330 residues of spike.

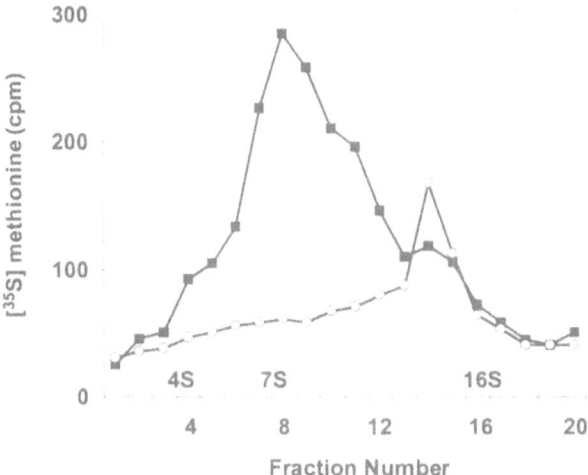

Figure 2. Capture of newly-synthesized [^{35}S] spikes after their sedimentation on sucrose gradients. HeLa cells were infected with recombinant vaccinia vectors encoding spike proteins. After pulse-labelling with [^{35}S] methionine from 7.5 to 8.0 hours postinfection, cells were dissolved with nonionic detergent and lysates were sedimented on sucrose gradients. Spikes in gradient fractions were immunoprecipitated with either anti-JHM antiserum (solid squares) or with sMHVR-Fc (open circles) and the captured radioactivity was counted by scintillation spectrometry. Markers for sedimentation rate were horseradish peroxidase (4S), immunoglobulin G (7S) and β-galactosidase (16S).

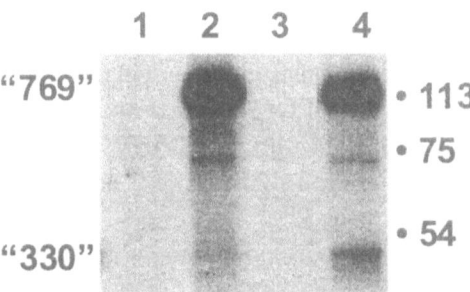

Figure 3. Co-immunoprecipitation of S_{769} and S_{330} fragments by mAb J.2.6. Spike fragments were synthesized in HeLa cells from vaccinia vectors and were metabolically radiolabelled with [^{35}S] methionine. After immunoprecipitation of proteins from lysates, SDS-polyacrylamide gel electrophoresis and flourography were performed to reveal the "769" and "330" fragments. The position of molecular weight markers is indicated at the right in kilodaltons. Lane 1: No S (negative control). Lane 2: S_{769}. Lane 3: S_{330}. Lane 4: S_{769} + S_{330}.

3.2   Cell-surface presentation of spike proteins: Stability of the S1-S2 interaction and its relationship to the membrane fusion reaction

Oligomerized S1-S2 complexes encoded by JHM are exceedingly thermolabile. At 37°C, the infectivity of JHM virus particles declines rapidly as spikes become denatured. Tissue-culture growth of JHM gives rise to more thermostable variants, some of which have fixed mutations into the spike gene. Some of these thermostability mutations are easy to identify as they are relatively large ~ 450 nucleotide, in-frame deletions within the S1-encoding region (Gallagher *et al.*, 1990). The "deletion-prone region" (DPR) of S1 is always within residues ~ 420 to ~ 600.

Mutations associated with increased thermostability are also associated with long-lived S1-S2 interactions. This was demonstrated by comparing the kinetics of S1-S2 formation and decay for S_{JHM} and one of its variants, $S_{\Delta DPR2}$ ($\Delta 437$-585). Pulse-chase radiolabelling, immunoprecipitation with sMHVR-Fc, and S1 / S2 detection after electrophoresis revealed that a stable, cell-associated S1-S2 complex was never observed for S_{JHM} (Fig. 4A). It appeared that an S1-S2 complex decayed quickly after it was formed. S1 was always found in culture media, while free S2 could not be identified in cells or media and was presumably rapidly degraded. In contrast, $S1_{\Delta DPR2}$ -S2 complexes were readily captured from cell lysates and relatively little free $S1_{\Delta DPR2}$ was found in culture media (Fig. 4B). These results depicting stable maintenance of $S1_{\Delta DPR2}$ on the plasma membrane were corroborated by our finding that cultures synthesizing $S_{\Delta DPR2}$ readily absorbed exogenously-added sMHVR-Ig at levels over 10 times that of S_{JHM} (data not shown). Thus the JHM strain of MHV can accumulate spike mutations that increase the duration of the S1-S2 association and thereby help to maintain virus infectivity at 37°C.

The JHM spike protein is far more potent at mediating membrane fusion than its more thermostable ΔDPR variants. This was demonstrated by performing a series of assays for membrane fusion-dependent cytoplasmic mixing between spike-bearing (effector) and MHVR-bearing (target) cells. To measure the extent of fusion-dependent cytoplasmic mixing, we used the assay of Nussbaum *et al.* (1994) which involves infecting target cells with vCB21R-lacZ (Alkhatib *et al.*, 1996), thereby introducing a transcriptionally-silent β galactosidase reporter gene under T7 promoter control. Since spike-bearing effector cells are infected with vTF7.3 (Fuerst *et al.*, 1987), which encodes T7 RNA polymerase, effector:target cell fusion brings T7 RNA polymerase in contact with the $T7_{pro}$-lacZ and β-galactosidase is synthesized.

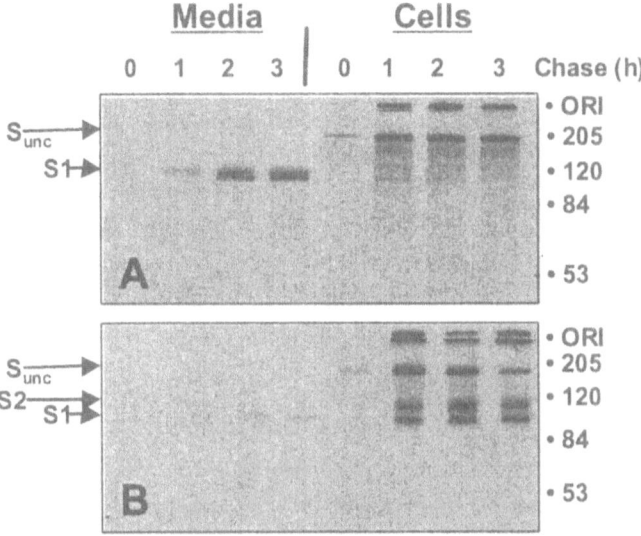

Figure 4. Evaluation of spike protein synthesis, proteolytic fragmentation and elution into media. HeLa cells producing spike proteins were radiolabelled with [^{35}S] methionine for 10 min and then chased for the indicated time periods before cell lysis and immunoprecipitation with sMHVR-Fc. Captured spike proteins were revealed after SDS-polyacrylamide gel electrophoresis and flourography. Molecular weight markers are indicated at the right in kilodaltons. S_{unc} denotes the uncleaved precursor of S1 and S2 fragments. Panel A: S_{JHM}. Panel B: $S_{\Delta DPR2}$. This figure was adapted from Krueger *et al.*, submitted.

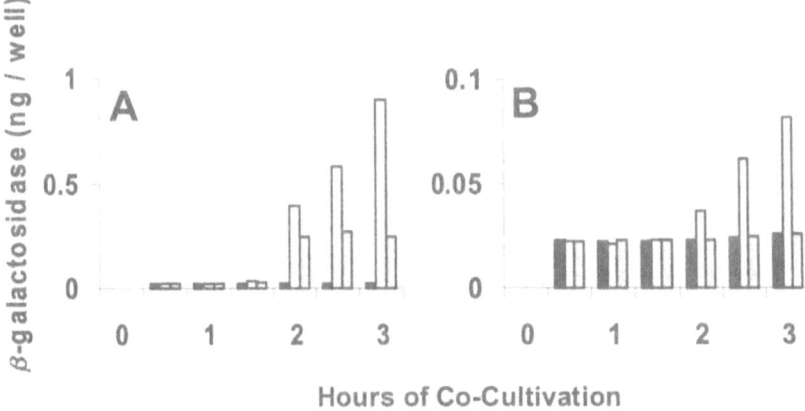

Figure 5. Quantitative assessment of intercellular fusion indicating that S_{JHM} is uniquely capable of MHV receptor-independent syncytium formation. Cytoplasmic mixing that occurs within syncytia of co-cultivated cells was measured by β-galactosidase production, as described in the text. Solid bars; no spikes on effector cells (negative control). Open bars; S_{JHM} effector cells. Shaded bars; $S_{\Delta DPR2}$ effector cells. Panel A: Hela-MHVR (+R) target cells were overlaid onto the designated S-bearing effector cell monolayers, and lysate-associated β-galactosidase activities were measured at the indicated times after co-cultivation. Panel B: HeLa (-R) target cells were used. This figure was adapted from Krueger *et al.*, submitted.

When target cells displayed the MHVR (Rao and Gallagher, 1998), the $S_{\Delta DPR2}$ stimulated β-galactosidase production that was about 25% the level of S_{JHM} (Fig. 5A). A more striking discovery was made from a parallel fusion assay in which target cells lacked the MHVR. Here the $S_{\Delta DPR2}$ did not stimulate any β-galactosidase production, indicating that the MHVR is required to generate a fusion-competent conformation. However, S_{JHM} could indeed fuse with cells lacking the MHVR, as assessed by β-galactosidase induction (Fig. 5B). A hyperactive membrane fusion activity that can occur even without prior binding of the MHVR is therefore correlated with the extremely labile $S1_{JHM}$-S2 interaction.

4.   DISCUSSION

The JHM spike is a very large glycoprotein of 1376 amino acids and 20 predicted asparagine-linked carbohydrates. Its folding in the ER is complex, likely requiring assistance from numerous ER chaperones. Using sMHVR-Fc as an oligomer-specific immunoadhesin, we found that full-length spikes are synthesized and then slowly folded into oligomers in a process requiring about one hour. It is interesting that spike proteins were not captured by soluble MHV receptors until they formed oligomers (Fig. 2), indicating that additional folding events take place during or shortly after oligomerization to form the MHVR-binding site.

The large spike protein can be functionally subdivided in that the MHVR-binding site can form when only about one-fourth of the protein (residues 1-330) are synthesized (Kubo *et al.*, 1994). We found that the S_{330} fragment contains region(s) conferring homo-oligomerization (Fig. 3). It is not yet clear whether each monomer of the spike homo-oligomer contains a separate and independent MHVR-binding site. It is also not clear whether the binding of MHVR to this amino-terminal region causes conformational changes in the carboxy-terminal S2 fragments that are thought to carry out the membrane fusion reaction. Such changes would have to occur however if the MHVR is to be considered a trigger for spike-mediated membrane fusion.

Limiting step(s) on the pathway to S1-S2 complexes on the plasma membrane occurs between formation of the uncleaved and cleaved spike oligomer. This is evident from the detection of uncleaved spikes throughout every 3 h chase period (see Fig. 4). However, small proportions of each spike population do progress through the exocytic pathway during the chase periods. For JHM, the spikes that reach the cell surface do not maintain their MHVR-binding sites. The S1 fragment, along with its site for receptor interaction, is rapidly jettisoned. The S2 that remains cell-associated is rapidly degraded. Tissue-culture adaptation selects for mutations that delay this process of S1-S2 elution. One example of this is depicted in this chapter – the deletion of S1 residues 437-585 (ΔDPR2) permits stable S1-S2 complexes (Fig. 4). Increased stability of the S1-S2 complex was correlated with reduced membrane fusion activity, as the $S_{\Delta DPR2}$ was far less potent in mediating intercellular fusion despite its continued accumulation on cell surfaces (Fig. 5).

It is clear that the coronaviruses are capable of encoding a spectrum of different spike structures. Some of these structures may be extremely unstable, thereby creating a relatively small free energy barrier between the so-called "native" and "fusion-active" conformations. The JHM spike is the prototypic example here. For JHM, the one-way conversion to the fusion-active conformation does not even require the free energy that is presumably released upon MHVR interaction, making it capable of "MHVR-independent" development of syncytia. Maintenance of virus infectivity at 37°C requires increased stability of the protruding spikes, and this occurs through mutation and selection. Increased stability of the native conformation necessarily places a higher energy barrier in front of the "fusion-active" state, and thus the presumed energy of MHVR interaction is required to activate membrane fusion by mutant spikes. Relative to the prototype JHM, viruses with stabilized spike proteins are associated with reduced *in vivo* virulence and with slower spread of infection (Fazakerley *et al.*, 1992). Perhaps this restricted dissemination is explained by *in vivo* environments that are unable to drive the tissue-culture adapted spikes through a high-energy barrier and into the fusion-active state.

ACKNOWLEDGMENTS

This work was supported by Public Health Service Research Grant NS31616 from the National Institutes of Health.

REFERENCES

Alkhatib, G., Broder, C.C. and Berger, E.A., 1996, Cell-type specific fusion cofactors determine human immunodeficiency virus type 1 tropism for T-cell lines versus primary macrophages. *J. Virol.* **70**: 5487-5494.

Beauchemin, N., Draber, P., Dveksler, G., Gold, P., Gray-Owen, S., Grunert, F., Hammarstrom, S., Holmes, K., Karlsson, A., Kuroki, M., Lin, S.H., Lucka, L., Najjar, S., Neumaier, M., Obrink, B., Shively, J., Skubitz, K., Stanners, C., and Thomas, P., 1999, Redefined nomenclature for members of the carcinoembryonic antigen family. *Exptal. Cell Research* **252**: 243-249.

Cavanagh, D., 1995, The coronavirus surface glycoprotein. In *The Coronaviridae* (S. G. Siddell, ed.), Plenum Press, New York and London, pp. 73-115.

Cheever, F.S., Daniels, J.B., Pappenheimer, A.M., and Bailey, O.T., 1949, A murine virus (JHM) causing disseminated encephalomyelitis with extensive destruction of myelin: Isolation and biological properties of the virus. *J. Exp. Med.* **90**:181-194.

Dalziel, R.G., Lampert, P.W., Talbot, P.J., and Buchmeier, M.J., 1986, Site-specific alteration of murine hepatitis virus type 4 peplomer glycoprotein E2 results in reduced neurovirulence. *J. Virol.* **59**: 463-471.

Dveksler, G.S., Pensiero, M.N., Dieffenbach, C.W., Cardellichio, C.B., Basile, A.A., Elia, P.E., and Holmes, K.V., 1993, Mouse hepatitis virus strain A59 and blocking antireceptor monoclonal antibody bind to the N-terminal domain of cellular receptor. *Proc. Natl. Acad. Sci. USA* **90**: 1716-1720.

Fazakerley, J. K., Parker, S.E., Bloom, F., and Buchmeier, M.J., 1992, The V5A13.1 envelope glycoprotein deletion mutant of mouse hepatitis virus type-4 is neuroattenuated by its reduced rate of spread in the central nervous system. *Virology* **187**:178-188.

Fleming, J. O., Stohlman, S.A., Harmon, R.C., Lai, M.M.C., Frelinger, J.A., and Weiner, L.P., 1983, Antigenic relationships of murine coronaviruses: analysis using monoclonal antibodies to JHM (MHV-4) virus. *Virology* **131**:296-307.

Fuerst, T. R., Earl, P.L., and Moss, B., 1987, Use of a hybrid vaccinia virus-T7 RNA polymerase system for expression of target genes. *Mol. Cell. Biol.* **7**:2538-2544.

Gallagher, T.M., Parker, S.E., and Buchmeier, M.J., 1990, Neutralization-resistant variants of a neurotropic coronavirus are generated by deletions within the amino-terminal half of the spike glycoprotein. *J. Virol.* **64**:731-741.

Gallagher, T.M.,1997, A role for naturally occurring variation of the murine coronavirus spike protein in stabilizing association with the cellular receptor. *J. Virol.* **71**:3129-3137.

Krueger, D.K., Kelly, S.M., Lewicki, D.N., Ruffolo, R., and Gallagher, T.M., 2000, The unique receptor-independent membrane fusion activity of MHV strain JHM is eliminated by mutations in disparate regions of the spike gene. *J. Virol.* Submitted.

Kubo, H., Yamada, Y.K., and Taguchi, F., 1994, Localization of neutralizing epitopes and the receptor-binding site within the amino-terminal 330 amino acids of the murine coronavirus spike protein. *J. Virol.* **68**: 5403-5410.

Leparc-Goffart, I., Hingley, S.T., Chua, M.M., Phillips, J., Lavi, E., and Weiss, S.R., 1998, Targeted recombination within the spike gene of murine coronavirus mouse hepatitis virus A59: Q159 is a determinant of hepatotropism. *J. Virol.* **72**: 9628-9636.

Nussbaum, O., Broder, C.C., and Berger, E.A., 1994, Fusogenic mechanisms of enveloped-virus glycoproteins analyzed by a novel recombinant vaccinia virus-based assay quantitating cell fusion-dependent reporter gene activation. *J. Virol.* **68**: 5411-5422.

Opstelten, D.-J.E., Raamsman, M.J.B., Wolfs, K., Horzinek, M.C., and Rottier, P.J.M., 1995, Envelope glycoprotein interactions in coronavirus assembly. *J. Cell Biol.* **131**: 339-349.

Rao, P.V., and Gallagher, T.M., 1998, Intracellular complexes of viral spike and cellular receptor accumulate during cytopathic murine coronavirus infections. *J. Virol.* **72**: 3278-3288.

Sanchez, C.M., Izeta, A., Sanchez-Morgado, J.M., Alonso, S., Sola, I., Balasch, M., Plana-Duran, J., and Enjuanes, L., 1999, Targeted recombination demonstrates that the spike gene of transmissible gastroenteritis coronavirus is a determinant of its enteric tropism and virulence. *J. Virol.* **73**: 7607-7618.

Sturman, L.S., Ricard, C.S., and Holmes, K.V., 1990, Conformational change of the coronavirus peplomer glycoprotein at pH 8.0 and 37C correlates with virus aggregation and virus-induced cell fusion. *J. Virol.* **64**: 3042-3050.

Vennema, H., Rottier, P.J.M., Heijnen, L., Godeke, G.J., Horzinek, M.C., and Spaan, W.J.M., 1990, Biosynthesis and function of the coronavirus spike protein. *Adv. Expt. Med. Biol.* **276**: 9-19.

Human Coronavirus HCoV-229E Enters Susceptible Cells via the Endocytic Pathway

DIANNA M. BLAU AND KATHRYN V. HOLMES
University of Colorado Health Sciences Center, Department of Microbiology, 4200 E 9th Avenue, Denver, CO, 80262, USA

1. INTRODUCTION

Human coronaviruses (HCoV) are important causes of upper respiratory infections in humans of all ages. In addition, HCoVs have occasionally been associated with pneumonia, meningitis and diarrhea. HCoV RNA has been detected by RT-PCR in up to 40% of adult human brains (Stewart et al., 1992). There are two serotypes of HCoV represented by HCoV-229E and HCoV-OC43. HCoV-229E is a member of coronavirus serogroup I, which also includes porcine transmissible gastroenteritis virus (TGEV) and feline and canine coronaviruses. The viral attachment protein S of HCoV-229E is unlike S proteins of many coronaviruses in serogroup II, in that it is not cleaved during virus assembly, nor does it cause syncytia formation.

In order to infect susceptible cells, the S glycoprotein of HCoV-229E binds to its receptor, human aminopeptidase N (hAPN) also known as CD13, a metalloprotease (Yeager et al., 1992). hAPN is expressed on a variety of cells including monocytes, granulocytes, neuronal cells and the apical surface of renal, lung and intestinal epithelial cells (Look et al., 1989, Lachance et al. 1998). After attachment of S to hAPN, the viral envelope must fuse with a cellular membrane, either the plasma membrane or an endocytic membrane. Most strains of mouse hepatitis virus (MHV) in serogroup II are believed to gain entry into cells by fusing at the plasma membrane. This is supported by the data that MHV causes cell fusion at neutral or alkaline pH. (Sturman et al., 1990). In addition it has been

The Nidoviruses (Coronaviruses and Arteriviruses).
Edited by Ehud Lavi *et al.*, Kluwer Academic/Plenum Publishers, 2001.

reported that internalization of MHV-A59 by endocytosis does not lead to a productive infection (Kooi et al., 1991). In contrast, porcine transmissible gastroenteritis virus (TGEV), after binding to its receptor, porcine APN (pAPN), is observed by electron microscopy in endocytic pits. TGEV infection is blocked by ammonium chloride or bafilomycin A1, agents that prevent the acidification of endosomes (Hansen et al. 1998). These data suggest that TGEV penetrates at the membrane of an acidic intracellular compartment, although MHV penetrates by fusion at the plasma membrane.

We have studied the entry of HCoV-229E into polarized human colon carcinoma cells (Caco-2) cultured on permeable filters. hAPN is expressed predominantly on the apical surfaces of these cells (LeBivic et al., 1990). In this report we describe the preferential entry of HCoV-229E at the apical surfaces of polarized Caco-2 cells. We also show that the entry of HCoV-229E into MRC-5 human lung epithelial cells is inhibited by drugs that block the acidification of endosomes. These findings suggest that HCoV-229E undergoes endocytosis after binding of S to hAPN at the plasma membrane, and the virion is then sorted into endosomes where fusion of the viral envelope and endocytic membrane occur.

2.   RESULTS

The primary site of replication of HCoV-229E is in human respiratory epithelial cells. To investigate the entry process of HCoV-229E into susceptible cells *in vitro*, we first assayed the interaction of virions with the polarized epithelial cells. Caco-2 cells, grown on permeable filters to allow access to apical and basal surfaces, were inoculated with HCoV-229E via either the apical or basal medium, and the amount of virus released into each medium was then determined. Table 1 shows that in polarized Caco-2 cells, HCoV-229E entered the cells more efficiently from the apical surface than from the basal surface. The virus was released from the polarized cells into both the apical and basal media. The inefficiency of the virus infecting at the basal surface is not due to a physical barrier presented by the filter because nonpolarized MRC-5 cells were effectively infected from either the apical or the basal membrane. Virus was also released from MRC-5 cells into both the apical and basal media. These results indicate that HCoV-229E enters polarized epithelial cells preferentially at the apical membrane.

Table 1. Virus yields released from polarized Caco-2 cells

Inoculation	Apical Medium pfu/ml	Basal Medium pfu/ml
Apical	2.3×10^7	2.0×10^7
Basal	7.0×10^3	4.5×10^4

To further investigate the route of entry of HCoV-229E virions into susceptible cells, we treated cells with agents that inhibit the acidification of the endosomes. If HCoV-229E enters by fusing with endocytic membranes, it is likely that these drugs will inhibit infection. Chloroquine, a weak base, and bafilomycin A1, a specific inhibitor of the vacuolar ATP-ase proton pump both block the acidification of endosomes. As seen in Table 2, treatment of MRC-5 cells with these drugs during virus adsorption inhibits viral infection, whereas treatment after viral replication has begun does not reduce the percent of cells expressing HCoV-229E antigen.

Table 2. Percent of cells expressing 229E antigen at 12 hpi

Treatment	-1 to 1 hpi	1 to 3 hpi	8 to 12 hpi
None	100	100	100
Chloroquine (50µM)	28 ±1.4	98 ±0.8	80 ±0.9
Bafilomycin A (500nM)	10 ±2.3	60 ±1.3	90 ±1.2

Coronaviruses acquire their envelopes and bud intracellularly in the intermediate compartment between the rough endoplastic reticulum and the Golgi apparatus (Tooze et al., 1984). Virions are then transported in vesicles to the plasma membrane and released by exocytosis (Tooze et al., 1987). It is thus possible that chloroquine and bafilomycin A1 affect not only the entry but also the release of the virus. Viral yields were compared from cells incubated with drugs either before virus inoculation or at a later time when virus is starting to be released (Figure 1). Incubation in chloroquine or bafilomycin A1 before and during virus inoculation resulted in a decrease in viral titers when compared to untreated, inoculated cells. In contrast, when these drugs were added 8-12 hours post inoculation, there was no significant decrease in viral yields compared to untreated, inoculated cells. These results show that the lysosomotrophic drugs inhibit early at virus entry but at later times do not affect the release of virus.

Nocodazole, a microtubule depolymerizing agent, blocks transport from early endosomes to late endosomes in certain cell types (Gruenberg and Howell 1989). To determine if endosomal transport was required for HCoV-229E infection, cells were treated during virus adsorption or continuously with nocodazole. Figure 2 shows the growth curves of infectious virus recovered from these and untreated cells. Cells treated with nocodazole yielded lower titers of HCoV-229E than untreated, HCoV-229E infected cells. These results show that microtubule-disrupting drugs inhibit the early stage of virus infection, supporting the hypothesis that HCoV-229E enters cells through the endocytic pathway.

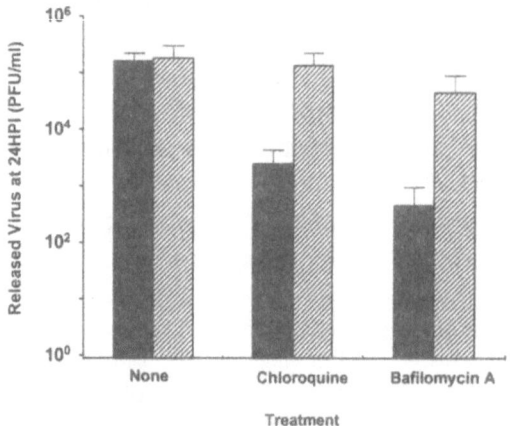

Figure 1. Lysosomotropic drugs inhibited HCoV-229E entry not virus maturation and release. MRC-5 cells were treated at -1 to 1 hpi (solid) or 8-12 hpi (hatched), either with medium alone or with chloroquine (50μM) or bafilomycin (500nM).

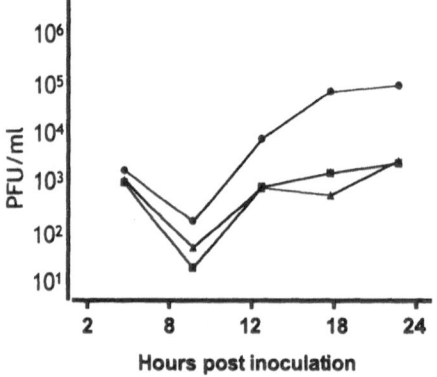

Figure 2. Nocodazole inhibited HCoV-229E infection. MRC-5 cells were treated with medium alone (circles), or nocodazole 6mg/ml either from -1 to 3 hpi (squares) or continuously (triangle) from 1 hpi until collection.

3.   DISCUSSION

In order to establish an infection, HCoV-229E must bind to a specific cell receptor, penetrate by fusion of the viral envelope with the cellular membrane and release its plus strand RNA genome to the cytoplasm. We previously showed that HCoV-229E uses hAPN as its cellular receptor

(Yeager et al., 1992). We have further investigated the entry process by determining the site of penetration of the virus. We used the Caco-2 cell line, which is a well-characterized human polarized epithelial cell line. The receptor, hAPN, is found predominately on the apical surface (LeBivic et al., 1990) and this correlates with the results presented here that HCoV-229E enters preferentially from the apical surface of these cells. Similar results were also observed when differentiated human airway epithelia were inoculated with HCoV-229E (Wang et al., 2000).

We also investigated the site of penetration of HCoV-229E in MRC-5 cells. Drugs that block the acidification of endosomes inhibited infection of cells by HCoV-229E. Influenza virus is a well-characterized enveloped virus that initiates infection by endocytosis. Influenza hemagglutinin (HA), its viral attachment protein, undergoes a conformational change at the low pH found in endosomes (for review Hernandez et al., 1996). This change in HA allows fusion of the viral envelope with the endosomal membrane to occur. HA glycoprotein is cleaved during virus maturation and is found on the virion envelope as two subunits HA1 and HA2. Unlike influenza, HCoV-229E S glycoprotein is not proteolytically cleaved. Incubation of HCoV-229E at low pH alone does not induce observable conformational changes in HCoV-229E S glycoprotein assayed by protease sensitivity and association with liposomes (data not shown). This correlates with the data that treatment of HCoV-229E virions at low pH alone does not neutralize the virus (Lamarre and Talbot, 1989, personal unpublished data). Similarly TGEV, which also enters through an endocytic pathway, retains its infectivity at low pH treatment (Laude et al., 1981). It may be that a conformational change occurs that is readily reversible. Another possibility is that viral infectivity requires incubation at low pH and with the receptor or another cofactor in order to undergo changes that are required for penetration of the virus to occur. Further work characterizing the steps needed for HCoV-229E entry will be valuable as it appears that this virus is using another route and mechanism of entry when compared to the better characterized coronavirus, MHV.

ACKNOWLEDGEMENTS

We thank Bruce Zelus for critical review of the manuscript. This work was supported by research grant AI26075 and T32 AI07537 (D.B.) from the National Institutes of Health.

REFERENCES

Hansen, G. H., Delmas, B., Besnardeau, L., Vogel, L. K., Laude, H., Sjostrom, H., and Noren, O., 1998, The coronavirus transmissible gastroenteritis virus causes infection after receptor-mediated endocytosis and acid-dependent fusion with an intracellular compartment. *J. Virol.* **72**:527-534.

Hernandez, L. D., Hoffman, L. R., Wolfsberg, T. G., and White, J. M., 1996, Virus-cell and cell-cell fusion. *Annu. Rev. Cell Dev. Biol.* **12**:627-661.

Gruenberg, J., and Howell, K. E., 1989, Membrane traffic in endocytosis: insights from cell-free assays. *Annu. Rev. Cell Biol.* **5**:453-481.

Kooi, C., Cervin, M., and Anderson R., 1991, Differentiation of acid-pH-dependent and -nondependent entry pathways for mouse hepatitis virus. *Virology.* **180**:108-119.

Lachance, C., Arbour, N., Cashman, N. R., and Talbot, P. J., 1998, Involvement of aminopeptidase N (CD13) in infection of human neural cells by human coronavirus 229E. *J. Virol.* **72**:6511-6519.

Lamarre, A., and Talbot, P. J., 1989, Effect of pH and temperature on the infectivity of human coronavirus 229E. *Can. J. Microbiol.* **35**:972-974.

Laude, H., Gelfi, J., and Aynaud, J. M., 1981, In vitro properties of low- and high- passage strains of gastroenteritis coronavirus of swine. *Am. J. Vet. Res.* **42**:447-449.

LeBivic, A., Quaroni, A., Nichols, B., and Rodriguez-Boulan, E., 1990, Biogenic pathways of plasma membrane proteins in Caco-2, a human intestinal epithelial cell line. *J. Cell Biol.* **111**:1351-1361.

Look, A. T., Ashmun, R. A., Shapiro, L. H., and Peiper, S. C., 1989, Human myeloid plasma membrane glycoprotein CD13 (gp150) is identical to aminopeptidase N. *J. Clin. Invest.* **83**:1299-1307.

Stewart, J. N., Mounir, S., and Talbot, P. J., 1992, Human coronavirus gene expression in the brains of multiple sclerosis patients. *Virology* **191**:502-505.

Sturman, L. S., Ricard, C. S., and Holmes, K. V., 1990, Conformational change of the coronavirus peplomer glycoprotein at pH 8.0 and 37° C correlates with virus aggregation and virus-induced cell fusion. *J. Virol.* **64**:3042-3050.

Tooze, J., Tooze, S. A., and Fuller, S. D., 1987, Sorting of progeny coronavirus from condensed secretory proteins at the exit from the trans-Golgi network of AtT20 cells. *J. Cell Biol.* **105**: 1215-26.

Tooze, J., Tooze, S., and Warren, G., 1984, Replication of coronavirus MHV-A59 in sac- cells: determination of the first site of budding of progeny virions. *Eur. J. Cell Biol.* **33**:281-293.

Wang, G., Deering, C., Macke, M., Shao, J., Burns, R., Blau, D. M., Holmes, K. V., Davidson, B. L., Perlman, S., and McCray Jr., P. B., 2000, Human coronavirus 229E infects polarized airway epithelia from the apical surface. Submitted to J. Virol.

Yeager, C. L., Ashmun, R. A., Williams, R. K., Cardellichio, C. B., Shapiro, L. H., Look, A. T., and Holmes, K. V., 1992, Human aminopeptidase N is a receptor for human coronavirus 229E. *Nature* **357**:420-422.

Addition of a Single Glycosylation Site to hAPN Blocks Human Coronavirus-229E Receptor Activity

DAVID E. WENTWORTH AND KATHRYN V. HOLMES

Department of Microbiology, University of Colorado Health Sciences Center, Denver, Colorado

1. INTRODUCTION

The cellular receptor for HCoV-229E is human aminopeptidase N (hAPN). Murine fibroblasts that are nonpermissive for HCoV-229E become susceptible after transfection with an hAPN expression plasmid (Yeager *et al.*, 1992). In addition, antibodies to hAPN block infection of human neural cells by HCoV-229E (Lachance *et al.*, 1998). hAPN, also called CD13, is a 150 kDa glycoprotein that is a membrane peptidase (Look *et al.*, 1989). APN is expressed by many cell types including epithelial cells of the kidney, intestine, respiratory tracts and at synaptic junctions in the CNS (Kenny and Maroux, 1982; Look *et al.*, 1989; Riemann *et al.*, 1999; Noren *et al.*, 1997).

Aminopeptidase N (APN) also serves as the major receptor for serogroup I coronaviruses that infect pigs, cats and dogs (Delmas *et al.*, 1992; Delmas *et al.*, 1993; Tresnan *et al.*, 1996). In general, APN is used in a species specific manner. However, feline APN (fAPN), serves as a receptor for many serogroup I coronaviruses including feline (FECoV and FIPV), porcine (TGEV), human (HCoV-229E), and canine (CCoV) (Tresnan *et al.*, 1996). Studies using chimeras between the human, feline, and porcine APN glycoproteins have identified two regions that are important in coronavirus receptor activity. Amino acids 717-813 of the pAPN are required for TGEV receptor activity (Delmas *et al.*, 1994) and amino acids 670-840 of fAPN are important for receptor activity of feline and porcine viruses (Hegyi and Kolb, 1998). Furthermore, substitution of amino acids 283-290 of pAPN

The Nidoviruses (Coronaviruses and Arteriviruses).
Edited by Ehud Lavi *et al.*, Kluwer Academic/Plenum Publishers, 2001.

199

with the corresponding amino acids 288-295 of hAPN resulted in a protein that serves as a receptor for both TGEV and HCoV-229E (Kolb *et al.*, 1997).

We used wild type and nine mutant hAPN proteins expressed in mammalian cell lines to demonstrate that the addition of a glycosylation signal in hAPN at N291 blocks HCoV-229E receptor activity.

2.   MATERIALS AND METHODS

IFA assays were done as described (Tresnan *et al.*, 1996) using MAbs specific for hAPN (WM47) or HCoV-229E spike glycoprotein (5-11H.6). BHK-21 cells were transfected using lipofectamine™ (GIBCO Laboratories) according to manufacturer's instructions. hAPN expression plasmids were mutated using primers encoding the desired nucleotide changes and PFU polymerase was used to extend the primers. This was followed by DpnI digestion and transformation of E. coli DH5α (GIBCO Laboratories). Clones were screened by restriction digestion and sequenced to verify the desired mutations.

3.   RESULTS

3.1   Substitution of hAPN amino acids 288-295 for the corresponding pAPN sequence blocked infection by HCoV-229E

BHK-21 cells were transfected with plasmids that express wild type hAPN (wt-hAPN) or a mutant hAPN containing nucleotide changes that encode 6 amino acid changes to the corresponding pAPN amino acids at 288, 289, 291, 292, 293, and 295 (p6-hAPN). Two days post-transfection the cells have similar levels of APN expression as determined by IFA with anti hAPN MAb WM47 (Table 1). Transfected cells were inoculated with HCoV-229E and 20 h. later the receptor activity was determined by IFA using anti spike MAb 5-11H.6. Cells expressing wt-hAPN are infected by HCoV-229E. In contrast, cells expressing p6-hAPN and those transfected by the empty expression plasmid (pCi-neo) are not infected by HCoV-229E (Table 1). The 6 amino acids from pAPN used to create p6-hAPN had a putative N-linked glycosylation site. Reversion of this site by mutagenesis of p6-hAPN to change asparagine 291 to glutamic acid (N291E) or threonine 293 to glutamine (T293Q), generated p6-hAPN/N291E and p6-hAPN/T293Q, respectively. Transfection of BHK-21 cells with wt-hAPN, p6-hAPN, p6-hAPN/N291E, p6-hAPN/T293Q or pCi-neo resulted in expression of hAPN in all of the cells except those transfected by empty vector as demonstrated by IFA (Table 1). HCoV-229E inoculation of these transfected cells resulted in infection of cells expressing wt-hAPN or the revertants (p6-hAPN/N291E and p6-hAPN/T293Q), but cells transfected with p6-hAPN or pCi-neo remained nonpermissive (Table 1). None of these APN expression constructs have receptor activity for the porcine coronavirus TGEV (Table 1).

3.2 Glycosylation of hAPN at amino acid 291 blocked infection by HCoV-229E

We used mutagenesis to insert a glycosylation signal at N291 of wt-hAPN to determine if glycosylation here was the only change needed to block HCoV-229E infection. The wt-hAPN expression plasmid was mutated at E291N and Q293T creating hAPN/N291KT. Then hAPN/N291KT was reverted to wild type at N291E or T293Q creating two other expression constructs, hAPN/E291KT and hAPN/N291KQ. Transfection of BHK-21 cells with wt-hAPN and each mutant hAPN expression plasmid resulted in hAPN expression as evidenced by IFA with MAb WM47 48 h. post-transfection. IFA of the transfected BHK-21 cells with anti-S glycoprotein MAb (5-11H.6) 20 h. p.i. with HCoV-229E illustrated that cells expressing wt-hAPN, hAPN/E291KT or hAPN/N291KQ were HCoV-229E permissive (Table 1). Cells transfected with hAPN/N291KT, which has the glycosylation signal, have markedly fewer IFA positive cells (Table 1).

Table 1.

APN	Amino Acids 288-295	hAPN Expression	Glycosylation of N291	HCoV-229E Permissive	TGEV Permissive
wt-hAPN	DYVEKQAS	Y	N	Y	N
pAPN	QSVNETAQ*	NA	Y	N	Y
p6-hAPN	QSVNETAQ	Y	Y	N	N
p6-hAPN/N291E	QSVEETAQ	Y	N	Y	N
p6-hAPN/T293Q	QSVNEQAQ	Y	N	Y	N
hAPN/N291KT	DYVNKTAS	Y	Y	N	ND
hAPN/E291KT	DYVEKTAS	Y	N	Y	ND
hAPN/N291KQ	DYVNKQAS	Y	N	Y	ND
pCi-neo		NA	NA	N	N

Amino acids are listed in single letter code, * indicates amino acids 283-290 in pAPN,
Y = yes, N = no, NA = not applicable and ND = not determined

3.3   Removal of a glycosylation signal at amino acid 818 of hAPN does not make hAPN a receptor for TGEV or FECoV

Feline APN has receptor activity for feline, canine, human and porcine coronaviruses. The fAPN has fewer predicted glycosylation sites than hAPN or pAPN. We found a potential N-linked glycosylation site in hAPN at amino acids 818-820 that is not in the analogous regions of pAPN or fAPN. To determine if the removal of this glycosylation signal converted hAPN into a functional receptor for porcine or feline coronaviruses, we removed it from wt-hAPN and p6-hAPN by creating hAPN/N818E, p6-hAPN/N818E and p6-hAPN/T820E. These plasmids were transfected into nonpermissive CMT93 cells. These cells were selected for G418 resistance and sorted for similar hAPN expression levels by FACS with MAb WM47. These stably transfected CMT93 cell lines and *Felis catus* cell whole fetus (FCWF) cells were inoculated with HCoV-229E, TGEV or FECoV (79-1683). The FCWF cells were used as a positive control because they express fAPN and are permissive to HCoV-229E, TGEV, and FCoV (Tresnan *et al.*, 1996). Immunobloting of proteins isolated 20 h. p.i. showed that removal of the potential glycosylation site at N818 of hAPN did not alter HCoV-229E receptor activity (Table 2). Although hAPN/N818E, like the analogous region of fAPN, lacks an N-linked glycosylation signal at amino acid 818, it lacked receptor activity for TGEV and FCoV (Table 2). Removal of the 818-820 glycosylation signal from p6hAPN yielded a protein predicted to be glycosylated at N291 and lack the signal at amino acid 818 which is similar to pAPN. Cells expressing proteins from p6-hAPN/N818E or p6-hAPN/T820E lacked receptor activity for TGEV and FECoV (Table 2).

Table 2.

APN	Glycosylation		Permissive For		
	N291	N818	HCoV-229E	TGEV	FECoV
fAPN	N	N	Y	Y	Y
pAPN	Y	N	N	Y	N
wt-hAPN	N	Y	Y	N	N
hAPN/N818E	N	N	Y	N	N
p6-hAPN	Y	Y	N	N	N
p6-hAPN/N818E	Y	N	N	N	N
p6-hAPN/T820E	Y	N	N	N	N
pCi-neo	NA	NA	N	N	N

Y = yes, N = no and NA = not applicable

4.   DISCUSSION

The species specificity and tissue tropism of viruses are, in large part, determined by the specificity of virus-receptor interactions. The p6-hAPN mutant that has pAPN amino acids 283-290 substituted for 288-295 of hAPN, completely lacked receptor activity for HCoV-229E. The most important change among these mutations was the introduction of a potential glycosylation signal from amino acids 291-293 of hAPN. To determine if the introduction of the glycan at N291 was responsible, two other mutants generated from p6-hAPN that changed N291E or T293Q but also had 5 amino acid changes in this region were tested for HCoV-229E receptor activity.   These two revertants were functional HCoV-229E receptors. Feline APN also has amino acid differences from hAPN at D288, K292, Q293 and S295.   This suggests that the block in receptor activity of p6-hAPN is due to glycosylation of N291 rather than the individual amino acid substitutions.   This is best demonstrated by the hAPN/N291KT construct that contained only this glycosylation signal and lacked receptor activity. However, revertants (hAPN/E291KT and hAPN/N291KQ) that lack this signal have HCoV-229E receptor function.   This data demonstrates that the glycosylation of hAPN at N291 blocks infection.

Studies of a region of APN essential for TGEV and FECoV receptor activity identified a potential glycosylation signal at N818 of hAPN that is not conserved in pAPN or fAPN.   However mutants of hAPN created to remove this potential glycosylation signal did not have receptor activity for TGEV or FECoV, but retained receptor activity for HCoV-229E.   These results show that removal of a potential glycosylation signal at N818 of hAPN is not sufficient to confer receptor activity for porcine and feline coronaviruses.

ACKNOWLEDGEMENTS

The authors would like to thank Pierre Talbot for the anti spike MAb 5-11H.6, Justin Hagee for his technical assistance, Karen Helm at the UCHSC Cancer Center for FACS and Dr. Bruce Zelus for his critique of this manuscript.   This work was supported by NIH grant AI26075 and Dr. Wentworth was supported by NIH Neurovirology-Molecular Biology Training Grant-T32 NS07321.

REFERENCES

Delmas, B., Gelfi, J., Kut, E., Sjostrom, H., Noren, O., and Laude, H.(1994). Determinants essential for the transmissible gastroenteritis virus-receptor interaction reside within a domain of aminopeptidase-N that is distinct from the enzymatic site. *Journal of Virology* **68**, 5216-5224.

Delmas, B., Gelfi, J., L'Haridon, R., Vogel, L.K., Sjostrom, H., Noren, O., Laude, and H.(1992). Aminopeptidase N is a major receptor for the entero-pathogenic coronavirus TGEV. *Nature* **357**, 417-420.

Delmas, B., Gelfi, J., Sjostrom, H., Noren, O., and Laude, H.(1993). Further characterization of aminopeptidase-N as a receptor for coronaviruses. *Advances in Experimental Medicine & Biology* **342**, 293-298.

Hegyi, A. and Kolb, A.F.(1998). Characterization of determinants involved in the feline infectious peritonitis virus receptor function of feline aminopeptidase N. *Journal of General Virology* **79**, 1387-1391.

Kenny, A.J. and Maroux, S.(1982). Topology of microvillar membrane hydrolases of kidney and intestine. [Review] [153 refs]. *Physiological Reviews* **62**, 91-128.

Kolb, A.F., Hegyi, A., and Siddell, S.G.(1997). Identification of residues critical for the human coronavirus 229E receptor function of human aminopeptidase N. *Journal of General Virology* **78**, 2795-2802.

Lachance, C., Arbour, N., Cashman, N.R., and Talbot, P.J.(1998). Involvement of aminopeptidase N (CD13) in infection of human neural cells by human coronavirus 229E. *Journal of Virology* **72**, 6511-6519.

Look, A.T., Ashmun, R.A., Shapiro, L.H., and Peiper, S.C.(1989). Human myeloid plasma membrane glycoprotein CD13 (gp150) is identical to aminopeptidase N. *Journal of Clinical Investigation* **83**, 1299-1307.

Noren, O., Sjostrom, H., and Olsen, J.(1997). *Cell-Surface Peptidases in Health and Disease* (Kenney, A.J. and Boustead, C.M., Eds.) BIOS Scientific Publishers, Oxford. 175-191.

Riemann, D., Kehlen, A., and Langner, J.(1999). CD13 - not just a marker in leukemia typing [Review]. *Immunology Today* **20**, 83-88.

Tresnan, D.B., Levis, R., and Holmes, K.V.(1996). Feline aminopeptidase N serves as a receptor for feline, canine, porcine, and human coronaviruses in serogroup I. *Journal of Virology* **70**, 8669-8674.

Yeager, C.L., Ashmun, R.A., Williams, R.K., Cardellichio, C.B., Shapiro, L.H., Look, AT, and Holmes, K.V.(1992). Human aminopeptidase N is a receptor for human coronavirus 229E. *Nature* **357**, 420-422.

Effects of Amino Acid Insertions in the Cysteine-Rich Domain of the MHV-A59 Spike Protein on Cell Fusion

KEVIN W. CHANG AND JAMES L. GOMBOLD
Department of Microbiology and Immunology, Louisiana State University Health Sciences Center, Shreveport, LA

1. INTRODUCTION

The spike protein is a key determinant in the pathogenesis of mouse hepatitis virus strain A59 (Hingley et al., 1994). The spike is a type I viral envelope glycoprotein that mediates receptor-binding and viral entry through virus-cell fusion. In addition, when present on the cell surface, the spike can also mediate cell-cell fusion between infected and neighboring cells. During maturation, the spike protein is post-translationally cleaved into two 90 kDa subunits, S1 and S2. Unlike other viral fusion proteins, fusion mediated by the spike protein does not require a cleavage step that liberates a fusion peptide on the N-terminus of the membrane bound subunit (Bos et al., 1997; Bos et al., 1995; Gombold et al., 1993). Instead, a portion of heptad repeat 1 has been suggested to serve as an internal fusion peptide (Luo and Weiss, 1998a; Luo and Weiss, 1998b). In the current model of fusion, fusion peptides are proposed to mediate lipid mixing between the outer leaflets of the viral envelope and cell membrane. This leads to an intermediate stage, termed hemifusion, in which the inner leaflets remain intact. Resolution of hemifusion is thought to be mediated by the transmembrane anchor. Among coronaviruses, there is significant sequence conservation within the spike transmembrane anchor and cysteine-rich (*cys*) domain. Both the transmembrane anchor and the *cys* domain of the spike protein have been shown to be necessary for cell-cell fusion activity (Bos et al., 1995; Chang et al., 2000). In this report, we examine how changes in the length of the

The Nidoviruses (Coronaviruses and Arteriviruses).
Edited by Ehud Lavi *et al.*, Kluwer Academic/Plenum Publishers, 2001.

transmembrane anchor and the consequent change in spacing of the *cys* domain affect cell fusion activity.

2.   MATERIAL AND METHODS

2.1   Cells and virus

DBT cells were grown in Dulbecco's Modified Eagles Medium (DMEM) containing 10% fetal bovine serum. Mouse hepatitis virus strain A59 was propagated and titrated by plaque assay on DBT cells. Vaccinia virus WR and the T7 RNA polymerase recombinant vTF7.3 (Fuerst et al., 1987; Fuerst et al., 1986) were prepared as described previously (Chang et al., 2000).

2.2   PCR mutagenesis

PCR mutagenesis was performed as described previously (Chang et al., 2000) using mutagenic primers to generate the desired mutations. The resulting PCR products were ligated into the cDNA of the wild-type spike protein under transcriptional control of the T7 RNA polymerase promoter.

2.3   β-galactosidase-fusion assay

A fusion assay based on the expression of β-galactosidase, described originally by Nussbaum et al. (1994), was modified for use with the MHV spike protein as previously described (Chang et al., 2000). Briefly, effector DBT cells were infected with VTF7.3, a recombinant vaccinia virus encoding and expressing the T7 RNA polymerase at 10 PFU per cell for 1 hr at 37°C. The cells were then washed once with PBS and transfected with plasmid DNA encoding the spike protein under the control of the T7 RNA polymerase promoter and EMC IRES sequence. Reporter cells were infected with wild-type vaccinia virus, strain WR, for 1 h at 37°C. This was followed by transfection with plasmid DNA encoding *lacZ* under the control of the T7 RNA polymerase promoter and EMC IRES sequence. At 4 h post-transfection the effector and reporter cells were removed with trypsin, mixed, and incubated for 4 hr at 37°C in a 96-well plate. Following the incubation, the cells were lysed by adding Triton X-100 in PBS to a final concentration of 1%. The level of β-galactosidase expression was measured by mixing aliquots of the cell lysate with the β-galactosidase substrate, CPRG (Boehringer Mannheim), and measuring the optical density at 570 nm.

2.4   Immunofluorescence

Cell surface expression of the wild type and mutant spike proteins was examined by immunofluorescence staining. At six hours post-transfection, the cells were incubated with polyclonal goat anti-spike serum AO4 (kindly provided by Dr. Kathryn Holmes, University of Colorado Health Sciences Center), washed, and then fixed in 4% paraformaldehyde in phosphate buffered saline. The cells were then incubated in fluorescein-conjugated rabbit anti-goat IgG, washed, and examined for fluorescence on an Olympus BX-50 fluorescence microscope.

3.   RESULTS AND DISCUSSION

3.1   Extension of the spike transmembrane anchor

Chimeric spike proteins that substitute the A59 transmembrane anchor with the anchor from glycoprotein D (gD) of herpes simplex virus type 1 are fusion defective (Chang et al., 2000). The *cys* domain of the spike, located immediately downstream of the membrane spanning domain, is also required for membrane fusion, but it cannot restore fusion activity to fusion defective A59:gD chimeric spike proteins. The region of the A59 transmembrane anchor upstream of the *cys* domain is four amino acids shorter than the gD anchor (Fig. 1A). To determine if the extension of the transmembrane anchor contributed to the fusion negative phenotypes of the chimeric spike proteins, PCR mutagenesis was used to generate insertion mutants of the wild type A59 spike protein that contained either one, two, or all four of the additional amino acids present at the C-terminus of gD anchor (Fig. 1A). The insertion of just one amino acid (valine, Ins-V) immediately upstream of the *cys* domain reduced the fusion activity by 98% while the insertion of two or four amino acids (Ins-VY or Ins-VYWM, respectively) reduced fusion activity by 99% or greater (Fig. 1B). Membrane fusion requires that the spike protein be transported to the cell surface. Consequently, expression of wild type or mutant spike proteins on the cell surface was examined by immunofluorescence using an anti-spike serum (Fig. 2). Each of the mutant spike proteins was present on the cell surface in levels comparable to wild type spike, indicating that the insertion of either one, two, or four amino acids between the hydrophobic membrane-spanning domain and the *cys* domain does not interfere with transport of the protein to the cell surface. The increased size of the transmembrane anchor that results from these mutations may affect fusion by altering the position of the cysteine-rich domain relative to the transmembrane anchor.

(A)

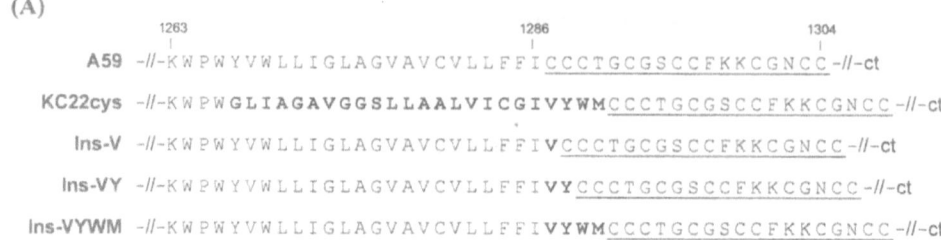

(B)

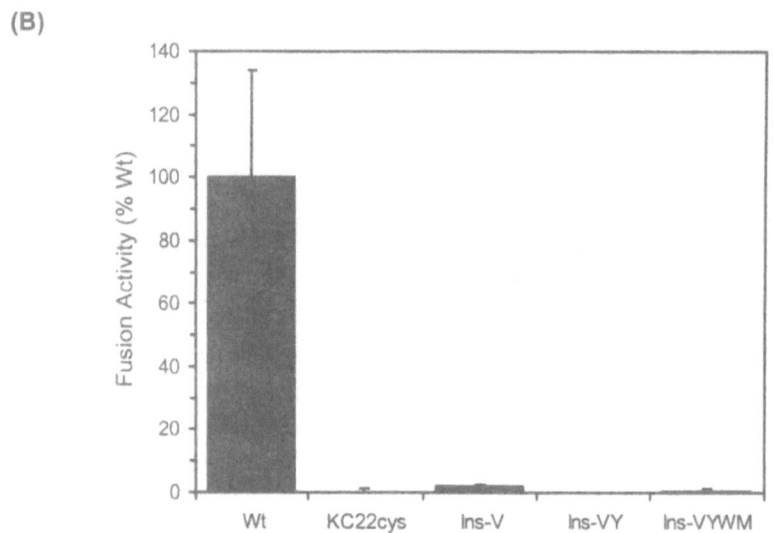

Figure 1. Amino acid insertions upstream of the spike cys domain prevents cell-cell fusion.
(A) Sequence of the hydrophobic region of the membrane-spanning domain (amino acids
1263 to 1286) and cysteine-rich domain (underlined, amino acids 1287 to 1304) of the wild
type A59 and chimeric spike proteins. KC22cys is chimeric spike protein possessing the
majority of the HSV gD transmembrane anchor (bold face) fused to the A59 cys domain and
cytoplasmic tail (ct) and is fusion defective as shown previously (Chang et al., 2000). Three
mutants were created using PCR mutagenesis techniques to insert one (Ins-V), two (Ins-VY)
or all four (Ins-VYWM) of the amino acids present at the C-terminus of the gD
transmembrane anchor. (B) Fusion activities of the wild type (Wt) and mutant spike proteins
were measured as described in Materials and Methods. The data were normalized to wild
type and plotted as the mean percent fusion ± standard deviation.

3.2   Single alanine insertions within the cysteine-rich domain

The insertion of a single valine residue immediately upstream of the
cys domain interfered strongly with fusion, but it was unclear if this resulted
from the position of the insertion or the nature of the amino acid that was

inserted. To address this question, single alanine residues, which have a smaller side group and are less hydrophobic than valine, were inserted immediately upstream of the *cys* domain or at various positions within the domain. When a single alanine was inserted immediately upstream of the *cys* domain at position 1287, fusion activity decreased 75% compared to wild type spike (Fig. 3). Even though the reduction in fusion was not as large as that observed with the valine insertion, this result implies that the position of the insertion is a critical factor contributing to the inhibition of fusion.

To further examine this issue, we constructed a series of mutants in which single alanine residues were inserted at several sites downstream of position 1287 but still within the *cys* domain. Measurement of cell-cell fusion activity for each of the mutants showed that insertion of alanine at position 1290 (A1290) reduced fusion nearly 80% (Fig. 3). In contrast, insertions at positions 1294, 1298, or 1303 did not negatively affect cell-cell fusion activity of the spike. Thus, fusion was affected when insertions were placed at positions up to and including amino acid 1290. Since we did not examine mutants with insertions between positions 1290 and 1294, we do not know if the region sensitive to insertion mutations extends farther downstream. These results suggest that the position of one or more amino acids in the cys domain upstream of and including threonine 1290 (and possible glycine 1293) may be crucial for fusion to occur properly. It is interesting to note that two glycine residues at positions 1291 and 1293 are found in the cysteine-rich domain of A59 and that both of these residues are conserved for the antigenic group II coronaviruses (Chang et al., 2000). In addition, glycine 1291 is conserved among all coronavirus spike proteins that we have examined. It will be necessary to examine additional mutants to determine if the position of either or both of these two glycines is important for fusion.

Glycines that lie within the transmembrane domain of viral fusion proteins have been suggested to function as "glycine hinges" (Cleverley and Lenard, 1998). These hinges are proposed to allow for flexibility within the transmembrane anchor and thereby promote disruption of the hemifusion diaphragm, an intermediate in the fusion process. If the glycines that are conserved in the A59 spike *cys* domain are functionally important during fusion, we would anticipate that insertion of alanines upstream of either or both would interfere with fusion similar to the alanine insertions at positions 1287 and 1290. In addition, mutation of glycine 1291 and/or glycine 1293 would be predicted to reduce cell-cell fusion activity of the spike. This would be consistent with our previous observation that deletion of the cysteine-rich domain, where both glycines reside, abrogates fusion activity but does not prevent hemifusion (Chang et al., 2000).

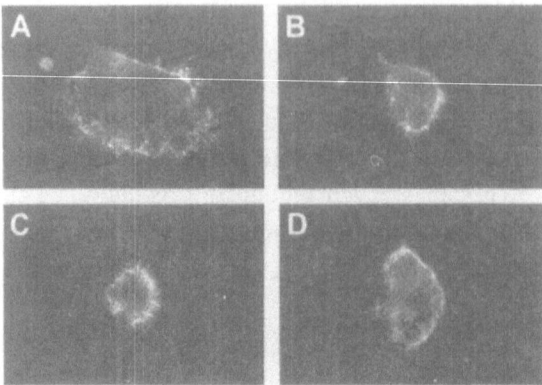

Figure 2. Mutant spike proteins are expressed on the cell surface. Mutant spike proteins were expressed in DBT cells and the cells were stained by immunofluorescence as described in Materials and Methods. (A) Wild type spike, (B) Ins-V, (C) Ins-VY, (D) Ins-VYWM.

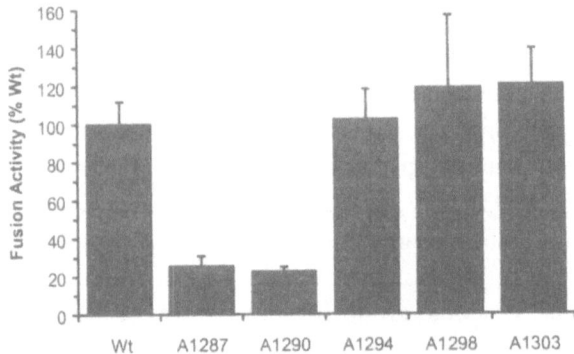

Figure 3. Alanine insertions upstream of position 1290 inhibit cell-cell fusion. PCR mutagenesis was used to generate a series of spike mutants in which a single alanine residue was inserted upstream of the cys domain (position 1287, A1287) or at various positions within the domain (A1290 to A1303). Fusion activities of the wild type (Wt) and mutant spike proteins were measured as described in Materials and Methods. The data were normalized to wild type and plotted as the mean percent fusion ± standard deviation.

4.   ACKNOWLEDGEMENTS

We thank Dr. Kay Holmes for providing the AO4 antiserum and David Ramage and Laura Perkins for the excellent technical assistance. This work was supported by grant LEQSF-RD-A-17 from the Louisiana Education Quality Support Fund and by the Centers for Excellence in Cancer Research and in Arthritis and Rheumatology at the Louisiana State University Health Sciences Center in Shreveport.

REFERENCES

Bos, E. C., Luytjes, W., and Spaan, W. J. M. (1997). The function of the spike protein of mouse hepatitis virus strain a59 can be studied on virus-like particles: cleavage is not required for infectivity. *J. Virol.* **71**(12), 9427-9433.

Bos, E. C. W., Heijnen, L., Luytjes, W., and Spaan, W. J. M. (1995). Mutational analysis of the murine coronavirus spike protein: effect on cell-to-cell fusion. *Virology* **214**, 453-463.

Chang, K. W., Sheng, Y., and Gombold, J. L. (2000). Coronavirus-induced membrane fusion requires the cysteine-rich domain in the spike protein. *Virology* **269**, 212-224.

Cleverley, D. Z., and Lenard, J. (1998). The transmembrane domain in viral fusion: Essential role for a conserved glycine residue in vesicular stomatitis virus G protein. *Proc. Natl. Acad. Sci. USA* **95**, 3425-3430.

Fuerst, T. R., Earl, P. L., and Moss, B. (1987). Use of a hybrid vaccinia virus-T7 RNA polymerase system for expression of target genes. *Mol. Cell Biol.* **7**(7), 2538-2544.

Fuerst, T. R., Niles, E. G., Studier, F. W., and Moss, B. (1986). Eukaryotic transient-expression system based on recombinant vaccinia virus that synthesizes bacteriophage T7 RNA polymerase. *Proc. Natl. Acad. Sci.* **83**, 8122-8126.

Gombold, J. L., Hingley, S. T., and Weiss, S. R. (1993). Fusion-defective mutants of mouse hepatitis virus A59 contain a mutation in the spike protein cleavage signal. *J. Virol.* **67**, 4504-4512.

Hingley, S. T., Gombold, J. L., Lavi, E., and Weiss, S. R. (1994). MHV-A59 fusion mutants are attenuated and display altered hepatotropism. *Virology* **200**, 1-10.

Luo, Z., and Weiss, S. R. (1998a). Roles in cell-to-cell fusion of two conserved hydrophobic regions in the murine coronavirus spike protein. *Virology* **244**, 483-494.

Luo, Z. L., and Weiss, S. R. (1998b). Mutational analysis of fusion peptide-like regions in the mouse hepatitis virus strain A59 spike protein. In *Coronaviruses and Arteriviruses* (L. Enjuanes, S.G. Siddel and W Spann, eds) Plenum Press, New York, pp17-23.

Nussbaum, O., Broder, C. C., and Berger, E. A. (1994). Fusogenic mechanisms of enveloped-virus glycoproteins analyzed by a novel recombinant vaccinia virus-based assay quantitating cell fusion-dependent reporter gene activation. *J.Virol.* **68**, 5411-5422.

Involvement in Fusion Activity of an Epitope in the S2 Subunit of Murine Coronavirus Spike Protein

[1]FUMIHIRO TAGUCHI AND [1,2]YOHKO K. SHIMAZAKI

[1]*National Institute of Neuroscience, NCNP, 4-1-1 Ogawahigashi, Kodaira, Tokyo 187-8502;*
[2]*National Veterinary Assay Laboratory, 1-15-1 Tokura, Kokubunji, Tokyo 185-8511 Japan*

1.   INTRODUCTION

The spike projecting from the virion of mouse hepatitis virus (MHV) is composed of a spike (S) glycoprotein with 180 to 200 kDa. The S protein is cleaved into N-terminal S1 and C-terminal S2 subunits. The S1 forms the globular part of the spike and the S2 its stalk portion. An important biological function of the S protein is binding to the virus receptor, which is mediated by a domain located in the N terminal 330 residues of the S1 (S1N330) (Kubo et al., 1994). Virus-cell fusion is also mediated by the S protein (Collins et al., 1982). Various regions in the S2 are involved in fusion activity. A candidate fusion peptide was reported to reside in the S2 (Luo and Weiss, 1998).

A monoclonal antibody (MAb) 5B19.2, which recognizses an epitope A (S2A) composed of nine hydrophobic amino acids in the S2 (Luytjes et al., 1989), has virus neutralizing (VN) and fusion inhibition (FI) activities (Collins et al., 1982). Antibodies to the S2A play a part for protection in mice (Koolen et al., 1990). These findings suggest that the S2A plays an important role in either virus-receptor interactions and/or viral entry into cells. In this study, we investigate the involvement of the S2A in the receptor-binding and fusion activities by using S2A antiserum with VN and FI activities as well as S proteins with mutated S2A.

The Nidoviruses (Coronaviruses and Arteriviruses).

Edited by Ehud Lavi *et al.*, Kluwer Academic/Plenum Publishers, 2001.

2.   MATERIALS AND METHODS

2.1   Cells, viruses and antibodies

DBT and RK13 cells as well as two MHV JHMV variants, cl-2 and sp-4 (Taguchi and Fleming, 1989), were used. The recombinant vaccinia virus vTF7.3 encoding the T7 RNA polymerase was used to express the JHMV S protein or MHV receptor protein. S2A antiserum (α-S2A) was made using rabbits by a synthetic peptide corresponding to amino acids 896 to 911 of JHMV cl-2 S protein (Taguchi et al., 1992). MAb 5B19.2 was provided by M. J. Buchmeier (Collins et al., 1982). S1-specific MAbs 3, 6, 7, 13, and 93 recognize conformational epitopes located in the S1N330 (Kubo et al., 1993).

2.2   Construction of S genes with mutations in the S2A and their expression

Mutations in the S2A shown in Fig. 1 were created by polymerase chain reaction principally as described previously (Taguchi, 1993) using a pair of primers containing the desired mutations. The S proteins with mutated S2A were expressed in DBT cells as described previously in vaccinia virus–T7 system (Feurst et al., 1986; Suzuki and Taguchi, 1996).

3.   RESULTS

3.1   VN and FI activities of S2A antiserum

MAb 5B19.2 recognizes the S2A and displays VN and FI activities, suggesting that the S2A is involved in receptor-binding and/or fusion activities. To confirm that anti-S2A serum has VN and FI activities and also to test above possibility, we have raised α-S2A by immunising rabbits with a synthetic peptide. We compared the VN and FI activities of the α-S2A with those of 5B19.2 and anti-S1 MAbs. Both α-S2A and 5B19.2 showed very low VN titers. Their specific VN values (VN titer/ELISA titer) were 10^3 to 10^4-fold lower than the VN values of anti-S1 MAbs. In contrast, their specific FI values (FI titer/ELISA titer) were the same or slightly higher than those of anti-S1 MAbs. These suggest the α-S2A and 5B19.2 neutralize JHMV by a different mechanism than anti-S1 MAbs.

										Fusion activity
S2A	L	L	G	C	I	G	S	T	C	++++
Mut -1	P	-	-	G	-	-	T	-	-	-
2	-	P	-	-	M	-	-	S	-	+
3	-	-	E	-	-	S	-	-	S	++++
4	R	-	-	R	-	-	R	-	-	-
5	-	H	-	-	R	-	-	R	-	++++
6	-	-	E	-	-	D	-	-	R	++++
41	R	-	-	-	-	-	-	-	-	+++
44	-	-	-	R	-	-	-	-	-	++++
47	-	-	-	-	-	-	R	-	-	++++

Figure 1. Amino acid sequences of the S2A and the fusogenicity of the S proteins. Mutations in the mutant S2A are shown. – indicates the residue identical to that of wt. Fusogenicity is shown as the proportion of cells involved in syncytia. ++++:75~100%, +++: 50~75%, ++:25~50%, +: 10~25%, -: <10%

3.2 Effects of S2A antiserum on receptor binding by JHMV

We examined whether the neutralization by α-S2A is due to the inhibition of the binding of virus to the receptor, CEACAM1[a]-2s (MHVR1). We treated 1-5 × 10^5 PFU of JHMV with 20 VN units of various antibodies at room temperature for 1 h. Twenty VN units neutralized more than 95% of 5 × 10^5 PFU JHMV. The treated viruses were examined by a viral overlay protein blot assay (VOPBA) using MAb 7 to determine whether they bound to MHVR1; neither the α-S2A nor the anti-S1 MAbs used in this study prevents the binding of MAb 7 to S protein. Bands of 40 to 50 kDa corresponding to the MHVR1 were visible when viruses were treated with either MAb 13, α-S2A, or normal rabbit serum, indicating that these antibodies fail to block the binding of virus to the receptor (Fig. 2). In contrast, no or only a faint band was produced after treatment with MAbs 3, 6, and 93, indicating these MAbs effectively block the receptor binding of the virus. Even 200 VN units of MAb 13 failed to block the binding (Fig. 2).

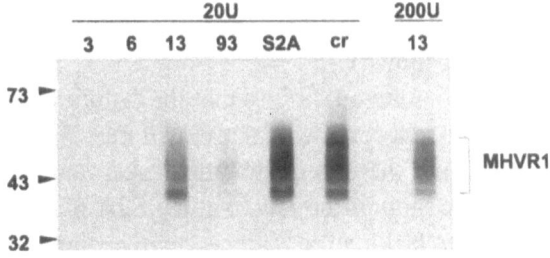

Figure 2. Inhibition of the receptor-binding of JHMV by various anti-MHV S antibodies as examined by VOPBA. Viruses were treated with 20 VN units (U) or 200U of α-S2A, MAbs 3, 6, 13, 93 or normal rabbit serum (cr) and allowed to bind to MHVR1 prepared on membrane paper.

These results suggest the failure of α-S2A to block the binding of virus to the receptor is not due to low antiviral titer but rather because the S2A is not involved in receptor binding.

3.3   Fusion activity of the S proteins with mutations in the S2A

To evaluate the role of the S2A in the process of virus-cell fusion, we have prepared a series of mutant S proteins with amino acid substitutions in the S2A (Fig. 1). In mut-1 to mut-3, three hydrophobic residues were replaced by different hydrophobic ones, while they were changed to hydrophilic ones in mut-4 to mut-47. These mutated S proteins were expressed in DBT cells by vaccinia virus-T7 system. Syncytia formation was monitored by staining with Giemsa solution (Kubo et al, 1993). There was no difference in transfection efficiency of the plasmids containing various S genes as examined using a vector, pG1NT7 β-gal, which encodes β-galactosidase downstream of the T7 promoter. It was shown that mut-1 and mut-4 almost completely lost the fusion activity and the fusogenicity of mut-2 and mut-41 was strongly and moderately reduced, whereas the other mutants displayed the fusion activity to the similar extent as wt S protein (Fig. 1 and 3). These results suggest that Leu_1, Cys_4 and Ser_7 influence the fusogenicity of the S protein, with Leu_1 being the most important. Other residues were also important for fusion activity, as shown by the reduced fusogenicity of mut-2. Furthermore, these results suggest that the hydrophobicity of the S2A is not critical for its fusion activity.

We have examined the mutant S proteins expressed in DBT cells by Western blot. Both uncleaved and cleaved forms of S proteins were detected, though cleavability differed among the mutants. Differences in cleavability did not correlate with differences in fusogenicity (data not shown), consistent with our previous finding (Taguchi, 1993). We further found that wt as well as mutant S proteins were similarly expressed on the cell membrane (data not shown).

These findings rule out the possibility that the failure of mut-1 and mut-4 to induce fusion and the reduced fusion activity of mut-2 and mut-41 are due to impaired transport onto the cell membrane. Taken together, our analysis of mutations in the S2A support the idea that the S2A plays a critical role in fusion formation, despite lacking the features of an ordinary fusion peptide.

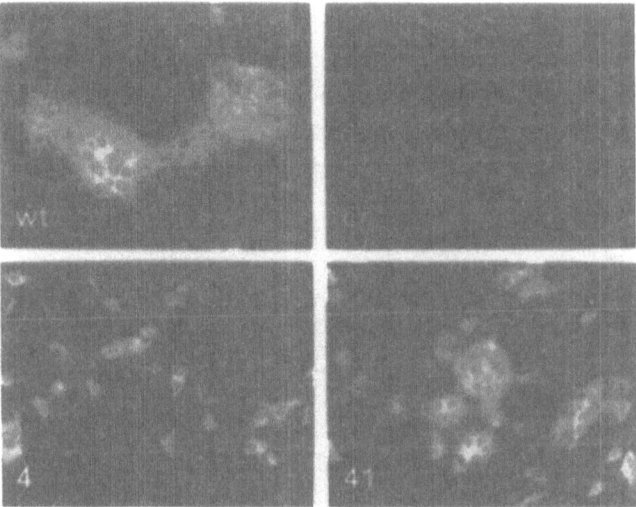

Figure 3. Syncytia formation by the S protein. DBT cells were infected with vTF7.3 and transfected with the S genes of wt, mut-4 and mut-41. The S proteins were detected by the S1 specific MAb at 6h posttransfection. Cr indicates DBT cells transfected with vector alone.

4.   DISCUSSION

The present study indicates that the S2A could be involved in fusion activity. In the N terminus of MHV S2, there is no hydrophobic amino acid cluster (Schmidt et al., 1987), which is called fusion peptide generally found in fusogenic enveloped RNA viruses (White, 1990). It was reported recently, however, that a region, PEP1, in heptad repeat 1 of MHV-A59 could work as an internal fusion peptide (Luo and Weiss, 1998). The fusion peptide has several common features, which are conserved in PEP1 but not in S2A. Completion of fusion events for MHV may require several distinct steps of interaction between the viral envelope protein and the cell membrane, with some functions performed by the PEP1 and others by the S2A. Further studies should be undertaken to determine the molecular mechanism by which the S2A participates in fusion events, as well as any specific cellular molecules that may cooperate with the S2A prior to or during fusion events.

5.   CONCLUSION

MAb 5B19.2 having VN and FI activities binds the S2A consisting of nine hydrophobic amino acids in the S2 of MHV S protein, suggesting the involvement of the S2A in receptor-binding and/or virus-cell fusion. While the binding of virus to the receptor was blocked by the anti-S1 MAbs that recognize the receptor binding domain, it was not blocked by the S2A antiserum, indicating that the S2A is not involved in receptor binding, nor it influences the receptor binding of the S1N330. S proteins with mutations in a few residues in the S2A were not fusogenic, even when they were replaced by hydrophobic ones, while most S proteins containing mutations in other residues retained the fusion activity. These results suggest that the S2A is important for the fusion activity of the MHV S protein during viral entry into cells.

REFERENCES

Collins, A. R.,. Knobler, R. L., Powell, H., and M. Buchmeier, M. J. (1982) Monoclonal antibodies to murine hepatitis virus-4 (strain JHM) define the viral glycoprotein responsible for attachment and cell fusion. *Virology* **119**, 358-371.

Fuerst, T. R., Niles, E. G., Studier, F. W., and Moss, B. (1986) Eukaryotic transient expression system based on recombinant vaccinia virus that synthesizes T7 RNA polymerase. *Proc. Natl. Acad. Sci. U.S.A.* **83**, 8122-8126.

Koolen, J. J. M., Borst, M. A., Horzinek, J. M C, and Spaan W. J. M. (1990) Immunogenic peptide comprising a mouse hepatitis virus A59 B-cell epitope and an influenza virus T-cell epitope protects against lethal infection. *J. Virol.* **64**, 6270-6273.

Kubo, H., Takase, S. Y., and Taguchi, F. (1993) Neutralization and fusion inhibition activities of monoclonal antibodies specific for the S1 subunit of the spike protein of neurovirulent murine coronavirus JHMV cl-2 variant. *J. Gen. Virol.* **74**, 1421-1425.

Kubo, H., Yamada, Y. K., and Taguchi, F. (1994) Localization of neutralizing epitopes and the receptor-binding site within the amino-terminal 330 amino acids of the murine coronavirus spike protein. *J. Virol.* **68**, 5403-5410.

Luo, Z., and Weiss, S. R. (1998) Roles in cell-cell fusion of two conserved hydrophobic regions in the murine coronavirus spike protein. *Virology* **244**, 483-494.

Luytjes, W., Geerts, D., Posthumus, W., Meloen, R., and Spaan, W. J. M. (1989) Amino acid sequence of a conserved neutralizing epitope of murine coronaviruses. *J. Virol.* **63**, 1408-1412.

Schmidt, I., Skinner, M., and Siddell, S. (1987) Nucleotide sequence of the gene encoding the surface projection glycoprotein of coronavirus MHV-JHM. *J. Gen. Virol.* **68**, 47-56.

Suzuki, H., and Taguchi, F. (1996) Analysis of the receptor binding site of murine coronavirus spike glycoprotein. *J. Virol.* **70**, 2632-2636.

Taguchi, F. (1993) Fusion formation by uncleaved spike protein of murine coronavirus JHMV variant cl-2. *J. Virol.* **67**, 1195-1202.

Taguchi, F., and Fleming, J. O. (1989) Comparison of six different murine coronavirus JHM variants by monoclonal antibodies against the E2 glycoprotein. *Virology* **169**, 233-235.

Taguchi, F., Ikeda, T., and Shida, H. (1992) Molecular cloning and expression of a spike protein of neurovirulent murine coronavirus JHMV variant cl-2. *J. Gen. Virol.* **73**, 1065-1072.

White, J. M. (1990) Viral and cellular membrane fusion proteins. *Annu. Rev. Physiol.* **52**, 675-697.

Are Intestinal Mucins Involved in the Pathogenicity of Transmissible Gastroenteritis Coronavirus?

CHRISTEL SCWEGMANN, GERT ZIMMER, AND GEORG HERRLER
Institut für Virologie, Tierärztliche Hochschule Hannover, Bünteweg 17, 30559 Hannover, Germany

1.   INTRODUCTION

Transmissible gastroenteritis virus (TGEV) is a prototype of enteropathogenic coronaviruses. The virus causes diarrhea in pigs of all ages. Infections are most severe in newborn piglets where letality can be as high as 100% (Pensaert *et al* 1993). The determinants of the enterotropism of TGEV are not known. A crucial factor appears to be the sialic acid binding activity of this virus (Schultze *et al* 1996). The ability of TGEV to attach to sialoglycoconjugates allows the virus to bind to erythrocytes. This interaction results in an agglutination reaction that probably has no physiological importance. However, hemagglutination provides a convenient assay for the sialic acid binding activity and allows quantitation of the virus. Mutants of TGEV that have lost their haemagglutinating activity because of a single point mutation in the S protein also had lost enteropathogenicity (Krempl *et al* 1997). Porcine respiratory coronavirus (PRCV), a respiratory variant of TGEV also lacks hemagglutinating activity. Both PRCV and the hemagglutination-deficient mutants are still able to grow in cell culture. Therefore, the sialic acid binding activity appears to be essential for enteropathogenicity but dispensable for growth in cultured cells. The sialic acid binding activity is located on the viral surface protein S. Another binding activity of this glycoprotein is the ability to interact with aminopeptidase N, the cellular receptor for TGEV (Delmas *et*

The Nidoviruses (Coronaviruses and Arteriviruses).
Edited by Ehud Lavi *et al.*, Kluwer Academic/Plenum Publishers, 2001.

al 1992). PRCV and the hemagglutination-deficient mutants have retained the ability to bind to aminopeptidase N. Therefore, the interaction with aminopeptidase N – though essential for the infection of cells – does not explain the enteropathogenicity of TGEV.

How the sialic acid binding activity contributes to the enteropathogenicity of TGEV is not known. One possibility is that it may facilitate the binding to and infection of enterocytes. Another possibility is that the sialic acid binding activity may protect the virus from detrimental effects encountered during passage through the gastrointestinal tract, such as the action of detergent-like bile salts. The latter view is supported by the finding that a concentration of the detergent octylglucoside that completely inactivates PRCV and hemagglutination-deficient mutants, only results in a partial reduction of the infectivity of TGEV (Krempl *et al* 2000). This protection may be achieved by binding of sialoglycoconjugates to the viral surface. Potential ligands encountered in the gastrointestinal tract are mucins because of their high content of sialic acids.

As there is an age-dependent difference in the severity of the disease caused by TGEV, we wondered whether the sialic acid binding activity may explain these differences. Therefore, it was of interest whether intestinal mucins obtained from piglets and pigs differ in the amount and in the type of sialic acid. To answer this question we isolated mucins from the small intestine and analyzed the sialic acids by high performance liquid chromatography (HPLC).

2.   METHODS

Mucins were isolated from two piglets (12 to 14 days old) and from two pigs (10 to 15 weeks old) as described previously (Enss *et al* 1996). Freeze-dried preparations of the mucins were suspended in water to determine the sialic acid content by HPLC-analysis (Krempl *et al* 2000).

3.   RESULTS

Intestinal mucins were obtained form two piglets that were in the age group where TGEV infections are most severe. These samples were compared to mucins from two older animals. Mucin preparation involved Sepharose chromatography. The sialic acid of the mucin fractions were analyzed by HPLC. The result is shown in Fig. 1. The two samples derived from piglets, both showed two sialic acid peaks, one representing N-glycolylneuraminic acid (Neu5Gc) and the other representing Nacetylneuraminic acid (Neu5Ac). The samples resembled each other in that the Neu5Gc peak was more prominent than the Neu5Ac peak. They differed in the total amount of sialic acid. This difference is most likely explained by the presence of nonsialylated substances in one of the samples. One of the samples derived from adult pigs was found to have a sialic acid profile similar to one of the piglet samples both in the Neu5Gc/Neu5Ac ratio and in the total amount of sialic acid. The other sample was quite different with the Neu5Ac peak being much more prominent than the Neu5Gc peak. This result indicates that there is a variation in the distribution of sialic acids present in the mucin samples and that there are no clearcut differences between the samples of young and adult animals.

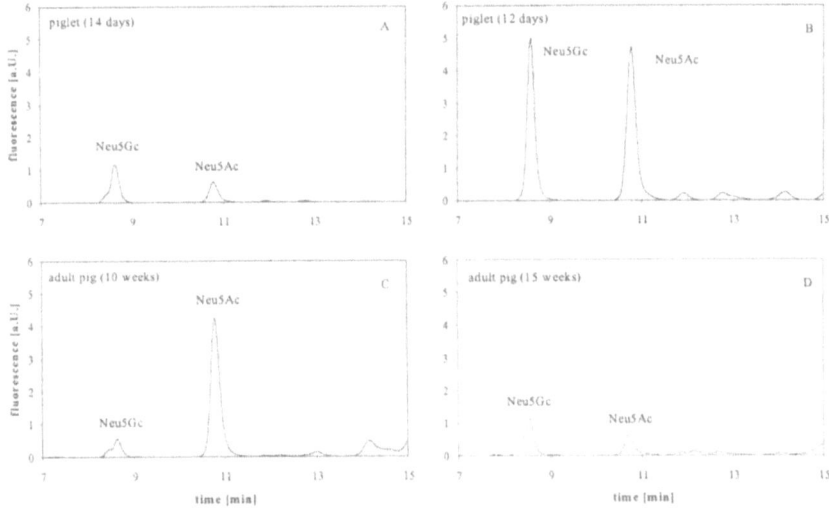

Figure 1. HPLC-analysis of the sialic acid content of mucins from two piglets and two pigs.

4. CONCLUSION

Most enteropathogenic viruses that enter the host via the gastrointestinal tract are nonenveloped viruses. Coronaviruses are unique among this group of viruses because they contain a lipid envelope. The predominance of nonenveloped viruses has been explained by the ability of nonenveloped viruses to resist the harsh conditions encountered during gastrointestinal passage, e.g. the low pH and the proteases. Enveloped viruses might be expected to be sensitive to such environmental factors, especially to the action of detergent-like bile salts. How coronaviruses survive under these conditions is unclear. An attractive idea is the binding of protective substances to the viral surface. Sialoglycoconjugates are potential candidates for ligands that may help coronaviruses to survive. Several coronaviruses have been shown to contain a sialic acid binding activity. Whereas bovine coronavirus and serologically related viruses use this binding activity in the process of infection for attachment to cellular surface receptors (Schultze *et al* 1992), TGEV does not require it to infect cultured cells. For several TGEV mutants as well as for the respiratory variant PRCV loss of the enteropathogenicity has been shown to be associated with the loss of the sialic acid binding activity. So, TGEV is ideal for analyzing whether the binding of protective ligands is required to survive in the gastrointestinal tract.

Mucins are prime candidates for sialoglycoconjugates that may bind to the surface of TGEV. Mucins are abundantly found in the gastrointestinal tract and they are rich in sialic acids. If intestinal mucins play a role in the enteropathogenicity of TGEV, one might expect age-dependent differences, because TGEV infections are much more severe in piglets than they are in adult animals. Our analysis showed that the mucins from piglets contain both Neu5Gc and Neu5Ac. The former type of sialic acid was present in higher amounts in the two samples analyzed. As TGEV has a preference for Neu5Gc over Neu5Ac – at least in the recognition of low amounts of sialic acid-, the sialic profile found in the piglet samples is in accordance with the expected profile of a ligand for the sialic acid binding activity of TGEV. In the mucin sample of one of the two adult pigs, the sialic acid profile was similar to that found in the two piglet samples. In the other sample of adult pigs, a predominance of Neu5Ac was detected. Our data do not indicate that there are differences between young and adult animals in the sialic acid content of intestinal mucins, that can be interpreted such that piglet mucins are less protective ligands. Therefore, other explanations have to be considered. One possibility that the sialic acid binding activity is required for binding to intestinal epithelial cells and that the sialoglycoconjugates expressed on the surface of piglet cells allow a more efficient binding compared to glycoconjugates of cells from adult animals. Work is in progress to find out whether this concept is correct.

ACKNOWLEDGMENTS

This work was supported by a grant from Deutsche Forschungsgemeinschaft (SFB280).

REFERENCES

Delmas, B., Gelfi, J., L'Haridon, R., Vogel, L.K., Sjöström, H., Noren, O. and Laude, H., 1992, Aminopeptidase N is a major receptor for the entero-pathogenic coronavirus TGEV. *Nature* **357**: 417-420.

Enss, M.-L., Schmidt-Wittig, U., Müller, H., Mai, U.E.H., Coenen, M. and Hedrich, H.J., 1996, Response of germfree rat colonic mucous cells to peroral endotoxin application. *European Journal of Cell Biology* **71**: 99-104.

Krempl, C., Schultze, B., Laude, H. and Herrler, G., 1997, Point mutations in the S protein connect the sialic acid binding activity with the enteropathogenicity of transmissible gastroenteritis coronavirus. *Journal of Virology* **71**: 3285-3287.

Krempl, C., Ballesteros, M.L., Zimmer, G., Enjuanes, L., Klenk, H.-D. and Herrler, G., 2000, Characterization of the sialic acid binding activity of transmissible gastroenteritis coronavirus by analysis of haemagglutination-deficient mutants. *Jounal of General Virology* **81**: 489-96.

Pensaert, M., Callebaut, P. and Cox, E., 1993, Enteric coronaviruses of animals. In *Viral Infections of the Gastrointestinal Tract.* Edited by A.Z. Kapikian. New York, Marcel Dekker, pp. 627-696.

Schultze, B. and Herrler, G., 1992, Bovine coronavirus uses N-acetyl-9-O-acetylneuraminic acid as a receptor determinant to initiate the infection of cultured cells. *Journal of General Virology* **73**:901-906.

Schultze, B., Krempl, C., Ballesteros, M.L., Shaw, L., Schauer, R., Enjuanes, L. and Herrler, G., 1996, Transmissible gastroenteritis coronavirus, but not the related porcine respiratory coronavirus, has a sialic acid (N-glycolylneuraminic acid) binding activity. *Journal of Virology* **70**:5634-5637.

Biochemical Properties and Processing of the Three Major Structural Proteins of PRRS Virus Expressed by Recombinant Adenoviruses
Structural, functional and community aspects

CARL A. GAGNON,[1,2,3] YVES LANGELIER,[3] BERNARD MASSIE,[2] AND SERGE DEA[1]

[1]INRS-Institut Armand-Frappier, Centre de Microbiologie et Biotechnologie, Laval, P.Q., Canada, H7V 1B7. [2]Biotechnologie Research Institute, CNRC, Montreal, P.Q., Canada. [3]Centre Hospitalier de l'Université de Montréal, Montreal, P.Q., Canada.

1. INTRODUCTION

The ORFs 5 to 7 encode for the three major structural proteins of porcine reproductive and respiratory syndrome virus (PRRSV) : the envelope glycoprotein GP_5 (25-26 kDa), the non-glycosylated membrane protein M (18-19 kDa) and the nucleocapsid protein N (14-15 kDa), respectively (Mardassi et al., 1995; Meulenberg et al., 1993). The latter structural proteins of PRRSV are closely associated both in the infected-cells and in the virion, the GP_5 and M proteins being associated in the form of heterodimers (Mardassi et al., 1996). The GP_5, which is highly glycosylated (Mardassi et al., 1996), have been found to play an important role in the induction of a protective immune response (Pirzadeh and Dea, 1998). Immunization experiments of mice with E. coli-expressed GST-ORF5 recombinant fusion protein, as well as with purified PRRSV, induced specific anti-GP_5 neutra-lizing MAbs (Pirzadeh and Dea, 1997). The antibodies in PRRSV-infected pigs antisera responsible for the viral neutralisation in cell culture have been also determined to be GP_5 specific (Gonin et al., 1999). In DNA immunization experiments, pigs injected with a plasmidic vector expressing the ORF5 gene not only produced neutralizing antibodies to PRRSV, but were also protected against development of clinical disease and lung lesions following an intratracheal challenge with a high

The Nidoviruses (Coronaviruses and Arteriviruses).
Edited by Ehud Lavi et al., Kluwer Academic/Plenum Publishers, 2001.

infectious dose of PRRSV. On the other hand, parenteral inoculation of SPF piglets with *E. coli*-expressed GST-ORF5 recombinant fusion protein rather resulted in an increased severity of lung lesions, despite the development of high titers (> 2048) of non-neutralizing antibody titers to GP$_5$ (Pirzadeh and Dea, 1998). Consequently, the use of an eucaryotic expression system should be preconized to produce large amounts of a recombinant protein that would have to preserve the major characteristics (glycosylation, conformational epitopes) of the native major envelope glycoprotein of PRRSV. Recently, a replication defective human type 5 adenovirus (Ad) has been used successfully for the eucaryotic expression and characterization of the ORF3 product of PRRSV (Mardassi et al., 1998). Previous investigators have also demonstrated that recombinant Ad (recAd) carrying the structural genes of different viruses, mainly the envelope or spike glycoproteins of murine and porcine coronaviruses, were effective for the induction of antibodies to the expressed rec glycoproteins that conferred protection (Both et al., 1993; Torres et al., 1995; Wesseling et al., 1993).

The main objectives of the present study were to construct recAds expressing the three major structural proteins of PRRSV and to assess if the recombinant (rec) expressed proteins have conserved the antigenicity, glycosylation properties and biological functions (cytotoxicity, proapoptotic phenotype, immunogenicity) of the native major structural proteins of PRRSV.

2.   METHODOLOGY AND RESULTS

The IAF-Klop PRRSV N, M and GP$_5$ coding sequences (ORF7, ORF6 and ORF5, respectively) (EMBL/Genbank accession No. U64928) were inserted into the unique *Bam*H1 site of the adenovirus transfer vectors pAdCMV5 or pAdTR5 (Massie et al., 1998) so that the ORFs 5, 6 and 7 coding sequences would be under the control of the constitutive human CMV immediate-early (IE) promoter/enhancer (pAdCMV5) or the tetracyclin regulatable promotor (pAdTR5). Rec plasmids were rescued into the genome of Ad/CMVlacZ by homologous recombination in 293 cells, as described elsewhere (Ascadi et al., 1994) and the following replication-defective recAds were generated: AdTR5/ORF5, AdCMV5/ORF6 and AdT/R5/ORF7. The recAdCMV/tTA permitted the constitutive expression of tTA in infected cells which is required for expression in recAd-infected cells of the foreign gene that has been cloned dowstream and under the control of the TR5 promotor. In the presence of doxycyclin (1 µg/ml), the expression of the foreign gene is inhibited in recAd-infected cells (Massie et al., 1998). Immunolo-gical identification of the native viral proteins, as well as of that of the 293- expressed rec proteins was confirmed by Western blot (WB) and radioimmuno-precipitation (RIPA) using the homologous hyperimmune porcine anti-PRRSV serum, rabbit monospecific antisera ($\alpha5$, $\alpha6$ and $\alpha7$) and MAbs. The fact that the rec proteins were recognized by antibodies specific for each of the three major structural proteins of PRRSV in both WB and RIPA, was indicative that at least the major antigenic determinants of the native viral structural proteins have been conserved (data not shown). Also the rec proteins displayed the same electro-phoretic profiles than that of the native proteins of PRRSV (data not shown).

At 18 to 24 h post infection (pi), following a starvation period of 60 min in methionine-deprived DMEM, the PRRSV-infected MARC-145 cells or the 293 recAds-infected cells were pulse-labelled with 250µCi of [^{35}S]methionine (sp activity of 1,120 Ci/mmole, Amersham Searle Co., Oakville, Ontario) for 5 h. Cells lysates were incubated with the $\alpha5$ rabbit monospecific antiserum and the immune complexes were collected by addition of protein A-sepharose CL4B beads (Pharmacia) and analyzed by 12% SDS-PAGE, as previously described (Mardassi et al., 1996). The immune complexes were then incubated with either Endo-β-N-acetylglucosaminidase H (Endo H), Endoglycosidase F/N-Glycosi-dase F (Glyco F) or Endo-β-Galactosidase (Endo β) to characterize the N-glycosylation type of the GP_5 (Fig. 1). The recGP_5 appeared different from the native protein in regards of its glycosylation process. The recGP_5 apparently possessed poly-*N*-acetyllactosamine residues because of its sensitivity to Endo β digestion (Fig. 1a and b, lane b). Also the recGP_5 did not possess oligosacharides of the complex type since it was totally sensitive to Endo H (Fig. 1b, lane h) contrary to its native counterpart (Fig.1a, lane h). Differences observed in the N-glycosylation was independent of the cell substrate since AdTR5/ORF5 propagated in MARC-145 cells have the same endoglycosidases digestion profiles than AdTR5/ORF5 propagated in 293 cells (Fig. 1b and c).

Coexpression of the three major structural proteins in recAds-infected 293 cells did not changed the fate of the glycosylation process of GP_5 (Fig. 2). Also, even if the N, M and GP_5 proteins were present in 293 cells, no formation of viral particles could be observed in the recAds-infected cells by electron microscopy and no formation of M-GP_5 heterodimers was observed in RIPA experiments (data not shown).

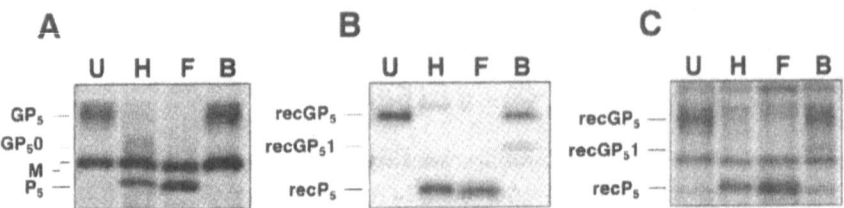

Figure 1. Glycosylation nature of the recGP₅ in 293 and MARC-145 cells. Immunoprecipitation profiles obtained with: A) lysate of PRRSV-infected MARC-145 cells; B) lysate of AdTR5/ORF5-infected 293 cells; C) lysate of AdTR5/ORF5 + AdCMV/tTA infected MARC-145 cells. In all cases, viral proteins were immunoprecipitated using α5 monospecific antiserum. The immuno-precipitated proteins were untreated (lane U) or treated with endoglycosidases; Glyco F (lane F), Endo H (lane H) and Endo β (lane B).

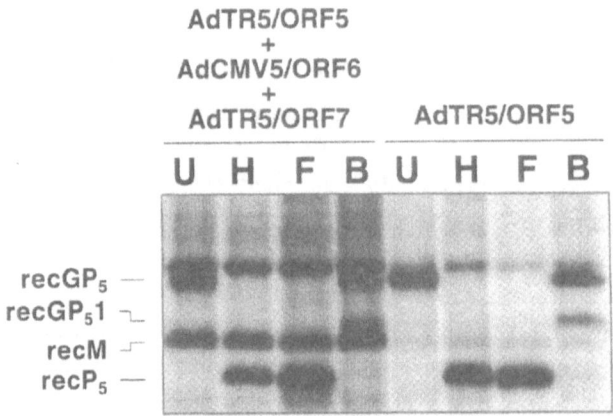

Figure 2. Glycosylation nature of the coexpressed recN, recM and recGP₅ proteins in 293 cells. RIPA experiments were conducted with cell lysates of recAds-infected 293 cells. Expressed rec proteins were immunoprecipitated with α5, α6 and α7 monospecific antisera, then electrophoresed after no treatment (U) or treatment with endoglycosidases Glyco F (lane F), Endo H (lane H) and Endo β (lane B).

Both genotypes of PRRSV have been reported to induce apoptosis in the infected cells (Suarez et al., 1996; Sur et al., 1997). On the other hand, only the GP₅ of the European genotype has been reported to induce apoptosis (Suarez et al., 1996). Since, there is only a 52% aa identity between the European and the North American genotypes, the apoptotic phetotype of the GP₅ of North American IAF-Klop strain was verified by infecting MARC-145 cells with the recAd AdTR5/ORF5. The cytopathic changes observed following expression of the recGP₅ in MARC-145 cells appeared almost identical to those observed in the case of a PRRSV infection (Kim et al., 1993). Indeed, scattered and enlarged cells with very small granular inclusions in their cytoplasm usually started to be detectable by 24 h pi, than infected cells showed a tendency to clump by foci that could be well delineated from the non-affected monolayer by 36 to 48 h pi. Numerous infected cells eventually detached from the plates, more than 75% of the cell monolayers being severely damaged by 72 h pi (data not shown). No effect on MARC-145 cells could be observed at 72 h pi when doxycyclin was added to the medium. To verify if the cytopathic effect was due in part to apoptosis, the level of procaspase 3 activation was measured by incubating cell lysates in the presence of a specific subtrate for caspase 3, the DEVD-AMC fluorogenic substrate (BIOMOL Research Laboratories, Inc.). When caspase 3 is present in the cell lysates, the AMC substrate is cleaved and the fluorescence increases with time. The results obtained were reported in Fig. 3 and expressed as constant fluorescence released (fluorescence units or FU) per second per μg of cell lysates. Fig. 3 demonstrated that the recGP$_5$ induced apoptosis in MARC-145 cells and that degenerative effects were not due to Ad itself because in the presence of doxycyclin the inhibition of the GP$_5$ synthesis correlated with inhibition of the procaspase 3 activation, the level of caspase 3 being 11 times higher when the GP$_5$ was expressed in the cells. Also, PRRSV infection induced an activation of the procaspase 3.

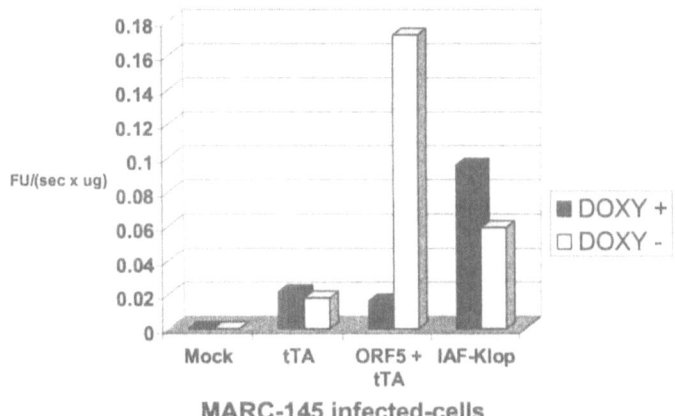

Figure 3. Procaspase 3 activation. MARC-145 cells were infected with only AdCMV/tTA or with both AdCMV/tTA and AdTR5/ORF5 at a MOI of 100 PFU/cell in the presence or absence of 1 μg/ml of doxycyclin.

Our main objective is to construct an efficient recombinant vaccine against PRRSV infection. Consequently , to study the immunogenicity of recAds, mice have been immunized with AdTR5/ORF7 and AdTR5/ORF5. The mice develop-ped specific antibodies against the N and GP$_5$ proteins, antibody titers of 64 to 128 being detected in their sera by indirect immunofluorescence (data not shown) 55 days post-inoculation.

3. DISCUSSION

We are not yet able to tell if the differences observed concerning the glycosylation type and processing of the GP_5 expressed individually by recAd may alter its biological and immunological functions. Nevertheless, recAds were established as useful tools to study the structural proteins of PRRSV *in vitro* and *in vivo*. Furthermore, since genetic immunization was previously found to trigger the immune system of pigs to produce neutralizing antibodies to PRRSV and induced protection against development of the clinical disease and histopathological lesions, but failed to eliminate the presence of the virus in lungs of infected animals, data suggested that mucosal immunity may have a major role to play in the protection against this infection. The use of human or porcine adenoviruses as viral vectors for the induction of mucosal immune reponses against enteric or respiratory viral diseases in pigs has been proposed by several investigators and found to be very effective because of their tropism for the BALT (bronchoalveolar-associated lymphoid tissues) and GALT (gut-associated lymphoid tissue). Such viral vector, either replicative-defective or replicative, may proved very efficient against PRRSV infection. However, more studies are needed in order to identify the viral proteins carrying antigenic determinants involved in the effective humoral and cellular immune response and whether these determinants are well preserved among different field strains.

ACKNOWLEDGMENTS

This work was partly supported by the National Science and Engineering Research Council of Canada, strategic grant #STP0202083 and Biovet Inc., St-Hyacinthe, Qc, Canada. C.A. Gagnon is a recipient of a fellowship from the Medical Research Council of Canada.

REFERENCES

Ascadi, G., Jani, A., Massie, B., Simoneau, M., Holland, P., Blaschuk, K. and Karpati, G. (1994). A differential efficiency of adenovirus-mediated in vivo gene transfer into skeletal muscle cells of different maturity. Human molecular genetic 3:578-584.

Both, G.W., Lockett, L.J., Janardhana, V., Edwards, S.J., Bellamy, A.R., Graham, F.L., Prevec, L. and Andrew, M.E. (1993). Protective immunity to rotavirus-induced diarrhoea is passively transferred to newborn mice from naive dams vaccinated with a single dose of recombinant adenovirus expressing rotavirus VP7sc. Virology 193:940-950.

Gonin, P., Pirzadeh, B., Gagnon, C.A. and Dea, S. (1999) Seroneutralization of porcine reproductive and respiratory syndrome virus correlates with antibody response to the GP5 major envelope glycoprotein. J. Vet. Diagn. Invest. 11: 20-26.

Kim, H. S., Kwang, J., Yoon, L.J., Joo, H.S. and Frey, M.L. (1993). Enhanced replication of porcine reproductive and respiratory syndrome (PRRS) virus in a homogeneous subpopulation of MA-104 cell line. Arch.Virol. 133:477-483.

Mardassi, H., Mounir, S. and Dea, S. (1995) Molecular analysis of the ORF3-7 of porcine reproductive and respiratory syndrome virus, Quebec reference strain. Arch. Virol. 140: 1405-1418.

Mardassi, H., Massie, B. and Dea, S. (1996) Intracellular synthesis, processing and transport of proteins encoded by ORFs 5 to 7 of porcine reproductive and respiratory syndrome virus. Virology 221: 98-112.

Mardassi, H., Gonin, P., Gagnon, C.A., Massie, B. and Dea, S. (1998). A subset of porcine reproductive and respiratory syndrome virus (PRRSV) GP3 glycoprotein is released into the culture medium of cells as a non-virion-associated and membrane-free (soluble) form. J. Virol. 72:6298-6306.

Massie, B., Couture, F., Lamoureux, L., Mosser, D.D., Guibault, C., Jolicoeur, P., Bélanger, F. and Langelier, Y. (1998). Inducible overexpression of a toxic protein by an adenovirus vector with a tetracycline-regulatable expression cassette. J. Virol. 72:2289-2296.

Meulenberg, J.J.M., Hulst, M.M., de Meijer, E.J., Moonen, P.L.J.M., den Besten, A., de Kluyver, E.P., Wensvoort, G. and Moormann, R.J.M. (1993). Lelystad virus, the causative agent of porcine epidemic abortion and respiratory syndrome (PEARS) is related to LDV and EAV. Virology 192:62-74.

Pirzadeh, B. and Dea, S. (1997). Monoclonal antibodies to the ORF5 product of porcine reproductive and respiratory syndrome virus define linear neutralizing determinants. J. Gen. Virol. 78:1867-1873.

Pirzadeh, B. and Dea, S. (1998). Immune response in pigs vaccinated with plasmid DNA encoding ORF5 of porcine reproductive and respiratory syndrome virus. Journal of General Virology 79:989-999.

Suárez, P., Díaz-Guerra, M., Prieto, C., Esteban, M., Castro, J.M., Nieto, A. and Ortín, J. (1996). Open reading frame 5 of porcine reproductive and respiratory syndrome virus as a cause of virus-induced apoptosis. J. Virol. 70:2876-2882.

Sur, J.H., Doster, A.R., Christian, J.C.S., Galeota, J.A., Wills, R.W., Zimmerman, J.J. and Osario, F.A. (1997). Porcine reproductive and respiratory syndrome virus replicates in testicular germ cells, alters spermatogenesis, and induces germ cell death by apoptosis. J. Virol.71:9170-9179.

Torres, J.M., Sanchez, C., Sune, C., Smerdou, C., Prevec, L., Graham F. and Enjuanes, L. (1995). Induction of antibodies protecting against Transmissible Gastroenteritis Coronavirus (TGEV) by recombinant adenovirus expressing TGEV spike protein. Virology 213:503-516.

Wesseling, J.G., Godeke, G.-J., Schijns, V.E.C.J., Prevec, L., Graham, F.L., Horzinek, M.C. and Rottier. P.J.M. (1993). Mouse hepatitis virus spike and nucleocapsid proteins expressed by adenovirus vectors protect mice against a lethal infection. J. Gen. Virol. 74:2061-2069.

Inefficient Infection of Soluble Receptor-Resistant Mutants of Murine Coronavirus in Cells Expressing MHVR2 Receptor

SHUTOKU MATSUYAMA AND FUMIHIRO TAGUCHI

National Institute of Neuroscience, NCNP 4-1-1 Ogawahigashi, Kodaira, Tokyo 187-8502, Japan

1. INTRODUCTION

We have isolated several soluble receptor-resistant (srr) mutants of mouse hepatitis virus (MHV) strain JHMV cl-2, which are resistant to neutralization with the soluble receptor CEACAM1[a] (MHVR1) derived from MHV-susceptible BALB/c mice (Saeki et al., 1997). We report here that srr mutants infected cultured cells expressing MHVR1 as efficiently as wt virus, yet failed to efficiently infect cells expressing CEACAM1[b] (MHVR2) derived from MHV-resistant SJL mice. Our results suggest that inefficient infection by srr mutants in these cells is due to defects in viral entry into cells.

2. MATERIALS AND METHODS

2.1 Viruses and cells

MHV strain JHMV cl-2 (wt cl-2) (Taguchi et al., 1985) and srr mutants (srr7, srr11, srr18) (Saeki et al., 1997) were used. Srr11 has an amino acid change in the S1 subunit at position 65 (Leu to His). Srr 7 and 18 have a

The Nidoviruses (Coronaviruses and Arteriviruses).
Edited by Ehud Lavi *et al.*, Kluwer Academic/Plenum Publishers, 2001.

change in the S2 subunit at position 1114 (Leu to Phe) and 1163 (Cys to Phe) respectively. BHK-21 (BHK), BHK-R1 and BHK-R2 cell lines were used for these studies. BHK-R1 and BHK-R2 cells constitutively express the MHVR1 and MHVR2, respectively (Saeki et al., 1997).

2.2   Virus Overlay Protein Blot Assay (VOPBA)

The virus overlay protein blot assay (VOPBA) was used to detect the binding of the S proteins to the receptor protein, as described previously (Saeki et al., 1997). A cross-linker BS^3 [Bis(sulfosuccinimidyl)substrate] was used to form crosslinks between S protein and MHVR2.

2.3   Construction of expression vectors containing the S genes and their expression by vTF7.3

DNA encoding the wt S protein was amplified by RT-PCR from the original wt S gene and ligated into the pTarget vector in the downstream the T7 promoter to create pTar-cl-2S. To make srr S constructs, we have substituted the segment that contains the srr mutation for the corresponding segment of pTar-cl-2S. The vectors containing various S genes were expressed in BHK-R1 and BHK-R2 cells using vTF7.3 (Fuerst et al., 1986).

3.   RESULTS AND DISCUSSION

MHV JHMV and its srr mutants, srr7, srr11 and srr18, grew and induced syncytia equally well in BHK-R1. In contrast, srr syncytia formations (Figure 1) and growth (data not shown) were drastically (srr7 and srr11) and moderately (srr18) reduced relative to wild type (wt) virus in BHK-R2.

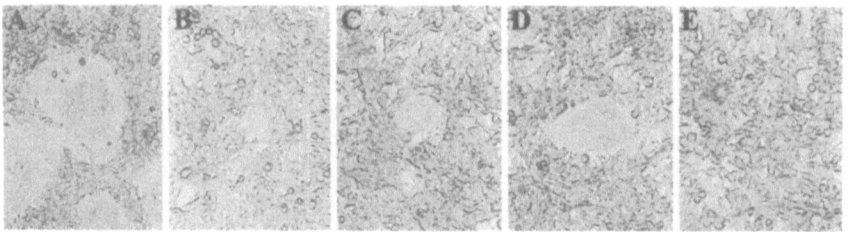

Figure 1. Cytopathic effects of wt and srr mutants in BHK-R2 cells.
BHK-R2 cells infected with wt cl-2(A), srr7(B), srr11(C), srr18(D) at an MOI=0.1, or mock(E) were photographed under the light microscope at 18h after infection.

Since the difference in infection in BHK-R2 cells could be related to their interaction with MHVR2, we examined the direct binding of these viruses to MHVR2 by a VOPBA in the presence of cross-linker, BS^3. There was no difference between srr7, srr18 and wt in their ability to bind to BHK-R2. The binding ability of srr11 was weaker relative to the wt ability. However, these binding features were very similar to those observed in the binding to MHVR1; srr7 and srr18 bound to MHVR1 in a same efficiency as wt, yet srr11 did less efficiently. (Figure 2).

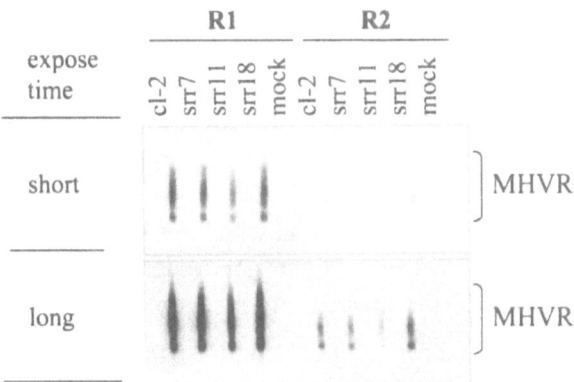

Figure 2. Analysis of virus-binding to the MHVR1 and MHVR2 by VOPBA.
Same amounts of S protein of wt cl-2 or srr mutants were incubated with the soluble MHVR1 and MHVR2 transferred to the membrane paper in the presence of cross-linker, BS^3. The binding of viruses were evaluated with the S1 specific monoclonal antibody No.7 conjugated with biotin and HRPO labeled avidin by enhanced chemiluminescence. X-ray film was exposed for 2 min (long) or 15 sec (short).

Table 1. Fusion activity of transfected S proteins of wt and srr in BHK-R1 and BHK-R2 cells.

S protein	Fusion index (%)		Reduction (%) [b]
	BHK-R1	BHK-R2	
cl-2	93± 4[a]	74±14	21
srr7	87±11	15± 9**	83
srr11	87±10	6± 8**	93
srr18	90± 8	27± 3**	70

[a] Mean fusion indices ± SD. Fused cells and nuclei were counted in four different fields, and fusion indices [1-(cells /nuclei)] ×100 were calculated.
[b] Relative reduction of fusion index in BHK-R2 compared to the value for BHK-R1.
** (P < 0.01), Significant reduction by Student's t test compared with the cl-2 fusion index in BHK-R2 cells.

Since the above data suggest that srr mutants bind to both MHVR1 and MHVR2 with similar efficiency as wt did, we have examined their entry into these cells. As a marker of viral entry potential, the fusogenicity of these srr

S proteins was examined. As shown in Table 1, wt and srr mutants induced syncytia with similar efficiencies in BHK-R1 cells. The fusion index of BHK-R2 cells, however, was remarkably low when srr S proteins, but not wt S proteins, were expressed. The finding that srr7 and srr11 fusion activity was reduced to a greater degree than that of srr18 or wt is compatible with the degree of their lesser efficiency of infection in BHK-R2 cells.

From the results obtained, we have concluded that srr mutants can enter into BHK-R1 cells more efficiectly than into BHK-R2 cells, though there is little difference in the binding of srr mutants to MHVR1 and to MHVR2. However, wt virus bound and entered into BHK-R1 and BHK-R2 cells to the same extent.

S proteins of srr mutants may become fusogenic when bound to a high-affinity receptor, but not when bound to a low-affinity receptor, while the wt S protein becomes fusogenic even when bound to a low-affinity receptor. It is likely that binding to either a high- or low-affinity receptor causes the conformational changes of wt S protein which result in virus-cell membrane fusion. Srr mutants, however, might not undergo such conformational changes after binding to the low-affinity receptor because of the mutations in their S protein. Studies are currently in progress to search the conformational changes of S protein.

4.   CONCLUSION

Srr mutants infected BHK-R1 as efficiently as wt, while their infection to BHK-R2 was less efficient than wt virus. The inefficient infection was not due to their low affinity to MHVR2, but due to the low fusogenicity. These results suggest that the srr inefficient infection in BHK-R2 cells results from an impaired entry into cells.

REFERENCES

Fuerst, T.R., E. G. Niles, F. W. Studier, and B. Moss. (1986). Eukaryotic transient expression system based on recombinant vaccinia virus that synthesizes T7 RNA polymerase. *Proc. Natl. Acad. Sci. U.S.A.* **83**: 8122-8126

Saeki, K., N. Ohtsuka, and F. Taguchi. (1997). Identification of spike protein residues of murine coronavirus responsible for receptor-binding activity by use of soluble receptor-resistant mutants. *J. Virol.* **71**: 9024-9031

Taguchi, F., S. G. Siddell, H. Wege, and V .ter Meulen. (1985). Characterization of a variant virus selected in rat brain after infection by coronavirus mouse hepatitis virus JHM. *J. Virol.* **54**: 429-435.

A Study on Mouse Hepatitis Virus Receptor Genotype in the Wild Mouse

[1,3]NOBUHISA OHTSUKA, [2]KIMIYUKI TSUCHIYA, [3]EIICHI HONDA, AND [1]FUMIHIRO TAGUCHI

[1]National Institute of Neuroscience, NCNP, 4-1-1 Ogawahigashi, Kodaira, Tokyo 187-8502, [2]Department of Physiology, Miyazaki Medical College, 5200 Kihara, Kiyotake, Miyazaki, Miyazaki 889-1692, [3]Department of Veterinary Medicine, Faculty of Agriculture, Tokyo University of Agriculture and Technology, 3-5-8 Saiwai, Fuchu, Tokyo 183-8509, Japan

1. INTRODUCTION

Whereas most of laboratory mouse strains are susceptible to mouse hepatitis virus (MHV) infection, only SJL mouse strain is resistant (Smith et al., 1984). MHV susceptible mouse strains have CEACAM1[a] (MHVR1) as a receptor for MHV, while SJL mice have CEACAM1[b] (MHVR2) (Ohtsuka and Taguchi, 1997). Although these molecules are functional receptors for MHV (Yokomori and Lai, 1992), there is a striking difference in receptor functionality between these two proteins. MHVR1 has 250~500 times higher receptor-binding activity relative to MHVR2 as examined by a virus overlay protein blot assay and 10~30 times higher receptor activity when expressed on cells without MHV receptor (Ohtsuka et al., 1996, Rao et al., 1997). These findings suggest that the difference in susceptibility to MHV infection between MHV-susceptible mouse strains and SJL could result from the difference in receptor functionality found between MHVR1 and MHVR2.

The Nidoviruses (Coronaviruses and Arteriviruses).
Edited by Ehud Lavi et al., Kluwer Academic/Plenum Publishers, 2001.

The susceptibility of mice to MHV is determined by a single gene located on chromosome 7 and the susceptibility is dominant over the resistance (Smith et al., 1984). MHV receptor gene is also mapped on chromosome 7 (Robins et al., 1991). These facts in combination with the difference in receptor functionality detected between MHVR1 and MHVR2 suggest the possibility that the gene controlling MHV susceptibility is identical to the MHV receptor gene. To investigate this possibility, we have generated the progeny mice between BALB/c and SJL. Analysis of these progeny showed that all mice with R2/R2 genotype were resistant, while those with R1 gene were susceptible, suggesting that MHV receptor gene determines the susceptibility of mice to MHV infection (Ohtsuka and Taguchi, 1997). In order to further test above possibility using the wild mouse, we first of all examined the MHV receptor genotype of several wild mouse subspecies distributed in Asia and Europe. The results indicated that both MHVR1 and MHVR2 are common to all of the wild mouse subspecies examined.

2.   MATERIALS AND METHODS

2.1   Wild mouse subspecies used in this study

Wild mice seized in Europe and Asia were used in the present study. *Mus. musculus. (M. m.) domesticus* lives in western Europe. *M. spicilegus* distributes from northern area of Black sea to the river Volga. *M. m. bactrianus* distributes from Pakistan to the west. *M. m. musculus* distributes from western Europe to northern area of China. *M. m. castaneus* distributes from Pakistan to east southern Asia. *M. m. molossinus* distributed in Japan is a hybrid between northern *M. m. musculus* and *M. m. castaneus*.

2.2   Genotyping

Genomic DNA samples were prepared from each mouse tail and N-terminal domain of MHVR was amplified from the genomic DNA by nested PCR. First PCR was performed with 5'-CCTCACTTTTAGCCTCCTGGAG -3' and 5'-ACATGAAATTGCACAGTCGC-3' primers, and second PCR was done with 5'-GCTGAAGTCACCATTGAGGC-3' and FITC labeled 5'-AGCAGGGATCCATTGCTGTA-3' primers. These products were denatured in formamide by heating at 94˚C for 2 minutes. Then single strand conformation polymorphism (SSCP) of these products were analysed by electrophoresis on 6% polyacrylamide gel for 1.5 hour at 26˚C.

3.   RESULTS

We investigated the MHV receptor genotype of various *Mus musculus* subspecies. Since remarkable difference in DNA sequence has been reported in the N-terminal domain of MHVR1 and MHVR2 genes, we amplified this region by nested PCR. The resultant DNA products were then analyzed by SSCP. As shown in figure 1, most of wild mouse subspecies retained both MHVR1 and MHVR2 genotypes. More than half of wild mice in each subspecies had MHVR2. These results suggested that MHVR2, which is very rare in laboratory mice, is common to the wild mouse population. Although some mice contained mutations in the MHV receptor genes, none of those mutations affected the function of each MHVR1 or MHVR2 (data not shown).

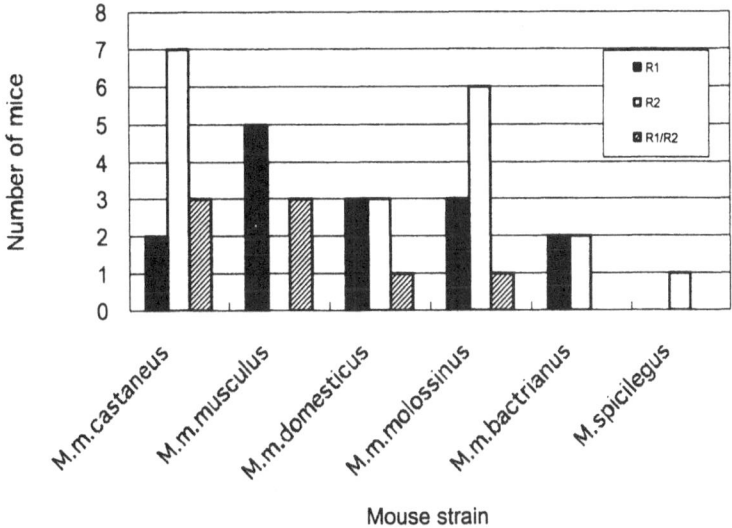

Figure 1. Typing of the MHV receptor gene in the wild mouse subspecies.

4.   DISCUSSION

The present study has clearly shown that MHVR2 is common to all *Mus* subspecies examined, though that is very rare in the laboratory mouse strains. So far, only SJL mouse strain is reported to have MHVR2.

MHV strains, MHV-A59 and JHMV, multiply in cells with MHVR1 receptor 10 to 100 times better than in cells with MHVR2 (Ohtsuka et al.,

1997, Rao et al., 1997). Their preferential growth in MHVR1 cells could result from the adaptation of the virus to MHVR1 during many repeated passages in DBT or 17cl-1 cells generally used for MHV propagation. DBT and 17cl-1 cells are derived from CDF1 mouse (BALB/c and DBA hybrid) (Kumanishi, 1967) and BALB/c mouse (Sturman and Takemoto, 1972), respectively, implying that these cell lines have only MHVR1. Their adaptation to MHVR1 could account for the higher virulence of these viruses in mice expressing MHVR1 relative to those expressing MHVR2. To test this possibility, the study of MHV isolated from the wild mouse is important.

5. CONCLUSION

It was demonstrated in the present study that all of wild mouse subspecies retained the MHVR2 gene. This is very different from the distribution of MHVR2 in the laboratory mouse. Studies are in progress using the wild mouse to see whether the resistance of mice to MHV links to the MHV receptor gene.

REFERENCES

Dveksler, G.S., Pensiero, M.N., Cardellichio, C.B., Williams, R.K., Jiang, G.S., Holmes, K.V., and Dieffenbach, C.W., 1991, Cloning of the mouse hepatitis virus (MHV) receptor: expression in human and hamster cell lines confers susceptibility to MHV. *J. Virol.* **65**: 6881-6891.

Kumanishi, T., 1967, Brain tumors induced with Rous sarcoma virus, Schmidt-Ruppin strain. I. Induction of brain tumors in adult mice with Rous chicken sarcoma cells. *Jpn. J. Exp. Med.* **37**: 461-474

Ohtsuka, N., Yamada, Y.K., and Taguchi, F., 1996, Difference in virus-binding activity of two distinct receptor proteins for mouse hepatitis virus. *J. Gen. Virol.* **77**: 1683-1692.

Ohtsuka, N., and Taguchi, F., 1997, Mouse susceptibility to mouse hepatitis virus infection is linked to viral receptor genotype. *J. Virol.* **71**: 8860-8863.

Rao, P.V., Kumari, S., and Gallagher, T.M., 1997, Identification of a contiguous 6-residue determinant in the MHV receptor that controls the level of virion binding to cells. *Virology* **229**: 336-348.

Robbins, J., Robbins, P.F., Kozak, C.A., and Callahan, R., 1991, The mouse biliary glycoprotein gene (*Bgp*): Partial nucleotide sequence, expression and chromosomal assignment. *Genomics* **10**: 583-587.

Smith, M.S., Click, R.E., and Plagemann, P.G., 1984, Control of mouse hepatitis virus replication in macrophages by a recessive gene on chromosome 7. *J Immunol* **133**: 428-432.

Sturman, L.S., and Takemoto, K.K., 1972, Enhanced growth of a murine coronavirus in transformed mouse cells. *Infect. Immun.* **6**: 501-507

Yokomori, K., and Lai, M.M., 1992, The receptor for mouse hepatitis virus in the resistant mouse strain SJL is functional: implications for the requirement of a second factor for viral infection. *J. Virol.* **66**: 6931-6938.

Arterivirus RNA Synthesis Dissected
Nucleotides, membranes, amino acids, and a bit of zinc...

ERIC J. SNIJDER

Department of Virology, Center of Infectious Diseases, Leiden University Medical Center, LUMC P4-26, P.O. Box 9600, 2300 RC Leiden, the Netherlands. E-mail: E.J.Snijder@LUMC.nl

1. INTRODUCTION

Over the past ten years, research on by far the largest nidovirus protein, the nonstructural gene 1 or replicase gene, has gone through a number of characteristic stages. First, cDNA copies of various nidovirus replicase genes were cloned and sequenced, with those of the coronaviruses infectious bronchitis virus (IBV) (Boursnell *et al* 1987) and mouse hepatitis virus (MHV) (Lee *et al* 1991) and the arterivirus equine arteritis virus (EAV) (den Boon *et al* 1991) being the first to be completed. Subsequently, extensive sequence comparisons revealed very interesting similarities/homologies within the nidovirus group (then still classified as corona-, toro-, and arteriviruses) and between nidoviruses and other positive-stranded RNA viruses. About five years later, the taxonomic debate on the position of the various nidovirus subgroups ended at the 1996 International Congress of Virology in Jerusalem. The family *Arteriviridae* and the order *Nidovirales* (containing the *Coronaviridae* and *Arteriviridae*) were formally established to acknowledge both the unique properties of arteriviruses and coronaviruses as well as their intriguing ancestral relationship at the level of replicase genes, genome organization, and replication strategy (Snijder and Spaan 1995, de Vries *et al* 1997).

In the meantime, the experimental characterization of the replicase gene products had mainly been focused on the identification of protease domains

The Nidoviruses (Coronaviruses and Arteriviruses).
Edited by Ehud Lavi *et al.*, Kluwer Academic/Plenum Publishers, 2001.

and the elucidation of the pathways used for proteolytic processing of the replicase ORF1a and ORF1ab polyproteins (for a review, see Ziebuhr *et al* 2000). A more recent point of interest is the membrane-association of the nidovirus replication complex. With the advent of reverse genetics systems for arteriviruses in 1996 (Meulenberg *et al* 1998, van Dinten *et al* 1997) and, more recently, for coronaviruses (Almazan *et al* 2000), the stage has now been set for the functional characterization of replicase gene products in the context of their main task in the viral life cycle: RNA synthesis. The size of the nidovirus nonstructural polyprotein varies between 3175 residues for the arterivirus EAV and approximately 7200 amino acids for the coronavirus MHV. This size difference of the replicase again illustrates how both conservation and variation have played a part in nidovirus evolution. In addition to a number of highly conserved domains/functions that can be considered to form the "core" of the nidovirus replicase, each virus (or cluster of viruses) appears to have developed its own set of accessory nonstructural functions.

2.   PROTEOLYTIC PROCESSING

Proteolytic processing of nonstructural proteins fulfils a key role in the life cycle of most viruses. In the course of virus evolution, highly specific virus-encoded proteases have evolved and their importance for the regulation of virus replication is becoming more and more evident (for reviews, see Dougherty and Semler 1993, Gorbalenya and Snijder 1996). A comprehensive review on the properties and activities of nidovirus proteases has been published recently in the Journal of General Virology (Ziebuhr *et al* 2000). A basic characterization of the various protease domains has been carried out and a cleavage site map has been obtained for prototypic coronavirus and arterivirus replicases. Still, given the size of the nidovirus replicase and the details that have emerged for the replicases of other groups of positive-stranded RNA viruses, it is quite likely that -in processing terms- we have merely revealed the tip of an iceberg. Some of the most challenging topics that remain to be investigated are clearly related to the role of proteolytic processing in the regulation of the nidovirus life cycle. That investigating these aspects is not a simple exercise of "knocking out" cleavage sites, has become clear from our work with the EAV infectious cDNA clone. The systematic mutagenesis of each of the EAV replicase cleavage sites (van Dinten *et al* 1999, M. A. Tijms and E. J. Snijder; unpublished data) has revealed that -basically- they are all essential for normal viral RNA synthesis. This conclusion includes sites that are part of the previously described "minor" pathway for processing of the C-terminal half of the ORF1a protein (Wassenaar *et al* 1997). This indicates that -as far as the *in vivo* analysis of proteolytic processing by mutagenesis is concerned- a more sophisticated approach is required, for example by using partially cleavable mutant sites and/or partially active protease mutants.

3.   MEMBRANE ASSOCIATION OF THE REPLICATION COMPLEX

Having raised antibodies against replicase subunits, using them in immunofluorescence assays to analyze infected cells was a logical next step that was taken by several nidovirus research groups. In Leiden, the initial results obtained for EAV were followed by a comparative analysis for MHV.

Possibly due to the number of labs working on either virus group, the results for arteriviruses seem to be more consistent than those for coronaviruses: all EAV replicase subunits (with the exception of a part of nsp1) and *de novo* RNA synthesis (visualized by BrUTP labeling) colocalize in the perinuclear region of the three cell types tested (Pedersen *et al* 1999, van der Meer *et al* 1998). Also a substantial part of the EAV N protein was found in this region of the cell, but its involvement in genome replication or subgenomic RNA transcription is highly unlikely, since N gene expression can be inactivated without a significant effect on either process (Molenkamp *et al* 2000).

For MHV, the situation is much less clear: different replicase subunits have been reported to localize to different membrane compartments in different cell types (Bost *et al* 2000, Shi *et al* 1999, Sims *et al* 2000, van der Meer *et al* 1999). Co-localization of RNA synthesis and N protein has been reported, but is not as complete as for EAV.

Two technical complications that are sometimes overlooked are (i) the fact that coronavirus-infected cells fuse into syncytia that are unsuitable for a reliable cell biological analysis, and (ii) (true for both virus groups) that replicase processing takes time and antisera do not discriminate between processing intermediates and end products.

There is no doubt that -in the end- we will need the power of the electron microscope to understand the details of nidoviral membrane-associated RNA synthesis. The advantages of a 100-fold higher resolution need hardly be explained, and also allows for the identification of cellular compartments on the basis of their morphology and the presence of well-defined markers. So far, however, only one replicase study for each nidovirus subgroup included electron microscopy (Pedersen *et al* 1999, van der Meer *et al* 1999) and therefore it is good to interpret these data with caution. For the moment, two main differences between arteri- and coronaviruses appear to exist: a replication complex associated with double ER-derived membranes for arteriviruses (EAV) versus a complex associated with single membranes of another origin, most likely endosomal/lysosomal, for coronaviruses (MHV). The general distribution in the cell of these two replication complexes is somewhat different (Fig. 1): the arterivirus signal is generally closer to the nucleus and a bit less punctate.

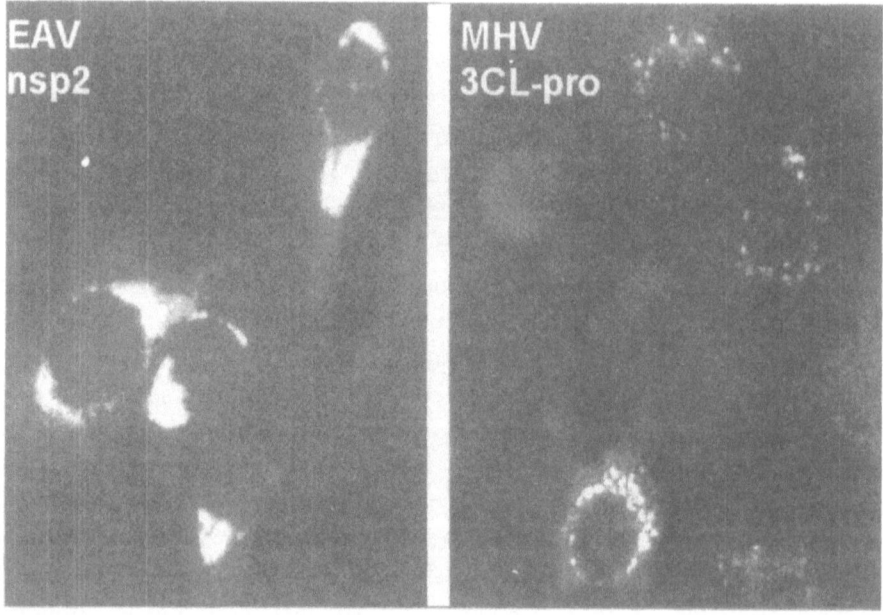

Figure 1. Immunofluorescence staining for specific replicase subunits of the arterivirus EAV (left) in BHK-21 cells and the coronavirus MHV (right) in mouse L cells at a time point around the peak of viral RNA synthesis. For both viruses, the staining is representative for the majority of replicase subunits and *de novo* RNA synthesis (see text).

The formation of paired membranes and double membrane vesicles (DMVs) is a typical feature of arterivirus replication, although the use of double membranes as matrix for RNA synthesis is not restricted to arteriviruses. Double membrane structures in arterivirus-infected cells were seen over 30 years ago and have now been convincingly linked to RNA synthesis. A number of suggestive EM pictures indicate that they originate from ER membranes (Pedersen *et al* 1999), although a thorough biochemical characterization remains to be carried out. We have previously shown that we can induce DMVs that strikingly resemble those in infected cells by expressing EAV nsp2-7, which comprises 1,400 residues of the ORF1a protein, including four major hydrophobic regions. Recently, we have narrowed this down to nsp2 and nsp3 (K. W. Pedersen and E. J. Snijder; unpublished data), which were already known to have a very strong interaction from our analysis of the proteolytic processing of the ORF1a protein (Snijder *et al* 1994). Co-expression of nsp2 and nsp3, but not the expression of either protein by itself, results in the generation of DMVs.

Clearly we have no well-defined ideas about the way in which the EAV ORF1a proteins associate with membranes. Computer predictions suggest up to 10 trans-membrane domains and furthermore a sort of "processing logic" can be applied (Fig. 2).

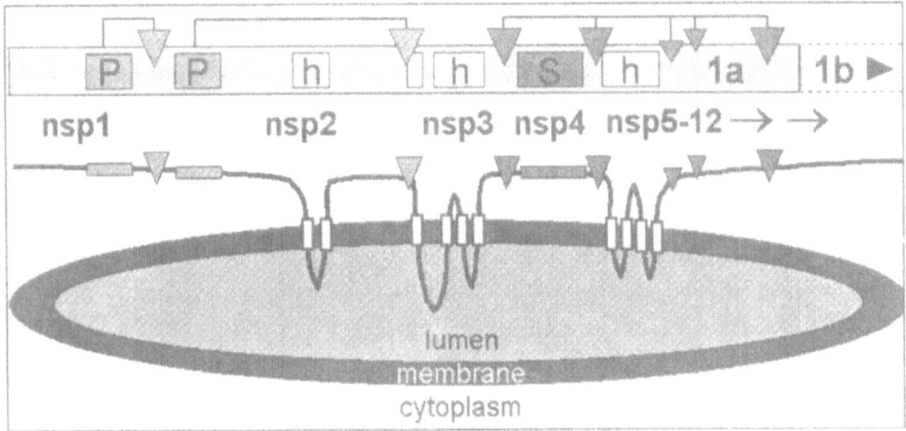

Figure 2. Tentative model for membrane association of the EAV ORF1a protein. The processing scheme of the ORF1a protein is depicted and the hydrophobic domains in nsp2, nsp3 and nsp5 are indicated. The model is based on computer predictions and processing information. Abbreviations: P, papain-like cysteine protease; S, chymotrypsin-like serine protease; h, hydrophobic domain.

For example, the nsp2 protease and the nsp2/3 cleavage site should obviously be on the same side of the membrane, making it logical that the nsp2 hydrophobic domain spans the membrane twice. Likewise, three (and not two or four) is the obvious number of transmembrane segments in the second (largest) hydrophobic domain of nsp3, because otherwise the nsp4 protease would end up in the lumen and not have access to the sites it is known to cleave.

Particularly interesting questions that remain to be addressed relate to (i) the mechanism of translocation, which is probably post-translational since there are no N-terminal signal sequences on the replicase polyprotein, (ii) the coordination of processing and membrane association, (iii) the role of established interactions, like the one between nsp2 and nsp3 (Snijder *et al* 1994) and the cofactor role of nsp2 for the nsp4 protease in cleaving the nsp4/5 site (Wassenaar *et al* 1997), and (iv) the mechanism of the formation of double membranes and DMVs.

4.   RNA SYNTHESIS

Following the generation of the EAV infectious cDNA clone in 1996 (van Dinten *et al* 1997), the RNA sequences thought to be involved in discontinuous mRNA transcription were an obvious primary target for site-directed mutagenesis. In our opinion, the discontinuous minus strand extension model (Fig. 3), proposed by Sawicki and Sawicki (Sawicki and Sawicki 1995), is the nidovirus transcription model that is best supported by the experimental data available for corona- and arteriviruses.

According to this model, nidovirus minus strand transcription is either continuous, yielding the genomic minus strand, or discontinuous. The latter process involves a jump, which partially depends on a base pairing interaction between conserved transcription-regulating sequences (TRSs), and yields subgenomic minus strands that serve as templates for the transcription of subgenomic mRNAs. An important aspect of the nidovirus transcription strategy is that the 3' end of the genome contains an array of body TRS sequences, all directing this jump with their own specific activity, and probably also influencing each other's activity in making subgenomic minus strands.

If we try to dissect the discontinuous step in transcription, which may strongly resemble an RNA recombination event (Brian and Spaan, 1997, Nagy and Simon, 1997), we can discriminate at least four steps, each of which may involve specific RNA-RNA and RNA-protein interactions. There is probably attenuation of the transcriptase, translocation of the nascent strand, base-pairing at the leader TRS, and reinitiation of transcription. Although, in our EAV reverse genetics system, we can analyze the impact of mutations only after conversion of the subgenomic minus strand into subgenomic mRNAs, a number of predictions could be made on the basis of the transcription model. First, leader and body TRS mutations were expected to affect base-pairing and identical mutations in leader and body TRSs should complement each other. Second, since not all body TRSs are completely identical, leader TRS mutations might have different effects on the transcription of different subgenomic mRNAs. Third, body TRS mutations may or may not affect attenuation: if they do not, the number of transcriptases reaching upstream body TRSs should remain the same; but if they do for example reduce attenuation, transcription from upstream TRSs may go up.

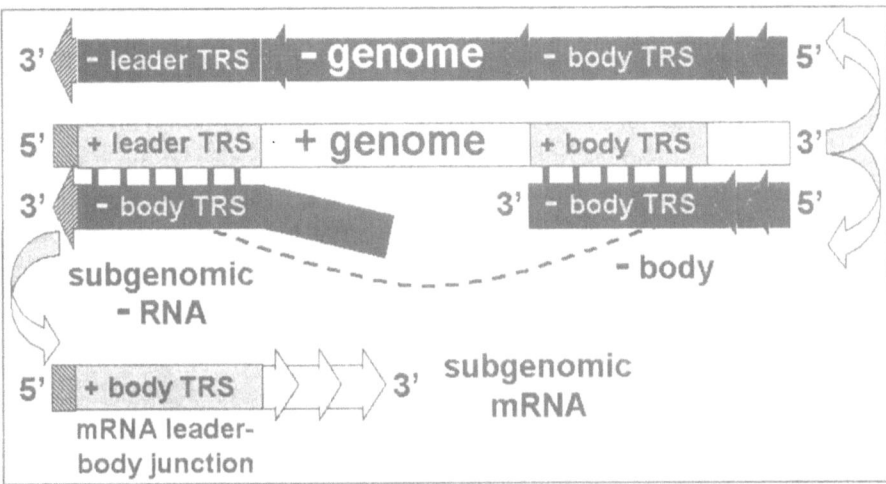

Figure 3. Discontinuous minus strand extension model for nidovirus transcription as proposed by Sawicki and Sawicki (1995). The genomic plus strand is copied into a genomic minus strand that serves as the template for genome replication. In addition, minus strand synthesis can be interrupted at a body TRS, after which the nascent strand is translocated and the TRS complement at the 3' end of the nascent minus strand base-pairs with the genomic leader TRS. Transcription is resumed to add the anti-leader sequence and the resulting subgenomic minus strand (containing the body TRS at its leader-body junction site) is the template for subgenomic mRNA transcription.

Evidence for the importance of the TRS and the existence of a base-pairing interaction between plus leader and minus body TRSs was previously published in a paper by van Marle *et al.* (1999a). However, in this study only the two C residues of the conserved 5' UCAACU 3' sequence that forms the EAV TRS were targeted by mutagenesis. A more recent, comprehensive mutagenesis study of the leader TRS and RNA7 body TRS (A. O. Pasternak, E. van den Born, W. J. M. Spaan, and E. J. Snijder; unpublished data) has revealed that, depending on the nucleotide that is changed, the results can vary quite dramatically. For example, mutants in which identical (leader and RNA7 body) TRS mutations do not restore transcription at all have been obtained. Likewise, body TRS mutations that appear to up-regulate the activity of more upstream body TRSs have been identified. A detailed characterization of these TRS mutants is now in progress.

In addition to RNA primary structure, RNA secondary structure is being studied as a factor that may be involved in the regulation of transcription. A conserved "leader TRS hairpin" has been predicted in the 5' end of all arterivirus genomes and seems a very good candidate to be involved in the base-pairing step of transcription (van Marle *et al* 1999a). At the other end of the genome, a recent analysis of the body TRS regions suggests a link between body TRS activity and its presence in a non-base-paired region of the predicted RNA structure (Pasternak *et al* 2000). Experiments to corroborate this prediction are now in progress.

5.   PROTEIN FUNCTIONS IN RNA TRANSCRIPTION

We have previously proposed that the nidovirus replicase may consist of a basic or "core" transcriptase, containing subunits like the RNA-dependent RNA polymerase and helicase found in other systems, that is supplemented with more specialized domains/proteins involved exclusively in subgenomic RNA synthesis (van Marle *et al* 1999a). Thus, one can envision multiple possibilities for discontinuous transcription-specific protein functions, e.g. proteins interacting with body TRSs, leader TRS, or the leader TRS hairpin (Fig. 4).

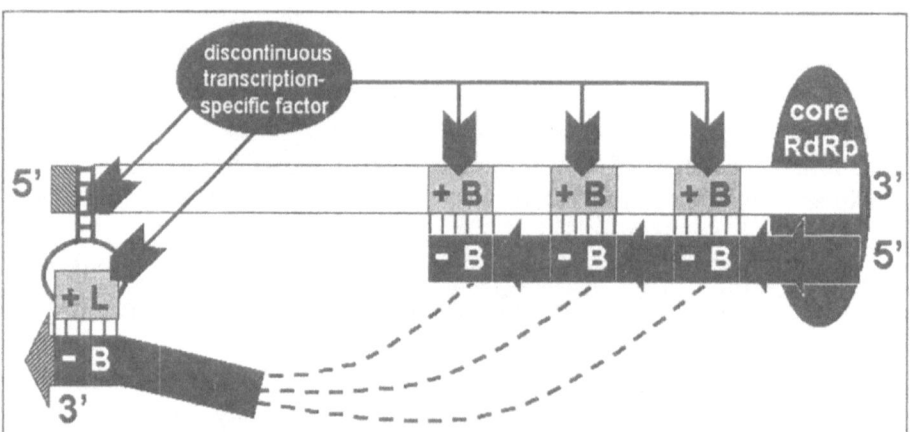

Figure 4. Potential sites for protein-RNA interactions during EAV discontinuous minus strand synthesis. The nidovirus RdRp complex may operate like a core transcriptase that is supplemented with specific protein factors to direct discontinuous minus strand synthesis.

We were confronted with such a transcription-specific function even before the full-length cDNA clone was generated (van Dinten *et al* 1997). Mutant EAV030F had a single replicase mutation in nsp10 (Ser-2429 to Pro) that selectively knocked out subgenomic RNA transcription, thereby rendering the clone noninfectious. The EAV030F mutation maps to nsp10, which can be considered the most conserved nidovirus replicase subunit, with an N-terminal predicted zinc finger domain (van Dinten *et al* 2000) and a C-terminal helicase, for which unique enzymatic properties were recently revealed that are shared by corona- and arteriviruses (Seybert *et al* 2000). The EAV030F mutation is located in a region that seems to connect zinc finger and helicase (Fig. 5).

We have recently shown that the EAV030F mutant is not completely defective in subgenomic RNA synthesis (van Marle *et al* 1999b). It was estimated that subgenomic plus and minus strand synthesis are about 500-fold reduced, which was supported by a construct in which we inserted the CAT gene at the position of the N protein gene and quantified mRNA7 transcription on the basis of reporter gene expression.

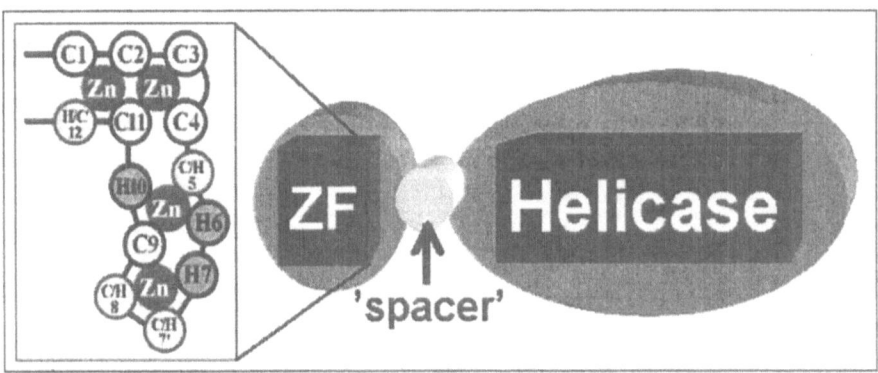

Figure 5. Tentative model for zinc binding by the N-terminal domain of the EAV nsp10 helicase protein (van Dinten *et al* 2000).

More recently, an extensive mutagenesis of the nsp10 zinc finger region and the region containing the original EAV030F mutation was performed, details of which can be found in a paper by van Dinten *et al* (2000). Briefly, all 13 conserved Cys and His residues, which are predicted to be part of an unusual zinc-binding domain (Fig. 5), were found to be important for both genome replication and subgenomic RNA synthesis. However, two additional mutants in the region downstream of the zinc finger displayed EAV030F-like phenotypes and suggested that this region may indeed function as a "spacer" between the zinc finger and helicase domains. The flexibility of this spacer may be essential for a function of nsp10 in subgenomic RNA transcription.

Complementation of the EAV030F defect was attempted using a replicon containing the EAV030F replicase and an IRES/nsp10 cassette in the genomic 3' end (van Dinten *et al* 2000). Expression of wild-type nsp10 from this locus in the genome was indeed possible, but unfortunately complementation did not occur. Immunofluor-escence studies suggested that the IRES-expressed nsp10 did not localize to the replication complex and may have been unable to force the mutant nsp10, which is still active in genome replication, out of the complex.

However, a similar approach for nsp1 was very successful (Fig. 6) and has basically allowed us to establish that nsp1 is a factor that is fully dispensable for genome replication, but essential for discontinuous transcription (M. A. Tijms, L. C. van Dinten, A. E. Gorbalenya, and E. J. Snijder; unpublished data).

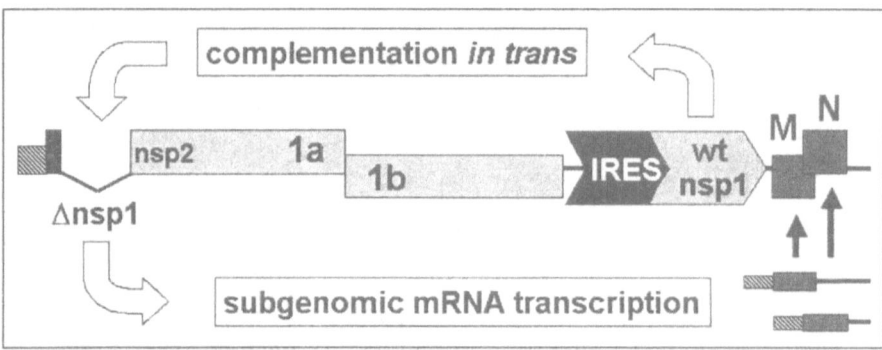

Figure 6. Complementation of the function of EAV nsp1 in subgenomic RNA transcription. First, the nsp1-coding region was deleted from the 5' end of the EAV genome, which resulted in a complete block of subgenomic mRNA transcription. Subsequently, a number of structural genes in the 3' end of the genome were replaced by an IRES element followed by the nsp1 gene. Upon transfection of this construct, nsp1 was expressed from the genomic RNA and subgenomic RNA synthesis (as monitored by the expression of the M and N genes) was found to be restored. These results established the role of nsp1 as a factor specifically involved in subgenomic mRNA synthesis.

Deletion of nsp1 inactivated mRNA synthesis completely, and in this case the IRES-approach yielded a complementation system in which subgenomic RNA synthesis was restored. In contrast to the EAV030F mutant, the block of transcription upon nsp1 deletion seems to be absolute. The identification of a putative zinc finger in the nsp1 N-terminus (M. A. Tijms, L. C. van Dinten, A. E. Gorbalenya, and E. J. Snijder; unpublished data) is an additional intriguing feature of nsp1, since zinc finger domains are widespread in for example transcription factors. Nevertheless, it remains to be established how directly or indirectly nsp1 interacts with the EAV transcription machinery. The fact that the protein does not only colocalize with the viral replication complex, but is also partially found in the nucleus of the infected cell (M. A. Tijms, Y. van der Meer, and E. J. Snijder;

unpublished data), suggests that it may be a multifunctional replicase subunit.

ACKNOWLEDGMENTS

A large number of people have contributed to the work described in this chapter, and to arterivirus/nidovirus research in Leiden as a whole. I would like to acknowledge the contributions of Ketil Pedersen (Oslo University), Jacomine Krijnse Locker (EMBL, Heidelberg), Yvonne van der Meer (LUMC), Johan den Boon (LUMC), Peter Bredenbeek (LUMC), Willem Luytjes (LUMC), Guido van Marle (LUMC), Richard Molenkamp (LUMC), Babette Rozier (LUMC), Sasha Pasternak (LUMC), Erwin van den Born (LUMC), Sasha Gultyaev (Department of Chemistry, Leiden University), Fred Wassenaar (LUMC), Leonie van Dinten (LUMC), Sietske Rensen (LUMC), Hans van Tol (LUMC), Marieke Tijms (LUMC), Sasha Gorbalenya (SAIC/NCI-FCRDC, Frederick), Jessika Dobbe (LUMC), Sophie Greve (LUMC), and (in many ways) Willy Spaan. This work was supported by several grants from the Council for Chemical Sciences of the Netherlands Organization for Scientific Research (CW-NWO).

REFERENCES

Almazan, F., Gonzalez, J.M., Penzes, Z., Izeta, A., Calvo, E., Plana-Duran, J., and Enjuanes, L., 2000, Engineering the largest RNA virus genome as an infectious bacterial artificial chromosome. *Proc. Natl. Acad. Sci. U.S.A.* **97**:5516-5521.

Bost, A.G., Carnahan, R.H., Lu, X.T., and Denison, M.R., 2000, Four proteins processed from the replicase gene polyprotein of mouse hepatitis virus colocalize in the cell periphery and adjacent to sites of virion assembly. *J. Virol.* **74**: 3379-3387.

Boursnell, M.E., Brown, T.D.K., Foulds, I.J., Green, P.F., Tomley, F.M., and Binns, M.M., 1987, Completion of the sequence of the genome of the coronavirus avian infectious bronchitis virus. *J. Gen. Virol.* **68**: 57-77.

Brian, D.A. and Spaan, W.J.M., 1997, Recombination and coronavirus defective interfering RNAs. *Semin. Virol.* **8**: 101-111.

den Boon, J.A., Snijder, E.J., Chirnside, E.D., de Vries, A.A.F., Horzinek, M.C., and Spaan, W.J.M., 1991, Equine arteritis virus is not a togavirus but belongs to the coronaviruslike superfamily. *J. Virol.* **65**: 2910-2920.

de Vries, A.A.F., Horzinek, M.C., Rottier, P.J.M., and de Groot, R.J., 1997, The genome organization of the Nidovirales: similarities and differences between arteri-, toro-, and coronaviruses. *Semin. Virol.* **8**: 33-47.

Dougherty, W.G. and Semler, B.L., 1993, Expression of virus-encoded proteinases: functional and structural similarities with cellular enzymes. Microbiol. Rev. 57: 781-822.

Gorbalenya, A.E. and Snijder, E.J., 1996, Viral cysteine proteases. *Persp. Drug Discov. Design* **6**: 64-86.

Lee, H.J., Shieh, C.K., Gorbalenya, A.E., Koonin, E.V., La Monica, N., Tuler, J., Bagdzhadzhyan, A., and Lai, M.M.C., 1991, The complete sequence (22 kilobases) of murine coronavirus gene 1 encoding the putative proteases and RNA polymerase. *Virology* **180**: 567-582.

Meulenberg, J.J.M., Bos-de Ruijter, J.N.A., Wensvoort, G., and Moormann, R.J.M., 1998, Infectious transcripts from cloned genome-length cDNA of porcine reproductive respiratory syndrome virus. *J. Virol.* **72**: 380-387.

Molenkamp, R., van Tol, H., Rozier, B.C.D., van der Meer, Y., Spaan, W.J.M., and Snijder, E.J., 2000, The arterivirus replicase is the only viral protein required for genome replication and subgenomic mRNA transcription. *J. Gen. Virol.* **81**: 2491-2496.

Nagy, P.D. and Simon, A.E., 1997, New insights into the mechanisms of RNA recombination. *Virology* **235**: 1-9.

Pasternak, A.O., Gultyaev, A.P., Spaan, W.J.M., and Snijder, E.J., 2000, Genetic manipulation of arterivirus alternative mRNA leader-body junction sites reveals tight regulation of structural protein expression. *J. Virol.* **74**, in press.

Pedersen, K.W., van der Meer, Y., Roos, N., and Snijder, E.J., 1999, Open reading frame 1a-encoded subunits of the arterivirus replicase induce endoplasmic reticulum-derived double-membrane vesicles which carry the viral replication complex. *J. Virol.* **73**: 2016-2026.

Sawicki, S.G. and Sawicki, D.L., 1995, Coronaviruses use discontinuous extension for synthesis of subgenome-length negative strands. *Adv. Exp. Biol. Med.* **380**: 499-506.

Seybert, A., van Dinten, L.C., Snijder, E.J., and Ziebuhr, J., 2000, The biochemical characterization of the equine arteritis virus helicase suggests a close functional relationship between arterivirus and coronavirus helicases. *J. Virol.* **74**: 9586-9593.

Shi, S.T., Schiller, J.J., Kanjanahaluethai, A., Baker, S.C., Oh, J.W., and Lai, M.M., 1999, Colocalization and membrane association of murine hepatitis virus gene 1 products and De novo-synthesized viral RNA in infected cells. *J. Virol.* **73**: 5957-5969.

Sims, A.C., Ostermann, J., and Denison, M.R., 2000, Mouse hepatitis virus replicase proteins associate with two distinct populations of intracellular membranes. *J. Virol.* **74**: 5647-5654.

Snijder, E.J., and Spaan, W.J.M., 1995, The coronaviruslike superfamily. In: The Coronaviridae (S.G.Siddell, ed), Plenum Press, New York, N.Y., pp. 239-255.

Snijder, E.J., Wassenaar, A.L.M., and Spaan, W.J.M., 1994, Proteolytic processing of the replicase ORF1a protein of equine arteritis virus. *J. Virol.* **68**: 5755-5764.

van der Meer, Y., Snijder, E.J., Dobbe, J.C., Schleich, S., Denison, M.R., Spaan, W.J.M., and Krijnse Locker, J., 1999, Localization of mouse hepatitis virus nonstructural proteins and RNA synthesis indicates a role for late endosomes in viral replication. *J. Virol.* **73**: 7641-7657.

van der Meer, Y., van Tol, H., Krijnse Locker, J., and Snijder, E.J., 1998, ORF1a-encoded replicase subunits are involved in the membrane association of the arterivirus replication complex. *J. Virol.* **72**: 6689-6698.

van Dinten, L.C., den Boon, J.A., Wassenaar, A.L.M., Spaan, W.J.M., and Snijder, E.J., 1997, An infectious arterivirus cDNA clone: identification of a replicase point mutation which abolishes discontinuous mRNA transcription. *Proc. Natl. Acad. Sci. U.S.A.* **94**: 991-996.

van Dinten, L.C., Rensen, S., Spaan, W.J.M., Gorbalenya, A.E., and Snijder, E.J., 1999, Proteolytic processing of the open reading frame 1b-encoded part of arterivirus replicase is mediated by nsp4 serine protease and is essential for virus replication. *J. Virol.* **73**: 2027-2037.

van Dinten, L.C., van Tol, H., Gorbalenya, A.E., and Snijder, E.J., 2000, The predicted metal-binding region of the arterivirus helicase protein is involved in subgenomic mRNA synthesis, genome replication, and virion biogenesis. *J. Virol.* **74**: 5213-5223.

van Marle, G., Dobbe, J.C., Gultyaev, A.P., Luytjes, W., Spaan, W.J.M., and Snijder, E.J., 1999a, Arterivirus discontinuous mRNA transcription is guided by base- pairing between sense and antisense transcription-regulating sequences. *Proc. Natl. Acad. Sci. U.S.A.* **96**: 12056-12061.

van Marle, G., van Dinten, L.C., Luytjes, W., Spaan, W.J.M., and Snijder, E.J., 1999b, Characterization of an equine arteritis virus replicase mutant defective in subgenomic mRNA synthesis. *J. Virol.* **73**: 5274-5281.

Wassenaar, A.L.M., Spaan, W.J.M., Gorbalenya, A.E., and Snijder, E.J., 1997, Alternative proteolytic processing of the arterivirus replicase ORF1a polyprotein: evidence that NSP2 acts as a cofactor for the NSP4 serine protease. *J. Virol.* **71**: 9313-9322.

Ziebuhr, J., Snijder, E.J., and Gorbalenya, A.E., 2000, Virus-encoded proteinases and proteolytic processing in the *Nidovirales*. *J. Gen. Virol.* **81**: 853-879.

Guanosine Triphosphatase Activity of the Human Coronavirus Helicase

ANJA SEYBERT AND JOHN ZIEBUHR
Institute of Virology, University of Würzburg, 97078 Würzburg, Germany

1. INTRODUCTION

RNA helicases are a diverse class of enzymes that use the energy of nucleoside triphosphate (NTP) hydrolysis to unwind duplex RNA structures. They are involved in virtually every aspect of RNA metabolism, including transcription, RNA splicing, translation, RNA export, ribosome biogenesis, mitochondrial gene expression and regulation of mRNA stability (Schmid and Linder 1992, Lohman and Bjornson 1996, Kadaré and Haenni 1997). On the basis of conserved sequence motifs, helicases have been divided into 3 large superfamilies (SF), SF1 to SF3, and two smaller families (Gorbalenya et al. 1989, Gorbalenya and Koonin 1993b), which contain both RNA and DNA helicases.

Apart from the RNA-dependent RNA polymerases (RdRp), helicases are the most conserved subunits of the replication machinery of (+)RNA viruses (Gorbalenya et al. 1988, Gorbalenya and Koonin 1993a, Koonin and Dolja, 1993). Thus, based on sequence analyses, putative helicase domains have been identified in most (+)RNA virus genomes and there is a large body of genetic and reverse-genetic information to suggest a key function of helicases in the life-cycle of (+)RNA viruses (Kadare and Haenni 1997). Remarkably, (–)RNA viruses and retroviruses do not encode helicases (Gorbalenya et al. 1988). While a number of (+)RNA virus helicases of SF2 have been characterized in considerable detail, there is nearly no information on the biochemical properties of RNA virus SF1 helicases. The latter superfamily includes putative helicases from more than 15 (+)RNA virus

The Nidoviruses (Coronaviruses and Arteriviruses).
Edited by Ehud Lavi *et al.*, Kluwer Academic/Plenum Publishers, 2001.

families but, to date, no convincing evidence for duplex-unwinding activity has been obtained for most of these proteins.

The human coronavirus 229E (HCoV) replicase gene encodes two large polyproteins that are extensively processed by virus-encoded proteinases (Ziebuhr *et al.* 2000). One of the mature processing products, p66^{HEL}, has previously been predicted to contain an SF1 helicase domain (Gorbalenya *et al.* 1988). Biochemical data to support these predictions have recently been reported (Seybert *et al.* 2000). Specifically, clear evidence was presented to show nucleic acid duplex-unwinding activity associated with this viral protein. Interestingly, the functional analysis of p66^{HEL} revealed a 5'-to-3' polarity of the unwinding reaction, whereas SF2 RNA virus helicases have been shown to operate in 3'-to-5' direction.

2. MATERIALS AND METHODS

2.1 Protein Expression and Purification Using Baculovirus Recombinants

The construction of baculovirus recombinants expressing recombinant forms of p66^{HEL} has been described previously (Seybert *et al.* 2000). Briefly, the coding sequence of the HCoV pp1ab amino acids 4998 to 5592 was inserted into pBlueBacHis2B DNA (Invitrogen; Groningen, Netherlands). The resultant plasmid, pBlueBacHis2B-Hel, essentially encodes the complete HCoV helicase domain fused to an amino-terminal histidine tag.

A recombination-PCR method was used to introduce a point mutation into the helicase-coding sequence of pBlueBacHis2B-Hel. In the resultant plasmid, pBlueBacHis2B-Hel-KA, the codon for the HCoV pp1ab amino acid Lys-5284, AAA, was substituted by GCA, which encodes Ala.

The plasmids pBlueBacHis2B-Hel and pBlueBacHis2B-Hel-KA were used to derive two recombinant baculoviruses, designated vBac-Hel and vBac–Hel–KA, respectively. Cell culture, transfections, isolation of recombinant baculoviruses, plaque purifications, protein expression in High Five™ insect cells (Invitrogen) and nickel-affinity chromatography purification were done as previously described (Seybert *et al.* 2000). The recombinant proteins purified from vBac–Hel- or vBac–Hel–KA-infected insect cells were designated HEL and HEL–KA, respectively.

2.2 Nucleoside Triphosphatase Assay

In the adenosine triphosphatase (ATPase) assay, HEL (3, 15 and 450 fmol, respectively) or HEL–KA (10 pmol) were incubated in a volume of

40µl containing 20 mM HEPES-KOH pH 7.4, 300 µM ATP, 5 mM magnesium acetate, 2 mM dithiothreitol, 25 µg/ml bovine serum albumin and 250 nCi of [γ-^{32}P] ATP (3000 Ci/mmol). In the guanosine triphosphatase (GTPase) assay, ATP and [γ-^{32}P] ATP were replaced by 300 µM GTP and 250 nCi [γ-^{32}P] GTP (3000 Ci/mmol), respectively. When included, poly(U) was at a concentration of 150 µg/ml. The reactions were incubated at 30°C for 30 min and stopped by adding EDTA to a final concentration of 100 mM. The samples were analyzed by polyethyleneimine-cellulose thin layer chromatography with 0.15 M formic acid-0.15 M LiCl (pH 3.0) as the liquid phase. The reaction products were analyzed by phosphorimaging of the dried chromatographic plates (ImageQuant software; Molecular Dynamics, Sunnyvale, CA).

2.3   Preparation of the Partial-Duplex DNA Substrate

The synthetic oligonucleotide 5'–GGTGCAGCCGCAGCGGTGCTCG–d(pT)$_{30}$–3' was labelled with [γ-^{32}P] ATP (3000 Ci/mmol) using T4 polynucleotide kinase and purified by phenol/chloroform extraction and gel filtration chromatography using Micro Bio-Spin 6 columns (Bio-Rad Laboratories, Munich, Germany). The DNA duplex was produced by annealing the labelled oligonucleotide to the oligonucleotide 5'–d(pT)$_{30}$–CGAGCACCGCTGCGGCTGCACC–3' in buffer E (25 mM HEPES-KOH pH 7.4, 500 mM NaCl, 1 mM EDTA, 0.1% (w/v) SDS). The annealing reaction, which contained a tenfold excess of unlabelled DNA over [γ-^{32}P] ATP-labelled DNA, was denatured for 5 min at 95°C and slowly cooled to room temperature. The resultant substrate was a twin-tailed ("forked") DNA duplex; i.e., it contained both 5' and 3' single-stranded regions on one end of the partial-duplex DNA.

2.4   Duplex-Unwinding Assay

HEL (30 fmol) or HEL–KA (210 fmol) were incubated in a volume of 40 µl with 15 fmol of the partial-duplex DNA substrate for 30 min at 30°C in a buffer containing 20 mM HEPES-KOH pH 7.4, 37.5 mM NaCl, 5 mM ATP or GTP, respectively, 10% glycerol, 5 mM magnesium acetate, 2 mM dithiothreitol and 0.1 mg/ml bovine serum albumin. The reactions were stopped by the addition of 10 µl of 5% (v/v) SDS, 15% (w/v) Ficoll and 100 mM EDTA. The reaction products were separated on 12% polyacrylamide-1×TBE gels (acrylamide/bis-acrylamide ratio of 19 to 1) at 4 W until the bromphenol blue dye approached the bottom of the gel. The gels were exposed to X-ray film at –70°C.

3.   RESULTS AND DISCUSSION

A recombinant form of the HCoV helicase, HEL, has recently been shown to have ATPase activity *in vitro* (Seybert *et al*. 2000). To examine if HEL is also able to hydrolyze GTP, we have incubated [γ-^{32}P] GTP with HEL and the ATPase-deficient control protein HEL–KA. As Figure 1 (lane 2) shows, HEL effectively hydrolyzes GTP. We also found that the substrate conversion by HEL can be significantly stimulated by poly(U). Thus, in the presence of 150 µg/ml poly(U), 15 fmol of HEL gave a similar rate of hydrolysis as 450 fmol HEL without the polynucleotide cofactor. Even very low amounts of enzyme gave significant GTPase activity if poly(U) was included (lane 6). The rate of GTP hydrolysis was found to be similar to that of ATP hydrolysis (Fig.1, cf. lanes 3, 5 and 4, 6). As expected, no significant substrate hydrolysis was observed in the control reactions using buffer or HEL–KA (Figure 1, lanes 7 to 10). We conclude from these data that the HCoV helicase is able to effectively hydrolyze GTP and, in agreement with the structural model for the nucleic acid-induced activation of the *Bacillus stearothermophilus* PcrA helicase-associated ATPase activity (Soultanas *et al*. 1999), we propose that the binding to single-stranded RNA induces a

ATP (300 µM)	+	−	+	−	+	−	+	−	+	−
GTP (300 µM)	−	+	−	+	−	+	−	+	−	+
HEL (fmol)	450	450	15	15	3	3	−	−	−	−
HEL–KA (pmol)	−	−	−	−	−	−	−	−	10	10
poly(U) (150 µg/ml)| −	−	+	+	+	+	+	+	+	+
	1	2	3	4	5	6	7	8	9	10

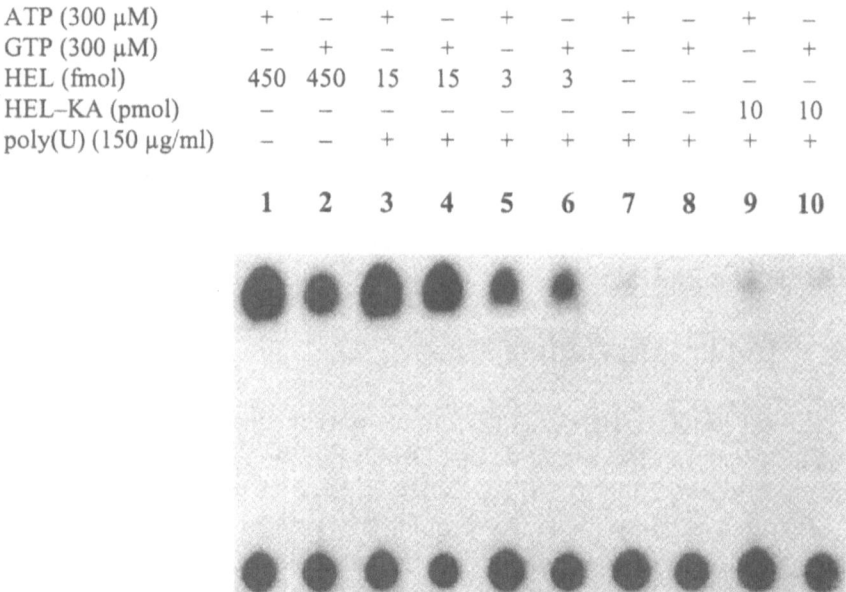

Figure 1. ATPase and GTPase activities of HEL. Reactions containing either 300 µM ATP (lanes 1, 3, 5, 7, and 9) or 300 µM GTP (lanes 2, 4, 6, 8, and 10) were incubated with (i) different amounts of HEL (lanes 1 to 6), (ii) buffer alone (lanes 7 and 8), or (iii) ATPase-deficient control protein HEL–KA (lanes 9 and 10). The reaction products were analyzed by thin-layer chromatography and phosphorimaging (see Materials and Methods for details). The composition of the individual reactions is indicated above.

conformational change in the NTPase active site of HEL, which stabilizes the bound NTP molecule in a conformation that is required for rapid substrate hydrolysis.

It is generally accepted that the strand separation of nucleic acid duplexes by helicases is an energy-dependent process. Thus, NTP is a necessary cofactor for duplex-unwinding activity. In agreement with this model, we recently found that HEL requires ATP for duplex-unwinding activity. Also, the substitution of a lysine residue (Lys-5284 to Ala), which is part of the Walker A motif (Walker *et al.* 1982) and is predicted to be involved in binding and/or hydrolysis of NTP, resulted in an ATPase-deficient and, hence, helicase-deficient protein (HEL–KA).

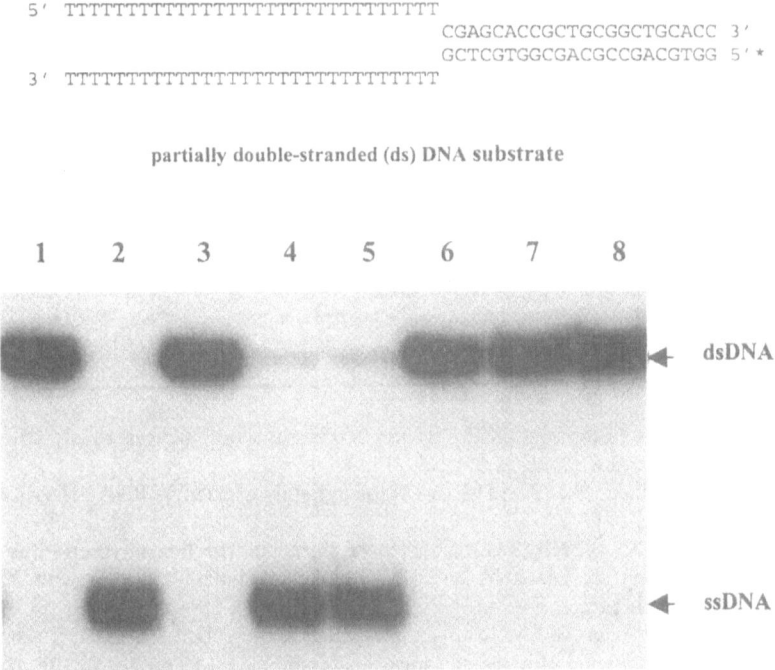

Figure 2. NTP-dependent DNA duplex-unwinding activity of HEL. Reaction conditions were as described in Materials and Methods with approximately 15 fmol of DNA substrate per reaction. The structure of the partial-duplex DNA substrate is shown schematically with the radiolabelled strand marked by an asterisk. The reaction products were separated on a non-denaturing 12% polyacrylamide gel and visualized by autoradiography. The positions of the partially double-stranded substrate (dsDNA) and the displaced, monomeric product (ssDNA) are indicated. Lanes: 1, reaction without protein; 2, heat-denatured DNA substrate; 3, reaction containing 30 fmol HEL in the absence of NTP; 4, reaction containing 30 fmol HEL in the presence of 5 mM ATP; 5, reaction containing 30 fmol HEL in the presence of 5 mM GTP; 6, reaction containing 210 fmol HEL–KA in the absence of NTP; 7, reaction containing 210 fmol HEL–KA in the presence of 5 mM ATP; 8, reaction containing 210 fmol HEL–KA in the presence of 5 mM GTP.

ACKNOWLEDGMENTS

This work was supported by grants from the Deutsche Forschungsgemeinschaft (SI 357/4-1) and the Fonds der chemischen Industrie (FCI).

REFERENCES

Gorbalenya, A.E., and Koonin, E.V., 1989, Viral proteins containing the purine NTP-binding sequence pattern. *Nucleic Acids Res.* **17:** 8413-8440.

Gorbalenya, A.E., and Koonin, E.V., 1993a, Comparative analysis of the amino acid sequences of the key enzymes of the replication and expression of positive-strand RNA viruses. Validity of the approach and functional and evolutionary implications. *Sov. Sci. Rev. D. Physicochem. Biol.* **11:** 1-84.

Gorbalenya, A.E., and Koonin, E.V., 1993b, Helicases: amino acid sequence comparisons and structure-function relationships. *Curr. Opin. Struct. Biol.* **3:** 419-429.

Gorbalenya, A.E., Koonin, E.V., Donchenko, A.P., and Blinov, V.M., 1988, A novel superfamily of nucleoside triphosphate-binding motif containing proteins which are probably involved in duplex unwinding in DNA and RNA replication and recombination. *FEBS Lett.* **235:** 16-24.

Gorbalenya, A.E., Koonin, E.V., Donchenko, A.P., and Blinov, V.M., 1989, Two related superfamilies of putative helicases involved in replication, recombination, repair and expression of DNA and RNA genomes. *Nucleic Acids Res.* **17:** 4713-30.

Kadaré, G., and Haenni, A.L., 1997, Virus-encoded RNA helicases. *J. Virol.* **71:** 2583-2590.

Koonin, E.V., and Dolja, V.V., 1993, Evolution and taxonomy of positive-strand RNA viruses: implications of comparative analysis of amino acid sequences. *Crit. Rev. Biochem. Mol. Biol.* **28:** 375-430.

Lohman, T.M., and Bjornson, K.P., 1996, Mechanisms of helicase-catalyzed DNA unwinding. *Annu. Rev. Biochem.* **65:** 169-214.

Schmid, S.R., and Linder, P., 1992, D-E-A-D protein family of putative RNA helicases. *Mol. Microbiol.* **6:** 283-291.

Seybert, A., Hegyi, A., Siddell, S.G., and Ziebuhr, J., 2000, The human coronavirus 229E superfamily 1 helicase has RNA and DNA duplex-unwinding activities with 5'-to-3' polarity. *RNA* **6:** in press.

Soultanas, P., Dillingham, M.S., Velankar, S.S., and Wigley, D.B., 1999, DNA binding mediates conformational changes and metal ion coordination in the active site of PcrA helicase. *J. Mol. Biol.* **290:** 137-148.

Walker, J.E., Saraste, M., Runswick, M.J., and Gay, N.J., 1982, Distantly related sequences in the alpha- and beta-subunits of ATP synthase, myosin, kinases and other ATP-requiring enzymes and a common nucleotide binding fold. *EMBO J.* **1:** 945-951.

Ziebuhr, J., Snijder, E.J., and Gorbalenya, A.E., 2000, Virus-encoded proteinases and proteolytic processing in the *Nidovirales*. *J. Gen. Virol.* **81:** 853-879.

A Strategy for the Generation of an Infectious Transmissible Gastroenteritis Coronavirus from Cloned cDNA

FERNANDO ALMAZAN, JOSE M. GONZALEZ, ZOLTAN PENZES, ANDER IZETA, ENRIQUE CALVO, AND LUIS ENJUANES
Centro Nacional de Biotecnología, CSIC, Department of Molecular and Cell Biology, Campus Universidad Autónoma, Cantoblanco, 28049 Madrid, Spain

1. INTRODUCTION

To date reverse genetics of coronavirus has been possible by targeted recombination following the procedure initially developed by Masters' group (Koetzner *et al.*, 1992). However, the construction of a full-length cDNA clone, from which infectious RNA may be transcribed, will considerably improve the genetic manipulation of coronaviruses. Unfortunately, the size of the coronavirus genome and the instability in bacteria of plasmids carrying coronavirus replicase sequences have hampered the construction of a full-length cDNA clone (Masters, 1999). To overcome these problems we have combined three strategies: (i) a two-step amplification system that couples transcription in the nucleus from the cytomegalovirus (CMV) promoter with a second amplification step in the cytoplasm by the viral polymerase; (ii) the construction of the full-length cDNA from a defective minigenome (DI) that was stably and efficiently replicated by the helper virus (Izeta *et al.*, 1999); and (iii) the full-length cDNA was cloned as a bacterial artificial chromosome (BAC), a low-copy number plasmid, which is present in one or two copies per cell.

In the present study, we report the recovery of infectious transmissible gastroenteritis virus (TGEV) from cloned cDNA and show that this procedure can be used to generate a genetically modified TGEV.

The Nidoviruses (Coronaviruses and Arteriviruses).
Edited by Ehud Lavi *et al.*, Kluwer Academic/Plenum Publishers, 2001.

2.   MATERIALS AND METHODS

2.1   Cells, Viruses, Plasmid and Bacteria Strains

Epithelial swine testis (ST) cells were kindly provided by L. Saif (Ohio State University, OH). The TGEV strain PUR46-MAD (PUR-MAD) and PUR46-C11 (PUR-C11) (Sánchez *et al.*, 1999) were grown and titered as described (Sánchez *et al.*, 1990).

Plasmid pBeloBAC11 (Wang *et al.*, 1997) was kindly provided by H. Shizuya and M. Simon (California Institute of Technology, Pasadena, CA). *E. coli* DH10B strain was obtained from GIBCO/BRL.

2.2   Construction of the TGEV Full-Length cDNA

As a backbone for the construction of a full-length cDNA clone of the TGEV, the minigenome DI-C, derived of the PUR-MAD strain, was used (Izeta *et al.*, 1999). DI-C RNA has three deletions ($\Delta 1$, $\Delta 2$, and $\Delta 3$) within ORFs 1a, 1b, and between genes S and 7, respectively. These deletions were restored by cloning a set of cDNA fragments generated by standard RT-PCR techniques, using the medium copy number plasmid pACNR1180. Deletions $\Delta 2$ and $\Delta 3$ were restored generating a stable plasmid. However, when the deletion $\Delta 1$ was completed, the resultant plasmid was unstable within the bacteria, and only mutated forms were recovered. To overcome this toxicity problem, TGEV cDNAs were cloned in pBeloBAC11 leading to plasmids pBAC-TGEV^{ClaI}, containing a 5.2 kb *ClaI-ClaI* fragment from nucleotides 4,417 to 9,615, and pBAC-TGEV$^{\Delta ClaI}$, encoding the rest of the TGEV genome (Almazán *et al.*, 2000). The last step for the generation of the full-length cDNA (pBAC-TGEV^{FL}) consisted of the insertion of de *ClaI-ClaI* fragment into *ClaI*-linearized pBAC-TGEV$^{\Delta ClaI}$ (Fig. 1).

The full-length cDNA was under the control of the CMV promoter and flanked at its 3' end by a 24 bp poly(A) tail followed by the hepatitis delta virus ribozyme and the bovine growth hormone termination and polyadenylation sequences (Izeta *et al.*, 1999). Details of this construction are presented in the report entitled "Cloning of a Transmissible Gastroenteritis Coronavirus Full-Length cDNA", also published in this book.

2.3   Recovery of Infectious TGEV from the cDNA

ST cells (10^6) were grown to 60% confluence and were transfected with 10 μg of either pBAC-TGEV^{FL} or pBAC-TGEV$^{FL-(ClaI)RS}$ (pBAC-

TGEV^{FL} plasmid carrying the *ClaI-ClaI* fragment in the reverse-sense) using 15 µg of Lipofectine (Life Technologies, GIBCO) according to the manufacture's specifications. After an incubation period of 2 days, the cell supernatant (referred to as passage 0) was passaged six times on fresh ST cells and the presence of virus was analyzed by plaque titration. After six passages the virus was cloned by three plaque purification steps.

2.4   Virulence Assay

The in vivo growth and virulence of TGEV isolates and the recovered virus were determined as described (Sánchez *et al.*, 1999). The virus titers in lung, jejunum, and ileum were determined 2 days after infection.

3.   RESULTS AND DISCUSSION

3.1   Rescue of an Infectious TGEV from a cDNA Clone

To obtain a cDNA encoding a full-length TGEV RNA, a cDNA encoding the TGEV derived DI-C was used as the starting point. The three deletions that DI-C has in relationship to the parental virus PUR-MAD were restored (see *Materials and Methods*) and the full-length

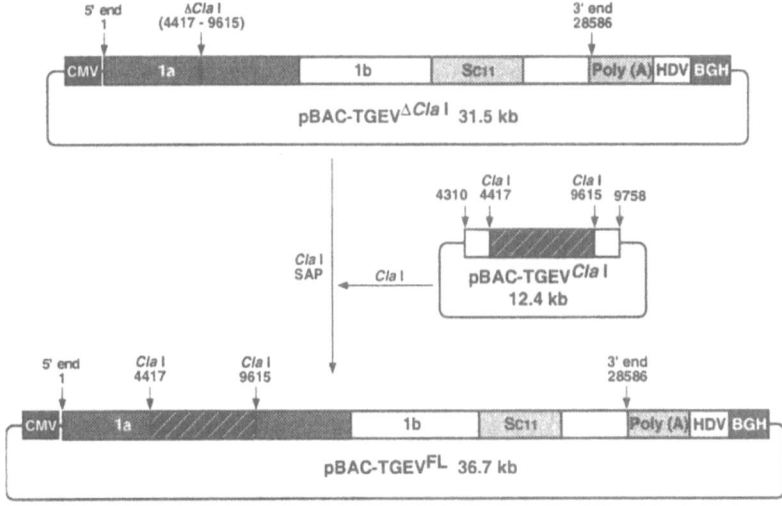

Figure 1. Cloning of the TGEV cDNA in pBeloBAC11. Plasmid pBAC-TGEV^{FL} were generated as described in *Materials and Methods*. CMV, cytomegalovirus immediate-early promoter; Poly(A), tail of 24 A residues; HDV, hepatitis delta virus ribozyme; BGH, bovine growth hormone termination and polyadenylation sequences; SC11, S gene of PUR-C11 strain; SAP, shrimp alkaline phosphatase.

cDNA was cloned as a BAC (Fig. 1) (Almazán *et al.*, 2000). The resulting plasmid, pBAC-TGEV^{FL}, was stable for at least 80 generations in DH10B cells. In addition, to generate a cDNA encoding a fully active TGEV that would replicate both within the enteric and the respiratory tracts and preserve the virulence of the original *in vivo* isolates, the spike (S) gene of the PUR-MAD strain, which replicates abundantly within the respiratory tract and scarcely in the enteric tract of swine, was replaced by the S gene of PUR-C11 strain, which replicates with high titers within both the respiratory and the enteric tracts (Sánchez *et al.*, 1999).

To recover an infectious TGEV from the cDNA clone, ST cells were transfected with plasmid pBAC-TGEV^{FL}, and the cell supernatant was passaged six additional times. Virus titers quickly increased with passage and were around 10^8 pfu/ml by passage 4. However, in the mock-transfected cultures or in cells transfected with the same plasmid but carrying the *Cla*I fragment in the reverse-sense (pBAC-TGEV$^{FL-(ClaI)RS}$) no virus was recovered (Fig. 2). After six passages, the virus was cloned, and the selected virus was named rPUR-MAD-SC11.

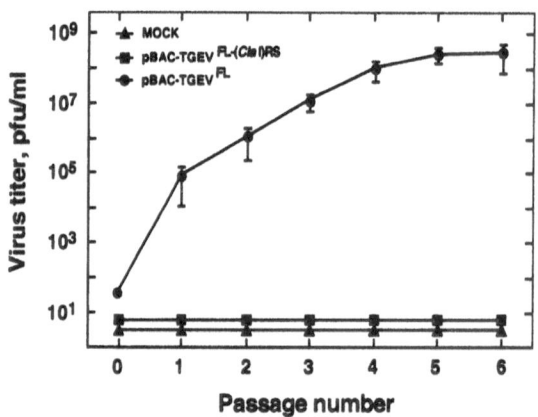

Figure 2. Infectious TGEV recovered from cDNA. After transfection, the recovered virus was passaged, and the culture supernatant were titrated on ST cells. Error bars represent standard deviations of the mean from six experiments.

The rescued virus conserved all the genetics markers introduced throughout the sequence and showed the antigenic characteristics expected for the synthetic virus. The cytopathic effects produced by the rescued virus included induction of cell fusion and formation of large size plaques. These characteristics are identical to those of the parental virus, which provided the S gene, and not to the PUR-MAD strain, which provided the rest of the genome, suggesting that the S gene is a determinant of cell fusion and plaque morphology.

Since the cDNA was transcribed within the nucleus, we investigated if there was splicing of the genomic RNA during its translocation from the nucleus to the cytoplasm. For this purpose, the sequences with the highest splicing potential along the TGEV sequence were determined, and the RNA fragments with the potential splice sites were amplified by RT-PCR by using as template the cytoplasmic RNA at passage 0 and 1. Splicing was observed in only one amplified fragment, and only 20% of the molecules were spliced. Interestingly, the genome with no splicing was favored by selection after one passage. To asses definitively whether splicing had taken place in the viral RNA selected during virus replication, we determined the full-length sequence of the cloned virus. Splicing was not detected and only five nucleotide differences were observed between the sequence of the rescued RNA genome and that of the cDNA clone.

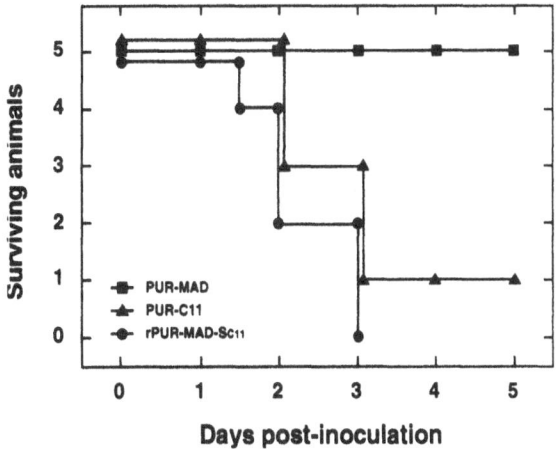

Figure 3. Surviving newborn minipigs infected with rPUR-MAD-SC11, PUR-MAD, or PUR-C11 at 48h after birth with 2×10^8 pfu per animal.

3.2 Tropism and Virulence of the Infectious cDNA

The virulence and tropism of clone rPUR-MAD-SC11 have been analyzed during infection of breast-fed newborn animals. rPUR-MAD-SC11 showed a mortality of 100%, similar to that of the parental virus that provided the S gene (PUR-C11). In contrast, the parental virus providing all of the genes except the S gene (PUR-MAD) produced no clinical signs, as expected (Fig. 3). On the other hand, the rPUR-MAD-SC11 virus grew in the jejunum and ileum of infected animals to titers as high as those of the parental enteric virus (PUR-C11), whereas the parental virus PUR-MAD produced low titers. Both parental viruses and the rescued one grew very well in the lungs. These data indicated that the S gene is a determinant of TGEV tropism and virulence.

4.   CONCLUSIONS

A fully functional infectious cDNA clone, leading to a virulent TGEV that replicates both in the enteric and the respiratory tracts, has been engineered. This cDNA clone will have an important impact on the study of mechanisms of coronavirus replication and transcription and provides an invaluable tool for the experimental investigation of virus-host interactions.

ACKNOWLEDGMENTS

This research was supported by grants from the Comisión Interministerial de Ciencia y Tecnología (Spain), the Community of Madrid, the European Community (Biotechnology, FAIR and Control of Infectious Diseases Programs), and Fort Dodge Veterinaria (Spain).

REFERENCES

Almazán, F., González, J. M., Pénzes, Z., Izeta, A., Calvo, E., Plana-Durán, J. and Enjuanes, L. 2000. Engineering the largest RNA virus genome as an infectious bacterial artificial chromosome. *Proc. Natl. Acad. Sci. U. S. A.* 97:5516-5521.

Izeta, A., Smerdou, C., Alonso, S., Penzes, Z., Méndez, A., Plana-Durán, J. and Enjuanes, L. 1996. Replication and packaging of transmissible gastroenteritis coronavirus-derived synthetic minigenomes. *J. Virol* 73: 1535-1545.

Koetzner, C. A., Parker, M. M., Ricard, C. S., Sturman, L. S. and Masters, P. S. 1992. Repair and mutagenesis of the genome of a deletion mutant of the coronavirus mouse hepatitis virus by targeted RNA recombination. *J. Virol.* 66: 1841-1848.

Masters, P. S. 1999. Reverse genetics of the largest RNA viruses. *Adv. Virus Res.* 53: 245-264.

Sánchez, C. M., Jiménez, G., Laviada, M. D., Correa, I., Suñé, C., Bullido, M. J., Gebaguer, F., Smerdou, C., Callebaut, P. et al. 1990. Antigenic homology among coronaviruses related to transmissible gastroenteritis virus. *Virology* 174: 410-417.

Sánchez, C. M., Izeta, A., Sánchez-Morgado, J. M., Alonso, S., Sola, I., Balasch, M., Plana-Durán, J. and Enjuanes, L. 1999. Targeted recombination demonstrates that the spike gene of transmissible gastroenteritis coronavirus is a determinant of its enteric tropism and virulence. *J. Virol.* 73: 7607-7618.

Wang, K., Boysen, C., Shizuya, H., Simon, M. I. and Hood, L. 1997. Complete nucleotide sequence of two generations of a bacterial artificial chromosome cloning vector. *Biotechniques* 23: 992-994.

Processing of the Replicase of Murine Coronavirus: Papain-like Proteinase 2 (PLP2) Acts to Generate p150 and p44

AMORNRAT KANJANAHALUETHAI AND SUSAN C. BAKER

Department of Microbiology and Immunology, Loyala University of Chicago, Stritch School of Medicine, Maywood, IL USA

1.   INTRODUCTION

For Nidoviruses, proteolytic processing of a large polyprotein translated from the 5'-end of the genomic RNA is required for the maturation and assembly of the viral replicase complex. The scheme used to process the arterivirus equine arteritis virus (EAV) replicase polyprotein has been experimentally determined (reviewed in Snijder and Meulenberg, 1998). The EAV replicase polyprotein is processed by 3 distinct proteinases to generate 12 mature protein products. Polyprotein intermediates have been identified as well as major and minor processing pathways. However, the role of the intermediates and the alternate pathways remains to be investigated. For the coronavirus mouse hepatitis virus (MHV), several laboratories are investigating the pathways used to process the replicase polyprotein. Two major processing models have been postulated and are shown in Figure 1. Studies from our laboratory showed that p150 is an intermediate to the 3C-like proteinase (3CLpro) product p27 and that p150 likely extends to include the putative membrane-spanning protein domain 1, MP1 (Schiller *et al* 1998). In contrast, other investigators have not detected the p150 precursor and postulate that a p240 product is adjacent to p27 (Denison *et al* 1992; 1995 and Lu *et al* 1998). In this study, we developed a specific antiserum to the MP1 domain (anti-D11) and determined that the

The Nidoviruses (Coronaviruses and Arteriviruses).
Edited by Ehud Lavi *et al.*, Kluwer Academic/Plenum Publishers, 2001.

MP1 domain is indeed part of the p150 intermediate. Furthermore, we show that MHV papain-like proteinase 2 (PLP2) is responsible for cleaving the polyprotein at the putative p150 cleavage site. These results show that PLP2 is an active enzyme.

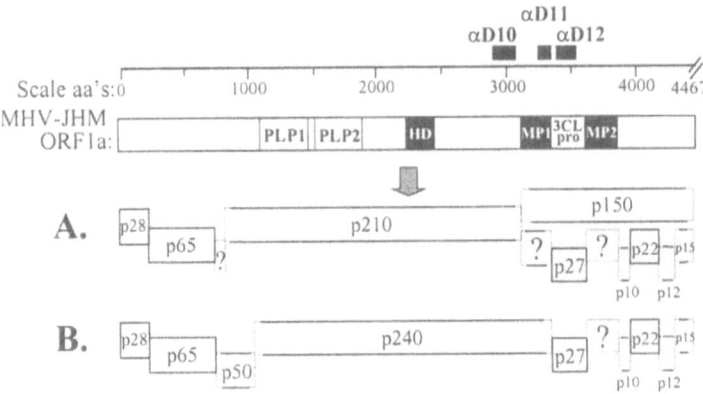

Figure 1. Two models for the proteolytic processing of the MHV ORF1a polyprotein. Model A is modified from Schiller *et al* (1998) and depicts the p150 precursor to the 3CLpro product. Model B is from the data of Denison *et al* (1992 and 1995) and Lu *et al* (1998) and indicates that the MP1 domain is part of the p240 protein.

2.   MATERIALS AND METHODS

2.1   Virus, cell line and immunoprecipitation

The MHV-JHM-X strain of coronavirus was used to infect HeLa-MHVR cells at moi 1. Infected cells were radiolabled with 100 μCi/ml (^{35}S)-translabel from 5 to 7 h post-infection (pi). Lysates were prepared and subjected to immunoprecipitation as previous described by our laboratory (Schiller *et al* 1998). The polyclonal rabbit antisera were generated after injection of GST-MHV fusion proteins and the specificity of the antisera was demonstrated as previously described (anti-D10 and anti-D12 in Schiller *et al* 1998; anti-D11 in Kanjanahaluethai and Baker, 2000). The anti-V5 monoclonal antibody (Invitrogen) is directed against an epitope tag built into the constructs.

2.2   Transfection of plasmid DNA and vTF7.3-mediated expression of MHV replicase products

Plasmid DNAs encoding the PLP2 domain and putative p150 cleavage site (substrate) under control of the T7 promoter were generated and expressed via the vaccinia-T7 system (Fuerst *et al* 1978) as described in detail by Kanajanahaluethai and Baker (2000). Newly synthesized proteins were radiolabeled with 50 µCi/ml (^{35}S)-translabel from 5.5 to 10.5 h pi and lysates were prepared and subjected to immuno-precipitation with the designated antisera. Products were analyzed by 5-10% SDS-PAGE.

3.   RESULTS AND DISCUSSION

Previously, we hypothesized that the MP1 domain was part of the p150 intermediate generated during the processing of the MHV replicase polyprotein (Schiller *et al* 1998). To test this hypothesis, we generated a specific antiserum, anti-D11, and used it to immunoprecipitate replicase products from MHV-infected cells. We compared the products immuno-precipitated with anti-D11 to products detected by antisera directed against a region upstream of MP1 (D10 region) and the region downstream of MP1 (the 3CLpro region precipitated by anti-D12). As shown in Figure 2, anti-D11 sera precipitates MHV specific products of >300 kilodaltons (kDa), 150 kDa and 44 kDa (lane 4). The p150 product migrates slightly faster than a product detected with all preimmune and immune sera tested (indicated by the open arrow), and co-migrates with the p150 product identified by anti-D12 (Schiller *et al* 1998). The p150 product is clearly distinct from the p210 and p290 proteins detected with anti-D10 (compare lane 2 with lanes 4 and 6). These results indicate that the p150 intermediate does indeed encompass the MP1 domain. Furthermore, pulse-chase analysis and proteinase inhibition studies indicate that p150 is a precursor to p44 and p27, and that the proteinase responsible for generating p150 is not sensitive to proteinase inhibitor E64d (Kanjanahaluethai and Baker, 2000).

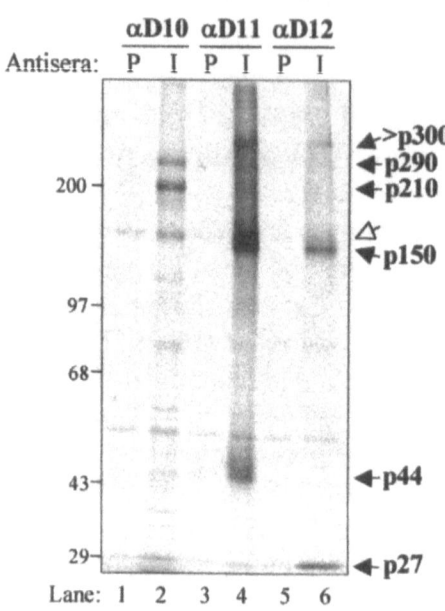

Figure 2. Detection of replicase gene products from MHV-infected HeLa-MHVR cells. Viral products detected by immunoprecipitation using preimmune (P) and immune (I) sera were separated by 5-10% SDS-PAGE. MHV replicase gene products >p300, p290, p210, p150, p44 and p27 are indicated by filled arrows. A non-specific product co-precipitated at low level by both preimmune and immune sera is indicated by the open arrow.

To identify the proteinase domain responsible for cleaving p150, we generated constructs encoding a single proteinase domain (PLP1, PLP2 or 3CLpro) and a construct encoding the putative p150 cleavage site (the substrate), and tested for the ability of the proteinase domain to cleave the substrate *in trans*. We found that the protein encoded by the PLP2 domain was indeed an active proteinase and rapidly cleaved the substrate to generate p44 (Figure 3). Neither the PLP1 nor the 3CLpro domains acted *in trans* to efficiently process the substrate (data not shown). These results show that the previously predicted PLP2 domain is an active proteinase and acts to process the precursor polyprotein to generate the p150 intermediate. P150 is likely then processed by the encoded 3CLpro domain to generate p44, p27, p10, p22, p12, p15 and the MP2 product (Figure 4A).

The p44 product identified in this study is analogus to the nsp3 replicase product of EAV (van der Meer et al 1998) and the 41 kDa product recently identified for avian infectious bronchitis virus (IBV) (Lim et al 2000). The 41 kDa IBV replicase product was shown to be modified by N-linked glycosylation. It will be interesting to determine if the MHV p44 replicase product is also glycosylated.

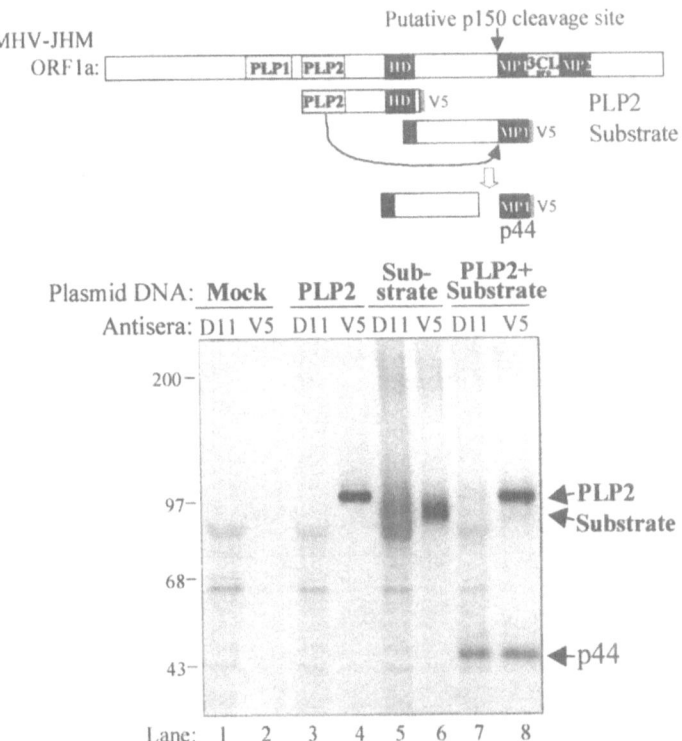

Figure 3. Detection of trans-cleavage activity by MHV PLP2. Plasmids encoding the PLP2 domain and the putative p150 cleavage site were transfected (either independently or together) into vTF7.3-infected HeLa-MHVR cells, newly synthesized proteins were radiolabeled and subjected to immunoprecipitation with the indicated antisera, and analyzed by 5-10% SDS-PAGE. The PLP2, full length substrate and p44 cleavage products are indicated by arrows.

The processing of p150 may be critical for releasing a hydrophobic sequence (MP1) that acts as an important membrane anchor for the MHV replicase complex (Figure 4B). We speculate that early processing events such as the cleavage of p28 by PLP1 and cleavage of p150 by PLP2 occur co-translationally *in cis* and are required to direct the intermediates encoding the hydrophobic domains to become membrane associated. This membrane association may drive the formation of the MHV double membrane vesicles (DMVs) (van der Meer *et al* 1998; Gosert *et al* 2000). The DMVs are likely the sites where MHV RNA synthesis occurs. The processing intermediates may serve as the replicase complex that mediates negative strand RNA synthesis, whereas further processing may be required to allow the complex to function in positive strand RNA synthesis. This scenario is based in part on our knowledge of the role of proteolytic processing in the switch from negative strand RNA synthesis to positive strand RNA synthesis in alphaviruses such as Sindbis virus (Lemm *et al* 1994; Shirako and Strauss, 1994). In that system, the nsP2 proteinase acts initially *in cis* and then *in trans* to process the replicase precursor from the P123-nsP4 complex responsible for negative strand RNA synthesis to the nsP1, nsP2, nsP3, nsP4 complex that mediates positive strand RNA synthesis. It will be interesting to determine if proteolytic processing of the MHV replicase complex is required for a similar switch from negative to positive strand RNA synthesis.

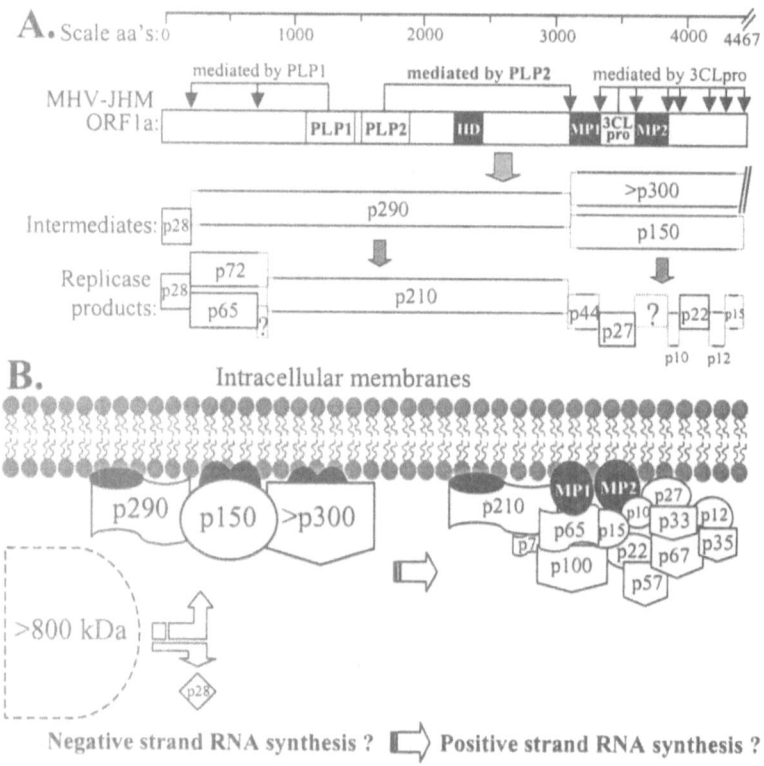

Figure 4. A) Schematic diagram of MHV ORF1a replicase intermediates and products. The proteinase domain that mediates each cleavage event is indicated. B) Working model for MHV replicase membrane association, assembly and function. All putative ORF1a and ORF1b products are indicated.

ACKNOWLEDGEMENTS

We thank David Axtell for his assistance in the production of GST-fusion protein and John Zaryczny for his assistance with rabbit injection and sera collection. This work was supported by Public Health Service Research Grant AI 32065.

REFERENCES

Denison, M.R., Hughes, S.A., and Weiss, S.R. 1995. Identification and characterization of a 65-kDa protein processed from the gene 1 polyprotein of the murine coronavirus MHV-A59. *Virology* **207**:316-320.

Denison, M.R., Zoltick, P.W., Hughes, S.A., Giangreco, B., Olson, A.L., Perlman, S., Leibowitz, L.L., and Weiss, S.R. 1992. Intracellular processing of the N-terminal ORF1a proteins of the coronavirus MHV-A59 requires multiple proteolytic events. *Virology* **189**:274-284.

Fuerst, T.R., Niles, E.G., Studier, F.W., and Moss, B. 1986. Eukaryotic transient-expression system based on recombinant vaccinia virus that synthesizes bacteriophage T7 RNA polymerase. *Proc. Natl. Acad. Sci. USA.* **83**:8122-8126.

Gosert, R., Kanjanahaluethai, A., Egger, D., Bienz, K., and Baker, S.C. 2000. Comparison of replicase localization in different types of mouse hepatitis virus (MHV)-infected cells. In *The Nidoviruses*. (E. Lavi, ed.), Plenum Press, New York (in press).

Kanjanahaluethai, A., and Baker, S.C. 2000. Identification of mouse hepatitis virus papain-like proteinase 2 activity. *J. Virol.* (in press)

Lemm, J.A., Rumenapf, T., Strauss, E.G., Strauss, J.H., and Rice, C.M. 1994. Polypeptide requirements for assembly of functional Sindbis virus replication complexes: A model for the temporal regulation of minus- and plus-strand RNA synthesis. *EMBO J.* **13**: 2925-2934.

Lim, K.P., Ng, L.F.P., and Liu, D.X. 2000. Identification of a novel cleavage activity of the first papain-like proteinase domain encoded by open reading frame 1a of the coronavirus avian infectious bronchitis virus and characterization of the cleavage products. *J. Virol.* **74**:1674-1685.

Lu, Y., Sims, A.C., and Denison, M.R. 1998. Mouse hepatitis virus 3C-like protease cleaves a 22-kilodalton protein from the open reading frame 1a polyprotein in virus-infected cells and in vitro. *J. Virol.* **72**:2265-2271.

Schiller, J.J., Kanjanahaluethai, A., and Baker, S.C. 1998. Processing of the coronavirus MHV-JHM polymerase polyprotein: Identification of precursors and proteolytic products spanning 400 kilodaltons of ORF1a. *Virology* **242**:288-302.

Skirako, Y., and Strauss, J.H. 1994. Regulation of Sindbis virus RNA replication: Uncleaved P123 and nsP4 function in minus-strand RNA synthesis, whereas cleaved products from P123 are required for efficient plus-strand RNA synthesis. *J. Virol.* **68**:1874-1885.

Snijder, E.J., and Meulenberg, J.J.M. 1998. The molecular biology of arteriviruses. *J. Gen. Virol.* **79**:961-979.

van der Meer, Y., van Tol, H., Locker, J.K., and Snijder, E.J. 1998. ORF1a-encoded replicase subunits are involved in the membrane association of the arterivirus replication complex. *J. Virol.* **72**:6689-6698.

Comparison of Replicase Localization in Different Types of Mouse Hepatitis Virus (MHV)-infected Cells

RAINER GOSERT[1], AMORNRAT KANJANAHALUETHAI[2], DENISE EGGER[1], KURT BIENZ,[1] AND SUSAN C. BAKER[2]

[1]Institute for Medical Microbiology, Basel, Switzerland. [2]Department of Microbiology and Immunology, Loyola University of Chicago, Stritch School of Medicine, Maywood, IL, USA

1. INTRODUCTION

The replication complex (RC) of virtually all positive strand RNA viruses has been shown to be intimately associated with cellular membranes. However, different RNA viruses seem to target or recruit distinct membranes for the assembly of their RCs. For example, poliovirus replicates its RNA on the surface of membranous vesicles which seem to evolve from the endoplasmic reticulum (ER). The vesicles form a rosette-like structure (Bienz *et al* 1992; Egger *et al* 1996). For tobacco etch potyvirus viral replication takes place at ER-derived vesicles and results in a collapse of the ER (Schaad *et al* 1997). In contrast, alphaviruses appear to use the cytosolic surface of endocytic organelles for the formation of their RCs (Froshauer *et al* 1988). For Nidoviruses such as arteriviruses and coronaviruses, the story is even more complex. Recently, it was shown that equine arteritis virus (EAV) generates ER-derived double-membrane vesicles (DMVs) (Pedersen *et al* 1999). In contrast, studies of the coronavirus mouse hepatitis virus (MHV) implicated a role for late endosomes in the formation of the RC (van der Meer *et al* 1999). Recently, we have shown that translation products of the MHV replicase gene localized to different membrane structures in different cell lines. For a human cell line, viral replicase products colocalized with golgi markers, but in a murine cell line, the viral products and ER-derived membranes colocalized (Shi *et al* 1999). To extend our studies, we wanted to determine the viral and cellular factor(s) that drive the generation of the MHV RC at distinct membranes in the different cell lines.

The Nidoviruses (Coronaviruses and Arteriviruses).
Edited by Ehud Lavi *et al.*, Kluwer Academic/Plenum Publishers, 2001.

To investigate the localization of the MHV RC, we initiated ultra-structural studies. Here, we present our preliminary analysis in the electron microscope (EM) and report that our results supported our earlier localization studies. We detected a difference in the distribution of MHV-induced DMVs in the human and murine cell lines. We also initiated biochemical studies to identify MHV proteins that allow the assembly of a membrane-associated RC. We found that the MHV replicase product p44 acts as an integral membrane protein, implicating it as a likely membrane-anchor protein for the MHV RC.

2.   MATERIALS AND METHODS

2.1   Cell lines

17Cl-1 cells were maintained and passaged as described by Sturman and Takemoto (1972). HeLa-MHVR cells, expressing high levels of the MHV receptor, were maintained and passaged as described by Gallagher (1996).

2.2   Triton X-114 extraction

HeLa-MHVR cells were infected with MHV-A59 at moi 1. Newly synthesized proteins were radiolabeled with 100 µCi/ml of (^{35}S)-translabel from 3.5-5.5 h post-infection (pi) and extractions were carried out essentially as described by Border (1981). Pellet and soluble fractions were subjected to immunoprecipitation as previously described (Schiller *et al* 1998).

2.3   Electron microscopy

Confluent monolayers of HeLa-MHVR or 17Cl-1 cells were infected at moi 1 with MHV-A59. At 5 h pi, cells were fixed with 2.5% glutaraldehyde and 2% OsO$_4$ and embedded in Poly/Bed 812 (Polysciences) according to standard protocols. The ultrathin sections were examined with a Philips CM 100 EM.

3.   RESULTS AND DISCUSSION

In this study, we describe our preliminary biochemical and ultrastructural analysis of the MHV RC. We wanted to determine which replicase subunit(s) may act as putative integral membrane proteins to anchor the RC to intracellular membranes and to determine if MHV-infection induces the formation of intracellular vesicles which are likely the sites for MHV replication.

Our first approach was to characterize MHV replicase products biochemically using Triton X-114 extraction (Bordier 1981). This procedure separates proteins principally by their hydrophobic properties, with hydrophobic, integral membrane proteins partitioning into the detergent phase, while hydrophilic proteins remain in the aqueous phase. As shown in Figure 1, the amino-terminal replicase product, p28, remains in the soluble fraction whereas the putative membrane-spanning protein 1, MP1 (p44), is detected in the membranous pellet fraction. As expected, MHV virion integral membrane matrix protein M, partitions to the membranous pellet fraction. Overall, these results show that the MHV replicase product MP1 (p44) does indeed act as an integral membrane protein and may serve as an important anchor protein for the RC.

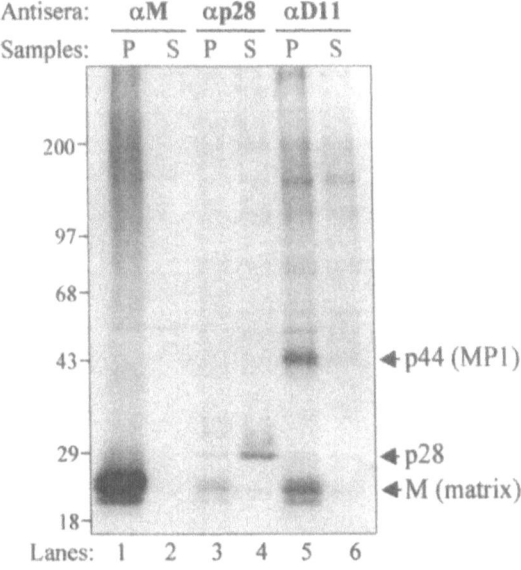

Figure 1. Triton X-114 extraction analysis of MHV replicase products. A schematic diagram depicting the region to which the specific antisera was generated and a map of the known MHV replicase products of ORF1a is shown at the top. MHV products detected after extraction with Triton X-114 and immunoprecipitation with specific antisera are shown below. The pelleted (P) fraction contains integral membrane proteins such as the matrix (M) protein of the MHV virion. The supernatant (S) fraction contains cytosolic proteins and non-integral membrane proteins.

Our second approach was to examine MHV-infected cells by EM to determine intracellular membrane alterations. We tested two distinct cell lines, human cells encoding the MHV receptor, HeLa-MHVR cells, and a mouse fibroblast cell line, 17Cl-1 cells. Our previous immunofluorescene results indicated that the MHV replicase products localized to different sites in these two cell lines (Shi *et al* 1999), therefore we wanted to investigate these distinct localization patterns at the EM level.

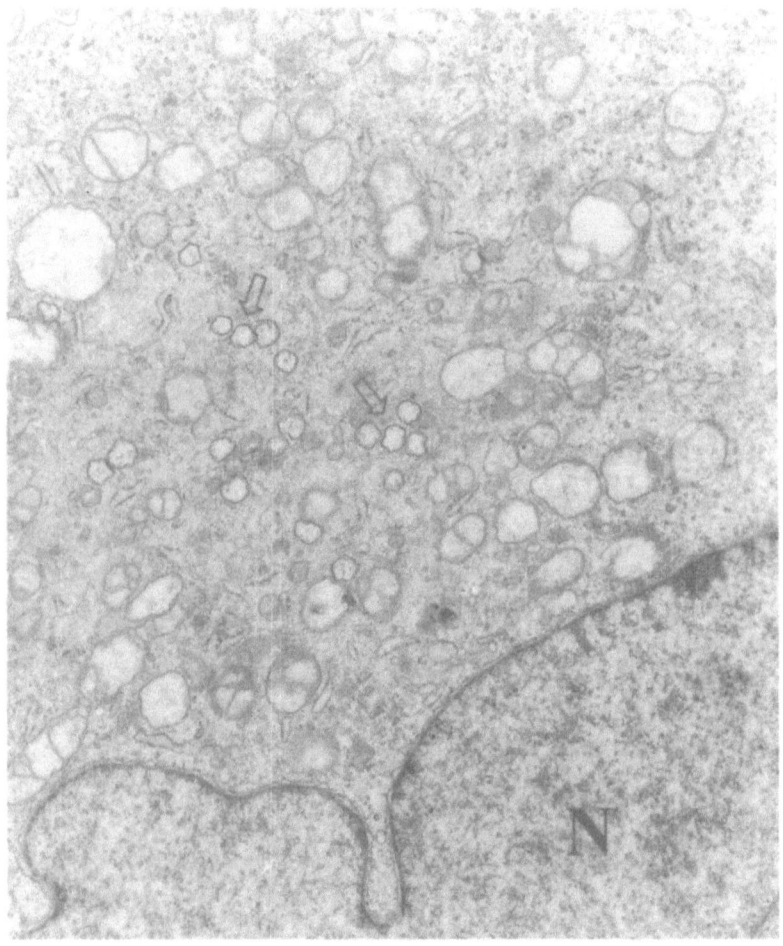

Figure 2. EM analysis of intracellular membrane structures in MHV-A59 infected HeLa-MHVR cells at 5 h pi. Magnification: 15,000x. N: Nucleus. MHV-induced vesicles are indicated by the open arrows.

MHV-infected HeLa-MHVR cells contained virus-induced vesicles in the cytoplasm that were found alone or in small clusters (Figure 2). The vesicle membranes appear as electron dense structures, with a thickness more consistent with double membranes. High resolution images indicate that the vesicle membranes consist of 2 lipid bilayers (data not shown). EM analysis of MHV-infected 17Cl-1 cells revealed similar, virus-induced vesicles (Figure 3). However, fewer vesicles were detected and the vesicles were found primarily as vesicle clusters. We also noted a clear dilatation of the ER in MHV-infected 17Cl-1 cells, while the golgi complex remained intact at 5 h pi (data not shown).

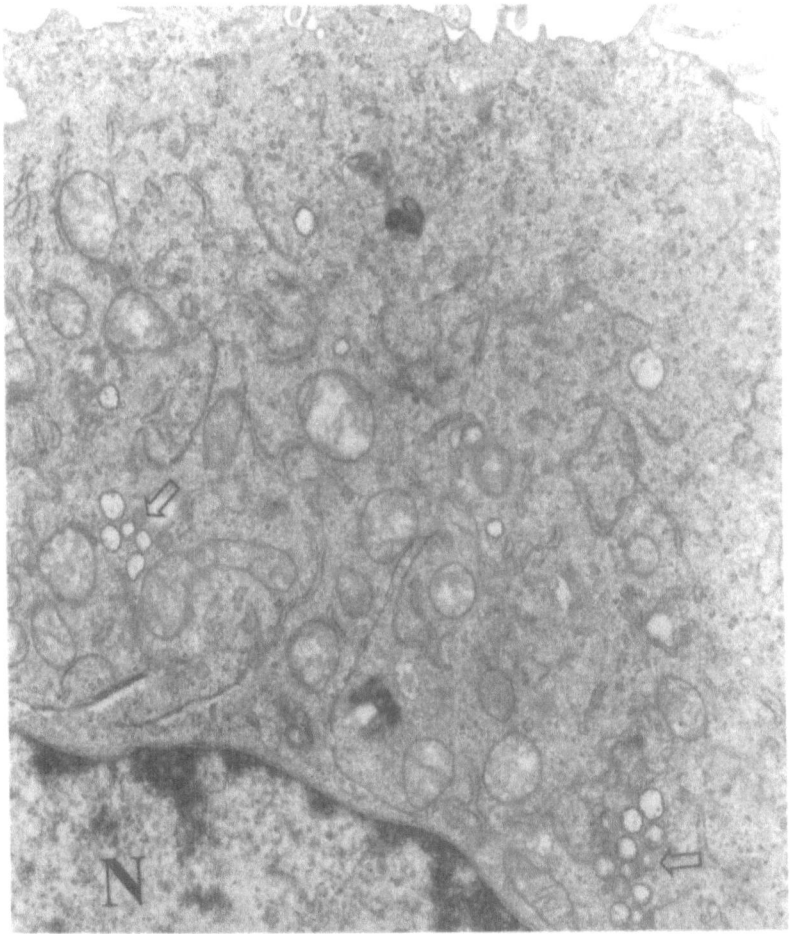

Figure 3. EM analysis of intracellular membrane structures in MHV-A59 infected 17Cl-1 cells at 5 h pi. Magnification: 15,000x. N: Nucleus. MHV-induced vesicle clusters are indicated by the open arrows. Swollen ER is seen in the central area of the picture.

In MHV-infected HeLa-MHVR cells, the large number of vesicles in a clustered area seen by EM compared well to the intense gene 1 translation product staining that colocalized with markers of the golgi complex shown by IF studies in the confocal microscope (Shi *et al* 1999). For MHV-infected 17Cl-1 cells, the dispersed vesicle clusters detected by EM correlate with a perinuclear punctuate signal of viral proteins. Further analysis including immuno-EM and in situ hybridization with MHV-specific riboprobes will be required to characterize the protein and RNA content of the DMVs. Detection of newly synthesized RNA by BrUTP incorporation at the ultrastructural level will identify the location of MHV replication. In addition, important questions remain concerning which viral and cellular factor(s) drive the formation of the vesicles and how these factors influence vesicle distribution in different cell lines. These studies should be helpful in defining the viral and cellular processes that drive MHV RNA synthesis.

ACKNOWLEDGEMENTS

This work was supported by Public Health Service Research Grant AI 32065 (to S.C.B.) and grant 31-055397.98/1 from the Swiss National Foundation (to K.B.).

REFERENCES

Bienz, K., Egger, D., Pfister, T., and Troxler, M.   1992. Structural and functional characterization of the poliovirus replication complex. *J. Virol.* **66**:2740-2747.

Bordier C. 1981. Phase separation of integral membrane proteins in Triton X-114 solution. *J. Biol. Chem.* **256**:1604-1607.

Egger, D., Pasamontes,L., Bolten,R., Boyko,V., and Bienz, K.   1996. Reversible dissociation of the poliovirus replication complex: Functions and interactions of its components in viral RNA synthesis. *J. Virol.* **70**:8675-8683.

Froshauer, S., Kartenbeck, J. and Helenius, A. 1988. Alphavirus RNA replicase is located on the cytoplasmic surface of endosomes and lysosomes. *J. Cell Biol.* **107**:2075-2086.

Gallagher, T.M. 1996. Murine coronavirus membrane fusion is blocked by modification of thiols buried within the spike protein. *J. Virol.* **70**:4683-4690.

Pedersen, K.W., van der Meer, Y., Roos, N., and Snijder, E.J. 1999. Open reading frame 1a-encoded subunits of the arterivirus replicase induce endoplasmic reticulum-derived double-membrane vesicles which carry the viral replication complex. *J. Virol.* **73**:2016-2026.

Schaad, M.C., Jensen, P.E., and Carrington, J.C.   1997. Formation of plant RNA virus replication complexes on membranes: Role of an endoplasmic reticulum-targeted viral protein. *EMBO J.* **16**:4049-4059.

Shi, S.T., Schiller, J.J., Kanjanahaluethai, A., Baker, S.C., Oh, J.-W., and Lai, M.M.C. 1999. Colocalization and membrane association of mouse hepatitis virus gene 1 products and de novo-synthesized viral RNA in infected cells. *J. Virol.* **73**:5957-5969.

Sturman, L.S., and Takemoto, K.K.   1972. Enhanced growth of a murine coronavirus in transformed mouse cells. *Infect. Immun.* **6**:501-507.

van der Meer, Y., Snijder, E.J., Dobbe, J.C., Schleich, S., Denison, M.R., Spaan, W.J.M., and Locker, J.K. 1999. Localization of mouse hepatitis virus nonstructural proteins and RNA synthesis indicates a role for late endosomes in viral replication. *J. Virol.* **73**:7641-7657.

Exploiting DNA Immunization to Generate Polyclonal Antisera to Coronavirus Replicase Proteins

SUSAN C. BAKER, AMORNRAT KANJANAHALUETHAI, NATHAN M. SHERER, DAVID D. AXTELL, AND JENNIFER J. SCHILLER
Department of Microbiology and Immunology, Loyola University of Chicago, Stritch School of Medicine, Maywood, IL, USA

1. INTRODUCTION

Over the last 10 years, a large number of studies have shown that immunization with plasmid DNA encoding an antigen can elicit specific immune responses in animals (reviewed in Donnelly *et al* 1997; Robinson and Torres 1997). The type of immune response generated depends on several factors, such as the route of inoculation, whether the DNA-encoded antigen remains intracellular or is secreted, and the stability and antigenicity of the protein products. DNA immunization is an attractive alternative to traditional purified protein immunizations because plasmid DNA is relatively easy and inexpensive to generate. However, additional studies are required to determine the optimal route of delivery of plasmid DNA and if additional factors may help "direct" the encoded antigen to elicit a specific type of immune response.

For our studies of the MHV replicase complex, we wanted to exploit DNA vaccine technology to generate polyclonal antisera to replicase products. To this end, we tested 3 different routes of DNA immunization and 2 expression constructs to determine the best conditions to stimulate a strong humoral immune response to our antigen of interest. We found that priming rabbits with plasmid DNA encoding a chimeric cytotoxic T lymphocyte antigen 4-MHV domain 11 fusion protein, termed CTLA4-D11, followed by a booster inoculation with GST-D11 protein elicited a high

The Nidoviruses (Coronaviruses and Arteriviruses).
Edited by Ehud Lavi *et al.*, Kluwer Academic/Plenum Publishers, 2001.

283

affinity antibody response and generated polyclonal antisera that recognized MHV replicase proteins by ELISA, Western blot and immunoprecipitation assays.

2.   MATERIALS AND METHODS

2.1   Plasmid DNA

Plasmid pD11 was generated by amplification of the D11 region (MHV-JHM nucleotides 9,941-10,204 numbering as per Lee *et al* 1991 as modified by Bonilla *et al* 1994) with oligonucleotides containing Not I sites followed by ligation into the pCMV-βgal vector (Invitrogen) backbone.   Plasmid pCTLA4-D11 was generated by amplification of the rabbit CTLA4 ectodomain (nucleotides 49-525, numbering according to Isono and Seto, 1995) from pCTLA4-Ig (a kind gift from Dr. David Dichek, UCSF, CA). The CTLA4 ectodomain PCR product was digested with Not I and EcoR I and ligated with an EcoR I-Not I D11 fragment and then ligated into the Not I digested pCMV vector.   The resultant plasmid was checked for orientation of the insert by digestion with EcoR I and designated pCTLA4-D11.

2.2   DNA immunization and analysis of antisera

DNA immunizations were performed essentially as described by Sundaram *et al* (1996) using an Helios gene gun (BioRad), or by directly injecting 1.0 mg/ml of plasmid DNA in PBS intramuscularly (IM).   The generation of the GST-D11 fusion protein and the resulting anti-D11 sera are described elsewhere (Kanjanahaluethai and Baker 2000).   Antibody titer of DNA immunized animals was determined by ELISA to GST-D11 fusion protein and by immunoprecipitation using methods previously described by our laboratory (Schiller *et al* 1998).

3.   RESULTS AND DISCUSSION

We wanted to exploit DNA immunization and develop a relatively inexpensive and non-labor intensive method to generate polyclonal antisera in rabbits.   Previous studies demonstrated that immunization of rabbits with DNA encoding β-galactosidase (β-gal) did indeed induce specific antibodies that recognized β-gal protein in an ELISA and Western blot assay (Sundaram *et al* 1996).   In this study, our antigen of interest was the D11 domain of the MHV-JHM replicase polyprotein. The D11 domain is located just upstream of the MHV 3C-like proteinase (3CLpro) domain. Previously, we showed that 3CLpro was generated from a p150 precusor polyprotein (Schiller *et al* 1998). We wanted to generate antibodies specific to the D11 domain to help us determine if p150 extended upstream of the 3CLpro region.

First, we generated pD11 DNA (Figure 1) as described in the Materials and Methods and immunized rabbits using three different routes of inoculation: 1) intramuscular (IM) inoculation of 1 mg plasmid DNA suspended in phosphate buffered saline (PBS) (rabbits 1 and 2); 2) IM administration of DNA in PBS with prior adminstration of bupivacaine and co-adminstration of cardiotoxin according to the method of Wells, 1993 (rabbits 3 and 4); and 3) gene gun adminstration of plasmid DNA according to Sundaram *et al* (1996) (rabbits 5 and 6). Each rabbit was inoculated with DNA at 3 week intervals and blood collected 10 days after each inoculation. As a control, 2 rabbits were injected with purified GST-D11 fusion protein (rabbits 7 and 8, described in detail in Kanjanahaluethai and Baker 2000). Specific antibody response was determined by ELISA to GST-D11 protein. As shown in Table 1, none of the pD11-inoculated rabbits generated a specific antibody response. This was likely because the D11 antigen was not secreted from transfected cells (data not shown). DNA vaccines expressing intracellular proteins are more likely to generate a strong cytotoxic T lymphocyte response and minimal if any, antibody response.

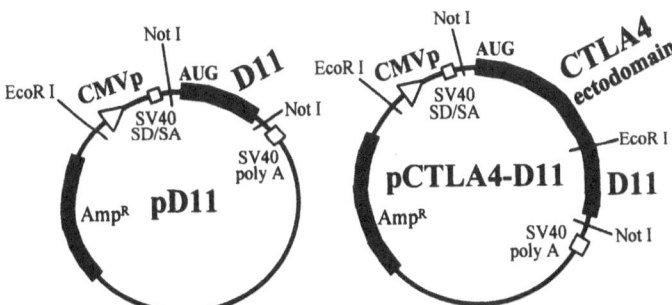

Figure 1. Schematic diagram of the constructs generated for DNA immunization of rabbits. The vector backbone is derived by digestion of pCMV-Bgal (Invitrogen) with Not I and isolation of the vector fragment. The D11 and CTLA4-D11 inserts were derived by PCR amplification as described in the methods section.

To determine if a plasmid DNA encoding a secreted form of D11 would stimulate a humoral immune response in rabbits, we constructed pCTLA4-D11 (Figure 1). This plasmid DNA encodes the ectodomain of rabbit CTLA4 in frame with the D11 domain. Studies by Boyle and co-workers showed that DNA vaccines encoding CTLA4-chimeric antigens generated a strong humoral immune response, likely because the CTLA4 ectodomain binds to B7 on antigen presenting cells (APCs) (Boyle *et al* 1998; Linsley *et al* 1991). Thus, the CTLA4 ectodomain facilitates the secretion of the chimeric antigen and likely directs it to bind APCs.

We tested DNA immunization via IM inoculation (rabbits 9 and 10) and gene gun inoculation (rabbits 11 and 12) for induction of specific antibodies to D11 fusion protein. As seen in Table 1, D11-specific antibodies were induced in all 4 rabbits, with higher titers detected in the gene gun inoculated animals. This is consistent with the results of other investigators who showed that plasmid DNA administered by a gene gun enters dendritic cells, the optimal site for induction of an immune response to the encoded antigen (Torres *et al* 1997; Akbari *et al* 1999; Takashima and Morita 1999). Overall, the results of the pD11 and pCTLA4-D11 inoculation studies show that DNA encoding a CTLA4-fusion protein administered via the gene gun stimulates the highest titers of specific antibody (Table 1).

Table 1. Specific antibody response detected in sera isolated from rabbits immunized with GST-D11 protein or with plasmid DNA encoding the MHV-D11 domain.

Rabbit	Antigen	Inj/Ad[b]	Titer to GST-D11 protein[a]			
			Pre	1st	2nd	3rd
1	pD11	IM	<10	n.d.[c]	<10	<10
2	pD11	IM	<10	n.d.	<10	<10
3	pD11	IM/bup+CT	<10	n.d.	<10	<10
4	pD11	IM/bup+CT	<10	n.d.	<10	<10
5	pD11	gg	<10	<10	<10	<10
6	pD11	gg	80	80	80	80
7	GST-D11	IM+SC/FA	<10	n.d.	256,000	512,000
8	GST-D11	IM+SC/FA	<10	n.d.	512,000	1,024,000
9	pCTLA4-D11	IM	<10	<10	160	320
10	pCTLA4-D11	IM	<10	<10	1,040	2,080
11	pCTLA4-D11	gg	<10	140	2,080	10,240
12	pCTLA4-D11	gg	280	1,040	10,240	40,960

[a] Preimmune sera (Pre) and sera isolated from blood taken 10-14 days after each of three injections were tested for specific antibody response by ELISA to GST-D11 protein
[b] Method of injection, IM = intramuscular; bup+CT = bupivacaine + cardiotoxin; gg = gene gun; SC = subcutaneous; FA = Freund's adjuvant
[c] n.d. = not determined

Our next step was to determine if the antibodies generated by DNA immunization were of sufficient affinity to immunoprecipitate viral antigens from MHV-infected cells. HeLa-MHVR cells were infected with MHV-A59 at a moi of 10 and radiolabeled with 100 μCi/ml (^{35}S)-translabel from 3.5-5.5 h pi. Lysates were prepared and subjected to immunoprecipitation with 5 μl of sera from rabbits 1-6 and 9-12, and products analyzed by 5-10% SDS-PAGE. However, no MHV-specific products were detected (data not shown). A review of recent literature revealed that boosting DNA immunized animals with soluble protein stimulates maturation of the antibody response to the specific antigen (Richmond *et al* 1998; Levtin *et al* 1997; Boyle *et al* 1997). We boosted rabbits 2, 4, 9, 10, 11 and 12 with 0.5 mg of purified GST-D11 protein in complete Freund's adjuvant, isolated sera 14 days after the boost and determined the titer to GST-D11 protein by ELISA. Rabbits 2 and 4 had low titer (40-120) consistent with a primary response to immunization (data not shown). In contrast, rabbits 9-12 exhibited 2-10 fold increases in specific antibody titer, consistent with a secondary response to the D11 antigen. We then tested the polyclonal antisera from DNA primed-protein boosted animals to determine if it was of sufficient affinity to immunoprecipitate replicase products from MHV-infected cells as described above. As shown in Figure 2, MHV replicase products p150 and >p300 were detected from lysates immunoprecipitated with sera from DNA primed and protein boosted animals (lanes 2 and 4).

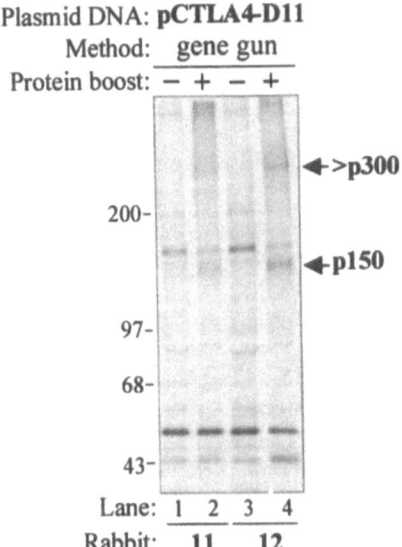

Figure 2. Detection of MHV-specific replicase products with antisera from pCTLA4-D11 primed, GST-D11 protein boosted rabbits. Radiolabeled MHV-infected cell lysates were prepared and subjected to immunoprecipitation as previously described (Schiller et al, 1998). Antisera were from rabbits 11 and 12 after 3 DNA immunizations (-) or after the DNA immunized rabbits had been boosted with 0.5 mg purified GST-D11 protein (+). Products of the immunoprecipitation were analyzed by 5-10% SDS-PAGE and subjected to autoradiography. MHV-specific products >p300 and p150 are indicated.

We noted that replicase product p44 was not resolved from background band on this gel. It should also be noted that the pre- and post-protein boosting sera were diluted to identical ELISA titers for this experiment. Thus, immunoprecipitation of the viral replicase product is likely due to affinity-boosting of the response, not simply to higher titer of antibody.

Overall, we present data demonstrating the utility of the DNA priming/protein boosting procedure for the rapid generation of high affinity polyclonal antisera in rabbits. This DNA priming/protein boosting approach is likely to facilate the generation of polyclonal antisera to viral or cellular proteins that are challenging to express in sufficient quantity for standard, protein immunization protocols. In addition, this procedure may elicit higher avidity antibodies or antibodies to epitopes that may not be revealed during protein immunization. We anticipate that this approach will help us in our efforts to generate polyclonal antisera to all MHV replicase products and to identify their localization and function in MHV RNA synthesis.

ACKNOWLEDGEMENTS

We thank John Zaryczny for his assistance with rabbit injection and sera collection. This work was supported by Public Health Service Research Grant AI 32065.

REFERENCES

Akbari, O., Panjwani, N., Garcia, S., Tascon, R., Lowrie, D., and Stockinger, B. 1999. DNA vaccination: Transfection and activation of dendritic cells as key events for immunity. *J. Exp. Med.* **189**:169-177.

Bonilla, P.J., Gorbalenya, A.E., and Weiss, S.R. 1994. Mouse hepatitis virus strain A59 RNA polymerase gene ORF 1a: Heterogeneity among MHV strains. *Virology* **198**:736-740.

Boyle, J.S., Brady, J.L., and Lew, A.M. 1998. Enhanced responses to a DNA vaccine encoding a fusion antigen that is directed to sites of immune induction. *Nature* **392**:408-411.

Boyle, J.S., Silva, A., Brady, J.L., and Lew, A.M. 1997. DNA immunization: Induction of higher avidity antibody and effect of route on T cell cytotoxicity. *Proc. Natl. Acad. Sci. USA* **94**:14626-14631.

Donnelly, J.J., Ulmer, J.B., Shiver, J.W., and Liu, M.A. 1997. DNA vaccines. *Annu. Rev. Immunol.* **15**:617-648.

Isono, T., and Seto, A. 1995. Cloning and sequencing of the rabbit gene encoding T-cell costimulatory molecules. *Immunogenetics* **42**:217-220.

Kanjanahaluethai, A., and Baker, S.C. 2000. Identification of activity of mouse hepatitis virus papain-like proteinase 2. *J. Virol.* (in press)

Lee, H.-J., Shieh, C.-K., Gorbalenya, A.E., Koonin, E.V., Monica, N.L., Tuler, J., Bagdzhadzhyan, A., and Lai, M.M.C. 1991. The complete sequence (22 kilobases) of murine coronavirus gene 1 encoding the putative proteases and RNA polymerase. *Virology* **180**:567-582.

Letvin, N.L., Montefiori, D.C., Yasutomi, Y., Perry, H.C., Davies, M.E., Lekutis, C., Alroy, M., Freed, D.C., Lord, C.I., Handt, L.K., Liu, M.A., and Shiver, J.W. 1997. Potent, protective anti-HIV immune responses generated by bimodal HIV envelope DNA plus protein vaccination. *Proc. Natl. Acad. Sci. USA* **94**:9378-9383.

Linsley, P.S., Brady, W., Urnes, M., Grosmaire, L.S., Damle, N.K., and Ledbetter, J.A. 1991. CTLA-4 is a second receptor for the B cell activation antigen B7. *J. Exp. Med.* **174**: 561-569.

Richmond, J.F.L., Lu, S., Santoro, J.C., Weng, J., Hu, S.L., Montefiori, D.C., and Robinson, H.L. 1998. Studies of the neutralizing activity and avidity of anti-human immunodeficiency virus type 1 env antibody elicited by DNA priming and protein boosting. *J. Virol.* **72**:9092-9100.

Robinson, H.L., and Torres, C.A T. 1997. DNA vaccines. *Semin. Immunol.* **9**:271-283.

Schiller, J.J., Kanjanahaluethai, A., and Baker, S.C. 1998. Processing of the coronavirus MHV-JHM polymerase polyprotein: Identification of precursors and proteolytic products spanning 400 kilodaltons of ORF1a. *Virology* **242**:288-302.

Sundaram, P., Xiao, W., and Brandsma, J.L. 1996. Particle-mediated delivery of recombinant expression vectors to rabbit skin induces high-titered polyclonal antisera (and circumvents purification of a protein immunogen). *Nucl. Acids Res.* **24**:1375-1377.

Takashima, A., and Morita, A. 1999. Dendritic cells in genetic immunization. *J. Leukoc. Biol.* **66**:350-356.

Torres, C.A.T., Iwasaki, A., Barber, B.H., and Robinson, H.L. 1997. Differential dependence on target site tissue for gene gun and intramuscular DNA immunizations. *J. Immunol.* **158**:4529-4532.

Wells, D.J. 1993. Improved gene transfer by direct plasmid injection associated with regeneration in mouse skeletal muscle. *FEBS* **332**:179-182.

Further Identification and Characterization of Products Processed from the Coronavirus Avian Infectious Bronchitis Virus (IBV) 1a Polyprotein by the 3C-like Proteinase

LISA F. P. NG, H. Y. XU, AND D. X. LIU
Institute of Molecular Agrobiology, The National University of Singapore, 1 Research Link, Singapore 117604

1.   INTRODUCTION

Proteolytic processing of the IBV 441-kDa 1a and 741-kDa 1a/1b fusion polyproteins to smaller mature products is mediated by two viral proteinases, namely, the papain-like proteinase and the 3C-like proteinase (Fig. 1).   The 3C-like proteinase was identified as a 33-kDa protein in IBV-infected Vero cells (Lim *et al.*, 2000).   Characterization of the proteinase activities has shown that it cleaves the bulk of the polyproteins at conserved Q-S (G and N) dipeptide bonds, resulting in the release of more than ten mature products (Liu *et al.*, 1994, 1997, 1998; Ng and Liu, 1998, 2000).   It was shown to be associated with the membrane fraction in virus-infected cells by both cellular and biochemical studies (Ng and Liu, 2000).

In this report, we confirm the presence of three more cleavage products of 34, 6 and 16 kDa.   Taken together with the previously identified 24- and 10-kDa proteins (Ng and Liu, 1998; Liu *et al.*, 1997; Lim *et al.*, 2000), five products were released from the 1a polyprotein by the 3C-like proteinase, in addition to the 33-kDa proteinase itself.   Subcellular localization of the five products in virus-infected cells shows fluorescence at the perinuclear region with granular and vesicular staining, but distinct staining patterns were observed when the proteins were expressed individually in intact cells.

The Nidoviruses (Coronaviruses and Arteriviruses).
Edited by Ehud Lavi *et al.*, Kluwer Academic/Plenum Publishers, 2001.

These results suggest that interaction of the individual cleavage products may occur in virus-infected cells in order to form the membrane-bound viral replication complexes.

2.　　MATERIALS AND METHODS

2.1　　Virus and cells

The egg-adapted Beaudette strain of IBV (ATCC VR-22), obtained from the American Type Culture Collection (ATCC), and was adapted to Vero cells and prepared as described previously (Ng and Liu, 1998, 2000).

Vero cells and Cos-7 cells were grown at 37°C in 5% CO_2 and maintained in Dulbecco's modified minimal essential medium (Gibco BRL, Life Technologies) supplemented with 10% newborn calf serum.

2.2　　Radiolabeling of IBV-infected and mock-infected Vero cells

Confluent monolayers of Vero cells grown on 60 mm dishes were infected with IBV at a multiplicity of infection of approximately 2 PFU/cell. The cells were labeled for 4 h with 25 μCi/ml [^{35}S] methionine-cysteine at 6 h postinfection before harvesting.

2.3　　Radioimmunoprecipitation

Radioimmunoprecipitation with polyclonal rabbit antisera was carried out as described previously (Liu *et al.*, 1994).

2.4　　SDS-polyacrylamide gel electrophoresis

SDS-PAGE was carried out with 15% and 17.5% polyacrylamide concentrations, and the labeled polypeptides were detected by autoradiography or fluorography of dried gels.

2.5　　Western Blot

SDS-PAGE of viral polypeptides was carried out as mentioned and viral proteins were transferred to nitrocellulose membrane by the semi-dry method (Bio-Rad). The membrane was blocked overnight at 4°C in blocking buffer (5% milk in 20 mM Tris-Cl 7.4, 150 mM NaCl, 0.1% Tween 20) and incubated with specific polyclonal rabbit antiserum (anti-16) diluted in blocking buffer (1:500 dilution) at RT°C for 2 h and washed with TBST (20 mM Tris-Cl 7.4, 150 mM NaCl, 0.1% Tween 20). Membrane was blocked in blocking buffer for another 20 min and incubated with anti-rabbit IgG conjugated with horseradish peroxidase (Dako) diluted in blocking buffer (1:2500) at RT°C for 1 h followed by colour development (ECL, Amersham).

2.6 Polymerase chain reaction (PCR)

Appropriate primers and template DNAs were used in amplification reactions with cloned PFU DNA polymerase (Stratagene) under standard buffer conditions with 2mM $MgCl_2$.

2.7 Indirect Immunofluorescence microscopy

Cells were grown on coverslips and infected with IBV or transfected with appropriate plasmid DNAs. After washed with PBS containing 1 mM $CaCl_2$ and 1 mM $MgCl_2$ (PBSCM), the cells were fixed with 4% paraformaldehyde (in PBSCM) for 30 min at room temperature and permeabilized with 0.1% saponin (in PBSCM), followed by incubation with specific antiserum at room temperature for 2 h. Antibodies were diluted in fluorescence dilution buffer (PBSCM with 5% normal goat serum, 5% newborn calf serum, and 2% bovine serum albumin, pH 7.6). The cells were then washed with PBSCM and incubated with anti-rabbit or anti-mouse IgG conjugated to fluorescein isothiocynate (Sigma) in the fluorescence dilution buffer at 4°C for 1 h before mounting.

2.8 Construction of plasmids

Plasmids pIBV9, pIBV3C and pBP5 were described before (Liu *et al.*, 1994, 1997 and Ng and Liu, 2000). The IBV sequences present in these constructs are nucleotides 8693 to 10925, 8865 to 9786 and 10752 to 12312 respectively.

Plasmids pT7tag34k and pT7tag16k cover nucleotides 9787 to 10662 and 11875 to 12309, respectively were constructed by ligation of PCR fragments covering the appropriate regions into *Nco* I-digested pT7tag (contains an 11-amino-acid T7tag: MASMTGGQQMG). pT7tag34k was generated using pIBV9 as the template and LN-76 (5'-AGATTACCATGGC TTCTTTTGTAAGA-3') and LN-75 (5'-CTTGCCATGGAAACTGTA GCAATAGG-3') as the cloning primers. pT7tag16k was cloned using pBP5 as the template and LN-9 (5'-TCTTACCATGGAGTCTAAAGGGCAT-3') and LN-70 (5'-AGCACCATGGAAACAG AAGATTTTGG-3') as the cloning primers.

3.   RESULTS AND DISCUSSION

3.1   Identification of two novel proteins in IBV-infected cells

Previously, we have identified the 33-kDa 3C-like proteinase and two other cleavage products, the 24- and 10-kDa proteins from the 1a polyprotein (Fig. 1; Lim *et al.*, 2000; Ng and Liu, 1998; Liu *et al.*, 1997). Recently, a Q^{3379}-A^{3380} dipeptide bond was demonstrated to be a cleavage site of the 3C-like proteinase (Ng and Liu *et al.*, 2000). Three more products with calculated molecular masses of 34, 9, and 16 kDa, respectively were expected to be released from cleavage by the 3C-like proteinase at the C-terminal region of the 1a polyprotein. Two region specific antisera were raised and similar approaches were used to study the protein expression. Anti-6, raised against the IBV sequence encoded between nucleotides 10666 and 10914 immunoprecipitated a specific product from IBV-infected Vero cells. The protein migrated on SDS-PAGE as a 6 kDa protein instead of the estimated size of 9 kDa (Fig. 2, lane 1). Western blot analysis with antiserum anti-16, raised against the IBV sequence encoded between nucleotides 11787 and 12312 resulted in the detection of a 16-kDa protein from IBV-infected cells from 6 h p.i. onwards (Fig. 2, lanes 5 to 8). Both proteins are released by the 3C-like proteinase at Q^{3379}-A^{3380} / Q^{3462}-S^{3463} and Q^{3783}-S^{3784} / Q^{3928}-S^{3929} dipeptide bonds, respectively (Fig. 1).

This data was consistent with that from MHV and HCV (Bost *et al.*, 2000; Ziebuhr and Siddell, 1999), thus the previously predicted Q^{3213}-G^{3214} dipeptide bond is not a real cleavage site of the 3C-like proteinase. Site-directed mutagenesis and *in vitro* studies confirmed that this site is not used (data not shown). Cleavage at the Q^{3086}-S^{3087} and Q^{3379}-A^{3380} dipeptide bond would result in the release of a mature product of approximately 34 kDa (Fig.1). Efforts to raise region specific antiserum to detect this product failed, possibly due to the presence of extremely hydrophobic amino acid residues.

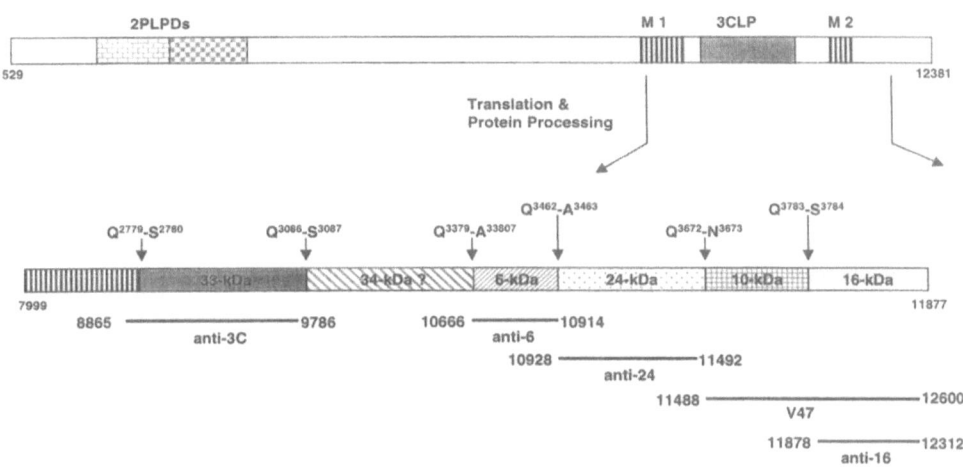

Figure 1. Diagram of ORF1a, illustrating the two overlapping papain-like proteinase domains (PLPDs) and the 3C-like proteinase domain (3CLP).

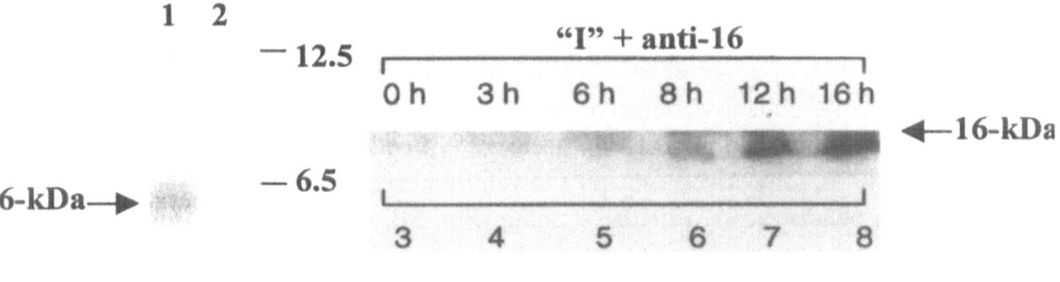

1 : "I" + anti-6
2 : "M" + anti-6

Figure 2. Detection of the 6- and 16-kDa proteins in IBV-infected Vero cells (I) with anti-6 (lanes 1 and 2) and anti-16 (lanes 3 to 8), respectively.

3.2 Intracellular localization of the five viral products

The subcellular localization of the five cleavage products was first studied by fusing the individual proteins with the green fluorescent protein

(GFP). Transfection of Cos-7 cells with plasmids expressing the 34-, 24-and 16-kDa proteins showed that the three proteins are localized to intracellular membrane structures (Fig. 3B, D and F). However, transfection of Cos-7 cells with plasmids expressing the 6- and 10-kDa proteins showed diffuse distribution patterns, which are similar to that of the GFP control (Fig. 3A, C and E).

Due to the absence of a specific antiserum, the 34-kDa protein was tagged to a T7 tag, and a highly specific anti-T7 monoclonal antibody (Novagen) was used to stain cells in indirect immunofluorescence assay. Cos-7 cells transfected with pT7tag29k showed perinuclear staining (Fig. 4A and B), confirming that this hydrophobic protein is associated with membrane compartments. Cos-7 cells expressing the 6- and 10-kDa proteins were stained with anti-6 and V47 respectively, showing cytoplasmic staining, suggesting that these two proteins are localized in the cytoplasm when expressed alone (Fig. 4C, D, G and H). Immunofluorescence of Cos-7 cells expressing the 24-kDa protein with anti-24 showed perinuclear staining (Fig. 4E and F). Positive staining was observed on nonpermeabilized Cos-7 cells transfected with pT7tag16k (Fig. 4J), suggesting that the 16-kDa protein may be transported to the cell surface. In permeabilized cells, the 16-kDa protein appeared to be localized in the perinuclear region (vesicles) (Fig. 4L). Immunofluorescence with the respective antisera on non-transfected Cos-7 cells gave negative staining (Fig. 4I and K).

Finally, immunostaining of IBV-infected Vero cells with anti-6, anti-24, V47 and anti-16, respectively, clearly showed fluorescence at the perinuclear region with punctate and vesicular staining (Fig. 5B, D, F and H). Mock-infected cells did not give similar profiles (Fig. 5A, C, E and G).

4. CONCLUSION

The 34-, 6-, 24-, 10- and 16-kDa proteins may interact in virus-infected cells together with the 3C-like proteinase to form the membrane-bound viral replication complexes.

REFERENCES

Bost, A. G., Carnahan, R., Lu, X. T., and Denison, M. R., 2000, Four proteins processed from the replicase gene polyprotein of mouse hepatitis virus colocalize in the cell periphery and adjacent to sites of virion assembly. *J. Virol.* **74**: 3379-3387.

3.

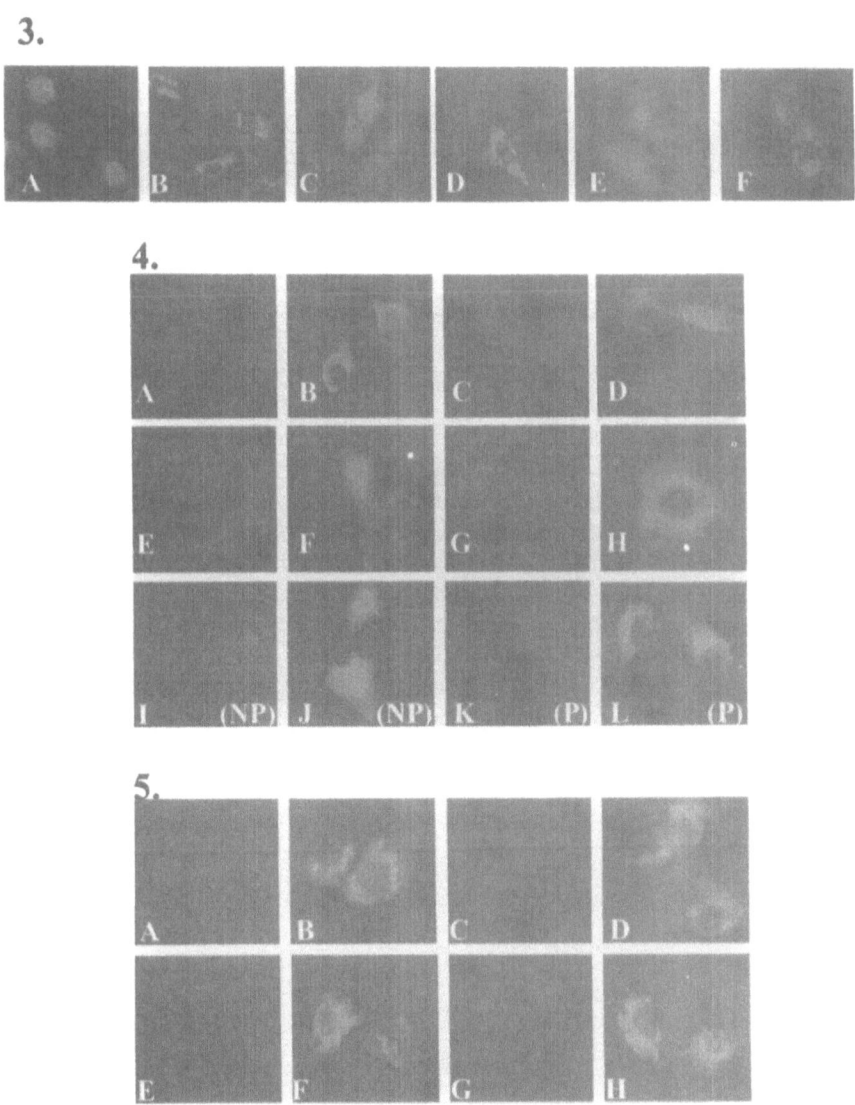

4.

5.

Figure 3. (A - F): Cos-7 cells overexpressing the GFP, 34-, 6-, 24-, 10- and 16-kDa proteins.

Figure 4. (A - L): Indirect immunofluorescence of Cos-7 cells overexpressing the 34-, 9-, 24-, 10- and 16-kDa proteins (NP: Non permeabilized; P: Permeabilized).

Figure 5. (A - H): Indirect immunofluorescence staining of Mock-infected (M) and IBV-infected (I) Vero cells with anti-6, anti-24, V47 and anti-16.

Lim, K. P., Ng, L. F. P., and Liu, D. X., 2000, Identification of a novel cleavage activity of the first papain-like proteinase domain encoded by ORF 1a of the coronavirus avian infectious bronchitis virus and characterization of the cleavage products. *J. Virol.* **74**: 1674-1685.

Liu, D. X., Brierley, I., Tibbles, K. W., and Brown, T. D. K., 1994, A 100-kilodalton polypeptide encoded by open reading frame (ORF) 1b of the coronavirus infectious bronchitis virus is processed by ORF 1a products. *J. Virol.* **68**: 5772-5780.

Liu, D. X., Xu, H. Y., and Brown, T. D. K., 1997, Proteolytic processing of the coronavirus infectious bronchitis virus 1a polyprotein: identification of a 10-kilodalton polypeptide and determination of its cleavage sites. *J. Virol.* **71**: 1814-1820.

Liu, D. X., Shen, S., Xu, H. Y., and Wang, S. F., 1998, Proteolytic mapping of the coronavirus infectious bronchitis virus 1b polyprotein: evidence for the presence of four cleavage sites of the 3C-like proteinase and identification of two novel cleavage products. *Virology* **246**: 288-297.

Ng, L. F. P., and Liu, D. X., 1998, Identification of a 24-kDa polypeptide processed from the coronavirus infectious bronchitis virus 1a polyprotein by the 3C-like proteinase and determination of its cleavage sites. *Virology* **243**: 388-395.

Ng, L. F. P., and Liu, D. X., 2000, Further characterization of the coronavirus infectious bronchitis virus 3C-like proteinase and determination of a new cleavage site. *Virology* **272**: (in press).

Ziebuhr, J., and Siddell, S. G., 1999, Processing of the human coronavirus 229E replicase polyproteins by the virus-encoded 3C-like proteinase: identification of proteolytic products and cleavage sites common to pp1a and pp1ab. *J. Virol.* **73**: 177-185.

CD8 T Cell Mediated Immunity to Neurotropic MHV Infection

[1,2]CORNELIA C. BERGMANN, [3]NORMAN W. MARTEN, [3]DAVID R. HINTON, [2]BEATRIZ PARRA, AND [1,2]STEPHEN A. STOHLMAN

Departments of [1]Neurology, [2]Molecular Microbiology and Immunology, and [3]Pathology University of Southern California, Keck School of Medicine, 1333 San Pablo Street, MCH142, Los Angeles, CA 90033

1.   INTRODUCTION

Neurological disease induced by neurotropic coronaviruses has received considerable attention not only as a model for virus induced demyelination, but also of immune regulation and viral persistence within the central nervous system (CNS). The neurotropic JHM strain (JHMV) of mouse hepatitis virus (MHV) produces an acute CNS infection characterized by encephalomyelitis and demyelination. Several strains and variants have been derived from the lethal, parental JHMV to better study the demyelinating process as a subacute disease resembling the human CNS disease, Multiple Sclerosis. A concise overview of historical aspects, derivation of neurotropic coronavirus strains and variants, together with types of neurological disease and immune responses associated with these viruses is provided in the previous volume of this series (Perlman, 1998) and in several other recent reviews (Houtman and Fleming 1996, Lane and Buchmeier 1997, Stohlman *et al* 1999). Advances in molecular immunology techniques, combined with the use of selective exclusion of immune components, either using antibody (Ab) mediated depletion or genetic disruption, has resulted in significant progress in dissecting the interactions between virus, CNS cells, and immune responses and their role in demyelination. This chapter therefore focuses on advances of the past 3-4 years in elucidating interactions between CNS infection and immune responses leading to persistence. As MHV variants

The Nidoviruses (Coronaviruses and Arteriviruses).
Edited by Ehud Lavi *et al.*, Kluwer Academic/Plenum Publishers, 2001.

differ vastly in tropism, spread, clinical symptoms and mortality, this review focuses largely on a murine model of infection using the neuroattenuated 2.2v-1 variant of JHMV (Fleming *et al* 1986). This mAb selected variant establishes a persistent infection associated with extensive, subacute, ongoing CNS demyelination in the absence of infectious virus. During acute infection this virus replicates primarily in microglia, astrocytes and oligodendrocytes. Although the immune response of immunocompetent hosts generally eliminates infectious virus within 14 days post infection (p.i.), survivors remain persistently infected as evidenced by the detection of viral RNA (vRNA). Despite suffering from varying degrees of encephalitis and paralysis during acute infection, mice generally recover from clinical symptoms and remain asymptomatic during viral CNS persistence.

2.   IMMUNITY DURING ACUTE INFECTION

2.1   CD8 T cell activation and recruitment into the CNS

Many acute infections are controlled by CD8$^+$ T cells, which are primed in the draining lymph nodes and home to the site of infection, where they execute effector functions (Ahmed and Gray 1996, Doherty *et al* 1994). However, the CNS provides a suitable environment for persistence due to the absence of classical lymphatic drainage, the presence of the blood brain barrier, low levels of MHC expression and relative resistance of resident CNS cells to apoptosis, factors all contributing to inefficient immunity (Cserr and Knopf 1997, Lipton and Gildon 1997). Acute JHMV infection of the CNS is nevertheless associated with potent T cell responses, which are critical for the clearance of infectious virus and survival of the host (Stohlman *et al* 1995, Williamson and Stohlman 1990). The strength of the CD8$^+$ T cell response is documented by class I restricted virus specific cytolytic activity of mononuclear cells isolated from the acutely infected CNS (Stohlman *et al* 1993, Castro and Perlman 1996). By contrast, cytolytic activity in peripheral lymphoid organs can only be found following *in vitro* stimulation with antigen (Castro and Perlman 1996), suggesting that most virus specific T cells are either recruited to the CNS or that effector function is predominantly acquired in the CNS due to high levels of viral antigen. The development of class I tetramer reagents has provided a powerful technology to phenotypically monitor virus specific CD8$^+$ T cell expansion and trafficking independent of functional differentiation (Flynn *et al* 1998, Murali-Krishna *et al* 1998). This technology takes advantage of tetramer formation via avidin binding to specifically biotinylated class I chains refolded with peptide and β₂m to detect antigen specific T cell receptors (TCR) on CD8⁺ T cells using flow cytometric analysis (Altman *et al* 1996). Analysis of CNS infiltrating cells during acute JHMV infection using class I tetramers comprising the immunodominant epitopes from the nucleocapsid (N) and spike (S) proteins, revealed that the CD8⁺ T cell compartment comprised up to 50-60% virus specific cells, independent of responder mouse strain (Bergmann *et al* 1999, Marten *et al* 2000a, Pewe *et al* 1999). However, tetramer staining in CD8⁺ splenocytes and lymph nodes was at most 2-5% (Bergmann *et al* 1999). Cytolytic function in the CNS thus correlated with higher frequencies of virus specific T cells within the CNS. This is supported by ~20 fold higher frequencies of CNS mononuclear cells secreting IFN-γ in response to peptide stimulation in ELISPOT assays compared to splenocytes or lymph node cells (Bergmann, unpublished). These results support CD8⁺ T cells as vital contributors in mediating viral clearance (Stohlman *et al* 1995). The inability to provide sterile immunity within the CNS can therefore not be attributed to poor induction, or inefficient recruitment of effector CD8⁺ T cells. Persistence may rather be due to resistance of some infected CNS cell types to CD8⁺ T cell effector functions.

2.2 Mechanisms of virus clearance

As demonstrated by adoptive transfer studies, virus specific CD8⁺ T cells limit replication in astrocytes and microglia, and to a lesser extent in oligodendrocytes; however, they do not inhibit persistent infection (Stohlman *et al* 1995). Infection of perforin deficient (P⁻ᐟ⁻) and IFN-γ deficient (IFN-γ⁻ᐟ⁻) mice with 2.2v-1 revealed that different effector mechanism are utilized to control replication in a cell type specific manner. Whereas CD8⁺ T cells eliminate virus from microglia and astrocytes via a perforin dependent mechanism (Lin *et al* 1997), IFN-γ is critical for controlling replication in oligodendroglia (Parra *et al* 1999). The lack of either component alone resulted in delayed viral clearance; this was more pronounced in IFN-γ⁻ᐟ⁻ mice, although perforin-dependent cytotoxicity was not compromised (Parra *et al* 1999). IFN-γ⁻ᐟ⁻ mice infected with 2.2v-1 also exhibited dramatically increased clinical disease and mortality compared to control mice. Similarly, 2.2v-1 infected P⁻ᐟ⁻ mice recovered poorly from acute disease compared to control mice and developed a progressive paralytic disease despite initial improvement. More extensive demyelination in P⁻ᐟ⁻ mice suggested that demyelination is independent of perforin mediated mechanisms (Lin *et al* 1997). Overall the pathogenesis in P⁻ᐟ⁻ and IFN-γ⁻ᐟ⁻ mice indicated that IFN-γ is more critical for virus clearance and disease recovery than perforin-mediated cytotoxicity. Interestingly, IFN-γ is dispensable for virus clearance from neurons as shown by elimination of the neuronotropic OBLV-60 JHMV variant from the CNS of IFN-$\gamma^{-/-}$ mice (Lane *et al* 1997). Nevertheless, the primary source and mechanism of protective IFN-γ is yet to be identified, as NK, CD4$^+$ and CD8$^+$ T cells are all found in the CNS during acute infection (Williamson *et al* 1991). The relative expression of class I and class II MHC molecules on distinct CNS cell types is also likely to determine differential susceptibility to T cell produced IFN-γ at distinct phases of the infection (Sedgwick and Hickey 1997).

Efforts to define a role of Fas-FasL interactions as a perforin independent lytic mechanism contributing to viral clearance or immunopathology were unsuccessful using Fas-deficient (lpr) mice (Parra *et al* 2000). Infection of lpr mice also indicated that Fas-mediated cytotoxicity did not contribute to virus induced inflammation, mononuclear cell infiltration, extent of demyelination or frequencies of apoptotic cells. However, uncontrolled virus replication in the simultaneous absence of perforin and Fas uncovered the potential use of alternate cytolytic pathways in viral clearance (Parra *et al* 2000). These results, obtained using chimeric mice generated by reconstitution of lethally irradiated wt or lpr mice with splenocytes from P$^{+/+}$ or P$^{-/-}$ mice, suggested the possibility of compensation for perforin via Fas under conditions in which perforin mediated cytotoxicity is diminished, i.e. after the majority of infectious virus is cleared.

In addition to CD8$^+$ T cells, CD4$^+$ T cells exert crucial anti-viral activity. Adoptive transfer and selective depletion of CD4$^+$ T cells indicate that the mechanisms by which CD4$^+$ T cells provide protection are also multifactorial. Although CD4$^+$ T cells were not required for induction of CD8$^+$ T cell responses following infection with lethal JHMV, they enhanced CD8$^+$ T cell expansion and/or activation in splenocytes (Stohlman *et al* 1998). More importantly, although the absence of CD4$^+$ T cells did not inhibit entry of CD8$^+$ T cells into the CNS parenchyma, their accumulation was reduced, coincident with an increase in apoptotic cells. Thus, despite the paucity of CD4$^+$ T cells to enter the CNS parenchyma during acute JHMV infection, they contribute to the maintenance of CD8$^+$ T cell function and viability in the CNS (Stohlman *et al* 1998). However, CD4$^+$ T cells are found within the CNS parenchyma following 2.2v-1 infection (Lin et al 1997). Although there is no evidence for CD4$^+$ T cell mediated cytotoxicity in the CNS, their ability to modify viral persistence via secretion of chemokines or cytokines, especially IFN-γ, is unknown. The notion that soluble mediators may contribute directly to protection is supported by adoptive transfer of CD4$^+$ T cells into infected recipients (Stohlman *et al* 1999). On the other hand CD4$^+$ T cells may be detrimental by accelerating CNS inflammation and enhancing demyelination (Lane *et al* 1997).

3. IMMUNITY DURING PERSISTENT INFECTION

3.1 Retention of T cells during persistence

CD4$^+$ and CD8$^+$ T cell infiltration into the CNS both peak between days 8 and 10 p.i. and gradually decline thereafter. However, even after infectious virus is eliminated, a high percentage of T cells remains in the CNS (Bergmann *et al* 1999, Marten *et al* 2000a,b). This observation, together with T cell retention in a model of neurotropic influenza virus infection (Hawke *et al* 1998) suggested that the CNS environment may retain T cells for prolonged periods even after initial viral clearance. Analysis of CD8$^+$ T cell functional activity revealed that cytolytic function is rapidly downregulated even before infectious JHMV is completely cleared. Furthermore, direct *ex vivo* cytotoxicity remained undetectable in persistently infected mice (Bergmann *et al* 1999, Marten *et al* 2000a). In contrast to impaired antigen responsiveness at the cytolytic level, IFN-γ secretion was maintained, indicating differential regulation of distinct effector functions (Bergmann *et al* 1999).

Interestingly, the relative percentage of virus specific cells within the CD8$^+$ population, as determined by tetramer staining, remains constant throughout infection (Bergmann *et al* 1999, Marten *et al* 2000a, b). This apparent lack of enrichment of virus specific cells indicated a stable steady state among T cell populations residing in the CNS. The specificity of non-tetramer staining CD8$^+$ T cells remains to be resolved. As T cell retention in the CNS is thought to require cognate antigen recognition (Hickey *et al* 1997), T cells specific for as yet unidentified JHMV epitopes, possibly located within the nonstructural proteins, cannot be ruled out. The potential existence of autoantigen-specific T cells is less likely as persistently infected mice show no symptoms of autoimmune disease. However, infections are generally carried out in mouse strains relatively resistant to CNS autoimmune disease. Host genetic factors may therefore contribute to the suppression of T cell reactivity or overt clinical symptoms associated with autoimmune disease.

T cell retention despite barely detectable vRNA levels, in addition to persistence of T cells in the influenza model, suggested that residual antigen is redundant for T cell maintenance in the CNS. This is indirectly supported by the abundant presence of D^b-S510 tetramer binding CD8$^+$ T cells in the CNS of infected mice, in which the wild type (wt) epitope encoding vRNA sequence is no longer detectable due to the exclusive emergence of CTL escape mutants (Pewe *et al* 1999). T cells may thus be trapped in the CNS or recruited by an antigen independent mechanism. The requirement of viral persistence for retention of T cells was addressed by taking advantage of a nonpathogenic 2.2v-1 variant designated 2.2/7.2v-2 (Fleming *et al* 1986). Despite a deletion in the hypervariable region comprising the immunodominant S510 epitope, infection was associated with similar replication patterns and peak virus titers compared to parental 2.2v-1 in both C57BL/6 (H-2^b) and BALB/c (H-2^d) mice (Fleming *et al* 1986, Marten *et al* 2000a). In contrast to subacute paralysis and extensive demyelination typical of 2.2v-1 infected mice, 2.2/7.2v-2 infection was clinically silent and caused little to no demyelination. Nevertheless, phenotypic and functional analysis of CNS infiltrating cells demonstrated that this infection induced potent CD8$^+$ T cell responses similar to those induced by the pathogenic variant (Marten *et al* 2000a, b). Furthermore, although spread of 2.2/7.2v-2 to the spinal cord was delayed, transient replication was associated with an influx of T cells similar in magnitude to 2.2v-1 infected mice. At 8-9 weeks p.i. vRNA was no longer detectable in 2.2/7.2v-2 infected mice, indicating complete resolution of infection, whereas vRNA was still evident in spinal cords from 2.2v-1 infected mice (Marten *et al* 2000b). Despite transient infiltration during the acute phase, the lack of vRNA coincided with the absence of both CD8$^+$ and CD4$^+$ T cells. T cell retention is thus tightly linked to the presence of vRNA and presumably antigen expression, indicating that viral persistence is a driving force for T cell maintenance within the CNS. Whether chronic T cell activation is a result of direct MHC-TCR dependent or cytokine/chemokine mediated bystander mechanisms remains to be elucidated. Potential candidates mediating ongoing recruitment may be the chemokines CRG-2 and RANTES (Lane *et al* 1998).

The functional role of CD8$^+$ T cells during persistence is unclear; however, several lines of evidence support dynamic T cell populations associated with persistence: 1) Unlike memory CD8$^+$ T cells characterized by the CD44^{hi}, CD62L^{lo}, CD69$^-$ activation/memory phenotype, the majority of CD8$^+$ T cells in the persistently infected CNS express the very early activation marker CD69, independent of specificity (Bergmann *et al* 1999). This marker is typically only transiently upregulated early during T cell activation; however, CD69 expression on CNS derived CD8$^+$ T cells peaks coincident with clearance of infectious virus and remains upregulated, indicating ongoing stimulation. 2) In H-2^{bxd} mice responding to both the immunodominant N and S epitopes, ongoing chronic activation was evident by a switch in immunodominance from N to S specific CD8$^+$ T cells during the course of infection (Bergmann *et al* 1999); 3) CD8$^+$ T cells isolated from persistently infected mice exhibited highly focused reactivity to the wt N epitope sequence, compared to broader specificity characteristic of the acute CD8$^+$ populations (Marten *et al* 1999). While a broad polyclonal TcR specificity can thus potentially accommodate mutations arising during acute replication, persistence is associated with recruitment or survival of T cell subsets with exquisite specificity for the wt epitope. This may serve to limit the potential for autoantigen crossreactive CD8[+] T cells.

The selective depletion of immune functions combined with highly sensitive methods to monitor individual T cell populations have thus been valuable tools to elucidate the mechanisms by which both CD8[+] and CD4[+] T cells provide protection during acute JHMV infection. However, their contribution in preventing viral recrudescence during persistence is an open question. Recent results from the analysis of JHMV infection in mice homozygous for disruption of the Ig μ gene (IgM[-/-]) and thus genetically deficient in B cells and Ab, clearly demonstrated increased mortality associated with recrudescence of infectious virus (Lin *et al* 1999). Although replication was initially controlled by cell mediated immunity in the absence of Ab, T cells alone did not suffice to effectively clear infectious virus to below detection resulting in subsequent uncontrolled replication. Passive transfer of anti-JHMV Ab following initial clearance prevented reactivation of infectious virus within the CNS of IgM[-/-] mice. However, as these mice lack B cells and Ab, it is unclear if B cells themselves play an additional Ab independent role in suppressing virus, e.g. by supporting T cell activation peripherally or maintaining them in the CNS. Defects in T cell function are evident during LCMV infection of IgM[-/-] mice (Homann *et al* 1998). Nevertheless, these data suggest that humoral immunity plays no role in controlling virus during acute infection but is crucial in establishing and maintaining CNS viral persistence.

4. CONCLUSIONS

Virus infections of the CNS frequently result in persistence associated with subclinical disease and/or disease after a long latent period. To guarantee survival of both the host and the virus, an equilibrium must be established between virus replication, spread and the immune system to minimize virus- or immune-induced pathology. Control of JHMV infection in the CNS involves a well orchestrated dynamic immune response, in which distinct players take on critical roles at various time points. Whereas cellular effector mechanisms control acute infection in a cell type specific manner, the humoral immune response plays a critical role in preventing viral recrudescence. Although distinct molecular effector mechanisms have been defined, a number of important issues relative to the regulation of immunopathology and mechanisms of persistence within the CNS require further investigation: Contribution of persisting T cell subsets to the demyelinating process, factors regulating T cell recruitment and effector function, direct antiviral role of CD4$^+$ T cells, and finally mechanisms via which antibody or B cells themselves suppress virus replication.

ACKNOWLEDGMENTS

This work was supported by National Institutes of Health Grants NS18146 and AI33314.

REFERENCES

Ahmed, R., and Gray, D., 1996, Immunological memory and protective immunity: understanding their relation. *Science* 272: 54-60.

Altman, J. D., Moss, P.A. H., Goulder, P. J. R., Barouch, D. H., McHeyzer-Williams, M. G., Bell, J. I., McMichael, A. J., and Davis, M. M., 1996, Phenotypic analysis of antigen-specific T lymphocytes. *Science* 274: 94-96.

Bergmann, C.C., Altman, J.D., Hinton, D.R., and Stohlman, S.A., 1999, Inverted immunodominance and impaired cytolytic function of CD8$^+$ T cells during viral persistence in the CNS. *J. Immunol.* 63: 3379-3387.

Castro, R.F., and Perlman, S., 1996, Differential antigen recognition by T cells from the spleen and central nervous system of coronavirus-infected mice. *Virology* 222: 247-251.

Cserr, H.F. and Knopf, P.M., 1997, Cervical lymphatics, the blood brain barrier, and immunoreactivity of the brain. In *Immunology of the Central Nervous System* (R.W. Keane and W.F. Hickey, eds.) Oxford University Press, Oxford pp. 134-152.

Doherty, P.C., Hou, S, and Tripp, R. A., 1994, CD8+ T-cell memory to viruses. *Cur. Opin. Immunol.* 6: 545-552.

Fleming, J. O., Trousdale, M., Zactarim, F.E., Stohlman, S.A., and Weiner, L.P., 1986, Pathogenicity of antigenic variants of murine coronavirus JHM selected with mAb. *J. Virol.* 58: 869-875.

Flynn, K..J., Belz, G.T., Altman J.D., Ahmed, R., Woodland, D.L., and Doherty P.C., 1998, Virus-specific CD8$^+$ T cells in primary and secondary influenza pneumonia. *Immunity* 8: 683-691.

Hawke, S., Stevenson, P.G., Freeman, S. and Bangham C.R.M., 1998, Long-term persistence of activated cytotoxic T lymphocytes after viral infection of the central nervous system. *J. Exp. Med.* 187: 1575-1582.

Hickey, W.F., Lassman, H., and Cross, A.H., 1997, Lymphocyte entry and the initiation of inflammation in the central nervous system. In *Immunology of the Central Nervous System* (R.W. Keane and W.F. Hickey, eds.) Oxford University Press, Oxford pp. 200-225.

Homann, D., Tishon, A., Berger, D.P. , Weigle, W.O., von Herrath, M.G., and Oldstone, M.B.A., 1998, Evidence of an underlying CD4 helper and CD8 T-cell defect in B-cell-deficient mice: Failure to clear persistent virus infection after adoptive immunotherapy with virus-specific memory cells from µMT/µMT mice. *J. Virol.* 72: 9208-9216.

Houtman, J.J., and Fleming, J.O., 1996, Pathogenesis of mouse hepatitis virus-induced demyelination. *J. Neurovirol.* 2: 361-376.

Lane, T.E., Asensio, V.C., Yu, N., Paoletti, A.D., Campbell, I.L., and Buchmeier, M.J., 1998, Dynamic regulation of alpha- and beta-chemokine expression in the central nervous system during mouse hepatitis virus-induced demyelinating disease. *J. Immunol.* **160**: 970-978.

Lane, T.E., Liu, M.T., Chen, B.P., Asensio, V.C., Samawi, R.M., Paoletti, A.D., Campbell, I.L., Kunkel, S.L., Fox, H.S., and Buchmeier, M.J., 2000, A central role for CD4 T cells and RANTES in virus-induced central nervous system inflammation and demyelination. *J. Virol.* **74**: 1415-1424.

Lane, T., and Buchmeier, M. J., 1997, Murine coronavirus infection: a paradigm for virus-induced demyelinating disease. *Trends in Microbiology* 1:9-14.

Lane, T. E., Paoletti, A.D., and Buchmeier M.J., 1997, Disassociation between the in vitro and in vivo effects of nitric oxide on a neurotropic murine coronavirus. *J. Virol.* **71**: 2202-2210.

Lin, M.T., Stohlman, S.A., and Hinton, D.R., 1997, Mouse hepatitis virus is cleared from the central nervous systems of mice lacking perforin-mediated cytolysis. *J. Virol.* **71**: 383-391.

Lin, M.T., Hinton, D.R., Marten, N.W., Bergmann, C.C., Stohlman, S.A., 1999, Antibody prevents virus reactivation within the central nervous system. J. Immunol. **162**: 7358-7368.

Lipton, H.L., and Gilden, D.H., 1997, Viral Diseases of the Nervous System: Persistent Infections. In *Viral Pathogenesis* (N. Nathanson et al., eds) Lippencott-Raven Publishers, Philadelphia, pp. 855-869.

Marten, N.W., Stohlman, S.A., Atkinson, R.D., Hinton, D.R., Fleming, J.O., and Bergmann, C.C., 2000a, Contributions of CD8$^+$ T cells and viral spread to demyelinationg disease. *J. Immunol.* **164**: 4080-4088

Marten, N.W., Stohlman, S.A., and Bergmann, C.C., 2000b, The role of viral persistence in retaining CD8$^+$ T cells within the central nervous system. *J. Virol.* In press.

Marten, N.W., Stohlman, S.A., Smith-Begolka, W., Miller, S.D., Dimacali, E., Yao, Q., Stohl, S., Goverman, J., and Bergmann C.C., 1999, Selection of CD8$^+$ T cells with a highly focused specificity during viral persistence in the central nervous system. *J. Immunol.* **162**: 3905-3914.

Murali-Krishna, K., Altman, J.D., Suresh, M., Sourdive, D.J.D., Zajac, A.J., Miller, J.S., Slansky, J., and Ahmed, R., 1998, Counting antigen-specific CD8 T cells: A reevaluation of bystander activation during viral infection. *Immunity* **8**: 177-187.

Parra, B., Hinton,D., Marten, N.W., Bergmann, C.C., Lin, M.T., Yang, C. S., Stohlman, .A., 1999 Gamma Interferon is required for viral clearance from central nervous system oligodendroglia *J. Immunol.* **162**: 1641-1647.

Parra, B., Lin, M.T., Stohlman, S.A., and Bergmann, C.C., Atkinson, R., and Hinton, D.A., 2000, Fas-Fas Ligand Interactions do not contribute to the pathogenesis of mouse hepatitis virus in the presence of perforin mediated cytolysis. *J.Virol.* **74**: 2447-2450.

Perlman, S. 1998. Pathogenesis of coronavirus-induced infections. *Adv. Exp. Med. Biol.* **440**: 503-513.

Pewe, L., Heard, S.B., Bergmann, C.C., Dailey, M.O., and Perlman, S., 1999, Selection of T cells escape mutants in mice infected with a neurotropic coronavirus: Quantitative estimate of TCR Diversity in the infected CNS. *J. Immunol.* **163**: 6106-6113.

Sedgwick, J.D., and Hickey, W.F., 1997, Antigen presentation in the central nervous system. In *Immunology of the Central Nervous System* (R.W. Keane and W.F. Hickey, eds.) Oxford University Press, Oxford, pp. 364-418.

Stohlman, S.A., Bergmann, C.C., Lin, M.T., Cua, D.J., and Hinton, D., 1998, CTL effector function within the central nervous system requires CD4$^+$ T cells. *J. Immunol.* **160**: 2896-2904.

Stohlman, S.A., Bergmann, C.C., and Perlman, S., 1999, Persistent mouse hepatitis viral infections. In *Persistent Viral Infections*. (R. Ahmed and I. Chen, eds.) John Wiley and Sons, pp. 537-557.

Stohlman, S.A., Bergmann, C.C., van der Veen, R., and Hinton, D., 1995, Mouse hepatitis virus-specific cytotoxic T lymphocytes protect from lethal infection without eliminating virus from the central nervous system. *J. Virol.* **69**: 684-694.

Stohlman, S.A., Kyuwa, S., Polo, J.M., Brady, D., Lai, M..M.C., and Bergmann, C.C., 1993, Characterization of mouse hepatitis virus-specific cytotoxic T cells derived from the central nervous system of mice infected with the JHM strain. *J. Virol.* **67**: 7050-7059.

Williamson, J.S.P., and Stohlman, S.A., 1990, Effective clearance of mouse hepatitis virus from the central nervous system requires both CD4[+] and CD8[+] T cells. *J. Virol.* **64**: 4589-4592

Williamson, J.S., Sykes, K.C., and Stohlman, S.A., 1991, Characterization of brain-infiltrating mononuclear cells during infection with mouse hepatitis virus strain JHM. *J. Neuroimmunol.* **32**: 199-207.

Coronavirus Derived Expression Systems
Progress and problems

[1]LUIS ENJUANES, ISABEL SOLA, FERNANDO ALMAZAN, ANDER IZETA, JOSE M. GONZALEZ, AND SARA ALONSO
[1]Centro Nacional de Biotecnología, CSIC, Department of Molecular and Cell Biology, Campus Universidad Autónoma, Cantoblanco, 28049 Madrid, Spain

1. INTRODUCTION

Coronaviruses have several advantages to be used as vectors over other viral expression systems: (i) coronaviruses are single-stranded RNA viruses that replicate within the cytoplasm without a DNA intermediary, making unlikely the integration of the virus genome into the host cell chromosome; (ii) these viruses have the largest RNA genome known having in principle room for the insertion of large foreign genes; (iii) since coronaviruses in general infect the mucosal surfaces, both respiratory and enteric, they may be used to induce a strong secretory immune response; (iv) the tropism of coronaviruses may be modified by the manipulation of the spike (S) protein allowing the engineering of the tropism of the vector; and, (v) non-pathogenic coronavirus strains infecting most species of interest are available to develop expression systems.

Two types of expression vectors have been developed based on coronavirus genomes (Fig. 1), one requires two components (helper dependent) and the other, a single genome that is modified either by targeted recombination or by engineering a cDNA encoding an infectious RNA.

The Nidoviruses (Coronaviruses and Arteriviruses).
Edited by Ehud Lavi *et al.*, Kluwer Academic/Plenum Publishers, 2001.

This review will focus on the advantages and limitations of these novel coronavirus expression systems, and the attempts to increase their expression levels by studying the influence of the transcription regulatory sequences (TRSs).

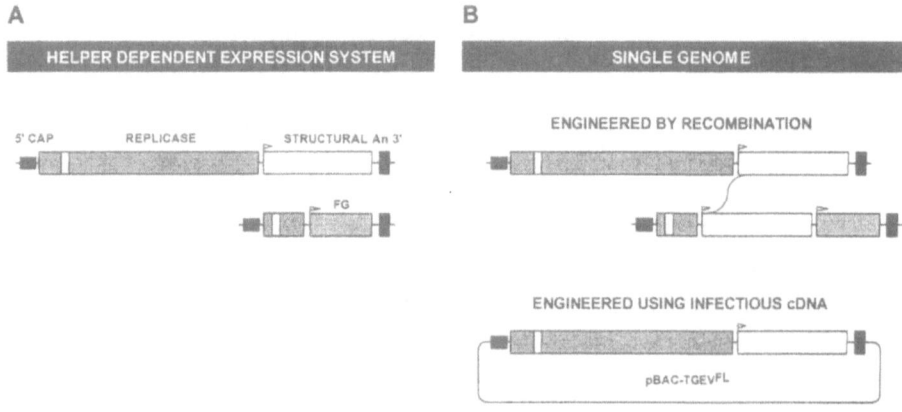

Figure 1. Coronavirus derived expression systems. A. Helper dependent expression system based in two components, the helper virus and a minigenome carrying the foreign gene (FG). An, poly A. B. Single genome engineered either by targeted recombination of by using an infectious coronavirus cDNA clone (pBAC-TGEV^FL) derived from TGEV genome.

2.   HELPER DEPENDENT EXPRESSION SYSTEMS

The helper dependent expression systems have been developed using members of the three groups of coronaviruses (Fig. 2). Coronavirus derived minigenomes have a theoretical cloning capacity close to 25 kb, since minigenome RNAs of about 3 kb are efficiently amplified and packaged by the helper virus and the virus genome has about 30 kb. This is in principle the largest cloning capacity for a vector based on RNA virus genomes.

Most of the initial work required for the development of helper dependent expression systems based on coronaviruses has been done with MHV defective RNAs (Fig. 2) (Liao, Zhang, and Lai, 1995; Lin and Lai, 1993; Zhang *et al.*, 1997). Three heterologous genes have been expressed using MHV system, chloramphenicol acetyltransferase (CAT), hemagglutinin-esterase (HE), and interferon γ (Fig. 2). After intracerebral inoculation of the virus vectors expressing CAT and HE into mice, HE- or CAT-specific subgenomic mRNAs were detected in the brain at days 1 and 2 p.i. but not later, indicating that the genes in the defective genome (DI) vector were expressed only in the early stage of viral infection (Zhang *et al.*, 1998). CAT expression has also been shown using an internal ribosomal entry site sequence (IRES) of encephalomyocarditis virus (Lin and Lai, 1993) (Fig. 2). The murine IFN-γ gene was secreted into culture medium as early as 6 hr post-transfection and reached a peak level at 12 hr post-transfection.

Infection of susceptible mice with DI RNA producing IFN-γ caused significantly milder disease, accompanied by less virus replication than that caused by virus containing a control DI vector (Lai *et al.*, 1997; Zhang *et al.*, 1997).

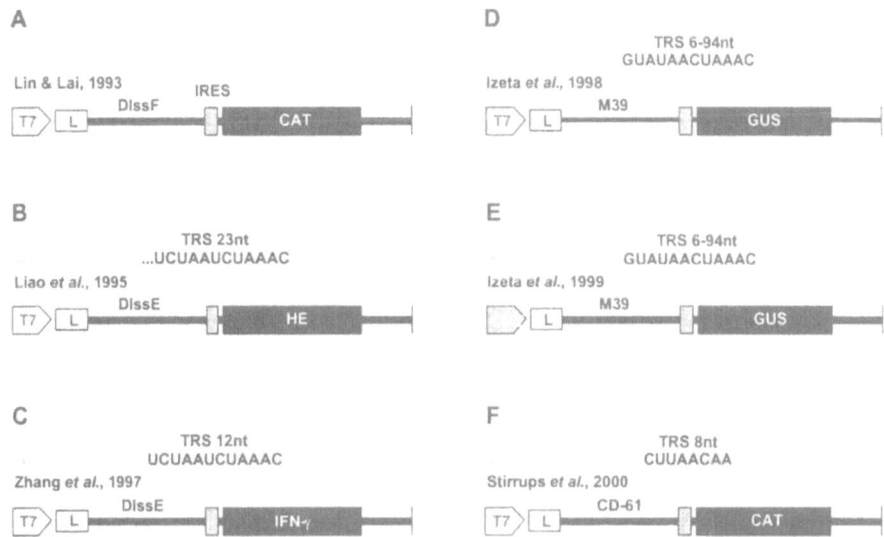

Figure 2. Summary of helper-dependent expression systems based on coronavirus derived minigenomes. A, B, and C. expression modules based MHV on minigenomes DIssF and DIssE cloned after the T7 bacteriophage polymerase (T7), used to express chloramphenicol acetyltransferase (CAT), hemagglutinin-esterase (HE) or interferon-γ using either an IRES (A), or transcription regulatory sequences (B and C). D, E. expression modules based on the TGEV derived minigenome M39 used to express the β-glucuronidase (GUS); the minigenome is cloned after the CMV promoter (Izeta *et al.*, 1998). F. Expression module based on the IBV derived minigenome CD-61 used to express CAT.

Group 1 coronaviruses such as transmissible gastroenteritis virus (TGEV) has also been used to express foreign proteins (Fig. 2). The vector included a two step amplification (Dubensky *et al.*, 1996), by cloning a cDNA copy of M39 minigenome after the CMV promoter. Minigenome RNAs are first amplified in the nucleus by the cellular RNA *pol II*, then the RNAs are translocated into the cytoplasm where they are amplified by the viral replicase of the helper virus. The smallest TGEV derived minigenome (M33) that was replicated by the helper virus and efficiently packaged was 3.3 kb (Izeta *et al.*, 1999; Méndez *et al.*, 1996). In addition to GUS, the ORF5 of the porcine respiratory and reproductive syndrome virus (PRRSV) has been expressed (Alonso *et al.*, 2000b).

The HCoV-229E has also been used to express new subgenomic mRNAs, although it has not been applied to the expression of a foreign protein (Thiel, Siddell, and Herold, 1998). In addition, a defective RNA (CD-61) derived from the Beaudette strain of the IBV virus (Penzes *et al.*,

1994; Penzes *et al.*, 1996) was used as an RNA vector for the expression of CAT (Fig. 2) (Stirrups *et al.*, 2000).

A helper dependent expression system has recently been described based on arteriviruses, closely related to coronaviruses (Molenkamp *et al.*, 2000). Using equine arteritis virus (EAV) minigenomes of 3.8 kb the CAT reporter gene has been produced. The smallest defective RNA replicated by the helper virus had a 3.0-kb length, but this RNA was not packaged.

The expression levels have not been quantified in terms of protein mass for MHV expression systems. Expression levels of CAT between 1-2 µg per 10^6 cells have been described using IBV minigenomes. The highest amount of protein (1 to 8 µg of GUS per 10^6 cells) have been obtained using a two step amplification system based on TGEV derived minigenomes with optimized TRSs (Alonso *et al.*, 2000a; Izeta *et al.*, 1999). These protein levels are similar to those described for vectors based on other positive strand RNA viruses such as poliovirus and the Venezuelan encephalitis virus (VEEV) (4 µg per 10^6), but still lower than the expression levels described for Sindbis virus: 50 µg per 10^6 (Agapov *et al.*, 1998) and SFV: 80-300 µg per 10^6 cells (DiCiommo and Bremner, 1998; Liljeström and Garoff, 1991).

3.   SINGLE GENOME CORONAVIRUS VECTORS

3.1.   Vectors constructed by targeted recombination

Reverse genetics has been possible by targeted recombination between a helper virus and either non-replicative or replicative coronavirus derived RNAs (Fig. 3A) (Masters, 1999). Targeted recombination has been mediated by one or two cross-overs. Changes were introduced within the S gene that modified MHV pathogenicity (Leparc-Goffart *et al.*, 1998). The gene encoding green fluorescent protein (GFP) was inserted into MHV between genes S and E by targeted recombination, resulting in the creation of the largest known RNA viral genome (Fischer *et al.*, 1997). Mutations have also been created by targeted mutagenesis within the E and the M genes showing the crucial role of these genes in assembly (de Haan *et al.*, 1998; Fisher and Goff, 1998).

Targeted recombination mediated by two cross-overs allowed the replacement of the S gene of a respiratory strain of TGEV by the S gene of enteric TGEV leading to the isolation of viruses with a modified tropism and virulence (Sánchez *et al.*, 1999). A new strategy for the selection of TGEV recombinants was based in the elimination of the parental replicative virus by its simultaneous neutralization with two mAbs (Fig. 3B) (Sola *et al.*, 2000).

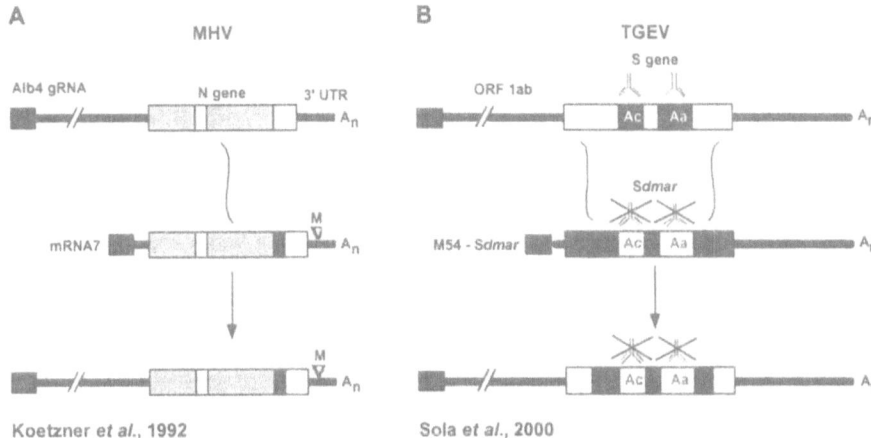

Figure 3. Single genome expression based on the engineering of coronavirus minigenomes by targeted recombination. A. Basic scheme of targeted recombination in MHV. The black boxe indicates the approximate location of the N gene region (87nt) that is deleted in the Alb4 mutant. M, insertion of 5 nt used as a genetic marker (M). B. Targeted recombination within the S gene of TGEV and a minigenome carrying the information for an S gene with three nucleotide mutations (*Sdmar*) that allow the escaping from the neutralization by two mAbs specific for antigenic sub-sites Ac and Aa of S protein.

The frequencies of the targeted recombination event in MHV and TGEV recombination were found higher than the standard prediction for the recombination frequency of a multiple cross-over. This frequency was expected to be the product of the frequencies of the individual recombination events. This suggests that the alignment of two templates is the rate-limiting event in recombination and, once this has been achieved, the barrier to multiple crossovers may be only marginally higher than that for single crossovers (Masters, 1999; Sola *et al.*, 2000).

3.2. Coronavirus vectors derived from an infectious cDNA clone

The construction of a full-length genomic cDNA clone could considerably improve the genetic manipulation of coronaviruses. Now, for the first time, the construction of an infectious TGEV cDNA clone has been possible (Almazan *et al.*, 2000). To obtain an infectious cDNA three strategies have been combined (i) the construction of the full-length cDNA was started from a DI that was stably and efficiently replicated by the helper virus (Izeta *et al.*, 1999; Méndez *et al.*, 1996). Using this DI, the full-length genome was completed and the performance of the enlarged genome was checked after each step. This approach allowed for the identification of a cDNA fragment that was toxic to the bacterial host. This finding was used

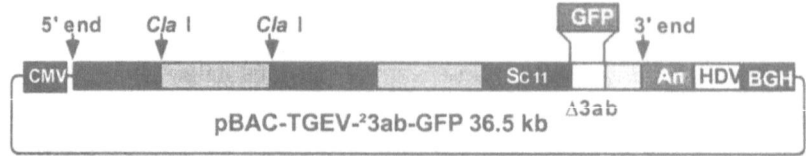

Figure 4. Expression of GFP using an infectious TGEV cDNA clone. Genes 3a and 3b were deleted in the TGEV infectious cDNA, cloned in a bacterial artificial chromosome (BAC), leading to a replication competent cDNA (pBAC-TGEV-Δ3ab-GFP). GFP (0.72 kb) was inserted within the position of the deleted genes after the TRS of gene 3a. High (>40 μg/10^6 cells) GFP expression levels were obtained. CMV, immediate-early cytomegalovirus promoter. GFP, green fluorescent protein. Sc11, S gene of PUR-C11 TGEV strain. An, poly A. HDV, hepatitis delta-virus ribozyme. BGH, bovine growth hormone termination and polyadenylation signals.

to advantage by reintroducing the toxic fragment into the viral cDNA in the last cloning step; (ii) in order to express the long coronavirus genome, and to add the 5' cap, a two-step amplification system that couples transcription in the nucleus from the CMV promoter, with a second amplification in the cytoplasm, driven by the viral replicase, was used; and, (iii) to increase viral cDNA stability within bacteria, the cDNA was cloned as a bacterial artificial chromosome (BAC), that produces only one or two plasmid copies per cell. The full-length cDNA was divided into two plasmids because their fusion into one reduced the stability of the cDNA. One plasmid contained all virus sequences except for a fragment Cla I to Cla I of about 5 kb that was included within a second BAC. A fully functional infectious cDNA clone, leading to a virulent virus able to infect both the enteric and respiratory tracts, was engineered by inserting the Cla I fragment into the rest of the TGEV cDNA sequence (Fig. 4). Using the TGEV cDNA, the GFP gene was cloned by replacing the non-essential 3a and 3b genes, leading to an engineered genome with high stability (Fig. 4) (Sola *et al.*, 2000).

The theoretical cloning capacity for an expression system based on a single coronavirus genome like TGEV may be around 3 kb taking into account that: (i) the non-essential 3a and 3b genes (~1.0 kb) have been deleted; (ii) the standard S gene can been replaced by that of PRCV mutants with a deletion of 0.67 kb; and (iii) both DNA and RNA viruses may accept genomes with sizes up to 105 % of the wild type genome (Afanasiev *et al.*, 1999; Bett, Prevec, and Graham, 1993; Parks and Graham, 1997). The present cloning capacity of the coronavirus vectors is within the range expected, since other RNA virus vectors, such as those derived from the Sindbis virus and VEEV, with a genome of around 12 kb, accept stable inserts of about 1 kb in size (Bredenbeek and Rice, 1992; Caley *et al.*, 1997).

4. REGULATION OF TRANSCRIPTION

4.1. Introduction

Coronavirus RNA synthesis occurs in the cytoplasm via a negative-strand RNA intermediate. Both genome-size and subgenomic negative-strand RNAs, which correspond in number of species and size to those of the virus-specific mRNAs have been detected. The two transcription models compatible with most of the experimental data are leader-primed transcription and discontinuous transcription during negative-strand RNA synthesis (Lai, 1998). Recently, more experimental evidence is being generated supporting the second model (Baric and Yount, 2000; Sawicki and Sawicki, 1990; Sethna, Hung, and Brian, 1989; van Marle *et al.*, 1999).

Viral RNA replication and transcription may involve cellular proteins taken from the translation machinery of host cells (Lai, 1998). Two cellular hnRNPs, polypyrimidine tract-binding protein (PTB) and hnRNP A1, bind to the transcription regulatory sequences (TRSs) of MHV RNA and may participate in its transcription (Li *et al.*, 1999; Li *et al.*, 1997).

Many factors including RNA primary and secondary structure, RNA-protein and protein-protein interactions could influence mRNA abundance. One of these factors, the nature of the TRSs and the extent of their complementarity with the 3' end sequence of the leader, may be the most relevant, and is discussed below.

The TRSs include the core sequence (CS), previously named intergenic sequence (IG), that is a short conserved sequence element upstream of the transcription units. Because the leader-mRNA junction occurs within the CS, this motif or its minus-sense counterpart (cCS) are considered to be crucial for mRNA synthesis. The nature of the cCS probably influences transcription throughout its potential basepairing with the leader 3' end. According to this model the cCS should act as a classical promoter where transcription is initiated. Alternatively this sequence may slow down or even detach the transcriptase complex, according to the discontinuous transcription during negative-strand RNA synthesis model.

Most of the information on coronavirus transcription has been generated using helper dependent expression systems based on minigenomes encoding new subgenomic mRNAs. The CSs of coronaviruses belonging to groups I (hexameric 5'-CUAAAC-3') and II (heptameric 5'-UCUAAAC-3') share homology, whereas the CS of coronaviruses belonging to group III, like that of IBV have the most divergent sequence (5'-UAACAA-3').

4.2. Extent of basepairing and mRNA levels

The potential basepairing between the 3' end of the leader and the cCS differs slightly among the different coronavirus genes. For MHV, the extent of the basepairing ranges from 9 to 18 basepairs and these CSs were sufficient to direct subgenomic DI RNA synthesis (Joo and Makino, 1992; van der Most, De Groot, and Spaan, 1994; van der Most and Spaan, 1995).

In MHV cCS strength is affected only slightly when a single nucleotide is mutated (Joo and Makino, 1992; van der Most, De Groot, andSpaan, 1994). Exceptionally, substitutions in some positions result in a more than ten-fold reduction of transcription. These data suggest that transcription initiation requires a duplex of a minimal stability. Extending this basepairing does not increase cCS strength.

Using TGEV derived RNA minigenomes, we have shown that the CS sequence 5'-CUAAAC-3' is required and sufficient for high expression levels (Alonso *et al.*, 2000a). Similarly, in IBV, expression of the reporter gene was under one canonical octameric IBV CS sequence 5'-CUUAACAA-3' (Stirrups *et al.*, 2000). In Arteriviruses, it has also been shown that subgenomic mRNA (sgmRNA) synthesis requires base-pairing between the leader 3'-end and the cCS. EAV CS consists of pentanucleotide 5'-UCAAC-3' (van Marle *et al.*, 1999). Thus, expression both in coronavirus and arteriviruses can be driven by a TRS with less than 18 nt in size.

The sequences flanking the consensus core sequence 3'-UCUAAAC-5' affected the efficiency of subgenomic DI RNA transcription (Joo and Makino, 1992; Makino and Joo, 1993; Makino, Joo, and Makino, 1991; van der Most, De Groot, andSpaan, 1994). The insertion of a 12 nt sequence including the 5'-UCUAAAC-3' CS at different locations of the DI RNA resulted in different efficiencies of subgenomic DI synthesis as a consequence of the flanking sequences in each position, and not due to the location of the 12 nt sequence on the DI genome.

In TGEV, the absence of the core CS (5'-CUAAAC-3') or the deletion of the U within ORF 3b CS led to the complete abrogation of mRNA transcription (Alonso *et al.*, 2000a). The insertion of the hexameric 5'-CUAAAC-3' CS restored expression levels more than 400-fold above background. The addition of 5' upstream sequences flanking the core CS from the TGEV N gene, led to an increase in transcription of up to 10-fold, indicating the benefit of TRSs of larger (88 nt) size.

The sequences 3' downstream to the core CS of seven viral genes (S, 3a, 3b, E, M, N, and 7) have sizes ranging from 3 to 37 nt. Expression modules in which the 5' flanking sequence was kept constant, and the 3' CS flanking sequences were provided by each of the seven 3' flanking sequences of the viral genes, led to similar expression levels, with the exception of that from ORF 3a. This construct gave expression levels 5- to 10-fold lower. Thus, there was no correlation between the length of the 3' flanking sequences and the expression level (Alonso *et al.*, 2000a), but careful selection of the 3' flanking sequences is recommended to optimize mRNA levels.

More data is required to clarify the role of baseparing between the leader 3' end and the CS, and also the relevance of the primary or secondary sequence of the TRSs.

4.3. CS copy number effect on transcription

Insertion of two to three CS copies within a defective RNA using MHV, BCoV and IBV resulted in the decreased transcription of the larger mRNA (Joo and Makino, 1995; van Marle *et al.*, 1995). In all cases a negative effect on the transcription of upstream CSs by the downstream ones was observed. When several CSs are inserted in tandem, transcription preferentially occurred at the 3'-most TRS (Krishnan, Chang, and Brian, 1996; Stirrups *et al.*, 2000).

4.4. Influence of the insertion site

Using a TGEV derived helper dependent expression system, the reporter gene GUS was inserted at different nucleotide distances from the 5'. The expression levels increased from the 5' to 3' end by one thousand-fold (Alonso *et al.*, 2000a).

In a systematic study using MHV, a 0.4 kb region including a TRS of 12 nt flanked by 0.2 kb from upstream and downstream was inserted throughout the sequence at seven different positions within a 2.2 kb minigenome (Jeong *et al.*, 1996). The position of the insert along the minigenome did not influence the mRNA expression level. In the experiments performed with TGEV, the insertion site close to the 5' end probably have affected essential primary or secondary structures required for minigenome replication, thus reducing the significance of this result in relationship to the insertion site. Furthermore, no difference in expression levels was observed with the MHV system, in which the flanking sequences were kept constant for all insertion sites, suggesting that the location of the insertion site *per se* does not necessarily affect transcription levels and that the differences observed with TGEV were mostly due to the CS-flanking sequences.

4.5. Expression system stability and insert size

Expression from MHV defective RNAs of CAT, HE and murine IFN-γ genes using was not observed beyond passages 2, 3 and 4, respectively. Using minigenomes derived from TGEV and IBV expression was more stable but highly dependent on the nature of the heterologous gene that was expressed. GUS or CAT expression with TGEV or IBV derived minigenomes, respectively, was observed for about ten passages (Alonso *et al.*, 2000a; Stirrups *et al.*, 2000). In general, the insertion of a heterologous gene such as GUS into TGEV derived minigenomes led to a 50-fold reduction in the levels of minigenome RNA replication (Alonso *et al.*, 2000a). The limited stability of the helper dependent expression systems is most likely due to the foreign gene, since TGEV minigenomes in the absence of the heterologous gene are efficiently rescued for at least 30 passages (Izeta *et al.*, 1999; Méndez *et al.*, 1996). The recombination frequency in MHV, TGEV, and IBV may be inversely proportional to the stability of the recombinants expressing a foreign gene. The stability of the MHV expression system is the lowest, probably because of the higher recombination frequency within MHV (Lai, 1996).

The stability of the expression system is conditioned by the type of polymerases involved in the amplification of the minigenome and in the transcription of the mRNA (Agapov *et al.*, 1998). For *in vitro* transcribed minigenome RNAs, the accumulation of mutations with T7 DNA-dependent RNA-polymerase is 10^{-4} to 10^{-5} (Boyer, Bebenek, and Kunkel, 1992). Minigenome RNAs transcribed by viral RNA-dependent RNA-polymerases will have an accumulation of mutations of 10^{-3} to 10^{-4} (de Mercoyrol *et al.*, 1992). An improvement in expression stability should be observed by using expression systems initiated by DNA transfection, such as those based on the expression of the minigenomes under CMV promoter since an eukaryotic RNA polymerase II has an estimated error frequency of 5×10^{-6} (de Mercoyrol *et al.*, 1992).

5.   CONCLUSIONS

Both helper-dependent expression systems, based on two components, and single genomes constructed by targeted recombination or by using an infectious cDNA have been developed. The sequences that regulate transcription have been characterized. Expression of high amounts of heterologous antigens (1 to 8 μg/10^6 cells) have been achieved, and the expression levels have been maintained for around 10 passages. These expression levels should be sufficient to elicite protective immune responses.

Single genome coronavirus vectors have been constructed efficiently expressing a foreign gene such as GFP. Thus, a new avenue with high potential has been opened for coronaviruses which have unique properties, such as a long genome size and enteric tropism, that makes them of high interest as expression vectors for vaccine development and gene therapy. The possibility of engineering the tissue and species tropism will make coronavirus very flexible expression systems, since the same vector could be modified to target expression to different organs and animal species, including humans.

ACKNOWLEDGEMENTS

This work has been supported by grants from the Comisión Interministerial de Ciencia y Tecnología (CICYT), La Consejería de Educación y Cultura de la Comunidad de Madrid, and Fort Dodge Veterinaria from Spain, and the European Communities (Key Action 2: Infectious Diseases). JO and JMG received fellowships from the Department of Research and Technology. ISG, FA and PB, received contracts from the European Union Biotechnology, FAIR and Key Action 2: Infectious Diseases. AI and SA received fellowships from the Department of Education, University and Research of the Gobierno Vasco; JMS, received a fellowship from the Veterinary College of the Community of Madrid, CR received a fellowship from the Spanish Department of Health.

REFERENCES

Afanasiev, B. N., Ward, T. W., Beaty, B. J., and Carlson, J. O., 1999, Transduction of *Aedes aegypti* mosquitoes with vectors derived from *Aedes* densovirus. *Virology* **257**: 62-72.

Agapov, E. V., Frolov, I., Lindenbach, B. D., Pragai, B. M., Schlesinger, S., and Rice, C. M., 1998, Noncytopathic Sindbis virus RNA vectors for heterologous gene expression. *Proc. Natl. Acad. Sci. USA* **95**: 12989-12994.

Almazan, F., González, J. M., Pénzes, Z., Izeta, A., Calvo, E., Plana-Durán, J., and Enjuanes, L., 2000, Engineering the largest RNA virus genome as an infectious bacterial artificial chromosome. *Proc. Natl. Acad. Sci. USA* **97**: 5516-5521.

Alonso, S., Izeta, A., Sola, I., and Enjuanes, L., 2000a, Transcription regulatory sequences in transmissible gastroenteritis coronavirus. *Submitted.*

Alonso, S., Sola, I., Wege, H., Teifke, J., and Enjuanes, L., 2000b, Heterologous gene expression in tissue culture and *in vivo* using a transmissible gastroenteritis coronavirus helper dependent system. *Submitted.*

Baric, R. S., and Yount, B., 2000, Subgenomic negative-strand RNA function during mouse hepatitis virus infection. *J. Virol.* **74**: 4039-4046.

Bett, A. J., Prevec, L., and Graham, F. L., 1993, Packaging capacity and stability of human adenovirus type 5 vectors. *J. Virol.* **67**: 5911-5921.

Boyer, J. C., Bebenek, K., and Kunkel, T. A., 1992, Unequal human immunodeficiency virus type 1 reverse transcriptase error rates with RNA and DNA templates. *Proc. Natl. Acad. Sci. USA* **89**: 6919-6923.

Bredenbeek, P. J., and Rice, C. M., 1992, Animal RNA virus expression systems. *Semin. Virol.* **3**: 297-310.

Caley, I. J., Betts, M. R., Irlebeck, D. M., Davis, N. L., Swanstrom, R., Frelinger, J. A., and Johnston, R. E., 1997, Humoral, mucosal, and cellular immunity in response to a human immunodeficiency virus type 1 immunogen expressed by a venezuelan equine encephalitis virus vaccine vector. *J. Virol.* **71**: 3031-3038.

de Haan, C. A. M., Kuo, L., Masters, P. S., Vennema, H., and Rottier, P. J. M., 1998, Coronavirus particle assembly: primary structure requirements of the membrane protein. *J. Virol.* **72**: 6838-6850.

de Mercoyrol, L., Corda, Y., Job, C., and Job, D., 1992, Accuracy of wheat-germ RNA polymerase II. General enzymatic properties and effect of template conformational transition from right-handed B- DNA to left-handed Z-DNA. *Eur. J. Biochem.* **206**: 49-58.

DiCiommo, D. P., and Bremner, R., 1998, Rapid, high level protein production using DNA-based Semliki Forest virus vectors. *J. Biol. Chem.* **17**: 18060-18066.

Dubensky, T. W., Driver, D. A., Polo, J. M., Belli, B. A., Latham, E. M., Ibanez, C. E., Chada, S., Brumm, D., Banks, T. A., Mento, S. J., Jolly, D. J., and Chang, S. M. W., 1996, Sindbis virus DNA-based expression vectors: utility for in vitro and in vivo gene transfer. *J. Virol.* **70**: 508-519.

Fischer, F., Stegen, C. F., Koetzner, C. A., and Masters, P. S., 1997, Analysis of a recombinant mouse hepatitis virus expressing a foreign gene reveals a novel aspect of coronavirus transcription. *J. Virol.* **71**: 5148-5160.

Fisher, J., and Goff, S. P., 1998, Mutational analysis of stem-loops in the RNA packaging signal of the Moloney murine leukemia virus. *Virology* **244**: 133-145.

Izeta, A., Sánchez, C. M., Smerdou, C., Méndez, A., Alonso, S., Balasch, M., Plana-Durán, J., and Enjuanes, L., 1998, The spike protein of transmissible gastroenteritis coronavirus controls the tropism of pseudorecombinant virions engineered using synthetic minigenomes. *Adv. Exp. Med. Biol.* **440**: 207-214.

Izeta, A., Smerdou, C., Alonso, S., Penzes, Z., Méndez, A., Plana-Durán, J., and Enjuanes, L., 1999, Replication and packaging of transmissible gastroenteritis coronavirus-derived synthetic minigenomes. *J. Virol.* **73**: 1535-1545.

Jeong, Y. S., Repass, J. F., Kim, Y.-N., Hwang, S.-M., and Makino, S., 1996, Coronavirus transcription mediated by sequences flanking the transcription consensus sequence. *Virology* **217**: 311-322.

Joo, M., and Makino, S., 1992, Mutagenic analysis of the coronavirus intergenic consensus sequence. *J. Virol.* **66**: 6330-6337.

Joo, M., and Makino, S., 1995, The effect of two closely inserted transcription consensus sequences on coronavirus transcription. *J. Virol.* **69**: 272-280.

Krishnan, R., Chang, R. Y., and Brian, D. A., 1996, Tandem placement of a coronavirus promoter results in enhanced mRNA synthesis from the downstream-most initiation site. *Virology* **218**: 400-405.

Lai, M. M. C., 1996, Recombination in large RNA viruses: coronaviruses. *Semin. Virol.* **7**: 381-388.

Lai, M. M. C., 1998, Cellular factors in the transcription and replication of viral RNA genomes: a parallel to DNA-dependent RNA transcription. *Virology* **244**: 1-12.

Lai, M. M. C., Zhang, X., Hinton, D., and Stohlman, S., 1997, Modulation of mouse hepatitis virus infection by defective-interfering RNA-mediated expression of viral proteins and cytokines. *J. Neurovirol.* **3**:(Supp. 1) S33-S34.

Leparc-Goffart, I., Hingley, S. T., Chua, M. M., Phillips, J., Lavi, E., and Weiss, S. R., 1998, Targeted recombination within the spike gene of murine coronavirus mouse hepatitis virus-A59: Q159 is a determinant of hepatotropism. *J. Virol.* **72**: 9628-9636.

Li, H.-P., Huang, P., Park, S., and Lai, M. M. C., 1999, Polypyrimidine tract-binding protein binds to the leader RNA of mouse hepatitis virus and serves as a regulator of viral transcription. *J. Virol.* **73**: 772-777.

Li, H.-P., Zhang, X., Duncan, R., Comai, L., and Lai, M. M. C., 1997, Heterogeneous nuclear ribonucleoprotein A1 binds to the transcription-regulatory region of mouse hepatitis virus RNA. *Proc. Natl. Acad. Sci. USA* **94**: 9544-9549.

Liao, C. L., Zhang, X., and Lai, M. M. C., 1995, Coronavirus defective-interfering RNA as an expression vector: the generation of a pseudorecombinant mouse hepatitis virus expressing hemagglutinin-esterase. *Virology* **208**: 319-327.

Liljeström, P., and Garoff, H., 1991, A new generation of animal cell expression vectors based on the Semliki Forest virus replicon. *Biotechnology* **9**: 1356-1361.

Lin, Y. J., and Lai, M. M. C., 1993, Deletion mapping of a mouse hepatitis virus defective interfering RNA reveals the requirement of an internal and discontiguous sequence for replication. *J. Virol.* **67**: 6110-6118.

Makino, S., and Joo, M., 1993, Effect of intergenic consensus sequence flanking sequences on coronavirus transcription. *J. Virol.* **67**: 3304-3311.

Makino, S., Joo, M., and Makino, J. K., 1991, A system for study of coronavirus messenger RNA synthesis: a regulated, expressed subgenomic defective interfering RNA results from intergenic site insertion. *J. Virol.* **65**: 6031-6041.

Masters, P. S., 1999, Reverse genetics of the largest RNA viruses. *Adv. Virus Res.* **53**: 245-264.

Méndez, A., Smerdou, C., Izeta, A., Gebauer, F., and Enjuanes, L., 1996, Molecular characterization of transmissible gastroenteritis coronavirus defective interfering genomes: packaging and heterogeneity. *Virology* **217**: 495-507.

Molenkamp, R., Rozier, B. C. D., Greve, S., Spaan, W. J. M., and Snijder, E. J., 2000, Isolation and characterization of an arterivirus defective interfering RNA genome. *J. Virol.* **74**: 3156-3165.

Parks, R. J., and Graham, F. L., 1997, A helper-dependent system for adenovirus vector production helps define a lower limit for efficient DNA packaging. *J. Virol.* **71**: 3293-3298.

Penzes, Z., Tibbles, K., Shaw, K., Britton, P., Brown, T. D. K., and Cavanagh, D., 1994, Characterization of a replicating and packaged defective RNA of avian coronavirus infectious bronchitis virus. *Virology* **203**: 286-293.

Penzes, Z., Wroe, C., Brown, T. D. K., Britton, P., and Cavanagh, D., 1996, Replication and packaging of coronavirus infectious bronchitis virus defective RNAs lacking a long open reading frame. *J. Virol.* **70**: 8660-8668.

Sánchez, C. M., Izeta, A., Sánchez-Morgado, J. M., Alonso, S., Sola, I., Balasch, M., Plana-Durán, J., and Enjuanes, L., 1999, Targeted recombination demonstrates that the spike gene of transmissible gastroenteritis coronavirus is a determinant of its enteric tropism and virulence. *J. Virol.* **73**: 7607-7618.

Sawicki, S. G., and Sawicki, D. L., 1990, Coronavirus transcription: subgenomic mouse hepatitis virus replicative intermediates function in RNA synthesis. *J. Virol.* **64**: 1050-1056.

Sethna, P. B., Hung, S.-L., and Brian, D. A., 1989, Coronavirus subgenomic minus-strand RNAs and the potential for mRNA replicons. *Proc. Natl. Acad. Sci. USA* **86**: 5626-5630.

Sola, I., Izeta, A., González, J. M., and Enjuanes, L., 2000, Tissue specific expression into the mucosal surface using a single genome vector based on recombinant coronaviruses. *Submitted*.

Stirrups, K., Shaw, K., Evans, S., Dalton, K., Casais, R., Cavanagh, D., and Britton, P., 2000, Expression of reporter genes from the coronavirus infectious bronchitis virus defective RNA CD-61. *J. Gen. Virol.* **In press**: 000-000.

Thiel, V., Siddell, S. G., and Herold, J., 1998, Replication and transcription of HCV 229E replicons. *Adv. Exp. Med. Biol.* **440**: 109-114.

van der Most, R. G., De Groot, R. J., and Spaan, W. J. M., 1994, Subgenomic RNA synthesis directed by a synthetic defective interfering RNA of mouse hepatitis virus: a study of coronavirus transcription initiation. *J. Virol.* **68**: 3656-3666.

van der Most, R. G., and Spaan, W. J. M., 1995, Coronavirus replication, transcription, and RNA recombination. *In* "The Coronaviridae" (S. G. Siddell, Ed.), pp. 11-31. Plenum Press, New York.

van Marle, G., Dobbe, J. C., Gultyaev, A. P., Luytjes, W., Spaan, W. J. M., and Snijder, E. J., 1999, Arterivirus discontinuous mRNA transcription is guided by base pairing between sense and antisense transcription-regulating sequences. *Proc. Nat. Acad. Sc. USA* **96**: 12056-12061.

van Marle, G., Luytjes, W., Van der Most, R. G., van der Straaten, T., and Spaan, W. J. M., 1995, Regulation of Coronavirus mRNA transcription. *J. Virol.* **69**: 7851-7856.

Zhang, X., Hinton, D. R., Cua, D. J., Stohlman, S. A., and Lai, M. M. C., 1997, Expression of interferon-γ by a coronavirus defective-interfering RNA vector and its effect on viral replication, spread, and pathogenicity. *Virology* **233**: 327-338.

Zhang, X., Hinton, D. R., Park, S., Parra, B., Liao, C.-L., and Lai, M. M. C., 1998, Expression of hemagglutinin/esterase by a mouse hepatitis virus coronavirus defective-interfering RNA alters viral pathogenesis. *Virology* **242**: 170-183.

The CXC Chemokines IP-10 and Mig are Essential in Host Defense Following Infection with a Neurotropic Coronavirus

M. T. LIU[1], B. P. CHEN[1], P. OERTEL[1], M. J. BUCHMEIER[2], T. A. HAMILTON[3], D. A. ARMSTRONG[3], AND T. E. LANE[1]

[1]Department of Molecular Biology and Biochemistry, University of California at Irvine, California, USA; [2] The Scripps Research Institute, La Jolla, California, USA; [3]Department of Immunology, The Lerner Research Institute, Cleveland, Ohio, USA.

1.   INTRODUCTION

Chemokines represent an ever-growing family of secreted proteins that function as potent mediators of inflammation (for review, see Luster, 1998). These molecules have been classified depending on the number and spacing of the first two conserved amino terminal cysteine residues into the C, CC, CXC, and CX_3C family. Studies have shown that chemokines target specific leukocyte populations during periods of inflammation (Luster, 1998; Lane et al., 2000; Biddison et al., 1998; Kolb et al., 1999). In addition, chemokines have been shown to be prominently expressed following viral infection of the CNS (Lane et al., 1998; Cheret et al., 1997, Asensio and Campbell, 1997; Hoffman et al., 1999). However, the functional significance of chemokine expression within this environment has not been fully defined.

1.1   The MHV model of CNS disease

Mouse hepatitis virus (MHV) is a positive strand RNA virus that is a member of the *Cornaviridae* family. Infection of susceptible strains of mice with MHV results in an acute encephalomyelitis followed by chronic

The Nidoviruses (Coronaviruses and Arteriviruses).

Edited by Ehud Lavi et al., Kluwer Academic/Plenum Publishers, 2001.

neurological disease. The acute stage is characterized by wide spread viral infection of neurons and glial cells (Houtman and Fleming, 1996; Buchmeier and Lane, 1999). Inflammatory CD4[+] and CD8[+] T cells as well as the antiviral cytokine IFN-γ are essential to host defense against MHV-induced neurological disease (Lane et al., 1997, 2000; Parra et al., 1999; Hoffman et al., 1999).

1.2   The CXC chemokines IP-10 and Mig

Intracranial infection of mice with MHV results in a dramatic increase in chemokine expression within the CNS (Lane et al., 1998). The present study focuses on the chemokines interferon inducible protein-10 (IP-10) and monokine induced by interferon gamma (Mig) which are prominently expressed during the acute stage of disease (Lane et al., 1998). IP-10 and Mig are non-ELR CXC chemokines (chemokines lacking the glutamic acid-leucine-arginine motif in the amino terminus) which have been shown to have a chemotactic effect on T cells by binding to the chemokine receptor CXCR3 (Biddison et al., 1998; Loetsher et al., 1996; Farber, 1997; Piali et al., 1998). The present study was undertaken to investigate the contribution of IP-10 and Mig in host defense against MHV infection of the CNS.

2.   IP-10 AND MIG IN HOST DEFENSE

IP-10 and Mig were selectively inhibited through intraperitoneal administration of rabbit polyclonal anti-IP-10 or anti-Mig antisera to MHV-infected mice. Control mice were treated with normal rabbit serum (NRS). Approximately 50% of control animals survived up to 12 days post infection and successfully cleared infectious virus (2.1 ± 0.2 $\log_{10}$ pfu/g tissue n=14). In contrast, mice treated with either anti-IP-10 or anti-Mig showed a decreased ability to clear virus (anti-IP-10: 5.7 ± 0.3 $\log_{10}$ pfu/g tissue n=15, anti-Mig: 5.4 ± -0.3 $\log_{10}$ pfu/g tissue n=4) which corresponded to an increase in mortality (Figure 1). These data indicate that both IP-10 and Mig play an essential role in host defense by contributing to viral clearance.

Previous studies have shown that both CD4[+] and CD8[+] T cells are essential for optimal host defense following viral infection (2). Mice treated with either anti-IP-10 or anti-Mig displayed a significant reduction in both CD4[+] and CD8[+] T cell infiltration when compared to NRS treated control animals (Figure 2) indicating that both IP-10 and Mig function in host defense through attraction of T cells which participate in viral clearance.

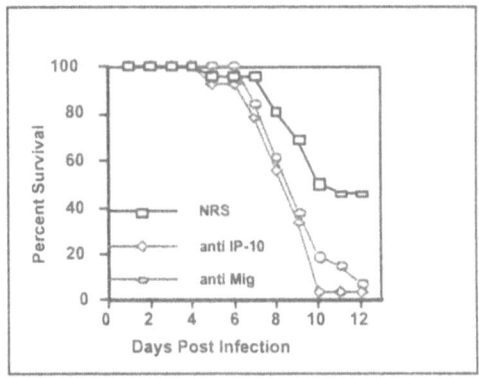

Figure 1. MHV infected mice treated with either anti-IP-10 or anti-Mig display increased mortality when compared to infected mice treated with NRS.

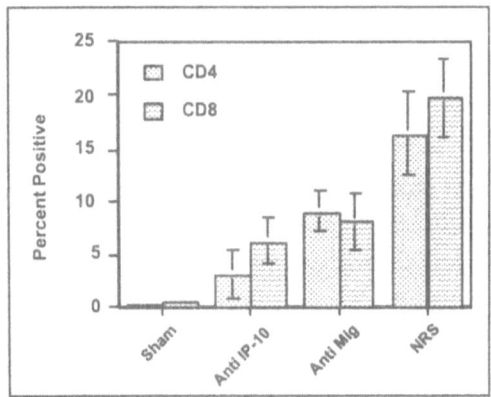

Figure 2. Mice treated with either anti-IP-10 or anti-Mig have decreased levels of CD4 and CD8 infiltration into the CNS as compared to NRS treated control animals. (82.4% and 70% for anti-IP-10 mice $P \leq 0.001$ and 45.4% and 60.2% for anti-Mig $P \leq 0.005$).

One mechanism by which T cells contribute to viral clearance is through release of the anti-viral cytokine IFN-γ (11,12). Examination of IFN-γ transcript levels in the CNS by ribonuclease protection assay revealed that anti-IP-10 and anti-Mig treated mice displayed a significant decrease ($P < 0.05$) (Data not shown) in transcript levels as compared to levels present in NRS control mice. Corresponding with the decrease in IFN-γ transcript levels was a significant decrease ($P \leq 0.05$) of IFN-γ protein levels in anti-IP-10 and anti-Mig treated mice (Figure 3).

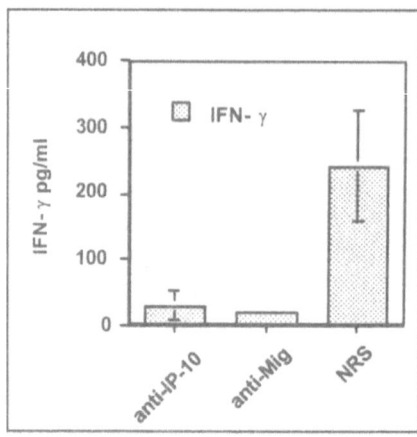

Figure 3. IFN-γ protein levels are significantly decreased in anti-IP-10 and anti-Mig treated mice. (P<0.05)

3.   CONCLUSION

Neutralization of IP-10 and Mig activity results in i) increased mortality, ii) delayed viral clearance, and iii) inhibition of a protective Th1 response characterized by infiltrating T cells and IFN-γ expression. Collectively, these data indicate that both IP-10 and Mig are essential in contributing to host defense following MHV infection of the CNS. Moreover, IP-10 has been demonstrated to be expressed within the CNS following infection with other viruses such as lymphocytic choriomeningitis virus (LCMV) and Theiler's virus suggesting that IP-10 exerts a protective effect in these models (Hoffman et al., 1999; Theil et al., 2000).

In addition to contributing to host defense during the acute stage of MHV infection, IP-10 is expressed during the chronic stage of disease almost exclusively within areas undergoing demyelination (Lane et al., 1998). A recent study has implicated a role for CD4[+] T cells in driving demyelination in persistently infected mice (Lane et al., 2000). Taken together, this suggests that persistent viral infection results in chronic IP-10 expression which serves to attract CD4[+] T cells to sites of MHV persistence. CD4[+] T cells contribute to demyelination by attracting macrophages which ultimately results in myelin destruction. Therefore, regulation of T cell entry into the CNS by targeting molecules such as IP-10 and Mig may represent promising targets for therapeutic interventions of human CNS inflammatory diseases.

REFERENCES

Asensio V.C., and I.L. Campbell. 1997. Chemokine gene expression in the brains of mice with lymphocytic choriomeningitis. *J. Virol. 71: 7832-7840.*

Biddison W.E., W.W. Cruikshank, D.M. Center, C.M. Pelfrey, D.D. Taub, and R.V. Turner. 1998. CD8+ myelin peptide-specific T cells can chemoattract CD4+ myelin peptide specific T cells: importance of IFN-inducible protein 10. *J. Immunol. 160: 444-448.*

Buchmeier M.J. and T.E. Lane. 1999. Viral-induced neurodegenerative disease. *Curr. Op. Micro. 2: 398-402.*

Cheret A., R. Le Grand, P. Caufour, O. Neildez, F. Matheux, F. theodoro, F. Boussin, B. Vaslin, and D. Dormont. 1997. Chemoattractant factors (IP-10, MIP-1alpha, IL-16) mRNA expression in mononuclear cells from diferent tissues during acute SIVmac251 infection of macaques. *J. Med. Primatol. 26:19-26.*

Farber J.M. 1997. Mig and IP-10: CXC chemokines that target lymphocytes. *J. Leukoc. Biol. 61: 246-257.*

Hoffman L.M. B.T. Fife, W.S. Begolka, S.D. Miller, and W.J. Karpus. 1999. Central nervous system chemokine expression during Theiler's virus-induced demyelinating disease. *J. Neurovirol. 5: 635-642.*

Houtman J.J., and J.O. Fleming. 1996. Pathogenesis of mouse hepatitis virus-induced demyelination. *J. Neurovirol. 2: 361-376.*

Kolb S.A., B. Sporer, F. Lahrtz, U. Koedel, H.W. Pfister, and A. Fontana. 1999. Identification of a T cell chemotactic factor in the cerebral spinal fluid of HIV-infected individuals as interferon gamma inducible protein 10. *J. Immunol. 163: 5686-5692.*

Lane T.E., A.D. Paoletti, and M.J. Buchmeier. 1997. Disassociation between the in vitro and in vivo effects of nitric oxide on a neurotropic murine coronavirus. *J. Virol. 71: 2202-2210.*

Lane T.E., V.C. Asensio, N.Yu, A.D. Paoletti, I.L. Campbell, and M.J. Buchmeier. 1998. Dynamic regulation of alpha and beta chemokine expression in the central nervous system during mouse hepatitis virus-induced demyelinating disease. *J. Immunol. 160: 970-978.*

Lane, T.E., M.T. Liu, B.P. Chen, V.C. Asensio, R.M. Samawi, A.D. Paoletti, I.L. Campbell, S.L. Kunkel, H.S. Fox, and M.J. Buchmeier. 2000. A central role for CD4[+] T cells and RANTES in virus-induced central nervous system inflammation and demyelination. *J. Virol. 74: 1415-1424.*

Loetsher M., B. Gerber, P. Loetscher, S.A. Jones, L. Piali, I.C. Lewis, M. Baggiolini, and B. Moser. 1996. Chemokine receptor specific for IP-10 and Mig: structure, function and expression in activated T-lymphocytes. *J. Exp. Med. 184: 963-969.*

Luster, A.D. 1998. Chemokines – chemotactic cytokines that mediate inflammation. *N. Engl. J. Med. 338:436-445.*

Parra B., D.R. Hinton, N.W. Marten, C.C. Bergmann, M.T. Lin, C.S. Yang, and S.A. Stohlman. 1999. IFN-gamma is required for viral clearance from central nervous system oligodendroglia. *J. Immunol. 162: 1641-1647.*

Piali L., C. Weber, G. LaRosa, C.R. Mackay, T.A. Springer, I. Clark-Lewis, and B. Moser. 1998. The chemokine receptor CXCR3 mediates rapid and shear-resistant adhesion-induction of effector T lymphocytes by the chemokines IP-10 and Mig. *Eur. J. Immunol. 28: 961-972.*

Theil D.J., I. Tsunoda, J.E. Libbey, T.J. Derfuss, and R.S. Fujinami. 2000. Alterations in cytokine but not chemokine mRNA expression during three distinct Theiler's virus infections. *J. Neuroimmunol. 104: 22-30.*

Regulation of Matrix Metalloproteinase (MMP) and Tissue Inhibitor of Matrix Metalloproteinase (TIMP) Genes During JHMV Infection of the Central Nervous System

JIEHAO ZHOU,[1] STEPHEN A. STOHLMAN,[2,3] NORMAN W. MARTEN,[1] AND DAVID R. HINTON[1]

Departments of [1]Pathology, [2]Neurology and [3]Molecular Microbiology and Immunology Keck School of Medicine, University of Southern California, 1333 San Pablo Street, Los Angeles, CA 90033

1.   INTRODUCTION

The central nervous system (CNS) is refractive to many aspects of the immune system primarily due to its limited ability to repair damage induced by the cytopathic mechanisms deployed by most immune cells. A primary obstacle to CNS inflammation is the blood-brain-barrier (BBB), which limits entry of immune cells into the CNS. To pass the BBB, inflammatory cells release matrix metalloproteinases (MMPs); a growing family of proteases with overlapping substrate specificities for components of the extracellular matrix (reveiwed by Kieseier et al. 1999). MMPs break down the extracellular matrix surrounding the endothelial layer of BBB thereby permitting peripheral immune cells to traverse the BBB and migrate through the parenchyma of the CNS in response to inflammatory signals. To limit potential damage resulting from infiltration of inflammatory cells, MMP activity is tightly regulated at both the level of gene expression and proenzyme activation as well as by expression of a second gene family, the tissue inhibitors of MMPs (TIMPs). TIMPs act as competitive inhibitors for the active sites of MMPs and thus limit inflammatory infiltrates.

The Nidoviruses (Coronaviruses and Arteriviruses).
Edited by Ehud Lavi *et al.*, Kluwer Academic/Plenum Publishers, 2001.

To evaluate the role of individual MMPs and TIMPS in facilitating mononuclear cell infiltration during acute CNS viral infection, male BALB/c mice were infected with JHMV, a neurotropic strain of mouse hepatitis virus. JHMV induces lesions similar to those associated with the human disease multiple sclerosis and pathology is associated with extensive infiltration by NK cells, CD4$^+$ and CD8$^+$ T cells and peripheral macrophages. These inflammatory cells appear as early as day 5 to 6 post infection (p.i.) and peak by day 7 p.i. (Williamson et al. 1991) Naïve and JHMV infected mice were assessed for MMP and TIMP gene expression by means of a RNAse protection assay (RPA). Results from RPA revealed that genes encoding MMP-2, -9, -10 and -14 as well as TIMP-2 and -3 were already expressed within the uninfected CNS, however their expression remained unchanged in response to JHMV infection. By contrast MMP-3 and -12, which were not detected in naïve mice, were rapidly induced following JHMV infection. TIMP-1 was similarly induced within the CNS of JHMV infected mice. These data suggest that MMP-3 and -12 may be involved in immune cell infiltration of the CNS and that their activity may be limited by expression of TIMP-1 during acute JHMV expression.

2.   MATERIALS AND METHODS

Animals and virus. Male BALB/c mice were purchased from National Cancer Institute and used between 6 to 8 weeks of age. Mice were infected with 1000 PFU of the 2.2-V-1 strain of JHMV (Fleming et al. 1986).

RNA isolation and RPA. Total RNA was prepared from brains of naïve and JHMV infected mice using TRIzol reagent in accordance with the manufacturer's protocol. Plasmids containing MMP and TIMP RPA probes were a gift from Iain Campbell (Pagenstecher et al. 1997). 10 μg of total brain RNA was hybridized with α-^{32}p labeled MMP or TIMP probe sets and RPA were performed as described (Pagenstecher et al. 1997). Protected probe fragments were separated on a 5% acrylamide/7 M urea denaturing gel. Intensities of bands were quantitated using a Molecular Dynamics phosphorimager.

3.   RESULTS

Infection with the neurotropic strain of JHMV results in extensive infiltration by peripheral mononuclear cells. To identify the role of individual MMPs involved in mononuclear cell infiltration of the CNS, a multi-probe RPA was used to assay MMP mRNA expression during acute

JHMV infection. Total RNA prepared from the brains of both naïve and infected mice sacrificed at the indicated time points was hybridized with a probe set specific for MMP-2, -3 and -9 (Fig. 1A). Both the MMP-2 and -9 genes were expressed on a constituative basis, with no apparent differences between naïve and infected mice. However, induction of MMP-3 expression, which was not detectable within the CNS of naïve mice, was induced quite rapidly following JHMV infection and was observed as early as day 2 p.i. MMP-3 expression was transient, however, peaking at day 4 p.i. and was not detected after day 10 p.i.

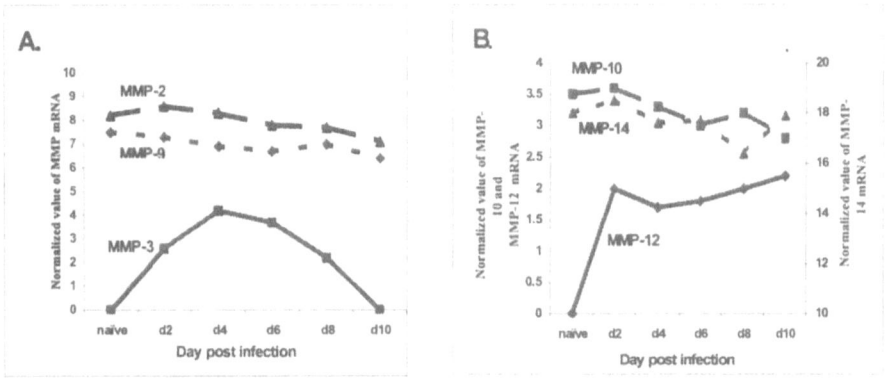

Figure 1. Expression of MMP RNA during acute JHMV infection. RPA were performed to determine the relative expression of MMP-2, -3 and -9 mRNA (Panel A) and MMP-10, -12 and -14 mRNA (Panel B) within brains of naïve or JHMV infected mice sacrificed at the indicated time points.

A second probe set was used to examine expression of MMP-10, -12 and -14 (Fig. 1B). MMP-14 mRNA (more commonly referred to as membrane type MMP-1 [MT1-MMP]), was expressed at a very high level compared to expression of other MMP genes, possibly due to its role as an activator of secreted MMP proenzymes. However, MMP-14 expression was constitutive and no induction of mRNA was observed following acute JHMV infection. Similarly, MMP-10 was expressed within the CNS of naïve mice, but was not induced within the CNS of JHMV infected mice. MMP-12 expression was not observed within the CNS of naïve mice, but was detected as early as day 2 p.i. in the brains of JHMV infected mice. In contrast to the transient expression of MMP-3, however, MMP-12 expression remained elevated after clearance of infectious virus and was detected as late as day 30 p.i.

TIMPs are a family of specific regulatory inhibitors of MMP activity and are critical for limiting the extent of immune cell infiltration. To determine whether expression of TIMPs were altered by JHMV infection, total brain

RNA was hybridized with a RPA probe set consisting of TIMP-1, -2 and -3 (Fig. 2). TIMP-1 mRNA was barely detectable in naïve mice, but expression was rapidly upregulated following JHMV infection. TIMP-1 mRNA expression peaked at day 6 p.i. and remained elevated compared to naïve mice as late as day 30 p.i. By contrast expression of TIMP genes 2 and 3, although detectable within the brains of naïve mice, was not altered by JHMV infection.

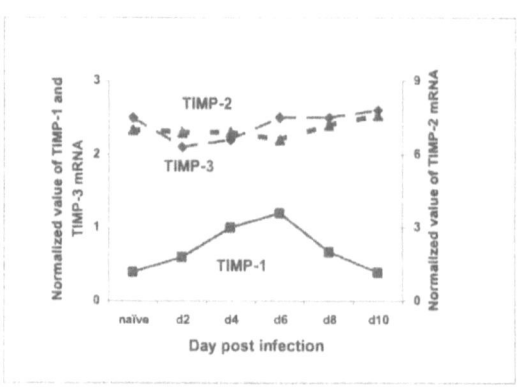

Figure 2. TIMP mRNA expression during acute JHMV infection. RPA were performed to determine the relative expression of TIMP-1, -2, and -3 mRNA within brains of naïve or JHMV infected mice sacrificed at the indicated time points.

4.   DISCUSSION

In order to elucidate the role of MMPs during CNS inflammation, MMP and TIMP gene expression was examined following acute JHMV infection. mRNA expression from the MMP-2, -9, -10 and -14 genes was detected in the brain of naïve BALB/c mice. However, no alterations in mRNA expression were observed from any of these genes following intracerebral JHMV infection. Expression of these MMPs within the CNS of naïve mice and subsequent lack of change in expression during inflammation suggests that these genes may be involved in normal remodeling of the extracellular matrix within CNS.

It has been reported that MMP-2 and -9 are the predominant MMPs produced by T cells upon binding with adhesion molecules (Esparza et al. 1999, Goetzl et al. 1996). Surprisingly, mRNA expression of both MMP-2 and –9, which were detected in the CNS of naïve mice, were not induced by JHMV infection despite extensive T cell infiltration (data not shown). Although, posttranscriptional regulation of these proteins cannot be ruled out, it appears likely that other MMPs may be required to make a contribution towards permeation of the extracellular matrix of the BBB during acute JHMV infection. MMP-3 and -12, which were not detected within the CNS of naïve mice, were both induced within 48 h of JHMV infection. MMP-3 was expressed in a transient fashion, whereas MMP-12 remained elevated following infection with JHMV. MMP-3 may be involved in the breakdown of the BBB as it is specific for several components of the BBB basal lamina. Identifying the role of MMP-12, which is an elastase, may be more problematic, although elastin has been shown to account for up to 4% of the ECM of brain microvessels (Faris et al. 1982). These data suggest that MMP-3 and –12, which have been linked to expression by mononuclear cells (Ozenci et al. 1999, Shapiro et al. 1999, Maeda and Sobel 1996) and are also up-regulated during EAE (Pagenstecher et al. 1998) may play an active role during mononuclear cell infiltration in response to CNS infection.

Examination of TIMP mRNA levels revealed that TIMP-2 and -3 are expressed in naïve mice and that expression remains constitutive following JHMV infection. By contrast, TIMP-1 expression was induced several fold following JHMV infection. These data on TIMP expression following JHMV infection are a quite similar to those reported for mice following induction of EAE (Pagenstecher et al. 1998) and suggest that TIMP-1 may be the primary inhibitor of MMPs released during either viral or autoimmune induced inflammation. To be noted, simultaneous up-regulation of MMP-3 and TIMP-1 in the JHMV model is consistent with the previous reports that TIMP-1 can form complexes with MMP-3 *in vitro* (Gomis-Ruth et al. 1997), suggesting specific inhibition of MMP-3 by TIMP-1. Studies of *in vitro* cultured cells and murine models of neuropathogenesis suggests that astrocytes may be a primary source of TIMP-1 (Giraudon et al. 1998 Pagenstecher et al. 1998). Taken together these data suggest that T cells and microglia/macrophages may function in a proinflamatory role through release of MMP-3 and -12 and that this action is controlled by expression of TIMP-1, possibly secreted by astrocytes.

REFERENCES

Esparza, J., C. Vilardell, J. Calvo, M. Juan, J. Vives, A. Urbano-Marquez, J. Yague and M. C. Cid. Fibronectin upregulates gelatinase B (MMP-9) and induces coordinate expression of gelatinase A (MMP-2) and its activator MT1-MMP (MMP-14) by human T lymphocyte cell lines. A process repressed through RAS/MAP kinase signalling pathways. *Blood*. 1999. **94**:2754-66

Faris B., P. Mozzicato, R. Ferrera, M. Glembourtt, P. Toselli and C. Franzblau. Collagen of Brain Microvessel Preparations. *Mircovasc. Res.* 1982. **23**:171-179

Fleming, J. O., M. Trousdale, F. E. Zactarim, S. A. Stohlman and L. P. Weiner. Pathogenicity of antigenic variants of murine coronavirus JHM selected with mAb. *J. Virol.* 1986. **58**:869-875.

Giraudon, P., R. Szymocha, S. Buart, A. Bernard, L. Cartier, M. F. Belin and H. Akaoka. T lymphocytes activated by persistent viral infection differentially modify the expression of metalloproteinases and their endogenous inhibitors, TIMPs, in human astrocytes: relevance to HTLV-I-induced neurological disease. *J. Immunol.* 2000. **164**:2718-2727.

Goetzl EJ. Banda MJ. Leppert D. Matrix metalloproteinases in immunity. *J. Immunol.* 1996. **156**:1-4.

Gomis-Ruth, F.-X., K. Maskos, M. Betz, A. Bergner, R. Huber, K. Suszuki, N. Yoshida, H. Nagase, K Brew, G. P. Bourenkov, H. Bartunik and W. Bode. Mechanism of inhibition of the human matrix metalloproteinase stromelysin-1 by TIMP-1. *Nature.* 1997. **389**:77-81.

Kieseier, B.C., T. Seifert, G. Giovannoni and H. P. Hartung. Matrix metalloproteinases in inflammatory demyelination: targets for treatment. *Neurology.* 1999. **53**:20-25.

Maeda, A. and R. A. Sobel. Matrix metalloproteinases in the normal human central nervous system, microglial nodules and multiple sclerosis lesions. *J. Neuropath. Exp. Neuro.* 1996. **55**:300-309.

Ozenci V., L. Rinaldi, N. Teleshova, D. Matusevicius, P. Kivisakk, M. Kouwenhoven and H. Link. Metalloproteinases and their tissue inhibitors in multiple sclerosis. *J. Autoimmun.* 1999. **12**:297-303.

Pagenstecher, A., A. K. Stalder and I. L. Campbell. RNAse protection assays for the simultaneous and semiquantitative analysis of multiple murine matrix metalloproteinase (MMP) and MMP inhibitor mRNAs. *J. Immunol. Meth.* 1997. **206**:1-9.

Pagenstecher, A., A. K. Stalder, C. L. Kincaid, S. D. Shapiro, I. L. Campbell. Differential expression of matrix metalloproteinase and tissue inhibitor of matrix metalloproteinase genes in the mouse central nervous system in normal and inflammatory states. *Am. J. Path.* 1998. **152**:729-741.

Shapiro, S.D. Diverse roles of macrophage matrix metalloproteinases in tissue destruction and tumor growth. *Thromb & Haemo.* 1999. **82**:846-849.

Williamson, J. S. P., K. Sykes and S. A. Stohlman. Characterization of brain infiltrating mononuclear cells during infection with mouse hepatitis virus strain JHM. *J. Neuroimmunol.* 1991. **32**:199-207.

IFN-γ Secreted by Virus-Specific CD8+ T Cells Contribute to CNS Viral Clearance

BEATRIZ PARRA,[1] CORNELIA C. BERGMANN,[1,2] DAVID R. HINTON,[3] ROSCOE ATKINSON,[2,3] AND STEPHEN A. STOHLMAN[1,2]
Departments of Molecular Microbiology and Immunology[1], Neurology[2], and Pathology[3] . Keck School of Medicine, University of Southern California, Los Angeles, CA 90033

1. INTRODUCTION

Replication of mouse hepatitis virus strain JHM (JHMV) in the central nervous system (CNS) is controlled by CD8+ T cells. However, persistent infection and subsequent chronic demyelination are established. Understanding effector mechanisms during acute infection may help understand viral persistence. CD8+ T cells respond to viral infections via two different cellular mechanisms, lysis of infected cells and secretion of anti-viral cytokines (Kaki *et al* 1995, Ruby *et al* 1991, Young *et al* 1995). JHMV replication in astrocytes and microglia is controlled by perforin-dependent cytolysis (Lin *et al* 1997) whereas replication in oligodendrocytes is controlled via IFN-γ (Parra *et al* 1999). The contributions of IFN-γ and perforin-dependent CD8+ T cell function in viral clearance were examined in SCID mice following adoptive transfer of CD8+ T cells deficient in IFN-γ secretion.

2. MATERIAL AND METHODS

Mice: BALB/c SCID and wild type (wt) BALB/c mice were obtained from NCI (Frederick, MD). Homozygous IFN-γ[-/-] BALB/c mice were provided by Robert Coffman, DNAX Research Corporation (Palo Alto, CA).

The Nidoviruses (Coronaviruses and Arteriviruses).
Edited by Ehud Lavi *et al.*, Kluwer Academic/Plenum Publishers, 2001.

Virus infection: SCID mice were infected with the 2.2v-1 strain of JHMV (Fleming *et al* 1986). Wt mice and IFN-$\gamma^{-/-}$ donors were immunized i.p. with 10^6 pfu of JHMV and 4 weeks later spleens were removed to purify CD8⁺ T cells.

CD8⁺ T cell purification: Spleen cells from immunize mice were depleted of B cells and macrophages by panning. CD4⁺T cells were depleted by adsorbance to magnetic beads (Miltenyi Biotec Inc. Auburn, CA.). Recoveries of CD8⁺ T cells were approximately 80% as determined by flow cytometry analysis.

Adoptive Transfers: Recipient SCID mice were adoptive transferred with CD8⁺ T cells from either IFNγ$^{-/-}$ or wt immunized mice intravenously with $1-2\times10^7$ CD8⁺ T cells and infected 5 h later. Viral replication and pathogenesis were determined at d 10 and 14 p.i., when the control SCID mice began to succumb.

CNS mononuclear cell populations: Mononuclear cells were isolated as previously described (Bergmann *et al* 1999). Virus specific CD8⁺ T cells were identified by anti-CD8 and a tetrameric L^d-N-318 reagent and assayed for *ex vivo* CTL activity as described (Bergmann *et al* 1999). Microglial cells (CD45 ^{low} CD11b⁺) were characterized by staining with anti-CD45 and anti-CD11b⁺ mAbs (Pharmingen, San Diego, CA). MHC class I L^d expression was determined by the mean fluorescence intensity with an anti-H2L^d specific mAb (PharMingen).

Histopathology: Tissues were prepared for paraffin and frozen sections as described (Parra *et al* 1999). Viral antigen (Ag) was detected with mAb J3.3. CD8⁺ cells were identified in frozen sections using rat anti-CD8a mAb (PharMingen). Sections were stained with mAb J3.3 and Ab specific for astrocytes (GFAP) or microglia (CD11b) as described (Parra *et al* 1999).

3.   RESULTS

3.1   IFN-$\gamma^{-/-}$ CD8⁺ T cells reduce virus replication less efficiently than IFN-$\gamma^{+/+}$ CD8⁺ T cells.

Immunodeficient SCID mice reconstituted with CD8⁺ T cells from immune wt (IFN-$\gamma^{+/+}$) donors reduced CNS viral replication at 10 and 14 days p.i. compared with control SCID mice (Fig. 1). CD8⁺ T cells derived from immune IFN-γ $^{-/-}$ donors were less effective than IFN-γ$^{+/+}$ CD8⁺ T cells (Fig 1). These results demonstrate that IFN-γ derived from CD8⁺ T cells alone influences JHMV clearance from the CNS. Nevertheless, virus was reduced by IFN-$\gamma^{-/-}$ CD8⁺ T cells compared to control SCID mice (Fig 1), consistent with expression of perforin mediated cytotoxicity. Viral Ag in recipients of IFN-$\gamma^{-/-}$ CD8$^+$ T cells was dramatically increased compared to IFN-$\gamma^{+/+}$ recipients at 14 days p.i. Viral Ag localized to multiple cell types in brain of both reconstituted groups; however, it was mainly localized to spinal cord oligodendrocytes (Fig. 3). This contrast to the multiple infected cell types in spinal cords of control SCID mice. In contrast to reduced viral Ag positive cells in the CNS of wt CD8$^+$ T cell recipients, an increase in infected oligodendrocytes was present in IFN-$\gamma^{-/-}$ CD8$^+$ T cell recipients (Fig. 2), consistent with CD8$^+$ T cells controlling JHMV replication in oligodendrocytes via IFN-γ (Parra *et al* 1999).

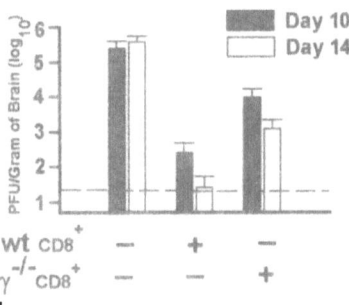

Figure 1. CD8$^+$ T cells reduce JHMV replication via IFN-γ. CD8$^+$ T cells from JHMV immune IFN-$\gamma^{-/-}$ or wt mice were transferred into SCID mice and infected with JHMV. Virus replication at 10 and 14 days p.i is expressed as Log$_{10}$ PFU/gm tissue. Titers are means of at least 4 mice/group. Dashed line is the assay detection limit. Data are representative of 2 experiments.

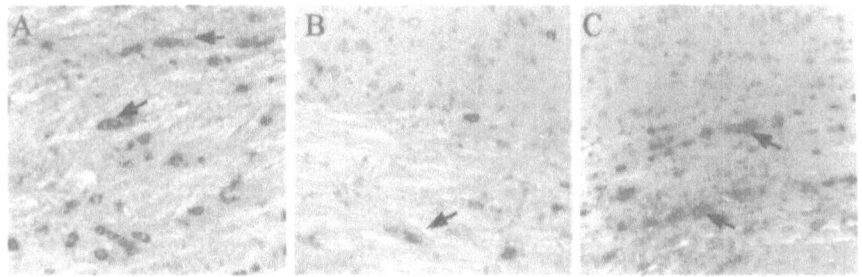

Figure 2. Viral Ag in oligodendroglia is reduced via CD8$^+$-derived IFN-γ. Viral Ag in spinal cords: A) unreconstituted; B) IFN-$\gamma^{+/+}$ CD8$^+$ reconstituted and; C) IFN-$\gamma^{-/-}$ CD8$^+$ reconstituted SCID mice at day 14 p.i. Arrowhead points indicate oligodendrocytes.

Functional virus specific IFN-$\gamma^{-/-}$ CD8$^+$ T cells are recruited to CNS. Brain infiltrating cells showed similar proportions of virus specific CD8$^+$ T cells recruited in both groups of reconstituted SCID mice (Fig. 3). IFN-$\gamma^{-/-}$ CD8$^+$ T cells are recruited by 4 d p.i. and accumulated with both the same

kinetic and to similar numbers in both groups (data not shown). Equal CD8⁺
T cells within the brain parenchyma and spinal cord white matter tracts were
in both groups (data not shown). CD8⁺ T cells deficient in IFN-γ secretion
within the CNS at d 10 p.i. retained *ex vivo* cytolysis (Fig. 3).

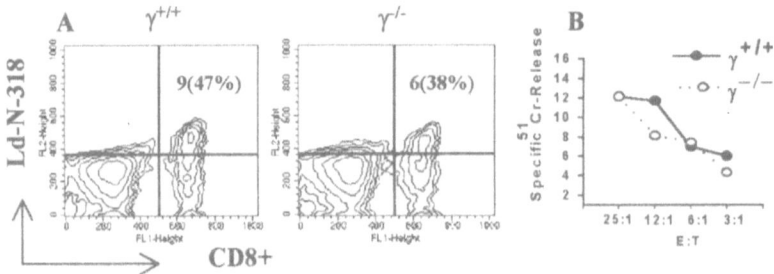

Figure 3. Virus-specific CD8⁺ T cells recruited into the CNS of CD8⁺ IFN-γ⁻/⁻ T cell
reconstituted SCID mice. CNS infiltrating cells were prepared from reconstituted mice at d
10 p.i. Cells were stained for CD8⁺ and virus-specific TcR (L^d -N-318 tetramer) (Panel **A**).
Numbers represent the percentage of total population. Data from two experiments. (Panel B).
Ex vivo ⁵¹Cr release from pN peptide coated targets. Nonspecific cytolysis was zero as
determined with uncoated targets.

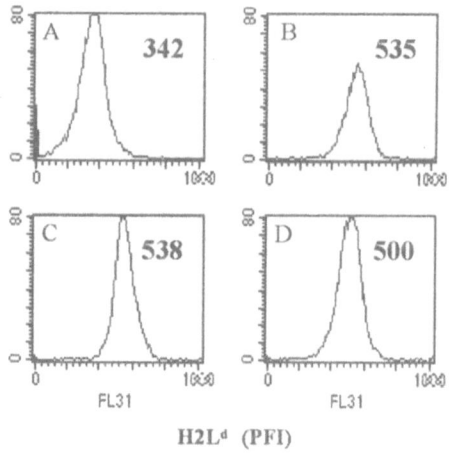

H2L^d (PFI)

Figure 4. MHC class I expression on microglia during JHMV infection is upregulated
independently of CD8-derived IFN-γ . CNS cells were isolated from (A) naïve, (B) JHMV
infected and infected (C) IFN-γ⁺/⁺ or (D) IFN-γ⁻/⁻ CD8⁺-reconstituted SCID mice. Cells were
gated on microglia (CD45^{low} and CD11b⁺). L^d expression was determined as peak mean
fluorescence intensity (PFI). Data represent two independent experiments.

CD8⁺ T cell derived IFN-γ independent increased MHC class I
expression. To determine if the lack of IFN-γ from CD8⁺ T cells influenced
JHMV clearance, MHC class I L^d, the immunodominant N epitope
restriction element was examined on microglia/macrophages. L^d expression

was found on microglia of control SCID mice after infection (Fig. 4A, 4B). Microglia in both CD8$^+$ T cell recipient groups have similar levels of increased L^d expression (Fig. 4C, 4D).

4. DISCUSSION

JHMV infection of mice with perforin or IFN-γ deficiencies suggested that both effector mechanisms played critical roles in viral clearance (Lin *et al* 1997, Parra *et al* 1999). Although the previous data suggested a predominant role of the CD8$^+$ T cell response in clearance, analysis of gene deleted mice is complicated by additional cell types. These experiments were conducted by reconstitution of immunodeficient SCID mice to eliminate contributions of CD4$^+$ T cells and B cells. The results demonstrate that in addition to cytotoxicity, IFN-γ is a key CD8$^+$ T cell effector important for the control of the acute CNS infection. JHMV is highly oligotropic and CD8$^+$ T cells may have an absent or diminished cytotoxic action on this low MHC class I expressing cell type. However, in wt mice virus is eliminated from oligodendrocytes by a vigorous CTL response (Stohlman *et al* 1995). CD8$^+$ T cells competent for IFN-γ secretion efficiently reduced virus from oligodendrocytes of infected SCID mice compared to partial elimination in recipients of IFN-γ$^{-/-}$ CD8$^+$ T cells. IFN-γ is not required for CTL induction (Parra *et al* 1999, Graham *et al* 1993) or for homing to the site(s) of infection (Nansen *et al* 1998). Indeed, IFN-γ$^{-/-}$ CD8$^+$ T cells with intact perforin-dependent cytotoxicity trafficked normally into the CNS and reduced virus from most MHC class I expressing cells.

It is possible that IFN-γ is required for increased expression of MHC class I on targets, thereby promoting effective recognition (Young *et al* 1995). Class I was expressed at similar levels on microglial of SCID mice reconstituted with CD8$^+$ T cells from both groups. Increased expression occurred after infection, independent of CD8$^+$ T cells, consistent with early NK cell recruitment (data not shown) and the IFN type I (α/β) mediated increase in MHC expression following viral infection (Njenja *et al* 1997). These data support the concept that CD8$^+$ T cells inhibit virus in oligodendrocytes via a IFN-γ dependent mechanism. However, IFN-γ may also inhibit viral spread from other cell types by limiting susceptible cells. Therefore, virus accumulation in oligodendrocytes during an infection in which CD8$^+$ T cells are unable to secrete IFN-γ could result from both increase viral spread, in addition to limited cytotoxicity. These data are consistent with the increase in CNS viral Ag in SCID mice reconstituted with IFN-γ$^{-/-}$ CD8$^+$ T cells. However, it is equally likely that an inability to lyse infected oligodendroglia, in addition to the lack of a direct IFN-γ mediated anti-viral activity, results in infection of other CNS cell types seeded by uncontrolled oligodendroglia infection. Although the precise role of $CD8^+$ T cell effector mechanisms is complicated by the dynamics of an ongoing infection, these data support the hypothesis that separate effector mechanisms are functioning at the single cell type level within the CNS.

REFERENCES

Kagi, D., Seiler, P., Pavlovic, P., Ledermann, B., Burki, K., Zinkernagel, R.M., and Hengatner, H., 1995, The roles of perforin and fas-dependent cytotoxicity in protection against cytopathic and non cytopathic viruses. *Eur. J. Immunol.* **25**:3256-3262.

Ruby, J., and Ramshaw, I., 1991, The antiviral activity of immune CD8+ T cells is dependent on interferon-γ. *Lymphokine Cytokine Res.* **10**:353-358.

Young, H.A., and Hardy, K.J., 1995, Role of interferon-γ in immune cell regulation. *J. Leukoc. Biol.* **58**:373-381.

Lin, M.T., Stohlman, S.A., and Hinton, D.R., 1997, Mouse hepatitis virus is cleared from the central nervous systems of mice lacking perforin-mediated cytolysis. *J. Virol.* **71**:383-391.

Parra, B., Hinton, D.R., Marten, N.W., Bergmann, C.C., Lin, M.T., Yang, C.S., and Stohlman, S.A., 1999, IFN-γ is required for viral clearance from central nervous system oligodendroglia. *J. Immunol.* **162**:1641-1647.

Fleming, J.O., Trousdale, M.D., el-Zaatari, F.A., Stohlman, S.A., and Weiner, L.P., 1986 Pathogenicity of antigenic variants of murine coronavirus JHM selected with monoclonal antibodies. *J. Virol.* **58**:869-875.

Bergmann, C.C., Altman, J.D., Hinton, D., and Stohlman, S.A., 1999, Inverted Immunodominance and Impaired Cytolytic Function of $CD8^+$ T Cells During Viral Persistence in the Central Nervous System. *J. Immunol.* **163**:3379-3387.

Stohlman, S.A., Bergmann, C.C., van der Veen, R., and Hinton, D.R., 1995 Mouse hepatitis virus-specific cytotoxic T lymphocytes protect from lethal infection without eliminating virus from oligodendroglia. *J. Virol.* **69**:684-694.

Graham, M.B., Dalton, D.K., Giltinan, D., Braciale, V.L., Stewart, T.A., and Braciale, T.J., 1993, Response to influenza infection in mice with a targeted disruption in the interferon γ gene. *J. Exp. Med.* **178**:1725-1732.

Nansen, A., Christensen, J.P., Ropke, C., Marker, O., Scheynius, A., and Thomsen, A.R., 1998, Role of interferon-γ in the pathogenesis of LCMV-induced meningitis: unimpaired leukocyte recruitment, but deficient macrophage activation in interferon-γ knock-out mice. *J. Neuroimmunol.* **86**:202-212.

Njenja, M.K., Pease, L.R., Wettstein, P., Mark, T., and Rodriguez, M., 1997, Interferon γ mediates early-virus induced expression of H2d and H2k in the central nervous system. *Lab. Invest.* **77**:71-84.

The Role of CD4 and CD8 T Cells in MHV-JHM-Induced Demyelination

[1]GREGORY F. WU, [2]AJAI A. DANDEKAR, [3]LECIA PEWE, AND [1,4]STANLEY PERLMAN

Interdisciplinary Programs in [1]Neuroscience and [2]Immunology and Departments of [3]Pediatrics and [4]Microbiology, University of Iowa, Iowa City, Iowa 52242

1.   INTRODUCTION

Demyelination following central nervous system (CNS) infection with the neurotropic coronavirus, mouse hepatitis virus, strain JHM (MHV-JHM) has been shown to be dependent on the immune response of the host (Perlman, 1998). Although controversial, both CD4 and CD8 T cells are thought to contribute to demyelination (Perlman, 1998). However, the mechanisms by which each T cell subset contributes to demyelination remains unknown. To address this issue, the following questions were raised. First, what is the magnitude of the T cell response during immune-mediated demyelination? Second, what is the timing of the immune response leading to demyelination? Third, what are the roles of CD4 and CD8 T cells - in particular virus-specific T cells - during MHV-JHM-induced demyelination? In order to address these questions, a previously described adoptive transfer system (Wu and Perlman, 1999) was utilized to measure the quantity of virus-specific T cells during the process of immune-mediated demyelination. Furthermore, the individual contributions of CD4 or CD8 T cells to MHV-JHM-induced demyelinating disease were determined by depletion studies.

The Nidoviruses (Coronaviruses and Arteriviruses).
Edited by Ehud Lavi *et al.*, Kluwer Academic/Plenum Publishers, 2001.

2.   MATERIALS AND METHODS

2.1   Virus

The neuroattenuated variant of MHV-JHM, strain J2.2-v1 (MHV-J2.2-v1), was generously provided by Dr. J. Fleming (University of Wisconsin, Madison).

2.2   Adoptive Transfer

Mice with genetic disruption of the recombination activating gene (RAG1-/-), obtained from Jackson Laboratories (Bar Harbor, ME), were infected with $1x10^3$ PFU MHV-J2.2-v1 in 30 μL by intracranial injection (Wu and Perlman, 1999). Donor splenocytes were isolated from C57Bl/6 (B6) mice, obtained from the National Cancer Institute (Bethesda, MD), that were immunized intraperitoneally (i.p.) with wild-type MHV-JHM. These cells were delivered to infected RAG1-/- mice 3 days post-inoculation (p.i.) as previously described (Wu and Perlman, 1999). In some experiments, CD4 or CD8 T cells were depleted prior to transfer as previously described (Wu *et al.*, 2000). In order to grade the clinical disease observed following adoptive transfer, the following scoring system was utilized: 0 – asymptomatic, 1 – limp tail, 2 – wobbly gait with righting difficulty, 3 – hind-limb weakness, 4 – hind-limb paralysis, 5 – moribund/dead. Brains and spinal cords from adoptive transfer recipients were harvested 7 to 15 days post-transfer (p.t.).

2.3   FACS Analysis

Antigen-specific T cells were identified by staining for intracellular IFN-γ using PE-conjugated anti-IFN-γ antibody (Pharmingen), as previously described (Wu *et al.*, 2000). Peptides corresponding to the CD4 (M-133-147; I-A^b restricted) and CD8 (S-510-518, H-2D^b restricted; S-598-605, H-2K^b restricted) T cell epitopes were used at a final concentration of 5 μM and 1 μM, respectively.

2.4   Histology

Preparation of spinal cord sections was done as previously described (Wu and Perlman, 1999). 8 μm sections were stained with luxol fast blue (LFB) and digitized. The percentage of demyelinated white matter was quantified as previously described (Xue *et al.*, 1999).

3. RESULTS

3.1 MHV-specific cells rapidly infiltrate the CNS following adoptive transfer

Infection of RAG1-/- mice, lacking B and T lymphocytes, with MHV-J2.2-v1 does not result in CNS demyelination. However, adoptive transfer of splenocytes from syngeneic B6 mice, immunized to MHV-JHM, into MHV-J2.2-v1-infected RAG1-/- mice consistently results in demyelinating disease (Wu and Perlman, 1999). In order to investigate the magnitude and timing of the immune response to MHV-JHM following adoptive transfer, intracellular cytokine staining was performed. The number of virus-specific CD4 and CD8 T cells was determined by quantifying the number of cells producing IFN-γ in response to peptides representing the known CD4 and CD8 T cell epitopes (Perlman, 1998), as described in Section 2.3. Analysis of the donor cell population, isolated from B6 mice 6 days after immunization with MHV-JHM, revealed a small fraction of CD4 T cells specific for M-133-147, the immunodominant CD4 T cell epitope (Figure 1). Similarly, a small fraction of CD8 T cells were found to produce IFN-γ in response to both the immunodominant CD8 T cell epitope, S-510-518, and the subdominant CD8 T cell epitope, S-598-605 (Figure 1). In contrast, analysis of lymphocytes harvested from the CNS of asymptomatic adoptive transfer recipients at day 4.5 p.t. demonstrated that a large fraction of both CD4 and CD8 T cells responded to the respective immunodominant T cell epitopes. Approximately 20% of CD4 T cells were specific for M-133-147, while approximately 30% of CD8 T cells responded to S-510-518 (Figure 1). However, the fraction of CD8 T cells that responded to S-598-605 remained low (less than 5%). At seven days p.t., a time at which mice have developed demyelination, the fraction of M-133-147-specific CD4 T cells remained around 20%. On the other hand, only 20% of the CD8 T cells isolated from adoptive transfer recipients on day 7 p.t. were specific for S-510-518. Nonetheless, due to the overall increase in CD8 T cells on day 7 p.t., the absolute number of CD8 T cells increased approximately 10-fold (data not shown). These data demonstrate that MHV-specific CD4 and CD8 T cells rapidly infiltrate the CNS following adoptive transfer.

3.2 CD4 and CD8 T cells make distinct contributions to MHV-JHM-induced demyelinating disease

In order to determine the specific contributions made by CD4 and CD8 T cells to the pathogenesis of demyelination following infection with MHV,

MHV-J2.2-v1-infected RAG1-/- mice were given donor cells depleted of each subset. A clinical difference was observed between recipients of CD4 T cell and CD8 T cell enriched donors. CD4 T cell enriched recipients developed a more rapid and severe course of disease, frequently becoming moribund by day 7 p.t. In contrast, CD8 T cell enriched recipients exhibited a less severe clinical phenotype, often developing clinical disease as late as 11 days p.t. (Figure 2). When adoptive transfer recipients were analysed according to the clinical scale described in Section 2.2, there was significantly more severe clinical disease observed in CD4 T cell enriched adoptive transfer recipients relative to undepleted recipients (p <0.5). Spinal cord sections from each group were analysed for demyelination as described in Section 2.5. The percentage of demyelinated white matter was found to be less in CD4 T cell enriched recipients in comparison to recipients of undepleted donor cells. No difference was observed in the level of infectious virus isolated from each adoptive transfer population, indicating that the more severe clinical disease observed in CD4 T cell enriched recipients was not due to inefficient viral clearance (data not shown).

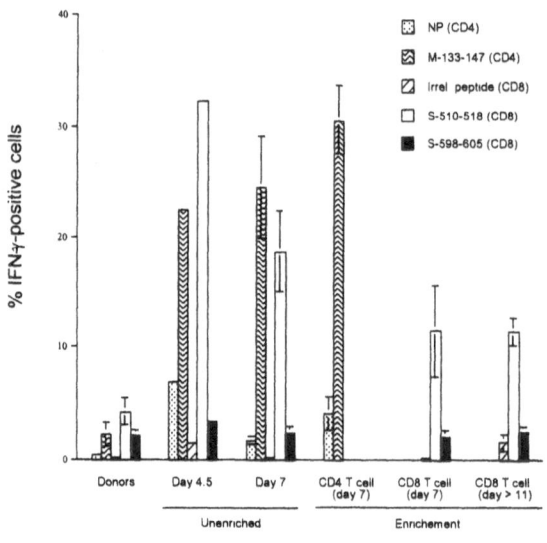

Figure 1. Quantity of MHV-specific CD4 and CD8 T cells before and after adoptive transfer. Donor splenocytes were isolated from B6 mice inoculated i.p. with MHV-JHM six days prior to adoptive transfer. The fraction of MHV-specific CD4 and CD8 T cells was determined by measuring the fraction of cells producing IFN-γ in response to peptides representing the respective epitopes (see section 2.3). Day 4.5 represents an average of 2 experiments, while all other values represent the average of at least 3 experiments.

No CD8 or CD4 T cells were detected in CD4 or CD8 T cell enriched recipients, respectively (Figure 1). Interestingly, the fraction of S-510-518-specific CD8 T cells was lower in the absence of CD4 T cells (Figure 1). Furthermore, CD8 T cell enriched recipients demonstrated an increase in the amount of demyelination and clinical disease at late timepoints (Figure 2), without a corresponding increase in the percentage of S-510-518 CD8 T cells (Figure 1).

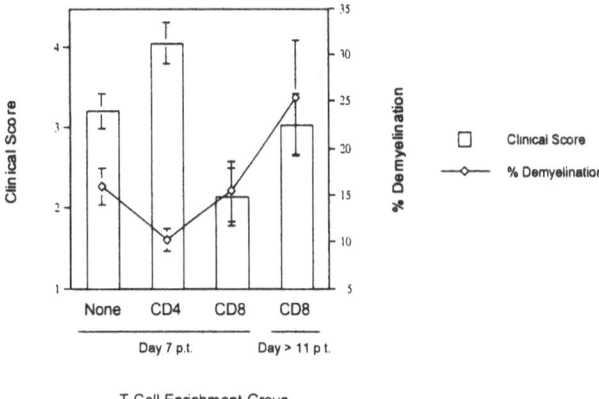

Figure 2. Clinical disease and demyelination in CD4 and CD8 T cell enriched recipients. Clinical scoring was performed using the scale described in Section 2.2. Percentage of demyelination within the spinal cord was determined as described in Section 2.5. A significantly lower percentage of demyelination was observed in CD4 T cell enriched recipients ($p < 0.05$) along with significantly more severe clinical disease ($p < 0.05$) than undepleted recipients.

4. CONCLUSIONS

1. A large percentage of MHV-specific T cells rapidly infiltrates the CNS following adoptive transfer. This rapid infiltration may be due to both the high levels of virus antigen present in the CNS of MHV-J2.2-v1-infected RAG1-/- mice prior to transfer and the activated state of donor cells. In addition, the expression of specific chemokines is likely to be important in the recruitment of T cells to the CNS. Previous reports have shown that a large fraction of CD8 T cells are specific for S-510-518 during acute and chronic MHV-induced CNS disease (Bergmann *et al.*, 1999; Pewe *et al.*, 1999).

2. In the absence of CD4 T cells, fewer MHV-specific T cells are detected within the CNS of adoptive transfer recipients. CD4 T cell help may be required for effective trafficking and/or survival of CD8 T cells

(Stohlman *et al.*, 1998). That these animals still developed demyelination, and that greater demyelination was observed without a corresponding increase in S-510-518-specific CD8 T cells at later timepoints, suggests that demyelination is induced after a certain threshold of cells is reached.

3. Neither CD4 nor CD8 T cells are essential for the development of MHV-induced demyelination. The process by which demyelination develops following adoptive transfer is therefore redundant to a degree. In contrast, CD4 T cells are critical to the development of clinical disease, inflammation and demyelination (Lane *et al.*, 2000). The studies reported herein specifically target the effector phase - rather than the induction phase - of the immune response, and therefore may be less susceptible to the depletion of CD4 T cells. The mechanisms by which CD4 and CD8 T cells induce demyelination may be unique or identical. Nonetheless, no single effector molecule has been shown to be essential. However, macrophages are most likely an important downstream mediator of demyelination (Wu and Perlman, 1999).

4. Greater clinical disease, accompanied by less demyelination, is observed in CD4 T cell-enriched adoptive transfer recipients. Macrophages may play a key role in the development of clinical disease, since a greater degree of F4/80-positive cell infiltration into the gray matter was observed in CD4 T cell enriched recipients (Wu *et al.*, 2000). Overall, these studies demonstrate the unique properties of CD4 and CD8 T cells in their contribution to MVH-JHM-induced demyelination and provide a foundation for further investigation into the mechanisms of virus-induced demyelination.

REFERENCES

Bergmann, C. C., Altman, J. D., Hinton, D., Stohlman, S. A. (1999). Inverted immunodominance and impaired cytolytic function of CD8+ T cells during viral persistence in the central nervous system. *J. Immunol.* **163**, 3379-3387.

Lane, T. E., Liu, M. T., Chen, B. P., Asensio, V. C., Samawi, R. M., Paoletti, A. D., Campbell, I. L., Kunkel, S. L., Fox, H. S., Buchmeier, M. J. (2000). A central role for CD4+ T-cells and RANTES in virus-induced central nervous system inflammation and demyelination. *J. Virol.* **74**, 1415-1424.

Perlman, S. (1998). Pathogenesis of coronavirus-induced infections: Review of pathological and immunological aspects. *Adv. Expt. Med. Biol.* **440**, 503-513.

Pewe, L., Heard, S. B., Bergmann, C. C., Dailey, M. O., Perlman, S. (1999). Selection of CTL escape mutants in mice infected with a neurotropic coronavirus: Quantitative estimate of TCR diversity in the infected CNS. *J. Immunol.* **163**, 6106-6113.

Stohlman, S. A., Bergmann, C. C., Lin, M. T., Cua, D. J., Hinton, D. R. (1998). CTL effector function within the central nervous system requires CD4+ T Cells. *J. Immunol.* **160**, 2896-2904.

Wu, G., Dandekar, A., Pewe, L., Perlman, S. CD4 and CD8 T cells have redundant but not identical roles in virus-induced demyelination. *J. Immunol.* in press.

Wu, G. F., Perlman, S. (1999). Macrophage infiltration, but not apoptosis, is correlated with immune-mediated demyelination following murine infection with a neurotropic coronavirus. *J. Virol.* **73**, 8771-8780.

Xue, S., Sun, N., van Rooijen, N., Perlman, S. (1999). Depletion of blood-borne macrophages does not reduce demyelination in mice infected with a neurotropic coronavirus. *J. Virol.* **73**, 6327-6334.

Acute CNS Infection is Insufficient to Mediate Chronic T Cell Retention

[1]NORMAN W. MARTEN, [2]MAUREEN HOHMAN, [2,3]STEPHEN A. STOHLMAN, [1]ROSCOE D. ATKINSON, [1]DAVID R. HINTON, AND [2,3]CORNELIA C. BERGMANN

Departments of [1]Pathology, [2]Molecular Microbiology and Immunology and [3]Neurology, Keck School of Medicine, University of Southern California, Los Angeles, CA 90033

1.   INTRODUCTION

CD8[+] T cells control infection by the neurotropic JHM strain of mouse hepatitis virus (JHMV) by eliminating infectious virus and reducing CNS pathology (Stohlman et al. 1995). Virus-specific CTL in BALB/c mice (H-2^d) respond almost exclusively to a single immunodominant epitope in the nucleocapsid (N) protein (Bergmann et al. 1993). During acute JHMV infection, CD8[+] T cells account for up to 40% of CNS mononuclear cells (MNC) (Bergmann et al. 1999). Following viral clearance, virus specific CD8[+] T cells remain in CNS for several weeks (Bergmann et al. 1999).

This report elucidates the requirement of persisting antigen (Ag) in retaining primary effector CD8[+] T cells within the CNS by comparing the responses to two antibody selected variants of JHMV, 2.2-V-1 and 2.2/7.2-V-2, abbreviated V-1 and V-2, respectively (Fleming et al. 1996 & 1997). V-1 forms a persistent CNS infection accompanied by demyelination (Fleming et al. 1996). By contrast, V-2 causes little to no demyelination (Fleming et al. 1997). Herein, we demonstrate that whereas V-1 persists in the form of vRNA within the spinal cord, infection with V-2 is transient and vRNA is reduced to below detectable limits. Analysis of CNS MNC revealed similar peaks of T cell infiltration during acute infection followed by clearance of both infectious viruses. However, in V-2 infected mice which clear vRNA, CNS T cells eventually decline to levels typical of naïve mice. By contrast T cells were maintained within the spinal cords of mice persistently infected with V-1. Virus mediated retention of infiltrating cells was specific for T cells, as other cell types including NK cells, B cells and peripheral macrophages, were not retained within the CNS of mice persistently infected with V-1.

2.   MATERIAL AND METHODS

Mice and viruses. Male BALB/c ($H-2^d$) mice were purchased from National Cancer Institute (Frederick, MD) at 6 weeks of age and infected within 1 week of arrival. Acute and persistent CNS infections were induced by intracranial injection of 1000 plaque forming units of JHMV variants 2.2-V-1 or 2.2/7.2-V-1 (Fleming et al. 1996 & 1997).

CNS MNC. MNC were derived separately from either spinal cords alone or both the brain and spinal cord of infected mice as described previously (Bergmann et al. 1999).

Determination of viral RNA. RNA isolation and RT-PCR amplification of viral N and host HPRT were performed as described (Marten et al. 2000b).

FACS analysis of CNS derived MNC. Surface markers were determined by staining with the following mAb: CD8 (53.67), CD4 (GK1.5), CD19 (1D3), CD45 (30-F11), panNK (DX5) (Pharmingen, San Diego, CA) and F4/80 (Cl:A3-1) (Serotech, Raleigh, NC) and the L^d-N318 MHC class-I tetramer has been described elsewhere (Bergmann et al. 1999).

3.   RESULTS

To confirm viral persistence following V-1 infection of BALB/c mice and establish that V-2 is cleared from the CNS, mice were analysed for the presence of vRNA from the viral N gene (Table 1) in brain and spinal cord. Following a single RT-PCR amplification, vRNA was detected in all RNA preparations from infected mice sacrificed at d 7 and 11 p.i. However, levels of vRNA fell sharply by d 33 p.i. and all samples from mice sacrificed at d 33 and 63 p.i. were reamplified with a nested N primer set. Following the second amplification, vRNA was detected in all but one spinal cord sample from V-2 infected mice at d 33 p.i. However, by d 63 p.i., vRNA was undetectable within the CNS of V-2 infected mice. Although vRNA was only detected in the brains of 2 out of 4 V-1 infected mice at d 63 p.i., vRNA was still present within spinal cords of all V-1 infected mice. No differences in product yield were observed following HPRT amplification nor was N RNA detected in naïve mice (data not shown). Although both virus strains replicate with similar efficiencies (Marten et al. 2000a), V-1 preferentially infects spinal cords during persistence, whereas V-2 vRNA was cleared from the CNS.

Table 1. Persistence of V-1 and V-2 vRNA within the CNS

Virus strain	Brain				S.C.			
	d 7	D 11	d 33	d 63	d 7	D 11	d 33	d 63
V-1	4/4	4/4	4/4	2/4	4/4	4/4	4/4	4/4
V-2	4/4	4/4	4/4	0/4	4/4	4/4	3/4	0/4

[a]Number of mice with persisting viral N RNA within the indicated CNS tissue as determined by RT-PCR.

Spinal cords were monitored for T cell infiltration to determine the role of potential class I Ag presentation associated with persisting vRNA in retaining CD8[+] T cells within the CNS. Mice infected with both V-1 and V-2 were sacrificed during and following resolution of acute infection. Peak infiltration was observed on d 10 p.i. when CD8[+] T cells accounted for 29 to 34 % of the MNC population in mice infected with both viruses (Table 2). 37 to 43 % of CNS CD8[+] T cells expressed T cell receptors specific for the JHMV-N epitope as indicated by tetramer staining. Although infectious virus has been cleared for several weeks by d 33 p.i., CD8[+] T cells were still present in spinal cords of mice infected with both viruses, albeit at two to three fold lower levels compared to peak frequencies. However, by d 70 p.i. at which time V-2 RNA was no longer detectable, the frequency of CD8[+] T cells within the CNS had dropped to levels typical of naïve mice. By contrast, both tetramer[+]CD8[+] T cells and tetramer[-]CD8[+] T cells remained within the spinal cords of mice persistently infected with V-1 at d 70 p.i. These data suggest that persisting vRNA within the CNS is required to maintain the continued presence of both N-specific CD8[+] T cells as well as CD8[+] T cells with yet undefined specificity.

Table 2. Frequency of MNC isolated from spinal cord[a]

Cell Types	Virus	d 7 p.i.	d 10 p.i.	d 33 p.i.	d 70 p.i.
CD8[+] T cells	V-1	30.0 % (35.9)[b]	35.0 % (42.9)	14.8 % (25.7)	8.5% (30.6)
	V-2	20.4 % (30.4)	29.3 % (36.9)	8.5 % (36.5)	0.3 % (33.3)
CD4[+] T cells	V-1	19.9 %	18.6 %	9.5 %	10.7 %
	V-2	8.3 %	21.8 %	2.6 %	1.6 %
B cells	V-1	2.3 %	1.7 %	1.4 %	0.6 %
	V-2	ND[c]	1.8 %	1.1 %	0.3 %
NK cells	V-1	11.9 %	3.3 %	1.9 %	1.1 %
	V-2	ND	1.7 %	ND	ND

[a]The percentages of MNC types isolated from spinal cord were determined by flow cytometry. [b]Numbers in parenthesis represent the percentage of tetramer[+]CD8[+] T cells among total CD8[+] T cells. [c]ND: not determined.

To determine if the influence of persisting vRNA was specific for retention of CD8[+] T cells, MNC from spinal cords of both V-1 and V-2 infected mice were also examined for CD4[+] T cells by flow cytometry. Both viruses induced similar peak infiltration of CD4[+] T cells, comprising 19 to 22 % of spinal cord MNC, at d 10 p.i. (Table 2). In contrast to CD8[+] T cells, the frequency of CD4[+] T cells in spinal cords of V-1 infected mice was at least twice that obtained from V-2 infected mice at both early (day 7 p.i.) and later time points (d 33 and 70 p.i.). By d 70 p.i. CD4[+] T cells comprised only 1.6% of the total spinal cord MNC population from V-2 infected mice. By contrast, CD4[+] T cells still accounted for greater than 10% of MNC isolated from spinal cords of mice persistently infected with V-1 at d 70 p.i. These data suggest that CD4[+] as well as CD8[+] T cells are retained in conjunction with persisting vRNA.

To assess retention or ongoing recruitment of other infiltrating cell types, spinal cord MNC were examined for B cells, NK cells and macrophages/microglia. CD19[+] B cells comprised a minor population of no more than 2.3% of spinal cord MNC at any time point examined (Table 2). NK cells accounted for approximately 12 % of MNC from V-1 infected mice at the peak of their infiltration at d 7 p.i. (Table 2). However, at all later points NK cells comprised < 3 % of the MNC population. Macrophages/ microglia make up approximately 30 to 35 % of the MNC population at d 7 p.i. but become the dominant population at later time points as the frequencies of other MNC types decline (data not shown). To distinguish parenchymal microglia and macrophages during infection, CNS MNC were analysed for expression of CD45, which marks bone marrow derived cells and F4/80. Parenchymal microglia are characterized by a CD45[lo]/F4/80[+] phenotype, whereas blood borne macrophages are CD45[hi]/F4/80[+] (Carson et al. 1998). At d 8 p.i., infiltrating macrophages represent approximately 30% of F4/80[+] (phagocytic) CNS cells (Table 3). However, by d 62 p.i., this population had diminished to only 2%; the remainder being represented by CD45[lo]/F4/80[+] microglia. These data suggest that unlike T cells, other MNC which infiltrate the CNS during acute V-1 infection do not remain within the CNS of persistently infected mice.

Table 3. Infiltrating macrophages are present during acute but not chronic stages of infection[a].

Cell Types	Naïve	d 8 p.i.	d 62 p.i
Microglia	99 %	70 %	98%
Macrophage	1 %	30%	2%

[a]MNC were isolated from CNS of naïve and V-1 infected mice sacrificed at d 8 and 62 p.i. Cells were stained with CD45 and F4/80 directly *ex vivo* to differentiate between resident microglia (CD45[lo]/F480[hi]) and infiltrating macrophages (CD45[hi]/F4/80[hi]).

4.   DISCUSSION

CNS infection with JHMV induces a vigorous CD8$^+$ CTL response, which rapidly clears infectious virus (Stohlman et al. 1995). However, CD8$^+$ T cells remain in the CNS for several months following resolution of acute JHMV infection (Bergmann et al. 1999). To identify mechanisms involved in T cell retention, particularly the role of persisting virus, the immune response to two distinct JHMV variants were analysed. V-1 initiates a persistent infection within the spinal cord whereas V-2 is cleared below detection following transient infection of the spinal cord. Infection with either virus induced similar levels of peak CD4$^+$ and CD8$^+$ T cell responses. However, during persistence both CD8$^+$ and CD4$^+$ T cells were retained in the spinal cords of V-1 infected mice in conjunction with persisting vRNA. By contrast, spinal cords of V-2 infected mice were essentially free of peripheral MNC infiltrates following clearance of vRNA. These data indicate that persisting vRNA is required to maintain T cells within the CNS, possibly acting as a source of protein for ongoing Ag presentation.

The ratio of tetramer$^-$CD8$^+$ T cells to tetramer$^+$CD8$^+$ T cells remained roughly constant throughout the course of infection with either strain of virus. Furthermore, tetramer$^-$CD8$^+$ T cells were retained along with virus specific CD8$^+$ T cells within the spinal cords of V-1 infected mice during persistence. This result might be explained by the inability of the pN-L^d tetramer to distinguish a large fraction of N-specific T cell pool due to low expression of TcR or the presence of CD8$^+$ T cells with specificity to as yet undefined virus epitopes. Since both tetramer$^+$CD8$^+$ and tetramer$^-$CD8$^+$ T cells remain in spinal cords of V-1 infected but not in V-2 infected mice, it is unlikely that tetramer$^-$CD8$^+$ T cells are specific for host epitope(s).

The mechanism for retention of peripheral MNC appears to be Ag specific as neither NK cells nor peripheral macrophages were retained within the CNS of V-1 infected mice. Determination of infiltration by B cells based on CD19 expression may provide an under-estimation due to the loss of distinctive surface markers during differentiation into plasma cells. Analysis of an influenza model of CNS infection has suggested that retention of CD8$^+$ T cells in the CNS is independent of viral Ag (Hawke et al. 1998). However, the CTL responding to the influenza infection were derived from a memory population previously activated and expanded by a peripheral immunization. Thus, whereas memory T cells may be retained within the CNS independent of chronic Ag, the data presented here suggests that retention of T cells responding to a primary infection within the CNS requires the presence of persisting virus.

REFERENCES

Bergmann, C. C. , M. McMillan, and S. A. Stohlman. Characterization of the L^d-restricted cytotoxic T lymphocyte epitope in the mouse hepatitis virus nucleocapsid protein. J. Virol. 1993;67:7041-7049.

Bergmann, C. C., J. D. Altman, D. Hinton and S. A. Stohlman. Inverted Immunodominance and impaired cytolytic function of CD8$^+$ T cells during viral persistence in the CNS. J. Immunol. 1999;163:3379-3387.

Carson, M. J., C. R. Reilly, J.G. Sutcliffe and D. Lo. Mature microglia resemble immature antigen-presenting cells. Glia. 1998. 22:72-85.

Fleming, J. O., M. Trousdale, F. E. Zactarim, S. A. Stohlman and L. P. Weiner. Pathogenicity of antigenic variants of murine coronavirus JHM selected with mAb. J. Virol. 1986;58:869-875.

Fleming, J. O., M. Trousdale, S. A. Stohlman and L. P. Weiner. Pathogenic characteristics of neutralization-resistant variants of JHM coronavirus (MHV-4). Adv. Exp. Med. Biol. 1987;218:333-342.

Hawke, S., P. G. Stevenson, S. Freeman, and C. R. M. Bangham. Long-term persistence of activated cytotoxic T lymphocytes after viral infection of the central nervous system. J. Exp. Med. 1998;187:1575-1582.

Marten, N. W., S. A. Stohlman, R. D. Atkinson, D. R. Hinton, J. O. Fleming and C. C. Bergmann. Contributions of CD8$^+$ T cells and viral spread to demyelinationg disease. 2000a. in press.

Marten, N. W., S. A. Stohlman and C. C. Bergmann. Role of viral persistence in retaining CD8+ T cells within the CNS. 2000b. in press.

Stohlman, S. A., C. C. Bergmann, R. van der Veen and D. Hinton. Mouse hepatitis virus-specific cytotoxic T lymphocytes protect from lethal infection without eliminating virus from the central nervous system. J. Virol. 1995;69:684-69.

Generation from Multiple Sclerosis Patients of Long-Term T-Cell Clones that are Activated by both Human Coronavirus and Myelin Antigens

[1]ANNIE BOUCHER, [1]FRANÇOIS DENIS, [2]PIERRE DUQUETTE, AND [1]PIERRE J. TALBOT

[1]Human Health Research Center, INRS-Institut Armand-Frappier, Université du Québec, Laval, Québec, CANADA H7V 1B7; [2]MS Clinic, Notre-Dame Hospital, Montréal, Québec, CANADA H2L 4K

1. INTRODUCTION

Multiple sclerosis (MS) is a chronic central nervous system (CNS) disorder characterized by inflammation and myelin destruction involving autoreactive lymphocytes. While precise disease etiology is still unknown, it appears to be multifactorial, involving both genetic (such as HLA haplotype) and environmental factors (Sadvonick *et al.* 1993, Ebers *et al.* 1993). Various microbial infections have been proposed to be associated with triggering the pathology, but no conclusive evidence has been drawn so far.

Human coronaviruses (HCoV), of which two serotypes, OC43 and 229E are known, cause 10 to 35% of common colds. It has been hypothesized that a natural infection by these ubiquitous respiratory pathogens may lead in some genetically predisposed individuals to a demyelinating pathology. This rests mainly on observations that experimental murine coronavirus infections of genetically susceptible mice and rats induces an MS-like disease: CNS inflammation, cycles of demyelination and remyelination, and activation of myelin-reactive T cells (Lane *et al.* 1997, Lampert *et al.* 1973, Weiner *et al.* 1973, Watanabe *et al.* 1983). Moreover, accumulating evidence is consistent with the neurotropic and neuroinvasive potential of

The Nidoviruses (Coronaviruses and Arteriviruses).
Edited by Ehud Lavi *et al.*, Kluwer Academic/Plenum Publishers, 2001.

HCoV: 1) acute and persistent infection of neural cell lines (Collins and Sorensen 1996, Arbour *et al.* 1999a,b); 2) infection of primary cultures of astrocytes, microglia and cerebral endothelial cells (Cabirac *et al.* 1995, Bonavia *et al.* 1997); 3) viral isolation from MS brains (Burks *et al.* 1980); 4) RT-PCR and *in situ* hybridization detection of both HCoV-229E and HCoV-OC43 in the brain, with a potential association with MS (Murray *et al.* 1992, Stewart *et al.* 1992, Arbour *et al.* 2000). Microbial pathogens might activate MS-associated autoreactive lymphocytes through molecular mimicry in individuals that are genetically predisposed by their HLA haplotype to recognize T-cell epitopes shared between the pathogen and myelin antigens (Oldstone 1997, Wucherpfennig and Strominger 1995, Cantor *et al.* 1998).

To evaluate whether molecular mimicry between HCoV and myelin antigens could participate in MS pathogenesis, T-cell cross-reactivity between myelin basic protein (MBP), a putative MS autoantigen and HCoV-229E was investigated in MS patients (Talbot *et al.* 1996). Results summarized in Table 1 show that of the numerous peripheral HCoV-229E or MBP-reactive T-cell lines generated, cross-reactivity involved 29% of T-cell lines from 10 of 16 MS patients and only 1.3% in 2 of 14 controls ($p<0.0001$). To ascertain whether this MS-associated T-cell cross-reactivity could be linked to molecular mimicry, it was mandatory to demonstrate activation of T cells by both virus and myelin antigens at the single-cell level. Our previous study was also extended to HCoV-OC43 and another major myelin component and putative MS autoantigen, proteolipid protein (PLP). Importantly, both MBP and PLP can induce experimental allergic encephalomyelitis (EAE) in rodents (Wekerle *et al.* 1994). We now describe the methodology used to produce and expand human T cell clones specific for coronaviral and myelin antigens and the production of cross-reactive T-cell clones.

Table 1. MS-associated cross-reactivity of peripheral T-cell lines for MBP and HCoV-229E (results from Talbot et al. 1996)

Donors	n	Cross-reactive T-cell lines		
		Lines	n	%
MS	16	39	10	29
Controls	14	2	2	1.3

2. MATERIALS AND METHODS

2.1 Donors

Patients diagnosed with either chronic progressive or relapsing remitting MS and who had not received treatment in the last six months were selected at random with prior informed consent and approval of the experimental protocol from institutional ethics committees.

2.2 Antigens

Human MBP was prepared from a normal adult brain (Montreal Brain Bank, Douglas Hospital, Verdun, Québec, Canada) as described (Talbot *et al.* 1996). Chromatography-purified PLP was a kind gift of Dr. Mario Moscarello (Hospital for Sick Children, Toronto, Ontario, Canada). HCoV-229E was propagated in the human embryonic lung cell line L132 (ATCC). Viral antigen was prepared from L132 cells infected with HCoV-229E at a MOI of 0.01 at 33°C for 43 hours. Cells were lysed and clarified by low speed centrifugation. Before use, infectious virus in the viral antigen preparation (10^6 $TCID_{50}$/mL) was inactivated by exposure to ultra-violet light for 20 minutes. HCoV-OC43 was propagated in the human rectal carcinoma cell line HRT-18 (ATCC), cultivated and harvested as described above. Control antigen was obtained from parallel cultures. HHV-6 antigens in the form of infected cell lysates were provided by Dr. Louis Flamand, Laval University, Québec, Canada.

2.3 Generation and Maintenance of T-Cell Clones

Human blood samples (300-400 mL) were collected from human donors and buffy coat prepared. Peripheral blood lymphocytes (PBL) were separated on Ficoll/Hypaque (Pharmacia) gradients and 1×10^6 cells/mL were seeded into 96-well microtiter plates (Talbot *et al* 1996). After 7 days of incubation, 100 U/mL of interleukin-2 (Hoffman La Roche) was added. At day 12, primary T-cell lines were tested for antigen specificity by a tritiated thymidine incorporation assay. On day 16, antigen-specific lines were cloned by limiting dilution into 60-microwell Terazaki plates. A concentration of 5×10^5 autologous irradiated antigen-presenting cells/ well (APC; PBL prepared as described) were added, in the presence of 30 µg/mL of either MBP or PLP (6 µg/well) or a 1/20 final dilution of viral antigens. Interleukin-2 (50 U/mL) was added to enhance cloning efficiency. After incubation for 10-13 days, each well was checked for cell growth. Cells from plates showing growth in less than 33% of the wells were selected. Clones were tested for specificity in a proliferation assay but were also expanded in the presence of autologous APC and the specific antigen (MBP, PLP, HCoV-229E or OC43).

2.4   Antigen-Specific Proliferation Assays

Proliferation assays were performed by adding 1×10^4 T cells per well into 96-well round bottom microtiter plates in the presence of irradiated APC and various antigens in complete culture medium, as described above (0.1 mL/well). These specific and control antigens were MBP and PLP (300 µg/well), HCoV-229E and -OC43 viral antigens in infected cell lysates (final dilution of 1/2), tetanus toxin (1 µg), appropriate dilutions of uninfected cell lysates, and serum-free complete culture medium. Cells were incubated for 72 hours, with 1 µCi/well of tritiated-thymidine (Amersham) added for the last 18 hours. Cells were harvested onto glass microfiber filters (Skatron) on a 96-well Skatron model 11050 Micro cell harvester and counted in 2 mL of Ultimagold scintillation fluid (Packard), using a Canberra Packard Tri-Carb 2200a scintillation counter. The stimulation index was calculated as the ratio of the radioactivity (cpm) incorporated in the presence of specific over control antigen and a ratio above 3.0, with at least 1000 cpm incorporated was considered positive.

3.   RESULTS

3.1   Human T-cell clones specific to coronaviral and/or myelin antigens

From 32 MS patients, the protocol described above led to the generation of 114 T-cell clones responding to viral antigens and 31 T-cell clones activated by myelin antigens, (Table 2). Positive clones showed proliferation indices between 3.0 and 19.0. A total of 10 cross-reactive T-cell clones were obtained from 6 patients that were diagnosed with either chronic progressive or relapsing remitting MS.

Table 2. T-cell clones generated from peripheral blood of MS patients

Donors	Total number of T-cell clones generated		
(MS)	HCoV Monospecific	MBP or PLP Monospecific	Virus-Myelin Cross-Reactive
32	114	31	10

3.2 Antigen specificity of the cross-reactive T-cell clone

The antigenic specificities of the ten coronavirus-myelin cross-reactive T-cell clones generated in this study from 6 MS patients are listed in Table 3. Using the same protocol, human herpesvirus type-6 (HHV-6), another virus potentially associated with MS (Challoner *et al.*, 1995), was also tested. Although virus-specific T-cell clones were generated, no HHV-6 and myelin antigen cross-reactive clones were obtained (data not shown).

Table 3. Antigen specificity of cross-reactive T-cell clones

Clone #	Selecting antigen	HCoV-OC43	HCoV-229E	PLP	MBP
1	OC43	+	-	-	+
2	229E	-	+	-	+
3	PLP	-	+	+	-
4	PLP	-	+	+	-
5	MBP	-	+	-	+
6	MBP	-	+	-	+
7	OC43	+	-	+	+
8	229E	-	+	-	+
9	229E	-	+	+	+
10	229E	-	+	-	+

4. DISCUSSION

Our results are consistent with clonal T-cell cross-reactivity between either of the two known HCoV serotypes (229E and OC43) and two major myelin antigens and putative MS autoantigens (MBP and PLP). Even though, following Poisson's distribution, a clone is statistically obtained when growth is observed in less then 33% of seeded wells after cloning by limiting dilution, clonality will be confirmed by molecular techniques as well.

We were successful in elaborating a protocol to generate and stimulate human T-cell clones after a single blood sampling, despite the technical challenges involved in studying primary T cells in cultures. Autologous antigen-presenting cells were frozen and later thawed and irradiated when needed. This technique was thus independent of the patient's clinical status or his/her availability for blood samplings. The preservation of the clone's antigenic specificity was of premier concern and we were able to freeze and thaw the clones with no loss of antigenic specificity. Nevertheless, we remain concerned that anergy or apoptosis of T cells can occur upon repeated stimulations. Our protocol does consider these limitations and allows for the rapid determination of antigenic specificity, in order to preserve the T cells in a state that will allow freezing and later recovery of viable and functional cells. Nevertheless, we are currently making use of techniques such as immortalization of T-cell clones by infection with *herpes saimiri* virus, as well as T-cell receptor (TCR) reconstitution into hybridomas to allow for the detailed molecular characterization of the putative molecular mimicry mechanism underlying the observed myelin-coronavirus cross-reactivities.

Human coronavirus-myelin cross-reactive T-cell clones represent precious tools to understand the triggering of autoimmunity by molecular mimicry in the context of a human disease. How susceptibility is determined could be related to peptide presentation. Indeed, we are in the process of typing the HLA haplotypes of donors to look for any DR2 associations. Molecular mimicry involving viruses represents a very attractive and unifying mechanism that can nicely explain T-cell activation against myelin determinants and the possible involvement of several microbial pathogens in MS. Moreover, cytokines and inflammation could also act as key players by increasing antigen presentation and cellular recruitment. Other viruses could be involved in MS through similar mechanims, although our preliminary study did not reveal T-cell cross-reactivity with HHV-6. Finally, we will search for the presence of coronavirus-myelin cross-reactive T-cell clones in the CNS of MS patients and their possible involvement in local pathology.

5. CONCLUSION

Our observation that a respiratory virus can activate myelin-reactive T cells would explain the influence of environmental factors in disease etiology and add much credence to the possible role of such viruses in the triggering of MS in individuals that are genetically predisposed to such putative molecular mimicry at the level of T-cell recognition of foreign and self-antigens. Activation of such cells could also explain why MS relapses often follow respiratory infections. Moreover, a persistent CNS infection could represent a precipitating factor favoring local inflammation and contribute to recurrent disease exacerbations.

ACKNOWLEDGMENTS

This work was funded by the Multiple Sclerosis Society of Canada. We thank Dr. Mario Moscarello (University of Toronto) for providing PLP, and Dr. Louis Flamand (Université Laval) for providing HHV-6 antigens.

REFERENCES

Arbour, N., Ekandé, S., Côté, G., Lachance, C., Chagnon, F., Tardieu, M., Cashman, N.R., and Talbot, P.J., 1999a, Persistent infection of human oligodendrocytic and neuroglial cell lines by human coronavirus 229E, *J. Virol.* **73**:3326-3337.

Arbour, N., Côté, G., Lachance, C., Tardieu, M., Cashman, N.R., and Talbot, P.J., 1999b, Acute and persistent infection of human neural cell lines by human coronavirus OC43, *J. Virol.* **73**:3338-3350.

Arbour, N., Day, R., Newcombe, J., and Talbot, P.J., 2000, Neuroinvasion by human respiratory coronaviruses, *J. Virol.,* in press.

Bonavia, A., Arbour, N., Yong, V.W., and Talbot, P.J., 1997, Infection of primary cultures of human neural cells by human coronaviruses 229E and OC43, *J. Virol.* **71**:800-806.

Burks, J.S., DeVald, B.L., Jankovsky, L.D., and Gerdes, J.C. (1980) Two coronaviruses isolated from central nervous system tissue of two multiple sclerosis patients, *Science* **209**:933-934.

Challoner, P.B., Smith, K.T., Parker, J.D., Macleod, D.L., Coulter, S.N., Rose, T.M., Schultz, E.R., Bennett, J.L., Garber, R.L., Chang, M., Schad, P.A., Sewart, P.M., Nowinski, R.C., Brown, J.P., and Burmer, G.C., 1995, Plaque-associated expression of human herpesvirus 6 in multiple sclerosis, *Proc. Natl Acad. Sci. USA* **92**:7440-7444.

Collins, A.R., and Sorensen, O., 1986, Regulation of viral persistence in human glioblastoma and rhabdomyosarcoma cells infected with coronavirus OC43, *Microb. Pathog.* **6**:573-82.

Ebers, G.C., and Sadovnick, A.D., 1993, The role of genetic factors in multiple sclerosis susceptibility, *J. Neuroimmunol.* **54**:1-17.

Lane, T.E., and Buchmeier, M.J., 1997, Murine coronavirus infection : a paradigm for virus-induced demyelinating disease, *Trends Microbiol.* **5**:9-14.

Murray, R.S., Brown, B., Brian, D., and Cabirac, G.F., 1992, Detection of coronavirus RNA and antigen in multiple sclerosis brain, *Ann. Neurol.* **31**:525-533.

Oldstone M.B.A., 1987, Molecular mimicry and autoimmune disease, *Cell* **50**:819-820.

Sadovnick, A.D., Armstrong, H., Rice, G.P., Bulman, D., Hashimoto, L., Paty, D.W., Warren, S., Hader, W., Murray T.J., *et al.,* 1993, A population-based study of multiple sclerosis in twins: update, *Ann. Neurol.* **33**:281-285.

Stewart, J.N., Mounir, S., and Talbot, P.J., 1992, Human coronavirus gene expression in the brains of multiple sclerosis patients, *Virology* **191**: 502-505.

Talbot, P.J., Paquette, J.-S., Ciurli, C., Antel, J.P., and Ouellet, F., 1996, Myelin basic protein and human coronavirus 229E cross-reactive T cells in multiple sclerosis, *Ann. Neurol.* **39**:233-240.

Watanabe, R., Wege, H., and ter Meulen, V., 1983, Adoptive transfer of EAE-like lesions from rats with coronavirus-induced demyelinating encephalomyelitis, *Nature* **5930**:150-153.

Weiner, L.P., 1973, Pathogenesis of demyelination induced by a mouse hepatitis, *Arch. Neurol.* **5**:298-303.

Wekerle, H., Kojima, K., Lannes-Vieira, J., Lassmann, H., and Linington, C., 1994, Animal models, *Ann. Neurol.* **36**:S47-53.

Wucherpfennig, K.W., and Strominger, J.L., 1995, Molecular mimicry in T cell-mediated autoimmunity: viral peptides activate human T cell clones specific for myelin basic protein, *Cell* **80**:695-705.

Zhao, Z.S., Granucci, F., Yeh, L., Schaffer, P.A., and Cantor, H., 1998, Molecular mimicry by herpes simplex virus-type 1: autoimmune disease after viral infection, *Science* **279**:1344-1347.

The Role of B Cells in Mouse Hepatitis Virus Infection and Pathology

[1]AMY E. MATTHEWS, [1]SUSAN R. WEISS, [2]EHUD LAVI, [3]MARK SHLOMCHIK, AND [1]YVONNE PATERSON

[1]Department of Microbiology and [2]Department of Pathology, University of Pennsylvania, Philadelphia, PA; [3]Department of Laboratory Medicine and Section of Immunobiology, School of Medicine, Yale University, New Haven, CT.

1. INTRODUCTION

Murine hepatitis virus (MHV) strain A59 infects the central nervous system (CNS) and liver of mice, causing encephalitis and hepatitis. Virus is cleared from both organs by 16 days post-infection (dpi) (Lavi et al, 1986; data not shown). Infection of the CNS by this virus is associated with demyelination. Interestingly, demyelination peaks at around 30 dpi, after infectious virus is cleared and only viral RNA and small amounts of protein remain (Lavi et al, 1984a, 1984b; Sutherland et al, 1997; data not shown).

Lymphocytes are necessary to control the viral infection, and are considered critical for the development of demyelination also (Wang et al, 1990; Williamson and Stohlman, 1990; Houtman and Fleming, 1996). Although much effort has been devoted to understanding the role of T cells in the control of viral infection and development of demyelination, the other major populations of lymphocytes, the B cells, have been more difficult to study. We used several recent knockout strains of mice to investigate whether B cells are important in immunological control of viral replication or in MHV-A59 induced demyelination.

The Nidoviruses (Coronaviruses and Arteriviruses).

Edited by Ehud Lavi et al., Kluwer Academic/Plenum Publishers, 2001.

2.   VIRAL PERSISTENCE IN THE CNS IN THE ABSENCE OF B CELLS

To determine whether B cells are important for clearance of virus from the CNS, we injected 10-50 pfu MHV-A59 intracerebrally (i.c.) into mice that have defects in various pathways of B cell activity (Table 1). Thirty days after infection, we collected brains from euthanized, PBS perfused animals.   Virus from brain homogenate was titered on L2 cells as previously described (Gombold et al, 1993).

Infectious virus was successfully cleared in various strains of immunocompetent wildtype mice, but virus persisted in mice that lacked B cells or the ability to secrete IgG (Jackson Laboratory; Chan et al, 1999; Mark Shlomchik, personal communication).   These results suggest that antibodies, especially IgG, are important for clearance of virus.   Mice that lack the ability to activate the complement cascade (Circolo al, 199) or lack the major receptors for IgG (FcR) (Taconic Laboratories) clear virus, suggesting that antibody clears MHV-A59 by a mechanism independent of complement activation or FcR signalling.

Table 1. Persistence of infectious virus in the brains of mice deficient either in B cells, antibodies, or mechanisms of antibody activity.

Mouse strain	Deficiency	Presence of infectious virus in the CNS
C57BL/6	none	-
Igh6 (μMT)/B6	No B cells	+
C.B-17	none	-
JhD/C.B-17	No B cells	+
Meg/C.B-17	No class switching	+
Bonnie/C.B-17	No secreted antibody	+
FcRγsubchain knockout/B6	No FcRγI or III	-
C3 knockout/B6	No complement cascade	-

3.   LATE MORTALITIES IN B CELL DEFICIENT MICE ARE ASSOCIATED WITH HIGH VIRAL TITERS

In wildtype C57BL/6 or C.B-17 mice, MHV-A59 associated mortality occurs within 30 days (Lavi et al, 1984a).   After i.c. infection of B cell deficient Igh6 mice with MHV-A59, we observed that while these mice sometimes died acutely, they were also dying at late time points. In comparison, uninfected Igh6 mice did not die as they grew older. When viral titers were measured from brains collected from moribund or recently dead animals, we discovered that mortality was associated with very high viral titers in the CNS (Table 2). These titers were comparable to titers we observed at the peak of acute infection in C57BL/6 mice (average of $\log_{10}6.1 \pm 0.3$ from 3 experiments in C57BL/6 mice infected with 50 pfu MHV-A59 i.c.).

Table 2. Brain viral titers in B cell deficient mice that died late after infection. Mice were infected with MHV-A59 at 4 weeks of age.

Days post-infection	Dose infected with (pfu)	Brain viral titer (pfu/g, log base 10)
21	5000	6.54
26	5000	6.68
48	5000	5.88
60	500	6.70
62	500	5.40
72	50	5.94
81	5000	5.58
85	500	6.28
91	50	5.68

4.   DEMYELINATION IN THE ABSENCE OF B CELLS

Data from β_2 microglobulin knockout mice who lack normal CD8+ T cells and from CD4+ and CD8+ depleted mice infected i.c. with MHV-A59 demonstrated that CNS demyelination can occur in the absence of the major T cell populations (Sutherland et al, 1997; Gombold et al, 1995). Therefore, we hypothesized that the other major lymphocyte population, the B cells, could be important in the process of demyelination. We checked for demyelination by collecting spinal cords 30 dpi, embedding them in Epon, and staining sections with toluidine blue (Lavi et al, 1984a). Primary demyelination with preservation of axonal structure was observed (Fig. 1).

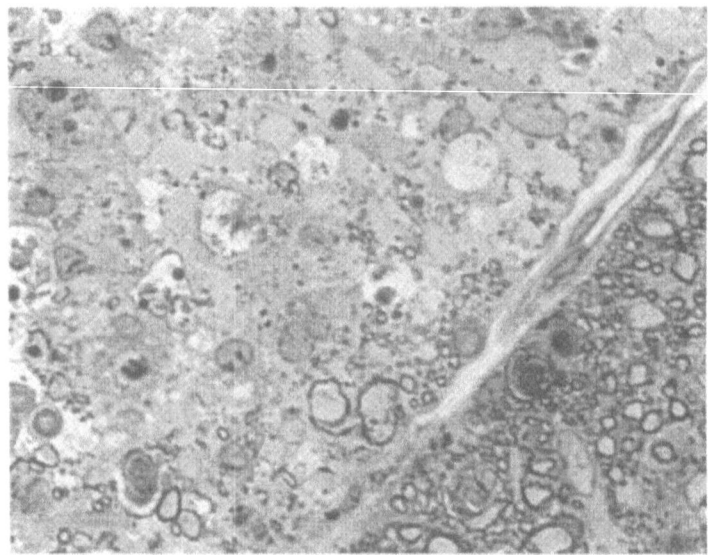

Figure 1. Demyelination in the white matter of the spinal cord of a B cell deficient mouse. Each dark ring represent the fatty myelin sheath coiled around a transverse section of a neuronal axon. These sheaths are common in the lower right corner. In the left side of the picture, very pale circles reveal the bare axons left behind after demyelination.

5. NORMAL T CELL ACTIVATION IN B CELL DEFICIENT MICE

Studies have demonstrated that T cell responses can be decreased in B cell deficient mice, although the requirement for B cells as antigen presenting cells for T cell activation varies depending on the antigen or organism used to prime for a response (Constant et al, 1995; Phillips et al, 1996). To determine whether CD4+ and CD8+ T cell responses are normal during MHV-A59 infection, we determined the expression of the activation marker CD69 on T cells in Igh6 and B6 mice infected with 10^5 pfu virus i.p. Splenocytes were collected 4 dpi during the peak of CD69 expression in the spleen, stained with fluorescently labelled antibodies against CD4, CD8, and/or CD69, and counted by flow cytometry. We observed no significant differences in the percentage of activated T cells in Igh6 mice as compared to wildtype mice.

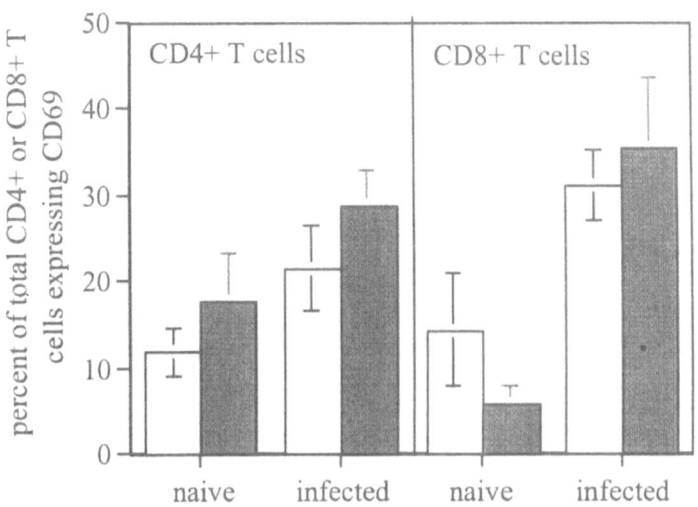

Figure 2. Activation of T cells in the absence of B cells as antigen presenting cells. White columns represent C57BL/6 mice; dark columns represent B cell deficient Igh6 mice. There were 6 mice per group. The percentage of activated cells was significantly increased (p<0.006) for each infected group as compared to the corresponding naïve group.

6. CONCLUSION

B cells are critical for the clearance of MHV-A59 from the CNS. In the absence of B cells and viral clearance, very high viral titers develop which may cause the deaths observed late after infection. Although B cells are required for viral clearance, they are not required for CNS demyelination. Finally, the persistence of viral titers is not due to defective T cell activation in the absence of B cells as antigen presenting cells, since early T cell activation as revealed by CD69 expression was normal in Igh6 mice.

ACKNOWLEDGMENTS

This work was supported by MS grant #RG2585 (SW), NIH-RO1-A143603 (MS), and GM31841 (YP).

REFERENCES

Chan O.T., Hannum L.G., Haberman A.M., Madaio M.P., and Shlomchik M.J. , 1999, A novel mouse with B cells but lacking serum antibody reveals an antibody-independent role for B cells in murine lupus. *J Exp Med.* **189**: 1639-1648.

Circolo A, Garnier G., Fukuda W., Wang X., Hidvegi T., Szalai A.J., Briles D.E., Volanakis J.E., Wetsel R.A., and Colten H.R. , 1999, Genetic disruption of the murine complement C3 promoter region generates deficient mice with extrahepatic expression of C3 mRNA. *Immunopharmacology.* **42**: 135-49.

Constant, S., Schweitzer, N., West, J., Ranney, P., and Bottomly, K., 1995, B lymphocytes can be competent antigen-presenting cells for priming CD4+ T cells to protein antigens in vivo. *J. Imm.* **155**: 3734-3741.

Gombold, J.L., Hingley, S.T., and Weiss, S.R., 1993, Fusion-defective mutants of mouse hepatitis virus A59 contain a mutation in the spike protein cleavage signal. *J. Virol.* **67**: 4505-4512.

Gombold, J.L., Sutherland, R.M., Lavi, E., Paterson, Y., and Weiss, S.R., 1995, Mouse hepatitis virus A59-induced demyelination can occur in the absence of CD8+ T cells. *Microbial Pathogenesis.* **18**: 211-221.

Houtman, J.J., and Fleming, J.O,, 1996, Dissociation of demyelination and viral clearance in congenitally immunodeficient mice infected with murine coronavirus JHM. *J. NeuroVirology.* **2**: 101-110.

Lavi, E., Gilden, D.H., Wroblewska, Z., Rorke, L.B., and Weiss, S.R., 1984a, Experimental demyelination produced by the A59 strain of mouse hepatitis virus. *Neurology.* **34**: 597-603.

Lavi, E., Gilden, D.H., Wroblewska, Z., Rorke, L.B., and Weiss, S.R., 1984b, Persistence of mouse hepatitis virus A59 RNA in a slow virus demyelinating infectionin mice as detected by in situ hybridization. *J. Virol.* **51**: 563-566.

Lavi, E., Gilden, D.H., Highkin, M.K., and Weiss, S.R., 1986, The organ tropism of mouse hepatitis virus A59 in mice is dependent on dose and route of inoculation. *Lab. Animal Sci.* **36**: 130-135.

Phillips, J.A., Romball, C.G., Hobbs, M.V., Ernst, D.N., Shultz, L., and Weigle, W.O., 1996, CD4+ T cell activation and tolerance induction in B cell knockout mice. *J. Exp. Med.* **183**: 1339-1344.

Sutherland, R.M., Chua, M-M, Lavi, E., Weiss, S.R., and Paterson, Y., 1997, CD4+ and CD8+ cells are not major effectors of mouse hepatitis virus A59-induced demyelinating disease. *J. Neurovirology.* **3**: 225-228.

Wang, F., Stohlman, S.A., and Fleming, J.O., 1990, Demyelination induced by murine hepatitis virus JHM strain (MHV-4) is immunologically mediated. *J. Neuroimmunol.* **30**: 31-41.

Williamson, J.S.P. and Stohlman, S.A., 1990, Effective clearance of murine hepatitis virus from the central nervous system requires both CD4+ and CD8+ T cells. *J. Virol.* **64**: 4589-4592.

B Cell Mediated Lysis of JHMV Infected Targets

BEATRIZ PARRA,[1] SHAWN MORALES,[1] RAMACHRISTNA
CHANDRAN,[2] AND STEPHEN A. STOHLMAN[1,2]
*Departments of Molecular Microbiology and Immunology[1] and Neurology[2], Keck School of
Medicine, University of Southern California, Los Angeles, CA 90033*

1.   INTRODUCTION

Humoral immunity has been implicated in various models of mouse
hepatitis virus (MHV) central nervous system (CNS) infection (Stohlman, *et
al.*, 1999). However, the lack of neutralizing Ab until the virus has been
cleared suggested that humoral immunity played little or no role in clearance
of JHMV from the CNS. Infection of mice lacking the ability to produce Ab
showed that following acute clearance, infectious virus reactivates within the
CNS (Lin *et al.*, 1999), demonstrating an essential role for humoral
immunity in preventing CNS virus reactivation. The absence of both B cells
and Ab in these mice leaves open the question of whether the B cells limit
CNS virus reactivation, a role potentially masked by Ab mediated
neutralization (Lin *et al.* 1999). B cells from naïve mice interact with MHV-
A59 infected cells resulting in lysis (Holmes *et al.* 1986, Welsh *et al.* 1986).
Lysis is blocked by neutralizing Ab (Wysocka *et al.* 1989), suggesting a role
of the S protein. B cells, although resistant to MHV infection, express high
levels of the MHV receptor (MHV-R) (Coutelier *et al.* 1994). Interactions
between the S protein and MHV-R appear to result in cytolysis via cell-cell
fusion (Wysocka *et al.* 1989) and is independent of Fas/FasL interactions
and TNF-α (Nishioka *et al.* 1993). These studies were initiated to explore
the potential role of B cell-MHV-R interactions in contributing to the
suppression of infectious virus within the CNS.

The Nidoviruses (Coronaviruses and Arteriviruses).
Edited by Ehud Lavi *et al.*. Kluwer Academic/Plenum Publishers, 2001.   369

1.1   Materials and Methods

Mice: BALB/c and C57/BL6 mice were purchased from the Jackson Laboratory, Bar Harbor MA. SJL mice were purchased from the National Institutes of Health (Fredrick, MD).

Viruses: JMHV was propagated and assayed described (Bergmann *et al.* 1993). Recombinant vaccinia viruses (rVV) expressing the JHMV S protein (vJS), the MHV-A59 S protein (vAS) and beta galactosidase (vSC8) were propagated as described (Bergmann *et al.* 1996).

Preparation of Effector cells: Single spleen cell suspensions were incubated on plastic petri plates at 37 C for 1 hour to remove adherent cells. This preparation contained 50 to 60% CD19$^+$ B cells. HeLa cells expressing the MHV-R (MHVR-HeLa) and a control line (HeLa+TA) were provided by Thomas Gallagher, Loyola University Medical Center.

Flow Cytometry: B cell were stained with anti-CD19 (ID3; PharMingen, San Diego CA). Biotinylated MHV-R specific mab CC-1 was supplied by Katheryn Holmes, University of Colorado Health Science Center.

Target cells: J774.1 cells were infected with rVV at a multiplicity of infection of 5 and with MHV at a multiplicity of 1 for 90 min. at 37C and incubated for 10 h prior to use.

Cytolytic assays: Targets were labeled with Na^{51}Cr. Labeled cells (1 x 10^4) were added to round bottomed 96 well plates and mixed with effector cells to achieve desired Effector/Target (E:T) ratios. Release was determined after 4 h incubation at 37C. Data are presented as % specific ^{51}Cr release = 100 x (experimental cpm – spontaneous cpm)/ (maximum cpm – spontaneous cpm).

2.   RESULTS

2.1   MHV-R Expression and B Cell Mediated Cytolysis

In contrast to CD19$^+$ B cells from C57BL/6 and BALB/c mice, which all express MHV-R, on only 30% of B cells from SJL mice express MHV-R (data not shown). B cells from BALB/c mice (Fig. 1) and C57BL/6 mice (data not shown) lyse targets infected with JHMV. These data confirm the lack of a requirement for MHC restriction in B cell mediated cytolysis (Holmes *et al.* 1986, Welsh *et al.* 1986). S protein-MHV-R interactions was confirmed by inhibition of lysis via anti-MHV-R mAb (Fig 1). B cells from SJL mice showed no demonstrable lysis of JHMV infected cells (Fig. 1).

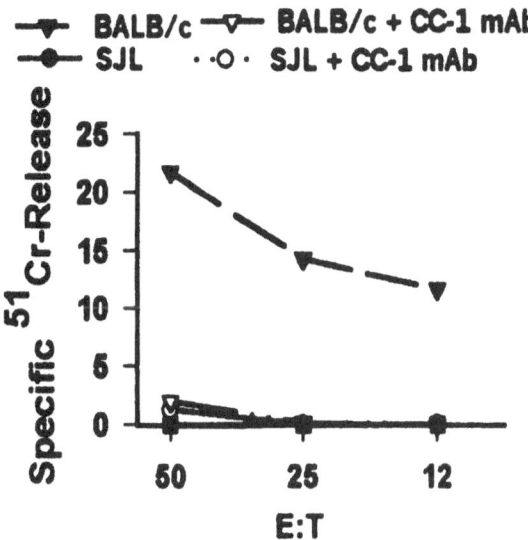

Figure 1. Receptor Dependence of B cells Mediated Cytolysis. B cells from BALB/c and SJL mice were tested for lyse of JHMV infected targets in the presence or absence of 1 ug/ml of anti-MHV-R CC-1 mAb.

2.1.1   S protein Fusion Dependent B Cell Cytolysis

To confirm that neutralizing Ab inhibits cytolysis and that lysis involves fusion, B cells were tested for lysis in the presence of anti-S protein mAb. J.2.6 (neutralization[+]/fusion inhibition[+]) and J.2.5 (neutralization[+]/fusion inhibition[-]) (Fleming *et al.* 1983), inhibit cytolysis (Fig. 2). M protein specific mAb J.3.9 did not inhibit cytolysis (Fig. 2), demonstrating that neutralizing anti-S protein mAb inhibit cytolysis. Cytolysis was also tested in the presence of the anti-S protein mAb J.2.5 (neutralization[-]/fusion inhibition[+]) (Fleming *et al.* 1983). J.2.5 inhibits B cell mediated cytolysis, consistent with a fusion dependent mechanism (Wysocka *et al.* 1989). In addition, cells infected with the nonfusogenic MHV-2 strain were also not susceptible to cytolysis, nor was cytolysis mediated by B cells from perforin deficient mice (data not shown). These data indicate that cytolysis requires expression of the S protein, involves cell-cell fusion, but is independent of other cytolytic pathways, i.e., Fas/FasL interactions, TNF-α or perforin mediated cytolysis.

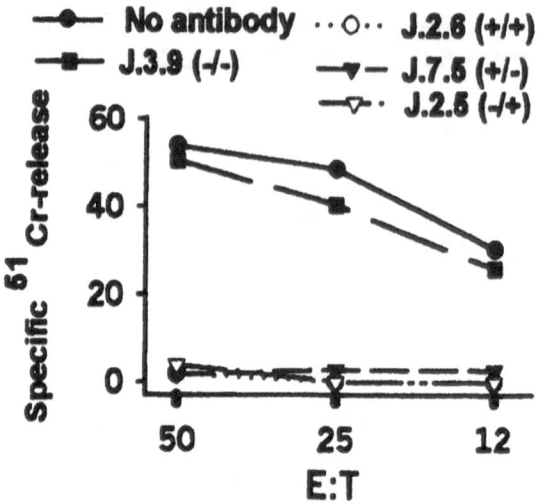

Figure 2. JHMV Neutralizing and Fusion Inhibiting mAb Inhibit Cytolysis. B cells from BALB/c mice were tested for lysis of JHMV infected targets in the presence of JHMV-specific mAb. The mAb were added at initiation at a final concentration of l ug/ml. Specific release was determined after 4 h incubation.

2.1.2   S Protein Expression is Sufficient for Cytolysis

To determine if cytolysis required other viral components, JHMV infected cells were compared to cells expressing only the MHV-A59 S protein (Vac-A59 S) or the JHMV S protein (Vac-JHMV S).   Cells expressing the S proteins from MHV-A59 or JHMV were lysed at levels comparable to those found with JHMV infected cells (Fig. 3). No cytolysis was detected using cells infected with rVV vSC8 (Vac-Control). These data demonstrate that MHV infection is not required for B cell mediated cytolysis and that expression of the MHV S protein is sufficient to initiate cytolysis.

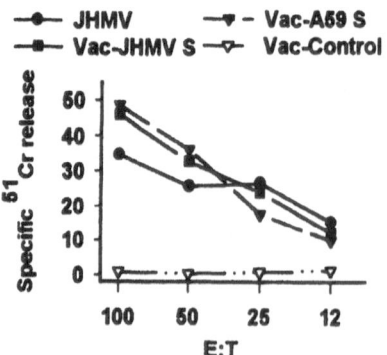

Figure 3. S Protein Expression is Sufficient for Cytolysis. Cells were infected with rVV expressing the MHV-A59 and JHMV S protein or control rVV. B cells were derived from BALB/c mice. Specific release was determined after 4 hours incubation.

2.1.2.1   MHV-R Expression is Sufficient for Cytolysis

To determine if cytolysis is dependent upon a unique B cell property, MHVR HeLa were examined for cytolysis. MHVR HeLa cells express similar levels of MHV-R compare to cytolytic CD19$^+$ B cells derived from naïve BALB/c mice (data not shown). JHMV infected cells were lysed by HeLa cells expressing MHV-R, but not control HeLa cells (HeLa+TA) (Fig. 4). Inhibiting S-MHV-R interactions via addition of 1 ug/ml anti-MHV-R mAb completely abolished cytolysis (data not shown). These data demonstrate that cytolysis resulting from interactions between S protein and cells expressing high levels of the MHV-R is not a unique to B cells.

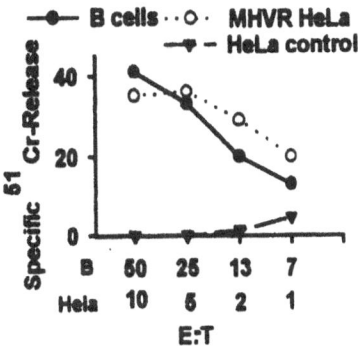

Figure 4. B Cells are not Required for Cytolysis. Cytolysis by B cells compared to HeLa cells expressing the MHV-R (MHVR-HeLa). B cells were added at E:T ratios beginning at 50:1. MHVR-HeLa and control HeLa+TA were added at E:T ratios beginning at 10:1. Specific release was determined after 4 h incubation.

3.   DISCUSSION

Reactivation of JHMV in the absence of Ab suggests that either Ab or B cells contribute to viral persistence. In addition to Ab secretion B cells could inhibit viral reactivation either via direct cytolysis or, since B cells are not susceptible to infection, by sequestering virus. To understand the potential role of B cells in preventing JHMV reactivation. The present study explored the mechanism(s) of B cell mediated cytolysis. The data demonstrate that B cells lyse cells expressing the JHMV S protein via a fusion dependent process. Lysis does not require infection; therefore, other components of the viral infection are not required for cytolysis. In addition MHV-R expression is sufficient and cytolysis is not dependent upon a unique B cell property.

A conceptual difficulty with a role for B cell cytolysis in preventing viral recrudescence is inhibition by anti-S protein Ab with either neutralizing or anti-fusogenic activity. During JHMV infection Ab can first be detected in serum at d 5 p. i. Neutralizing Ab is detected at approx. 10 d p. i. It is unclear how rapidly anti-fusogenic Ab are secreted or rapidly Ab infiltrates the CNS. It is also unclear how rapidly B cells secreting anti-S protein Ab are recruited into the CNS or if they express MHV-R. One possibility is that MHV-R+ B cells are lysed contributing a minimal amount to the reduction of infectious virus due to its intracellular site of replication but providing a plausible mechanism for the somewhat delayed appearance of the Ab response. Understanding the role of B cells during JHMV infection will await an examination of JHMV pathogenesis in mice which possess MHV-R^+ B cells, but are unable to secrete Ab.

REFERENCES

Bergmann, C., McMillan, M., and Stohlman, S.A., 1993, Characterization of the L^d restricted CTL epitope in the mouse hepatitis virus nucleocapsid protein. *J. Virol.* **67**:7041-7049.

Bergmann, C.C., Yao, Q., Lin, M., and Stohlman, S.A., 1996, The JHM strain of mouse hepatitis virus induces a spike protein-specific D^b-restricted cytotoxic T cell response. *J. Gen. Virol.* **77**:315-325.

Coutelier, J.-P., Godfraind, C., Dveksler, G.S., Wysocka, M., Cardellichio, C.B., Noel, H., and Holmes, K.V., 1994, B lymphocyte and macrophage expression of carcinoembryonic antigen-related adhesion molecules that serve as receptors for murine coronavirus. *Eur. J. Immunol.* **24**:1383-1390.

Fleming, J.O., Stohlman, S.A., Harmon, R.C., Lai, M.M.C., Frelinger, J.A., and Weiner, L.P., 1983, Antigenic relationship of murine coronaviruses: Analysis using monoclonal antibodies to JHM (MHV-4) virus. *Virology* **131**:296-307.

Holmes, K.V., and Welsh, R.M., 1986, Natural cytotoxicity against mouse hepatitis virus-infected target cells. I. Correlation of cytotoxicity with virus binding to leukocytes. *J. Immunol.* **136**:1446-1453.

Lin, M.T., Hinton, D.R., Marten, N.W., Bergmann, C.C., and Stohlman, S.A., 1999, Antibody prevents virus reactivation within the central nervous system. *J. Immunol.* **162**:7358-7368.

Nishioka, W.K., and Welsh, R.M., 1993, B cells induce apoptosis via a novel mechanism in fibroblasts infected with mouse hepatitis virus. *Nat. Immun.* **12**:113-127.

Stohlman, S.A., Bergmann, C.C., and Perlman, S., 1999, Mouse hepatitis virus. In *Persistent Viral Infections.* (R. Ahmed and I. Chen, eds.), John Wiley & Sons, p. 537-557.

Welsh, R.M., Haspel, M.V., Parker, D.C., and Holmes, K.V., 1986, Natural cytotoxicity against mouse hepatitis virus-infected cells. *J. Immunol.* **136**:1454-1460.

Williams, R.K., Jiang, G.-S., Snyder, S.W., Frana, M.F., and Holmes, K.V., 1990, Purification of the 110-kilodalton glycoprotein receptor for mouse hepatitis virus (MHV)-A59 from mouse liver and identification of a nonfunctional, homologous protein in MHV-resistant SJL/J mice. *J. Virol.* **64**:3817-3823.

Wysocka, M., Korngold, R., Yewdell, J. and Bennink, J., 1989, Target and effector cell fusion accounts for B lymphocyte-mediated lysis of mouse hepatitis virus-infected cells. *J. Gen. Virol.* **70**:1465-1472.

Polyclonal Activation of B Cells by Lactate Dehydrogenase-Elevating Virus is Mediated by N-Glycans on the Short Ectodomain of the Primary Envelope Glycoprotein

PETER G.W. PLAGEMANN,[1] QUENTIN A. JONES,[2] AND WILLIAM A. CAFRUNY[2]

[1]*Department of Microbiology, University of Minnesota, Minneapolis, MN, USA;* [2]*Division of Basic Biomedical Sciences, Molecular Microbiology & Immunology Group, University of South Dakota School of Medicine, Vermillion, SD, USA*

1. INTRODUCTION

The common strains of lactate dehydrogenase-elevating virus (LDV-P and LD-vx) are primary examples for viruses that cause a permanent polyclonal activation of B cells that results in IgG2a hyper-gammaglobulinemia and the generation of autoantibodies and circulating immune complexes in their host, the mouse (Notkins et al., 1966; Coutelier and van Snick, 1985; Li et al, 1990). Plasma IgG2a levels increase from generally below 0.5 mg/ml to 2-6 mg/ml by two weeks post infection (p.i.) and remain elevated thereafter (see later). LDV-P/vx cause life-long persistent viremic infections (see Fig. 1A) which are maintained by continuous rounds of replication in a renewable subpopulation of macrophages and resistance to host immune responses (Plagemann, 1996). Previous results have shown that the single neutralization epitope located on the short (about 30 amino acids long) ectodomain of the primary envelope glycoprotein, VP-3P, carries three large N-glycan chains in LDV-P and LDV-vx (see Fig. 1A insert) that suppress the immunogenicity of the epitope and impair antibody neutralization of the virions of these quasispecies (Chen et al., 2000; Plagemann et al., 1999). The present results indicate that the

The Nidoviruses (Coronaviruses and Arteriviruses).
Edited by Ehud Lavi *et al.*, Kluwer Academic/Plenum Publishers, 2001.

three N-glycans on the VP-3P ectodomains of LDV-P/vx also play a critical role in the polyclonal activation of B cells by these LDVs.

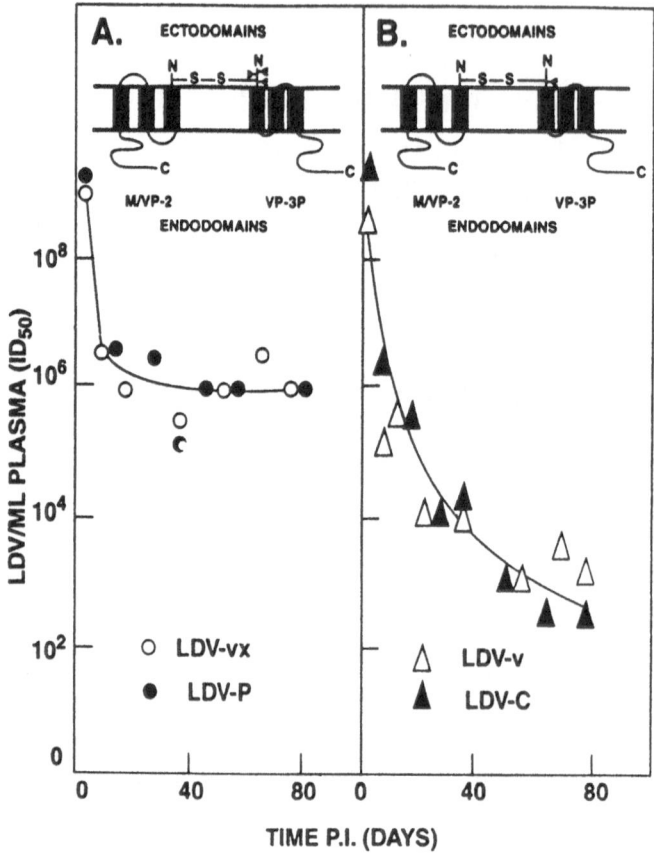

Figure 1. Time courses of viremia in FVB mice injected intraperitoneally with (A) biologically cloned LDV-P or LDV-vx and (B) with biologically cloned LDV-C or LDV-v (data are redrawn from Plagemann et al., 1999) and the proposed topography of VP-3P and M/VP-2 in the LDV envelope (inserts).

2.   RESULTS

The conclusion that the N-glycans on the VP-3P ectodomain play a critical role in the polyclonal activation of B cells by LDV is provided by comparing the IgG2a hypergammaglobulinemia and immune complex formation in mice infected with biologically cloned LDV-P/vx with those in mice infected by two biologically cloned variants of LDV, LDV-C and LDV-v, which have gained neuropathogenicity for C58 and AKR mice, but at the same time have lost the ability for high viremic persistent infection (Fig. 1B; Chen et al., 2000). Previous studies have shown that the VP-3P ectodomains of LDV-C/v lack the two N-terminal N-glycosylation sites (see Fig. 1B, insert) present on the VP-3P ectodomains of LDV-P/vx (Fig. 1A insert) which greatly increases the immunogenicity of the neutralization epitope and the sensitivity of LDV-C/v to antibody neutralization.

This results in rapid suppression of their replication in mice (Fig. 1B; Chen et al., 2000; Plagemann et al., 1999). We have now found that infection of C57BL/6 and FVB mice with LDV-C or LDV-v consistently induces a IgG2a hypergammaglobulinemia that was only one half or less as high as that induced by infection with LDV-P/vx (Fig. 2). Plasma IgG2a levels were quantified by capture ELISA as described previously (Li et al.; 1990). In order to allow valid comparisons between the two classes of LDV, we generally compared the time courses of IgG2a formation (4 time points each) in mice infected with at least one non-neuropathogenic and one neuropathogenic LDV in the same ELISA (2-fold plasma dilutions from 1:4000 to 1: 128,000) and have used the same standard solution of IgG2a for establishing standard curves (0.5 ng/ml to 1 mg/ml) in all assays. Similar results were obtained with duplicate mice assayed in the same experiment (Fig. 2C and D), and with additional mice infected with these LDVs (at least two other mice for each, data not shown) and no increase in plasma IgG2a was observed in an uninfected companion mouse (see Fig. 2C) consistent with previous results (Li et al., 1990).

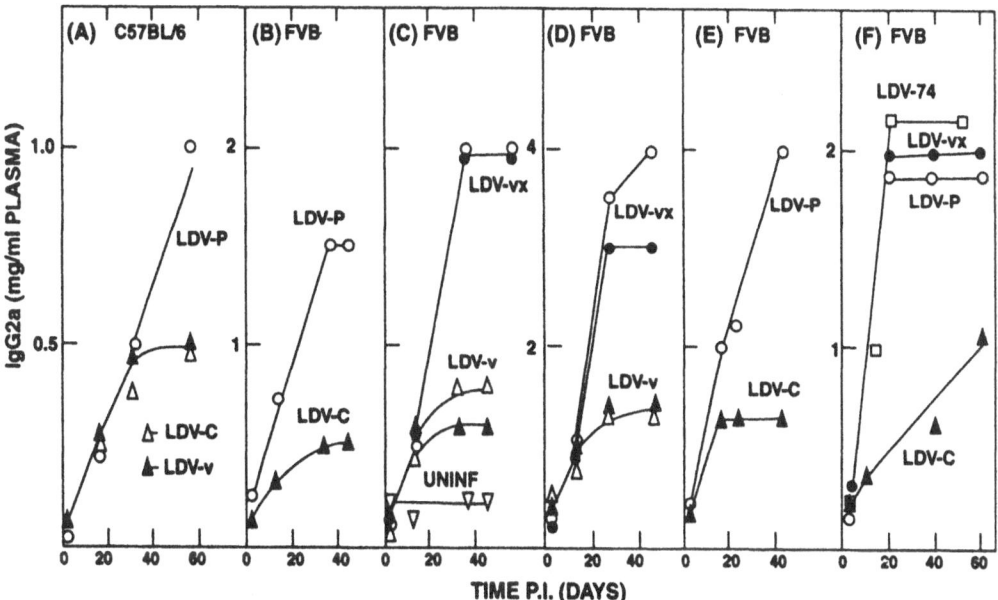

Figure 2. Time courses of plasma IgG2a elevation in C57BL/6 and FVB mice after infection with the indicated LDV quasispecies. All time courses are for single mice, but in (C) and (D) results for two companion mice are presented. Results in each frame come from a single ELISA, and were estimated from the same IgG2a standard curve. Comparable results were obtained in repeat ELISAs of the plasma samples (data not shown).

Since LDV-v is a genetic recombinant of LDV-vx that has, by a double recombination, specifically acquired the 5' end of ORF5 of LDV-C (~400 nt) that encodes the ectodomain of the primary envelope glycoprotein VP-3P (Fig. 3A; Li et al., 1999) the difference in polyclonal B cell activation between LDV-P/vx, on the one hand, and LDV-C/v, on the other hand, must reside in this segment of VP-3P. The most likely molecular structure responsible for this difference in the polyclonal B cell activation is the number of N-glycans associated with the VP-3P ectodomain (Fig. 1 and 3B) just as they are responsible for the other phenotypic differences between LDV-P/vx and LDV-C/v discussed already. There are no other amino acid differences in this VP-3P segment that correlate with the various differences in phenotypic properties of the two classes of LDV (Chen et al., 1998; Li et al., 1999).

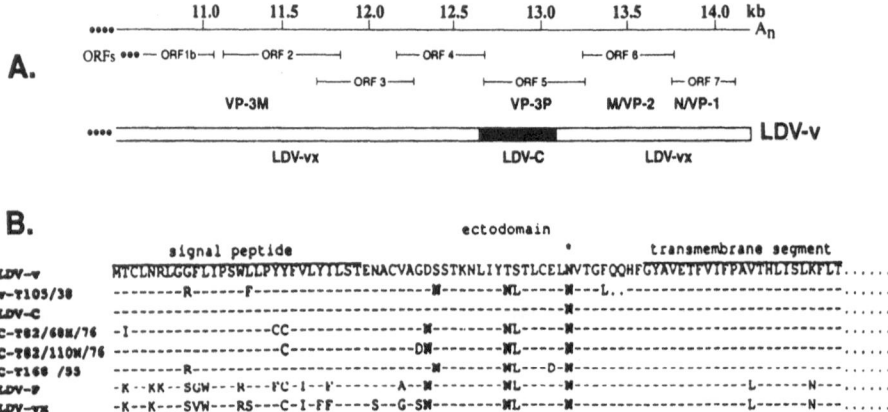

Figure 3. Organization of ORFs 2-7 of LDV and origins of the 3' and genome segments of the neuropathogenic genetic recombinant LDV-v (A) and amino acid sequences of the N-terminal ends of the VP-3Ps of various LDV quasispecies and of neutralization escape mutants of LDV-C/v (B). For the origins of the mutants of LDV-C/v see Chen et al., 2000. The signal peptide and the first transmembrane segment are overlined and the potential N-glycosylation sites are in bold face letters (* indicates the N-glycosylation site that is conserved in all LDV isolates).

This hypothesis was further explored by examining plasma IgG2a levels in FVB mice after infection with three neutralization escape mutants of LDV-C and LDV-v (LDV-v-T105/38, LDV-C-T168/55, LDV-C-T82/68N/76) whose VP-3P ectodomains had regained three N-glycosylation sites (see Fig. 3B). The plasma IgG2a elevation in duplicate mice infected with each of the three neutralization escape mutants was comparable to that in an LDV-P infected mouse and much higher than that in an LDV-C infected mouse (Fig. 4). Similarly, in a mouse infected with an LDV isolated from a wild house mouse in Montana (LDV-74), whose VP-3P ectodomain

possesses all three N-glycosylation sites (Li et al., 2000), plasma IgG2a
levels increased similarly as in LDV-P and LDV-vx infected mice (Fig. 2F).

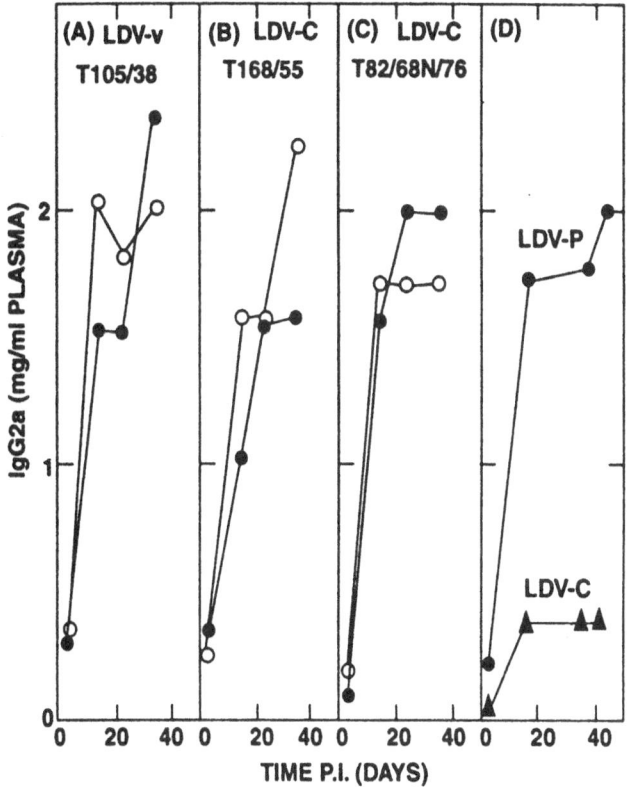

Figure 4. Comparison of the time courses of plasma IgG2a elevation in duplicate FVB mice
infected with three neutralization escape mutants of LDV-v and LDV-C (A-C) with those in
LDV-P and LDV-C infected mice (D). All values were obtained in the same ELISA and
estimated from the same IgG2a standard curve. Comparable results were obtained in a repeat
ELISA of the plasma samples (data not shown).

An even more drastic difference between mice infected with the common
non-neuropathogenic LDVs and the neuropathogenic mutants was observed
in the formation of circulating 150-300 kDa immune complexes that have
been found to become generated concomitant with an LDV induced
polyclonal activation of B cells (Cafruny et al., 1986; Hu et al., 1992). These
complexes contain IgG2a which most likely represents autoantibodies that
are generated as a result of the polyclonal activation of B cells and bound to
their auto antigens (Hu et al., 1992). They are recognized by binding in PBS-
Tween to ELISA plates that are not coated with antigen. They differ in size
and IgG isotype specificity from infectious virion-antibody complexes that
are also generated and persist in LDV-infected mice (Hu et al., 1992). Fig.
5A-C illustrates the appearance of plate binding immune complexes in mice

after infection with cloned LDV-P and LDV-vx. The time courses of immune complex formation were very similar for the two cloned quasispecies, just as was the case for the IgG2a hypergammaglobulinemia induced by them (Fig. 2) and similar to those previously reported for mice infected with original LDV isolates (Cafruny et al., 1986; Hu et al., 1992). In contrast, in our standard assay, which involves a 1 h incubation of the ELISA plates with alkaline phosphatase substrate, we detected little or no plate binding IgG in the plasmas of FVB or C57BL/6 mice infected with the neuropathogenic LDV-C or LDV-v (Fig. 5E-H, and data not shown).

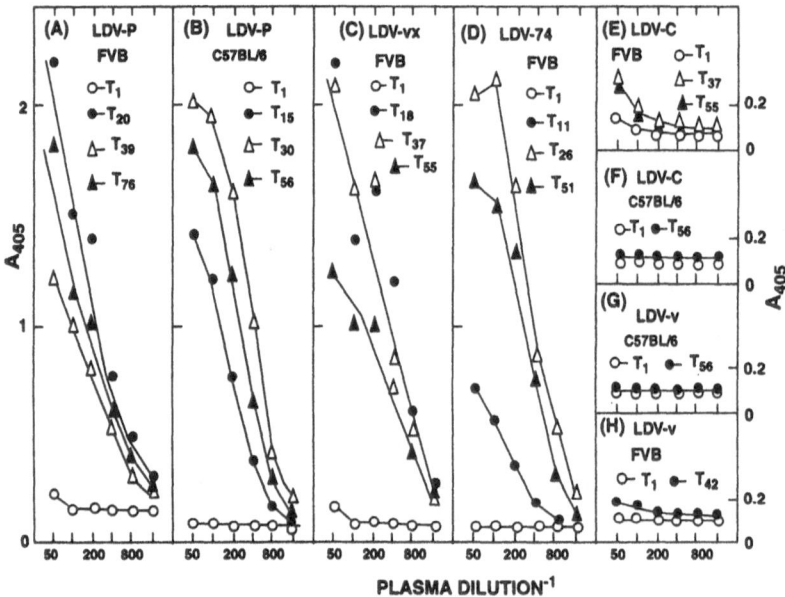

Figure 5. Time courses of appearance of ELISA plate binding activity in plasma of C57BL/6 and FVB mice after infection with the indicated LDVs. A405 is plotted against 2-fold dilutions of each plasma sample (T=time p.i. in days). The results in (B), (E), and (G) come from the same plate-binding assay and so do the results in (A) and (E) and the results in (C) and (H). The results are for single mice, which generally have also been assayed for plasma IgG2a levels (Fig. 2). Comparable results were obtained in repeat assays of the plasma samples and in duplicate mice where available.

On the other hand, immune complex formation in mice infected with the three neutralization escape mutants of LDV-C and LDV-v, whose VP-3P ectodomains had regained three N-glycosylation sites (see Fig. 3B), was comparable to that in mice infected with LDV-P and LDV-vx rather than to that in mice infected with the LDV-v (or LDV-C) parent (Fig. 6). The same was the case for the immune complex formation in a mouse infected with the wild house mouse LDV-74 (Fig. 5D), whose VP-3P ectodomain possesses three N-glycosylation sites, just like those of LDV-P and LDV-vx.

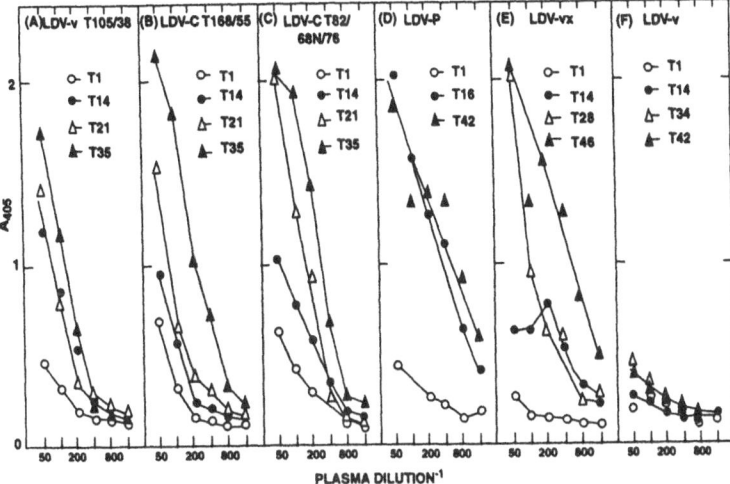

Figure 6. Comparison of the appearance of ELISA plate binding activity in FVB mice infected with three neutralization escape mutants of LDV-v and LDV-C (A-C) with that in mice infected with LDV-P, LDV-vx, and LDV-v (D-F) (T= time p.i. in days). All values were obtained in a single assay. The results are for single mice, but comparable results were observed in duplicate mice which were infected with the neutralization escape mutants and assayed along with plasma samples from LDV-P and LDV-C infected mice (data not shown).

3.   DISCUSSION

Previous studies have shown that the hypergammaglobulinemia caused by LDV is only partly dependent on T cells. It occurs to some extent in nude and T-cell depleted mice (Li et al., 1990; Coutelier et al., 1990; Hu et al., 1992) and in the complete absence of an anti-LDV antibody response (Rowland et al., 1994). Thus, it is not simply an indirect result of the antiviral immune response (bystander effect, Ahmed & Oldstone, 1984). It has also been shown that B cells from T cell depleted and non-depleted mice after an LDV infection exhibit a similar elevated proliferative response in vitro, but that the former do not produce IgG2a in vitro unless treated with lipopolysaccharide and IFNγ (Coutelier et al., 1990). Furthermore, T cells from LDV infected mice as early as 4 days p.i. produce IFNγ much more rapidly and to higher levels in vitro after exposure to concanavalin A than T cells from uninfected mice and IFNγ mRNA is detected in the spleen (Plagemann et al., 1995). The results suggested that an LDV protein may function as a direct B cell mitogen and that the generation of IgG2a producing plasma cells is then mediated by IFNγ produced by T cells largely generated in the course of the anti-LDV immune response (Plagemann et al., 1995). Taken together, the present results are consistent with this view and show that the N-glycans associated with the very short ectodomain of the primary envelope glycoprotein VP-3P of LDV-P/vx (Fig. 1) seem to represent the direct B cell mitogen of these LDVs behaving like a TI-1 antigen similar to bacterial polysaccharides, lipopolysaccharides and polymeric proteins (Janeway et al., 1999).

It is unclear, however, why the N-glycans on the VP-3P ectodomain play the major role in B-cell activation, since the LDV genome encodes three other glycoproteins (encoded by ORFs 2, 3 and 4; Plagemann, 1996). Perhaps the VP-3P ectodomain with its N-glycans exposed on the surface of LDV virions forms a rather rigid structure that is required for its B cell mitogenic activity (Fig. 1A). In addition, or alternatively, B cell activation may require cross linkage of receptor sites on the B cells, or at least multiple interactions between sites on the inducer and sites on a B cell as may be typical for TI-1 antigens (Janeway et al., 1999). Such multiple interactions can probably only be accomplished by the high density of VP-3P ectodomains covering intact LDV virions, since the ORF 2 protein seems to be only a very minor envelope glycoprotein and the ORF 2 and ORF 3 proteins are non-structural glycoproteins (Faaberg and Plagemann, 1995; 1997).

If the N-glycans of the VP-3P ectodomain are solely responsible for the B cell mitogenic activity of LDV, one would predict that loss of all three N-glycosylation sites on VP-3P would abolish all B cell mitogenic activity of LDV. We have not been able to test this hypothesis since we have not been able to strip LDV virions of the N-glycans without loss of infectivity and have not found an LDV mutant that lacks all N-glycosylation sites in VP-3P. Thus, we cannot rule out that some other LDV glycoprotein may contribute to the B cell mitogenic activity of LDV.

Another finding that requires an explanation is that the formation of plate binding immune complexes is reduced in LDV-C/v infected mice, as compared to LDV-P/vx infected mice, much more that the IgG2a hypergammaglobulinemia (Fig. 2 and 5). These results suggest that the increased production of polyclonal IgG2a does not necessarily lead to the generation of immune complexes containing IgG2a autoantibodies. An explanation may be derived from the fact that only a very minor fraction of the total IgG2a that is produced in LDV-infected mice is sequestered in immune complexes (Hu et al., 1992) as well as the likelihood that the autoantibodies in these complexes are produced by a subset of B cells, such as CD5 B cells (B1 cells; Janeway et al., 1999) that may respond differently to the mitogenic activity of LDV than the bulk of the B cells. On the other hand, it is possible that factors other than autoantibodies and their antigens are involved in the formation of the ELISA plate binding immune complexes since we have not been able to artificially generate plate binding immune complexes in vitro or to regenerate the immune complexes once they have been dissociated (Hu et al., 1992).

To our knowledge, this is the first report implicating N-glycans exposed on the surface of an enveloped virus in the polyclonal activation of B cells during an acute infection. It seems likely that other enveloped viruses that cause a polyclonal activation of B cells possess similar N-glycan containing structures associated with their envelope glycoproteins that function in a similar manner.

ACKNOWLEDGEMENTS

We thank Ron Jemmerson for editorial comments and Sara Veglahn and Patricia Nelson for excellent secretarial assistance.

REFERENCES

Ahmed, R., and Oldstone, M.B.A., 1984, Mechanisms and biological implications of virus-induced B-cell activation. In *Concepts in Viral Pathogenesis* (A.L. Notkins and M.B.A. Oldstone. eds.), Springer Verlag. New York, pp. 231-238.

Cafruny, W.A., Heruth, D.P., Jaqua, M.J., and Plagemann, P.G.W., 1986, Immunoglobulins that bind to uncoated ELISA plates: appearance in mice during infection with lactate dehydrogenase-elevating virus and in human anti-nuclear antibody positive sera. *J. Med. Virol.* **19**: 175-186.

Chen, Z., Li, K., and Plagemann, P.G.W., 2000, Neuropathogenicity and sensitivity to antibody neutralization of lactate dehydrogenase-elevating virus are determined by polylactosaminoglycan chains on the primary envelope glycoprotein. *Virology.* **266**: 88-98.

Chen, Z., Li, K., Rowland, R.R.R., Anderson, G.W., and Plagemann, P.G.W., 1998, Lactate dehydrogenase-elevating virus variants: cosegregation of neuropathogenicity and impaired ability for high viremic persistent infection. *J. Neurovirol.* **4**: 560-568.

Coutelier, J.-P., Coulie, P.G., Wauters, P., Heremans, H., and van der Logt, J.T.M., 1990, In vivo polyclonal B-lymphocyte activation elicited by murine viruses. *J. Virology.* **64**: 5383-5388.

Coutelier, J.-P., and van Snick, J., 1985, Isotypically restricted activation of B lymphocytes by lactic dehydrogenase virus. *Eur. J. Immunol.* **15**: 250-255.

Faaberg, K.S., and Plagemann, P.G.W., 1995, The envelope proteins of lactate dehydrogenase-elevating virus and their membrane topography. *Virology* **212**: 512-525.

Faaberg, K.S., and Plagemann, P.G.W., 1997, ORF3 of lactate dehydrogenase-elevating virus encodes a soluble, nonstructural, highly glycosylated, and antigenic protein. *Virology* **227**: 245-251.

Hu, B., Even, C., and Plagemann, P.G.W., 1992, Immune complexes that bind to ELISA plates not coated with antigen in mice infected with lactate dehydrogenase-elevating virus: Relationship to IgG2a and IgG2b-specific polyclonal activation of B cells. *Viral Immunol.* **5**: 27-38.

Janeway, C.A., Travers, P., Walport, M., and Capra, J.D., 1999, Immunobiology. 4th edn., pp. 3231-323. Garland Publishing.

Li, K., Chen, Z., and Plagemann, P.G.W., 1999, High frequency genetic recombination of an arterivirus, lactate dehydrogenase-elevating virus, in mice and evolution of neuropathogenic variants. *Virology* **258**: 73-83.

Li, K., Schuler, T., Chen, Z., Glass, G.E.G., Childs, J.E., and Plagemann, P.G.W., 2000, Isolation of lactate dehydrogenase-elevating viruses from wild house mice and their biological and molecular characterization. *Virus. Res.,* In press.

Li, X., Hu, B., Harty, J.T., Even, C., and Plagemann, P.G.W., 1990, Polyclonal B cell activation of IgG2a and IgG2b production by infection of mice with lactate dehydrogenase-elevating virus is partly dependent on CD4[+] lymphocytes. *Viral. Immunol.* **3**: 273-288.

Notkins, A.L., Mergenhagen, S.E., Rizzo, A.A., Scheele, C., and Waldmann, T.A., 1966, Elevated γ globulin and increased antibody production in mice infected with lactic dehydrogenase virus. *J. Exper. Med.* **123**: 347-356.

Plagemann, P.G.W., 1996, Lactate dehydrogenase-elevating virus and related viruses. In *Virology*, 3rd edn, (B.N. Fields, D.M. Knipe, & P.M. Howley, eds) Raven Press, New York, pp. 1105-1120.

Plagemann, P.G.W., Chen, Z., and Li., K., 1999, Polylactosaminoglycan chains on the ectodomain of the primary envelope glycoprotein of an arterivirus determine its neuropathogenicity, sensitivity to antibody neutralization and immunogenicity of the neutralization epitope. *Curr.Top. Virol.* **1**: 27-43.

Plagemann, P.G.W., Rowland, R.R.R., Even, C., and Faaberg, K.S., 1995, Lactate dehydrogenase-elevating virus--an ideal persistent virus? *Seminars in Immunopathobiol.* **17**: 167-186.

Rowland, R.R.R., Even, C., Anderson, G.W., Chen, Z., Hu, B., and Plagemann, P.G.W., 1994, Neonatal infection of mice with lactate dehydrogenase-elevating virus results in suppression of humoral antiviral immune response but does not alter the course of viremia or the polyclonal activation of B cells and immune complex formation. *J. Gen. Virol.* **75**: 1071-1081.

Influence of Changes in the Population of Target Cells and Appearance of Specific Antibodies on the Replication of Porcine Reproductive and Respiratory Syndrome Virus in the Lungs of Pigs

H.J. NAUWYNCK, G.G. LABARQUE, K. VAN REETH, AND M.B. PENSAERT

Laboratory of Virology, Faculty of Veterinary Medicine, Ghent University, Salisburylaan 133, B-9820 Merelbeke, Belgium

1. INTRODUCTION

Porcine reproductive and respiratory syndrome virus (PRRSV) is a new arterivirus that caused an epidemic of reproductive failure and respiratory problems in pigs, all over the world in the late eighties and early nineties.

Studies on organ pathogenesis showed that PRRSV primarily replicates in tonsils, lungs and draining lymph nodes and subsequently enters the blood (Duan *et al.*, 1997a). A cell-free viremia allows the virus to reach internal replication sites such as spleen, different internal lymph nodes and genital tract. Despite the onset of a humoral immune response at one week post inoculation (PI), the virus persists in tonsils until 21 days PI, in lungs until 35 to 49 days PI, in semen until 92 days PI and in serum until 150 days PI (Mengeling *et al.*, 1995, Christopher-Hennings *et al.*, 1995, Duan *et al.*, 1997a, Wills *et al.*, 1997).

PRRSV has a specific tropism for cells of the monocyte/macrophage lineage. However, there are clear restrictions. Well-differentiated macrophages in the lungs and lymphoid tissues are susceptible whereas well-differentiated peritoneal macrophages, macrophage progenitor cells in bone marrow and peripheral blood monocytes are refractory (Duan *et al.*, 1997a

The Nidoviruses (Coronaviruses and Arteriviruses).
Edited by Ehud Lavi *et al.*, Kluwer Academic/Plenum Publishers, 2001.

and b). PRRSV is only able to replicate when it passes successfully through a cascade of processes. It starts with the attachment, followed by entry, endosome formation, pH drop and release of the genome in the cytosol. Only a few steps have already been fully elucidated. A 210 kDa protein together with a heparin-like molecule have been shown to be of major importance in the early virus-cell interaction (Duan *et al.*, 1998, Vanderheijden *et al.*, 2000) and a receptor-mediated endocytosis, by which the virus enters the cell, has been documented (Nauwynck *et al.*, 1999). When one of these steps becomes blocked, virus replication will be halted.

In the present study, it was the purpose to relate replication of PRRSV with changes of cells of the monocyte/macrophage lineage expressing the putative receptor (the 41D3-reactive, 210 kDa protein) and the onset of total and specific neutralizing PRRSV-antibodies.

2. MATERIALS AND METHODS

2.1 Pigs, virus inoculation and collection of material

Twenty-nine gnotobiotic pigs were used at 4 to 5 weeks of age. Twenty-two pigs were inoculated intranasally with $10^{6.0}$ TCID$_{50}$ PRRSV (Lelystad). The remaining seven pigs were left uninoculated. One to three of the inoculated pigs were euthanatized at 1, 3, 5, 7, 9, 14, 20, 25, 30, 35, 40 and 52 days PI. Control pigs were euthanatized between 4 and 10 weeks of age. Upon euthanasia, lungs were collected for further cellular and virological examinations.

2.2 Cell analysis

The right lung was lavaged with 60 to 120 ml phosphate buffered saline (PBS) and the recovered lavage fluid was centrifuged (400xg, 4°C, 10 minutes). The supernatant was collected for virus isolation and determination of anti-PRRSV antibodies. The cell pellet was resuspended in PBS and the total number of BAL cells was determined. Cytocentrifuge preparations of BAL cells were made for a cytological staining to differentiate into mononuclear and polymorphonuclear cells and for an indirect immunofluorescence (IIF) staining to determine the percentage of viral antigen-positive cells by using the monoclonal antibodies WBE1 and WBE4-6, directed against the nucleocapsid of PRRSV (Drew *et al.*, 1995), as primary antibodies, biotinylated goat-anti-mouse IgG as secondary antibodies and streptavidin-FITC. The remaining BAL cells were analyzed with the flow cytometer (Becton Dickinson, FACSCalibur). For the phenotypic identification of neutrophils and cells of the monocyte/macrophage lineage, cells were stained with an IIF technique, as described above, using MAb 74-22-15 as primary antibody (Pescovitz *et al.*, 1984). The percentage of cells of the monocyte/macrophage lineage was determined by subtracting the percentage of neutrophils determined by a cytological staining from the 74-22-15$^+$ cells. Cells from the monocyte/macrophage lineage, expressing the 210kDa, putative PRRSV-receptor, were identified by an IIF technique, as described above, using MAb 41D3 as primary antibody (Duan *et al.*, 1998).

2.3 Virus titration and detection of viral antigen-positive cells

20% suspensions of lung tissues were made with PBS. Titrations were performed by inoculating 1-day-cultivated porcine alveolar macrophages (PAM) with tenfold dilutions of lung suspensions or BAL fluids. After three days cultivation, the inoculated PAMs were stained using an immunoperoxidase monolayer assay (IPMA), as described by Wensvoort *et al.* (1991).

Tissue samples from the lungs were embedded in methylcellulose medium and frozen at $-70°C$. Cryostat sections (5 to 8 μm) were made and fixed in acetone. An indirect streptavidin-biotin immunofluorescence technique, similar to that described for BAL cells, was used to stain and localize the viral antigen-positive cells in the lung tissue.

2.4 Serological examination

Anti-PRRSV antibody titers were determined in sera and BAL fluids using the IPMA technique described by Wensvoort *et al.* (1991). Neutralizing antibodies were determined in sera and BAL fluids using a virus neutralizing test on MARC-145 cells as described by Swenson *et al.* (1994). A MARC-145-adapted Lelystad strain was used in this assay.

3. RESULTS

3.1 BAL cell quantification and differentiation

Total BAL cell numbers of uninoculated control pigs ranged between 114 and 383 x10^6. The cells consisted of 97 to 98% of cells of the monocyte/macrophage lineage, 1% of neutrophils and 1 to 2% of non-phagocytes, presumably lymphocytes (74-22-15⁻ cells).

BAL cell populations in PRRSV-infected pigs are shown in Fig. 1. Their number and composition were quite similar to those of the uninoculated control pigs during the first 5 days PI. Mean BAL cell numbers increased from 140 x 10^6 at 5 days PI to 948 x 10^6 at 25 days PI and then remained at high levels until the end of the experiment with numbers ranging between 642 and 782 x 10^6. BAL cells consisted of 55 to 92% of cells of the monocyte/macrophage lineage, 1 to 15% of neutrophils (33% in one pig euthanatized at 9 days PI) and 6 to 31% of non-phagocytes.

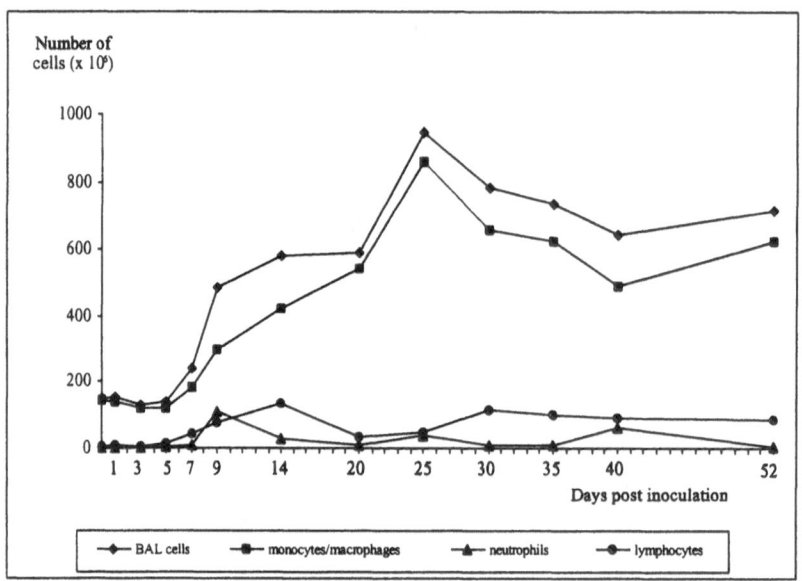

Figure 1. Quantification and differentiation of BAL cells throughout a PRRSV infection.

3.2   Flow cytometric analysis

In the uninoculated control pigs, the majority of BAL cells were flow-cytometrically recognized as large (population P1, high FSC value) and small (population P2, low FSC value) cells with a strong granularity (high SSC value). Based on their light-scattering properties and their reactivity with MAbs 74-22-15 and 41D3, these two populations were characterized as well-differentiated macrophages that express the putative PRRSV receptor. The particles with the lowest light scattering properties (population P3) were characterized as non-phagocytes, probably lymphocytes (74-22-15⁻, 41D3⁻) and fragments of cells of the monocyte/macrophage lineage (74-22-15⁺, 41D3⁺).

In PRRSV-infected pigs, the light-scattering characteristics of the BAL cell population were similar to those of the uninoculated control pigs during the first 3 days PI. In population P1, the absolute number of cells of the monocyte/macrophage lineage (74-22-15⁺) decreased to 25-50% of the original number (143 x10⁶) between 3 and 9 days PI, increased 2- to 3-fold between 14 and 30 days PI and remained at that level till the end of the study at 52 days PI. The 74-22-15⁺ cells were all 41D3⁺ with the exception of 25-50% between 9 and 25 days PI. In the smaller population P2 (original number: 33 x10⁶), a similar pattern was found. A small number of cells were present in the population P3 before 7 days PI (<38 x10⁶), with the same composition as in the uninoculated pigs; afterwards an influx of cells was observed (up to 284 x10⁶) with approximately one third of the cells 74-22-15⁻, 41D3⁻ (probably lymphocytes), one third 74-22-15⁺, 41D3⁻ (probably blood monocytes) and one third 74-22-15⁺, 41D3⁺ (probably fragmented macrophages).

3.3 Virological examinations of lung tissue and BAL fluid

The pattern of virus titres in lung tissue resembled that in BAL fluid. The highest virus titres were reached in BAL fluid at 7 days PI ($10^{7.3}$ TCID$_{50}$/ml) and in lung tissue at 9 days PI ($10^{7.1}$ TCID$_{50}$/g). Afterwards, virus titres decreased slowly during the next 5 weeks. Virus was not detected in the pig euthanatized at 52 days PI.

The results of the analysis of lung tissue and BAL cells by immunofluorescence for the detection of viral antigen-positive cells are presented in Fig. 2. Viral antigen-positive BAL cells were first observed at 1 day PI, increased to a maximum of 3% at 9 days PI, decreased to 0.5% at 14 days PI and remained at levels of 0.1 to 0.2% until 40 days PI. Single viral antigen-positive cells were observed from 3 until 35 days PI with a maximal number of 45 cells/100mm² lung tissue at 7 days PI. Viral antigen-positive foci were defined as areas in lung tissue consisting of groups of viral antigen-positive cells and cellular debris. Viral antigen-positive foci were found from 3 until 14 days PI with a maximal number of 37 foci/100mm² lung tissue at 9 days PI.

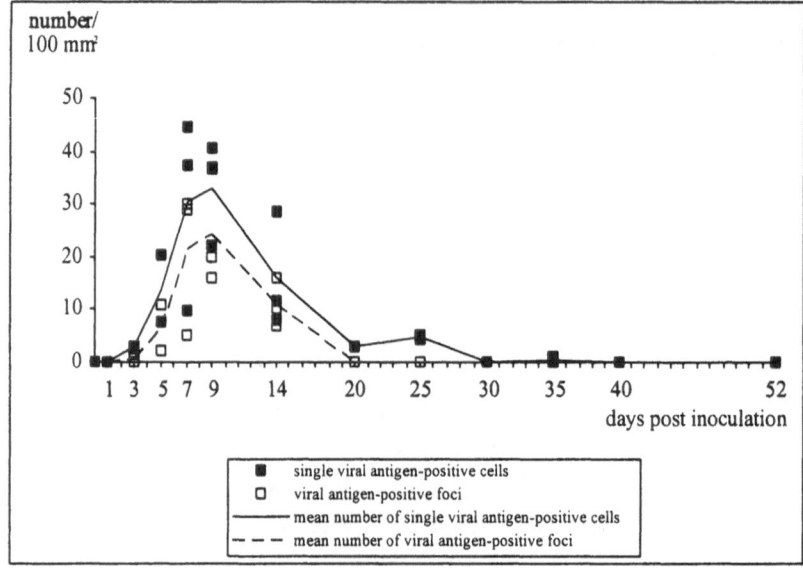

Figure 2. Quantification of single viral antigen-positive cells and viral antigen-positive foci in lung tissue (/100 mm^2) throughout a PRRSV infection

3.4   Antibodies in sera and BAL fluids

The antibody titers against PRRSV in sera and BAL fluids of PRRSV-inoculated pigs are presented in Fig. 3. PRRSV-specific antibodies in sera and BAL fluids were first detected by the IPMA at 9 days PI. Maximal titres were reached at 20 days PI in serum and at 25 days in BAL fluid.

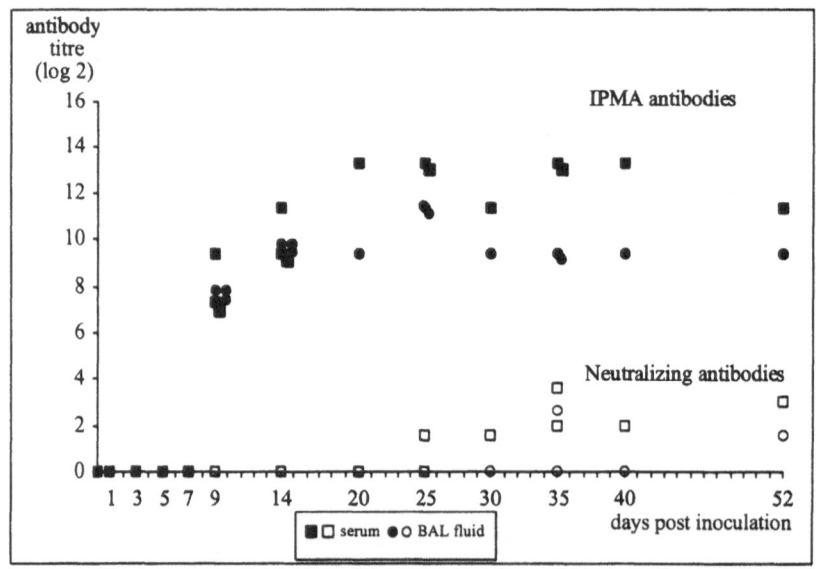

Figure 3. Antibody titres against PRRSV in sera and BAL fluids of PRRSV-inoculated pigs.

Neutralizing antibodies in sera were detected from 25 days PI. The titres remained at a low level ($2^{1-3.6}$) until the end of the study. Neutralizing antibodies in BAL fluids were only detected in two pigs, one euthanatized at 35 days and one at 52 days PI.

4. DISCUSSION

From 3 until 9 days after PRRSV inoculation, the total number of well-differentiated macrophages that express the putative PRRSV receptor was reduced. The reduction of this specific cell population is probably caused by a combination of cell lysis due to virus replication and apoptosis. The highest virus titres and number of viral antigen-positive cells in lungs and BAL fluids were indeed detected at 7 and 9 days PI and during the same period the first peaks of apoptosis have been detected in another study (Labarque *et al.*, 2000).

The total number of BAL cells increased starting from 9 days PI until 25 days PI due to an influx of mainly 74-22-15$^+$, 41D3$^-$ cells, probably blood monocytes, and to a less extent of 74-22-15$^-$, 41D3$^-$ cells, probably lymphocytes. The mechanism by which PRRSV induces a rather specific influx of monocytes is not yet known, but Van Reeth *et al.* (1999) suggested that chemotactic cytokines produced by PRRSV-infected macrophages mediate the influx of new cells of the monocyte/macrophage lineage. After 25 days PI, all 74-22-15$^+$ cells were 41D3 positive which is indicative for a quick change of newly infiltrated blood monocytes into well-differentiated macrophages that express the putative PRRSV receptor.

Despite the continuous increase of the number of cells of the monocyte/macrophage lineage from 9 until 25 days PI, the number of viral antigen-positive cells in lung tissue and BAL fluid decreased from 9 days PI. This is probably the result of (i) lack of susceptibility of the newly infiltrating blood monocytes (Duan *et al.*, 1997b) and (ii) appearance of anti-PRRSV antibodies in the lungs. It is remarkable that viral antigen-positive cells and cellular debris were mainly localized in foci in lung tissue until 14 days PI and that, thereafter, only single viral antigen-positive cells were observed. It is possible that most infected cells in foci become destroyed by antibody-dependent cell lysis. Single viral antigen-positive cells, morphologically recognized as macrophage-like cells, were observed until 35 days and 40 days PI in lung tissue and BAL fluid respectively. These single viral antigen-positive cells were the source of the virus detected in lung tissue and BAL fluid until 40 days PI. Why these single viral antigen-positive cells are able to persist despite the presence of the humoral immunity needs to be clarified.

Clearance of PRRSV from the lungs coincided with the appearance of neutralizing antibodies in sera and BAL fluids. However, since low amounts of PRRSV remain in the lungs of some pigs in spite of the presence of neutralizing antibodies in sera and BAL fluids, other immune factors or mechanisms such as cell-mediated immunity are probably involved as well in the complete elimination of the virus at this site. Why neutralizing antibodies appear so late in infection and remain at rather low levels is not yet known.

ACKNOWLEDGMENTS

The authors would like to thank Fernand De Backer, Tini De Lausnay, Chris Bracke, Lieve Sys and Chantal Van Marcke for their excellent technical assistance. The authors would also like to thank Dr. T. Drew for his gift of the monoclonal antibodies WBE1 and WBE4-6 and Dr. G. Wensvoort for the supply of the Lelystad isolate of PRRSV.

REFERENCES

Christopher-Hennings, J., Nelson, E.A., Hines, R.J., Nelson, J.K., Swenson, S.L., Zimmerman, J.J., Chase, C.C.L., Yaeger, M.J., and Benfield, D.A., 1995, Persistence of porcine reproductive and respiratory syndrome virus in serum and semen of adult boars, *Journal of Veterinary Diagnostic Investigation*, **7**: 456-464.

Drew, T.W., Meulenberg, J.J.M., Sands, J.J. and Paton, D.J., 1995, Production, characterization and reactivity of monoclonal antibodies to porcine reproductive and respiratory syndrome virus, *Journal of General Virology*, **76**: 1361-1369.

Duan, X., Nauwynck, H.J., and Pensaert, M.B., 1997a, Virus quantification and identification of cellular targets in the lungs and lymphoid tissues of pigs at different time intervals after inoculation with porcine reproductive and respiratory syndrome virus, *Veterinary Microbiology*, **56**: 9-19.

Duan, X., Nauwynck, H.J., and Pensaert, M.B., 1997b, Effects of origin and state of differentiation and activation of monocytes/macrophages on their susceptibility to PRRSV, *Archives of Virology*, **142**: 2483-2497.

Duan, X., Nauwynck, H.J., Favoreel, H.W., and Pensaert, M.B., 1998, Identification of a putative virus receptor for porcine reproductive and respiratory syndrome virus on porcine alveolar macrophages, *Journal of Virology*, **72**: 4520-4523.

Labarque, G.G., Nauwynck, H.J., Van Reeth, K. and Pensaert, M.B., 2000, Apoptosis in the lungs of pigs during an infection with a European strain of porcine reproductive and respiratory syndrome virus, *Advances in Experimental Medicine and biology*, (see elsewhere in this volume).

Mengeling, W.L., Lager, K.M., and Vorwald, A.C., 1995, Diagnosis of porcine reproductive and respiratory syndrome, *Journal of Veterinary Diagnostic Investigation*, **7**: 3-16.

Nauwynck, H.J., Duan, X., Favoreel, H.W., Van Oostveldt, P., and Pensaert, M.B., 1999, Entry of porcine reproductive and respiratory syndrome virus into porcine alveolar macrophages via receptor-mediated endocytosis, *Journal of General Virology* **80**: 297-305.

Pescovitz, M.D., Lunney, J.K., and Sachs, D.H., 1984, Preparation and characterization of monoclonal antibodies reacting with porcine PBL, *Journal of Immunology,* **133**: 368-375.

Swenson, S.L., Hill, H.T., Zimmerman, J.J., Evans, L.E., Landgraf, J.G. Wills, R.W., Sanderson, T.P., McGinley, M.J., Brevik, A.K., Ciszewski, D.K. and Frey, M.L., 1994, Excretion of porcine reproductive and respiratory syndrome virus in semen after experimentally induced infection in boars, *Journal of the American Veterinary Medical Association,* **204**: 1943-194.

Vanderheijden, N., Delputte, P., Nauwynck, H.J., and Pensaert, M.B., 2000, Effect of heparin on the entry of porcine reproductive and respiratory syndrome virus into alveolar macrophages, *Advances in Experimental Medicine and biology,* (see elsewhere in this volume).

Van Reeth, K., Labarque, G., Nauwynck H.J., and Pensaert, M.B., 1999, Differential production of proinflammatory cytokines in the pig lung during different respiratory virus infections: correlations with pathogenicity, *Research in Veterinary Science,* **67**: 47-52.

Wensvoort, B., Terpstra, C., Pol, J.M.A., ter Laak, E.A., Bloemraad, M., de Kluyver, E.P., Kragten, C., van Buiten, L., den Besten, A., Wagenaar, F., Broekhuijsen, J.M., Moonen, P.L.J., Zetstra, T., de Boer, E.A., Tibben, H.J., de Jong, M.F., van't Veld, P., Groenland, G.J.R., van Gennep, J.A., Voets, M.T., Verheijden, J.H.M., and Braamskamp, J., 1991, Mystery swine disease in the Netherlands: the isolation of Lelystad virus, *Veterinary Quarterly,* **13**: 121-130.

Wills, R.W., Zimmerman, J.J., Yoon, K.-J., Swenson, S.L., McGinley, M.J., Hill, H.T., Platt, K.B., Christopher-Hennings, J., and Nelson, E.A., 1997, Porcine reproductive and respiratory syndrome virus: a persistent infection, *Veterinary Microbiology,* **55**: 231-240.

Monoclonal Antibody Directed against a Membranous Protein of MARC-145 Cells Blocks Infection by PRRSV

DOMINIC THERRIEN AND SERGE DEA
Centre de Recherche en Microbiologie et Biotechnologie, INRS-Institut Armand-Frappier, Université du Québec, Laval, Québec, H7V 1B7

1.   INTRODUCTION

In its natural host, infection by the porcine reproductive and respiratory syndrome virus (PRRSV) is restricted to cells of the macrophage/monocyte lineage and testicular germ cells (Plagemann 1996; Sur et al., 1996). *In vitro,* only cell subclones derived from the established monkey kidney epithelial cell line MA-104, such as MARC-145 and CL2621, and primary cultures of porcine alveolar macrophages (PAMs) efficiently replicated the porcine arterivirus (Bautista et al., 1993). Factors implicated in this restricted tropism have not yet been elucidated, but PRRSV enters in PAMs and in MARC-145 cells through a mechanism of receptor-mediated endocytosis (Kreutz and Ackermann, 1996; Nauwynck et al., 1999). Recently, a 210 kDa protein has been proposed as a putative receptor for PRRSV on PAMs, since a MAb directed against this protein blocked virus infection of PAMs (Duan et al., 1998). However, this MAb failed to react with MARC-145 and porcine monocytes, suggesting that PRRSV may enter in these cells by interacting with other cellular proteins. MAbs to Gp4 and Gp5 enveloped glycoproteins of PRRSV neutralize viral infectivity, but the viral structural protein responsible for viral attachment is still to be identified (Meulenberg et al., 1997; Pirzadeh and Dea, 1997). In this study, the capacity of PRRSV to attach and infect different cell lines was first determined. Subsequently, two MAbs directed against a 60 to 66 kDa membranous protein of MARC-145

The Nidoviruses (Coronaviruses and Arteriviruses).
Edited by Ehud Lavi *et al.*, Kluwer Academic/Plenum Publishers, 2001.

cells were found to interfere with viral infection, thus representing a putative receptor or co-receptor for PRRSV.

2.   METHODOLOGY AND RESULTS

Virus binding to the host cells is a crucial step in the infection process. To study cell susceptibility to PRRSV, a binding assay was previously described to follow attachment of PRRSV on the surface of different cell lines by flow cytometry (Therrien et al., 2000). Infection was also performed on these cell lines to verify their capacity to replicate PRRSV. The cells were infected at a MOI of 0,01 with the Quebec strain, IAF-Klop. After 48 h, cells were fixed with acetone to perform immunofluorescence (IIF) with anti-N MAb IAF-K8 or harvested to verify the presence of PRRSV genome by RT-PCR using oligonucleotide primers 1010 PLS and 1011PLR, designed to amplify a 434 bp DNA fragment encompassing the entire ORF7 gene and a portion of the 3' terminal non-coding region of PRRSV (Mardassi et al., 1994). Only MARC-145 cells, PAMs, and a small population of porcine peripheral blood leukocytes which has been previously characterised as monocytes (Therrien et al., 2000) were infected efficiently by PRRSV. However, other cell lines tested (PT, PK-15, BHK-21, 293A, and RK-13), apparently resistant to PRRSV infection, were also able to bind PRRSV on their surface.

To further investigate on the nature of the restricted tropism, MAbs were raised against cell surface proteins. Three injections of 50:g of cell lysates of MARC-145 and PAMs cells, or PRRSV semi-purified virus, were given to Balb/C mice at 2 week-intervals, prior fusion experiments of spleen cells with Sp2/O myeloma cells. Two hybridomas, 18-A7 and 18-F6, were obtained that secreted MAbs to epitopes on the surface proteins of MARC-145 cells, but not on the surface of PAMs. Both MAbs blocked efficiently PRRSV infection of MARC-145 cells. The capacity of PRRSV to bind to MARC-145 cells was reduced by about 60 % following preincubation of MAb 18-F6 (Fig1.D). Western blotting analyses were performed to identified membranous protein reacting with MAb 18-A7. MARC-145 cells were harvested with 20 mM of EGTA pH8, 30 min at 37°C, then solubilized with LB-2 buffer (20 mM Tris-HCl, pH7.5, 150 mM NaCl, 5 mM EDTA, 0.5 % NP-40, 0.1 % Na-deoxycholate, 0.1 % SDS) (Mardassi et al. 1996). The proteins were separated by electrophoresis on SDS-PAGE gel and transferred on nitrocellulose membrane. A protein with a size ranging from 60 to 66 kDa was obtained.

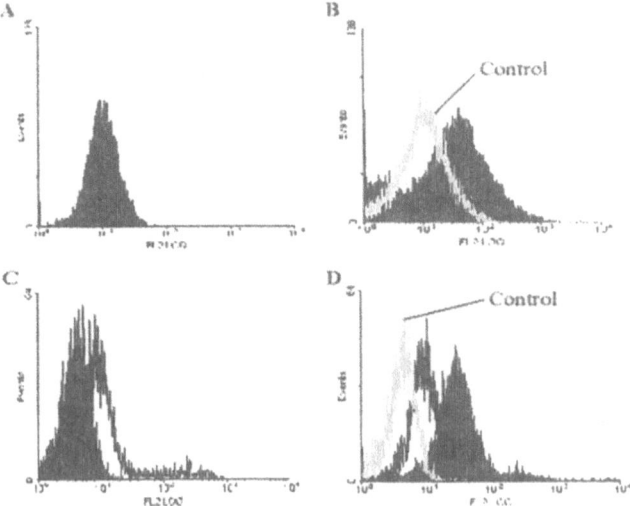

Figure 1. Binding of MAb 18-F6 and inhibition of PRRSV attachment. Suspensions of PAMs (A and B) or MARC-145 cells (C and D) were inoculated with the IAF-Klop strain of PRRSV at a MOI of 10 $TCID_{50}$ per cell. Following an adsorption period of 1 h at 4° C, infected cells were washed with PBS and incubated with biotinylated MAb IAF-K8 for 30 min at 37°. The surface bound immune complexes were labelled with streptavidin-phycoerythrin, and fluorescent level was analyzed by flow cytometry. Filled patterns represent control (A and C) cells incubated in the presence of biotinylated MAb IAF-K8 and streptavidin-phycoerythrin. B and D. Cells were incubated with the Mab 18-F6 for 30 min at room temperature prior to the binding assay (black line). In this case filled patterns represent cells incubated with the virus.

Figure 2. Reactivity of Mab 18-F6 against a cellular protein of 60 to 66 kDa. Western blot analysis of MARC-145 solubilized with LB-2 buffer (20 mM Tris-HCl, pH7.5, 150 mM NaCl, 5 mM EDTA, 0.5 % NP-40, 0.1 % sodium deoxycholate, 0.1 % SDS) (Mardassi et al. 1996). Following electrophoresis in 12% SDS-PAGE, proteins were electrophoretically trans-fered on nitrocellulose membrane, saturated with PBS containing 5 % of fat milk and incubated with Mab 18-F6. The reactions was revealed following an incubation in the presence of horsedish peroxidase labelled caprine IgG directed against mouse IgG, and staining of the immune complexes was obtained by a final incubation of the nitrocellulose membrane in a chloronaphtol solution containing 3% H_2O_2.

3.   DISCUSSION

Interaction of viral particles to specific receptors on the surface of permissive cells is the primary step of a viral infection. In this study, it was shown that the capacity of a cell line to bind PRRSV was not sufficient to allow a productive infection. In fact, it has been demonstrated that some cell lines known to be resistant to PRRSV infection can also bind the virus. Transfection of PRRSV genome into BHK-21 or Vero cells, thus bypassing surface cellular receptors, results in a productive infection (Kreutz, 1998), suggesting that PRRSV infection is probably restricted by its capacity to attach or penetrate the cytoplasmic membrane of the host cell. Data obtained raise the hypothesis that more than one receptor may be involved in PRRSV infection. The MAb, 18-F6, directed against a protein of 60 to 66 kDa on the surface of MARC-145 cells, was shown to block PRRSV infection. Since the MAb blocked only partially PRRSV binding, it could not be determined whether this protein is the initial binding factor or a co-receptor involved after binding of the virus. MAb 18-F6 failed to recognize an epitope on the surface of PAMs. Since a protein of 210 kDa has been proposed to be a potential receptor for PRRSV on the surface of PAMs, results of this study indicate that PRRSV enters in the MARC-145 cells through the interaction with a different receptor. EAV and SHFV, but not LDV, have previously demonstrated their ability to infect MA-104 cells, from which MARC-145 cells have been established (Plagemann 1996). This raises the hypothesis that PRRSV, EAV and SHFV may enter by the same pathway in these epithelial cells via a common receptor. This common receptor could be the protein recognized by the MAb 18-F6, but this remains to be further investigated.

ACKNOWLEDGMENTS

This work was supported by the National Science and Engineering Research Council of Canada, strategic grant #STP0202083. D. Therrien is a recipient of a PhD grant from the FCAR, Québec.

REFERENCES

Bautista, E.M., Goyal, S.M., Yoon, I..J, Joo, H.S., Collins, J.E. (1993) Comparison of porcine alveolar macrophages and CL2621 for the detection of porcine reproductive and respiratory syndrome (PRRS) virus and anti-PRRS antibody. *J Vet Diagn Invest* **5**, 163-165.

Duan, X., Nauwynck, H.J., Favoreel, H.W., Pensaert, M.B. (1998) Identification of a putative receptor for porcine reproductive and respiratory syndrome virus on porcine alveolar macrophages. *J Virol* **72,** 4520-4523.

Kreutz, L.C., Ackermann, M.R. (1996) Porcine reproductive and respiratory syndrome virus enters cells through a low pH-dependent endocytic pathway. *Virus Res* **42,** 137-147.

Kreutz, L.C. (1998) Cellular membrane factors are the major determinants of porcine reproductive and respiratory syndrome virus tropism. *Virus Res* **53,** 121-128

Mardassi H, Wilson L, Mounir S, Dea S (1994) Detection of porcine reproductive and respiratory syndrome virus and efficient differentiation between Canadian and European strains by reverse transcription and PCR amplification. J Clin Microbiol **32,** 2197-2203,

Mardassi, H., Massie, B., Dea, S. (1996) Intracellular synthesis, processing and transport of proteins encoded by ORFs 5 to 7 of porcine reproductive and respiratory syndrome virus *Virology* **221,** 98-112,

Meulenberg, J.J.M., Van Nieuwstadt, A.P., Van Essen-Zandbergen ,A., Langeveld, J.P.M. (1997) Posttranslational processing and identification of a neutralization domain of the GP4 protein encoded by ORF4 of Lelystad virus. *J Virol* **71,** 6061-6067,

Nauwynck, H.J., Duan, X., Favoreel, H.W., Van Oostveldt, P., Pensaert, M.B. (1999) Entry of porcine reproductive and respiratory syndrome virus into porcine alveolar macrophages via receptor-mediated endocytosis. *J Gen Virol* **80,** 297-305.

Pirzadeh, B., Dea, S. (1997) Monoclonal antibodies to the ORF5 product of porcine reproductive and respiratory syndrome virus define linear neutralizing determinants. *J Gen Virol* **78,** 1867-1873,

Plagemann, P.G.W. (1996) Lactate dehydrogenase-elevating virus and related viruses. Fields Virology 3rd ed., Raven publishers, Philadelphia, pp. 1105-1120.

Sur, J.H., Doster, A.R., Christian, J.S., Galeota,J.A., et al. (1997) Porcine reproductive and respiratory syndrome virus replicates in testicular germ cells, alters spermatogenesis, and induces germ cell death by apoptosis. *J Virol* **71,** 9170-9179.

Therrien, D., St-Pierre, Y., Dea, S. (2000) Preliminary characterization of protein binding factor for porcine reproductive and respiratory syndrome virus on the surface of permissive and non-permissive cells. *Arch. Virol.* (in press)

Detection of Antibodies to the Nucleocapsid Protein of PRRS Virus by a Competitive ELISA

[1]SERGE DEA, [1]LOUISE WILSON, [1]DOMINIC THERRIEN, AND
[2]ESTELA CORNAGLIA
[1]*Centre de Microbiologie et Biotechnologie, INRS-Institut Armand-Frappier, Université du Québec, Laval, Qc, Canada;* [2]*Biovet Inc., St-Hyacinthe, Qc, Canada*

1.   INTRODUCTION

The mature virions of the porcine reproductive and respiratory syndrome virus (PRRSV), a new porcine arterivirus, is made of three major structural proteins: a 25 kDa envelope glycoprotein (GP_5), an 18-19 kDa unglyco-sylated membrane protein (M), and a 15 kDa nucleocapsid (N) protein (Mardassi et al., 1995; Meulenberg et al., 1995). The N protein is the more abundant protein of the virion and is highly antigenic, which therefore makes it a suitable candidate for the detection of virus-specific antibodies and diagnosis of the disease (Loemba et al, 1996). It is also encoded by a relatively well conserved region of the viral genome, since a high degree of amino acid (aa) sequence identity has been observed among the N protein of North American (96-100%) and European (94-99%) strains (Meng et al., 1995; Suarez et al., 1994). Four to five domains of antigenic importance have been identified for the N protein, a common conformational antigenic site for European and North American strains being localized in the central region of the protein (Meulenberg et al., 1998; Wootton et al., 1998).

In this study, the genomic region encoding the N protein of a North American reference strain of PRRSV was cloned, expressed in *Escherichia coli*, purified and used as antigen in an indirect ELISA for detection of antibodies against PRRSV. Two MoAbs directed to highly conserved

The Nidoviruses (Coronaviruses and Arteriviruses).
Edited by Ehud Lavi *et al.*, Kluwer Academic/Plenum Publishers, 2001.

epitopes of North American PRRSV isolates were used in a competitive assay to improve the specificity of the test.

2.   METHODOLOGY AND RESULTS

The entire ORF7 gene of the Quebec cytopathogenic IAF-Klop strain of PRRSV was amplified by RT-PCR using primer pairs ORF7-S (5'-CTAAA-TATGCCAAATAACAAC-3') and ORF7.AS (5'-CTCAAGAATGCCAGCTCA-3') (Gonin et al., 1999). The primers were designed according to the sequence of the Quebec reference strain (EMBL/ GeneBank accession number U64928) and contained two restriction sites for *Eco*RI (sense primers) and *Bam*HI (anti-sense primers) at their 5' end for directional cloning. The complementary DNA corresponding to the entire ORF7 gene of the IAF-Klop strain was inserted into procaryotic expression vector pGEX-4T-1 (Pharmacia) to yield plasmid pGEX-7 (Mardassi et al., 1996).

Following incubation in the presence of IPTG, competent *E. coli* strain BL21 (DE3) cells that were transformed with pGEX-7 expressed the GST-N recombinant fusion protein, essentially in the form of inclusion bodies. The molecular mass (M_r) of the GST-N recombinant fusion protein, determined following SDS-PAGE analysis of bacterial cell lysates, was estimated to 39.6 kDa, in accordance with the value determined previously from the aa sequence of the native N protein of the Quebec reference strain (Mardassi et al., 1995). The GST-N recombinant fusion protein could be recovered following solubilization of the pelleted inclusion bodies in the presence of lysozyme, triton X-100 and 8 M urea, then enriched and purified by affinity chromatography on glutathione-Sepharose 4B. Three to four additional protein bands were revealed following SDS-PAGE analysis of the second and third eluates with apparent M_r s of 35.9, 32.5, 31.5 and 29 kDa. Only the 39.6 and 31.5 species could be revealed following Western immunoblotting, using homologous porcine anti-PRRSV serum and rabbit anti-GST-N monospecific hyperimmune serum.

The optimal dilutions of the antigen and the test serum in indirect ELISA were determined by checkerboard titration. Using the homologous porcine hyperimmune serum, the optimal dilutions were found to be 1:50 for the serum and 1:1000 for the antigen, which corresponded to a final concentration of 0.1 to 0.5 µg of protein per well. The highest P/N ratios were obtained with PBS containing 0.05% Tween 80 and 5% goat serum as the blocking (saturation) and dilution buffer. Comparable results were obtained by incubating the plates at room T° or 37°C, with incubation periods of 45 min for the test serum and the anti-porcine IgG conjugate. Data obtained by Western immunoblotting suggested that part of the background obtained with sow or adult sera was attributed to their reactivity with residual *E. coli* proteins (data not shown) that co-eluted with the GST-N recombinant usion protein following affinity chromatography. To avoid false-positive results due to bacterial proteins, a competitive ELISA was set-up where the capacity of clinical sera to interfere with the binding of anti-N MAbs (IAF-K8 and IAF-2B4), to the GST-N recombinant fusion protein was determined. Both MAbs were previously found to be directed against highly preserved conformational epitopes of North American isolates of PRRSV (Dea et al., 1996). Data for each test sera were expressed in percent competition, calculated by the following formula : % competition (sample) = $1 - (A_{450}$ (sample + MAb anti-N) / A_{450} (MAb alone)) X 100.

The evaluation of the competitive ELISA was carried out by first testing 95 pig sera obtained from previous experimental inoculation studies. A total of 20 of these pig sera were considered as negative by IIF (antibody titers < 16) and HerdCheck® ELISA (P/N values < 0.4). For both, the Herd-Check®ELISA and K8-ELISA, a linear correlation was obtained between P/N ratio or % competition values, with antibody titers determined by IIF. Comparison with HerdCheck® ELISA and IIF allowed the definition of a threshold range between 20 and 30% competition. In general, clinical sera given P/N values of > 2.0 in HerdCheck® ELISA showed >85% competition in the K8-ELISA corresponding to IIF titers > 1:512. Those sera for which P/N values varied between 1.0 and 2.0 in HerdCheck® ELISA yielded >50% competition in the K8-ELISA corresponding to IIF titers ranging from 1:64 and 1:256. Finally, sera with IIF titers ranging from 1:16 to ≤ 1:64 displayed P/N values of >0.4 < 1.0 in HerdCheck® ELISA, with percent competition values in K8-ELISA that varied from ≥ 20 to ≤ 50 %.

Comparison of data obtained in K8-ELISA with those obtained in IIF and HerdCheck® ELISA also allowed a comparison of the sensitivity and the specificity of these tests (Table 1). The data obtained with sera from experimentally-infected pigs showed that the K8-ELISA was capable of detecting anti-PRRSV antibodies in 86.7% (65/75) and 92.6 % (63/68) of pig sera that were considered as seropositive in IIF (titers > 16) and in HerdCheck® ELISA (P/N ratio > 0.4), with specificity values of 100% and 96.2 %, respectively. If a cut-off value of 1:32 rather than 1:16 was considered for IIF, the sensitivity and specificity of the K8-ELISA increased to 95.5 % and 96.5 %, respectively, but the performance of the test remained unchanged when compared to the HerdCheck® ELISA. Compa-rable results were obtained when using MAb IAF-2B4 in the competitive ELISA (data not shown). When tested on clinical samples (542 sera) from 28 positive and 28 negative pig herds, the K8-ELISA performed in a similar way to HerdCheck® and IIF tests as shown by Kappa values of 0.762 and 0.803. The sensitivity and specificity of K8-ELISA were 100% on a herd basis, whereas sensitivity a specificity of 98.7% were determined on an individual basis in comparison to HerdCheck[®]and IIF tests.

Table 1. Comparison of sensitivity and specificity of the 3 serological tests used for the validation of the K8-ELISA, $n= 95$

Test X	Cut-off values	IIF		IDEXX® ELISA		K8-ELISA	
		Se[a]	Sp[b]	Se	Sp	Se	Spe
IIF	1/16			100	74.1	100	66.7
IDEXX ELISA	P/N = 0.4	90.7	100			96.2	86.7
K8-ELISA	20%	86.7	100	92.6	96.2		
IIF	1/32			97.0	100	96.9	93.3
IDEXX ELISA	P/N = 0.4	100	93.1			96.3	86.7
K8-ELISA	20%	95.5	96.5	92.6	96.2		

[a] Sensitivity (test X) = 100 x number of positives in both X and reference test / total number of positives in the reference test.
[b] Specificity (test X) = 100 x number of negatives in both X and reference test / total number of negatives in the reference test.

3. DISCUSSION

Using sera from experimentally infected pigs, linear correlations were obtained by comparing the % of competition values to P/N ratios determined by Herdcheck[®] ELISA, as well as with IIF antibody titers, with relatively small standard deviation values. None of the sera that scored negative by Herdcheck [®] ELISA and by IIF exhibited reactivities > 30% in the K8-ELISA which therefore confirms the high specificity of this competitive ELISA. Furthermore, practically 100% correlation (sensitivity and specificity) was obtained with K8-ELISA if P/N ratios of > 0.5 were considered as the threshold of the Herdcheck[®] ELISA rather than > 0.4.

Although a study of the kinetics of antibody production was not per se using K8-ELISA, this new competitive ELISA should permit identification of pigs as early as 7 to 10 day after infection, since it has been previously demonstrated that IIF antibody titers of > 1:16 were usually obtained at that time post-infection using the same experimental pig sera (Loemba et al., 1996). Since MAb IAF-K8 was found to be directed to a well preserved epitope of the North American and European strains of PRRSV (Dea et al. 1996), we expect that a similar efficacy of the K8-ELISA should be obtained with clinical sera from European pig herds.

ACKNOWLEDGMENTS

This work was partly supported by the Conseil des Recherches en Pêcheries et AgroAlimentaire du Québec (grant # 4358) and Biovet Research Inc, St-Hyacinthe, Québec, Canada.

REFERENCES

Dea, S., Gagnon, C.A., Mardassi, H. and Milane, G. (1996) Antigenic variability among North-American and European strains of PRRSV as defined by monoclonal antibodies to the matrix protein. J Clin Microbiol 34:1488-1493.

Gonin, P., Pirzadeh, B., Gagnon, C.A. and Dea, S. (1999) Seroneutralization of porcine reproductive and respiratory syndrome virus correlates with antibody response to the GP5 major envelope glycoprotein. J. Vet. Diagn. Invest. 11: 20-26.

Loemba, H.D., Mounir, S., Mardassi, H., Archambault, D. and Dea, S. (1996) Kinetics of humoral immune response to the major structural proteins of the porcine reproductive and respiratory syndrome virus. Arch. Virol. 141: 751-761

Mardassi, H., Mounir, S. and Dea, S. (1995) Molecular analysis of the ORF3-7 of porcine reproductive and respiratory syndrome virus, Quebec reference strain. Arch. Virol. 140: 1405-1418.

Mardassi, H., Massie, B. and Dea, S. (1996) Intracellular synthesis, processing and transport of proteins encoded by ORFs 5 to 7 of porcine reproductive and respiratory syndrome virus. Virology 221: 98-112.

Meng, X-J., Paul, P.S., Halbur, P.G. and Lum, M.A. (1995) Phylogenetic analyses of the putative M (ORF6) and N (ORF 7) genes of porcine reproductive and respiratory syndrome virus (PRRSV): Implication for the existence of two genotypes of PRRSV in the USA and Europe. Arch. Virol. 140: 745-755.

Meulenberg, J.J.M., Petersen-Den Besten, A., de Kluyer, E.P., Moormann, R.J.M., Schaaper, W.M.M. and Wensvoort, G. (1995) Characterization of proteins encoded by ORFs 2 to 7 of Lelystad virus. Virology 206: 155-163.

Meulenberg, J.J.M., van Nieuwstadt, A.P., van Essen-Zandbergen, A., Bos-de-Ruijter, J.N.A., Langeveld, J.P.M. and Meloen, R.H. (1998) Localization and fine mapping of antigenic sites on the nucleocapsid protein of porcine reproductive and respiratory syndrome virus with monoclonal antibodies. Virology 252: 106-114.

Wootton, S.K., Nelson, E.A. and Yoo, D. (1998) Antigenic structure of the nucleocapsid protein of porcine reproductive and respiratory syndrome virus. Clin. Diagn. Lab. Immunol. 5: 773-779.

Enhancement of Defective RNA Expression Vectors as Potential Vaccine Delivery Systems for Avian Infectious Bronchitis Virus

BRIAN DOVE, KATHY SHAW, JULIAN HISCOX, DAVID CAVANAGH, AND PAUL BRITTON
Division of Molecular Biology, Institute for Animal Health, Compton Laboratory, Compton, Newbury, Berkshire, RG20 7NN, UK

1. INTRODUCTION

IBV defective RNA (D-RNA) CD-61, a deletion mutant of the naturally occurring D-RNA CD-91, is capable of being utilised as an RNA expression vector for heterologous genes (Pénzes *et al.*, 1996; Stirrups *et al.*, 2000a; Stirrups *et al.*, 2000b). The reporter genes chloramphenicol acetyltransferase (CAT) and luciferase, under the control of IBV gene 5 transcription associated sequence (TAS), are capable of expression from within CD-61 during D-RNA rescue. However the modification of CD-61 by the insertion of foreign gene sequences has an adverse effect on D-RNA rescue efficiency resulting in an eventual loss of foreign gene expression (Stirrups *et al.*, 2000b). We attempted to enhance heterologous gene expression in CD-61 by utilising the encephalomyocarditis virus (EMCV) internal ribosome entry site (IRES) as an alternative to gene 5 TAS to control CAT gene expression.

2. RESULTS AND DISCUSSION

CD-61 was modified to contain the CAT reporter gene under IRES control in the *Stu*I site present in domain II of CD-61 (IRES-CAT-*Stu*I) or in the *Pma*CI site present in domain III of CD-61 (IRES-CAT-*Pma*CI). CD-61-

The Nidoviruses (Coronaviruses and Arteriviruses).
Edited by Ehud Lavi *et al.*, Kluwer Academic/Plenum Publishers, 2001.

CAT contains the reporter gene CAT under the control of the IBV gene 5 TAS inserted into the *Pma*CI site in CD-61 (Stirrups *et al.*, 2000b) (Fig.1).

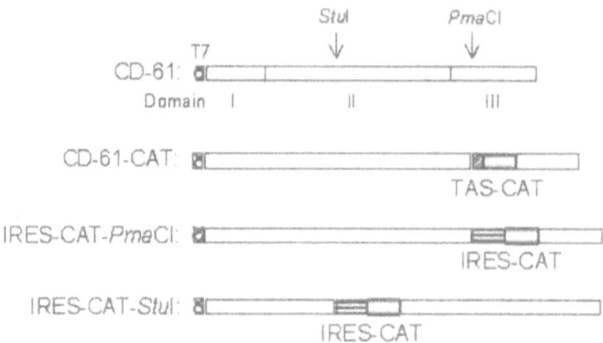

Figure 1. Modification of IBV D-RNA CD-61. All constructs were under the control of a T7 promoter.

To test the viability of the constructs, IRES-controlled expression of CAT in CD-61 was initially assessed *in vitro* using a rabbit reticulocyte lysate coupled transcription/translation cell free system. CAT protein production was detected and quantified by a CAT ELISA. IRES-CAT-*Stu*I and IRES-CAT-*Pma*CI showed a 45- to 65- fold higher production of CAT protein, respectively, under IRES control in a cell free system compared to CD-61-CAT (Fig. 2).

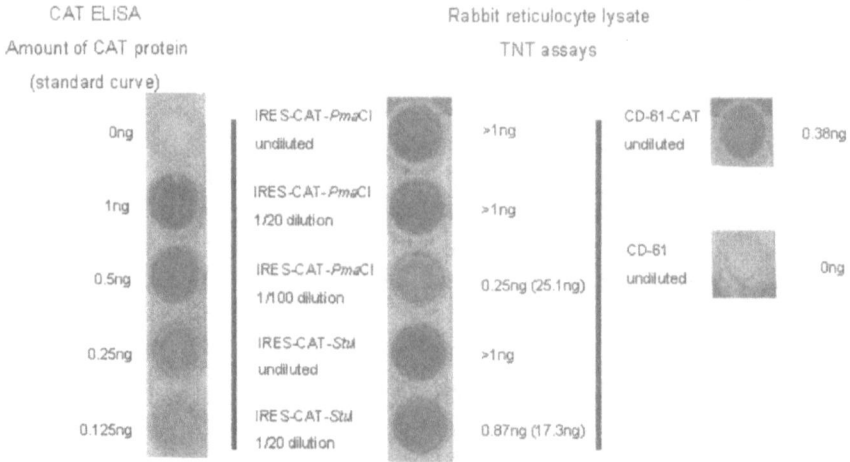

Figure 2. Expression of the CAT reporter gene from CD-61-CAT, IRES-CAT-*Stu* and IRES-CAT-*Pma*CI D-RNAs using a T7 transcription/translation coupled rabbit reticulocyte lysate cell free system. Samples with CAT protein values exceeding the calibration curve were diluted, with the total amount of CAT protein shown in brackets.

To determine the effectiveness of IRES-controlled expression in cell culture, T7 *in vitro*-generated RNA transcripts of CD-61-CAT, IRES-CAT-*Stu* and IRES-CAT-*Pma*CI were electroporated into either IBV-infected or mock-infected chick kidney (CK) cells. CAT protein from cell lysates was detected by CAT ELISA. Only passage 0 (P_0) CK cells infected with IBV and electroporated with CD-61-CAT showed the presence of CAT protein. No CAT protein was detected from the IRES constructs in either infected or mock-infected CK cells. Therefore, the preliminary results indicate that IRES-controlled CAT gene expression in D-RNAs during rescue by IBV in CK cells is ineffective. Further work is required to ascertain if the presence of an IRES sequence inhibits D-RNA replication.

ACKNOWLEDGEMENTS

This work was supported by the Ministry of Agriculture, Fisheries and Food, UK, Project code OD1905 and by grant number CT950064 of the Fourth RTD Framework Programme of the European Commission (EC). B. Dove was the recipient of an IAH studentship.

REFERENCES

Pénzes, Z., Wroe, C., Brown, T. D., Britton, P. & Cavanagh, D. 1996. Replication and packaging of coronavirus infectious bronchitis virus defective RNAs lacking a long open reading frame. *J. Virol.* **70**, 8660-8668.

Stirrups, K., Shaw, K., Evans, S., Dalton, K., Cavanagh, D. & Britton, P. 2000a. Leader switching occurs during the rescue of defective RNAs by heterologous strains of the coronavirus infectious bronchitis virus (IBV). *J. Gen. Virol.* **81**, 791-801.

Stirrups, K., Shaw, K., Evans, S., Dalton, K., Casais, R., Cavanagh, D. & Britton, P. 2000b. Expression of reporter genes from the defective RNA CD-61 of the coronavirus infectious bronchitis virus. *J. Gen. Virol.* **81**, 1687-1698.

A Virus-Neutralising Monoclonal Antibody Expressed in the Milk of Transgenic Mice

[1,2]A.F. KOLB, [3]J. WEBSTER, [3]C.B.A. WHITELAW, AND [2]S.G. SIDDELL
[1]Cell Physiology Group, Hannah Research Institute, Ayr, UK; [2]Institute of Virology, University of Würzburg, Würzburg, Germany; [3]Division of Molecular Biology, Roslin Institute, Roslin, UK

1. INTRODUCTION

Coronaviruses are frequently associated with respiratory and gastrointestinal disorders in both animals and man. In adult animals, coronavirus infections are generally mild, however, in neonates they often cause severe and sometimes lethal diseases (Enjuanes and van der Zeijst, 1995). Maternal antibodies supplied via the placenta and milk efficiently protect new-born animals against the fatal consequences of acute coronaviral infections during this critical phase (Homberger, 1992).

Vaccination against coronavirus infections have been employed with varying success (Saif and Wheeler, 1998) but the vaccines are usually highly strain specific (Homberger et al., 1992), dependent on specific routes of infection and often short-lived. Live vaccines are also associated with the danger of in vivo recombination leading to novel viruses with increased pathogenicity. As coronaviral infections cause a high mortality only during a short time period (up to 20 days post partum in mice), which largely coincides with the suckling period, transgenic animals producing a neutralising antibody as a recombinant protein in milk may provide an effective strategy to protect animals during this critical phase (Castilla et al., 1998; Sola et al., 1998). In order to test this hypothesis in vivo, we have generated transgenic mice expressing a highly neutralising monoclonal antibody (mab) directed against the neurotropic MHV strain JHM and

The Nidoviruses (Coronaviruses and Arteriviruses).
Edited by Ehud Lavi et al., Kluwer Academic/Plenum Publishers, 2001.

demonstrate in here that the recombinant antibody is secreted into the milk at yields up to 0.7mg/ml.

2.   RESULTS

2.1   Generation of transgenic mice

Neutralising antibodies can effectively prevent MHV-JHM infections in vitro and in vivo (Wege et al., 1984). Mab A1 (Wege et al., 1984), which binds to the S1 subunit of the MHV-JHM S-protein, is one of the most potent antibodies with regard to virus neutralisation and the inhibition of virus-induced cell-to-cell fusion. Chimeric open reading frames consisting of the variable regions of the mab A1 (Kolb and Siddell, 1997) and human immunoglobulin G (IgG) constant regions were inserted into the mammary gland specific expression vector pBJ41 (Sola et al., 1998), which is based on the ovine ß-lactoglobulin (ß-LG) gene. The two resulting expression vectors pBJ41-A1L (carrying the A1 light chain variable region) and pBJ41-A1H (carrying the A1 heavy chain variable region) were microinjected together with the ß-LG expression construct pSS1tgXS in a ratio of 1:1:1 into fertilised oocytes. pSS1tgXS is expressed in a copy-number dependent manner in transgenic animals (Whitelaw et al., 1992) and has been shown to increase the expression of co-linked transgenes (Clark et al., 1992). The molecular basis of this "transgene rescue effect" is not known, but it is assumed that the ß-LG construct is able to generate independent chromatin domains which escape the silencing effect of neighbouring chromatin structures.

Table 1. Transgene copy number. Transgene copy number was determined by Southern blot analysis as described (Whitelaw et al., 1992). Aliquots (20µg) of DNA were digested with BamHI, separated on a 1% agarose gel and blotted to a nylon membrane. The blot was subsequently probed with a 0.5kb SacI fragment of the ß-LG promoter, which is present in all 3 transgene constructs.

transgenic line	pSS1tgXS	copy number pBJ41-A1L	pBJ41-A1H
HEP3	16	2	5
HEP10	20	2	7
HEP17	nd	nd	nd
HEP20	12	1	6
HEP30	6	1	4
HEP36	10	2	4
HEP38	30	10	15
HEP50	2	2	10

The transgene copy number was determined in 7 of the 8 lines obtained (Table 1). Although the transgenes encoding the antibody light chain and the heavy chain were micro-injected into fertilised oocytes at equimolar levels, the transgene copy numbers seem to indicate that consistently more copies of the light than heavy chain construct were incorporated into the host genome. The reason for this is unknown but we believe it may be due to a leaky expression of the transgenes during early developmental stages which, in turn, leads to a cytotoxic over-production of heavy chain protein and the subsequent loss of embryos.

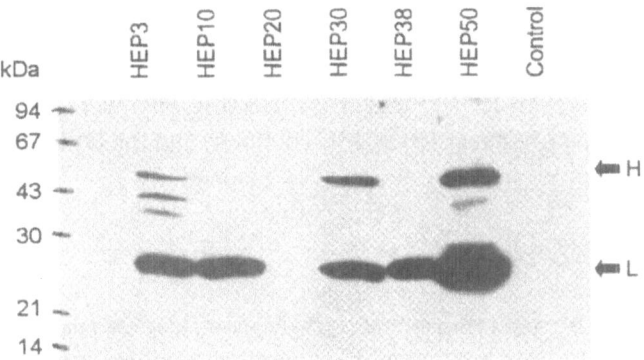

Figure 1. Western blot analysis. Aliquots of defatted/diluted milk samples (corresponding to 1 µl of milk) isolated from transgenic lines were separated by polyacrylamide gel electrophoresis. The human IgG portion of the chimeric recombinant antibody was detected using a peroxidase-linked rabbit-anti human IgG antiserum. The blot was developed using a chemiluminescent detection system (Pierce, UK). The positions of the antibody chains are indicated.

2.2   Protein analysis of milk derived from transgenic mice

Milk of transgenic dams was analysed for the expression of recombinant antibody by Western blotting using a rabbit anti-human IgG peroxidase-linked antiserum (Fig. 1). Of the 8 transgenic lines generated, 7 expressed heavy and light chain of the recombinant antibody. The light chain was always expressed in excess of the heavy chain. The concentrations of the heavy and light chain proteins were quantified by densitometry in comparison to IgG standards. A maximum concentration of heavy and light chain protein of 0.5mg/ml and 2.9mg/ml, respectively, was obtained in mouse line HEP50. This corresponds to a total recombinant IgG concentration of 0.7mg/ml. No correlation between the transgene copy numbers and the levels of antibody expression could be detected, indicating that the site of transgene integration had a dominant effect on gene expression. The transgenic mice described in here will provide a useful model system to validate whether transgenic animals producing a recombinant, neutralising antibody in milk can protect their offspring against virus infection.

ACKNOWLEDGEMENTS

We acknowledge the technical assistance of Claire Miller, Monika Lechermaier, Angelien Heister and Atiye Toksoy. This work was supported by the EC Bridge program (ERBSC1*CT000684) and the DFG (Si 357/1-1).

REFERENCES

Castilla, J., Pintado, B., Sola, I., Sanchez, M. J. and Enjuanes, L., 1998, Engineering passive immunity in transgenic mice secreting virus- neutralizing antibodies in milk. *Nat. Biotechnol.* 16, 349-54

Clark, A. J., Cowper, A., Wallace, R., Wright, G. and Simons, J. P., 1992, Rescuing transgene expression by co-integration. *Biotechnology* 10, 1450-1454

Enjuanes L. and van der Zeijst B.A.M., 1995, Molecular basis of transmissible gastroenteritis virus epidemiology. In *The Coronaviridae* (Siddell, S. G. ed.), Plenum Press, New York, pp. 337-376

Homberger, F.R., 1992, Maternally-derived passive immunity to enterotropic mouse hepatitis virus. *Arch. Virol.* 122, 133-141

Homberger, F.R., Barthold, S.W. and Smith, A.L., 1992, Duration and strain-specificity of immunity to enterotropic mouse hepatitis virus. *Lab. Anim. Sci.* 42, 347-351

Kolb, A.F. and Siddell, S.G., 1997, Expression of a recombinant monoclonal antibody from a bicistronic mRNA. *Hybridoma* 16, 421-426

Saif, L. and Wheeler, M.B., 1998, WAPing gastroenteristis with transgenic antibodies. *Nat. Biotechnol.* 16, 334-335

Sola, I., Castilla, J., Pintado, B., Sanchez, M. J., Whitelaw, C. B., Clark, A. J. and Enjuanes, L., 1998, Transgenic mice secreting coronavirus neutralizing antibodies into the milk. *J. Virol.* 72, 3762-3772

Wege, H., Dörries, R. and Wege, H., 1984, Hybridoma antibodies to the murine coronavirus JHM: characterization of epitopes on the peplomer protein E2. *J. of Gen. Virol.* 65, 1913-1941

Whitelaw, C.B.A., Harris, S., McClenaghan, M., Simons, J.P. and Clark, A.J., 1992. Position-independent expression of the ovine beta-lactoglobulin gene in transgenic mice. *Biochem. J.* 286, 31-39

Nidovirus Genome Replication and Subgenomic mRNA Synthesis
Pathways followed and cis-acting elements required

DAVID A. BRIAN
Department of Microbiology, University of Tennessee, Knoxville, TN

1. INTRODUCTION

Nidoviruses, specifically coronaviruses and arteriviruses, are recognized as unique among RNA viruses for utilizating a discontinuous transcription step during sgmRNA synthesis. Details of this pathway and of genome replication remain enigmatic, and identification of cis-acting elements and trans-acting factors involved is far from complete. The apparent absence of a common leader in toroviruses and differences in genome size among the viruses hint that fundamental differences in pathways of RNA synthesis might still be revealed. What are the current views?

2. PATHWAY OF NIDOVIRUS GENOME REPLICATION

A consensus view of nidovirus genome replication might be drawn as in Fig 1. This scheme depicts a unit-length plus-strand genome replicating through a unit-length minus-strand antigenome by an RdRp activity. The scheme is tentative since many details of coronavirus, torovirus and arterivirus genome replication are unknown (reviewed in de Vries et al 1997). Some unanswered questions follow.

The Nidoviruses (Coronaviruses and Arteriviruses).
Edited by Ehud Lavi *et al.*, Kluwer Academic/Plenum Publishers, 2001.

415

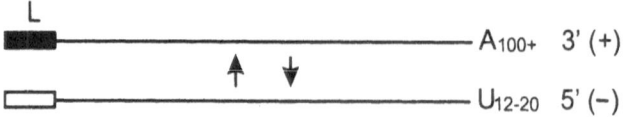

Figure 1. Proposed pathway for nidovirus genome replication. L=leader.

(1) Are only unit-length molecules made as intermediates? Published information only hints that nidovirus genomes replicate via anything other than unit-length molecules, and this for the minus strands of coronaviruses of which some appeared larger than unit length (Sethna et al 1989). (2) Does the leader play a special role in genome replication? Genomes of coronaviruses and arteriviruses possess a 5' leader common to all sgmRNAs (indicated by the filled box in Fig. 1) while toroviruses apparently do not although a leader of 8 nt or so, the size of an intergenic promoter sequence, has not been rigorously ruled out (Snijder et al 1991). Although the 5'-terminal sequence on all three genomes could behave similarly with regard to signals for initiation of translation and replication, the leader may impart a special, possibly optional, function. Genomic leaders on coronavirus DI RNAs show the remarkable property of 'leader switching' (Makino and Lai 1989, Chang et al 1996, Stirrups et al 2000). That is, the DI RNA leader within 24 h posttransfection becomes replaced by the leader of the helper virus when marked by natural sequence differences or by internal mutations. This has led some to postulate a possible discontinuous transcription step during genome synthesis (Zhang and Lai 1996). Does this occur between virus genome molecules too? If so, what advantages are served the virus by this behavior? Does leader switching occur in arteriviruses as well? Unless toroviruses have a leader that behaves similarly they cannot conform on this point. (3) Are the 5' ends of all nidovirus genomes and sgmRNAs capped? If so, what is the mechanism of capping in the absence of a documented nidoviral methyltransferase gene? (4) How is the poly(A) tail generated? Although it awaits confirmation, 5'-terminal oligo(U) sequences have been described on coronavirus minus-strand RNAs indicating that the antigenome is not quite unit length (Hofmann and Brian 1991). Is this the same story for toroviruses and arteriviruses? In coronaviruses, how is the poly(A) tail generated from an oligo(U) template? Is polyadenylation a virus or host-encoded function? Is oligo(U) length the same on genomic and sgmRNA minus strands? If not, might this be a regulatory factor allowing replication of one species (e.g. genome) but not the other (e.g. sgmRNA)?

3.   CIS-ACTING SIGNALS FOR GENOME REPLICATION

Cis-acting signals for genome replication might be functionally defined as those imparting replication competence to an extensively internally deleted genome or to a foreign RNA sequence (Levis et al 1986). Cis-acting signals presumably interact directly with trans-acting viral polymerase and associated proteins and perhaps cellular proteins to form the active replication complex. (1) What are the cis-acting signals for nidovirus genome replication? Initial identification of these has come from analyses of natural DI RNAs with further refinement from deletion analyses of replicating DI RNAs carrying a foreign sequence (reviewed in Brian and Spaan 1997) and from direct measurements of RNA synthesis (Liu et al 1994). DI RNAs have been described for four coronaviruses, one toroviruses, and one arteriviruses (Fig 2). It can be seen that the cis-acting sequences range from around 3% of vgRNA in MHV and BEV to around 18% in EAV. For MHV, IBV and EAV, minimal replication signals were deduced from the amounts of CAT produced by translation of the CAT reporter gene, and for TGEV, the amount of GUS produced. In many studies with reporter genes, replication of the DI RNA was deemed low. Do the amounts of translated product directly reflect genome (or even transcript) levels? Inasmuch as potent enhancers of translation exist in the plus-strand molecules of plant viruses (Miller et al 1997), caution would seem warranted when using quantitative analyses of translation products to deduce the strength of a promoter for RNA synthesis.

(2) What higher order RNA elements have been shown to act in cis for replication? To date only two have been, and these are in coronaviruses. The first is a cis-acting stem loop and the second a phylogenetically conserved cis-acting pseudoknot, both located just downstream of the N ORF in the 3' UTR (Hsue and Masters 1999, Williams et al 1999). Both were shown to function as higher order structures by mutational analyses. It remains to be shown whether these represent separate cis-acting elements or whether together they make up a single complex element. The latter seems likely since the UCUCU sequence at the base of the MHV stem loop (bases 230-234) is also part of the pseudoknot (bases 226-230). It also remains to be shown how these structures function in RNA replication but precedents from other higher order regulatory RNA elements would suggest a role for protein binding. To date, no protein has been identified as interacting with these elements. Do these or analogous structures also exist in toroviruses and arteriviruses?

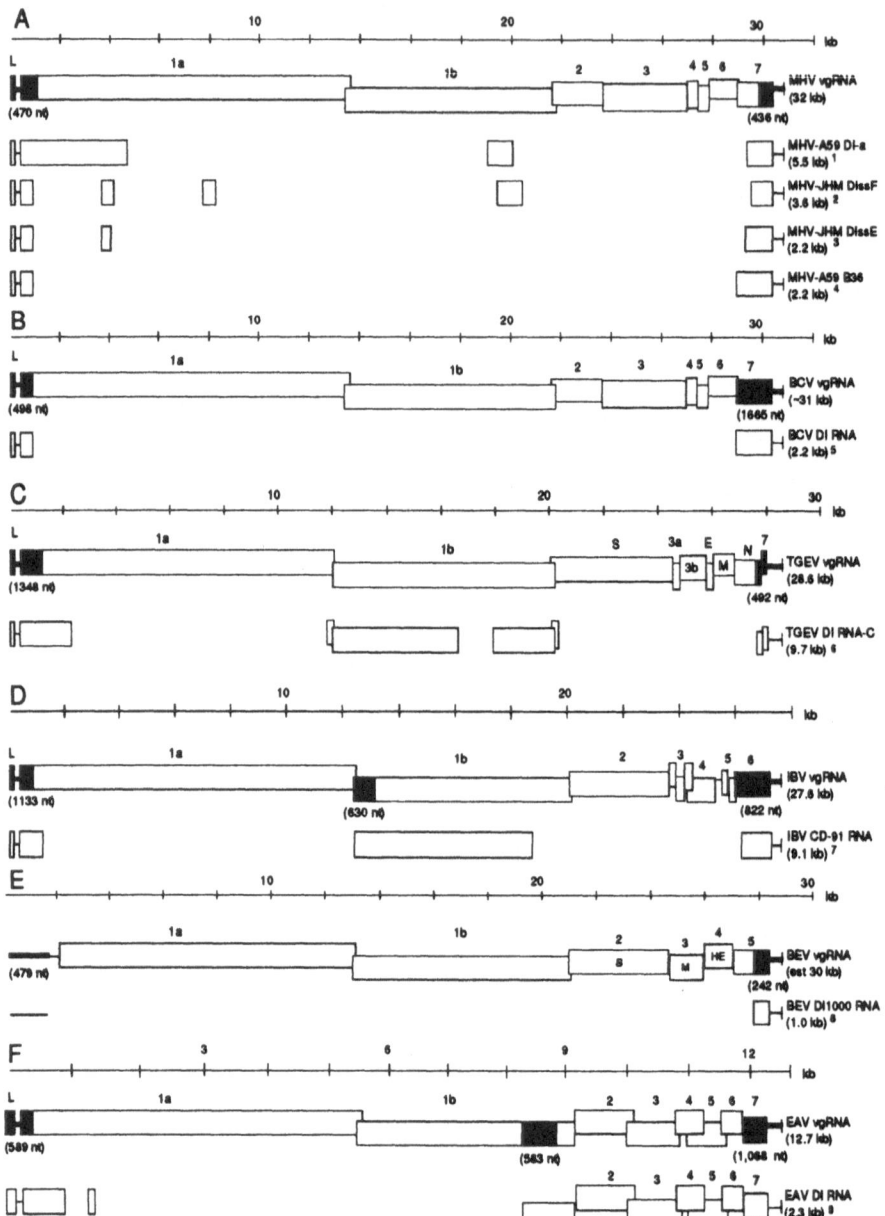

Figure 2. Initial estimates of minimal cis-acting sequence requirements for coronavirus, torovirus and arterivirus vgRNA replication as determined from analysis of DI RNAs. In some cases, deletion analysis, with or without an inserted exogenous sequence, was used. Genome regions shown to be cis-acting are represented by filled areas or drawn as heavy lines representing regions of the 5' and 3' UTRs and the number of nucleotides within these regions are shown. Filled regions represent 2.8% of the vgRNA for MHV, 6.9% for BCV, 6.4% for TGEV, 9.3% for IBV, 2.4% for BEV, and 17.6% for EAV. [1]van der Most et al 1991; [2]Lin and Lai 1993; [3]Kim et al 1993; [4]Masters et al 1994; [5]Chang et al 1994; [6]Izeta et al 1999; [7]Dalton et al 1998; [7]Snijder et al 1991; [8]Molenkamp et al 2000.

3) What protein factors, and hence potential transacting factors in the replication complex, have been shown interact with sequences mapping within the cis-acting regions of viral RNA? Twenty one have been, and of these, 17 are identified only by molecular weight (Table 1). Whether any of these function as transacting factors in genome replication remains to be shown.

Table 1. Proteins found to bind to replication-important cis-acting regions in nidovirus genomic or antigenomic RNA. They are thus potential trans-acting factors for genome replication.

Protein identity	Binding region
[1]hnRNP A1 (35-38kDa)	3' end of leader (-), IS (-), in MHV
[2,3]Polypyrimidine tract-binding protein (PTB) (55kDa)	5' end of leader (+) in MHV; complementary strand of the genome 3' UTR, nt 53-149 (strong) and nt 270-307 (weak).
[4]Poly(A)-binding protein (73 kDa)	3' poly(A) tail in MHV, BCV
[4] Proteins of 99, 95, 40-50, and 30 kDa	3'UTR of MHV and BCV
[5]Proteins of 120, 55 (2), 40 (2), and 25 kDa	nt 129-166 from the 3' end of MHV
[6]Proteins of 142, 120, 100, 55, and 33 kDa	nt 85-487 from the 3' end of MHV
[6]Proteins of 120, 103, 81, 70, and 55 kDa	nt 1-42 from the 3' end of MHV
[7]Proteins of 103, 86, 55, and 36 kDa	nt 116-184 from the 3' end of SHFV (-), and to a similar region in the 3' end of LDHV and EAV.
[8]Viral N protein	3' end of leader (+) in MHV
[9]Viral N protein	3' UTR in IBV

[1]Li et al 1997; [2]Li et al 1999; [3]Huang and Lai 1999; [4]Spangnolo and Hogue 2000; [5]Liu et al 1997; [6]Yu and Leibowitz 1995; [7]Hwang and Brinton 1998; [8]Nelson et al. 2000; [9]Zhou et al 1996

(4) What explains the curious cis-acting translation requirement (or strong preference) for coronavirus DI RNA replication (Chang and Brian 1996, de Groot et al 1992, Kim et al 1993, van der Most et al 1995)? How can this be a requirement for some DI RNAs but not others? If translation in cis is a requirement for replication of the virus genome, then which ORF suffices? If it is the 3'-terminal ORF, then how is initiation of translation on this ORF accomplished? It has been suggested that the linkage between translation and replication for DI RNAs may explain the requirement for the 3' poly(A) tail in replication, since 5' and 3' end interactions involving the poly(A)-binding protein are known to enhance translation in eucaryotic mRNAs (Spagnolo and Hogue 2000 and references therein).

4.   PATHWAY OF NIDOVIRUS SUBGEOMIC MESSENGER RNA SYNTHESIS (TRANSCRIPTION)

Historically, there have been two major concerns with regard to sgmRNA synthesis by coronaviruses and arteriviruses: 1) How does the leader get put on, and, 2) what is the general pathway of sgmRNA accumulation?

To the first question, two widely considered models of discontinuous RdRp synthesis have been put forward (discussed further in Lai and Cavanagh 1997, van der Most and Spaan 1995, Sawicki and Sawicki, 1998) (Fig. 3).   The first is the leader priming model in which there is a discontinuous step, an RdRp jump, during plus-strand synthesis of *each and every* sgmRNA molecule. This model postulates (i) existence of a free leader (80-140 nt in length) in excess amounts, (ii) annealing of the leader, with polymerase attached, to intergenic sites on the minus-strand antigenome, (iii) 3'->5' exonuclease activity (possibly a property of the RdRp) for trimming the excessively long leader back to the region of the intergenic sequence, (iv) priming, then completion of transcription.   Recent modifications of this model incorporate the concept of long-range interactions (Liao and Lai 1994, Lin et al 1996) and the binding of host protein hnRNP A1 to the AGAUUUG motifs in the antileader and antigenome in such a way that the two minus-strand templates are brought into close proximity thus facilitating the RdRp jump (Li et al 1997).   The second model postulates an RdRp jump during minus-strand synthesis at intergenic sites.   In this model, intergenic sites serve as attenuators of RdRp synthesis resulting in strand transfer of the polymerase at the intergenic site to the 3' end of the genomic leader producing a minus-strand template for sgmRNA.   This model requires the genomic leader be arranged such that the two intergenic sequences are in close proximity.   The minus-strand of the sgmRNA thus generated would then serve as template *for more than one round* of synthesis of sgmRNA. SgmRNA minus strands and ds RI molecules of sgmRNA length have been described, and metabolic labeling studies have shown these to be active in sgmRNA synthesis (Sawicki and Sawicki 1990; Schaad and Baric 1994; Sethna et al 1989).   In this model, no 3'->5' exonuclease activity is required for leader trimming.

If toroviruses have no common leader, then a discontinuous transcription process as observed for coronaviruses and arteriviruses cannot be a common nidovirus feature.   Elements of the transcription step, however, might be shared even in the absence of a leader.   For example, a novel transcription mechanism in dianthoviruses of plants has been described by Sit and Lommel (1998) in which elements are shared with the second transcription model described above.   In it, the RdRp is induced to terminate prematurely during minus-strand synthesis and thereby generates a minus-strand sgRNA product with a putative 3' terminal promoter-like sequence. This molecule is the putative template of sgmRNA synthesis. In the dianthovirus system a trans-acting RNA molecule functions as the attenuating element. Might there be an analogous mechanism in toroviruses? If so, can one find features in common between this and coronavirus and arterivirus transcription?

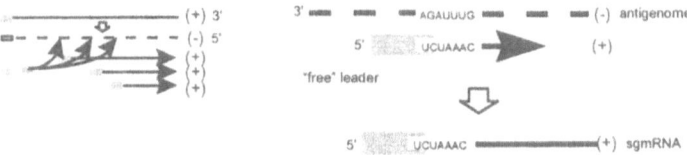

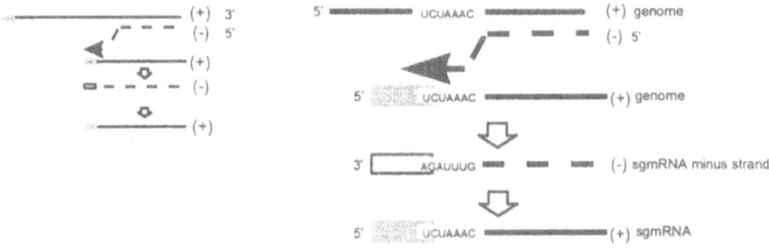

Figure 3. Two primary models used to explain addition of leader to sgmRNAs in coronaviruses. The second model seems to explain arteivirus data too (van Marle et al 1999).

It has been proposed that, once made, sgmRNAs can amplify via a replication mechanism (as sgmRNA replicons) (Sethna et al 1989). It remains to be demonstrated that sgmRNAs can be templates for anti-sgmRNAs and thereby contribute to sgmRNA amplification via a replication mechanism. If amplification does not occur by this route, then what restricts initiation of minus-strand synthesis to only genome-length (or DI genome-length) RNA?

5. CIS-ACTING ELEMENTS FOR SUBGENOMIC MESSENGER RNA SYNTHESIS

Until recently, the cis-acting elements for sgmRNA synthesis have been studied for only coronaviruses, and these in variously engineered DI RNAs. Thus, at this time it is not possible to depict a consensus nidovirus model of the cis-acting elements governing sgmRNA synthesis or even to state the rules by which transcription can be predicted. Four features of coronavirus (and in the first case, also arterivirus) transcription have been noted. (i) Base pairing between genomic leader-associated IS and the IS positioned within the minus-strand antigenome is necessary. Several coronavirus papers have demonstrated that minimal intergenic sequences of 9 to 18 nt in length is sufficient for generating sgmRNAs during replication by its insertion into the coronavirus genome or DI RNAs (Hsue and Masters 1999, Jeong and Makino 1992, Joo and Makino 1995, Krishnan et al 1996, Makino et al 1991, van der Most et al 1994, van Marle et al 1995). With the recently acquired full-length infectious clone for equine arteritis virus it has been elegantly demonstrated that base pairing between the genome 5' proximal and intergenc heptameric sequences is necessary for arterivirus sgmRNA synthesis (van Dinten et al 1997, van Marle et al 1999). (ii) The structural context within which the IS sequence is found has a profound but unpredictable effect on the strength of the sg promoter (An and Makino 1998, Fischer et al 1997, Hiscox et al 1995, Hsue and Masters 1999, Jeong and Makino 1992, Jeong et al 1996, Makino and Jeong 1993, van Marle et al 1995). (iii) The sequences flanking the IS sequence have an effect on promoter strength, but the rules predicting this have not been established. (iv) When two promoters are adjacent (say within 100 to 1000 bases), the downstream promoter has a powerfully suppressing effect on the upstream promoter.

New data from our laboratory also support the notion of a base pairing requirement for determining the site of the RdRp jump and suggest, in addition, that downstream sequences can play a major role in this decision (Ozdarendeli, et al, unpublished). The data are most consistent with RdRp template transfer taking place during minus-strand synthesis in a manner reminiscent of similarity-assisted recombination. The leader mRNA fusion site (most frequently) used for synthesis of the BCV 12.7 mRNA is GGUAGAC which occurs 14 nt downstream from a conforming UCCAAAC sequence (rarely used in this position) as determined by sequencing of asymmetrically-amplified RT-PCR products of leader-body junctions (Hofmann et al 1993). Secondary structure predictions indicated that the UCCAAAC would be buried within the stem of a stable (17.2 kCal/mole) stem-loop (Fig. 4) and perhaps this structure was preventing its use as a promoter. A 199-nt section of RNA containing this region of the genome and a 92-nt reporter sequence faithfully reproduced this expression from the BCV DI RNA pDrep1 when transfected into helper virus-infected cells (Chang et al 1994, Ozdarendelli et al unpublished) (Fig. 5A). To test the role of the stem-loop in the decision of RdRp to use site #2 rather than site #1, three extensive mutations each designed to unfold the stem-loop were made. These formed mutants m1, m2 and m4 as indicated in Fig. 4. With none of the mutants, including a 4th mutant in which UCUAAAC replaced

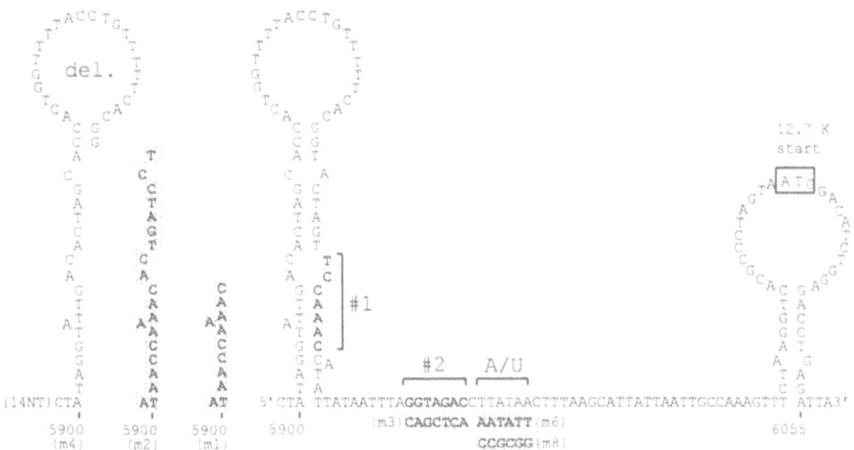

Figure 4. Context of the conforming (#1, rarely used) and nonconforming (#2, heavily used) intergenic sequences just upstream of the start site of the gene for the 12.7 kDa protein in the wild type bovine coronavirus genome and a description of the mutants used to alter this context in an experimental system (Ozdarendeli et al in preparation). Base positions noted are from a numbering of the viral genomic 3' proximal sequence of 8955 nt (excluding the poly(A) tail). Mutations made in the DI RNA (pDrepIS12.7) in attempts to influence the site of the RdRp jump are indicated as m (mutant) 1, in which bases shown were deleted, and m2, m3, m6, and m8, in which the bases (shaded) were substituted. Only m6 and m8 of those shown caused a switch in promoter usage from #2 to #1.

site #1, was site #1 used. All used site #2. When site #2 was made into a totally nonconforming CAGCTCA (m3) in the wt context, no subgenomic transcripts were made. To test the requirements by which the RdRp chooses site #2, two mutants were made in which the degree of base pairing between the leader-containing template and the DI RNA genome region just downstream of site #2 was altered. Since it had been noticed with BCV that leader switching occurred in a highly base-paired A/U-rich region just downstream of the IS region (Chang et al 1996), the A/U-rich region AUUAAU was changed to UAAUUA in m6 in order to decrease base-pairing but keep A/U richness. With mutant 6, sgmRNAs were made from site #1. Site #1 was also used with m8 in which base pairing and A/U richness were both diminished. Thus, downstream basepairing critically affected the site of the RdRp jump. These results are consistent with the earlier postulate that base-pairing downstream of the leader intergenic sequence strongly influences the replacement of a mutated leader with wild type by leader switching on DI RNA, even in the absence of an intergenic sequence on the donor strand. The region of base pairing was in the A/U-rich 9-nt region shown to be a required region for leader switching in MHV. To what extent are the mechanisms of leader switching and leader fusion during sgmRNA synthesis the same? Certainly, the genomic leader associated IS is not required for leader switching, but can adjacent sequences supplant the need for IS base pairing?

Thus, the full extent of the requirements for RNA-RNA base pairing, higher order RNA structures, and RNA-protein interactions remain to be clarified before a complete set of rules governing nidovirus sgRNA synthesis can be stated.

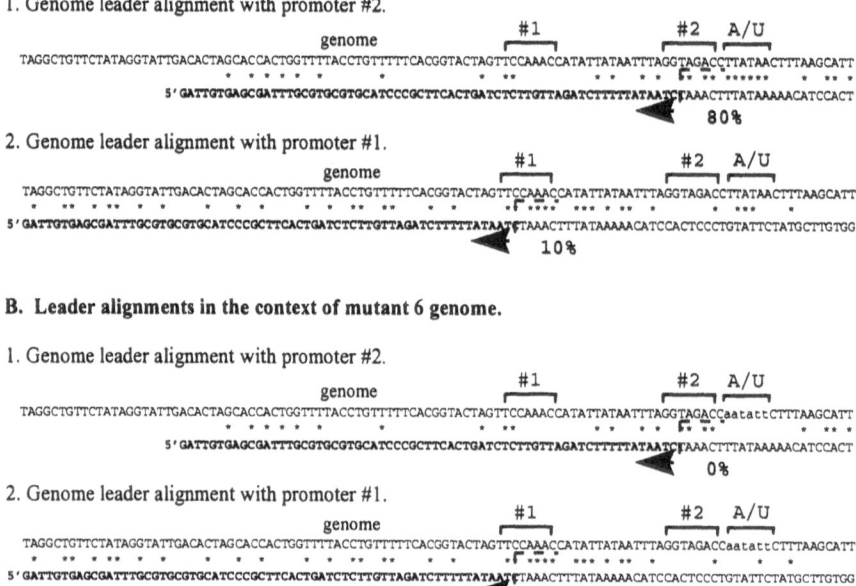

Figure 5. Base identities between the genomic leader intergenic sequence region and the upstream and downstream intergenic sequence regions for the 12.7 kDa protein gene in the bovine coronavirus. A positions of base identity (indicated by asterisks) are positions of potential base pairing between the leader-containing genomic plus-strand RNA and the nascent minus-strand RNA. Note that with mutant 6 in which the promoter used is #1, base pairings in the A/U-rich region just downstream of promoter #2 are diminished whereas base pairings just downstream of promoter #1 remain high. Retention of base pairing downstream of the potential intergenic sequence, therefore, correlates positively with promoter usage.

REFERENCES

An, S., and Makino, S., 1998, Characterizations of coronavirus cis-acting RNA elements and the transcription step affecting its transcription efficiency. *Virology* **243**: 198-207.

Brian, D.A., and Spaan, W.J.M, 1997, Recombination and coronavirus defective interfering RNAs. *Seminars in Virology* **8**: 101-111.

Chang, R.Y, and Brian, D.A., 1996, cis-requirement for N-specific protein sequence in bovine coronavirus defective interfering RNA replication. *J. Virol.* **70**: 2201-2207.

Chang, R.Y., Hofmann, M.A., Sethna, P.B., and Brian, D.A., 1994, A cis-acting function for the coronavirus leader in defective interfering RNA replication. *J. Virol.* **68**: 8223-8231.

Chang, R.-Y., Krishnan, R., and Brian, D.A., 1996, The UCUAAAC promoter motif is not required for high-frequency leader recombination in bovine coronavirus defective interfering RNA. *J. Virol.* **70**: 2720-2729.

Dalton, K., Penzes, Z., Wroe, C., Stirrups, K., Evans, S., Shaw, K., Brown, T.D.K., Britton, P., and Cavanagh, D., 1998, Sequence elements involved in the rescue of IBV defective RNA CD-91. *Adv. Exp. Med. Biol.* **440**: 253-258.

de Groot, R.J., van der Most, R.G., and Spaan, W.J.M., 1992, The fitness of defective interfering murine coronavirus DI-a and its derivatives is decreased by nonsense and frameshift mutations. *J. Virol.* **66**: 5898-5905.

de Vries, A.A.F., Horzinek, M.C., Rottier, P.J.M., and de Groot, R.J., 1997, The genome organization of the Nidovirales: similarities and differences between arteri-, toro-, and coronaviruses. *Semin. Virol.* **8**: 33-47.

Fischer, F., Stegen, C.F., Koetzner, C.A., and Masters, P.S., 1997, Analysis of a recombinant mouse hepatitis virus expressing a foreign gene reveals a novel aspect of coronavirus transcription. *J. Virol.* **71**: 5148-5160.

Hiscox, J.A., Mawditt, K.L., Cavanagh, D. and Britton, P., 1995, Investigation of the control of coronavirus subgenomic mRNA transcription by using T7-generated negative-sense RNA transcripts. *J. Virol.* **69**: 6219-6227.

Hofmann, M.A., and Brian, D.A., 1991, The 5-prime end of coronavirus minus-strand RNAs contain a short poly(U) tract. .*J. Virol.* **65**: 6331-6333.

Hofmann, M.A., Chang, R.-Y., Ku, S., and Brian, D.A., 1993, Leader-mRNA junction sequences are unique for each subgenomic mRNA species in the bovine coronavirus and remain so throughout persistent infection. *Virology* **196**:163-171.

Hsue, B., and Masters, P.S., 1997, A bulged stem-loop structure in the 3' untranslated region of the genome of the coronavirus mouse hepatitis virus is essential for replication. *J. Virol.* **71**: 7567-7578.

Hsue, B., and Masters, P.S., 1999, Insertion of a new transcriptional unit into the genome of mouse hepatitis virus. *J. Virol.* **73**: 6128-6135.

Huang, P., and Lai, M.M.C., 1999, Polypyrimidine tract-binding protein binds to the complementary strand of the mouse hepatitis virus 3' untranslated region, thereby altering RNA conformation. *J. Virol.* **73**: 9110-9116.

Hwang, Y.K., and Brinton, M.A., 1998, A 68-nucleotide sequence within the 3' noncoding region of simian hemorrhagic fever virus negative-strand RNA binds to four MA104 cell proteins. *J. Virol.* **72**: 4341-4351.

Izeta, A., Smerdou, C, Alonso, S., Penzes, Z., Mendez, A., Plana-Duran, J., and Enjuanes, L., 1999. Replication and packaging of transmissible gastroenteritis coronavirus-derived synthetic minigenomes. *J. Virol.* **73**: 1535-1545.

Jeong, Y.S., and Makino, S., 1992, Mechanism of coronavirus transcription: duration of primary transcription initiation activity and effect of subgenomic RNA transcription on RNA replication. *J. Virol.* **66**: 3339-3346.

Jeong, Y.S., Repass, J.F., Kim, Y.N., Hwang, S.M., and Makino, S., 1996, Coronavirus transcription mediated by sequences flanking the transcription consensus sequence. *Virology* **217**: 311-322.

Joo, M., and Makino, S., 1995, The effect of two closely inserted transcription consensus sequences on coronavirus transcription. *J. Virol.* **69**: 272-280.

Kim, Y.N., Jeong, Y.S., and Makino, S., 1993. Analysis of cis-acting sequences essential for coronavirus defective interfering RNA replication. *Virology* **197**: 53-63.

Kim, Y.N., Lai, M.M.C., and Makino, S., 1993, Generation and selection of coronavirus defective interfering RNA with large open reading frame by RNA recombination and possible editing. *Virology* **194**: 244-253.

Kim, Y.N. and Makino, S., 1995, Characterization of a murine coronavirus defective interfering RNA internal cis-acting replication signal. *J. Virol.* **69**: 4963-4971.

Krishnan, R., Chang, R.-Y., and Brian, D.A., 1996, Tandem placement of a coronavirus promoter results in enhanced mRNA synthesis from the downstream-most initiation site. *Virology* **218**: 400-405.

Lai, M.M.C., and Cavanagh, D., 1997, The molecular biology of coronaviruses. *Adv. Virus Res.* **48**: 1-100.

Levis, R., Weiss, B.G., Tsiang, M., Huang, H.V., and Schlesinger, A., 1986, Deletion mapping of Sindbis virus DI RNAs derived from cDNAs defines the sequences essential for replication and packaging. *Cell* **44**: 137-145.

Li, H.-P., Huang, P., Park, S., and Lai, M.M.C., 1999, Polypyrimidine tract-binding protein binds to the leader RNA of mouse hepatitis virus and serves as a regulator of viral transcription. *J. Virol.* **73**: 772-777.

Li, H.-P., Xhang, X., Duncan, R., Comai, L., and Lai, M.M.C., 1997, Heterogeneous nuclear ribonucleoprotein A1 binds to the transcription-regulatory region of mouse hepatitis virus RNA. *Proc. Natl. Acad. Sci. USA* **94**: 9544-9549.

Liao, C.L., and Lai, M.M.C., 1994, Requirement of the 5'-end genomic sequence as an upstream cis-acting element for coronavirus subgenomic mRNA transcription. *J. Virol.* **68**: 4727-4737.

Lin, Y.J., and Lai, M.M.C., 1993, Deletion mapping of a mouse hepatitis virus defective interfering RNA reveals the requirement of an internal and discontiguous sequence for replication. *J. Virol.* **67**: 6110-6118.

Lin, Y.J., Liao, C.L., and Lai, M.M.C., 1994, Identification of the cis-acting signal for minus-strand RNA synthesis of a murine coronavirus: implications for the role of minus-strand RNA in RNA replication and transcription. *J. Virol.* **68**: 8131-8140.

Lin, Y.J., Zhang, X., Wu, R.-C., and Lai, M.M.C., 1996, The 3' untranslated region of coronavirus RNA is required for subgenomic mRNA transcription from a defective interfering RNA. *J. Virol.* **70**: 7236-7240.

Liu, Q., Yu, W., and Leibowitz, J.L., 1997, A specific host cellular protein binding element near the 3' end of mouse hepatitis genomic RNA. *Virology* **232**: 74-85.

Makino, S., and Joo, M., 1993, Effect of intergenic consensus sequence flanking sequences on coronavirus transcription. *J. Virol.* **67**: 3304-3311.

Makino, S., Joo, M., and Makino, J.K., 1991, A system for study of coronavirus mRNA synthesis: A regulated, expressed subgenomic defective interfering RNA results from intergenic site insertion. *J. Virol.* **65**: 6031-6041.

Makino, S., and Lai, M.M.C., 1989, High-frequency leader sequence switching during coronavirus defective interfering RNA replication. *J. Virol.* **63**: 5285-5292.

Makino, S., Shieh, C.K., Soe, L.H., Baker, S.C., and Lai, M.M.C., 1988, Primary structure and translation of a defective interfering RNA of murine coronavirus. *Virology* **166**:1-11.

Masters, P.S., Koetzner, C.A., Kerr, C.A., and Heo, Y., 1994, Optimization of targeted RNA recombination and mapping of a novel nucleocapsid gene mutation in the coronavirus mouse hepatitis virus. *J. Virol.* **68**: 328-337.

Masters, P.S., 1999, Reverse genetics of the largest RNA viruses. *Adv. Virus Res.* **53**: 245-264.

Mendez, A., Smerdou, C., Izeta, A., Gebauer, F., and Enjuanes., L., 1996, Molecular characterization of transmissible gastroenteritis coronavirus defective interfering genomes: packaging and heterogeneity. *Virology* **217**: 495-507.

Miller, A.W., Brown, C.M., and Wang, S., 1997, New Punctuation for the genetic code: luteovirus gene expression. *Seminars in Virology* **8**: 3-13.

Molenkamp, R., Rozier, C.D., Greve, S., Spaan, W.J.M., and Snijder, E.J., 2000, Isolation and characterization of an arterivirus defective interfering RNA genome. *J. Virol.* **74**: 3156-3165.

Nelson, G.W., Stohlman, S.A., and Tahara, S.M., 2000, High affinity interaction between nucleocapsid protein and leader/intergenic sequence of mouse hepatitis virus RNA. *J. Gen. Virol.* **81**:181-188.

Ozdarendeli, A., Ku, S., Rochat, S., Williams, G.D., Senanayake, S.D., and Brian, D. A., unpublished data.

Penzes, Z., Wroe, C., Brown, T.D.K., Britton, P., and Cavanagh, D., 1996, Replication and packaging of coronavirus infectious bronchitiis virus defective RNAs lacking a long open reading frame. *J. Virol.* **70**: 8660-8668.

Sawicki, S.G., and Sawicki, D.L., 1990, Coronavirus transcription: subgenomic mouse hepatitis virus replicative intermediates function in mRNA synthesis. *J. Virol.* **64**: 1050-1056.

Sawicki, S.G., and Sawicki, D.L., 1995, Coronaviruses use discontinuous extension for synthesis of subgenome-lenght negative strands. *Adv. Exp. Med. Biol.* **380**: 499-506.

Schaad, M.C., and R.S. Baric, 1994, Genetics of mouse hepatitis virus transcription: evidence that subgenomic negative strands are functional templates. *J. Virol.* **68**:8169-8179.

Sethna, P.B., S.-L. Hung, and D.A. Brian, 1989, Coronavirus subgenomic minus-strand RNA and the potential for mRNA replicons. Proc. Natl. Acad. Sci. USA **86**: 5626-5630.

Sit, T.L., Vaewhongs, A.A., and Lommel, S.A., 1998, RNA-mediated trans-activation of transcription from a viral RNA. *Science* **281**:829-832.

Snijder, E.J, Den Boon, J.A., Horzinek, M.C., and Spaan, W.J.M., 1991, Characterization of defective interfering Berne virus RNAs. *J. Gen. Virol.* **72**: 1635-1643.

Snijder, E.J., and Meulenberg, J.J.M., 1998, The molecular biology of arteriviruses. *J. Gen. Virol.* **79**: 961-979.

Spagnolo, J.F., and Hogue, B.G., 2000, Host protein interactions with the 3' end of bovine coronavirus RNA and the requirement of the poly(A) tail for coronavirus defective genome replication. *J. Virol* **74**: 5053-5065.

Stirrups, K., Shaw, K., Evans, S., Dalton, K., Cananagh, D., and Britton, P., 2000, Leader switching occurs during the rescue of defective RNAs by heterologous strains of the coronavirus infectious bronchitis virus. *J. Gen. Virol.* **81**: 791-801.

van der Most, R.G., Bredenbeek, P.J., and Spaan, W.J.M., 1991, A domain at the 3' end of the polymerase gene is essential for encapsidation of coronavirus defective interfering RNAs. *J. Virol.* **65**: 3219-3226.

Van der Most, R.G., de Groot, R.J., and Spaan, W.J.M., 1994, Subgenomic RNA synthesis directed by a synthetic defective interfering RNA of mouse hepatitis virus: a study of coronavirus transcription initiation. *J. Virol.* **65**: 3656-3666.

van der Most, R.G., Luytjes, W., Rutjes, S., and Spaan W.J.M., 1995, Translation but not the encoded sequence is essential for the efficient propagation of the defective interfering RNAs of the coronavirus mouse hepatitis virus. *J. Virol.* **69**: 3744-3751.

van der Most, R.G., and Spaan W.J.M., 1995, Coronavirus replication, transcription, and RNA recombination. In The Coronaviridae (S.G. Siddell, ed.), Plenum Press, New York and London, pp. 11-31.

van Dinten, L.C., den Boon, J.A., Wassenaar, A.L.M., Spaan, W.J.M., and Snijder, E.J., 1997, An infectious arterivirus cDNA clone: identification of a replicase point mutation which abolishes discontinuous mRNA transcription. *Proc. Natl. Acad. Sci. USA* **94**: 991-996.

van Marle, G., Dobbe, J.D., Gultyaev, A.P., Luytjes, W., Spaan, W.J.M., and Snijder, E.J., 1999, Arterivirus discontinuous mRNA transcription is guided by base-pairing between sense and antisense transcription-regulating sequences. *Proc. Natl. Acad. Sci. USA* **96**: 12056-12061.

van Marle, G., Luytjes., W., van der Most, R.G., van der Straaten, T., and Spaan, W.J.M., 1995, Regulation of coronavirus mRNA transcription. *J. Virol.* **69**: 7851-7856.

van Marle, G., van Dinten, L.C., Spaan, W.J.M., Luytjes, W., and Snijder, E.J., 1999, Characterization of an equine arteritis virus replicase mutant defective in subgenomic mRNA synthesis. *J. Virol.* **73**: 5274-5281.

Williams, G.D., Chang, R.-Y., and Brian, D.A., 1999, A phylogenetically conserved hairpin-type 3''untranslated region pseudoknot functions in coronavirus RNA replication. *J. Virol.* **73**: 8349-8355.

Yu, W., and Leibowitz, J.L., 1995, Specific binding of host cellular proteins to multiple sites within the 3' end of mouse hepatitis virus genomic RNA. *J. Virol.* **69**: 2016-2023.

Zhang, X., and Lai, M.M.C., 1996, A 5'-proximal RNA sequence of murine coronavirus as a potential initiation site for genomic-length mRNA transcription. *J. Virol.* **70**: 705-711.

Zhou M., Williams. A.K., Chung, S.-I., Wang, L., and Collisson, E.W., 1996, The infectious bronchitis virus nucleocapsid protein binds RNA sequences in the 3' terminus of the genome. *Virology* **217**: 191-199.

Regulation of Mouse Hepatitis Virus RNA synthesis by Heterogeneous Nuclear Ribonucleoprotein A1

[1]STEPHANIE T. SHI, [1]PEIYONG HUANG, HSIN-PAI LI, AND [1,2]MICHAEL M. C. LAI

[1]Department of Molecular Microbiology and Immunology and [2]Howard Hughes Medical Institute, University of Southern California School of Medicine, Los Angeles, CA

1. INTRODUCTION

Mouse hepatitis virus (MHV) RNA synthesis, including replication of viral genome and transcription of subgenomic mRNAs, has been shown to be regulated by several viral RNA elements, including 5'-untranslated region (5'-UTR), *cis*- and *trans*-acting leader RNA, intergenic (IG) sequence, and 3'-UTR (Lai and Cavanagh 1997). Biochemical evidence suggests that these regulatory sequences likely interact with each other either directly or indirectly, probably through protein-RNA and protein-protein interactions involving both viral and cellular proteins (Zhang and Lai 1995). Several cellular proteins have been found to bind these regulatory sequences by UV-crosslinking experiments (Furuya and Lai 1993, Huang and Lai 1999); three of these proteins have been identified to be heterogeneous nuclear ribonucleoprotein A1 (hnRNP A1) (Li *et al* 1997), polypyrimidine-tract-binding protein (PTB) (Huang and Lai 1999, Li *et al* 1999), and poly(A)-binding protein (PABP) (Spagnolo and Hogue 2000).

hnRNP A1 is involved in pre-mRNA splicing and transport of cellular RNAs (Dreyfuss *et al* 1993). It is predominantly a nuclear protein, but also shuttles between the nucleus and the cytoplasm (Piñol-Roma and Dreyfuss 1992). The cytoplasmic hnRNP A1 has been implicated in the regulation of mRNA stability (Hamilton *et al* 1993, Henics *et al* 1994) and translation (Svitkin *et al* 1996). hnRNP A1 binds MHV (-)-strand leader and IG sequences (Fig. 1) (Furuya and Lai 1993). The extent of binding of hnRNP A1 to the (-)-strand IG sequences correlates with the efficiency of

The Nidoviruses (Coronaviruses and Arteriviruses).
Edited by Ehud Lavi *et al.*, Kluwer Academic/Plenum Publishers, 2001.

transcription from the IG site (Zhang and Lai 1995). In MHV-infected cells, hnRNP A1 relocates to the cytoplasm, where viral RNA synthesis occurs (Li *et al* 1997). hnRNP A1 also interacts with the MHV nucleocapsid (N) protein (Wang and Zhang 1999), which is required for MHV RNA synthesis (Compton *et al* 1987). Furthermore, hnRNP A1 mediates the formation of a ribonucleoprotein complex containing the MHV (-)-strand leader and IG sequences (Zhang *et al* 1999). hnRNP A1 and PTB bind to the precisely complementary sites on the (-)- and (+)-strand RNA, respectively, of the leader region of MHV RNA (Li *et al* 1997, Li *et al* 1999), and also the 5'- and 3'-ends of both the (+)- and (-)-strand RNAs (Huang and Lai 1999, unpublished) (Fig. 1). The complementarity of their binding sites on both ends of MHV RNA suggests that hnRNP A1 and PTB may mediate the interaction between these regions to form a ribonucleoprotein complex, which functions in MHV RNA synthesis.

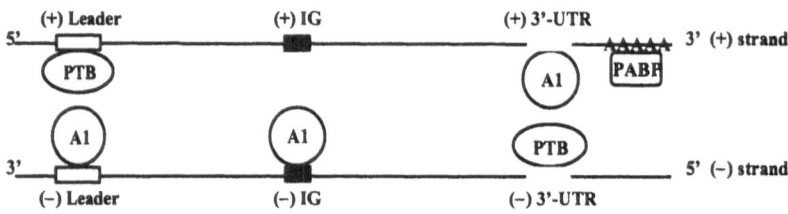

Figure 1. Schematic drawings of the cellular proteins that interact with MHV RNA.

2.   MATERIALS AND METHODS

2.1   Cells

Permanent DBT cell lines were established by transfecting pcDNA3.1 alone or the plasmid containing an open reading frame for the Flag-tagged hnRNP A1 or hnRNP A1ΔC, under the control of a CMV immediate-early gene promoter, into DBT cells and selected with 0.5 mg/ml G418.

2.2   Electrophoretic mobility shift and UV-crosslinking assays

Different amounts of GST-fusion proteins and the ^{32}P-labeled (-)-strand MHV 5'-end RNA (182 nt) were incubated for 10 min at 30°C. The protein-

RNA complexes were then separated on a 4% nondenaturing polyacrylamide gel according to the published procedures (Furuya and Lai 1993). UV-crosslinking assay was performed as described (Huang and Lai 1999).

3. RESULTS

3.1 The effects of the wt and mutant hnRNP A1 on MHV RNA synthesis

We established murine DBT cell lines stably expressing the Flag-tagged wt hnRNP A1 (DBT-A1) or a mutant hnRNP A1, which has a 75-amino acid deletion from the C terminus (DBT-A1ΔC) (Fig. 2). This mutant lacks part of the glycine-rich domain and the M9 sequence responsible for shuttling hnRNP A1 between the nucleus and the cytoplasm (Siomi and Dreyfuss 1995, Weighardt *et al* 1995). The Flag-tagged wt hnRNP A1 was localized almost exclusively in the nucleus, whereas the mutant hnRNP A1 was localized predominantly in the cytoplasm (data not shown) as predicted.

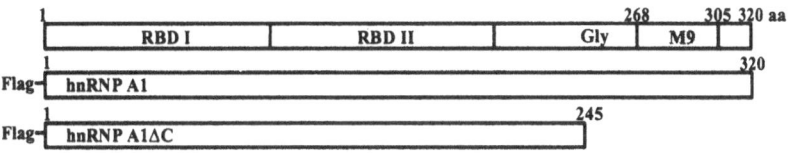

Figure 2. Diagrammatic structure of the wt and mutant hnRNP A1. RBD, RNA-binding domain; Gly, glycine-rich region.

We assessed the production of genomic and subgenomic MHV RNAs in these cell lines by Northern blot analysis. At 8 h p.i., there were significantly higher steady-state levels of all of the viral RNA species in DBT-A1 cells than in DBT-VEC cells, which were transfected with the empty vector (Fig. 3A). In contrast, no viral RNA was detected in DBT-A1ΔC cells at that time point. At 16 h p.i., viral RNA levels of DBT-VEC and DBT-A1 cells decreased generally because of the loss of dead cells, while the smaller subgenomic RNAs became detectable in DBT-A1ΔC cells (data not shown). By 24 h p.i., most viral RNA species became detectable in DBT-A1ΔC cells, while most of the DBT-A1 cells had already been completely lysed. Therefore, the wild-type hnRNP A1 accelerated MHV RNA synthesis, whereas its mutant significantly delayed it.

We also detected a defective-interfering (DI) RNA species (arrow in Fig. 3A), which appeared to be inhibited to a greater extent than other RNA species in DBT-A1ΔC cells. To confirm this result, we further studied replication of another DI RNA. DBT cells were infected with MHV-A59 and transfected with DIssE RNA of JHM (Makino and Lai 1989); the virus released was passaged twice in DBT cells and used to infect various cell lines. Similar to the A59 DI RNA, the replication of DIssE RNA was much more strongly inhibited than that of MHV genomic and subgenomic RNAs in DBT-A1ΔC cells (Fig. 3B). Our results thus suggest that MHV DI RNA replication is more dependent on the function of cytoplasmic hnRNP A1.

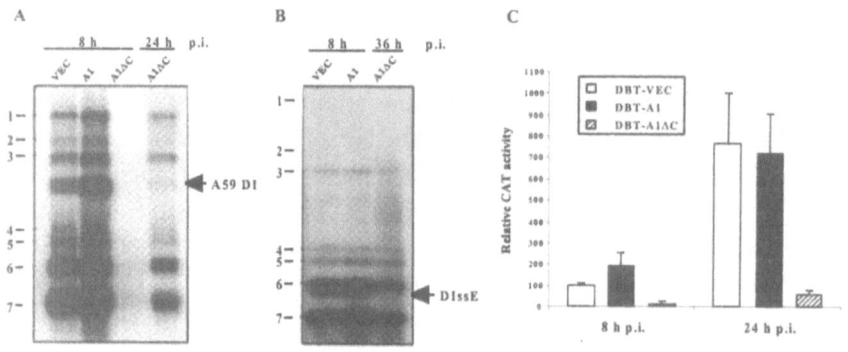

Figure 3. MHV RNA replication (A, B) and transcription (C) in DBT cells. Northern blot analysis of MHV RNA (A) or DIssE RNA (B). (C) CAT assay of A59-infected DBT cells transfected with the 25CAT DI RNA. The values represent averages of three independent experiments.

3.2   hnRNP A1ΔC inhibits transcription of MHV DI RNAs

To demonstrate that MHV RNA transcription machinery is defective in cells expressing the mutant hnRNP A1, we studied transcription of an MHV DI RNA, 25CAT (Liao and Lai 1994). The 25CAT RNA was transfected into MHV-A59-infected cells 1 h after infection. At 8 h p.i., CAT activity in DBT-A1 cells was higher than that in DBT-VEC cells (Fig. 3C). On the other hand, the CAT activity was very low in DBT-A1ΔC cells. At 24 h p.i., CAT activity in DBT-A1 cells became less than that in DBT-VEC cells because of the loss of dead DBT-A1 cells. CAT activity in DBT-A1ΔC remained significantly lower than that in DBT-VEC cells. Thus, there was an inhibition of mRNA transcription from the DI RNA, consistent with the results observed for the mRNA transcription from the wt viral genome.

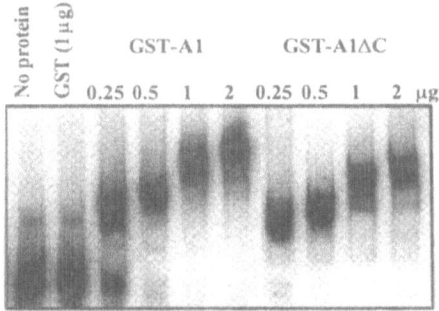

Figure 4. Electrophoretic mobility shift assay for the RNA-binding properties of the wt and mutant hnRNP A1 using the ^{32}P-labeled (-)-strand MHV leader RNA.

3.3 Characterization of the interaction between hnRNP A1 and MHV RNA

To explore the mechanism of the inhibitory effects of hnRNP A1ΔC, we examined the MHV RNA-binding ability of this mutant by electrophoretic mobility shift assay. GST-hnRNP A1 protein, but not GST, efficiently binds the (-)-strand MHV leader RNA to form an RNA-protein complex (Fig. 4). The complex increased in size with the increasing amounts of GST-hnRNP A1 (Fig. 4). A similar RNA-binding pattern was also observed for hnRNP A1ΔC. Furthermore, UV-crosslinking experiments showed that increasing amounts of purified GST-hnRNP A1ΔC efficiently competed with the cellular endogenous hnRNP A1 for the binding of (-)-strand MHV leader RNA (data not shown), indicating that the binding of hnRNP A1ΔC to RNA was not affected. Also, the ability of hnRNP A1ΔC to bind viral proteins N and gene 1 protein was not affected; however, this mutant protein failed to bind a cellular protein, p250, which was associated with the wild-type hnRNP A1 (data not shown).

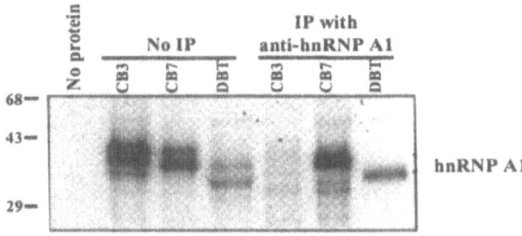

Figure 5. Detection of MHV RNA-binding proteins in CB3 and CB7 cells by UV-crosslinking.

3.4   hnRNP A1-related proteins bind to MHV RNA in CB3 cells

A mouse erythroleukemia cell line, CB3, was reported to lack detectable hnRNP A1 expression (Ben-David *et al* 1992). However, MHV was still able to replicate efficiently in these cells. Since hnRNP A1 protein is involved in a variety of important cellular functions, it is conceivable that other gene products may substitute for the function of hnRNP A1 in CB3 cells. We performed UV-crosslinking assays to detect MHV RNA-binding proteins present in CB3 and its parental cell line, CB7. hnRNP A1 was present only in DBT and CB7 cells as determined by immunoprecipitation with the monoclonal antibody against hnRNP A1 (Fig. 5). Three proteins comparable to hnRNP A1 in size that could interact with the MHV (-)-strand leader RNA were detected in CB3 cells, but none of them reacted with the antibody against hnRNP A1. The size of these proteins was slightly larger than hnRNP A1 in the DBT cell extracts. These proteins may represent hnRNP A1-related proteins. Therefore, multiple cellular proteins may have the capacity to be involved in MHV RNA synthesis.

4.   DISCUSSION

There is increasing evidence that RNA viruses subvert cellular factors for replication and transcription of viral RNAs (Lai 1998). In the present study, we showed that MHV RNA transcription and replication were enhanced by the overexpression of the wt hnRNP A1 protein, but inhibited by the expression of a dominant-negative hnRNP A1 mutant in DBT cell lines. Our results suggest that hnRNP A1 is a host protein involved in the formation of a cytoplasmic transcription/replication complex for viral RNA synthesis.

hnRNP A1ΔC caused a preferential inhibition of at least two DI RNA species, suggesting that the inhibition of MHV replication by the hnRNP A1 mutant was most likely a direct effect on viral RNA synthesis rather than an indirect effect on other aspects of cellular or viral functions. Since hnRNP A1 binds directly to the *cis*-acting MHV RNA sequences critical for MHV RNA transcription and replication, it is most likely that hnRNP A1 may participate in the formation of the transcription/replication complex. hnRNP A1 may modulate MHV RNA transcription or replication by participating in the processing, transport and controlling the stability of viral RNAs. Alternatively, hnRNP A1 may participate more directly in viral RNA synthesis in a similar role to that of transcription factors in DNA-dependent RNA synthesis, such as maintaining favorable RNA conformation for RNA synthesis.

The mechanism of the dominant-negative effects of hnRNP A1ΔC is still not clear. hnRNP A1ΔC retains the MHV RNA-binding and self-association ability (Fig. 4) and is capable of binding the viral proteins that are required for RNA replication. Preliminary data, however, suggest that the mutant protein is unable to bind certain cellular proteins, which may be involved in MHV RNA synthesis.

In summary, our data provide evidence that hnRNP A1 is directly or indirectly involved in MHV RNA transcription and replication. Our results also suggest that other related cellular proteins may substitute for the role of hnRNP A1 in MHV RNA synthesis. For example, other hnRNP proteins, such as hnRNP A2/B1 and hnRNP B2, are potential candidates because of their similarity in size, sequence (especially in RNA-binding domains), structure and function to hnRNP A1 (Dreyfuss *et al* 1993). Identification of the proteins in hnRNP A1-deficient CB3 cells will unveil more cellular factors that regulate MHV RNA synthesis.

ACKNOWLEDGMENTS

This work was partially supported by a National Institutes of Health research grant (to MMCL) and a postdoctoral fellowship (to STS). MMCL is an investigator of the Howard Hughes Medical Institute.

REFERENCES

Ben-David, Y., Bani, M.R., Chabot, B., De Koven, A., and Bernstein, A., 1992, Retroviral insertions downstream of the heterogeneous nuclear ribonucleoprotein A1 gene in erythroleukemia cells: evidence that A1 is not essential for cell growth. *Mol Cell Biol* **12:** 4449-4455.

Compton, S.R., Rogers, D.B., Holmes, K.V., Fertsch, D., Remenick, J., and McGowan, J.J., 1987, In vitro replication of mouse hepatitis virus strain A59. *J Virol* **61:** 1814-1820.

Dreyfuss, G., Matunis, M.J., Piñol-Roma, S., and Burd, C.G., 1993, hnRNP proteins and the biogenesis of mRNA. *Annu Rev Biochem* **62:** 289-321.

Furuya, T., and Lai, M.M.C., 1993, Three different cellular proteins bind to complementary sites on the 5'- end-positive and 3'-end-negative strands of mouse hepatitis virus RNA. *J Virol* **67:** 7215-7222.

Hamilton, B.J., Nagy, E., Malter, J.S., Arrick, B.A., and Rigby, W.F.C., 1993, Association of heterogeneous nuclear ribonucleoprotein A1 and C proteins with reiterated AUUUA sequences. *J Biol Chem* **268:** 8881-8887.

Henics, T., Sanfridson, A., Hamilton, B.J., Nagy, E., and Rigby, W.F.C., 1994, Enhanced stability of interleukin-2 mRNA in MLA 144 cells. Possible role of cytoplasmic AU-rich sequence-binding proteins. *J Biol Chem* **269:** 5377-5383.

Huang, P., and Lai, M.M.C., 1999, Polypyrimidine tract-binding protein binds to the complementary strand of the mouse hepatitis virus 3' untranslated region, thereby altering RNA conformation. *J Virol* **73**: 9110-9116.

Lai, M.M.C., and Cavanagh, D., 1997, The molecular biology of coronaviruses. *Adv Virus Res* **48**: 1-100.

Lai, M.M.C., 1998, Cellular factors in the transcription and replication of viral RNA genomes: a parallel to DNA-dependent RNA transcription. *Virology* **244**: 1-12.

Li, H.P., Zhang, X., Duncan, R., Comai, L., and Lai, M.M.C., 1997, Heterogeneous nuclear ribonucleoprotein A1 binds to the transcription- regulatory region of mouse hepatitis virus RNA. *Proc Natl Acad Sci U S A* **94**: 9544-9549.

Li, H.P., Huang, P., Park, S., and Lai, M.M.C., 1999, Polypyrimidine tract-binding protein binds to the leader RNA of mouse hepatitis virus and serves as a regulator of viral transcription. *J Virol* **73**: 772-777.

Liao, C.L., and Lai, M.M.C., 1994, Requirement of the 5'-end genomic sequence as an upstream cis-acting element for coronavirus subgenomic mRNA transcription. *J Virol* **68**: 4727-4737.

Makino, S., and Lai, M.M., 1989, High-frequency leader sequence switching during coronavirus defective interfering RNA replication. *J Virol* **63**: 5285-5292.

Piñol-Roma, S., and Dreyfuss, G., 1992, Shuttling of pre-mRNA binding proteins between nucleus and cytoplasm. *Nature* **355**: 730-732.

Siomi, H., and Dreyfuss, G., 1995, A nuclear localization domain in the hnRNP A1 protein. *J Cell Biol* **129**: 551-560.

Spagnolo, J.F., and Hogue, B.G., 2000, Host protein interactions with the 3' end of bovine coronavirus RNA and the requirement of the poly(A) tail for coronavirus defective genome replication. *J Virol* **74**: 5053-5065.

Svitkin, Y.V., Ovchinnikov, L.P., Dreyfuss, G., and Sonenberg, N., 1996, General RNA binding proteins render translation cap dependent. *EMBO J* **15**: 7147-7155.

Wang, Y., and Zhang, X., 1999, The nucleocapsid protein of coronavirus mouse hepatitis virus interacts with the cellular heterogeneous nuclear ribonucleoprotein A1 in vitro and in vivo. *Virology* **265**: 96-109.

Weighardt, F., Biamonti, G., and Riva, S., 1995, Nucleo-cytoplasmic distribution of human hnRNP proteins: a search for the targeting domains in hnRNP A1. *J Cell Sci* **108**: 545-555.

Zhang, X., and Lai, M.M.C., 1995, Interactions between the cytoplasmic proteins and the intergenic (promoter) sequence of mouse hepatitis virus RNA: correlation with the amounts of subgenomic mRNA transcribed. *J Virol* **69**: 1637-1644.

Zhang, X., Li, H.P., Xue, W., and Lai, M.M.C., 1999, Formation of a ribonucleoprotein complex of mouse hepatitis virus involving heterogeneous nuclear ribonucleoprotein A1 and transcription- regulatory elements of viral RNA. *Virology* **264**: 115-124.

Role of hnRNP A1 in Coronavirus RNA Synthesis

XUMING ZHANG, CHRISTOPHER LYLE, YICHENG WANG, AND LIN
ZENG
*Department of Microbiology and Immunology, University of Arkansas for Medical Sciences,
Little Rock, Arkansas 72205, USA*

1. INTRODUCTION

Several host proteins have been identified that bind to the cis-acting sequences of mouse hepatitis virus (MHV) RNAs (Furuya and Lai, 1993; Yu and Leibowitz, 1995; Zhang and Lai, 1995; Li et al., 1997, 1999). The first among those is the heterogeneous nuclear ribonucleoprotein (hnRNP) A1. In normal mammalian cells, hnRNP-A1 is predominantly localized in the nucleus, but it shuttles between the nucleus and the cytoplasm (Dreyfuss et al., 1993). It interacts with a number of RNAs, particularly pre-mRNA and snRNAs, and a number of proteins such as itself, other hnRNPs and some serine-arginine (SR)-rich proteins through its amino terminal RNA-binding domains and the carboxyl terminal glycine-rich domain, respectively, and thus is associated with the RNP complex (Dreyfuss et al. 1993; Buvoli et al., 1988; Cartegni et al., 1996). hnRNP A1 has been implicated in promoting RNA-reannealing (Buvoli et al., 1992) and in regulating alternative RNA-splicing (Dreyfuss et al., 1993). Its involvement in regulation of viral RNA processing of human immunodeficiency virus (HIV) and human T-cell leukemia virus (HTLV)-2 has also been suggested (Black et al., 1995, 1996). Recent evidence indicates that hnRNP A1 is required for inhibition of HIV-1 pre-mRNA splicing (Caputi et al., 1999). Previously, we found that hnRNP A1 specifically binds to the *cis*-acting sequences (the negative-strand leader and intergenic [IG] sequence) of MHV RNA (Furuya and Lai, 1993; Zhang and Lai, 1995). Site-directed mutagenesis of the IG sequence indicated that

The Nidoviruses (Coronaviruses and Arteriviruses).
Edited by Ehud Lavi *et al.*, Kluwer Academic/Plenum Publishers, 2001.

the extent of mRNA transcription correlates with the efficiency of hnRNP A1-binding to the IG sequence (Zhang and Lai, 1995). In MHV-infected cells, hnRNP A1 is translocated from the nucleus to the cytoplasm (Li et al., 1997). These findings suggest that hnRNP A1 may play a role in regulation of MHV RNA replication and transcription. However, to date, no direct evidence for a functional role of hnRNP A1 in MHV RNA synthesis has been reported.

In this study, we employed a dominant-negative approach by expressing mutant hnRNP A1 proteins in cells and then determining the effects of the mutants on viral replication. We found that the mutant hnRNP-A1 proteins indeed exerted inhibitory effects on viral cytopathogenicity, cell fusion, viral growth and viral RNA synthesis. Thus, this study provides the direct *in vivo* evidence demonstrating a functional role for hnRNP A1 in MHV RNA synthesis, establishing that hnRNP-A1 is a *bona fide* host factor in regulation of MHV RNA synthesis.

2.   MATERIALS AND METHODS

2.1   Virus, Cells and Antibodies

MHV strain JHM (Makino and Lai, 1989) was used exclusively throughout this study. The murine astrocytoma cell line DBT (Hirano et al., 1974) was used for virus growth, infection and cell lysate preparation. A monoclonal antibody specific to the glycine-rich domain of hnRNP-A1 was kindly provided by Dr. Gideon Dreyfuss (University of Pennsylvania). The monoclonal antibody M2 specific to the FLAG was purchased from Strategene (San Diego, CA).

2.2   Construction of Plasmids

A cDNA clone of the full-length murine hnRNP A1 gene was previously constructed in the pBluescript vector (Stratagene), termed pBS-mA1 (Zhang et al., 1999). For expression of the wild-type and mutant A1 proteins, pBS-mA1 DNA was used as a template for PCR amplification with a pair of primers specific for each construct. For the wild-type A1, ΔIX, ΔM9, and 2xRBD, the 5'-sense primer is 5'-mA1-ORF (5'-TTT <u>GGA TCC</u> ATG TCT AAG TCC GAG TC-3'), which contains a *Bam* HI site (underlined) at the 5'-end and the first 17 nt of the open reading frame (ORF). The antisense 3'-primers were 3'-mA1-ORF (5'-GAA CCT CCT GCC AC-3') for A1, 3'-mA1-VIII (5'-TTG GTT CCG TGG TTT-3') for

ΔIX, 3'-mA1-RBDII (5'-GCG ACC TCT CTG ACT-3') for 2xRBD, and 3'-mA1-ΔM9 (5'-GTT GTA ATT GCC-3') for ΔM9. All 3'-primers contain an additional 12-nt sequence (5'-TTT GAA TTC TTA-3) which includes an *Eco* RI site (underlined) and a stop codon, and a sequence encoding the 8 amino acids flag (5'-CTT GTC ATC GTC GTC CTT GTA GTC-3'). PCR was performed in a DNAsys Thermocycler (M.J. Researches Inc.) for 25 cycles under the following conditions: 95 °C for 30 sec. 56 °C for 1 min and 72 °C for 1 min. The PCR products were digested with *Bam* HI and blunt-ended with T4 DNA polymerase and then digested with *Eco*RI. The digested PCR fragments were cloned into the *Eco*RV and *Eco*RI sites of DI vector p25CAT to replace the CAT gene, generating pDE-A1, pDE-A1-ΔIX, pDE-A1-ΔM9, and pDE-A1-2xRBD, respectively. For expression of hnRNP A1 proteins by the recombinant vaccinia virus, the first ORF in the DI was removed by digesting the DNAs of pDE-A1 and mutants with *Sna*BI and *Spe*I. The smaller *Sna*BI-*Spe*I fragments (1.5 kb, representing the DI ORF) were eliminated by agarose gel electrophoresis, while the large *Sna*BI-*Spe*I fragments were excised from the gel and purified by the gel elution kit (Qiagen). The purified DNA fragments were blunt-ended with T4 DNA polymerase and self-ligated with T4 DNA ligase. The resultant constructs pT7-A1, pT7-A1-ΔIX, pT7-A1-ΔM9, and pT7-A1-2xRBD contain a T7 RNA polymerase promoter, a 20-nt sequence derived from the DI leader and a 15-nt from the intergenic sequence upstream of the ORF.

2.3　In Vitro Transcription, Transfection, and CAT Assay

Plasmid DNAs were linearized with *Xba*I and DI RNAs were transcribed in vitro with the MegaScript kit (Ambion Inc.). RNA and DNA transfection was carried out with the DOTAP transfection kit (Boehringer Mannheim). Cell lysates were prepared at 8 h post-transfection and subjected to CAT assay with the procedure as described previously (Zhang et al., 1994).

2.4　Radiolabeling of Proteins and Immunoprecipitation

For detection of the protein expression, cells were starved in methionine-free medium for half hour and labeled with ^{35}S-methionine at 100 μCi/ml (Amersham-Pharmacia-Biotech) in the presence of Actinomycin D for 3 hrs. Cells were then lysed with a radioimmunoprecipitation assay (RIPA) buffer (50 mM Tris, pH7.4, 150 mM NaCl, 0.5% NP-40, 0.1% SDS, 1 mM PMSF). Immunoprecipitation of the lysate was carried out on a rocking platform at 4 °C for 2 h to overnight with the M2 anti-FLAG Mab (Stratagene). The antibody-antigen complexes were then precipitated with protein A-agarose beads at 4 °C for 2-4 h. Agarose beads were washed 3 times with RIPA buffer. Proteins complexes were denatured by boiling for 3 min in Lammeli's sample loading buffer (100 mM Tris, pH6.8, 200 mM DTT, 4% SDS, 0.2% bromophenol blue, 20% glycerol) and analyzed by sodium dodecyl sulfate-polyacrylamide gel electrophoresis (SDS-PAGE). The gels were dried and exposed to X-ray film and autoradiographed.

2.5.   Radiolabeling and Analysis of Intracellular Viral RNAs

Cells were infected with MHV at an m.o.i. of 1. Intracellular viral RNAs were metabolically labeled with ^{3}H-uridine at 100 µCi/ml (Amersham-Pharmacia- BRL-Gibco) in the presence of actinomycin D (10µg/ml) at 4 h post infection for 4 h, and isolated with the Trizol reagent (Gibco-BRL). One µl of RNAs of each sample (from a total of 5×10^6 cells) were precipitated by 5 % trichloroacetic acid (TCA). The incorporation of ^{3}H-uridine into the TCA-precipitated materials on the fiberglass filters was then determined by a liquid scintillation counter (Beckman 2000). Radiolabeled RNAs were also denatured with 6 M glyoxyl, and analyzed by electrophoresis on 1% agarose gel.   Gels were treated with Amplifier (Amersham-Phamacia-BRL-Gibco), dried and exposed to X-ray film.

3.   RESULTS AND DISCUSSION

3.1   Expression of Mutant A1 Proteins by an MHV DI Expression System

In the defective-interfering RNA expression system, the gene of interest is placed under the control of an MHV-specific promoter (IG sequence); expression of the inserted gene depends on the presence of helper MHV functions and the transcription of a subgenomic mRNA containing the inserted gene. Here, we used this system for expression of mutant A1 proteins in MHV-infected cells. Because this homologous expression system is devoid of interference by foreign sequences, any effects of the expressed gene on MHV or DI RNA synthesis can be directly assessed. Cells were infected with MHV and transfected with DI RNAs containing the mutant gene. At 4 h post transfection, intracellular proteins were labeled with ^{35}S-methionine. Immunoprecipitation of the cell lysate was carried out with a monoclonal antibody (MAb) M2 specific to the flag (Statagene), which were fused to the C-terminus of each protein (Fig. 1). It was found that the full-length and deletion mutants of A1 protein were expressed. The size of each protein appeared to correspond to its estimated molecular weight. These proteins are specific because the M2 antibody also detected HE-flag (Liao et al., 1995) in the positive control, but did not detect endogenous hnRNP A1 protein or GFP in DE-GFP-transfected cell lysate.

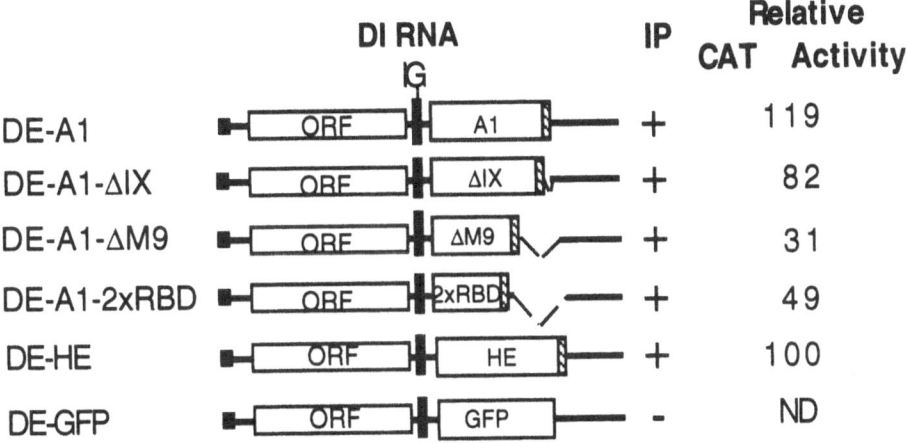

Figure 1. Effects of hnRNP-A1 mutants on the expression of a reporter gene. The names and structures of individual DI RNAs are indicated on the left. IG, intergenic sequence. The flag at the C-terminus is indicated by a striped box. IP, immunoprecipitation with anti-flag mAb. CAT activities in lysate from a 60 mm-Petri dish (5×10^6 cells) were determined by a CAT assay. Data shown are the averages from three individual experiments. The absolute CAT activities (cpm) are converted to percentages relative to the CAT activities of MHV-infected and mock-transfected cell lysate, in which the CAT activity is set to 100%. ND, not done.

3.2 Dominant Negative Effects of Mutant hnRNP A1 Proteins on MHV RNA Synthesis

We next determined the effects of A1 deletion mutants on MHV RNA synthesis using the DI RNA CAT reporter system. Following the expression of these proteins, a CAT reporter DI RNA was super-transfected. Three hours after the reporter DI RNA transfection, cell lysate was extracted and CAT activities were determined. As shown in Fig. 1, CAT activities decreased by approximately 70, 50, and 20% in ΔM9-, 2xRBD-, and ΔIX-expressing cells, respectively, as compared to those of the control (DE-HE-flag). CAT activity was slightly higher in cells expressing the wild-type A1 (119% of the control). We conclude that the deletion mutants of hnRNP A1 exhibited dominant-negative effects on CAT expression. Because CAT activity in this DI system reflects the synthesis of a subgenomic DI RNA (Liao and Lai, 1994), this result suggests that these mutants have dominant-negative effects on MHV RNA synthesis.

We then determined the effects of the mutant proteins on MHV RNA synthesis by directly measuring the amounts of de novo-synthesized viral RNAs. Cells were infected with MHV at an m.o.i. of 1, and transfected with DI RNAs containing the wild-type and mutant A1 genes. Newly-synthesized intracellular RNAs were metabolically labeled with ^{3}H-uridine in the presence of actinomycin D at 4 h post transfection for 4 h, and quantified after TCA precipitation. Fig. 2 shows that the amounts of total TCA-precipitable materials in ΔIX-, ΔM9-, 2xRBD-expressing cells were approximately 10, 70, and 60% less than that in the control, in which the DE-CAT was expressed. Consistent with the CAT assay (Fig. 1), TCA-precipitable materials in wild-type A1-transfected cells were slightly more than that in the CAT control (Fig. 2). Because actinomycin D was added to inhibit host DNA-dependent RNA transcription 1 h prior to and throughout the labeling period, the radiolabeled materials most likely are de novo-synthesized virus-specific RNAs. This interpretation is consistent with the result of Fig. 2, last column that only a background level of TCA-precipitable materials was detected from mock-infected and DE-CAT RNA-transfected cells, which also indicates that replication of DI RNAs depends on helper virus function. These results thus indicate that the mutants have dominant-negative effects on MHV RNA synthesis.

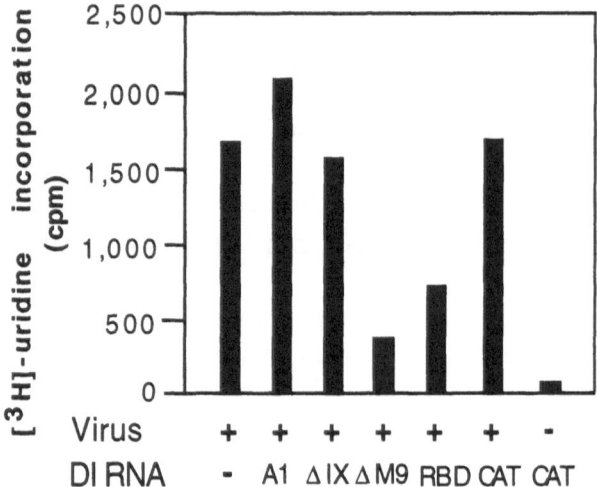

Figure 2. Effects of hnRNP-A1 mutants on MHV RNA synthesis. DBT cells were infected without (-) or with (+) MHV JHM strain at a multiplicity of infection of 1. At 1 h postinfection, cells were transfected without (-) or with (+) 5 µg of DI RNAs containing various inserted genes as indicated. Actinomycin D (10 µg/ml) was added to the cells at 3 h posttransfection. One hour later, cells were labeled with [^{3}H]-uridine (100 µCi/ml) for 3 h. Intracellular RNAs were isolated. One µl of each RNA sample was precipitated by 5% TCA, and radioactivity in the precipitates were counted in a liquid scintillation counter. Data indicate the mean of a duplicate experiment for each sample.

3.3 Expression of Mutant hnRNP A1 Proteins by a Recombinant Vaccinia Virus and Their Effects on MHV RNA Synthesis

We then used the recombinant vaccinia virus expression system to express mutant A1 proteins. Cells were infected with the recombinant vaccinia virus vTF7-3. At 1 h after vTF7-3 infection, *Xba*I-linearized plasmid DNAs, which contain a wild-type or mutant hnRNP A1 genes behind the T7 promoter (Fig. 3), were transfected. Following incubation at 37 °C for 10 h, cells were superinfected with MHV JHM strain to provide helper function necessary for DI RNA replication. At 1 h after MHV-infection, in vitro-transcribed CAT-reporter DI RNAs were transfected. Cell extracts were isolated at 4 h after reporter-transfection, and CAT activities in the lysate were then determined. The CAT activity measured in this system reflects the total effects of replication and transcription of both DI RNA and helper virus. As shown in Fig. 3, CAT activities from ΔIX-, ΔM9- and 2xRBD-expressing cells were 16, 71, and 57%, respectively, lower than that from the mock-transfected cells. In cells expressing the wild-type A1, the CAT activity was slightly higher. The CAT activity from GFP-expressing cells was also lower, for which the reason was unclear.

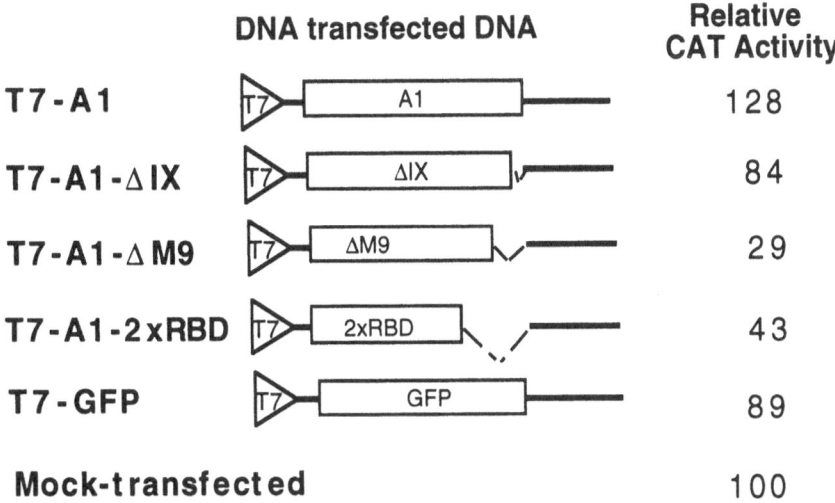

Figure 3. Effects of hnRNP-A1 mutants expressed by a recombinant vaccinia virus on the expression of a reporter gene. The names and structures of individual DNA constructs are indicated on the left. CAT activities in lysate from a 60 mm-Petri dish were determined by a CAT assay. Data shown are the mean value from a triplicate experiment. The absolute CAT activities (cpm) are converted to percentages relative to the CAT activities of vTF7-3-infected and mock-transfected cell lysate, in which the CAT activity is set to 100%.

In this study, we employed two different approaches to transiently express the wild-type and various deletion mutants of A1 protein in DBT cells and to determine their effects on MHV RNA synthesis. The advantage of using the DI RNA expression system is that it is devoid of potential interference by foreign genes. Its limitation is the dependence of DI gene expression on helper virus function. Therefore, on one hand, the more the helper virus replicates, the better the DI RNA gene expresses. On the other hand, when a large amount of helper virus RNA accumulate, dominant-negative effects may not be readily detected. This feature might have accounted for the low dominant-negative effect observed with the DI expression system (Fig. 1-2). The recombinant vaccinia virus expression system has the advantage in that the expression of the A1 mutants is independent of MHV function; once expressed, the effects of the mutants on MHV RNA synthesis can be observed early in MHV replication. Nevertheless, both approaches yielded similar results. Because both approaches express proteins transiently, the level of expression depends on the efficiency of transfection. We routinely achieved ≈10% transfection efficiency for RNA and much higher for DNA. Since the transfection efficiency was low, the number of cells doubly transfected with the expression DI and the reporter DI must have be extremely low. If so, one may wonder how we could observe any negative-effect of the mutant proteins. We speculate that, because most of the cells were fused to form polykaryons or even the whole monolayer was fused together at the time when the second DI (reporter) was transfected, the expressed proteins may have traveled between cells to exert their function. Indeed, when the GFP was expressed in the DI system, the fluorescence intensity was strong in a single cell prior to fusion and became weaker and more diffused when cells began to fuse (Zhang, unpublished observation).

We found that ΔM9 mutant exhibited the strongest inhibitory effect on the reporter gene expression and MHV RNA synthesis, while the ΔIX mutant had little such effect. Although both mutants contain RNA- and protein-binding activities (Zhang et al., 1999), ΔM9 lacks the M9 domain, which is known to be involved in protein-transport between the nucleus and the cytoplasm, while the ΔIX contains this domain. Thus, our data may suggest that the M9 domain may be involved in viral RNA synthesis. The lack of the functional M9 domain in the ΔM9 mutant possibly leads to an inhibitory effect on RNA synthesis. However, this does explain why 2xRBD, which has a larger deletion than ΔM9, had lower negative-effects than ΔM9. Apparently, some sequences other than the M9 domain may be also involved in regulation. Nevertheless, our present data indicate that cellular protein hnRNP A1 is functionally involved in coronavirus RNA synthesis. Whether such involvement is direct or indirect through other factors, and how it regulates viral RNA synthesis remain to be further investigated.

ACKNOWLEDGEMENTS

This work was supported by Research Project Grant RPG-98-090-01-MBC from the American Cancer Society.

REFERENCES

Black, A.C., Luo, J., Chun, S., Baker, A., Faser, J.K., and Rosenblatt, J.D. 1996. Specific binding of polypyrimidine tract binding protein and hnRNP A1 to HIV-1 CRS elements. Virus Genes 12: 275-285.

Black, A.C., Luo, J., Watanabe, C., Chun, S., Baker, A., Faser, J.K., Morgan, J.P., and Rosenblatt, J.D. 1995. Polypyrimidine tract-binding protein and heterogeneous nuclear ribonucleoprotein A1 bind to human T-cell leukemia virus type 2 RNA regulatory elements. J. Virol. 69: 6852-6858.

Buvoli, M., Biamonti, G., Tsoulfas, P., Bassi, M.T., Ghetti, A., Riva, S., and Morandi C. 1988. cDNA cloning of human hnRNP protein A1 reveals the existence of multiple mRNA isoforms. Nucl. Acids Res. 16: 3751-3770.

Buvoli, M., Cobianchi, F., and Riva, S. 1992. Interaction of hnRNP A1 with snRNPs and pre-mRNAs: evidence for a possible role of A1 RNA annealing activity in the first steps of spliceosome assembly. Nucl. Acids Res. 20: 5017-5025.

Caputi, M., Mayeda, A., Krainer, A.R., and Zahler, A.M. 1999. HnRNPA/B proteins are required for inhibition of HIV-1 pre-mRNA splicing. EMBO J. **18**:4060-4067.

Cartegni, L., Maconi, M., Morandi, E., Cobianchi, F., Riva, S., and Biamonti, G. 1996. hnRNP A1 selectively interacts through its Gly-rich domain with different RNA-binding proteins. J. Mol. Biol. 259: 337-348.

Dreyfuss, G., Matunis, M.J., Pinol-Roma, S., and Burd, C.G. 1993. hnRNP proteins and the biogenesis of mRNA. Annu. Rev. Biochem. 62: 289-321.

Furuya, T., and Lai, M.M.C. 1993. Three different cellular proteins bind to the complementary sites on the 5'-end positive- and 3'-end negative-strands of mouse hepatitis virus RNA. J. Virol. 67: 7215-7222.

Hirano, N., Fujiwara, K., Hino, S., and Matsumoto, M. 1974. Replication and plaque formation of mouse hepatitis virus (MHV-2) in mouse cell line DBT culture. Arch. Gesamte Virusforsch. 44: 298-302.

Li, H.P., Zhang, X.M., Duncan, R., Comai, L., and Lai, M.M.C. 1997. Heterogeneous nuclear ribonucleoprotein A1 binds to the transcription-regulatory region of mouse hepatitis virus RNA. Proc. Natl. Acad. Sci. USA 94: 9544-9549.

Li, H.P., Huang, P., Park, S., and Lai, M.M.C. 1999. Polypyrimidine tract-binding protein binds to the leader RNA of mouse hepatitis virus and serves as a regulator of viral transcription. J. Virol. **73**:772-777.

Liao, C.L., and Lai, M.M.C. 1994. Requirement of the 5'-end genomic sequence as an upstream cis-acting element for coronavirus subgenomic mRNA transcription. J. Virol. 68: 4727-4737.

Makino, S., and Lai, M.M.C. 1989. Evolution of the 5'-end of genomic RNA of murine coronaviruses during passages in vitro. Virology 169: 227-232.

Yu, W., and Leibowitz, J.L. 1995. Specific binding of host cellular proteins to multiple sites within the 3'-end of mouse hepatitis virus genomic RNA. J. Virol. 69:2016-2023.

Zhang, X.M., Liao, C.L., and Lai, M.M.C. 1994. Coronavirus leader RNA regulates and initiates subgenomic mRNA transcription both in trans and in cis. J. Virol. 68: 4738-4746.

Zhang, X.M., and Lai, M.M.C. 1995. Interactions between the cytoplasmic proteins and the intergenic (promoter) sequence of murine hepatitis virus RNAs: Correlation with the amounts of subgenomic mRNA transcribed. J. Virol. 69: 1637-1644.

Zhang, X.M., Li, H.P., Xue, W., and Lai, M.M.C. 1999. Formation of a ribonucleoprotein complex of mouse hepatitis virus involving heterogeneous nuclear ribonucleoprotein A1 and transcription-regulatory elements of viral RNA. Virology 264:115-124.

Expression of Transcriptional Units Using Transmissible Gastroenteritis Coronavirus Derived Minigenomes and Full-length cDNA Clones

ISABEL SOLA, SARA ALONSO, CARLOS SANCHEZ, J. MANUEL SANCHEZ-MORGADO, AND LUIS ENJUANES
Department of Molecular and Cell Biology, Centro Nacional de Biotecnología, CSIC. Campus Universidad Autónoma. 28049 Madrid. Spain

1. INTRODUCTION

Many factors including RNA primary and secondary structure and protein-RNA interactions may regulate mRNA abundance. The nature of the transcription regulatory sequences (TRSs) and the extent of their complementarity to the leader 3' end may be the most relevant and is discussed below. The TRSs include the core sequence (CS), previously named intergenic sequence (IG), that is a short conserved sequence element upstream the transcription units, and flanking sequences located upstream and downstream the CS. A helper dependent expression system based on transmissible gastroenteritis coronavirus (TGEV) derived minigenomes, encoding new subgenomic mRNAs, has been used to study the elements that regulate transcription in coronavirus. The optimization of TRSs can lead to the improvement of mRNA levels using both minigenomes and full-length cDNA clones.

The Nidoviruses (Coronaviruses and Arteriviruses).
Edited by Ehud Lavi *et al.*, Kluwer Academic/Plenum Publishers, 2001.

447

2.   OPTIMIZATION OF TRANSCRIPTION

2.1   Effect on mRNA levels of TGEV 5' and 3' core flanking sequences

Using TGEV derived minigenomes including the heterologous gene ß-glucuronidase (GUS) cloned under different TRSs, we have shown that the core CS sequence 5'-CUAAAC-3' is necessary and sufficient for mRNA expression. The absence of the CS (5'-CUAAAC-3') led to the complete abrogation of mRNA transcription. The insertion of the hexameric 5'-CUAAAC-3' restored expression levels more than 400-fold above the background. The addition of 5' upstream sequences flanking the core CS sequence from the TGEV N gene led to an increase in transcription up to 10-fold. The expression in coronavirus can be driven by a TRS with 20 nt in size, leading to 80% of the highest expression levels, produced by a TRSs of larger (88 nt) size (Alonso *et al.*, 2000) (Figure 1).

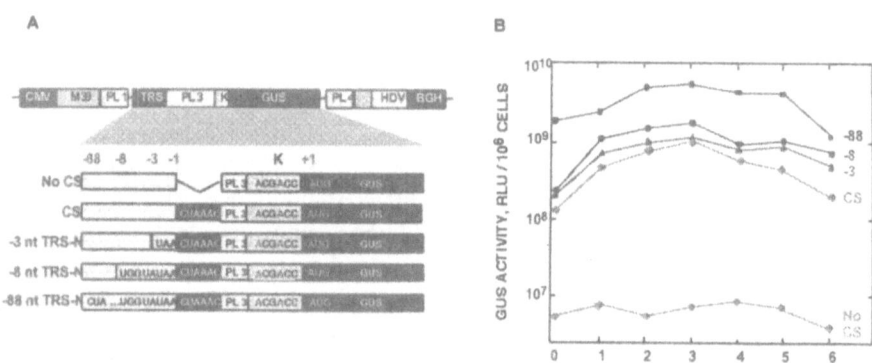

Figure 1. Effect of 5' flanking sequences on GUS expression. A. Expression cassettes including the GUS gene under different TRSs. TRSs include the core sequence 5'-CUAAAC-3' and 5' flanking sequences derived from TGEV N gene with different lengths. The minigenome M39 is cloned behind the cytomegalovirus promoter (CMV). B. GUS activity produced by constructs shown in A during six passages in cell culture. PL, polylinker. K, Kozak type sequence. An, poly A. HDV, hepatitis delta-virus ribozyme. BGH, bovine growth hormone termination and polyadenilation signals. RLU, relative luminometric units.

The sequences downstream the core CS of seven viral genes (S, 3a, 3b, E, M, N, and 7) have sizes ranging from 3 to 37 nt. Expression modules in which the 5' flanking sequence was kept constant, and the 3' CS flanking sequences were provided by each of the seven viral genes, led to similar expression levels of the heterologous mRNA, with the exception of that from ORF 3a. This construct gave expression levels 5- to 10-fold lower. Thus, there was no correlation between the length of the 3' flanking sequences and the expression level. Two non viral 3' flanking sequences were analyzed for mRNA expression. The sequence including an optimized Kozak motif produced mRNA levels as high as the best viral derived sequence. In contrast, when the sequences flanking the 3' end of the CS were engineered extending with 12 nt their potential basepairing with the 3' of the leader, mRNA levels were abrogated (Alonso *et al.*, 2000). The optimization of TRSs led to the expression of high amounts of heterologous protein (1 to 8 μg/10^6 cells), maintained for around 10 passages. These expression levels should be sufficient to elicite protective immune responses.

2.2   Influence of the insertion site

Using a TGEV derived helper dependent expression system, the expression cassette containing the GUS gene was inserted at different nucleotide distances from the 5' end. The expression levels increased from the 5' to 3' end by one thousand-fold (Alonso *et al.*, 2000). The insertion site close to the 5' end probably has affected essential primary or secondary structures required for minigenome replication, thus reducing the significance of this result in relationship to the insertion site. Furthermore, no difference in expression levels was observed with the MHV system, when the flanking sequences were kept constant for all insertion sites, suggesting that the location of the insertion site *per se* may not affect transcription levels and that the differences observed with TGEV were mostly due to the CS-flanking sequences.

2.3   Stability of TGEV helper dependent expresssion systems

GUS expression using TGEV derived minigenomes was observed for about ten passages (Alonso *et al.*, 2000). In general, the insertion of a heterologous gene such as GUS into TGEV derived minigenomes led to a 50-fold reduction in the levels of the minigenome RNA replication. The limited stability of the helper dependent expression systems is most likely due to the foreign gene, since TGEV minigenomes, in the absence of the heterologous gene, are efficiently rescued for at least 30 passages (Izeta *et al.*, 1999). The recombination frequency in MHV, TGEV, and IBV may be inversely proportional to the stability of the recombinants expressing a foreign gene.

3.   SINGLE GENOME TGEV VECTORS

To overcome the limited stability associated to RNA minigenomes, single genome TGEV vectors have been developed either by targeted recombination or by engineering an infectious cDNA encoding TGEV.

3.1   Vectors constructed by targeted recombination

Targeted recombination mediated by two cross-overs allowed the replacement of the S gene of a respiratory strain of TGEV by the S gene of enteric TGEV, leading to the isolation of viruses with a modified tropism and virulence (Sánchez *et al.*, 1999). A new strategy for the selection of TGEV recombinants was based in the elimination of the parental replicative virus by its simultaneous neutralization with two mAbs. The frequencies of the targeted recombination event in recombination for MHV and TGEV were found higher than the standard prediction for the recombination frequency of a multiple cross-over, suggesting that the alignment of two templates is the rate-limiting event in this recombination.

3.2   Coronavirus vectors derived from an infectious cDNA clone

Using the TGEV infectious cDNA (Almazan *et al.*, 2000), the green fluorescent protein (GFP) gene was cloned by replacing the non-essential 3a and 3b genes, leading to an engineered genome with high stability and expression levels around 40 µg/10^6 cells (Figure 2) (Sola *et al.*, 2000).

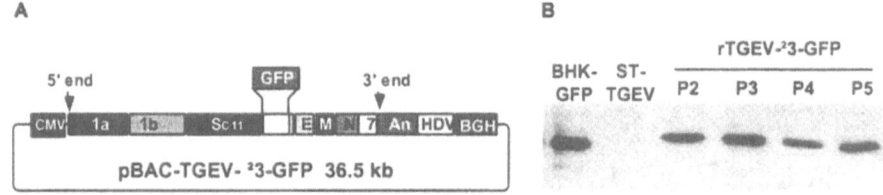

Figure 2. Expression of GFP using an infectious TGEV cDNA clone. **A**. Genes 3a and 3b were deleted in the TGEV infectious cDNA cloned in a bacterial artificial chromosome (BAC). GFP (0.72 kb) was inserted within the position of the deleted genes after the TRS of gene 3a, leading to a replication competent cDNA, rTGEV-Δ3-GFP. **B**. Western-blot analysis of GFP expression in cell culture along passages P2, P3, P4, P5. CMV, immediate-early cytomegalovirus promoter. Sc11, S gene of PUR-C11 TGEV strain. An, poly A. HDV, hepatitis delta-virus ribozyme. BGH, bovine growth hormone termination and polyadenylation signals. BHK-GFP, positive control of BHK cells expressing GFP. ST-TGEV, negative control of ST cells infected with *wt* TGEV.

4. CONCLUSION

Both helper-dependent expression systems, based on two components, and single genomes constructed by targeted recombination or by using an infectious cDNA have been developed. The sequences that regulate transcription have been characterized. Single genome coronavirus vectors have been constructed efficiently expressing a foreign gene such as GFP. Thus, a new avenue with high potential has been opened for coronaviruses which have unique properties, such as a long genome size and enteric tropism, that makes them of high interest as expression vectors for vaccine development and gene therapy. The possibility of engineering the tissue and species tropism will make coronavirus very flexible expression systems, since the same vector could be modified to target expression to different organs and animal species, including humans.

ACKNOWLEDGMENTS

The work was supported by grants from the Comisión Interministerial de Ciencia y Tecnología (CICYT), la Consejería de Educación y Cultura de la Comunidad de Madrid, and the European Community (Key Action 2: Infectious Diseases).

REFERENCES

Almazan, F., González, J. M., Pénzes, Z., Izeta, A., Calvo, E., Plana-Durán, J., and Enjuanes, L. (2000). Engineering the largest RNA virus genome as an infectious bacterial artificial chromosome. *Proc. Natl. Acad. Sci. USA* **97**, 5516-5521.

Alonso, S., Izeta, A., Sola, I., and Enjuanes, L. (2000). Transcription regulatory sequences in transmissible gastroenteritis coronavirus. *Submitted*.

Izeta, A., Smerdou, C., Alonso, S., Penzes, Z., Méndez, A., Plana-Durán, J., and Enjuanes, L. (1999). Replication and packaging of transmissible gastroenteritis coronavirus-derived synthetic minigenomes. *J. Virol.* **73**, 1535-1545.

Sánchez, C. M., Izeta, A., Sánchez-Morgado, J. M., Alonso, S., Sola, I., Balasch, M., Plana-Durán, J., and Enjuanes, L. (1999). Targeted recombination demonstrates that the spike gene of transmissible gastroenteritis coronavirus is a determinant of its enteric tropism and virulence. *J. Virol.* **73**, 7607-7618.

Sola, I., Alonso, S., Plana-Durán, J., and Enjuanes, L. (2000). Heterologous gene expression with a single genome derived from transmissible gastroenteritis coronavirus. *J. Virol.*, Submitted for publication.

Identification of the Mutations Responsible for the Phenotype of Three MHV RNA-negative ts Mutants

[1]S. SIDDELL, [2]D. SAWICKI, [1]Y. MEYER, [1]V.THIEL, AND [2]S. SAWICKI
[1]*Institute of Virology and Immunology, University of Würzburg, 97078, Germany:* [2]*Department of Microbiology and Immunology, Medical College of Ohio, Toledo*

1. INTRODUCTION

A number of laboratories have isolated and characterized conditionally lethal, temperature-sensitive (*ts*) mutants of the A59 strain of murine hepatitis virus (MHV) (Koolen *et al* 1983; Sturman *et al* 1987; Schaad *et al* 1990). By and large, these studies have focused on *ts* mutants with an RNA-negative phenotype, i.e. mutants that are unable to synthesize viral RNA at the restrictive temperature. RNA positive mutants that are able to synthesize viral RNA but fail to produce plaques at the restrictive temperature have also been isolated and analyzed (Masters *et al* 1994; Ricard *et al* 1995; Luytjes *et al* 1997).

We decided to initiate a systematic analysis of the phenotypes and genotypes of a collection of MHV-A59 *ts* mutants assembled from laboratories in Albany (Sturman *et al* 1987) Los Angeles and North Carolina (Schaad *et al* 1990), Utrecht (Koolen *et al* 1983) and Würzburg (Siddell, unpublished). Our intention is to characterize this set of mutants with respect to their phenotypes, i.e.

- growth characteristics at permissive and restrictive temperatures
- reversion or back mutation frequency
- ability to synthesize plus and minus strand RNA at the restrictive temperature and

The Nidoviruses (Coronaviruses and Arteriviruses).
Edited by Ehud Lavi *et al.*, Kluwer Academic/Plenum Publishers, 2001.

- relative replication and transcription rates (genomic RNA synthesis versus subgenomic mRNA synthesis) at the permissive temperature and after shifting to the restrictive temperature.
- and genotypes, i.e.complementation group of each mutant and causal mutation, identified by sequence analysis of *ts* mutant and revertant pairs.

In the long term, our goal is to provide further insights into the role of individual proteins encoded by the replicase gene in the assembly and function of the coronavirus replication-transcription complex

2.   METHODS

Complementation analysis was done as described (Schaad *et al* 1990) and Sawicki *et al* (in preparation). The distinct phenotypes of Alb *ts*22, LA *ts*6 and Alb *ts*16 were determined as described (Younker *et al* 1998 and in preparation). The replicase gene sequence of wild-type MHV-A59, mutants, and revertants was determined by cycle sequencing performed on DNA fragments produced by RT-PCR of genomic RNA that was isolated from plaque-purified virus (Sawicki *et al*, in preparation).

3.   RESULTS

Complementation analysis placed twenty-three mutants of MHV-59 into four complementation groups, numbered I-IV (Fig 1).

The mutations of three mutants, LA *ts*6 and Alb *ts*16 in cistron I and Alb *ts*22 in cistron II, were mapped by recombination and sequence analyses. Recombination analysis showed that mutations in LA *ts*6 and Alb *ts*16 were close to one another and distant from Albany *ts*22. Using Alb *ts*18 that has a mutation in the S gene (Ricard *et al* 1995) as a recombination marker, Alb *ts*16 was predicted to have a mutation close to and 5' of LA *ts*6, and Alb *ts*22 to have a mutation distant from and 3' of both LA *ts*6 and Alb *ts*16. We sequenced the entire replicase gene of Alb *ts*16 and *ts*22 and of LA *ts*6, revertants of each of the mutants and the wild-type parental MHV-A59 virus. The results of this analysis are summarized in Table 1.

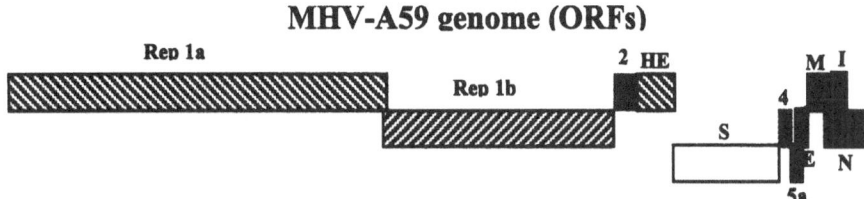

ts, RNA-negative mutants with mutant ORF 1ab Rep genes

Cistron I	Cistron II	Cistron III	Cistron IV
Alb *ts2*	Alb *ts22*	Wü *ts18*	Alb *ts17*
Alb *ts6*	Ut *ts261**	Wü *ts* 38	Wü *ts36*
Alb *ts8*	LA *ts18* ?	Ut *ts145*	LA *ts18* ?
Alb *ts9*		LA *ts18* ?	
Alb *ts16*			
Alb *ts19**			
LA *ts3*			
LA *ts6*			
LA*ts10*			
NC *ts2*			
NC *ts3*			
NC *ts11**			
Ut *ts88*			
Ut *ts261**			
Ut *ts329*			
Wü ts21*			

Figure 1. MHV-A59 RNA-negative mutants with replicase gene mutations. (*) mutants not producing revertants

Table 1. Identification of the mutations responsible for the phenotype of three RNA-negative-MHV ts mutants.

Mutant	Mutated Protein	Wild-type	ts	Revertant
Alb ts16	p33^{3Clpro}	Phe$_{219}$	Leu$_{219}$	Phe$_{219}$
LA ts6	p15	Gln$_{65}$	Glu$_{65}$	Gln$_{65}$
Alb ts22	p102^{POL}	His$_{868}$	Arg$_{868}$	His$_{868}$

In addition to being RNA-negative *ts* mutants, i.e., mutants unable to cause the synthesis of viral RNA when the infection is initiated and maintained at 37°C or higher, LA *ts*6, Alb *ts*16, and Alb *ts*22 have distinct phenotypes. These distinct phenotypes are observed when the infection is initiated at the permissive temperature (34°C or lower) and the infected cultures shifted to the restrictive temperature (40°C) after viral RNA synthesis has commenced. With Alb *ts*22, both plus and minus strand synthesis ceased immediately. With LA *ts*6, minus strand synthesis stopped immediately but plus strand synthesis continued at the same level as was occurring at the time of temperature shift; and, 30-60 min after shift to 40°C, plus strand synthesis began to diminish and by 3-4 hours was undetectable. With Alb *ts*16, both minus and plus strand synthesis was not greatly affected by a shift to 40°C. However, the rate of plus strand synthesis did not increase when Alb *ts*16 infected cultures were shifted to 40°C early, when the rate of viral RNA synthesis was increasing, i.e., between 5 and 8 hours at 30°C. Adding cycloheximide at the time of shift to 40°C, which inhibits minus strand synthesis within 15 min, caused plus strand synthesis to decline after 30-60 min (Younker *et al* 1998).

The phenotype of Alb *ts*22 is consistent with a mutation in the polymerase gene that makes elongation of nascent viral RNA temperature sensitive. Interestingly, we found that even at 30-34°C, Alb *ts*22 caused 4-5 fold less viral RNA synthesis compared to revertants of Alb *ts*22, the parental MHV-A59 virus, Alb *ts*16 or LA *ts*6. Moreover, Alb *ts*22 produced relatively more genomic than subgenomic mRNA (Younker *et al*, in preparation). It will be necessary to explain how a single amino acid change in the polymerase gene product, p102, that changed His_{868} to Arg_{868} results in both a *ts* phentype and an altered phenotype at permissive temperature.

The phenotype of LA *ts*6 would result from the failure to synthesize minus strand templates at 40°C. Eventually, the minus strands synthesized at permissive temperature are turned over (see Tao, Sawicki and Sawicki, these Proceedings) and plus strand synthesis declines coincidentally. The single amino acid change in the p15 protein that changes Gln_{65} to Glu_{65} apparently prevents the formation of minus strand polymerase activity at the restrictive temperature.

The phenotype of Alb *ts*16 results from a single amino acid change in the $p33^{3CLpro}$ protein that changes Phe_{219} to Leu_{219}. The mutated 3C-like protease activity would affect the rate of proteolytic cleavage of the replicase polyproteins, pp1a and pp1ab, at the restrictive temperature. Only enough minus strand polymerase activity would be formed at restrictive temperature to allow for the replenishment of minus strand templates being lost due to turnover. However, not enough minus strand templates would be produced at restrictive temperature to raise the rate of plus strand synthesis.

Alternatively, the mutation in the 3C-protease might specifically prevent the formation of plus strand polymerase activity, or the conversion of the minus strand polymerase activity to plus strand polymerase activity. As with alphaviruses, a precursor polyprotein in MHV might function in minus strand activity and proteolytic processing would unveil the plus strand polymerase activity (Wang *et al* 1994).

4. CONCLUSIONS

Our analysis of three MHV-A59 RNA-negative *ts* mutants allows us to make a number of interesting conclusions. First, the mutants Alb *ts*16 and LA *ts*6 both fall into the same complementation group (cistron I) and yet they have very distinct phenotypes. Indeed, the mutations responsible for these phenotypes are located in two different replicase polyprotein processed products, $p33^{3CLpro}$ and p15, present in pp1a and in pp1ab. One possible interpretation of this data is that the complementation group defining cistron I is in fact equivalent to the pp1a or pp1ab polyprotein precursor and $p33^{3CLpro}$ and p15 are cis-acting, i.e. the cleavage events that define these proteins occur after they have associated into a functional complex. This situation would be analogous to the relationships found for the Sindbis virus nsP2/nsP3 proteins (Wang *et al* 1994).

Second, our results demonstrate that the replicase gene products encoded in ORF 1a are involved directly in coronavirus RNA synthesis.

Third, LA *ts*6 was put into an "A" complementation group (Schaad *et al* 1990) and mapped to the 5' end of the genome by recombination analysis (Baric *et al* 1990). Based on our sequencing of LA *ts*6, the causal mutation is at the C-end of pp1a, close to where Fu and Baric (1994) suggested mutants belonging to their "C" complementation group would map. At this time we have no explanation for this discrepancy, which calls at least for a re-examination of past "cistron" assignments as causal mutations are identified.

ACKNOWLEDGEMENTS

We acknowledge the technical assistance of Barbara Schelle. Grants from the German Research Council (SGSi) and the National Institutes of Health (AI28506 to SGSa) supported this research.

REFERENCES

Baric, R. S., K. Fu, M. C. Schaad and S. A. Stohlman 1990. Establishing a genetic recombination map for murine coronavirus strain A59 complementation groups. *Virology* 177:646-56.

Fu, K. and R. S. Baric 1994. Map locations of mouse hepatitis virus temperature-sensitive mutants: confirmation of variable rates of recombination. *J Virol* 68:7458-66.

Koolen, M. J., A. D. Osterhaus, G. Van Steenis, M. C. Horzinek and B. A. Van der Zeijst 1983. Temperature-sensitive mutants of mouse hepatitis virus strain A59: isolation, characterization and neuropathogenic properties. *Virology* 125:393-402.

Luytjes, W., H. Gerritsma, E. Bos and W. Spaan 1997. Characterization of two temperature-sensitive mutants of coronavirus mouse hepatitis virus strain A59 with maturation defects in the spike protein. *J Virol* 71:949-55.

Masters, P. S., C. A. Koetzner, C. A. Kerr and Y. Heo 1994. Optimization of targeted RNA recombination and mapping of a novel nucleocapsid gene mutation in the coronavirus mouse hepatitis virus. *J Virol* 68:328-37.

Ricard, C. S., C. A. Koetzner, L. S. Sturman and P. S. Masters 1995. A conditional-lethal murine coronavirus mutant that fails to incorporate the spike glycoprotein into assembled virions. *Virus Res* 39:261-76.

Schaad, M. C., S. A. Stohlman, J. Egbert, K. Lum, K. Fu, T. Wei, Jr. and R. S. Baric 1990. Genetics of mouse hepatitis virus transcription: identification of cistrons which may function in positive and negative strand RNA synthesis. *Virology* 177:634-45.

Sturman, L. S., C. Eastwood, M. F. Frana, C. Duchala, F. Baker, C. S. Ricard, S. G. Sawicki and K. V. Holmes 1987. Temperature-sensitive mutants of MHV-A59. *Adv Exp Med Biol* 218:159-68.

Wang, Y. F., S. G. Sawicki and D. L. Sawicki 1994. Alphavirus nsP3 functions to form replication complexes transcribing negative-strand RNA. *J Virol* 68:6466-75.

Younker, D. R. and S. G. Sawicki 1998. Negative strand RNA synthesis by temperature-sensitive mutants of mouse hepatitis virus. *AdvExp Med Biol* 440:221-6.

MHV Subgenomic Negative Strand Function

R.S. BARIC[1,2], K. M. CURTIS[2], AND B. YOUNT[1]

[1]*Department of Epidemiology, Division of Infectious Diseases,School of Public Health,*
[2]*Department of Microbiology and Immunology, School of Medicine, University of North Carolina at Chapel Hill, Chapel Hill, North Carolina, 27599*

1.   INTRODUCTION

Mouse hepatitis virus (MHV), a member of *Nidovirales*, contains a ~32 KB linear, single-stranded, positive polarity RNA genome. Upon entry into the cell, the viral genome is transcribed into 7-8 subgenomic mRNAs ranging in size from ~1.0-32.0 KB. The positive strand mRNAs are arranged in a 3' co-terminal nested set, and each contains a 5' end ~72 nucleotide (nt) leader RNA sequence which is derived from the 5' end of the genome. Leader RNA sequences are joined to body sequences of each subgenomic length mRNA at highly conserved transcriptional start (TSE) sites located just upstream from the coding sequences of each viral gene (Baric et al., 1983; Sethna et al., 1989). In addition to the viral mRNAs, both full-length as well as subgenomic length negative strand RNAs and replicative form (RF) RNAs have been detected in porcine transmissible gastro-enteritis virus (TGEV), bovine coronavirus (BCV) and MHV-infected cells (Baric et al., 1983, Sethna et al., 1989; 1991; Sawicki and Sawicki, 1990). The subgenomic length negative strand RNA contains antileader RNA sequences (Sethna et al., 1991). It has been suggested that the subgenomic length negative strands function as dead end products of transcription (Yokomori et al., 1992; Jeong and Makino, 1992). In this manuscript, we provide direct evidence that the subgenomic negative strands function as the principle templates for the synthesis of each corresponding mRNA.

The Nidoviruses (Coronaviruses and Arteriviruses).
Edited by Ehud Lavi *et al.*, Kluwer Academic/Plenum Publishers, 2001.

2.   MATERIALS AND METHODS

2.1   Virus, Cells and Analysis of RNA

MHV-A59 was grown on DBT and 17Cl1 cells. Radiolabeling of viral RNAs was performed in 17Cl-1 cells. After infections with MHV-A59, the virus inoculums were removed and the cultures overlaid with ΔMEM medium (pH 6.8) containing 2% fetal clone II, 0.05 µg/ml gentamicin and 0.25 µg/ml kanamycin. Cultures were radiolabeled with 300 µCi/ml ^{32}P-orthophosphate for various times to radiolabel virus-specific mRNA and RI RNAs. The RNA was isolated as described previously (Schaad and Baric, 1994).

Cultures of 17Cl-1 cells were infected with MHV-A59 at a MOI of 10. At 6.0 hrs postinfection, cultures were radiolabeled with 1000 µCi/ml of ^{32}P-orthophosphate for 5, 15, 30, 45 and 60 min. Total RNA was isolated and purified. Viral mRNA and RF RNA were analysed according to the protocol described by Sawicki and Sawicki, 1990. To calculate the % molar ratios of mRNA and RF RNA, the % CPM in each viral RNA was divided by the molecular weight of the respective mRNA/RF RNA and normalised to 100% (Schaad and Baric, 1994).

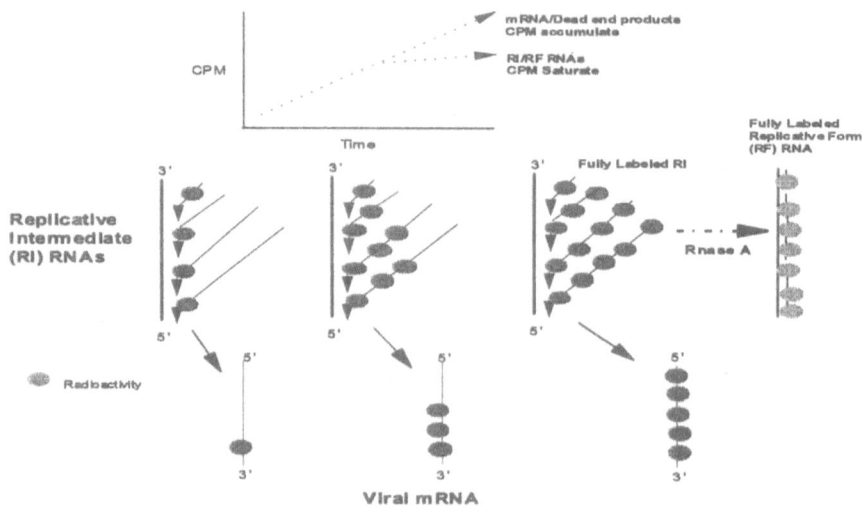

Figure 1.   Saturation Kinetics of Full Length and Subgenomic Length Replicative Intermediate RNAs and mRNAs.

3. RESULTS

For positive strand RNA viruses, it is understood that viral mRNA is transcribed from a complementary negative strand RNA. Nascent viral RNAs are rapidly radiolabeled and located in a partially single-stranded, partially double-stranded RI RNA (Simons and Strauss, 1972). Following RNase digestion, much of the recently transcribed RNA remains affiliated with double-stranded RF RNA cores (Figure 1). Previous studies by Sethna et al., 1989, 1991 used probe hybridization to identify the presence of full length and subgenomic length negative strand RNAs during transmissible gastroenteritis virus infection. These experiments could not address whether these negative strands were templates for mRNA synthesis. These experiments, however, were criticised because the subgenomic length negative strand RNAs could be dead-end products of transcription engaged in the synthesis of a single positive strand RNA (Jeong and Makino, 1992, Yokomori et al., 1992). If negative strands were actively engaged in continual nascent plus strand synthesis, both full length and subgenomic length RI RNAs should rapidly incorporate radiolabel and then saturate as the nascent positive strands become maximally radiolabeled and then dissociate from the RI complex (Simons and Strauss, 1972). In contrast, mRNA levels should increase steadily until degradation and transcription rates become equal.

To study the saturation kinetics of full length and subgenomic length RI RNAs, cultures of 17Cl-1 cells were infected with MHV-A59 at a MOI of 10 and radiolabeled with 1000 μCi/ml ^{32}P-orthophosphate for 5, 15, 30, 45 and 60 min. Under these conditions, nascent positive strands are preferentially labelled as rates of negative strand synthesis are reduced at later times postinfection. The intracellular RNAs were isolated and analysed for mRNA and RF RNA. Similar to findings reported by Sawicki and Sawicki, 1990, that increasing amounts of radiolabel were incorporated into all viral mRNAs following longer labelling periods (data not shown). In agreement with previous reports (Sawicki and Sawicki, 1990), full length and subgenomic length RF RNAs were rapidly labelled after a 5-min pulse. Increasing amounts of radiolabel were evident in the RF RNAs with longer labelling periods (15-30 min); afterwhich time the amount of radiolabel slowly increased in the RF RNAs (data not shown). For example, rapid increases in the levels of RF RNA 4 and RF RNA 5 were noted during the first 30 min pulse labelling period, afterwhich total counts in these RF RNAs became relatively constant. In contrast total counts in mRNA 4 and 5 increased steadily over the same labelling period (Figure 2). Such findings were consistent with the hypothesis that the nascent plus strands on full length and subgenomic length RI RNAs rapidly saturated with label and that all negative strands remained actively engaged in the synthesis of many new mRNAs.

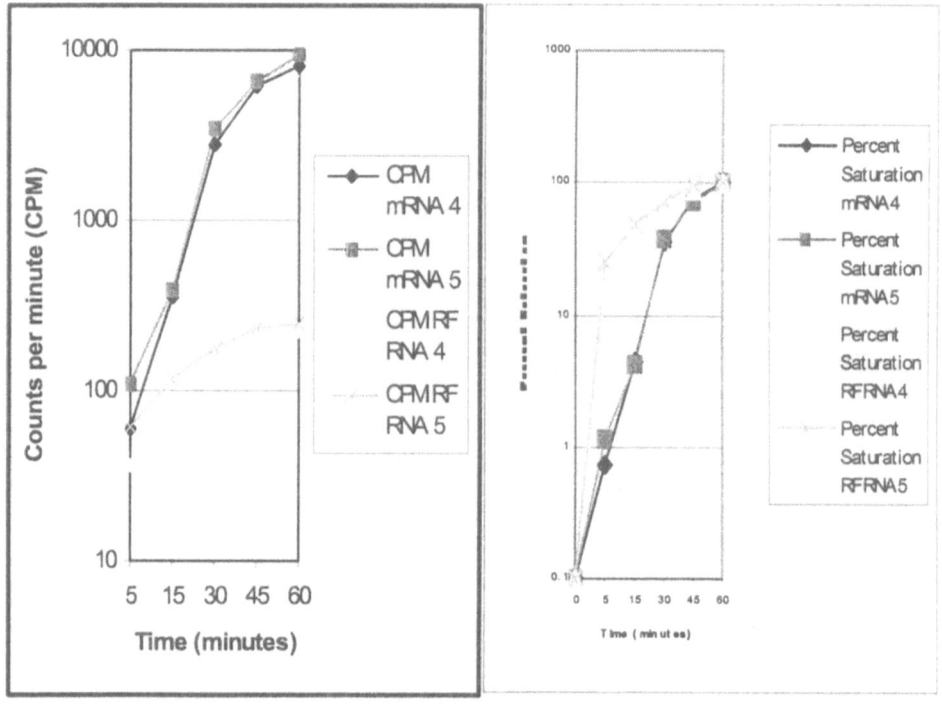

Figure 2. Radiolabeling of mRNA and RF RNA 4 and 5.

Figure 3. Saturation Kinetics of mRNA and RF RNA 4 and 5.

3.1   Saturation Kinetics of Genomic and Subgenomic Length RI RNAs

Labelling kinetics of individual MHV full length and subgenomic length RF RNAs reflected the kinetics of total RF RNA labelling. To provide additional evidence that the full length and subgenomic length negative strand RNAs were in transcriptionally active RI structures, we compared the saturation kinetics of mRNAs and RF RNAs 4 and 5 over the 1 hr time period (Figure 3). Importantly, while steady increases in mRNA 4 and 5 were noted throughout the labelling period, their respective RF RNAs rapidly saturated with label. By comparing the percent label incorporated as a function of total mRNA or RF RNA, saturation kinetics revealed that over 70% of the radiolabel was incorporated into full length and subgenomic length RF RNAs within 30 min of the addition of label (Figure 3). Only slight increases in the amount of RF RNA 4 or 5 were detected after this time. In contrast under identical conditions, <10% of the total label were incorporated into mRNA 4 or 5 within the first 30 minutes. These data were consistent with the hypothesis that both full length and subgenomic length negative strands remained in transcription active RI complexes engaged in the continual synthesis of numerous positive strand RNAs.

While the slight increase in total levels of the RF RNAs noted after 30 min was most likely due to new negative strand synthesis and hotter pools of nucleotide precursors, this could also represent RF RNA structures which had "burned out" and were accumulating during infection. If this were the case, then the relative percent molar ratio's of the transcriptionally-active full length RF RNA should decrease over time in proportion to the slow rate of increase in the molar ratio's of the slowly accumulating "burned out" subgenomic RF RNAs. To address this possibility, we calculated the percent molar ratio of each RF RNA and mRNA during the labelling period. Relative percent molar ratios of the viral mRNAs reflected the relative abundance of each RF RNA and remained relatively constant throughout the labelling period as well (data not shown).

4.   DISCUSSION

Several discontinuous transcription models have been proposed to explain the presence of leader RNA sequences on positive strand RNAs and antileader RNA sequences on full-length and subgenomic length negative strand RNAs (Baric et al., 1983; Sethna et al., 1989; Sawick and Sawicki, 1990). It is generally agreed that these smaller RNAs do not originate from larger precursors. The leader-primed transcription model proposed that a free leader RNA was synthesised from the 3' end of the full length minus strand (figure 4). Free leader RNA then acts in trans and binds with the different intergenic start elements (TSE) in a full length minus strand template to prime transcription of each of the subgenomic mRNAs (Baric et al., 1983). This model was heavily based on the finding of a single full length negative strand RNA and RF RNA in infected cells and the presence of mRNA 7 nascent plus strands on purified full length RI RNA (Lai et al., 1982; Baric et al., 1983). The former observation is clearly incorrect. The latter observation can no longer be viewed as definitive evidence for "leader-primed transcription" as such RNAs may represent template switching between full length and subgenomic length RI RNAs associated with high frequency RNA recombination (Baric et al., 1990).

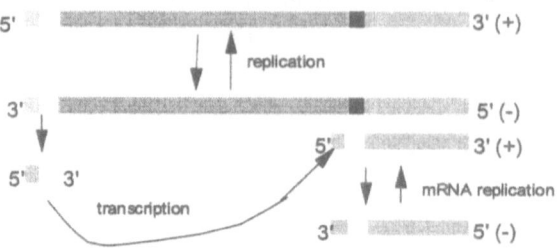

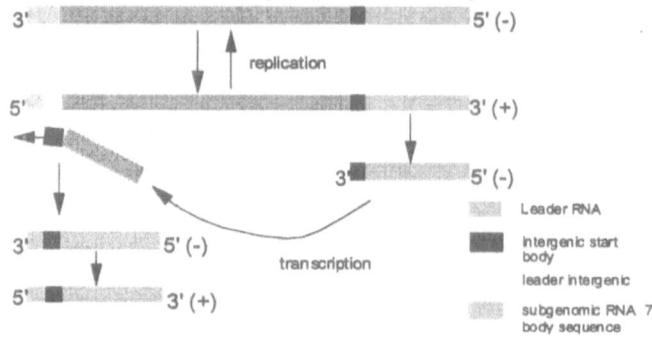

Figure 4. Models of Discontinuous Transcription. Panel A. Leader Primed Transcription. Panel B: transcription attenuation model.

All other data referenced in support of leader primed transcription will also support the transcription attenuation model for Nidovirales transcription (Makino et al., 1986; Li et al., 1997). In this model, the body TSE elements in the genome length template act as discontinuous elongation sites that allow for the polymerase/nascent minus strand to detach and reinitiate transcription (in cis or in trans) at the 5' end of the genome. This model predicts that all subgenomic negative strands originate from the genomic RNA directly, rather than from mRNA. In support of this hypothesis, tandem placement of a coronavirus promoter resulted in enhanced mRNA synthesis from the downstream most initiation sites (Krishman et al., 1996). During EAV transcription, the intergenic TSE element in subgenomic mRNA is derived from the TSE sequences encoded in the genomic RNA during negative strand synthesis rather than from leader sequences encoded at the 5' end of the genome. These studies support a mechanism of discontinuous negative strand synthesis reminiscent of the process of copy choice RNA recombination (Van Marle et al., 1999). The finding of a EAV replicase mutant deficient in subgenomic negative strand synthesis further supports the role of these RNA templates in mRNA synthesis (Van Marle et al., 1999). In light of these findings, clearly, transcription attenuation is the most attractive mechanism of Nidovirus discontinuous transcription of the subgenomic RNAs.

REFERENCES

Baric, R.S., S.A. Stohlman, and M.M.C. Lai. 1983. Characterization of replicative intermediate RNA of mouse hepatitis virus: presence of leader RNA sequences on nascent chains. J. Virol. **48**:633-640.

Baric, R.S., K.S. Fu, M.C. Schaad, and S.A. Stohlman. 1990. Establishing a genetic recombination map for MHV-A59 complementation groups. Virology 177: 646-656.

Jeong, Y.S. and Makino, S. 1992. Mechanism of coronavirus transcription: duration of primary transcription initiation activity and effects of subgenomic RNA transcription on RNA replication. J. Virol. 66:3339-3346.

Krishnan, R., Chang, R.-Y., and Brian, D.A. 1996. Tandem placement of a coronavirus promoter results in enhanced mRNA synthesis from the downstream-most initiation site. Virology, 218-400-405

Lai, M.M.C., Patton, C.D. and Stohlman, S.A. 1982. Replication of mouse hepatitis virus: negative-stranded RNA and replicative form RNA are of genome length. J. Virol. 44:487-492.

Li, H.-P., Zhang, X.M., Duncan, R., Comai, L., and Lai, M.M.C. (1997). Heterogeneous nuclear ribonuclleoprotein A1 binds to the transcription-regulatory region of mouse hepatitis virus RNA. Proc. Natl. Acad. Sci. USA 94, 9544-9549.

Makino, S., Stohlman, S.A. and Lai, M.M.C. 1986. Leader sequences of murine coronavirus mRNAs can be freely reassorted: evidence for the role of free leader RNA in transcription. Proc. Natl. Acad. Sci. USA 83:4204-4208.

Sawicki, S.G. and Sawicki, D.L. 1990. Coronavirus transcription: subgenomic mouse hepatitis virus replicative intermediates function in RNA synthesis. J. Virol. 64:1050-1056.

Schaad, M.C. and Baric, R.S. 1994. Genetics of mouse hepatitis virus transcription: Evidence that subgenomic negative strands are functional templates. J. Virol. 68, 8169-8197.

Sethna, P.B., Hung, S.-L. and Brian, D.A. 1989. Coronavirus subgenomic minus-strand RNAs and the potential for mRNA replicons. Proc. Natl. Acad. Sci. USA. 86:5626-5630.

Sethna, P.B., Hoffman, M.A. and Brian, D.A. 1991. Minus-strand copies of replicating coronavirus mRNAs contain antileaders. J. Virol. 65:320-325.

Simmons, D.T. and Strauss, J.H. 1972. Replication of sindbis virus. II. Multiple forms of double-stranded RNA isolated from infected cells. J. Mol. Biol. 71:615-631.

Van Marle, G., van Dinten, L.C., Spaan, W.J.M., Luytjes, W., and Snijder, E.J. 1999. Characterization of an equine arteritis virus replicase mutant defective in subgenomic mRNA synthesis. J. Virol. 71, 5274-5281.

Van Marle, G., Dobbe, J.C., Gultyaev, A.P., Luytjes, W., Spaan, W.J.M. and Snijder, E..J. 1999. Arteriovirus discontinuous mRNA transcription is guided by base pairing between sense and antisense transcription regulating sequences. PNAS USA 96, 12056-12061.

Yokomori, K. Banner, L.R. and Lai, M.M.C. 1992. Coronavirus mRNA transcription: UV light transcriptional mapping studies suggest an early requirement for a genomic-length template. J. Virol. 66:4671-4678.

Requirement of the Poly(A) Tail in Coronavirus Genome Replication

JEANNIE F. SPAGNOLO AND BRENDA G. HOGUE

Department of Molecular Virology and Microbiology, Baylor College of Medicine, Houston, Texas

1. INTRODUCTION

The 3' poly(A) tail plays an important, but as yet undefined role in coronavirus genome replication. To further examine the requirement for the coronavirus poly(A) tail, we created truncated poly(A) mutant defective interfering (DI) RNAs and observed the effects on replication. Bovine coronavirus (BCV) and mouse hepatitis coronavirus A59 (MHV-A59) DI RNAs with tails of 5 or 10 A residues were replicated, albeit at delayed kinetics as compared to DI RNAs with wild type tail lengths (>50 A residues). A BCV DI RNA lacking a poly(A) tail was unable to replicate; however, a MHV DI lacking a tail did replicate following multiple virus passages. Poly(A) tail extension/repair was concurrent with robust replication of the tail mutants. Binding of the host factor poly(A)- binding protein (PABP) appeared to correlate with the ability of DI RNAs to be replicated. Poly(A) tail mutants that were compromised for replication, or that were unable to replicate at all exhibited less *in vitro* PABP interaction. The data support the importance of the poly(A) tail in coronavirus replication and further delineate the minimal requirements for viral genome propagation.

The Nidoviruses (Coronaviruses and Arteriviruses).
Edited by Ehud Lavi *et al.*, Kluwer Academic/Plenum Publishers, 2001.   467

2.   BACKGROUND

The coronavirus 5' and 3' untranslated regions (UTRs) are necessary for genome replication. The 3' terminal 55 nucleotides of the 3' UTR and the poly(A) tail comprise the cis-acting signal for negative strand synthesis (Lin *et al.*, 1994). Defective genome positive strand synthesis requires UTRs from both the 5' and 3' ends of the genome (Kim *et al.*, 1993; Lin and Lai, 1993). The precise roles of these UTRs during replication are unknown, but it is likely that they function to recruit viral factors and possibly cellular proteins for formation of the viral replication complex.

The poly(A) tail has been shown to be important for replication of several positive-strand viruses including polio (Sarnow, 1989; Spector and Baltimore, 1974), EMCV (Cui *et al.*, 1993), and Sindbis (Hill *et al.*, 1997). MHV RNAs contain poly(A) tails of approximately 100-130 nt (Lai *et al.*, 1981). The mechanism by which the tail is added to coronavirus RNA is not known, though the presence of a 9-26 nt poly(U) tract at the 5' end of BCV negative-strand RNAs alludes to the likelihood that negative-strand synthesis initiates within the poly(A) tail (Hofmann and Brain, 1991). It remains to be determined how the tail is extended beyond the templated 9-26 nt. Irrespective of the mechanism of tail addition, the presence of a poly(A) tail of undetermined length is required for negative strand synthesis, the initial step of genome replication (Lin *et al.*, 1994).

To further investigate the requirement for the poly(A) tail in coronavirus replication, we sought to determine the effect of poly(A) tail truncation on DI RNA replication. DI replication was monitored over several virus passages and visualized by Northern blotting of total intracellular RNA. Repair of truncated poly(A) tails over the course of the experiment was examined by RT-PCR. We recently showed that PABP interacts with the poly(A) tail on group II coronavirus RNA (Spagnolo and Hogue, 2000). To determine whether binding of the host factor PABP correlated with the replication phenotypes of the mutants, a streptavidin-capture assay was developed to measure interactions with PABP *in vitro*.

3.   MATERIALS AND METHODS

Defective genome replication, PABP-3' UTR binding, and RT-PRC/poly(A) tail repair assays were all described in detail previously (Spagnolo and Hogue, 2000).

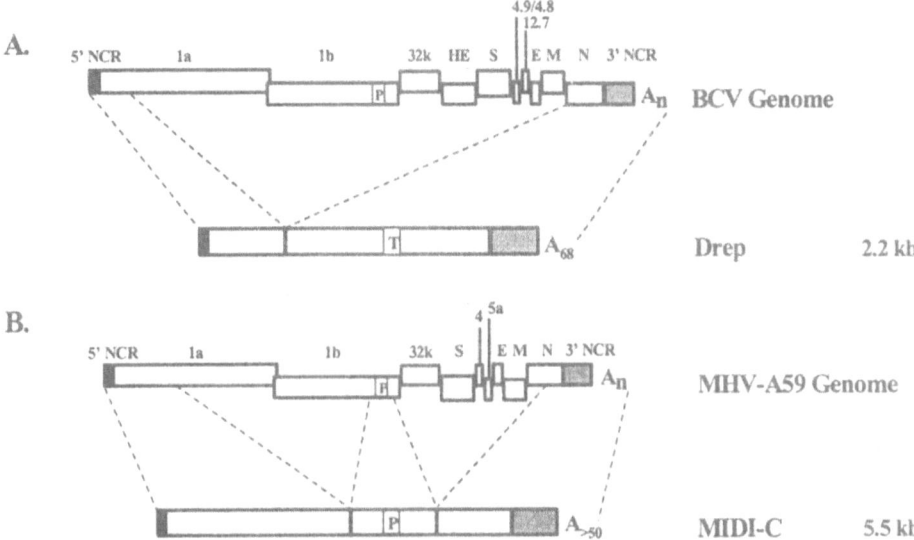

Figure 1. Schematic of BCV and MHV-A59 wt and defective genomes. The genes encoded by the open reading frames (white boxes) are labeled. The leader and 5' UTR are shown in black and the 3' UTR is shown crosshatched. Sequences found in BCV Drep (Chang *et al.*, 1994) and MHV MIDI-C (de Groot *et al.*, 1992) defective genomes are aligned with the position of origin in the parental genome and are indicated by dotted lines. Drep contains a 30 nt TGEV (T) reporter sequence. "P" indicates the position of the MHV and BCV packaging signals.

To determine the effect of poly(A) tail truncation on DI RNA replication, Drep and MIDI-C RNAs were constructed that contained poly(A) tails of 0/1, 5, or 10 A residues. *In vitro* transcribed, capped RNAs were transfected into mock- and BCV-infected HCT cells or MHV-infected 17Cl1 cells. Passage of virus supernatants onto new cell monolayers was performed at 24 h timepoints for BCV, and 12 h timepoints for MHV. DI replication was monitored over four virus passages by Northern blot analysis. Little to none of the transfected DI RNAs persisted in mock- infected HCT (Fig. 2A lanes 2-3) or 17Cl1 cells (Fig. 2B lanes 3-4). All Drep RNAs were present at 24 h for BCV (Fig. 2A lane 5); however, only Drep A_5, A_{10}, and A_{wt} were replicated upon virus passage (Fig. 2A lower panels, lanes 6-9). Drep A_1 was not replicated, or underwent replication so inefficiently that accumulation of this RNA was not detected by Northern blotting (Fig. 2A, upper panel, lanes 6-9). As was seen for Drep, MIDI-C A_5, A_{10}, and A_{wt} were clearly replicated and amplified upon virus passage (Fig. 2B, lower panels, lanes 7-10). Interestingly, MIDI-C A_0 accumulation was also observed by P3 or P4 (Fig 2B, upper panel, lanes 7-10).

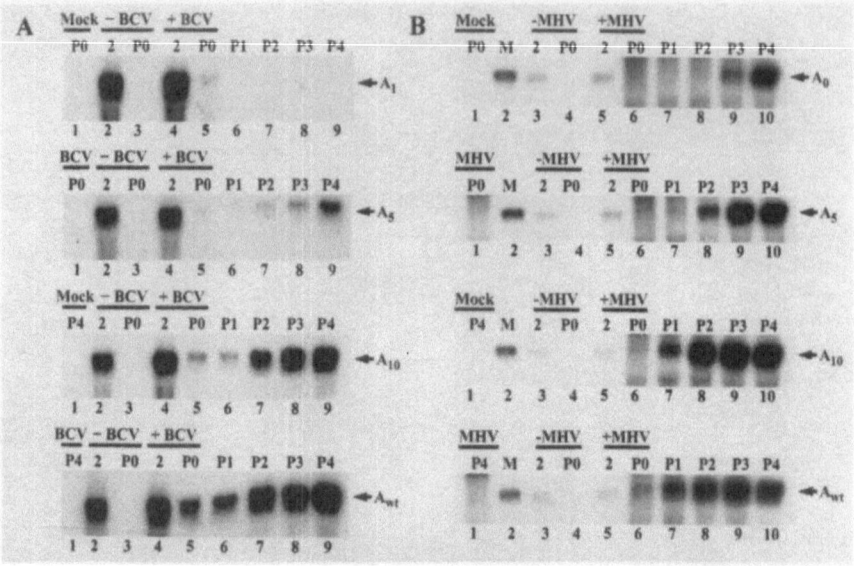

Figure 2. Replication of BCV and MHV defective genomes containing varying lengths of poly(A) tail. Mock- and BCV-infected HCT cells (A) or MHV-infected 17Cl1 cells (B) were transfected with Drep (A) or MIDI-C (B) RNAs. Total intracellular RNA was extracted, resolved on denaturing agarose gels, and vacuum blotted onto nylon membranes. Blots were probed with a 5'-end-labeled oligonucleotide complementary to the TGEV reporter sequence in Drep (A) or an MHV N riboprobe (B). P1-P4 indicates virus passage numbers. Virus passages were performed at 24 h intervals for BCV and 12 h for MHV. M denotes marker RNA (B).

To rule out any possibility that the observed differences in replication could be due to differences in RNA decay rates, stability of the transfected DI RNAs was monitored over a 12-24 h period following transfection by Northern blotting. All RNAs were comparable in stability and exhibited similar decay rates (Spagnolo and Hogue, 2000). Drep RNAs had a half-life of 10 h, and MIDI-C RNAs had a half-life of approximately 5.5 h. Quantitation of this data was performed by phosphorimaging. It therefore did not appear that any striking difference in RNA stability could account for the differences in replication competency of the mutant DI RNAs.

To determine whether binding of the host factor PABP better correlated with DI replication phenotypes, an assay was developed to detect PABP interactions with the BCV 3' or MHV3' UTRs containing various lengths of poly(A) tail. ^{35}S labeled PABP, or luciferase as a negative control, was incubated with biotinylated probes in the presence of nonspecific competitors. Immobilized streptavidin was added to recover the biotinylated RNA-protein complexes. Half of each sample was analyzed for recovery of radiolabeled protein by SDS-PAGE. The other half was monitored for comparable recovery of all biotinylated RNAs. No binding above background was detected between luciferase and any of the 3' UTR poly(A) tail RNAs. Binding of PABP to BCV 3' UTR A_5, A_{10}, and A_{wt}, and to MHV 3' UTR A_{10} and A_{wt} was detected (Fig. 3). Folding of the 3' UTR A_5 may account for lack of PABP binding. This data demonstrates that binding of PABP to 3' UTR RNAs correlates with the ability of the DI poly(A) tail mutants to be replicated.

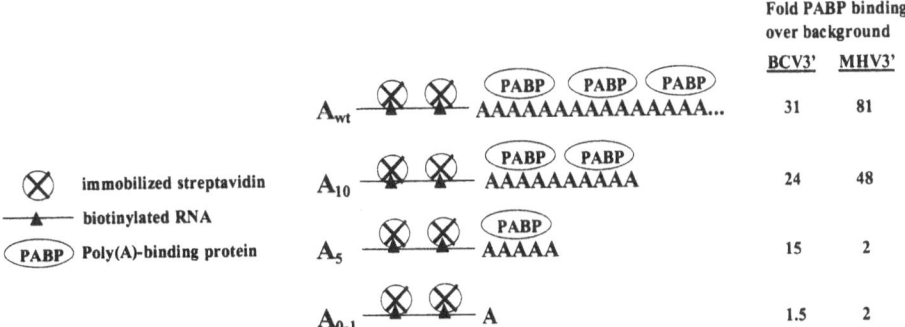

Figure 3. Schematic representation of *in vitro* binding of PABP to coronavirus 3' UTR RNAs. *In vitro* translated PABP was incubated with biotinylated BCV3'UTR or MHV3'UTR RNAs containing poly(A) tails of 0-1, 5, 10, or wt (>50) A-residues. Immobilized streptavidin was added to recover biotinylated RNA complexes, and samples were washed to remove any unbound RNA or protein. Samples were analyzed on 8% SDS-PAGE. PABP binding was quantitated by phosphorimaging.

The observation that MIDI-C A_0 was competent for replication by P3 or P4 led us to speculate that MIDI-C A_0 had undergone poly(A) tail repair over the duration of the experiment and thus became a suitable template for replication. To determine whether this was indeed the case, an RT-PCR assay was utilized. DI RNAs containing poly(A) tails of >10 A residues produce an 850 bp PCR product, whereas RNAs with tails $\leq$10 A residues are not amplified. Intracellular RNAs from the experiment shown in Fig. 2 were subject to RT-PCR analysis. Table 1 represents the results of these experiments. "+" denotes the appearance of the predicted 850 bp RT-PCR product, "-" indicates failure to amplify this product. MIDI-C A_{wt}, whose poly(A) tail is well over 10 A residues, yielded the 850 bp product from all virus passages, as expected. RT-PCR products for the poly(A) tail mutants were observed simultaneously with the appearance of robust replication of the DI RNAs (compare Table 1 with Fig. 2). These data demonstrate that *in vivo*, MIDI-C A_0, A_5, and $_{10}$ underwent poly(A) tail repair and were replicated.

Table 1. Appearance of RT-PCR product indicating poly(A) repair

Transcript	P0	P1	P2	P3	P4
MIDI-C A_0	--	--	--	++	+++
MIDI-C A_5	--	-/+	++	+++	+++
MIDI-C A_{10}	--	++	+++	+++	+++
MIDI-C A_{wt}	++	++	+++	+++	+++

5.   CONCLUSION

Our understanding of the molecular interactions required for coronavirus genome replication remains largely incomplete. In an effort to further elucidate the requirements for this process, we have begun to more closely examine the coronavirus 3' UTR and the poly(A) tail. In this report, we confirmed previous results indicating that the poly(A) tail is a cis-acting signal for coronavirus replication, and further showed that a poly(A) tail of 5 A residues is sufficient for initial DI replication. *In vitro* binding of the host factor PABP appeared to correlate with the ability of DI RNAs to be replicated.

Our data showed that poly(A) tail truncation compromised the ability of DI RNAs to be replicated, as DIs with truncated tails did replicate, but exhibited a decrease in the overall amount of replication. Poly(A) tail deletion resulted in failure of BCV Drep A_1 RNA to be replicated, whereas MHV MIDI-C A_0 was able to overcome any block to replication by passage three or four. RNA stability did not account for the differences in replication. It appears that BCV and MHV differ in their ability to deal with poly(A)-lacking DIs. MIDI-C A_0 underwent poly(A) tail extension during the course of the replication experiment and regained replication competency, whereas Drep A_1 became extinct. If repair occurs by a virus-specific mechanism, then MHV is able to repair MIDI-C A_0's defect in the four virus passages we have examined, whereas BCV is unable to do so for Drep A_1. We are currently trying to determine how repair occurs. It is likely due to recombination with helper virus, and if this is the case, would suggest that MHV may undergo recombination more readily than BCV.

Binding of the host factor PABP appears to correlate with the ability of the mutant DI RNAs to be replicated. What roles may PABP and the coronavirus poly(A) tail be playing during coronavirus replication? PABP and the poly(A) tail likely exert an initial effect on genome translation immediately after infection. PABP interactions with the coronavirus poly(A) tail and eIF4G, which is presumably bound to the coronavirus genome 5' cap, likely mediate 5'-3' end interactions and may facilitate genome translation. As translation appears to be required for efficient genome replication (Chang and Brian, 1996; van der Most *et al.*, 1995), poly(A) lacking DIs may be compromised in translation efficiency, and thereby do not replicate as well, or at all.

The coronavirus poly(A) tail likely plays an additional, direct role in genome replication by serving as part of the negative strand promoter. BCV negative strand RNA contains a 9-26 poly(U) tract, suggesting that negative strand RNA synthesis likely initiates within the poly(A) tail (Hofmann and Brain, 1991). DIs with shorter or nonexistent tails may therefore not be as efficient at or capable of being templates for negative strand RNA synthesis. Finally, initial interactions mediated in part by PABP that presumably juxtapose the 5' and 3' ends of coronavirus RNA may be important for building the replicase complex and may explain the requirement of sequences from both 5' and 3' ends of the RNA for positive strand synthesis.

ACKNOWLEDGMENTS

This work was supported by Public Health Service NIH grant AI33500 to B.G.H. from the National Institute of Allergy and Infectious Diseases. J.F.S. was supported in part by training grant AI07471 from the National Institutes of Health.

REFERENCES

Chang, R.Y. and Brian, D.A.(1996). cis Requirement for N-specific protein sequence in bovine coronavirus defective interfering RNA replication. *J Virol* **70**, 2201-2207.

Chang, R.Y., Hofmann, M.A., Sethna, P.B., and Brian, D.A.(1994). A cis-acting function for the coronavirus leader in defective interfering RNA replication. *J Virol* **68**, 8223-8231.

Cui, T., Sankar, S., and Porter, A.G.(1993). Binding of encephalomyocarditis virus RNA polymerase to the 3'-noncoding region of the viral RNA is specific and requires the 3'-poly(A) tail. *J Biol Chem* **268**, 26093-26098.

de Groot, R.J., van der Most, R.G., and Spaan, W.J.(1992). The fitness of defective interfering murine coronavirus DI-a and its derivatives is decreased by nonsense and frameshift mutations. *J Virol* **66**, 5898-5905.

Hill, K.R., Hajjou, M., Hu, J.Y., and Raju, R.(1997). RNA-RNA recombination in Sindbis virus: roles of the 3' conserved motif, poly(A) tail, and nonviral sequences of template RNAs in polymerase recognition and template switching. *J Virol* **71**, 2693-2704.

Hofmann, M.A. and Brian, D.A.(1991). The 5' end of coronavirus minus-strand RNAs contains a short poly(U) tract. *J Virol* **65**, 6331-6333.

Kim, Y.N., Jeong, Y.S., and Makino, S.(1993). Analysis of cis-acting sequences essential for coronavirus defective interfering RNA replication. *Virology* **197**, 53-63.

Lai, M.M., Brayton, P.R., Armen, R.C., Patton, C.D., Pugh, C., and Stohlman, S.A.(1981). Mouse hepatitis virus A59: mRNA structure and genetic localization of the sequence divergence from hepatotropic strain MHV-3. *J Virol* **39**, 823-834.

Lin, Y.J. and Lai, M.M.(1993). Deletion mapping of a mouse hepatitis virus defective interfering RNA reveals the requirement of an internal and discontiguous sequence for replication. *J Virol* **67**, 6110-6118.

Lin, Y.J., Liao, C.L., and Lai, M.M.(1994). Identification of the cis-acting signal for minus-strand RNA synthesis of a murine coronavirus: implications for the role of minus-strand RNA in RNA replication and transcription. *J Virol* **68**, 8131-8140.

Sarnow, P.(1989). Role of 3'-end sequences in infectivity of poliovirus transcripts made in vitro. *J Virol* **63**, 467-470.

Spagnolo, J.F. and Hogue, B.G.(2000). Host protein interactions with the 3' end of bovine coronavirus RNA and the requirement of the Poly(A) tail for coronavirus defective genome replication. *J Virol* **74**, 5053-5065.

Spector, D.H. and Baltimore, D.(1974). Requirement of 3'-terminal poly(adenylic acid) for the infectivity of poliovirus RNA. *Proc Natl Acad Sci U S A* **71**, 2983-2987.

van der Most, R.G., Luytjes, W., Rutjes, S., and Spaan, W.J.(1995). Translation but not the encoded sequence is essential for the efficient propagation of the defective interfering RNAs of the coronavirus mouse hepatitis virus. *J Virol* **69**, 3744-3751.

A Simple Strategy to Assemble Infectious RNA and DNA Clones

KRISTOPHER M. CURTIS[1], BOYD YOUNT[2], AND RALPH S. BARIC[1,2]

[1]Department of Microbiology and Immunology, School of Medicine, [2]Department of Epidemiology, Program of Infectious Diseases, School of Public Health, University of North Carolina at Chapel Hill, Chapel Hill, North Carolina, 27599

1. INTRODUCTION

The availability of infectious full-length cDNA clones is important for the molecular genetic analysis of the structure and function of RNA virus genomes (Ahlquist et al., 1984; Boursnell et al., 1987). Infectious cDNA clones for a number of positive-stranded RNA viruses have been developed, advancing our understanding of the molecular mechanisms of viral replication and pathogenesis, and has resulted in novel approaches for heterologous gene expression and vaccine development. Clearly, a full-length infectious construct of TGEV would enhance our understanding of all aspects of TGEV biology by providing a means for reverse genetic analysis.

2. METHODS AND MATERIALS

2.1 Mutagenesis, Cloning, and Assembly of a Full Length TGE Infectious Clone

The TGEV genome (Purdue strain, ATCC VR-763) was cloned from infected swine testicular cells (ST) and PCR mutagenesis was utilized to

The Nidoviruses (Coronaviruses and Arteriviruses).
Edited by Ehud Lavi et al., Kluwer Academic/Plenum Publishers, 2001.

generate unique junction sites at the 5' and 3' end of each cloned fragment (Figure 1). Standard recombinant DNA techniques were used to remove unwanted amino acid changes associated with reverse transcription or naturally occurring quasispecies variation and a consensus sequence was determined (Yount et al., submitted).

We isolated five cDNA subclones spanning the entire TGEV genome (designated TGEV A, B, C, DE and F) (Figure 1). The A fragment contains a unique 5' T7 promoter and the F fragment contains a 3' poly-T tail. Appropriately sized cloned fragments were isolated as described by Yount et al.

The pTGEV A, C, DE and F clones were stable in plasmid DNA's in *E.coli.* However, the B fragment was unstable and only a few slow growing isolates were obtained, all of which contained deletions or insertions in the wildtype sequence at a region consistent with those noted by other investigators (position 9973) (Denison et al., 1999; Eleouet et al., 1995). A unique BstxI site was engineered into the unstable region of the B fragment by primer-mediated mutagenesis (Yount et al., submitted) and used to bisect this fragment into B1 and B2 fragments (Figure 1). The region of instability, as suggested by locations of deletions and insertions, maps in or near the TGEV poliovirus 3C-like protease (3-Clpro) motif and may be bactericidal or unstable in microbial cloning vectors.

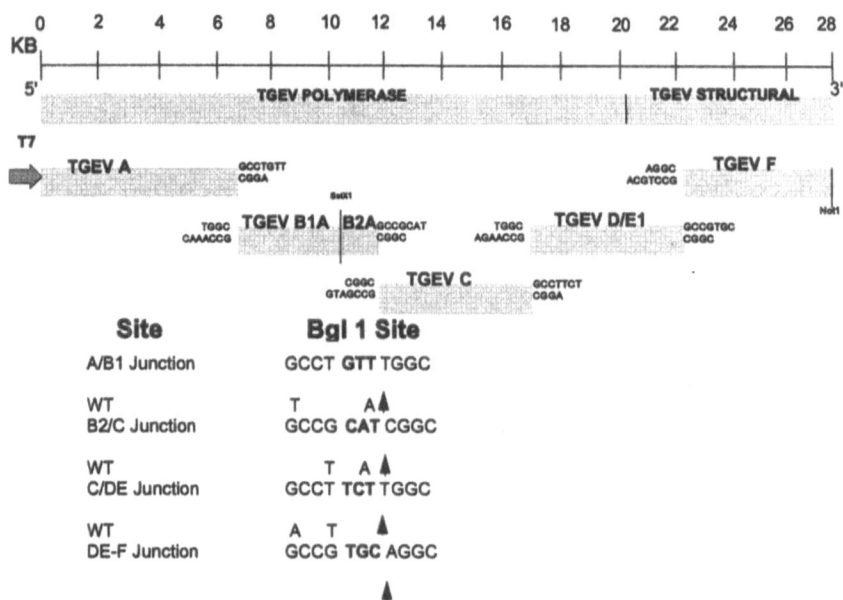

Figure 1. Assembly Strategy of TGEV Clone. Using RT-PCR and unique oligonucleotide primer mutagenesis, five clones spanning the entire TGEV genome were isolated using standard recombinant DNA techniques.

A full-length cDNA construct was assembled in two ligation steps. Each cDNA clone is flanked by unique Bgl1 sites and will only assemble with the appropriate adjacent subclone. The TGEV A and B1, B2 and C, and the DE-1 and F fragments were ligated overnight in the presence of T4 DNA ligase (Sambrook et al., 1989). Directionally assembled products were isolated from agarose gels and the TGE A/B1, B2/C3-2, DE-1/F fragments religated overnight. The final products were purified and capped-T7 transcripts were synthesized and analyzed as described by Yount et al.

2.2 Transfection, Virus Isolation, and Identification of Marker Mutations

Baby Hamster Kidney (BHK) cells were transfected with TGEV transcripts alone, TGEV + TGEV N gene transcripts, or just TGEV N gene transcripts and co-cultured with ST cells. Viral progeny were isolated as described by Yount et al.

Mutations introduced into the TGEV genome in order to engineer unique Bgl1 and BstX1 restrictions sites were identified in TGEV 1000 derived virus by standard molecular biology techniques as described by Yount et al.

3. DIRECTIONAL ASSEMBLY OF LARGE GENOMES AND CHROMOSOMES

Conventional restriction enzymes, such as Not1 and EcoR1, cleave identical DNA on average every 65,000 bp (Sambrook et al., 1989) and leave sticky ends that assemble with similarly cut DNA fragments in the presence of DNA ligase. Because of this, they rarely are appropriate choices for assembling large intact genomes or chromosomes. However, a second subclass of restriction enzymes (i.e. Bgl1, BstX1) also recognize palindrome sequences but leave random sticky ends of one to four nucleotides in length that are not complementary to most other sticky ends generated with the same enzyme at other sites in the DNA. Because a three nucleotide variable overhang is generated following cleavage with Bgl1, 64 different variable ends can be generated, which assemble only with the appropriate 3 nucleotide complementary overhang generated at an identical Bgl1 site. We hypothesized that a sequential series of smaller DNA subclones that are flanked by unique Bgl1 junctions could be directionally assembled into an intact, full-length infectious cDNA clone. To test this hypothesis, we assembled a full-length infectious clone of a coronavirus, thereby demonstrating its potential applications for assembling other large genomes or chromosomes *in vitro*.

4.   TRANSFECTION AND RECOVERY OF INFECTIOUS VIRUS

The efficiency of electroporation of full length TGEV transcripts into BHK cells was low, as indicated by indirect fluorescent assay (IFA), and only evident when electroporated with TGEV N gene transcripts (Figure 2). Supernatants from transfected BHK/ST cell co-cultures were passaged onto ST cells and CPE was observed at each passage only with supernatants derived from the TGEV+N transfected cultures (Figure 3). Virus isolated from these cultures formed plaques in agar (Figure 4). In addition, no significant differences in the replication of wildtype TGEV or TGEV 1000 derived viruses were noted in ST cells, and all viruses replicated to titers that approached 1×10^8 PFU/ml within 28 hrs (data not shown).

Figure 2. BHK Cell Electroporation Efficiency as Indicated by IFA. BHK cells were electroporated with TGEV transcripts alone, TGEV + TGEV N gene transcripts, or TGEV N gene transcripts alone.

4.1   Identification of Marker Mutations

Infectious virus derived from transfected cultures should each contain the four unique interconnecting junction sequences used in the construction of the infectious construct (Figure 1). Results from restriction fragment length polymorphism (data not shown) and sequencing analysis (Figure 5) of virus derived from passage one (icTGEV-1) and passage three (icTGEV-3) demonstrated that the appropriate marker mutations were absent from the wildtype TGEV genome and present in TGEV 1000 derived virus.

Figure 3. CPE typical of TGEV infection is evident at each passage. (A) uninfected ST cells, (B) passage 1, (C) passage 2, (D) passage 3, (E) passage 4

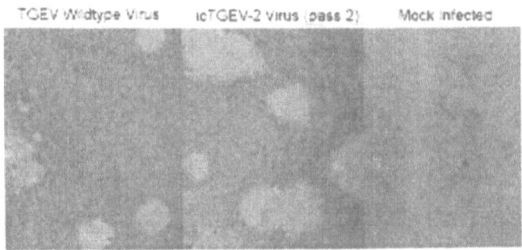

Figure 4. Plaque Assay. TGEV wildtype virus and TGEV 1000 derived virus were plaque assayed on ST cells. Mock infected cells were included as a negative control.

5.   CONCLUSION

Until recently, full length infectious clones have not been assembled because of size constraints in bacterial cloning vectors, regions of coronavirus chromosomal instability/toxicity in bacterial vectors, the requirement for a vector system which allows for simple reverse genetic applications, and the inability to drive full length transcripts *in vitro*. In this manuscript, we describe a simple and rapid approach to systematically assemble a full-length infectious coronavirus cDNA from a panel of six smaller subclones. Directional assembly methods should be appropriate for constructing full-length infectious constructs of other large RNA viruses, such as filoviruses and other coronaviruses, DNA genomes of herpesviruses, adenoviruses, and poxviruses, and those viral genomes which are unstable in prokaryotic vectors (Boursnell et al., 1987; Delmas et al., 1992; Peters et al.,

1996). In addition, the systematic assembly of large chromosomes or minichromosomes approaching several million base pairs in length is theoretically possible.

Clone	Junction	Sequence
icTGE-1	B2/C	TCCAGC<u>C</u>GCATC<u>G</u>GCTACA
wildtype		T A
icTGE-1	C/DE	CAAGGCCTT<u>C</u>TT<u>G</u>GCACAT
wildtype		T A
icTGE-1	DE/F	AGTAGC<u>C</u>GT<u>G</u>CAGGCTAGA
wildtype		A T
icTGE-3	B2/C	TCCAGC<u>C</u>GCATC<u>G</u>GCTACA
wildtype		T A
icTGE-3	C/DE	CAAGGCCTT<u>C</u>TT<u>G</u>GCACAT
wildtype		T A
icTGE-3	DE/F	AGTAGC<u>C</u>GT<u>G</u>CAGGCTAGA
wildtype		A T

Figure 5. Sequence analysis of marker mutations at fragment junctions of clones at passage 1 and 3. Marker mutations are underlined and wildtype nucleotides are shown.

ACKNOWLEDGMENTS

We would like to thank Robert E. Johnston, Nancy Davis, Patrick Harrington, Mary Schaad and Larwence Parks for helpful discussion and encouragement. This work was supported by a research grant from the National Institutes of Health (AI 23946).

REFERENCES

Ahlquist, P., French, R., Janda, M., and Loesch-Fries, L.S. 1984. Multicomponent RNA plant virus infection derived from cloned viral cDNA. Proc. Natl. Acad. Sci. USA 81, 7066-7070.

Boursnell, M.E., Brown, T.D., Foulds, I.J., Green, P.F., Tomley, F.M., and Binns, M.M. 1987. Completion of the sequence of the genome of the coronavirus avian infectious bronchitis virus. J. Gen. Virol. 68, 57-77.

Delmas, B., Gelfi, J., L'Haridon, R., Vogel, L.K., Sjostrom, H., Noren, O., and Laude, H.

1992. Aminopeptidase N is a major receptor for the enteropathogenic coronavirus TGEV. Nature 357:417-420.

Denison, M.R., Spaan, w.J., van der Meer, Y., Gibson, C.A., Sims, A.C., Prentice, B., and Lu, X.T. 1999. The putative helicase of the coronavirus mouse hepatitis virus is processed from the replicase gene polyprotein and localizes in complexes that are active in viral RNA synthesis. J. Virol 73: 6862-6871.

Eleouet, J.F., Rasschaert, D., Lambert, P., Levy, L., Vende, P., and Laude, H. 1995. The complete sequence (20kb) of the polyprotein-encoding gene 1 of transmissible gastroenteritis virus. Virology 206, 817-822.

Lee, H.J., Shieh, C.K., Gorbalenya, A.E. et al., 1991. The complete nucleotide sequence of murine coronavirus gene 1 encoding the putative proteases and RNA polymerase. Virology 180:567-582.

Peters, C.J., Sanchez, A., Rollin, P.E., Ksiazek, T.G. and Murphy, F.A. 1996. Filoviridae: Marburg and Ebola Viruses. In: Fields Virology, eds. Fields, B.N., Knipe, D.M. and Howley, P.M. Lippincott Williams and Wilkens, Philadelphia, Pa. pgs, 1161-1176.

Sambrook, J., Fritsch, E.F. and Maniatis, T. 1989. Enzymes used in Molecular Cloning. In: Molecular Cloning, A laboratory Manual, 2nd edition. Cold Spring Harbor Laboratory Press, Plainview, New York. pgs 5.1-5.31

Siddell, S.G. 1995. The coronaviridae, An introduction. In: The coronaviridae, eds. S.G. Siddell, Plenum Press, New York. Pgs 1-10.

Yount, B., K. M. Curtis, and R. S. Baric. A simple strategy for the systematic assembly of large RNA and DNA genomes: the transmissible gastroenteritis virus model. submitted

Chromatography of Mouse Hepatitis Virus Replicative Intermediate and Replicative Form RNA

DOROTHEA L. SAWICKI AND STANLEY G. SAWICKI

Department of Microbiology and Immunology, Medical College of Ohio, 3055 Arlington Avenue, Toledo, OH

1.   INTRODUCTION

One of our goals is to obtain large amounts of selected size classes of the coronavirus replicative intermediates (RI) and native replicative form (RF) RNA that contain the templates for viral genome and messenger RNA syntheses. The two replicative molecules differ in their relative amounts of single-stranded and double-stranded character, with the RF RNA considered essentially or completely double-stranded. It is known that there are seven species of RI RNA, each of which is proportional in size and abundance to one of the seven species of viral messenger RNA produced in infected cells (Sawicki and Sawicki 1990; 1998; Schaad and Baric 1994; Baric and Yount 2000). The presence in coronavirus-infected cells of seven viral minus strand templates was first reported by Sethna and Brian (1989). To assist in this effort, we undertook an analysis of how mouse hepatitis virus (MHV-A59) RI and RF RNA fractionate on three gel filtration matrices, Sepharose 2B, Superose 6 and Sephacryl S-1000. A second goal was to attempt to explain the failure by others to find subgenomic RI and their RF cores after ribonuclease treatment and gel filtration chromatography.

The Nidoviruses (Coronaviruses and Arteriviruses).
Edited by Ehud Lavi *et al.*, Kluwer Academic/Plenum Publishers, 2001.

2.   RESULTS

17Cl-1 cells were infected with MHV-A59 at an MOI of 65-80 and continuously labeled in the presence of 20 µg of actinomycin D/ml with ^{32}P-orthophosphate in phosphate-reduced medium (containing 1% normal levels) from 1.5-5 h p.i. The cells were harvested into buffer containing 5% lithium dodecyl sulfate and 200 µg of proteinase K/ml.   The RNA was deproteinized with acid phenol and chloroform, and collected by ethanol precipitation.   The total infected cell RNA was applied to a 1.5 x 90 cm column of Sepharose 2B (Pharmacia), as described in earlier studies (Sawicki and Gomatos 1976). Viral RNA was both excluded from the column, indicating these were equal to or larger than the exclusion limit of 1.3 kilobase-pairs of DNA or about 1 x 10^6 Daltons, and included or within the fractionation range of the column (Fig 1).   RNA in each fraction was collected by ethanol precipitation, digested with 10-30 units of ribonuclease T1/sample and analyzed on 1% agarose/TBE gels. Seven species of double-stranded RI cores or RF were found (Fig 1). The results indicate the largest three RIs are excluded by this matrix and the four smallest fractionate by size as they elute in order.   However, they elute broadly, over a large number of fractions, and different RI species elute together (overlap).

In a second approach, we combined velocity sedimentation and chromatography.   Total infected cell RNA was applied to a 15 to 30% sucrose gradient and sedimented to pellet single-stranded RNA 1 and place 28S ribosomal RNA in fraction 6 of 30 fractions and 18S ribosomal RNA in fraction 15 of 30, the middle of the gradient.   RNA in each fraction was collected, digested with ribonuclease T1 and then electrophoresed on agarose gels.   The results indicate velocity sedimentation separated genomic RF 1 from smaller RF cores, which sedimented according to size but again overlapped extensively (Sawicki, manuscript in preparation).   These were collected into 4 pools, numbered starting from the bottom of the gradient, and pools 2 and 4 were individually chromatographed on Sephacryl S-1000 columns (data not shown). Sephacryl S-1000 is a hydrophobic, rigid, allyl dextran/N-N'-methylenebisacrylamide matrix (Pharmacia) with a 20 kilobase-pair exclusion limit. To summarize the results, pool 2 RNA (~40S to 30S) was totally excluded by the column and contained the three largest RI cores, RF I-III. Early fractions contained only RF I; later fractions were enriched in RF II and RF III. Pool 4 RNA (16S to 10S) gave 2 peaks: a minor, excluded one containing small amounts of the three largest RF cores and a major, included peak containing the four smallest cores, RF IV-VII. Sephacryl S-1000 gave a better profile for the four smallest RF molecules, which eluted very narrowly and by size, than Sepharose 2B. But, even on this matrix, elution of each RF overlapped with that of at least one other.

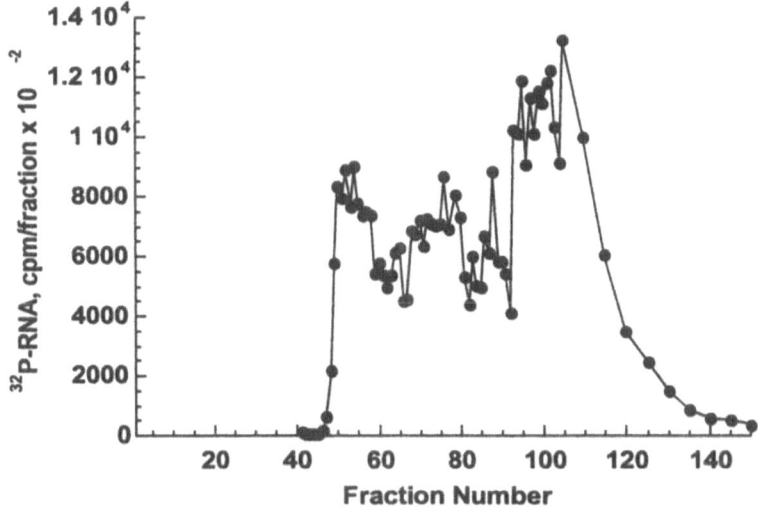

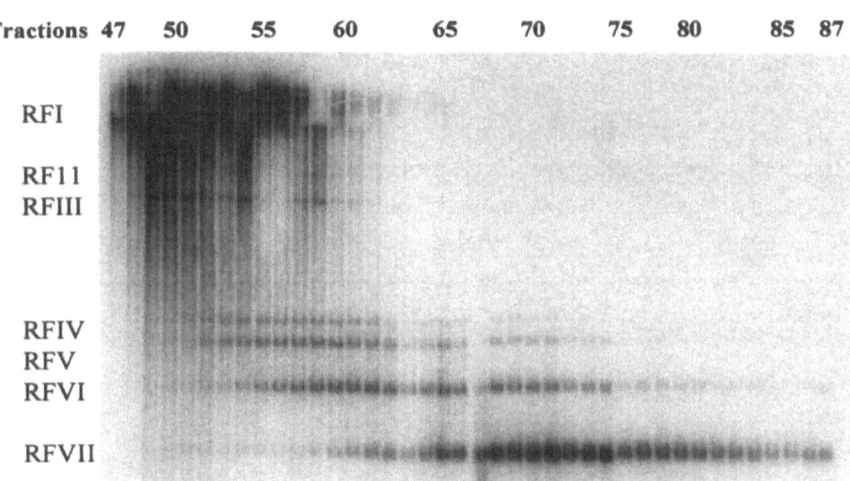

Figure 1. Chromatography on Sepharose 2B of MHV-infected cell RNA labeled with [32]-P-orthophosphate from 1.5-5 h p.i.. RNA larger than 3 million Daltons eluded in fractions 47-60 (excluded region). RNA in fractions 47-87 was analyzed by gel electrophoresis.

The elution profile of these molecules by FPLC on a Superose 6 column, a cross-linked agarose matrix similar to Sepharose 2B but having a particle size of 10-13 microns versus 60-200 microns for Sepharose 2B, resembled that of Sepharose 2B (Fig 2). The shorter column length, higher pressure during FPLC chromatography and/or smaller particle size may contribute to the observed retention of the three largest RF, especially RF I, and its elution even in fractions beyond those containing RF VII molecules that are 5% the size of RF I. Only Sephacryl S-1000 of the three matrices tested was able to include and fractionate all subgenomic RF. Of special note, comparison of the gel filtration results indicated coronavirus RI and RF RNA chromatographed close to what was predicted from the fractionation properties of each matrix.

2.1   There are Seven Native RI and Seven Native RF RNA

We next obtained native RF RNA. Native RF are soluble in 2M LiCl from their having no or very little single-stranded character. In contrast, native RIs have extensive single-stranded character and precipitate in high salt, as do single-stranded mRNA and ribosomal RNA. Infected cells were incubated continuously in medium containing 200 µCi of ^{3}H-uridine/ml and actinomycin D from 1.5-5.5 h p.i. to obtain RF and RI RNA labeled in both their plus and minus strand components. After deproteinization with acid-phenol, the total infected cell RNA was adjusted to 2M with LiCl, placed on ice for 18 hours, and centrifuged at 10,000 rpm for 1 hour to obtain the supernatant (LiCl soluble fraction) and precipitated (LiCl insoluble) fractions. The LiCl soluble RNA was collected by ethanol precipitation and directly chromatographed on Sepharose 2B. To summarize, three peaks of labeled RNA were found, the largest of which was excluded. Fractions were pooled and each pool was analyzed on gels without nuclease treatment. Pool 1 was the excluded peak, and pools 3 and 4 were the leading and trailing regions of the included first peak, respectively. Pool 1 was found to contain all of the largest three RF classes and small amounts of RF IV-VI. Pool 3 was enriched in the four smallest RF classes, and pool 4 contained predominantly RF VII. Thus, infected cells contain seven species of native RF RNA.

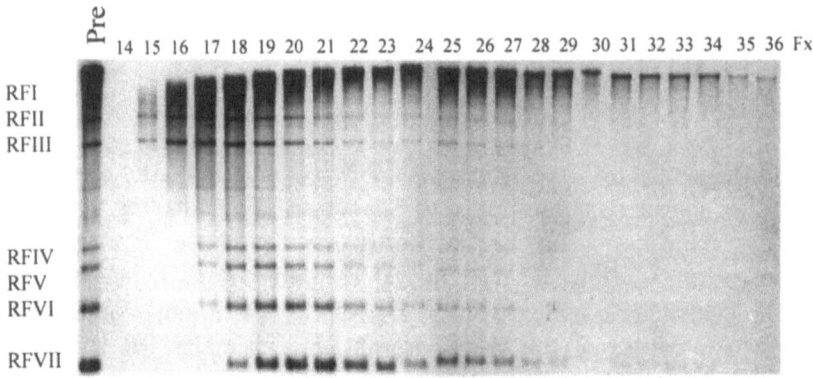

Figure 2. FPLC Chromatography of MHV-A59 RF RNA on Superose 6 gel filtration.

The LiCl insoluble fraction was also chromatographed on Sepharose 2B. Its amount of excluded RNA was >10 times greater than in the LiCl soluble fraction. The included RNA peak displayed a large and heterogeneous range of sizes, as expected for the seven species of viral mRNAs. RNA in alternate fractions was analysed on agarose gels directly to identify the elution position of each viral RNA species (fractions of 0.65, half the size of the LiCl soluble analysis, were collected for the LiCl insoluble RNA to attempt to obtain better resolution). Ethidium bromide staining of the gel indicated where 28S and 18S ribosomal RNA eluted. The results showed that genome or single-stranded RNA 1 eluted in the excluded volume as did the largest native RI, which was visible at the top of the gel in fractions 60-65. It was especially noted that RI 1 migrates slower than its single-stranded product on agarose gels, as do all of the RI or RF RNA relative to their single-stranded counterparts (Sawicki and Sawicki 1990). Eluting in an overlapping pattern, according to their relative sizes, and after the genome RNA were single-stranded mRNA 2-7. Viral mRNA 2-5 overlapped the elution of 28S ribosomal RNA, and mRNA 4-7 overlapped the 18S ribosomal RNA (Sawicki, in preparation).

About every 5 fractions of the Sepharose 2B fractionated LiCl insoluble RNA were pooled (numbered 1-20 beginning with the excluded fractions), treated with ribonuclease T1 and analyzed on gels to identify the elution profile of individual native RI classes. Pool 1 RNA contained most of the three largest RI classes and some single-stranded RNA, accounting for its high (49%) resistance to digestion when assayed for acid-precipitability after digestion with 5 µg of pancreatic ribonuclease/ml in 0.3 M NaCl or 2 x SSC. Pools 6-7 were enriched in the smallest, RI VII, and the other RI eluted in pools 2-5. We found the seven, native RI classes separated similarly to native RF RNA on Sepharose 2B and each eluted ahead of its single-stranded RNA counterpart, as expected.

2.2   Ribonuclease-resistance of Native RF and RI RNA

We also determined that native RF were fully double-stranded and that native RI were partially double-stranded. Digestion with 10-100 units of ribonuclease T1 left all seven RF species intact as judged by their identical migration on gels to untreated RF species. And, essentially 100% of labeled, native RF RNA was acid-precipitable after digestion with either ribonuclease T1 (30 units/sample) in 0.3 M NaCl at 30°C for 30 min or ribonuclease A (5 µg/ml) in 0.3 M NaCl or 2xSSC at 37°C for 30 min. While ribonuclease T1 treatment left native RF intact or generated RF cores from native RI RNA, overdigestion with ribonuclease A degraded coronavirus RI and RF molecules (Sawicki, in preparation). Ribonuclease in excess of 1 µg/ml destroyed large and midsize RF or RI RNA; higher amounts degraded small RF or RI species.

3.   CONCLUSIONS

Seven RF and RI molecules are readily detected after velocity sedimentation or after chromatography.   Sepharose 2B excluded RI I-III and fractionated the others by size; Sephacryl S-1000 excluded mainly RI/RF I and gave the best fractionation of all subgenomic RI and RF RNA. However, even Sephacryl S-1000 did not eliminate the overlap of the two mid-sized molecules with each other, and of the four smallest RI/RF molecules with each other.   Thus, electrophoresis on gels remains the best method to purify individual size-classes of native RF or RI RNA (Sawicki and Sawicki, 1990). Combining sedimentation and gel filtration provided enriched populations of large (I or I-III) size-classes or small (IV-VII) size-classes of RF and RI RNA.

The initial failure to find subgenomic RI RNA led to the leader–primed transcription model (Baric *et al* 1983).   Our results indicate this failure was likely due to technical error and not to any aberrant behavior of these molecules on gel filtration chromatography. Concentrations of ribonuclease A used by others of 10 µg/ml (Baric *et al* 1983) and 20 µg/ml (Lai *et al* 1982) destroyed viral RI and native RF molecules. Moreover, RF I migrates on gels slower than genome RNA 1, not faster as reported (Baric *et al* 1983), indicating that genome RF I was not recovered under conditions used by Lai and coworkers in 1982 and 1983.   Evidence arguing against the leader-primed transcription model is now extensive and includes work by our lab and that of Eric Snijder and Willy Spaan (van Marle *et al* 1999), of David Brian (Sethna *et al* 1989) and of Ralph Baric (Baric and Yount 2000); and this proceedings).   This evidence and the ability to explain the initial failure to find subgenomic RI and RF RNA lead us to state that the leader-primed model has been disproved.

REFERENCES

Baric, R.S., Stohlman, S.A., and Lai, M.M.C., 1983, Characterization of replicative
     intermediate RNA of mouse hepatitis virus: presence of leader RNA sequences on nascent
     chains. *J. Virology* 48:633-40.
Baric, R.S., and Yount, B., 2000, Subgenomic negative-strand RNA function during mouse
     hepatitis virus infection. *J. Virology* 74:4039-46.
Lai, M.M.C., Patton, C.D., and Stohlman, S.A., 1982, Replication of mouse hepatitis virus:
     negative-stranded RNA and replicative form RNA are of genome length. *J. Virology*
     44:487-92.

Sawicki, D.L, and Gomatos, P.J., 1976, Replication of Semliki Forest virus: Polyadenylate in plus strand RNA and polyuridylate in minus strand RNA. *J. Virology* 20:446-64.

Sawicki, S.G., and Sawicki, D.L., 1993, Coronavirus transcription: subgenomic mouse hepatitis virus replicative intermediates function in RNA synthesis. *J. Virology* 64:1050-56.

Sawicki, S.G., and Sawicki, D.L., 1998, A new model for coronavirus transcription. *Adv. Exp. Biol. Med.* 280:215-18.

Schaad, M.C., and Baric, R.S., 1994, Genetics of mouse hepatitis virus transcription: evidence that subgenomic negative strands are functional templates. *J. Virology* 68:8169-79

Sethna, P.B., Hung, S.-L., and Brian, D.A., 1989, Coronavirus subgenomic minus-strand RNAs and the potential for mRNA replicons. *Proc. Natl. Acad. Sci USA.* 86:5626-30.

van Marle, G., van Dinten, L.C., Spaan, W.J.M., Luytjes,W. and Snijder, E.J., 1999. Characterization of an equine arteritis virus replicase mutant defective in subgenomic mRNA synthesis. *J. Virology* 73: 5274-81.

Mouse Hepatitis Virus Minus-strand Templates are Unstable and Turnover During Viral Replication

TAO WANG AND STANLEY G. SAWICKI

Department of Microbiology and Immunology, Medical College of Ohio, 3055 Arlington Avenue, Toledo, OH

1.   INTRODUCTION

During the first 6 hours after infection at 37°C, MHV-A59 infected cells accumulate seven species of minus strands, one equal in length to the viral genome RNA and six that are subgenomic and with sizes equal to those of the six subgenomic viral mRNA (Sawicki and Sawicki 1990).   These templates are found in double-stranded replicative intermediates (RIs and RFs) that together with viral polymerases are active in transcription of the viral plus strand species.   Thus, replication of MHV-A59 conforms to the general rules discovered for plus strand RNA viruses.   It is unique in having a genome of ~32 kb and in utilizing a novel discontinuous transcription mechanism to form the templates for its 3' nest set of viral mRNA (Sawicki and Sawicki 1990, 1995, 1998;). Our studies have been verified by Schaad and Baric (1994) and Baric and Yount (2000).

Our studies of the kinetics of MHV-A59 RNA synthesis found both plus and minus strand syntheses occurred throughout the infectious cycle (Sawicki and Sawicki 1986). Minus strand synthesis (as determined by the incorporation of ^{3}H-uridine) slowed after reaching a maximum rate, which was at about the same time plus strand synthesis reached its maximum rate. Plus strand synthesis then remained at a constant or slowly decreasing rate for several hours. Minus strand synthesis was inhibited almost immediately after stopping translation, while plus strand synthesis continued for about 30-60 min and then declined. Because stopping minus strand synthesis early by

The Nidoviruses (Coronaviruses and Arteriviruses).
Edited by Ehud Lavi *et al.*, Kluwer Academic/Plenum Publishers, 2001.

491

blocking protein synthesis stopped plus strand synthesis from increasing, the number of minus strands appears to determine the rate of plus strand synthesis. These studies form the bases for our perspective on coronavirus transcription. Replication complexes (RCs) are formed when newly translated viral polymerase copies the genome into minus strands: The formation of RCs requires de novo protein synthesis. The minus strands are then converted into transcription complexes (TCs) producing plus strands. MHV TCs contain both genome and subgenome-length templates (Sawicki and Sawicki 1990). TCs producing genomes have similar, if not identical, properties as TCs making subgenomic mRNA because both show the same sensitivity to translation inhibition (Sawicki and Sawicki 1986). We had performed these analyses at times after plus strand synthesis had reached its maximum and minus strand synthesis was at a low rate. Therefore, the ^{3}H-uridine we detected in RF's was almost exclusively (>90%) in plus strands. From those experiments it was apparent that subgenomic mRNA was produced from anti-subgenomes, whereas TCs with anti-genomes were producing exclusively genomes. We began a study to investigate why overall rates of transcription decreased late in infection and yet minus strand synthesis continued.

2.   RESULTS

17Cl-1 cells infected with MHV-A59 at an MOI of 100 and continuously labeled in the presence of 20 μg of actinomycin D/ml with ^{3}H-uridine beginning at 1 h p.i. accumulated radiolabeled RNA that was viral mRNA and viral RI/RF RNA. Surprisingly, late in infection there was less accumulated RI/RF RNA as detected by either ethidium bromide staining (Fig 1) or autoradiography than had been in these infected cells earlier. Thus, levels of viral RI/RF RNA previously made by infected cells up to 6 h p.i. was found to decline significantly after 6 h p.i. This suggested the viral RI/RF RNA was not stable but rather turned over during replication. Contrary to the RI/RF RNA, viral plus strands were stable and accumulated over time (data not shown).

2.1   MHV RI/RF RNA and their minus strand templates are unstable

To further investigate this phenomenon, infected cells were treated with cycloheximide (CHI) at different times after infection. We included the early periods, when plus and minus strand syntheses were exponentially

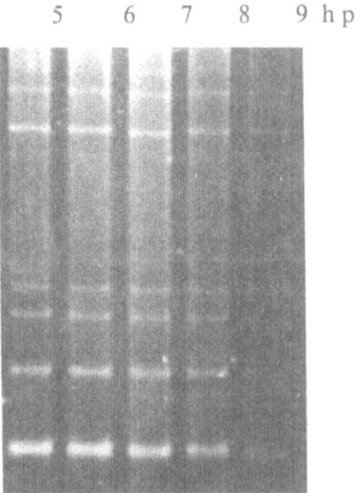

Figure 1. MHV-infected 17Cl-1 cells were harvested at the times indicated, the RNA was treated with RNase T1(1U/µg RNA), run on 0.8% agarose-TBE gels and stained with ethidium bromide.

increasing, and later periods, when plus strand synthesis was in its linear phase and minus strand synthesis was occurring at low rates. Treatment with CHI led to an abrupt cessation in minus strand synthesis and to a steady and linear decline in the rate of plus strand synthesis after a lag period of about 60 minutes. This pattern was the same whether cells were treated at 4 h p.i., an early period in infection, or at 5 h p.i or at 6 h p.i., confirming and extending our earlier results (Sawicki and Sawicki 1986). Thus, the ability of infected cells to continue linear RNA transcription late in infection before overall rates began to decline requires new protein synthesis. We investigated whether the synthesis of new minus strands were also needed during the late period because coronaviruses do not shut off minus strand transcription late but continue to synthesize new templates. We added ^{3}H-uridine to infected cells beginning at 6 h p.i. Radiolabel accumulated in RI/RF RNA and in new minus strands since the percentage of labeled RI/RF RNA that was in minus strands increased with time from a low of 5-10% at the start of the labeling period (6.5 h p.i.) to over 35% at 8 h p.i. This meant that at least 70% of minus strands in RI/RF RNA made during the first 6 hours were replaced by minus strands made after 6 h p.i. If verified, this indicated coronavirus minus strands are unstable templates and only function for a limited time and they must be continuously replaced by newly made ones for plus strand transcription rates to continue. It also suggested the loss of plus strand synthesis after CHI-treatment might result from the loss of the minus strand templates and not necessarily from the turnover of the plus

strand replicases or transcriptases.   Such template turnover may also be
under the control of a unique viral function.

2.2   Loss of minus strands is reversible but requires new protein synthesis

Removal of the CHI after a 4 h treatment period resulted in a burst of new
minus strand synthesis and a return of plus strand synthesis to normal levels
ongoing in untreated cultures (Fig 2). The newly synthesized minus strands
accounted for 70% or more of the total minus strand templates in viral RI
and RF RNA. This result indicates that essentially all the old templates have
been replaced by newly synthesized ones.

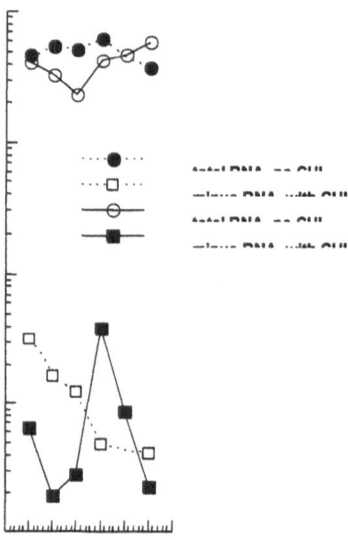

Figure 2. MHV-infected cells were pulse-labeled for 1 hour periods with ^{3}H-uridine in the
presence of actinomycin D and harvested at the times indicated. One set of cultures, treated
with CHI from 5-9 hr p.i., was washed free of CHI and labeled in the absence of CHI.
Incorporation/50,000 cells is shown.

2.3   Turnover is specific for coronavirus RI/RF RNA

Was the turnover of minus strands specific to MHV or did MHV infected
cells produce or activate a nuclease that would degrade any viral RI/RF
RNA? To investigate this issue, cells were infected with both MHV-A59
and the alphavirus Semliki Forest virus (SFV). Unlike MHV, alphavirus

RI/RF RNA is stable and does not turnover normally either late in infection or after treatment of cells with protein synthesis inhibitors. In cells infected with MHV and superinfected with SFV 1-2 hours later, which allowed the more slowly growing MHV to start transcription and accumulate TCs, only the MHV-A59 RI/RF RNA was lost over time. Therefore, turnover was specific to MHV RI/RF and their minus strands.

2.4   RI/RF minus strands are released in single-stranded form before they are degraded

We determined whether turnover of minus strands in MHV infected cells involved their release from double-stranded into a single-stranded form. If so, infected cells labeled continuously from 1 h p.i. would contain radiolabel in single-stranded minus strands. RNA from infected cells was either digested directly with RNase T1 to obtain the nuclease-resistant RF cores of the viral RIs or first allowed to hybridize by incubation at 68°C for 30 minutes and then at 25°C for another 30 minutes before RNase T1 treatment. Both samples were then passed through CF-11 columns to obtain the nuclease-resistant cores and the amount of labeled RF RNA was determined. The results (Table 1) indicated that after 6 h p.i. MHV-infected cells had less and less total RI/RF RNA with time but they also had a percentage of their total minus strands in single-stranded form. Hybridization before RNase digestion allowed single-stranded minus strands to form hybrids with the excess of viral plus strands (genomes and mRNA). Hybridization led to the recoverey of an additional, equal amount of RF RNA compared to lysates that were digested with RNase directly (Table 1). With time, however, the absolute amount of both RI/RF RNA and recoverable single-stranded minus strands decreased, explaining the steady decline of overall plus strand synthesis observed late in infection.

Table 1. Pre-hybridization recovered free minus strands late in infection. 17CL-1 cells were infected with MHV-A59 at an MOI of 100 and maintained at 37°C. Cultures were labeled continuously with ^{3}H-uridine (200 µCi/ml) in the presence of actinomycin D from 1 hr p.i. and harvested at the times indicated. Half of each infected cell total RNA sample was pre-hybridized by heat-fast cooling, followed by incubating at 68°C for 30 minutes and then at 25°C for 30 minutes. Viral RNA untreated or pre-hybridized was digested with RNase T1 and chromatographed on CF-11 cellulose to purify the RFs. The maximum RF cpm was at 6 h p.i.

| | RI/RF RNA cpm/50,000 cells | | % of maximal | |
h pi	No hybridization	hybridization	No hybridization	hybridization
4	7,883	7,475	57	53
6	13,801	14,122	100	100
7.5	3,246	6,617	24	47
9	1,236	2,357	9	17

3. CONCLUSIONS AND MODEL

Coronavirus infected cells synthesize minus strand templates that function for only a limited time as components of transcription complexes active in viral genome and subgenomic mRNA production. Thereafter, the minus strand are released from the transcription or the entire transcription complex dissociated; in either case, the minus strands were found in single-stranded form before they were degraded. This provides an explanation for why coronavirus infected cells continue to synthesize minus strand RNA after the maximum rate of plus strand transcription has been achieved. Thus, coronaviruses, and possibly other nidoviruses, employ a strategy for replication that requires constant production of both polyprotein forms of the viral polymerase components and minus strand templates. In poliovirus infected cells, templates for viral plus strand synthesis are also replaced at a slow, constant rate (Baltimore 1969). Paralleling but different from coronaviruses, the poliovirus minus strand components of RIs either were converted to inactive RFs after about 10-20 initiation cycles or were degraded. They were not found in single-stranded form in infected cells (reviewed in Koch and Koch 1985). The mechanism that specifically targets the MHV templates for degradation is of interest.

ACKNOWLEDGMENTS

This work was supported by the NIH (AI 28506) to SGS.

REFERENCES

Baltimore, D. "The Replication of Picornaviruses." In *The Biochemistry of Viruses*, H.B. Levy, ed. New York: Marcel Dekker, 1969

Baric, R.S., and Yount, B., 2000, Subgenomic minus-strand RNA function during mouse hepatitis virus infection. *J. Virology* 74:4039-46.

Koch, F. and Koch, G., 1985, *The Molecular Biology of Poliovirus*. New York: Springer-Verlag

Sawicki, D.L, and P.J. Gomatos., 1976, Replication of Semliki Forest virus: Polyadenylate in plus strand RNA and polyuridylate in minus strand RNA. *J. Virology* 20:446-64.

Sawicki, S.G., and Sawicki, D.L. 1986, Coronavirus minus-strand RNA synthesis and effect of cycloheximide on coronavirus RNA synthesis. *J. Virology* 57:328-34.

Sawicki, S.G., and Sawicki, D.L., 1990, Coronavirus transcription: subgenomic mouse hepatitis virus replicative intermediates function in RNA synthesis. *J. Virology* 64:1050-56.

Sawicki, S.G., and Sawicki, D.L., 1995, Coronaviruses use discontinuous extension for synthesis of subgenome-length negative strands. *Adv. Exp. Biol. Med.* 380:499-506.

Sawicki, S.G., and Sawicki, D.L., 1998, A new model for coronavirus transcription. *Adv. Exp. Biol. Med.* 440:215-19.

Schaad, M.C., and Baric, R.S., 1994, Genetics of mouse hepatitis virus transcription: evidence that subgenomic minus strands are functional templates. *J. Virology* 68:8169-79

Mutagenesis of the 3'42 Nucleotide Host Protein Binding Element of the MHV 3'UTR

REED F. JOHNSON AND JULIAN L. LEIBOWITZ

Department of Pathology and Laboratory Medicine, The Texas A&M University System Health Science Center, College Station, Texas 77802-1114

1.   INTRODUCTION

The mouse hepatitis virus (MHV) genome is a 32kb message sense single stranded RNA. Studies with defective interfering RNAs have shown that the 3' terminal 447nt are required for replication (Kim et al., 1993; Lin et al., 1993). Previous studies have identified two host protein binding elements within the 3' terminal 166nt of the 3'UTR. The first of these is located at nucleotides 154-129 [assigning position 1 to the first nucleotide 5' of the poly (A)] tail and the second is located within the 3' terminal 42 nucleotides (Liu et al., 1997; Yu and Leibowitz, 1995). Both elements contain the conserved motif UGAARNGAAGUU which is required for host protein binding and for DI RNA replication (Yu and Leibowitz, 1995; Liu et al., 1997). In the present study we further identify nucleotides involved in host protein binding and explore the role of RNA secondary structure in this process.

The Nidoviruses (Coronaviruses and Arteriviruses).
Edited by Ehud Lavi *et al.*, Kluwer Academic/Plenum Publishers, 2001.

2.   MATERIALS AND METHODS

2.1   Cells and Lysate preparation

17Cl-1 cells and lysates were prepared as described (Yu and Leibowitz, 1995).

2.2   Mutagenic PCR and Transcription

DE25 served as a template for amplifying and incorporating a T7 promoter into the 3' terminal 42 nucleotide host protein binding element by PCR. The resulting template was used for *in vitro* transcription reactions containing $[\alpha^{32}P]$-UTP. A portion of each transcription reaction was analyzed by electrophoresis.

2.3   Gel Mobility Shift RNase T1 Protection Assays

Gel mobility shift RNase T_1 protection assays were performed as described, except that 1.4 pmoles of each probe was used (Yu and Leibowitz, 1995). Competitive concentrations of unlabelled wild type probe as well as tRNA were used at 10, 25, 50, 100 fold molar excess to ensure specificity. Assays were quantitated by phosphorimagery. The binding activity of wild type RNA was set at 100%, and after arithmetically correcting for differences in base composition the binding activities of mutants were calculated as a percentage of wild type.

2.4   Computer Modelling

Computer modelling of the 3'42nt host protein binding element was carried out with Mfold, Version 3.0, available at http://www.ibc.wustl.edu/~zuker/rna/node3.html (Zuker et al. 1999).

3.   RESULTS

3.1   Gel Mobility Shift RNase T_1 Protection Assays

Mutagenesis of the 11nt motif contained in the last 42nts confirmed this motif's importance for host protein binding and DI replication (Yu and

Leibowitz, 1995). Computer models of the 3' (+)42 protein binding element predicted the formation of a small stem loop (Yu and Leibowitz, 1995). To determine if RNA secondary structure is important for host protein binding to the 3'(+)42 element we performed a series of gel mobility shift RNase T_1 protection assays with RNA probes containing mutations designed to disrupt this predicted stem loop structure. Mutants MT1A, MT2A, MT3C, MT5A, and MD10 were generated by mutagenic PCR and of these mutants, MT3C had the greatest effect on host protein binding. Mutants MT1A, MT2A, MT3C, and MT5A all carry mutations in the 11nt motif (Table 1). The mutations in MT1A and MT2A decreased their respective binding efficiencies to 68.5% and 64% relative to wild type RNA. Mfold, version 3.0, predicted that these mutations would not change the RNA secondary structure (Fig. 1). Mutant MT3C had a binding activity only 6.3% of wild type, while the mutation in MT5A resulted in a binding activity 25.2% of wild type RNA (Table 1). Both of these mutant RNAs were predicted to have dramatic changes in secondary structure from wild type (Fig. 2). For mutant MD10 the last 10 nucleotides were deleted. Deletion of these nucleotides decreased protein binding activity to 36.1% of wild type. Yu and Leibowitz's computer model suggested that the 10 terminal nucleotides were single stranded (Yu and Leibowitz, 1995). Secondary structure modelling with the current Mfold, version 3.0, suggests that the 10 terminal nucleotides are involved in secondary structure formation which may explain the decrease in binding.

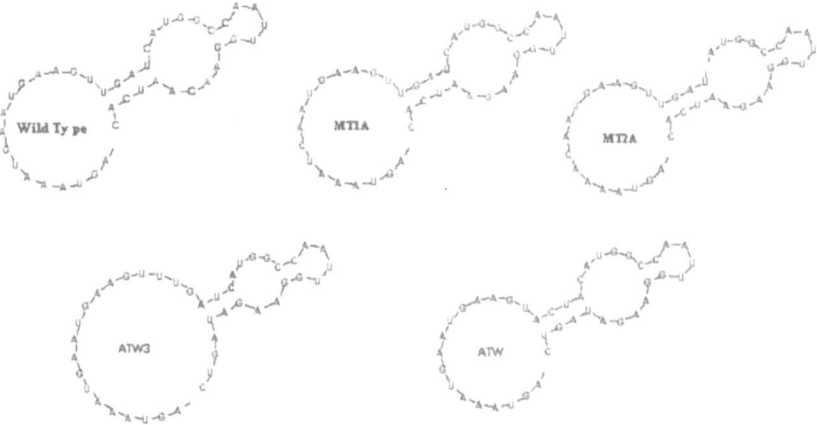

Figure 1. Secondary Structure models for RNA probes with minimal loss in binding activity or increased binding activity. The black dot indicates the 5' terminus of each molecule.

Table 1. Summary of binding activity for all probes.

Probe	Sequence	% Binding	Std Dev	Effect on 2° Structure
Wild type	AGUAAAUGAAUGAAGUUGAUC AUGGCCAAUUGGAAGAAUCAC	100	N/A	N/A
MT1A	AGUAAA**C**UAAUGAAGUUGAUC AUGGCCAAUUGGAAGAAUCAC	68.5	11.3	NONE
MT2A	AGUAAA**AC**AAUGAAGUUGAUC AUGGCCAAUUGGAAGAAUCAC	64	21.7	NONE
MT3C	AGUAAA**ACU**AUGAAGUUGAUC AUGGCCAAUUGGAAGAAUCAC	6.3	6.1	DISRUPTS
MT5A	AGUAAA**ACUUA**GAAGUUGAUC AUGGCCAAUUGGAAGAAUCAC	25.2	5.5	DISRUPTS
MD10	AGUAAAUGAAUGAAGUUGAUC AUGGCCAAUUG	44.1	2.5	DISRUPTS
M24C	AGUAAA**ACU**AUGAAGUUGAUC AU**C**GCCAAUUGGAAGAAUCAC	74.1	8.6	RESTORES
24C	AGUAAAUGAAUGAAGUUGAUC AU**C**GCCAAUUGGAAGAAUCAC	73.5	25.4	NONE
ATW5'	AGUAAAUGAAUGAAGU**ACUA**C AUGGCCAAUUGGAAGAAUCAC	7.2	8.7	DISRUPTS
ATW3'	AGUAAAUGAAUGAAGUUGAUC AUGGCCAAUUGGAAGA**UAGU**C	400	84.4	DISRUPTS
ATW	AGUAAAUGAAUGAAGU**ACUA**C AUGGCCAAUUGGAAGA**UAGU**C	185.8	34.7	RESTORES

If primary structure of the 11nt host protein binding motif is all that is required for protein binding, then additional mutations outside of the 11nt motif should have minimal effects on protein binding. If nucleotides outside of the motif also play some role in the RNA:protein interaction, mutating these nucleotides should effect binding. If a specific secondary structure is required for protein binding, then restoration of secondary structure in the presence of a deleterious mutation in the conserved motif should restore host protein binding to near wild type levels. Computer modelling generated compensatory sequence alterations that would restore the wild type secondary structure in the presence of the original deleterious mutations. A compensatory mutation for MT3C was found; changing nucleotide 19 from G to C restored predicted wild type secondary structure (Figs. 1 and 2). This probe, M24C, had a host protein binding activity 74.1% of wild type compared to 6.3% of wild type for MT3C. For probe 24C, a 19 G to C mutation was introduced alone. This mutation lies outside of the host protein binding motif, did not disrupt secondary structure, and had a binding activity 73.5% of wild type (Table 1, Fig. 1).

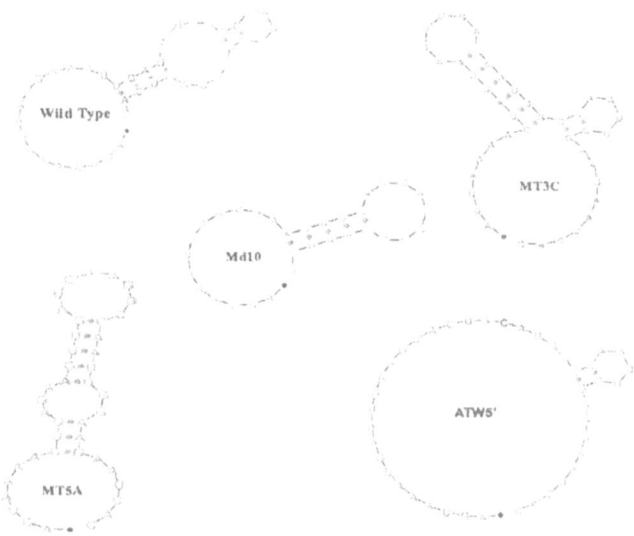

Figure 2. Secondary structure models of RNA probes that did not bind near wild type levels. The black dot indicates the 5' terminus of each molecule.

The computer predicted wild type secondary structure indicated base paring in two regions. Base pairing occurs between nucleotides 2-5 and 23-26, with a second base pairing occurring between nucleotides 10-11 and 16-17. To investigate the role of the base pairings between 2-5 and 23-26 we created the mutants ATW5', ATW3', and ATW (Figs. 1 and 2). When these structures were evaluated for changes in secondary structure, ATW maintained the wild type structure, ATW3' slightly altered its base pairing; ATW5' was almost completely single stranded except for the 10-11 to 16-17 base pairing. Surprisingly, ATW3' host protein binding activity was 4-fold greater than wild type RNA. ATW probe host protein binding ability also increased relative to wild type but not as dramatically, 1.85 fold (Table 1).

Base pairing at positions 10-11 with 16-17 was maintained in wild type, MT1A, MT2A, ATW, ATW3, M24C, and MT3C mutants. Mutations in the 10-11 and 16-17 region were examined with Mfold but compensatory mutations which restored wild type secondary structure could not be identified. Base pairings at positions 10-11 with positions 16-17 in wild type probe, probes with little binding activity, and probes with increased binding activity suggests that this pairing while conserved, is not sufficient for host-protein binding.

4.   DISCUSSION

We have shown that secondary structure plays an important role in binding of host proteins to the 3'(+)42 host protein binding element. Eleven probes were analyzed, and four of these probes, MT3C, MT5A, ATW5', and MD10, had a 60% or greater reduction in binding activity and also had significant alterations in their predicted secondary structures. Three probes MT1A, MT2A, and 24C had no predicted changes in secondary structure compared to wild type and had slightly decreased levels of host protein binding activity. M24C, contained the deleterious MT3C mutations along plus compensatory mutation which restored the predicted wild type secondary structure and protein binding activity, thus highlighting the role of secondary structure in host proteins binding to 3'(+)42 RNA.

ATW5' RNA has a predicted secondary structure radically different from the wild type structure and little protein binding activity. The predicted secondary structure of the ATW3' RNA is a variation of the wild type RNA structure (Fig. 1) and its host cell protein binding activity is four-fold greater than wild type. When the ATW3' and ATW5' mutations are placed in the same probe, the secondary structure of the RNA is predicted to match the wild type structure. The protein binding activity of the double mutant RNA is only 1.85-fold greater than wild type, an activity intermediate to that of the single mutations. These data are consistent with the view that secondary structure is a major determinant of protein binding activity of the 3'(+)42 RNA, although primary structure also makes an important contribution.

Non-translated sequences in the MHV genome such as the 5' leader and 3'UTR are essential for viral replication. Deletion studies have shown that *cis*-acting sequences in the last 447 nucleotides of the genome plus the poly (A) tail are essential for replication (Kim et al, 1993; Lin et al., 1993). The 3'(+)42 RNA lies within this region and is contained within the 3' 55 nucleotides required for minus strand RNA synthesis (Lin et al., 1994). Studies in our lab have shown that mutagenesis of the 11nt UGAARNGAAGUU motif within the 3'(+) 42 RNA which disrupt RNA-protein interaction also have a deleterious effect on DI replication(Yu and Leibowitz, 1995). The data presented here indicate that RNA secondary structure plays an essential role in the interaction of the 3'(+) 42 RNA with host proteins. Replication studies with the mutants generated in this work would test the hypothesis that these RNA secondary structures in the 3' UTR are important for replication. Studies of the two mutants with enhanced protein binding activity are likely to be particularly interesting. Such studies are currently under way in our lab.

ACKNOWLEDGEMENTS

This work was supported by National Multiple Sclerosis Society grant RG2203-B-6 and a gift from the Stearman family. The authors would like to thank Santosh Nanda, Qi Liu, and Elena Belyavskaya for all their help.

REFERENCES

Y.-N.Kim, Y.S.Jeong, and S.Makino, Analysis of cis-acting sequences essential for coronavirus defective interfering RNA replication., Virology 197:53 (1993).

Y.-J.Lin and M.M.C.Lai, Deletion mapping of a mouse hepatitis virus defective interfering RNA reveals the requirement of an internal and discontinous sequence for replication., J.Virol. 67:6110 (1993).

Y.Lin, C.Liao, and M.M.C.Lai, Identification of the cis-acting signal for minus-strand RNA synthesis of a murine coronavirus: implications for the role of minus-strand RNA in RNA replication and transcription , J.Virol. 68:8131 (1994).

Q.Liu, W.Yu, and J.L.Leibowitz, A specific host cellular protein binding element near the 3' end of mouse hepatitis virus genomic RNA., Virology 232:74 (1997).

W.Yu and J.L.Leibowitz, A conserved motif at the 3' end of mouse hepatitis virus genomic RNA required for host protein binding and viral RNA replication., Virology 214: 128 (1995).

W.Yu and J.L.Leibowitz, Specific binding of host cellular proteins to multiple sites within the 3' end of mouse hepatitis virus genomic RNA., J.Virol. 69:5033 (1995).

M.Zuker, D.H.Mathews, and D.H.Turner, Algorithms and Thermodynamics for RNA Secondary Structure Prediction: A Practical Guide, *in:* "RNA Biochemistry and Biotechnology", J.Barciszewski and B.F.C.Clark, eds., Kluwer Academic Publishers, New York (1999).

Use of an Infectious Bronchitis Virus D-RNA as an RNA Vector

PAUL BRITTON, KATHLEEN STIRRUPS, KEVIN DALTON, KATHLEEN
SHAW, SHARON EVANS, BENJAMIN NEUMAN, BRIAN DOVE, ROSA
CASAIS, AND DAVE CAVANAGH
*Division of Molecular Biology, Institute for Animal Health, Compton Laboratory, Compton,
Newbury, Berkshire, RG20 7NN, UK*

1.   INTRODUCTION

In the absence of a complete infectious IBV cDNA we have been
developing an alternative strategy for the production of recombinant IBVs,
utilising a defective RNA (D-RNA), CD-61. Coronavirus D-RNAs function
like a minigenome and are useful as RNA vectors for the expression of
heterologous genes and for targeted recombination. IBV D-RNA CD-61
(Pénzes *et al.*, 1996) was derived by deletion mutagenesis from a natural D-
RNA, CD-91, produced by multiple passage of high titre IBV Beaudette in
chick kidney (CK) cells (Pénzes *et al.*, 1994). CD-61 lacks internal parts of
the genome but contains the sequences required for replication and for
packaging into virus particles and can therefore be replicated and packaged
(rescued) in a helper virus-dependent manner.

2.   RESULTS AND DISCUSSION

We have developed two systems for the rescue of IBV D-RNA CD-61.
The first system relies on the electroporation of *in vitro* T7-generated CD-61
transcripts into IBV-infected cells. The second system involves co-infection
of IBV-infected cells with two recombinant fowlpox viruses (rFPV), one

The Nidoviruses (Coronaviruses and Arteriviruses).
Edited by Ehud Lavi *et al.*, Kluwer Academic/Plenum Publishers, 2001.

containing the cDNA corresponding to CD-61 (rFPV-CD-61) and the second rFPV expressing T7 RNA polymerase. Following co-infection of the IBV-infected cells with the two rFPVs CD-61 RNA is initially transcribed *in situ* using T7 RNA polymerase and then replicated and rescued on serial passage by the helper IBV (Fig. 1).

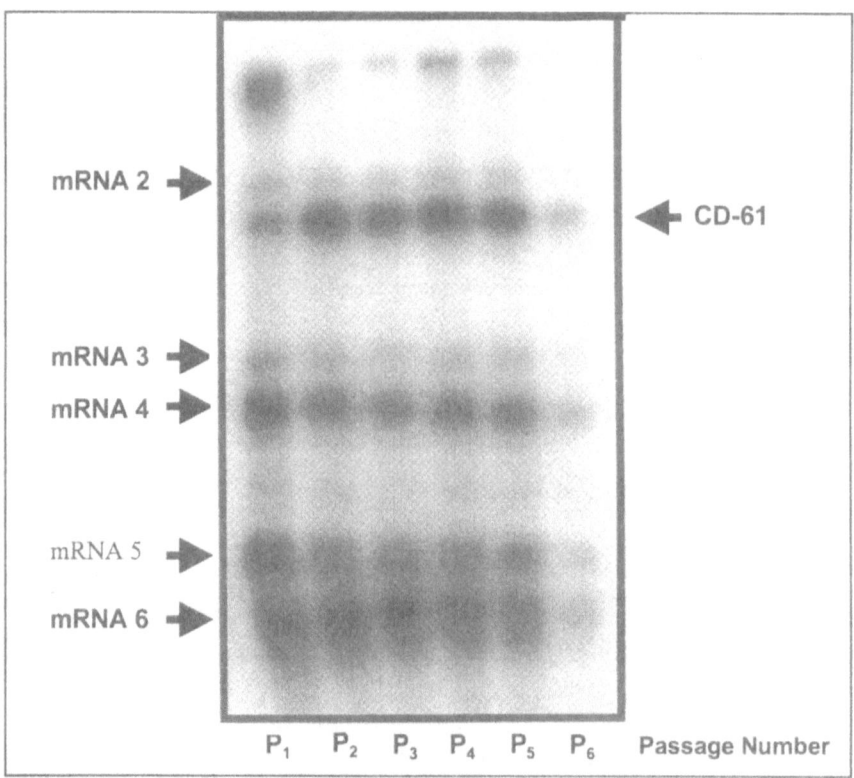

Figure 1. Northern blot analysis of IBV-derived RNAs showing rescue of CD-61 on serial passage (P_1-P_6) in IBV-infected CK cells. The IBV-derived RNAs, following electrophoresis in denaturing formaldehyde-agarose gels and transfer to nitrocellulose membrane, were detected by hybridisation with a ^{32}P-labelled IBV 3' UTR probe. CD-61 was derived from rFPV-CD-61.

IBV D-RNA CD-61 was investigated as a potential RNA vector for the expression of heterologous genes, using the luciferase (Luc) and chloramphenicol acetyltransferase (CAT) reporter genes. Expression of the genes was under the control of a transcription-associated sequence (TAS; Hiscox *et al.*, 1995) derived from the Beaudette gene 5, responsible for transcription of subgenomic mRNA 5 (Stirrups *et al.*, 2000b). However, following electroporation and subsequent rescue of *in vitro* T7-transcribed D-RNA CD-61-Luc (CD-61 containing the luc gene) the D-RNA was found to be unstable and incapable of expressing luciferase to the activity initially detected in the electroporated cells. In contrast, following rescue of D-RNA CD-61-CAT (CD-61 expressing the CAT gene) the D-RNA was shown to be capable of producing CAT protein in larger amounts than initially detected in the electroporated cells. This observation indicated that CD-61-CAT was more stable than CD-61-Luc. In some cases the amount of CAT protein detected following rescue of CD-61-CAT reached 1.6 ㎍/10^6 cells. The reporter genes were expressed from two different sites within CD-61, *Sna*BI in Domain II which interrupted the CD-61 specific ORF and *Pma*CI in Domain III which did not interrupt the CD-61 specific ORF (Pénzes *et al.*, 1996). The rescue of CD-61-CAT and expression of CAT was not affected by interruption of the CD-61-specific ORF (Fig. 2).

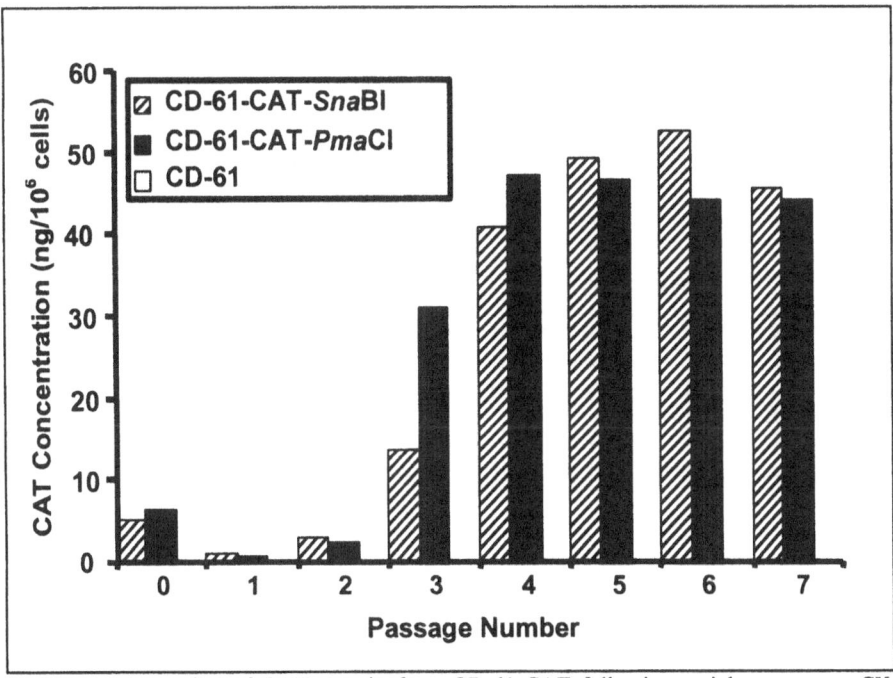

Figure 2. Expression of CAT protein from CD-61-CAT following serial passage on CK cells. The CAT gene was inserted in two different sites, *Sna*BI and *Pma*CI, with in the D-RNA. ELISA was used to detect CAT protein in cell lysates derived from CK cells infected with IBV and CD-61-CAT.

The IBV gene 5 TAS is composed of two tandem repeats of the IBV canonical consensus sequence CTTAACAA (Stirrups *et al.*, 2000b). We have demonstrated that only one sequence is required for expression of an mRNA, although both sequences can function as acceptor sites for acquisition of the leader sequence (Stirrups *et al.*, 2000b).

To determine whether CD-61-CAT could be rescued *in vivo*, cell supernatants containing helper IBV and CD-61-CAT were used to infect 11 day old embryonated eggs. Allantoic fluid from the infected embryonated eggs was then used to infect CK cells and cell lysates were analysed for the presence of CAT protein. Analysis of CK cell lysates, following infection with virus in allantoic fluid derived from the infected eggs, showed that CD-61-CAT was replicated *in ovo* (Fig. 3).

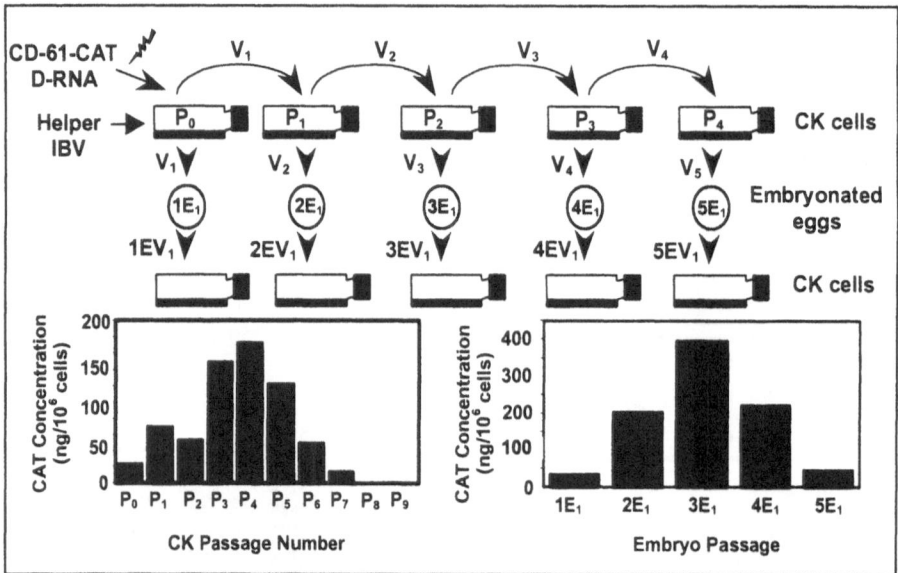

Figure 3. Analysis of CAT protein following infection of embryonated eggs with D-RNA CD-61-CAT. *In vitro* T7-transcribed CD-61-CAT was electroporated into IBV-infected CK cells. The rescued D-RNA was serially passaged on CK cells and cell lysates analysed for the presence of CAT protein. Virus (V_1-V_5) was either used to infect CK cells or 11 day old embryonated eggs. Virus ($1EV_1$-$5EV_1$) derived from the embryonated eggs, $1E_1$-$5E_1$ was used to infect CK cells to detect the presence of D-RNA CD-61-CAT via the expression of CAT protein.

However, serial passage of CD-61-CAT in embryonated eggs resulted in loss of the D-RNA (Fig. 4). Previous experiments had shown that IBV-derived D-RNAs containing heterologous genes eventually lost the ability to express the foreign gene (Stirrups *et al.*, 2000b). Therefore the loss of CAT expression from CD-61-CAT was not unexpected. Overall our results demonstrated that an IBV D-RNA containing a foreign gene was replicated *in ovo*.

We have demonstrated that CD-61 can be rescued by heterologous strains of IBV (Stirrups *et al.*, 2000a). The 5' ends of the heterologous IBV genomes were analysed and found to contain a variety of nucleotide substitutions when compared to each other and the Beaudette sequence from which CD-61 was derived. Analysis of the 5' ends of CD-61 rescued by the heterologous helper IBV identified that the Beaudette-derived leader sequence initially present on the electroporated *in vitro* T7-generated D-RNA transcript had been replaced with the leader sequence corresponding to

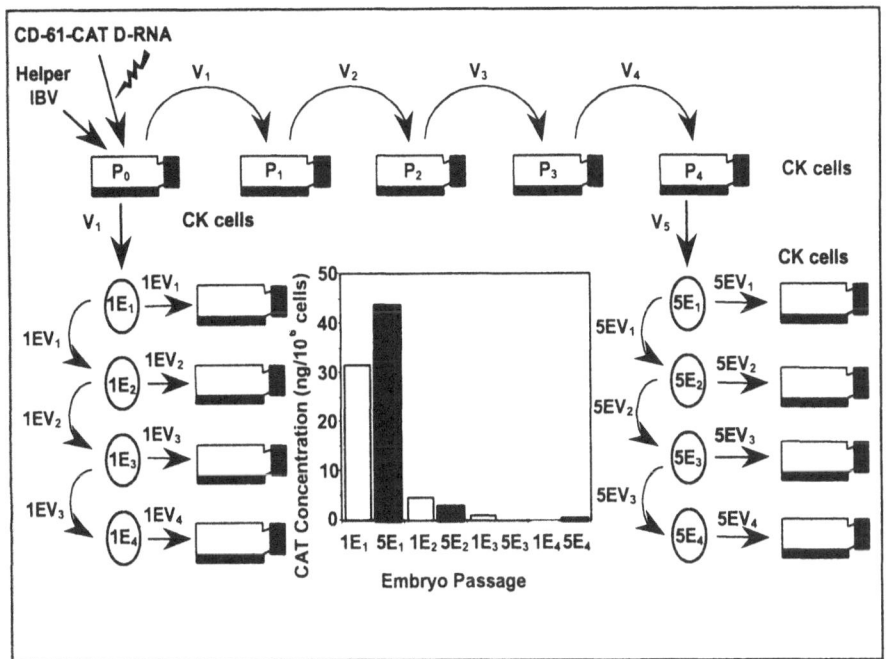

Figure 4. Analysis of CAT protein following serial passage of CD-61-CAT in embryonated eggs. IBV containing CD-61-CAT (V_1 and V_5) derived from CK cells was used to infect embryonated eggs. Virus ($1EV_1$ and $5EV_1$) derived from the allantoic fluid of infected eggs ($1E_1$ and $5E_1$) was serially passaged in eggs, $1E_1$-$1E_4$ and $5E_1$-$5E_4$. Virus derived ($1EV_1$-$1EV_4$ and $5EV_1$-$5EV_4$) from the infected eggs was used to infect CK cells to detect the presence of D-RNA CD-61-CAT via the expression of CAT protein.

the 5' end of the helper virus genome. In contrast, the adjacent 5' UTR sequence corresponded to the original CD-61 Beaudette sequence demonstrating that rescue of the unmodified CD-61 resulted in the phenomenon of leader switching (Makino & Lai, 1989). Three predicted stem-loop structures were identified within the 5' UTR on the various IBV strains (Stirrups *et al.*, 2000a). Stem-loop I showed a high degree of covariance amongst the IBV strains providing phylogenetic evidence that this structure exists and is potentially involved in replication (Stirrups *et al.*, 2000a), supporting previous observations that a bovine coronavirus (BCoV) stem-loop homologue was essential for replication of BCoV D-RNAs (Chang *et al.*, 1994).

ACKNOWLEDGEMENTS

This work was supported by the Ministry of Agriculture, Fisheries and Food, UK, Project code OD1905 and by grant number CT950064 of the Fourth RTD Framework Programme of the European Commission (EC). K.

Stirrups and K. Dalton were the holders of Research Studentships from the Biotechnology and Biological Sciences Research Council (BBSRC). B. Neuman and B. Dove were recipients of IAH studentships. S. Evans was supported by a BBSRC Realising Our Potential Award (ROPA). R. Casais was the recipient of a EC TMR Marie Curie Research Training Grant.

REFERENCES

Chang, R. Y., Hofmann, M. A., Sethna, P. B. & Brian, D. A. 1994. A *cis*-acting function for the coronavirus leader in defective interfering RNA replication. *J. Virol.* **68**, 8223-8231.

Hiscox, J. A., Mawditt, K. L., Cavanagh, D. & Britton, P. 1995. Investigation of the control of coronavirus subgenomic mRNA transcription by using T7-generated negative-sense RNA transcripts. *J. Virol.* **69**, 6219-6227.

Makino, S. & Lai, M. M. C. 1989. High-frequency leader sequence switching during coronavirus defective interfering RNA replication. *J. Virol.* **63**, 5285-5292.

Pénzes, Z., Tibbles, K., Shaw, K., Britton, P., Brown, T. D. K. & Cavanagh, D. 1994. Characterization of a replicating and packaged defective RNA of avian coronavirus infectious bronchitis virus. *Virology* **203**, 286-293.

Pénzes, Z., Wroe, C., Brown, T. D., Britton, P. & Cavanagh, D. 1996. Replication and packaging of coronavirus infectious bronchitis virus defective RNAs lacking a long open reading frame. *J. Virol.* **70**, 8660-8668.

Stirrups, K., Shaw, K., Evans, S., Dalton, K., Cavanagh, D. & Britton, P. 2000a. Leader switching occurs during the rescue of defective RNAs by heterologous strains of the coronavirus infectious bronchitis virus (IBV). *J. Gen. Virol.* **81**, 791-801.

Stirrups, K., Shaw, S., Evans, S., Dalton, K., Casais, R., Cavanagh, D. & Britton, P. 2000b. Expression of reporter genes from the defective RNA CD-61 of the coronavirus infectious bronchitis virus. *J. Gen. Virol.* **81**, 1687-1698.

Use of Defective RNAs Containing Reporter Genes to Investigate Targeted Recombination for Avian Infectious Bronchitis Virus

BENJAMIN NEUMAN, DAVE CAVANAGH, AND PAUL BRITTON
Division of Molecular Biology, Institute for Animal Health, Compton Laboratory, Compton, Newbury, Berkshire, RG20 7NN, UK

1.   INTRODUCTION

At present there is no infectious cDNA of IBV and the robustness of coronavirus cDNA reverse genetic systems remains undetermined. We have been investigating an alternate strategy for reverse genetic manipulation of IBV using defective RNAs (D-RNAs) to exchange genetic material with the IBV genome, based on efficient recombination systems for other coronaviruses. It has been previously shown that two strains of IBV can recombine *in ovo* (Kottier *et al.*, 1995) though recombination has not yet been demonstrated in cell culture. IBV D-RNA CD-91 (Pénzes *et al.*, 1994) consisting of regions from the 5'-end, open reading frame 1b and the 3'-end including most of the N gene was isolated and cloned after high multiplicity passage of IBV Beaudette. Internal sequences were deleted from CD-91 to generate CD-61 (Pénzes *et al.*, 1996). Both D-RNAs contain all the necessary signals for replication and packaging (rescue) by helper IBV.

2.   RESULTS AND DISCUSSION

We have developed several reporter gene-containing vectors based on CD-61 and CD-61Δ*Xba*I, which differs from CD-61 by a 2.2 kb internal deletion. Both D-RNAs were transcribed *in vitro* by T7 polymerase and electroporated into IBV-infected chick kidney (CK) cells. Cell culture

The Nidoviruses (Coronaviruses and Arteriviruses).
Edited by Ehud Lavi *et al.*, Kluwer Academic/Plenum Publishers, 2001.

medium containing virus and D-RNA particles was then passaged on fresh CK cells at 24 h intervals. RNA was extracted from cell lysates and analysed by Northern blot with ^{32}P-labelled DNA probes and by RT-PCR with primers which yield D-RNA-specific products. Both modified D-RNA constructs were rescued by IBV Beaudette (data not shown).

The chloramphenicol acetyltransferase (CAT) gene was placed under the control of the IBV gene 5 transcription associated sequence (TAS) element and inserted into pCD-61 (Stirrups *et al.*, 2000) and pCD-61Δ*Xba*I at the *Pma*CI site located in the N gene-derived region. The gene 5 TAS element consists of two tandem canonical TAS repeats (ACTTAACAA). CAT protein concentrations were compared up to passage 4 (P4) for each D-RNA (Figure 1). CAT expression from CD-61 CAT in N gene was better in later passages. In contrast, CAT expression from CD-61Δ*Xba*I CAT in N gene was highest in P0. D-RNAs with CAT in the N gene sequence could not be used for homologous recombination with IBV because the resulting recombinant IBV would have an interrupted N gene. The CAT gene was also inserted into pCD-61Δ*Xba*I at a unique *Hin*dIII site located in the sequence corresponding to the variable region (Williams *et al.*, 1993). of the Beaudette 3'-untranslated region (3'-UTR). CAT was expressed at a stable moderate level when inserted into the 3'-UTR region of CD-61Δ*Xba*I. Since we expect recombination to occur during P0, when expression from the 3'-UTR was similar to expression from CD-61 CAT in N gene, the 3'-UTR insertion site was deemed suitable for further reporter gene insertion and recombination studies.

The enhanced green fluorescent protein (EGFP; Clontech) gene was cloned as a gene cassette under control of the gene 5 TAS element into the *Hin*dIII (3'-UTR) or *Pma*CI (N gene) site of pCD-61Δ*Xba*I. Similarly, a hygromycin phosphotransferase – EGFP (HygEGFP; Clontech) fusion gene cassette or hygromycin phosphotransferase (Hyg) gene cassette was cloned into the *Hin*dIII site (3'-UTR) of pCD-61Δ*Xba*I. In P0, 0.1 to 1% of cells fluoresced and the number of fluorescent cells decreased with each passage (Figure 2). A few fluorescent cells were visible until P4 or P5. To determine whether these D-RNAs could be rescued by a heterologous strain of IBV, we attempted to rescue the EGFP gene-containing D-RNAs with IBV M41, a strain closely related to IBV Beaudette but lacking almost all of the variable region of the 3'-UTR. However, the number of fluorescent cells in each passage for M41 and Beaudette was similar. We were most interested in P0 for recombination purposes, so RNA samples were examined for evidence of recombination by RT-PCR using oligonucleotide primers from virus sequence not found in the D-RNA and complementary to the EGFP gene. However, no PCR products of the expected size were detected.

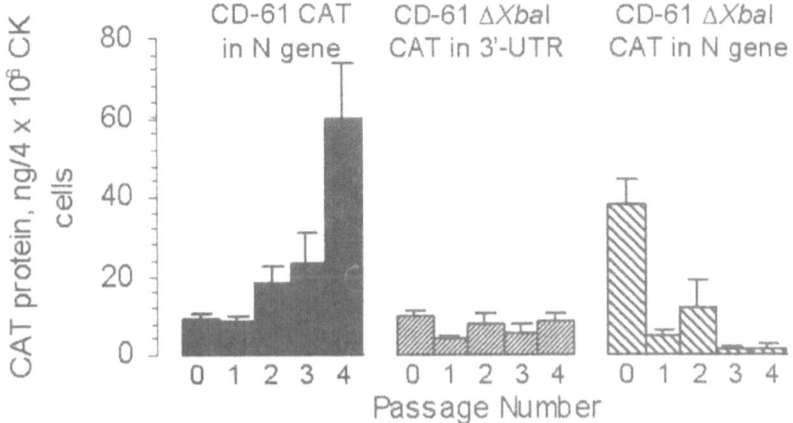

Figure 1. Expression of CAT protein from IBV D-RNAs after rescue by IBV Beaudette. The CAT gene cassette was inserted into regions of the D-RNA derived from the N gene or 3'-UTR. CAT protein in cell lysates was quantified by ELISA and adjusted for number of cells. The mean of six replicates +/- standard error of the mean is shown.

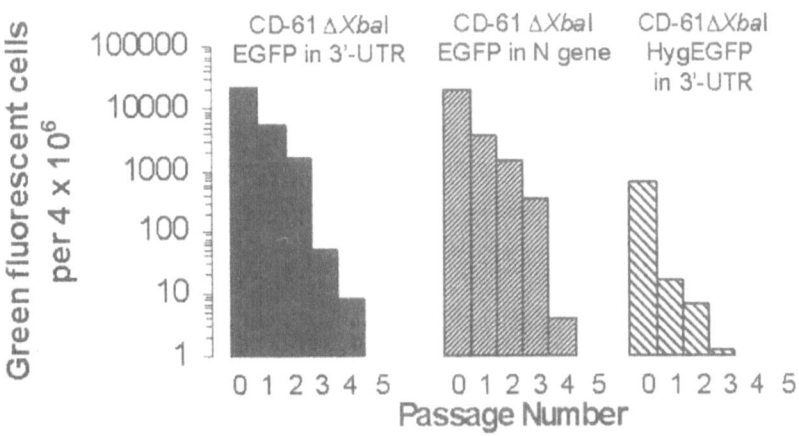

Figure 2. Number of cells expressing the EGFP and HygEGFP genes after serial 24 h passages. The EGFP gene cassette was inserted into the N gene or 3'-UTR sequence of pCD-61ΔXbaI and the HygEGFP gene cassette was inserted into the 3'-UTR. A 1 cm² or 5 cm² counting window was attached to the culture flask and bright green cells were counted using a mercury vapour illuminated Leitz Fluovert inverted microscope. The result was adjusted for number of cells as necessary and multiplied to obtain the total for each flask of CK cells.

Insertion of reporter genes into IBV D-RNAs seems to limit their accumulation during serial passage, and of the above constructs only CD-61 CAT in N gene was faintly visible on Northern blots. To confirm that each of the reporter genes was being transcribed from a D-RNA-derived reporter-specific mRNA, total RNA isolated from various passage points of rescue experiments was reverse transcribed with oligonucleotide primers complementary to each reporter gene. The resulting cDNAs were amplified by polymerase chain reaction using reporter-specific and leader primers and sequenced using the reporter-specific primers. IBV Beaudette mRNA 5 was reverse transcribed, amplified and sequenced from the same RNA extracts as a control. CAT gene mRNAs were detected from P1 and P2 (CD-61 CAT in N gene), EGFP gene mRNAs from P0 and P1 (CD-61Δ*Xba*I EGFP in N and CD-61Δ*Xba*I EGFP in 3'-UTR) and Hyg gene mRNAs from P0 to P5 (CD-61Δ*Xba*I Hyg in 3'-UTR). For each reporter gene mRNA and IBV Beaudette mRNA 5 the sequence obtained contained the IBV Beaudette leader fused to the 3'-most consensus TAS sequence of the gene 5 TAS element, consistent with the coronavirus transcription model of discontinuous synthesis during subgenomic (-) strand production (Sawicki and Sawicki, 1990).

We have optimised a potential selection system for Hyg gene or HygEGFP gene-containing recombinant IBV. Hygromycin B is an aminoglycoside antibiotic that inhibits translation by stabilising the peptidyl tRNA at the ribosomal acceptor site (Cabanas *et al.*, 1978; Gonzalez *et al.*, 1978). Uninfected cells are relatively impermeable to Hygromycin B, but cells infected with a number of viruses become permeable (Benedetto *et al.*, 1980). CK or Vero cells treated with hygromycin B were examined at 8 h intervals for three days. We determined the concentration at which only mild cytopathic effect was observed at 24 h (the duration of one viral passage) or 72 h (the duration of an IBV plaque assay; data not shown). Vero cells could tolerate 10-30 fold more hygromycin B than CK cells.

The effect of hygromycin B on virus titre during a 24 h passage of IBV Beaudette or IBV M41 was determined (Figure 3A). The highest concentrations of hygromycin B tolerated for 24 h gave 3-5 $\log_{10}$ reduction in virus titre. One or more passages in the presence of hygromycin B should enable us to enrich the surviving virus population with recombinant viruses. Relative hygromycin B resistance has been used to discriminate between BCoV field isolates (Kapil *et al.*, 1999), but we believe that Hyg-containing recombinant IBV would have an advantage over any resistant IBV that may emerge since similar systems have been successfully used to select recombinant retroviruses (Zhang *et al.*, 1999 and references therein). We also determined the effect of incubation with hygromycin during plaque assay and found that the size of IBV Beaudette plaques was reduced by about half (Figure 3B). As long as reporter gene insertion in the variable region of the 3'-UTR does not give a small-plaque phenotype we should be able to identify recombinants by plaque size.

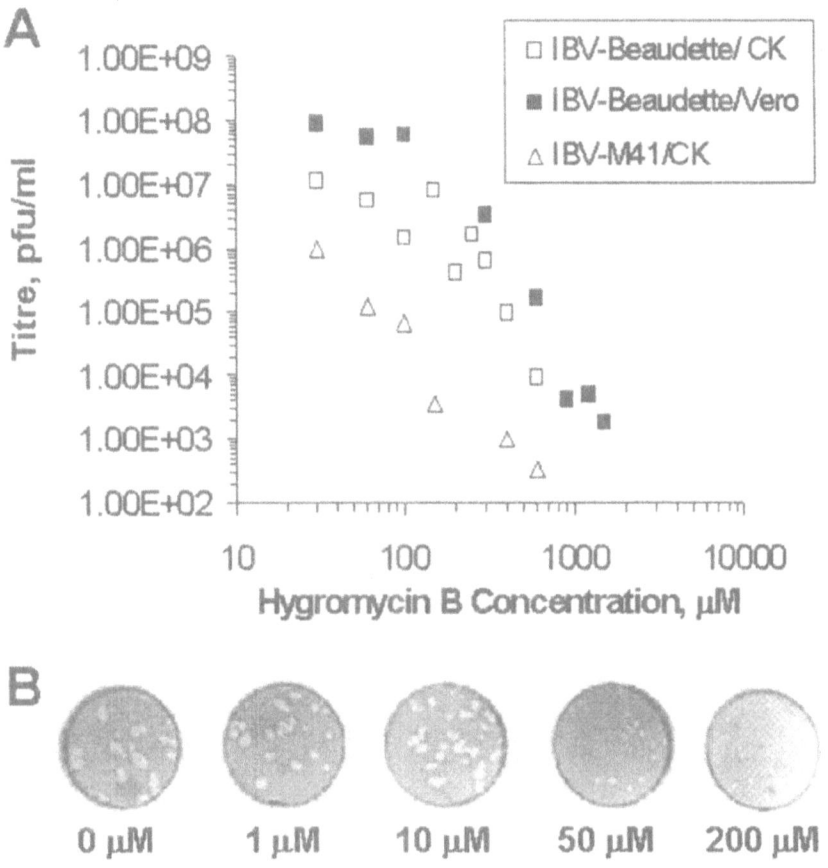

Figure 3. Effect of Hygromycin B on IBV titre and plaque size. (A) CK or Vero cells were incubated with hygromycin B for 6 hours before being washed and inoculated with IBV Beaudette or IBV M41 (approx. 10 p.f.u./cell) for 1 h. Cells were rinsed and incubated a further 24 h in medium containing the same concentration of hygromycin B. Titre was determined by plaque assay on CK cells. (B) Plaque assays on Vero cells inoculated with IBV Beaudette as usual and overlaid with agarose containing various concentrations of hygromycin B. Cells were fixed after 72 h and stained with 1% crystal violet to reveal plaques.

ACKNOWLEDGEMENTS

This work was supported by the Ministry of Agriculture, Fisheries and Food, UK, Project code OD1905 and by grant number CT950064 of the Fourth RTD Framework Programme of the European Commission (EC). B. Neuman was recipient of an IAH studentship.

REFERENCES

Benedetto, A., Rossi, G. B., Amici, C., Belardelli, F., Cioé, L., Carruba, G. & Carrasco, L. 1980. Inhibition of animal virus production by means of translation inhibitors unable to penetrate normal cells. *Virology* **106**, 123-132.

Cabanas, M. J., Vazquez, D., and Modolell, J. (1978). Dual interference of hygromycin B with ribosomal translocation and with aminoacyl-tRNA recognition. *Eur. J. Biochem.* **87**, 21-27.

Gonzalez, A., Jimenez, A., Vazquez, D., Davies, J. E., and Schindler, D. (1978). Studies on the mode of action of hygromycin B, an inhibitor of translocation in eukaryotes. *Biochim. Biophys. Acta* **521**, 459-469.

Kapil, S., Richardson, K. L., Maag, T. R. & Goyal, S. M. 1999. Characterization of bovine coronavirus isolates/from eight different states in the USA. *Vet. Microbiol.* **67**, 221-230.

Kottier, S. A., Cavanagh, D., and Britton, P. (1995). Experimental evidence of recombination in coronavirus infectious bronchitis virus. *Virology* **213**(2), 569-580.

Pénzes, Z., Tibbles, K., Shaw, K., Britton, P., Brown, T. D. K., and Cavanagh, D. (1994). Characterization of a replicating and packaged defective RNA of avian coronavirus infectious bronchitis virus. *Virology* **203**, 286-293.

Pénzes, Z., Wroe, C., Brown, T. D., Britton, P., and Cavanagh, D. (1996). Replication and packaging of coronavirus infectious bronchitis virus defective RNAs lacking a long open reading frame. *J. Virol.* **70**(12), 8660-8668.

Sawicki, S. G., and Sawicki, D. L. (1990). Coronavirus transcription - subgenomic mouse hepatitis virus replicative intermediates function in RNA synthesis. *J. Virol.* **64**, 1050-1056.

Stirrups, K. Shaw, K., Evans, S., Dalton, K., Casais, R., Cavanagh, D. & Britton, P. Expression of reporter genes from the defective RNA CD-61 of the coronavirus infectious bronchitis virus. *J. Gen. Virol.* **81**, 1687-1698.

Williams, A. K., Wang, L., Sneed, L. W., and Collisson, E. W. (1993). Analysis of a hypervariable region in the 3' non-coding end of the infectious bronchitis virus genome. *Virus Res.* **28**, 19-27.

Zhang J. & Sapp, C. M. Recombination between Two Identical Sequences within the Same Retroviral RNA Molecule. 1999. *J. Virol.* **73**, 5912-5917.

Characterization of an Arterivirus Defective Interfering RNA

Replication and homologous recombination

RICHARD MOLENKAMP, BABETTE C.D. ROZIER, SOPHIE GREVE, WILLY J.M. SPAAN, AND ERIC J. SNIJDER
Department of Virology, Center of Infectious Diseases, Leiden University Medical Center, Leiden, The Netherlands

1. INTRODUCTION

Recently, we have described the generation of DI-b, a natural equine arteritis virus (EAV) defective interfering (DI) RNA of 5.6 kb, and we have reported the construction of pEDI, a full-length cDNA copy of EAV DI-b RNA from which replication-competent RNA can be transcribed *in vitro* (Molenkamp *et al.*, 2000a, Molenkamp et al., 2000b). EDI RNA consists of three noncontiguous parts of the EAV genome fused in frame with respect to the replicase gene (Fig. 1). As a result the EDI RNA contains a truncated replicase open reading frame (ORF), which we will refer to as EDI-ORF. The importance of such a translation unit in coronavirus DI RNA propagation has been shown by a number of researchers (de Groot *et al.*, 1992; van der Most et al.,1995; Liao and Lai, 1995).

In this study we have analyzed the importance of the EDI-ORF in EDI RNA replication. The EDI-ORF was disrupted at different positions by the introduction of frameshift mutations. These were found either to block DI RNA replication completely or to be removed during virus passaging, probably due to homologous recombination with the helper virus genome. Using recombination assays based on EDI RNA and full-length EAV genomes containing specific mutations, the rate of homologous RNA recombination in the 3'- and 5'-proximal regions of the EAV genome was

The Nidoviruses (Coronaviruses and Arteriviruses).
Edited by Ehud Lavi *et al.*, Kluwer Academic/Plenum Publishers, 2001.

studied. Remarkably, the recombination frequency in the 5'-proximal region was found to be approximately 100-fold lower than that in the 3'-proximal part of the genome.

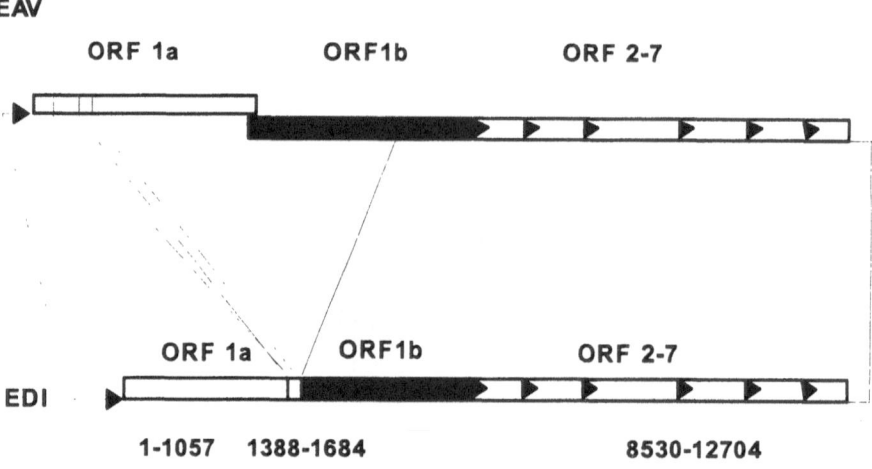

Figure 1. Schematic representation of the EAV genome and EDI RNA. Leader and body TRSs (see below) are depicted by arrowheads. The three noncontigous parts of the EAV genome of which EDI consists are indicated

2.   RESULTS

In view of the presence of a large ORF in many natural nidovirus DI RNAs, we have investigated the significance of the EDI-ORF in EAV DI RNA propagation. To this end, we engineered a set of pEDI derivatives containing C-terminally truncated EDI-ORFs of different sizes. By using a number of convenient restriction sites, we introduced frameshift mutations into the EDI-ORF. Furthermore, in construct pEDI-UAC, the EDI-ORF

Table 1. Overview of EDI-ORF frameshift mutants.

Construct	Restriction site	Position (nt)[a]	Position stop codon (nt)[a]	Length ORF
pEDI			2576	784
pEDI-UAC	N.A.	224	224	0
pEDI-127	*Mlu*I	589	605	127
pEDI-215	*Sst*I	857	869	215
pEDI-404	*Nco*I	1391	1436	404
pEDI-550	*Sal*I	1821	1874	550
pEDI-616	*Bam*HI	1975	2072	616

translation initiation codon was mutated. An overview of the EDI-ORF mutants is presented in Table 1. In vitro-transcribed RNA of the EDI-ORF mutants was transfected into EAV-infected baby hamster kidney (BHK-21) cells, and standard (undiluted) virus passaging experiments (Molenkamp *et al.*, 2000a) were performed. Intracellular RNA was isolated after P2 and analyzed by denaturing agarose gel electrophoresis and hybridization (Fig. 2). The positive control, EDI RNA containing the full-length EDI-ORF (784 aa), was replicated and passaged efficiently.Remarkably, RNA derived from the EDI-UAC translation initiation codon mutant could not be detected in P2 RNA samples. Likewise, the EDI-ORF mutants containing a relatively short reading frame (EDI-127 and EDI-215) could not be rescued. These results suggest that either translation of the 5'-terminal half of EDI-ORF or the N-terminal part of the EDI-protein is required for passaging of this EAV DI RNA.

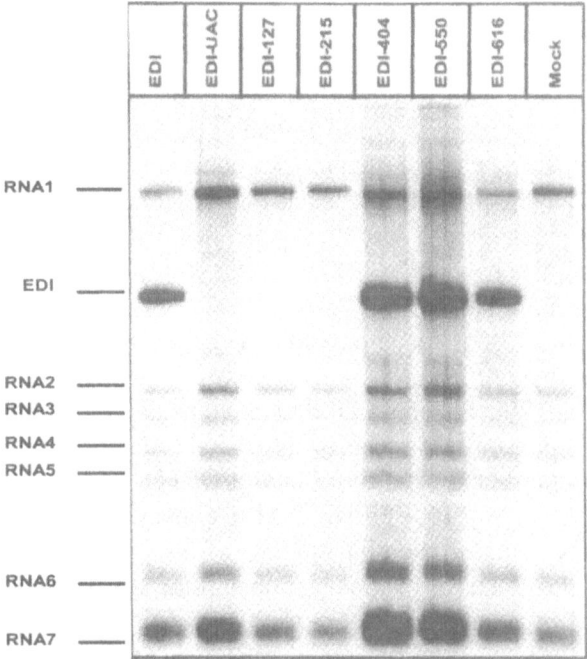

Figure 2. Analysis of the replication of EDI-ORF frameshift mutants. RNA from pEDI and EDI-ORF frameshift mutants was transfected into EAV-infected BHK-21 cells. Virus was harvested at 16 h posttransfection and passaged twice. P2 RNA was isolated at 12 h p.i. and subjected to gel electrophoresis and hybridization with an oligonucleotide recognizing the 3' ends of all viral mRNAs. The mock lane represents cells that were EAV infected but not transfected.

To analyze whether escape mutants containing a restored EDI-ORF were generated, intracellular P2 RNA of all EDI-ORF mutants was analyzed. The region corresponding to the EDI-ORF was amplified by using an EDI RNA-specific RT-PCR. PCR products were analyzed on agarose gel (data not shown) and the region containing the frameshift mutation was sequenced for each of the mutants that yielded a product. To our surprise, in all EDI-ORF frameshift mutants that were rescued efficiently (EDI-404, EDI-550, and EDI-616), the restriction site used to generate the frameshift mutation had been repaired and the flanking sequences in this region were completely identical to the wild-type (wt) EAV sequence. This strongly suggested that these mutants had undergone homologous recombination with the helper virus genome.

The results obtained with our EDI-ORF mutants strongly suggested that the rates of recombination may vary in different regions of the genome. To characterize this phenomenon in more detail, we investigated the relative rates of RNA recombination in the 3'- and 5'-proximal regions of the EAV genome. We designed recombination assays in which we made use of two previously described non-infectious but replication-competent full-length cDNA clones, mutants L3 and B3 (van Marle *et al.*, 1999). In these constructs, either the leader transcription-regulating sequence (TRS) (L3 mutant) or the RNA7 body TRS (B3 mutant) has been changed from 5'-UCAAC-3' to 5'-UGAAG-3'. In mutant L3 sg mRNA synthesis is completely abolished, whereas mutant B3 does not generate sg mRNA7. Thus, these mutants could serve as a potential partner in RNA recombination, which would yield an infectious EAV genome when the region containing the L3 or B3 TRS mutation would be exchanged for the corresponding region of an EDI RNA.

To analyze the rate of recombination in the 3'-proximal region of the EAV genome, mutant B3 was cotransfected with EDIC2-4150 RNA (Molenkamp *et al.*, 2000a). This EDI derivative contained the CAT reporter gene at the position normally occupied by ORF2b. In addition, EDIC2-4150 contains a deletion ranging from the 3' end of ORF3 (EAV nt 10723) to the 3' end of ORF5 (EAV nt 11636). EDIC2-4150 contained the wt RNA7 body TRS, and hence a single recombination with the B3 RNA in the 616-nt region between the 3' border of the deletion and the RNA7 TRS of EDIC2-4150 could yield a full-length, infectious EAV genome. Likewise, the rate of recombination in the 5'-proximal region of the genome was analyzed by cotransfection of the L3 mutant and EDIC2-4150. In this case, a single recombination event occurring in the 846-nt region between the leader TRS and the fusion of the first and second EDI segments of EDIC2-4150 (fusion site A) could generate a full-length, infectious RNA molecule. CAT expression was used to determine the number of double-transfected cells in an immunofluorescence assay (+/- 10% for both transfections). Medium from double-transfected cells was harvested at 12 h and the presence of infectious virus particles was determined by plaque assays (Fig. 3). Surprisingly, medium harvested from cotransfections of mutant B3 and

EDIC2-4150 (Fig. 3) contained substantially more recombinant virus particles than medium harvested from cotransfections of mutant L3 and EDIC2-4150 (Fig. 3). Only when a large amount of a 24-h harvest of the L3/EDIC2-4150 double transfection was used, a single plaque was observed (Fig. 3). Upon titration of the 24-h harvests from single transfections of mutant B3, mutant L3, or EDIC2-4150, no plaques were detected (Fig. 3), indicating that reversion of the L3 and B3 TRS mutations did not occur during the 24-h incubation period. We estimated that recombination in the 5'-proximal region of the genome was at least 100-fold less efficient than recombination in the 3'-terminal region.

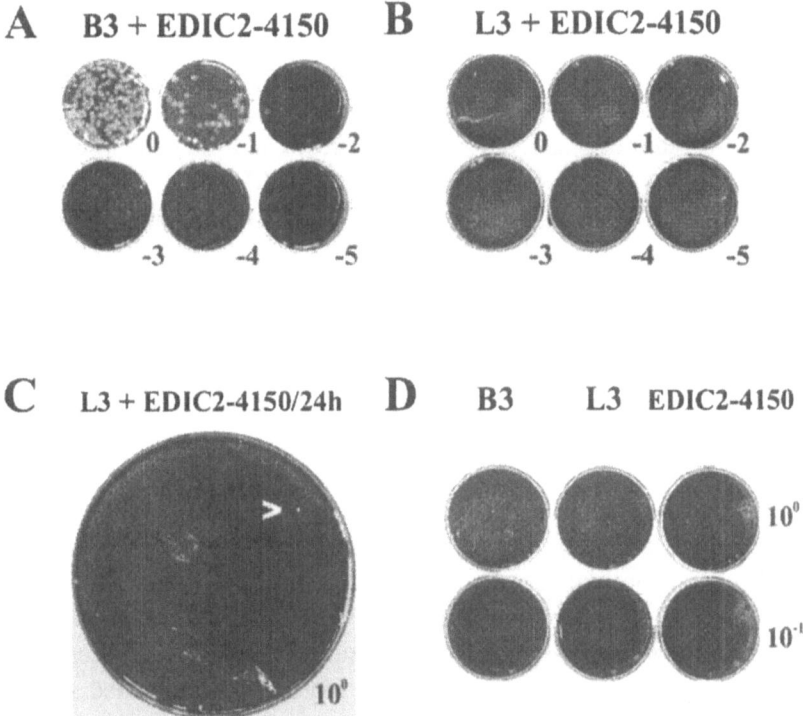

Figure 3. Titration of recombinant viruses. Medium from cotransfections of the B3 mutant (A) or L3 mutant (B) and EDIC2-4150 was harvested at 12 h posttransfection and titrated in plaque assays. Dilutions ranged from undiluted (upper left well) to 10^{-5} (lower right well). (C) Undiluted medium from a cotransfection of mutant L3 and EDIC2-4150, harvested at 24 h posttransfection, was used for a large-scale plaque assay. The single plaque observed is indicated with an arrow. (D) Control experiment. Medium isolated from single transfections of either the B3 mutant, the L3 mutant, or EDIC-4150 was harvested at 24 h posttransfection and used in a plaque assay. No plaques were detected.

3.   DISCUSSION

In this paper we demonstrate the importance of the EDI-ORF for the propagation of EDI RNA. As proposed by de Groot *et al.* (1992) and supported by van der Most *et al.* (1995), translation of the ORF in coronavirus DI RNAs, which often spans almost the entire DI genome, could enhance DI RNA stability. Consequently, the fitness of DI RNAs with an interrupted or truncated ORF might be reduced. In the case of EDI, the ORF covers only half of the RNA molecule (Molenkamp *et al.*, 2000a). However, it is interesting to note that in the region downstream of the truncated replicase ORF, in contrast to other nidovirus DI RNAs, EDI has retained the genomic 3'-terminal region that includes the *cis*-acting sequences required for the synthesis of all sg mRNAs. Therefore, we speculate that the nontranslated 3'-terminal part of the EDI genome might be stabilized by sg mRNA transcription rather than by translation. An alternative explanation for the requirement for a DI ORF is that one or more of the proteins it encodes is needed in *cis* for the replication of the DI RNA.

In this study, we compared the relative rates of recombination in the 5'- and 3'-proximal regions of an EAV DI RNA and mutant full-length genomes. In addition to recombinants, pseudotype virions could have been produced, containing both the DI RNA and the mutant genome, which might complement each other at the level of protein expression. However, our plaque assays and the analysis of the viral RNA isolated from individual plaques revealed that pseudotype virions were not generated (data not shown). Interestingly, our data showed that the relative rates of recombination in different regions of the EAV genome can vary. The relative recombination frequency in the 5'-proximal region of the genome was found to be approximately 100-fold lower than that in the 3'-terminal region. The high rate of recombination in the 3' end of the genome might be a direct result of the discontinuous nature of nidovirus sg RNA transcription. It has been proposed that minus-strand RNA synthesis is attenuated at body TRSs, a step which should be followed by base pairing with the leader TRS and reinitiation of minus-strand synthesis to add the antileader sequence (Konings *et al.*, 1988; Sawicki and Sawicki, 1995; van Marle et al., 1999; van Marle et al., 1995). In a similar manner, attenuation of minus strand synthesis might facilitate RNA recombination. The polymerase complex could reinitiate transcription of the nascent minus-strand at the same body TRS of a different template, and thus in our case continue to transcribe a recombinant full-length genome after initiating minus strand RNA synthesis from an EDI template. At the 5' end of the EAV genome, minus-strand synthesis might be attenuated much less frequently, which could explain the much lower rate of recombination in this region.

ACKNOWLEDGMENTS

We thank Marieke Tijms and Sasha Pasternak for helpful comments and discussions. We are grateful to Guido van Marle and Jessika Dobbe for the EAV mutant clones L3 and B3 and to Marieke Tijms for the PCR product containing the ORF1a translation initiation codon mutant. R.M. was supported by grant 700-31-020 from the Council for Chemical Sciences of the Netherlands Organization for Scientific Research (CW-NWO).

REFERENCES

de Groot, R. J., R. G. van der Most, and W. J. M. Spaan. 1992. The fitness of defective interfering murine coronavirus DI-a and its derivatives is decreased by nonsense and frameshift mutations. *J. Virol.* **66**:5898-5905.

Konings, D. A., P. J. Bredenbeek, J. F. Noten, P. Hogeweg, and W. J. M. Spaan. 1988. Differential premature termination of transcription as a proposed mechanism for the regulation of coronavirus gene expression. *Nucleic Acids Res.* **16**:10849-10860.

Liao, C. L., and M. M. C. Lai. 1995. A cis-acting viral protein is not required for the replication of a coronavirus defective-interfering RNA. *Virology* **209**:428-436.

Molenkamp, R., B. C. D. Rozier, S. Greve, W. J. M. Spaan, and E. J. Snijder. 2000a. Isolation and characterization of an arterivirus defective interfering RNA genome. *J. Virol.* **74**:3156-3165

Molenkamp, R., S. Greve, W.J.M. Spaan, and E.J. Snijder. 2000b. Efficient homologous RNA recombination and requirement for an open reading frame during replication of equine arteritis virus defective interfering RNAs. *J. Virol.* **74**:9062-9070

Sawicki, S. G., and D. L. Sawicki. 1995. Coronaviruses use discontinuous extension for synthesis of subgenome-length negative strands. *Adv. Exp. Biol. Med.* **380**:499-506.

van der Most, R. G., W. Luytjes, S. Rutjes, and W. J. M. Spaan. 1995. Translation but not the encoded sequence is essential for the efficient propagation of the defective interfering RNAs of the coronavirus mouse hepatitis virus. *J. Virol.* **69**:3744-3751.

van Marle, G., J. C. Dobbe, A. P. Gultyaev, W. Luytjes, W. J. M. Spaan, and E. J. Snijder. 1999. Arterivirus discontinuous mRNA transcription is guided by base-pairing between sense and antisense transcription-regulating sequences. *Proc. Natl. Acad. Sci. USA* **96**:12056-12061.

van Marle, G., W. Luytjes, R. G. van der Most, T. van der Straaten, and W. J. M. Spaan. 1995. Regulation of coronavirus mRNA transcription. *J. Virol.* **69**:7851-7856.

Packaged Heteroclite Subgenomic RNAs of PRRSV

SHISHAN YUAN, MICHAEL P. MURTAUGH, AND KAY S. FAABERG
Department of Veterinary PathoBiology, University of Minnesota, 1971 Commonwealth Avenue, St. Paul, MN 55108

1. INTRODUCTION

Porcine reproductive and respiratory syndrome virus (PRRSV) is a member of the Arteriviridae family of the Nidovirales (Meulenberg et al., 1993b; Nelsen et al., 1999). While the complete genomes of three PRRSV strains are known (Allende et al., 1999; Meulenberg et al., 1993; Nelsen et al., 1999), main questions remains to be elucidated concerning viral replication and transcription processes.

Coronaviruses, another Nidovirales, undergo high frequency RNA recombination (Lai, 1996) and produce DI RNAs (Makino et al., 1985; van der Most et al., 1991; Chang et al., 1994), which have been instrumental in determining regulatory cis-acting sequences for coronavirus replication and transcription. All known coronaviral DI RNAs were selected after repeated high multiplicity passage and were found to contain mostly multiple discontinuous genomic regions, while retaining similar viral 5' and 3' ends (Chang et al., 1994; Makino et al., 1985; Mendez et al., 1996).

We have demonstrated that PRRSV also undergoes homologous recombination thereby generating recombinant viruses (Yuan et al., 1999). There is yet no report about the appearance of DI RNAs. Since the production of defective interfering viral particles is a well-recognized consequence of passaging virus at high multiplicity of infection (m.o.i.), we hypothesized that PRRSV might also undergo non-homologous RNA recombination to generate DI RNAs. We found nine abundant subgenomic RNAs in cells and also in purified virions following high m.o.i. infection of

The Nidoviruses (Coronaviruses and Arteriviruses).

Edited by Ehud Lavi *et al.*, Kluwer Academic/Plenum Publishers, 2001.

MA-104 cells. However, the subgenomic RNAs are not typical with respect to other nidovirus DI RNAs, and required nucleotide direct repeats to mediate the non-homologous recombination. We suggest the term "heteroclite RNA" (meaning "deviating from common forms or rules") that accurately describes these novel PRRSV RNA species.

2.   MATERIALS AND METHODS

For serial consecutive passaging, VR-2332 was used to infect MA-104 cells at low m.o.i. (0.001) and high m.o.i. (2.5) and incubated in EMEM medium containing 10% fetal bovine serum (FBS).

At 36 h p.i., total cellular RNAs from MA-104 cells, mock infected or infected with PRRSV at low or high m.o.i., were isolated using an RNAeasy Midi Kit (QIAgen Inc.).

The supernatants were subjected to differential centrifugation, followed by pelleting through 0.5 M sucrose / 10 mM Tris (pH7.5) / 10 mM NaCl / 1 mM ethylenediaminetetraacetic acid (EDTA) cushion in a Beckman SW 27 tube and centrifuged at 24,000 rpm for 16 hoursto pellet virions. Viral RNA was purified as described (Nelsen et al., 1999).

RNA blotting was performed as described previously (Nelsen et al., 1999). Oligonucleotide probes complementary to individual genomic regions of PRRSV RNA were synthesized, 3'-end radiolabeled with [^{32}P]dATP (Amersham Life Sciences), and used for hybridization to the membrane-bound RNA in QuikHyb buffer (Stratagene).

Rapid amplification of cDNA ends (3'-RACE PCR) was performed on purified viral RNA templates or PRRSV-infected total cellular RNA essentially as described by Frohman (1994). The PCR products were separated on a 1% agarose gel and each DNA band was purified using QIAgen gel purification kit. The purified RT-PCR fragments were cloned into pGEM-T vector and sequenced by using the ABI 377 automatic sequencer.

3.   RESULTS

3.1   Northern blot analysis of PRRSV- infected total cellular RNA

When total RNA from infected MA-104 cells was analysed by northern blot, a probe specific for ORF 7 sequence (7-p14890) detected numerous extra subgenomic RNA species in addition to the viral genomic RNA and subgenomic mRNAs 2-7. When identical lanes were probed for a sequence in ORF1b (1b-p11982), the majority of the RNAs are negative except the genomic RNA. The third set of lanes, probed with an oligonucleotide specific to the 5' end of ORF 1a (1a-p222), revealed nine abundant subgenomic length RNA species. Unexpecetedly, the plaque purified virus produced the same RNA species to those seen in high m.o.i. viral passage.

To gain additional evidence on wether the existence of these RNAs was dependent on multiplicities of infection or passage number, we did similar analyses of total RNA from MA-104 cells infected with VR-2332 at low (0.001) MOI. Using the ORF1a-specific probe (1a-p222), the same size and pattern subgenomic RNAs were detected under all culture conditions. These results suggested that the production of subgenomic length ORF1a-specific RNAs may be a general property of PRRSV growth on MA-104 cells.

3.2 PRRSV atypical subgenomic RNAs are packaged into virions

Nested PCR was conducted to determine if intracellular atypical RNAs were packaged into virions, as described above. Both intracellular and extracellular RNA templates yielded the same PCR products. The pattern and sizes of the PRRSV-infected intracellular RNA products were consistent with the ORF1a-specific atypical RNA bands detected in northern blot analysis. Furthermore, the RNA harvested from purified virions produced comparably-sized products. These products were designated S-1 through S-7 (S=supernatant). Increased time of ultraviolet light exposure revealed two additional RT-PCR bands, S-8 and S-9. We previously demonstrated that PRRSV viral RNA, only when fully enclosed by the viral capsid, can escape RNase degradation (Yuan et al., 1999). Therefore, the results indicated that atypical PRRSV RNA species were packaged into virions. We arbitrarily designated these RNAs as I-1 through I-9 (I=intracellular), based on the sizes of the RNAs.

3.3 Nucleotide sequence of atypical subgenomic RNAs

Individual atypical RNA species were PCR amplified, cloned and sequenced to obtain the primary structure of the novel RNAs. In all cases, the nucleotide sequences revealed that the atypical subgenomic RNAs were composed of only two genomic regions of PRRSV, reflecting the joining of 5'-ORF1a nucleotides with downstream sequences in the 3'-end of the genome. The junction site was unique in most cases and existed at sites of short regions (3-6 nucleotides) of sequence identity between the specific 5' and 3' PRRSV genomic regions (Figure 1). In addition, we found an intracellular atypical RNA species (I-7) which was identical in sequence to a virion-derived atypical RNA (S-7b), consistent with the packaging of intracellular atypical subgenomic RNA into virions.

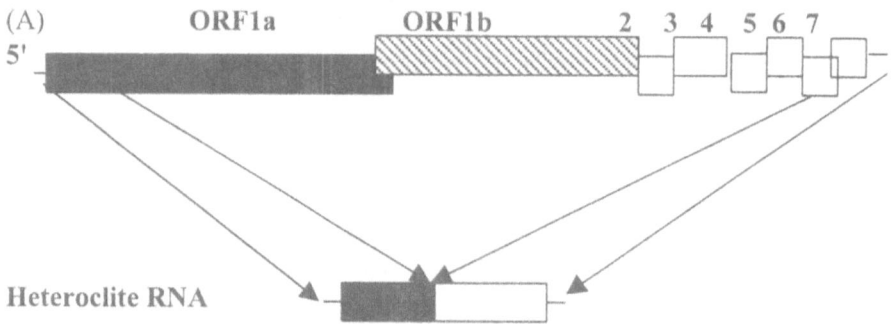

(B)

Heteroclite RNA	Direct repeat sequence	Junction position 5'	3'	size of sg RNA (kb)
S-9	CAAC	467	14342	1.53
S-8	TTCGC	878	14556	1.72
S-7	TCAGAG	1077	14567	1.91
H2S-7	TTTGGC	976	14429	1.95
I-7	TTTGGC	976	14429	1.95
I-9	GGC	471	14432	1.43

Figure 1. Primary structure of heteroclite RNAs. A. The overall structure of hetroclite RNAs and their relationship with that of the PRRSV genome. The arrows indicate the 5' and 3' regions from which the heteroclite RNAs originated. B. Nucleotide direct repeat (3 to 6 nt) forms the junction site between the two halves of the heteroclite RNAs.

4.   DISCUSSION

We identified nine novel PRRSV subgenomic RNAs that were constitutively present at all times during serial passage at high and low multiplicities of infection in MA-104 cells and were demonstrated in several PRRSV field strains that had not been cultured *in vitro* (data not shown). These atypical transcript species are different from typical defective interfering RNAs in that plaque purification did not elimnate these RNAs from viral stocks, interference specific to repeated high multiplicity passage was not observed (data not shown), and the subgenomic RNAs were present in virus isolated directly from the field. In contrast to the majority of coronavirus defective RNAs that inhibit the transcription of all (Mendez et al., 1996) or all but ORF 7 mRNA (Makino et al., 1990), PRRSV atypical subgenomic RNAs do not interfere with the genomic RNA and major subgenomic mRNAs. Furthermore, each of the atypical subgenomic RNAs was found to contain only contiguous sequences from the 5' and 3' ends of the viral genome, with a single large internal deletion. No ORF1b sequences which contains packaging signal of other Nidoviral genome was present. We propose that the PRRSV atypical RNA species described in this report be termed "heteroclite" subgenomic mRNAs because they result from single RNA recombination events similar to the proposed formation of the major subgenomic mRNAs, do not contain regions of ORF 1B shown to be critical in the formation of most coronaviral defective RNAs, seem to be constitutively present in infected cells, and are formed by a process of RNA similarity-assisted homologous recombination (Nagy P. 1997). These subgenomic RNAs thus appear to be part of the natural PRRSV life cycle.

ACKNOWLEDGMENTS

The authors would like to thank Dennis Foss, Daniel Mickelson, and Faith Klebs for their insightful discussions. The work was supported by grants from PIC USA, the National Pork Producers Council and Boehringer Ingelheim Vetmedica, Inc.

REFERENCES

Allende, R., Lewis, T. L., Lu, Z., Rock, D. L., Kutish, G. F., Ali, A., Doster, A. R., and Osorio, F. A. (1999). North American and European porcine reproductive and respiratory syndrome viruses differ in non-structural protein coding regions. J. Gen. Virol. 80, 307-315.

Brian, D. A. and Spaan, W. J. M. (1997). Recombination and coronavirus defective interfering RNAs. Sem. Virol. 8, 101-111.

Chang, R. Y., Hofmann, M. A., Sethna, P. B., and Brian, D. A. (1994). A cis-acting function for the coronavirus leader in defective-interfering RNA replication. J. Virol. 68, 8223–8231.

Frohman, M. A. (1994). On beyond classic RACE (Rapid Amplification of cDNA ends). PCR Methods and Applications 4, S40-S58.

Lai, M. M. C. (1996). Recombination in large RNA viruses: coronaviruses. Sem. Virol. 7, 381-388.

Makino, S., Taguchi, F., and Fujiwara, K. (1984). Defective interfering particles of mouse hepatitis virus. Virology 133, 9-17.

Makino, S., Yokomori, K., and Lai, M.M. (1990). Analysis of efficiently packaged defective interfering RNAs of murine coronavirus: localization of a possible RNA-packaging signal. J. Virol. 64, 6045-53.

Mendez, A., Smerdou, C., Izeta, A., Gebauer, F. and Enjuanes, L. (1996). Molecular characterization of transmissible gastroenteritis coronavirus defective interfering genomes: packaging and heterogeneity. Virology 217, 495–507.

Meulenberg, J. J. M., Hulst, M. M., de Meijer, E. J., Moonen, P. L. J. M., den Besten, A., de Kluyver, E. P., Wensvoort, G., and Moormann, R. J. M. (1993). Lelystad virus, the causative agent of porcine epidemic abortion and respiratory syndrome (PEARS), is related to LDV and EAV. Virology 192, 62-72.

Nagy, P. D. and Simon, A. E. (1997). New insights into the mechanism of RNA recombination. Virology 235, 1-9.

Nelsen, C. J., Murtaugh, M. P., and Faaberg, K. S. (1999). Porcine reproductive and respiratory syndrome virus comparison: divergent evolution on two continents. J. Virol. 73, 270-280.

van der Most, R. G., Bredenbeek, P. J. and Spaan, W. J. M. (1991). A domain at the 3' end of the polymerase gene is essential for encapsidation of coronavirus defective interfering RNAs. J. Virol. 65, 3219–3226.

Yuan, S., Nelsen, C.J., Murtaugh, M.P., Schmitt, B. J., and Faaberg, K. S. (1999). Recombination between North American strains of porcine reproductive and respiratory syndrome virus. Virus Res. 61, 87-98.

Cloning Of A Transmissible Gastroenteritis Coronavirus Full-Length cDNA

JOSE M. GONZALEZ, FERNANDO ALMAZAN, ZOLTAN PENZES, ENRIQUE CALVO, AND LUIS ENJUANES
Department of Molecular and Cell Biology, Centro Nacional de Biotecnologia, CSIC, Campus Universidad Autónoma. Cantoblanco, 28049 Madrid, Spain

1. INTRODUCTION

To understand gene function and expression in coronaviruses it would be of interest to obtain a cDNA encoding a full-length infectious RNA. This has not yet been possible due to the large size of the coronavirus genome and the instability of plasmids carrying coronavirus replicase sequences. In this report we describe the construction of a transmissible gastroenteritis virus (TGEV) full-length cDNA. For this purpose, we started from a cDNA encoding the defective RNA DI-C, that was stably and efficiently replicated by the helper virus (Izeta *et al.*, 1999). During the completion of the cDNA an ORF 1a fragment that was toxic to the bacteria was identified. Advantage of this finding was taken by reintroducing the toxic fragment into the viral cDNA in the last cloning step. To enhance the stability of the viral sequence, the cDNA was cloned as a bacterial artificial chromosome (BAC), a system that has been useful to stably clone large DNA from a variety of complex genomic sources into bacteria. The cytomegalovirus (CMV) immediate-early promoter was placed upstream of the cDNA to make use of a two-step amplification system that couples RNA pol II-driven transcription in the nucleus with the amplification by the viral replicase in the cytoplasm.

The Nidoviruses (Coronaviruses and Arteriviruses).
Edited by Ehud Lavi *et al.*, Kluwer Academic/Plenum Publishers, 2001.

2.   MATERIALS AND METHODS

2.1   Cells, Viruses, Plasmids and Bacteria Strains

The TGEV strains PUR46-MAD and PUR46-C11 were grown and titered as described (Sánchez *et al.*, 1999). Plasmid pACNR1180 was kindly provided by J.D. Tratschin (Institute of Virology and Immunoprophylaxis, Mittelhaüsern, Switzerland). Plasmid pBeloBAC11 was kindly provided by H. Shizuya and M. Simon (California Institute of Technology, Pasadena, CA). Plasmid pRL-CMV was obtained from Promega. *E. coli* DH10B strain was obtained from GIBCO/BRL.

2.2   Plasmid DNA Preparation, Sequence Analysis and Plasmid Stability

The preparation of plasmid DNA, the sequence analysis and the study of pBeloBAC11-based plasmids stability were performed as described (Almazán *et al.*, 2000).

3.   RESULTS AND DISCUSSION

3.1   Cloning and Assembly of a TGEV Genome cDNA

As a backbone for the assembly of a TGEV full-length cDNA, the plasmid pDI-C, encoding a TGEV-derived defective RNA was used. The DI-C cDNA was flanked by the CMV immediate early promoter at the 5' end and a synthetic poly-A tail of 24 bp followed by the hepatitis delta virus ribozyme and the bovine growth hormone termination and polyadenylation sequences at the 3' end. DI-C RNA has three deletions: Δ1 of 10 kb, Δ2 of 1.1 kb and Δ3 of 7.7 kb. A set of cDNAs encoding the missing sequences was generated from the respiratory PUR46-MAD strain RNA by standard RT-PCR techniques and assembled as described (Fig. 1). To make a fully functional virus, regions corresponding to the S gene were obtained from the virulent PUR46-C11 strain, which replicated to high titers in both the enteric and the respiratory tracts.

D1-5' fragment (positions 9349-11173) could not be cloned stably in *E. coli* cells, as DNA underwent sequence rearrangements (González *et al.*, 2000). Interestingly, D1-5' sequence was split into two overlapping clones spanning positions 8978-9758 and 9581-11173, respectively, that were fully stable. These results indicated that none of the individual fragments, but their joint sequences, were the source of toxicity. In the last step, the $ClaI^{4417}$-$ClaI^{9615}$ fragment was inserted into the plasmid encoding the rest of the TGEV genome (pAC-$TGEV^{\Delta ClaI}$), previously transferred to the low-copy number vector pACNR1180 (10-12 copies/cell) to minimize instability. However, *E. coli* cells did not tolerate the full-length cDNA.

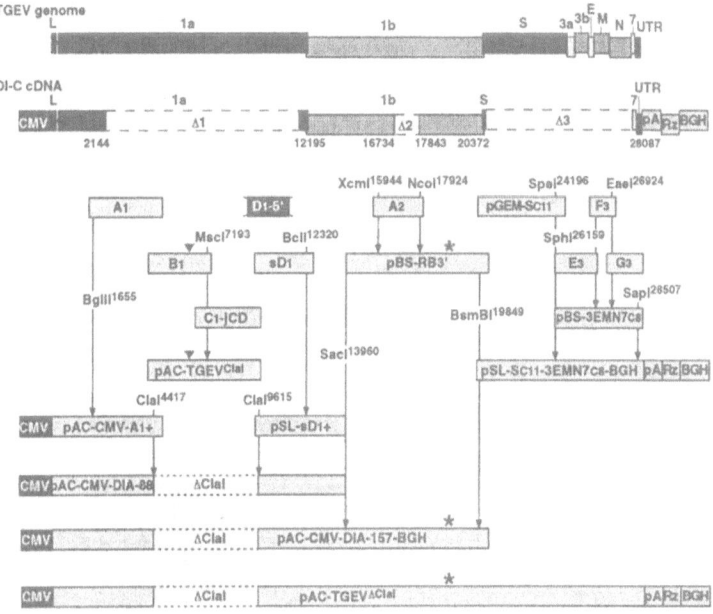

Figure 1. Strategy for the assembly of a TGEV full-length cDNA starting from the DI-C cDNA. CMV, cytomegalovirus immediate-early promoter; L, leader sequence; UTR, untranslated region; pA, tail of 24 A residues; Rz, hepatitis delta virus ribozyme; BGH, bovine growth hormone termination and polyadenylation sequences; ▼, marker at position 6752; *, marker at position 18997.

To improve definitively the stability of TGEV sequences in *E. coli* cells, the last step of the assembly was performed in the very-low-copy number vector pBeloBAC11 (1-2 copies/cell), allowing the construction of a full-length TGEV cDNA in *E. coli* DH10B cells.

3.2   Stability of the TGEV Full-Length cDNA as a BAC

The stability of the BAC-based plasmids was checked by restriction endonuclease analysis (González *et al.*, 2000). The plasmid encoding the full-length TGEV cDNA was stable in DH10B cells for at least 80 generations. Afterwards, it began to undergo rearrangements, mostly insertions from 0.7 to 1.6 kb in length throughout the ORF 1a (data not shown). The sequence of a ~0.7 kb insertion located at position 9480 was identical to that of an *E. coli* insertion sequence IS1, a transposon of 768 bp that is randomly inserted, generating 9-bp direct repeats in the target DNA.

To disrupt the toxic sequence and increase the stability of the TGEV full-length cDNA, a 133-nt intron from the plasmid pRL-CMV was inserted between the positions 9595 and 9596 to generate 5' and 3' intron splice sites matching the consensus sequence of mammalian introns. The intron-containing TGEV full-length cDNA was completely stable after 120 generations. *In vivo* transcription of the intron-containing full-length clones from the CMV promoter should assure splicing of the intron in the nucleus of the transfected cells..

4.   CONCLUSION

During the assembly of a TGEV full-length cDNA a ORF 1a region was determined as the cause of instability. The TGEV full-length cDNA was cloned as a bacterial artificial chromosome (BAC) in *E. coli* DH10B cells, where it was stable for, at least, 80 generations. The introduction of an intron in the ORF 1a completely stabilized the full-length cDNA.

ACKNOWLEDGMENTS

This research was supported by grants from the Comisión Interministerial de Ciencia y Tecnología (Spain), the Community of Madrid, the European Community (Biotechnology, FAIR and Control of Infectious Diseases Programs), and Fort Dodge Veterinaria (Spain).

REFERENCES

Almazán, F., González, J. M., Pénzes, Z., Izeta, A., Calvo, E., Plana-Durán, J. and Enjuanes, L. (2000). Engineering the largest RNA virus genome as an infectious bacterial artificial chromosome. *Proc. Natl. Acad. Sci. USA* **97**, 5516-5521.

González, J. M., Almazán, F., Pénzes, Z., Calvo, E. and Enjuanes, L. (2000). Construction of a stable transmissible gastroenteritis coronavirus full-length cDNA. Submitted.

Izeta, A., Smerdou, C., Alonso, S., Pénzes, Z., Méndez, A., Plana-Durán, J. y Enjuanes, L. (1999). Replication and packaging of transmissible gastroenteritis coronavirus-derived synthetic minigenomes. *J. Virol.* **73**, 1535-1545.

Sánchez, C. M., Izeta, A., Sánchez-Morgado, J. M., Alonso, S., Sola, I., Balasch, M., Plana-Durán, J. y Enjuanes, L. (1999). Targeted recombination demonstrates that the spike gene of transmissible gastroenteritis coronavirus is a determinant of its enteric tropism and virulence. *J. Virol.* **73**, 7607-7618.

Functional IBV Minigenomes Generated by Recombinant Fowl Pox Viruses for use in IBV-Targeted Recombination Studies

SHARON EVANS, DAVID CAVANAGH, AND PAUL BRITTON
Institute for Animal Health, Compton, Newbury, Berkshire, England

1.   INTRODUCTION

Having previously demonstrated that the IBV defective (D)-RNA CD-61, can be used as an RNA vector for the expression of heterologous genes via electroporation of transcripts into infected cells (Stirrups *et.al.,* 2000b) we wished to improve this procedure with another strategy. Hence, IBV D-RNA sequences (minigenomes) were introduced into the fowl pox virus (FPV) genome for use in a FPV/T7 system. Modified CD-61 D-RNA minigenomes were generated *in situ*, in a helper virus dependent manner, from cells co-infected with IBV, recombinant FPV containing CD-61 directed by the T7 promoter and rFPV expressing T7 RNA polymerase. Efficient rescue of CD-61 D-RNAs by helper IBV was demonstrated upon serial passage in tissue culture.

For recombination purposes, minigenomes (D-RNAs) require stable expression of extraneous sequences. Therefore, CD-61 was modified to contain the reporter gene luciferase (LUC). Recombinant FPV containing CD-61LUC was assessed in rFPV/T7 rescue studies for production of the D-RNA and expression of luciferase. In addition, we demonstrated the rescue of rFPV-generated CD-61 using the heterologous IBV strain M41.

The Nidoviruses (Coronaviruses and Arteriviruses).
Edited by Ehud Lavi *et al.,* Kluwer Academic/Plenum Publishers, 2001.

2. RESULTS AND DISCUSSION

We have already demonstrated that *in vitro* T7-derived D-RNA transcripts of CD-61 can be replicated and packaged (rescued) in a helper virus-dependent manner (Penzes *et al.*, 1994). Hence, CD-61 cDNA was modified by the addition of the hepatitis delta virus ribozyme plus T7 terminator, downstream of the 3' UTR, prior to cloning within the FPV transfer vector pEFL10 (Fig. 1). FPV-specific flanking sequences allow for homologous recombination of the CD-61 cassette into the genome of FPV. The LUC gene, under the control of IBV gene 5 transcription associated sequence (TAS; Hiscox *et.al.*, 1995) was cloned into the *Pma* CI site in domain III of CD-61.

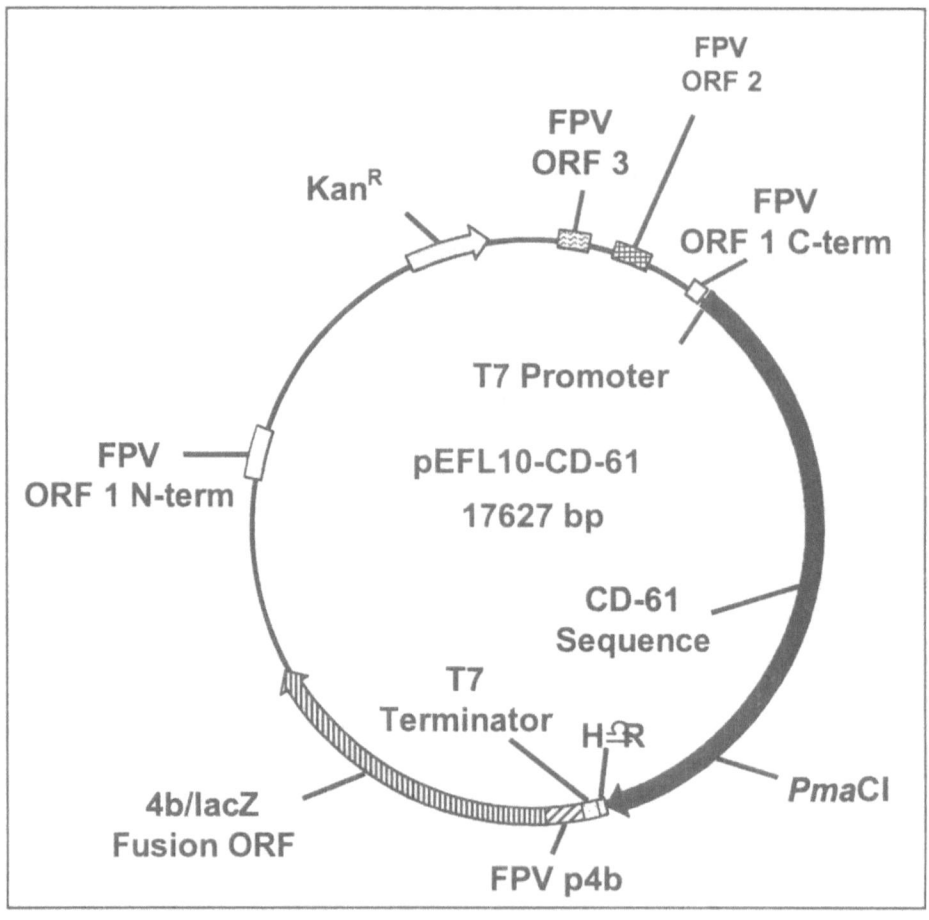

Figure 1. Schematic showing salient features of FPV transfer vector containing modified CD-61 D-RNA.

For rescue studies, chick kidney (CK) cells were co-infected with IBV (Beaudette), rFPV expressing T7 RNA polymerase and rFPV/CD-61. Serial passage of virus supernatants was follwed by total RNA extractions. Northern blot analysis of IBV-specific RNAs showed CD-61 D-RNA, which was rescued from passage 1 (P1) onwards. Similar experiments using Vero cells also demonstrated rescue of CD-61 D-RNA. Mammalian Vero cells do not support productive FPV infection, which demonstrates that the helper IBV has replicated and packaged the D-RNA; the presence of CD-61 in vero cell lysates is not the result of rFPV transmission.

Following efficient rescue of the IBV D-RNA in the rFPV/T7 system, we wished to assess the potential of rFPV/CD-61 for expression of heterologous genes. Recombinant FPV/CD-61LUC (containing the luciferase reporter gene) was isolated and analysed in rescue experiments. Extracted RNAs from CK cells co-infected with IBV, rFPV/T7 and rFPV/CD-61LUC were Northern blotted and hybridised to ^{32}P-labelled LUC-specific DNA. CD-61LUC D-RNA was detected from P1 and a potential LUC-specific mRNA was also present.

Infected cell lysates from rFPV/CD-61LUC rescue passages were assayed for expression of luciferase. Luciferase activity was observed from P0, showing a marked increase to P3/P4. Therafter, the activity declined to background levels by P7/P8. In previous studies, *in vitro* T7 transcripts of CD-61LUC were introduced into cells by electroporation prior to rescue. In this case luciferase levels were 200-fold less than for rFPV/CD-61LUC rescue. This demonstrates that the rFPV system supports expression of a reporter gene from the IBV D-RNA and is significantly more efficient than electroporation techniques.

The LUC-specific D-RNA-derived mRNA was amplified by RT.PCR prior to direct sequencing of the 5' end. Results identified the IBV leader sequence fused to the TAS site of gene 5, proximal to the LUC gene. Moreover, confirming the presence of an IBV-transcribed LUC-specific mRNA, as shown by Northern blotting.

Previous work has demonstrated rescue of CD-61, following electroporation of T7 transcripts, by heterologous IBV strains (Stirrups *et.al.,* 2000a). Consequently, we used IBV strain M41 as helper virus in the rFPV system. Total RNA from co-infection of CK cells with M41, rFPV/T7 and rFPV/CD-61 was assessed by Northern blotting. Results indicated that CD-61 D-RNA, derived from rFPV/CD-61, was rescued by IBV M41 from P2.

RT.PCR analysis of the D-RNA was used to sequence the 5' ends of the D-RNA rescued by heterologous helper IBV. The M41-rescued CD-61 had acquired the genomic leader sequence corresponding to the 5' of the M41 helper virus. Sequences downstream of the leader junction corresponded to the original CD-61 Beaudette sequence, confirming that leader switching

(Makino & Lai, 1989) occurred during rescue of CD-61 by M41. Furthermore, the leader switching event proves that rFPV/CD-61-derived D-RNA is transcribed and replicated by M41 IBV.

We have shown that fowl pox virus recombinants can improve delivery of D-RNA minigenomes to IBV-infected cells, which are replicated and packaged by the helper virus. This is expected to improve the opportunity for recombination between D-RNA containing a target gene and the helper IBV genome.

ACKNOWLEDGEMENTS

This work was supported by the Ministry of Agriculture, Fisheries and Food, UK, Project code OD1905 and by grant number CT950064 of the Fourth RTD Framework Programme of the European Commission (EC). S. Evans was supported by a BBSRC Realising Our Potential Award (ROPA).

REFERENCES

Hiscox, J. A.., Mawditt, K. L., Cavanagh, D. & Britton, P. (1995). Investigation of the control of coronavirus subgenomic mRNA transcription by using T7-generated negative sense RNA transcripts. *J. Virol*.**69**, 6219-6227.

Makino, S. & Lai, M. M. C. (1989). High-frequency leader sequence switching during coronavirus defective interfering RNA replication. *J. Virol*.**63**, 5285-5292.

Penzes, Z., Tibbles, K., Shaw, K., Britton, P., Brown, T. D. K. & Cavanagh, D. (1994). Characterisation of a replicating and passaged defective RNA of avian coronavirus infectious bronchitis virus. *Virology*. **203**, 286-293.

Stirrups, K., Shaw, K., Evans, S., Dalton, K., Cavanagh, D. & Britton, P. (2000a). Leader switching occurs during the rescue of defective RNAs by heterologous strains of the coronavirus infectious bronchitis virus (IBV). *J. Gen. Virol*.**81**, 791-801.

Stirrups, K., Shaw, K., Evans, S., Dalton, K., Casais, R., Cavanagh, D. & Britton, P. (2000b). Expression of reporter genes from the defective RNA CD-61 of the coronavirus infectious bronchitis virus. *J. Gen. Virol.* **81**, 1687-1698.

Heterogeneity of Subgenomic mRNAs of a Mutant Mouse Hepatitis Virus Strain JHM2C

XUMING ZHANG

Department of Microbiology and Immunology, University of Arkansas for Medical Sciences, Little Rock, Arkansas 72205

1.   INTRODUCTION

Mouse hepatitis virus (MHV), a prototype of murine coronavirus, contains a single-strand, positive-sense RNA genome of ≈32 kb in length (Lee et al., 1991; Pachuk et al. 1989). Upon viral infection into susceptible cells, the viral genomic RNA serves both as an mRNA for translation of the RNA-dependent RNA polymerase polyprotein, which is required for subsequent RNA transcription and replication and as a template for the synthesis of the negative-strand RNA (Lai and Cavanagh, 1997). The genome-length, negative-strand RNA, in turn, is used for the synthesis of the viral genome. Six to seven subgenomic mRNAs (mRNAs 2 to 7) are found in MHV-infected cells; they are co-nested at the 3'-ends. Each mRNA contains a leader sequence of approximately 70 nucleotides (nt) at the 5'-end, which is identical to the leader of the genomic RNA. Depending on MHV strains, there are two to four consensus UCUAA repeats with the last repeat being UCUAAAC, at the 3'-end of the leader. An identical or similar consensus sequence is present between each gene, termed intergenic (IG) sequence, which serves as a transcription initiation signal (promoter) for subgenomic mRNA synthesis (based on the leader-primed transcription model) or a termination signal for subgenomic negative-strand RNA synthesis (based on the discontinuous transcription on the negative-strand RNA) (Lai and Cavanagh, 1997, and ref. therein). Regardless of which transcription model coronavirus actually utilizes, the IG is the *cis*-acting sequence absolutely required for subgenomic RNA transcription; it serves as a joining point between the leader (or antileader) and the remaining body part of each subgenomic RNA. This structural feature lead to propose that coronavirus RNA transcription results from RNA-RNA base-pairing between the consensus sequences of the leader and the IG template.

Previously, we analyzed the subgenomic mRNA2-1 of a mutant MHV, JHM2c and found that mRNA2-1 is heterogeneous in the leader-body fusion site (Zhang and Lai, 1994). However, it was not known whether mRNA2-1 is unique among various mRNA species of JHM2c with respect to the leader-body fusion site. In the present study, the structures of all other subgenomic mRNA species of JHM2c were analyzed. It was found that the leader-body fusion sites in all subgenomic mRNA species were heterogeneous in JHM2c, thus supporting our previous hypothesis that the leader-body fusion during transcription does not require strict RNA base-pairing between the leader and the IG.

2.   MATERIALS AND METHODS

2.1   Cells and Virus

The murine astrocytoma cell line DBT (Hirano et al., 1974) was used for virus growth and infection. The naturally occurring small plaque mutant JHM2c of MHV (Makino et al. 1984) was used throughout this study.

2.2   Reverse Transcription and Polymerase Chain Reaction (RT-PCR) and Cloning of Viral Subgenomic mRNAs

For detection of viral subgenomic mRNAs, DBT cells were infected with JHM2c at a multiplicity of infection (m.o.i.) of 5. Virus grew in the presence of actinomycin D (10µg/ml). Intracellular RNAs were isolated at 7 h postinfection (p.i.) as described previously (Zhang et al., 1994). CDNAs were synthesized by RT with an antisense primer complementary to a sequence located approximately 300-nt downstream of each IG consensus sequence. The RT reaction was carried out at 42 °C for 90 min, and the PCR was performed in a thermocycler (DNA Engine, M.J. Research Inc.) for 30 cycles. The condition for each cycle was: denaturation at 95 °C for 30 seconds, annealing at 62 °C for 1 minute, and extension at 72 °C for 1 minute. PCR products were analyzed by agarose gel electrophoresis and cloned into the pTOPO2.1 TA cloning vector (In Vitrogen).

2.3   Analysis and Sequencing of cDNA Clones

All cDNA clones in the plasmid vector were analyzed by restriction enzyme digestion and agarose gel electrophoresis. Their sequences were determined with the automatic DNA sequencer (Model Prism 377, ABI) in the core facility of the Department of Microbiology and Immunology, UAMS. Either the T7 promoter primer or M13 reverse primer was used for DNA sequencing.

3.   RESULTS AND DISCUSSION

It has been shown previously that mRNA2-1 of JHM2c is heterogeneous at the leader-body fusion site. To determine whether such heterogeneity is a general phenomenon in subgenomic mRNA transcription of JHM2c, DBT cells were infected with MHV JHM2c. Intracellular viral RNAs were isolated. cDNAs were synthesized with an antisense primer to each subgenomic mRNA approximately 300-nt downstream of the consensus IG site, and were amplified with a sense primer specific to the leader. PCR fragments were directly cloned into the pTOPO2.1 TA cloning vector. Inserts were analyzed by agarose gel electrophoresis following restriction enzyme digestion with EcoRI. If each subgenomic mRNA species initiates from its respective IG consensus sequence, it would be expected that all cDNAs, which represent the 5'-end of each mRNAs, would be ≈370-nt in size. However, it was found that, in addition to the major fragment of ≈370-nt, longer or shorter inserts were identified in each subgenomic mRNA species (see Fig. 1 for an example, and further data not shown). Combined with the data on mRNA2-1 (Zhang and Lai, 1994), this result suggests that each sub-genomic mRNA species of JHM2c is heterogeneous in the initiation site.

To confirm that the difference in various mRNA species is due to the difference in transcription initiation site, DNA sequencing was performed to determine the sequence of all cDNA clones at the leader-body fusion site. Sequence results revealed that the majority of the cDNA clones contained the leader-body fusion site at the consensus IG region for each subgenomic mRNA species. Interestingly, some leader-body fusion sites occurred either upstream or downstream of the respective IG consensus sequence (Fig. 2 and data not shown), thus confirming that the observed size difference of the cDNA (see Fig. 1) represents the heterogeneous initiation sites of the subgenomic mRNA species. It is possible that some initiation sites, which are located at more than 300-nt downstream of the consensus IG, would have been overlooked, because any leader-body fusion occurring further downstream could not be amplified with the 3'-primers used in the RT-PCR

(Fig. 1). However, this procedure would not exclude the mRNAs initiated upstream of the IG. Nevertheless, it is noticeable that the majority of the leader-body fusion sites are located within a few hundred nucleotides upstream or downstream of the consensus IG sequence. This suggests that the IG region including the consensus sequence provides signal(s) for initiation. It is important to note that the leader-body fusion sites of these heterogeneous mRNAs contain little sequence homology with the templates, suggesting that, even if base-pairing between the leader and the IG consensus sequence plays an important regulatory role, it is not absolutely required for subgenomic mRNA transcription. Instead, other mechanisms are likely involved in such heterogeneous mRNA initiation, i.e. RNA secondary structure or protein-RNA and protein-protein interactions (Zhang et al., 1994; Zhang and Lai, 1994). Although the present data do not provide any direct evidence in supporting the latter potential mechanism, they are consistent with, and extend the previous finding that heterogeneous mRNA initiations is a general phenomenon for JHM2c. Further experiments are needed to determine whether mRNA heterogeneity occurs in other coronaviruses (Fisher et al., 1997).

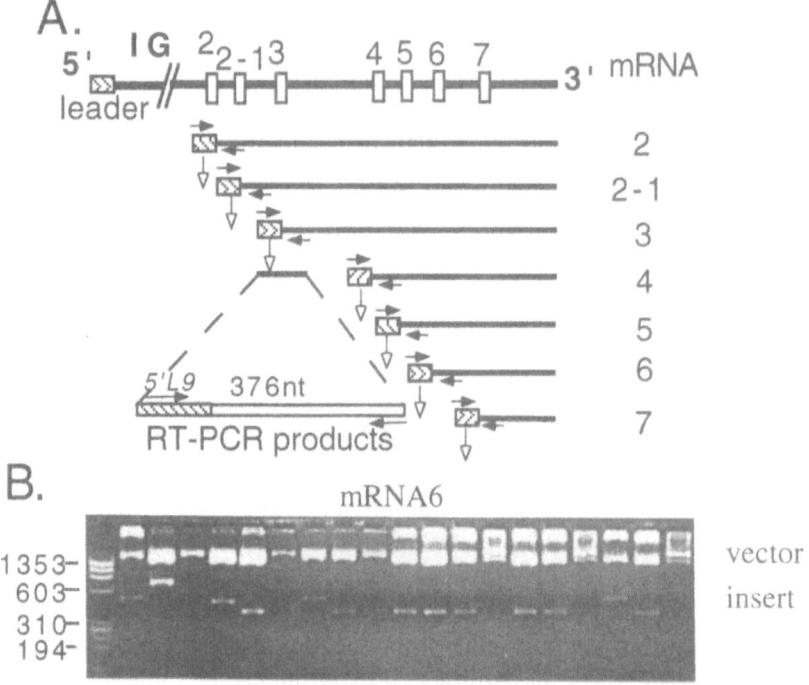

Figure 1. Strategy for RT-PCR amplification of subgenomic mRNAs of JHM2c. (A) Schematic diagram of the structure of subgenomic mRNAs and the locations of the primers. The size of the PCR fragments is indicated. (B) Heterogeneity of subgenomic mRNA 6. All clones in the pTOPO2.1 TA vector were digested with *Eco*R I, and analyzed by agarose gel electrophoresis.

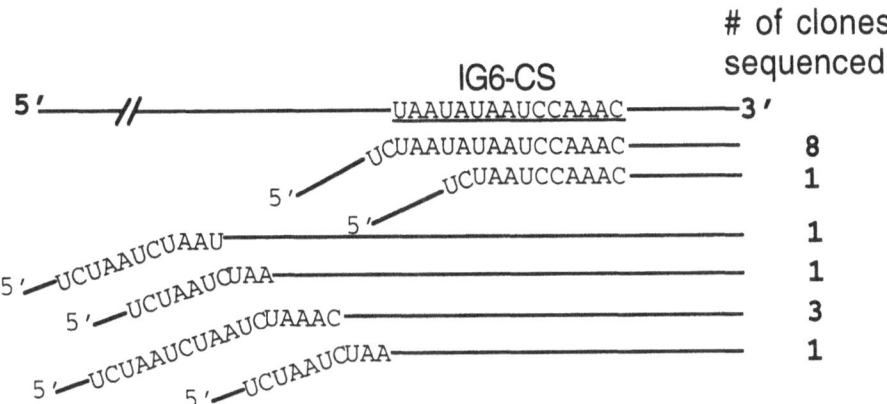

Figure 2. Summary of DNA sequencing results on the subgenomic mRNA 6. Template is shown in positive sense. Only the homology sequence between the template and mRNA is shown. Number of clones sequenced is indicated on the right.

ACKNOWLEDGMENTS

This work was supported by Public Health Services grants AI 41515 and AI 47188 from the National Institutes of Health, U.S.A.(to X.M.Z.).

REFERENCES

Fisher, F., Stegen, C.F., Koetzner, C.A., and Masters, P.S., 1997, Analysis of a recombinant mouse hepatitis virus expressing a foreign gene reveals a novel aspect of coronavirus transcription. *J. Virol.* **71**: 5148-5160.

Hirano, N., Fujiwara, K., Hino, S., and Matsumoto, M., 1974, Replication and plaque formation of mouse hepatitis virus (MHV-2) in mouse cell line DBT culture. *Arch. Gesamte Virusforsch.* **44:** 298-302.

Lai, M.M.C., and Cavanagh, D., 1997, The molecular biology of coronaviruses. *Adv. Virus Res.* **48**:1-100.

Lee, H.J., Shieh, C.K. Gorbalenya, A.E., Koonin, E.V., La Monica, N., Tuler, J., Bagdzhadzhyan, A., and Lai, M.M.C., 1991, The complete sequence (22 kilobases) of murine coronavirus gene 1 encoding the putative proteases and RNA polymerase. *Virology* **180:** 567-582.

Makino, S., Taguchi, F., Hirano, N., and Fujiwara, K., 1984, Analysis of genomic and intracellular viral RNAs of small plaque mutants of mouse hepatitis virus, JHM strain. *Virology* **39**:138-151.

Pachuk, C., Bredenbeek, P.J., Zoltick, P.W., Spaan, W.J.M., and Weiss, S.R., 1989, Molecular cloning of the gene encoding the putative polymerase of mouse hepatitis coronavirus strain A59. *Virology* **171**:141-148.

Zhang, X. M., and Lai, M.M.C., 1994, Unusual heterogeneity of leader-mRNA fusion in a murine coronavirus: Implications for the mechanism of RNA transcription and recombination. *J. Virol.* **68**:6626-6633.

Zhang, X. M., Liao, C.L., and Lai, M.M.C., 1994, Coronavirus leader RNA regulates and initiates subgenomic mRNA transcription both *in trans* and *in cis*. *J. Virol.* **68**:4738-4746.

Mapping of the RNA-Binding Domain of the Porcine Reproductive and Respiratory Syndrome Virus Nucleocapsid Protein

GIRISH C. DAGINAKATTE AND SANJAY KAPIL

Department of Diagnostic Medicine-Pathobiology, College of Veterinary Medicine, 1800 Denison Avenue,Manhattan, KS 66506

1. INTRODUCTION

The nucleocapsid (N) protein of PRRSV is a small basic protein of 15-kDa, constituting 20-40% of the protein content of virion (Casal *et al.*, 1998; Snijder and Meulenberg, 1998). It is 123 amino acids in length and approximately 20% of the residues are basic (Snijder and Meulenberg, 1998). The stretches of basic amino acid regions in the PRRSV N protein are conserved and these regions have been identified as nucleolar localization signals. The related coronavirus mouse hepatitis virus (MHV) protein interacts with RNA sequences in the leader RNA, the intergenic regions, and encapsidation signal, and this interaction is suggested to be important for transcription and encapsidation (Lai and Cavanagh, 1997).

The regions of the PRRSV RNA genome and the domains of the PRRSV N protein important for replication and RNA binding have not been defined. In this study, we constructed deletion mutants of PRRSV N protein and studied their interaction with the 5N leader and 3N NTR RNA of PRRSV genome by a northwestern blot assay. We identified a domain of N protein (residues 34 to 53) that interacts with specific domains of PRRSV RNA, and also defined the minimal binding regions of RNAs as the first 91 nt of 5N(+) leader RNA, and to the first 78 nt of 3N (+) NTR RNA.

The Nidoviruses (Coronaviruses and Arteriviruses).
Edited by Ehud Lavi *et al.*, Kluwer Academic/Plenum Publishers, 2001.

2.   MATERIALS AND METHODS

Full-length PRRSV N cDNA was synthesized from PRRS viral RNA by reverse transcription using primer N7 and was amplified by PCR using primer pair N1 and N7. The ORF 7 was subcloned into the pQE-30 expression vector. This resulting pQE-30-N was used as the parent plasmid to generate carboxy and amino terminal deletion mutants. All the recombinant proteins were expressed and purified according to the manufacturer's instructions (Qiagen Inc. Valencia, CA). The templates of different sizes for *in vitro* synthesis of 5N and 3N RNAs were generated by PCR from cDNAs using the specific primers. Purified recombinant mutant PRRSV N proteins were separated on a 16.5% T, 6% C Tricine-SDS-PAGE gel [8] and transferred to nitrocellulose membranes by electroblotting. Blots were hybridized with 1 X 10^5 cpm of [α^{32}P] UTP-radiolabeled RNA in binding buffer (50mM NaCl, 10mM Tris, 1mM EDTA, 0.02% bovine serum albumin, 0.02% Ficoll, and 0.02% polyvinyl pyrrrolidone, pH 7.0) for 6 h with gentle agitation. Unbound probe was removed by washing with binding buffer. Blots were dried, and bound RNA was visualized by autoradiography.

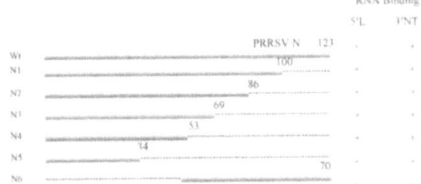

Figure 1. Deletion constructs of N protein. Deleted regions are indicated as dashed lines. Wt is the full length N protein. Length of each construct is indicated at the end as no. of amino acid residues. RNA binding is indicated as +/-.

3.   RESULTS

3.1   Identification of the RNA-binding domain of PRRSV N protein and protein binding domains of 5` and 3` NTR RNAs

To identify the RNA-binding domain, different mutants of PRRSV N protein were constructed as described above. Each deletion mutant differed by approximately 20 amino acid residues. To determine the amino acid residues of N protein involved in RNA-binding, we did northwestern blot assay with deletion mutants of N protein. The blots were probed with ^{32}P-labeled positive sense 5` leader (170 nt) and 3`(150 nt) NTR RNAs. All the N mutants except N5 mutant reacted similarly, with both the 5` leader and 3` NTR RNAs (Fig 2a and 2b). Quantification of radioactivity by scintillation counting from excised bands indicated that the interaction between N4 mutant and the PRRSV 5' leader and 3' NTR was greater than that between full length protein (Wt) and NTR RNAs. These observations indicated that, the RNA binding domain of PRRSV N protein is located between amino acid residues 34 and 53.

To identify the protein binding regions on the PRRSV 5` (+) leader 168 and 3` (+) 149 RNAs, different deletion constructs of NTRs were generated by PCR using specific primers to synthesize different RNA probes. 5' leader (+) 91 and 3' NTR (+) 78 RNAs interacted with N mutants in similar manner as full length RNAs. Thus, we have defined the minimal binding regions to first 91 nt of 5' leader and to first 78 nt of 3' NTR RNA.

A). Wt N1   N2 N3 N4 N5   B). Wt   N1 N2 N3 N4   N5

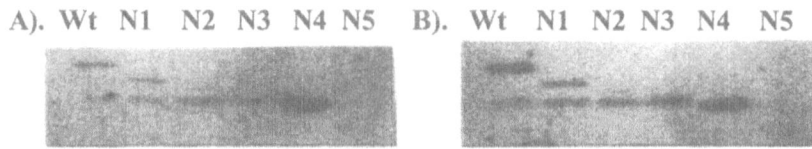

Figure 2. Northwestern blot analysis of recombinant deletion mutants of PRRSV N protein. The blots was probed with 5' (+) 91leader RNA (A) and 3' (+)78 NTR RNA (B). Lane 1,Wt N protein; lanes 2 to 6, PRRSV N deletion mutants N1, N2, N3, N4, N5, respectively.

3.2   Specificity of RNA-Protein Interactions

To confirm the above results, the PRRSV N construct lacking the RNA-binding domain (N6) (Fig 1) was constructed and checked for RNA binding. Along with N6 mutant, non specific mouse dihydro folate reductase (DHFR) and BCV N proteins having histidine tag were also checked for interaction with PRRSV 5' leader and 3' NTR RNAs. The PRRSV RNAs interacted only with N4 mutant but not with DHFR and BCV N, thus establishing the specificity of interaction. We also examined the binding of N protein in presence of unlabeled specific and nonspecific RNAs. Yeast t-RNA was used as nonspecific competitor. Blots were first incubated with unlabeled *in vitro* transcripts and then treated with labeled NTR RNAs. Results confirmed that the 5` and 3` NTR RNAs interact specifically with N protein.

Figure 3. Northwestern blot analysis confirming the specificity of PRRSV RNA-Protein interaction. The blots were probed with 5' (+) 168 leader RNA (A), and with 3' (+) 149 NTR RNA (B). Lane 1, BCV N protein; lanes 2 and 3 are PRRSN N mutants N6 and N5, respectively. Lane 4 is the DHFR protein.

4.   DISCUSSION

This is the first report demonstrating the interaction of the PRRSV nucleocapsid protein with 5N leader and 3N NTR RNAs in arteriviruses. In this study, we studied interactions of PRRSV N protein with 5N and 3N NTR RNAs by northwestern blot assay. The reaction of recombinant N fusion protein with NTR RNAs was similar to that of N protein from the purified virus. Therefore, we used prokaryotic system expressed fusion proteins for our mapping studies. The results presented here show that there is a specific interaction between N protein and NTR RNAs. We mapped the RNA-binding of PRRSV N protein to amino acid residues 34 to 53 (N4:NQSRGKGPGKKNKKKNPDKP) and also characterized the protein binding domain of NTR RNAs at the first 91 nt of 5Nleader and to first 78 nt of 3N NTR respectively.

The RNA-binding domain of the N protein has been characterized in many positive stranded RNA viruses (Nelson *et al.*, 2000). Recently in mouse hepatitis virus (MHV), a prototype member of Coronaviruses, the RNA-binding domain of N protein has been mapped to a 55 amino acids peptide containing residues 177-231 (Nelson *et al.*, 2000). The majority of amino acids residues involved in RNA-binding in all these RNA viruses were similar to those involved in our identified domain for PRRSV N protein. This RNA binding domain is rich in basic amino acids and is conserved among different isolates of PRRSV, however there is no homology with other arteriviruses. The predicted amino acid sequence of PRRSV N protein suggests, that the interactions between PRRSV N protein and RNAs is not only due to the high density of basic residues in this region. This was evident by our results in which other regions of N protein rich in basic amino acids did not bind to the RNA. The N4 mutant of PRRSV N protein has higher RNA-binding affinity compared with that of intact protein and, the N6 mutant lacking this region did not bind to RNA (Fig 3). This observation demonstrates that this region (34 to 53) alone imparts stronger RNA binding activity. This domain is the most immunoreactive region of

PRRSV N protein 10 and posses a unique epitope recognized by a monoclonal antibody SDOW (Wooton *et al.*, 1998). Our identified domain corresponds to a recently identified second nucleolar localization signal (NoLS) sequence of PRRSV N protein (Rowland *et al.*, 1999). These observations further demonstrate the importance and role of this domain of N protein in replication and encapsidation.

In arteriviruses regions of the RNA genome interacting with N protein have not been characterized. We have made an effort to study the regions of PRRSV RNA genome involved in interactions with N protein. In MHV, a sequence (AAUCYAAAC) present in 5` ends of mRNAs has been identified as the potential minimum specific ligand interacting with N protein (Nelson *et al.*, 2000). In our study we found that both 5N leader and 3N NTR RNAs interact similarly with PRRSV N. In 5N leader sequence of PRRSV, a 12 nucleotide stretch has been identified to form a potential RNA binding motif (Oleksiewicz *et al.*, 1999). This motif contains a CACCC site, which is present in both American and European 5N leader RNAs . This motif was present in our identified protein-binding domain of 5N leader RNA, but was not present in 3N NTR RNA. These observations indicate that CACCC motif alone might not responsible for binding with N protein. There was no sequence similarity between these RNAs. On comparison of predicted secondary structures of minimum binding regions of 5N leader and 3N NTR RNAs there was no similarity. Unlike in bovine coronavirus (BCV) in which the N protein of BCV interacts with the packaging signal of MHV (Cologna and Hogue, 2000), the PRRSV N protein did not bind to BCV 3N UTR RNA. We found that the denatured RNA probes reacted similarly as the non-denatured probes in northwestern blot assay (data not shown). These observations lead us to believe that binding of the PRRSV N protein to RNA might be sequence specific and further studies need to be done to identify the minimal aptamers interacting with PRRSV N protein. The interaction studies involving other regions of PRRSV RNA genome would help in identifying packaging signal of PRRSV. We are now constructing phage display libraries to study the role of each amino acid residue in this domain and are also studying ligand-protein interaction using multi-dimensional NMR spectroscopy to identify the secondary and tertiary structures of PRRS N protein involved in RNA binding.

REFERENCES

Casal, J. I., M. J. Rodriguez, J. Sarraseca, J. Garcia, J. Plana-Duran, and A. Sanz. 1998. Identification of a common antigenic site in the nucleocapsid protein of European and North American isolates of porcine reproductive and respiratory syndrome virus. Adv Exp Med Biol. **440**:469-77.

Cologna, R., and B. G. Hogue. 2000. Identification of a bovine coronavirus packaging signal. J Virol. **74**:580-3.

Corpet, F. 1988. Multiple sequence alignment with hierarchical clustering. Nucl. Acids Res., **16**(22), 10881-10890.

Lai, M. M., and D. Cavanagh. 1997. The molecular biology of coronaviruses. Adv Virus Res. **48**:1-100.

Nelson, G. W., S. A. Stohlman, and S. M. Tahara. 2000. High affinity interaction between nucleocapsid protein and leader/intergenic sequence of mouse hepatitis virus RNA. J Gen Virol. **81**:181-88.

Oleksiewicz, M. B., A. Botner, J. Nielsen, and T. Storgaard. 1999. Determination of 5'-leader sequences from radically disparate strains of porcine reproductive and respiratory syndrome virus reveals the presence of highly conserved sequence motifs. Arch Virol. **144**:981-7.

Rowland, R. R., R. Kervin, C. Kuckleburg, A. Sperlich, and D. A. Benfield. 1999. The localization of porcine reproductive and respiratory syndrome virus nucleocapsid protein to the nucleolus of infected cells and identification of a potential nucleolar localization signal sequence. Virus Res. **64**:1-12.

Schagger, H., and G. von Jagow. 1987. Tricine-sodium dodecyl sulfate-polyacrylamide gel electrophoresis for the separation of proteins in the range from 1 to 100 kDa. Anal Biochem. **166**:368-79.

Snijder, E. J., and J. J. Meulenberg. 1998. The molecular biology of arteriviruses. J Gen Virol. **79**:961-79.

Wootton, S. K., E. A. Nelson, and D. Yoo. 1998. Antigenic structure of the nucleocapsid protein of porcine reproductive and respiratory syndrome virus. Clin Diagn Lab Immunol. **5**:773-9.

Sequences Required for Replication and Packaging of IBV RNA

KEVIN DALTON,[1] ROSA CASAIS,[1] KATHLEEN SHAW,[1] KATHLEEN STIRRUPS,[1] SHARON EVANS,[1] T DAVID K BROWN,[2] PAUL BRITTON,[1] AND DAVE CAVANAGH[1]

[1]Institute for Animal Health, Compton Laboratory, Compton, Newbury RG20 7NN, UK: [2]Department of Pathology, Division of Virology, Unversity of Cambridge, Tennis Court Road, Cambridge CB2 1QP, UK.

1.   INTRODUCTION

We have used a naturally occurring IBV defective RNA (D-RNA), CD-91 (9.1 kb; Penzes et al., 1994) to investigate IBV replication and packaging signals. This D-RNA comprised the 5'-most 1133 nucleotides (domain I, including the 528 nucleotide 5' untranslated region, UTR) and 3'-most 1626 nucleotides (domain III) of the genome and a middle section (domain II) comprising discontinuous parts (6322 nucleotides in total) of the 1b open reading frame (ORF) of gene 1, the polymerase gene. In a limited study (Penzes et al., 1996) 3 kb of the ORF 1b sequence was removed to produce CD-61 which was replicated and packaged as efficiently as CD-91. Removal of a further 1.4 kb from ORF 1b produced D-RNA CD-44 which was not rescued (replicated and packaged; detected by Northern blot analysis) by helper IBV. This led us to propose that the 1.4 kb sequence was essential either for replication or, more likely, for packaging of the RNA. We have extended that investigation to an additional 26 modified D-RNAs. Previously we relied upon several passages of the rescued D-RNAs to produce sufficient RNA to be detectable by Northern blot analysis. The incorporation of a chloramphenicol acetyltransferase (CAT) reporter gene into the D-RNAs (Stirrups et al., 2000b) has allowed us to detect replication

The Nidoviruses (Coronaviruses and Arteriviruses).
Edited by Ehud Lavi et al.. Kluwer Academic/Plenum Publishers. 2001.

of IBV D-RNA constructs in transfected cells, without reliance on packaging to indicate that replication had occurred. Thus we have been able to distinguish the processes of replication and packaging.

2. METHODS

A CAT reporter gene was inserted at one of two positions in CD-61, under the control of the transcription associated sequence of the IBV gene 5 (Stirrups et al., 2000b). D-RNAs were electroporated into IBV-Beaudette-infected cells (passage 0; Stirrups et al., 2000a). CAT (measured by an ELISA) in P0 was indicative of *replication* of a D-RNA. The progeny of P0 were passaged up to five times (P1-P5). CAT or Northern blot detection showed that a D-RNA had been *rescued* – indicative of *packaging*. Deletion mutants of CD-61 (± CAT) were mostly made using existing restriction sites.

3. RESULTS

Only a few of the constructs are illustrated here. A D-RNA with the first 544, but not as few as 338, nucleotides of the 5'-terminus was replicated (Figure 1).

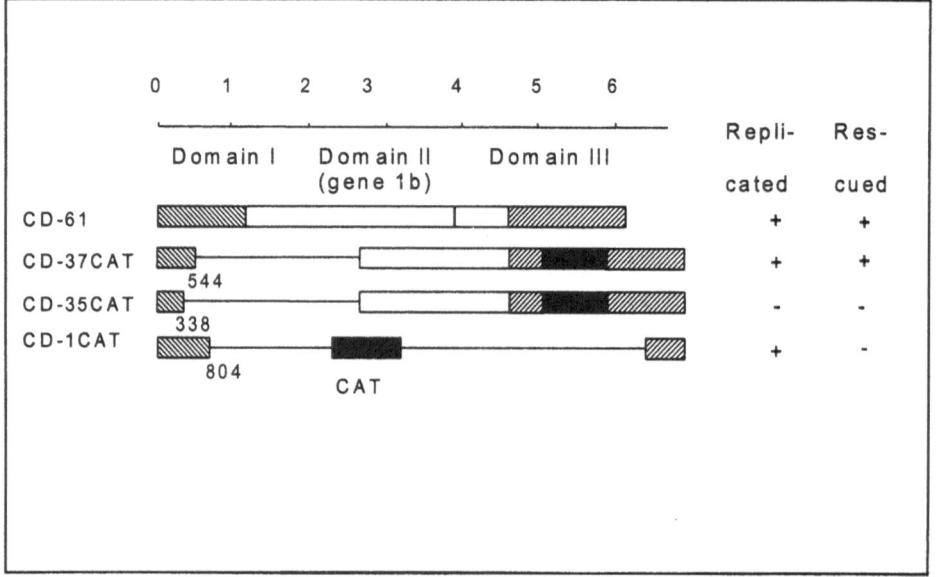

Figure 1. 5' sequences required for replication and packaging.

Region I of the 3' UTR, adjacent to the N gene, comprised 212 nucleotides and could be removed without impairment of replication or rescue of D-RNAs. A D-RNA with the final 338 nucleotides (CD-38CATstem+), including the 293 nucleotides in the highly conserved region II of the 3' UTR, was replicated (Figure 2). Thus the 5' terminal 544 and 3' terminal 338 nucleotides contained the necessary signals for RNA replication.

Phylogenetic analysis of 19 strains of IBV identified a conserved stem-loop at the 5' end of region II. Removal of the predicted stem loop (CD-38CATstem-) abolished replication (Figure 2).

D-RNAs in which the polymerase gene 1b sequence had been removed or replaced with all of the downstream genes were replicated well but rescued poorly, suggesting inefficient packaging. However, no *specific* part of the 1b gene was required for packaging.

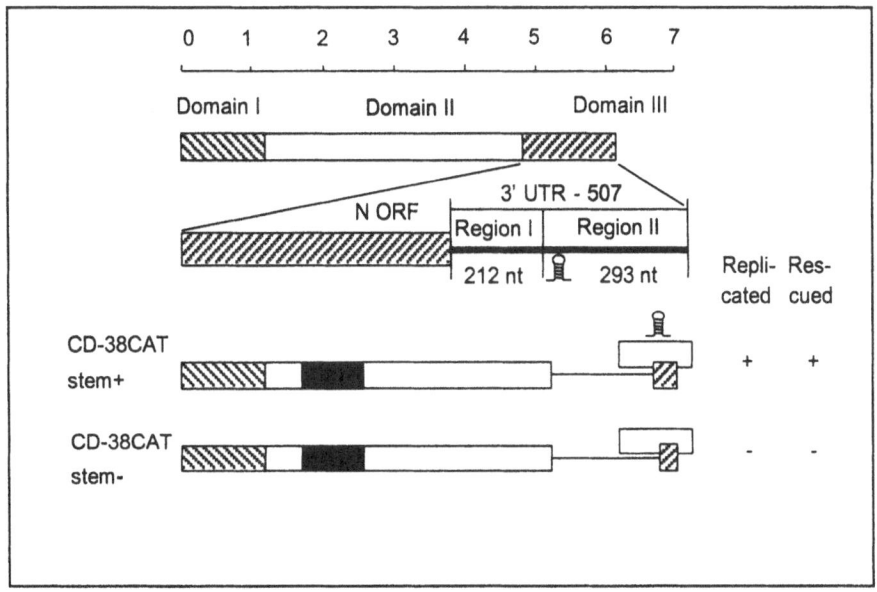

Figure 2. 3' sequences required for replication and packaging

4. DISCUSSION

The 5' UTR and region II of the 3' UTR contained the signals necessary for RNA replication; further deletions were lethal. Some gene 1b sequence, but not a specific part, was required for efficient packaging.

ACKNOWLEDGEMENTS

This work was supported by the Ministry of Agriculture, Fisheries and Food, UK (project code OD1904) and by grant number CT950064 of the Fourth RTD Framework Programme of the European Commission. Kevin Dalton and K Stirrups were the holders of Research Studentships from the Biotechnology and Biological Sciences Research Council (BBSRC). Sharon Evans was supported by a BBSRC Realising our Potential Award. Rosa Casais was the recipient of an EU TMR Marie Curie Research Training Grant.

REFERENCES

Penzes, Z., Tibbles, K., Shaw, K., Britton, P., Brown, T.D.K. and Cavanagh, D., 1994, Characterization of a replicating and packaged defective RNA of avian coronavirus infectious bronchitis virus (IBV). *Virology* **203**: 286-293.

Penzes, Z., Wroe C, Brown, T.D.K., Britton, P. and Cavanagh, D., 1996, Replication and packaging of coronavirus infectious bronchitis virus defective RNAs lacking a long open reading frame. *J. Virol.* **70**: 8660-8668.

Stirrups, K., Shaw, K., Evans, S., Dalton, K., Cavanagh, D. and Britton, P., 2000a, Leader switching occurs during the rescue of defective RNAs by heterologous strains of the coronavirus infectious bronchitis virus. *J. Gen. Virol.* **81**: 791-801.

Stirrups, K., Shaw, K., Evans, S., Dalton, K., Casais, R., Cavanagh, D. and Britton, P., 2000b, Expression of reporter genes from the coronavirus infectious bronchitis virus defective RNA CD-61. *J. Gen. Virol.*, **81**: 1687-1698.

Characterization of Temperature-sensitive (ts) Mutants of Coronavirus Infectious Bronchitis Virus (IBV)

SHUO SHEN AND DING XIANG LIU

Institute of Molecular Agrobiology, 1 Research Linke, National University of Singapore, Singapore 117604

1. INTRODUCTION

Coronavirus infectious bronchitis virus is a positive-stranded RNA virus, which synthesises a 3'-coterminal nested set of six subgenomic RNAs. Subgenomic RNA transcription and genome replication are directed by the viral replicase, which is expressed in the form of polyproteins 1a and 1a/b and subsequently processed into smaller nonstructural proteins by the virus-encoded proteinases. Major structural proteins S (spike protein), E (envelope protein), M (membrane protein) and N (nucleoprotein) are translated from subgenomic RNAs 2, 3, 4 and 6 and involved in nucleocapsid and virion assembling (Lai & Cavanagh 1998). Previous studies demonstrated that mutations and deletions on the structural proteins conferred the coronavirus mouse hepatitis virus ts phenotypes (Masters *et al* 1994 & Luytjes *et al* 1997). However, little is known about the mechanisms involved. In this study, ts mutants were generated by growing wild type virus at progressively lower temperatures from 35^0C to 28^0C on Vero cells. Two ts mutants were isolated from passages grown at 29^0C (ts291602) and 28^0C (ts282902). Sequence analysis reveals that mutations emerged in the S protein and an insertion occurred in the N protein. Biological and biochemical properties of ts mutants were characterised.

The Nidoviruses (Coronaviruses and Arteriviruses).
Edited by Ehud Lavi *et al.*, Kluwer Academic/Plenum Publishers, 2001.

2.   MATERIALS AND METHODS

2.1   Virus and cells

An IBV strain, Beaudette, was first adapted to grow on Vero cells at 37°C in our laboratory. In this study, the virus was plaque-purified and then passaged on Vero cells at progressively lower temperatures from 35°C to 28°C. Cells were maintained in complete DMEM medium (GIBCO BRL) supplemented with 10% newborn calf serum and infected with viruses at low multiplicity of infection (0.1). Plaque assays were performed as described previously (Shen *et al* 1994).

2.2   RT-PCR and sequencing

Viral RNA was extracted from partially purified viruses using the RNeasy Mini Kit (Qiagen) according to the manufacturer's instructions. Reverse transcription and polymerase chain reaction (RT-PCR) were performed using the Expand Reverse Transcription and High Fidelity PCR Kits (Boehringer Mannheim). Annealing and extension times of PCR were optimised for amplification of PCR products with different sizes using different primers. More than 100 specific primers were used for amplification, sequencing and cloning. Automated sequencing was carried out using PCR products and specific primers as previously described (Shen & Liu 2000). Sequence analyses were carried out using the GCG suite of programs.

2.3   Northern blot

Plasmid pGEMN1, covering nucleotides 26539 to 27608 under control of the SP6 promoter at the 3'-end of the IBV genome, was linearised with restriction enzyme PvuII (at position 27034). Transcription was performed using Dig Labelling Kit (Boehringer Mannheim) and the linearised template. The transcripts were treated with 20U/ml of RNase-free DNase (Boehringer) for 15 min at 37^{0}C, extracted with phenol/chloroform and precipitated with equal volume of 4M ammonium acetate and 6 volumes of ethanol. The pellet was washed once with cold 70% ethanol, resuspended in 50µl of H_2O and stored at -20^{0}C. Northern blot was performed as described previously (Shen *et al* 1999).

2.4 Analysis of viral structural proteins

Viral structural proteins were labelled, immunoprecipitated and separated on polyacrylamide gels as described previously (Liu *et al* 1998).

3. RESULTS

3.1 Isolation of ts mutants of IBV

An IBV strain Beaudette, grown on Vero cells at 37^0C, was plaque-purified and adapted to progressively lower temperatures by growing at 35^0C, 34^0C, 33^0C, 32^0C, 30^0C, 29^0C and 28^0C for 4, 5, 7, 44, 12, 21 and 29 times, respectively. Plaque assays of virus stocks were performed at both 32^0C and 40^0C. As shown in Table 1, viruses from early passages could form plaques at both temperatures at almost the same levels, but those from late passages only produced plaques at 32^0C. The results suggested that ts mutants emerged in the cold-adaptation process and dominated quickly when temperature was shifted to 28^0C. Two ts mutant clones, designated ts291602 and ts282902, were plaque-purified from passage 16 (29^0C), and passage 29 (28^0C), respectively. Plaque assays showed that ts291602 produced at least 1,000 time less plaques at nonpermissive temperature (40^0C) than at permissive temperature (32^0C), while ts282902 formed plaques only at permissive temperature. We believed that partially lethal and lethal ts mutants were generated through cold adaptation of the wild type on Vero cells.

3.2 Analysis of subgenomic RNA synthesis

Vero cells were infected with viruses at different temperatures. Viral subgenomic RNAs were extracted and detected with a probe complementary to the 3'-end of the positive-sense viral RNA from nucleotides 27034 to 27608 (nucleotide and amino acid numbering refer to those described by Boursnell *et al*, 1987). As shown in Figure 1, synthesis of viral subgenomic RNAs of ts mutant 28P29 (passage 29 at 28^0C) was hardly detectable under these conditions at 40^0C. On the other hand, normal amounts of the subgenomic RNAs were detected when the ts mutant 28P29 was grown at 32^0C and the wild type was grown at both 37^0C and 40^0C.

Table 1. Plaque assays of passages grown at different temperatures. Selected passages were assayed at both 32°C and 40°C.

Temperature	Passages	Passages assayed	Titre[#]	
			32°C	40°C
37°C	Wild type	wt63	5.0×10^1	2.0×10^6
32°C	P1 to P44	P44	2.7×10^5	2.0×10^6
30°C	P1 to P12	P12	1.1×10^5	1.4×10^5
29°C	P1 to P21	P10	1.5×10^4	5.0×10^3
		P13	2.5×10^5	2.5×10^2
		P21	5.0×10^5	1.3×10^3
28°C	P1 to P29	P1	1.0×10^5	5.0×10^1
		P5	3.3×10^6	NP*
		P29	2.9×10^6	NP

indicated as PFU/ml; * no plaques formed.

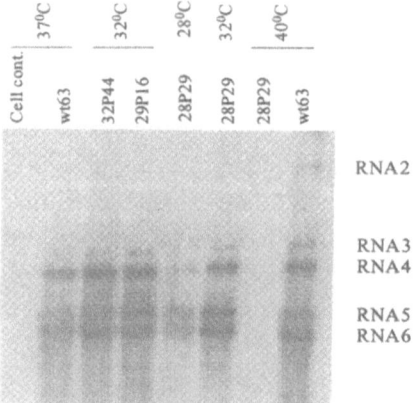

Figure 1. Subgenomic RNA synthesis. Northern blot analysis was carried out to detect viral subgenomic RNA synthesis of ts mutant 28P29 (passage 29 at 28°C) at 28°C, 32°C and 40°C (as indicated at the top). Cultures infected wild type wt63, passages 32P44 (passage 44 at 32°C) and 29P16 (passage 16 at 29°C) were used as controls and mock-infected culture as negative control. Subgenomic RNAs 2 to 6 are indicated.

3.3 Analysis of viral structural proteins

Cells were infected with wild type virus and ts mutants and viral proteins were labelled with ^{35}S-methionine. Viral structural proteins were immunoprecipitated with anti-IBV and antisera against the M and N proteins (Fig 2). The results revealed that the N proteins of the two mutants migrated slightly more slowly than that of the wild type, indicating mutation or insertion in the N protein of the ts mutants.

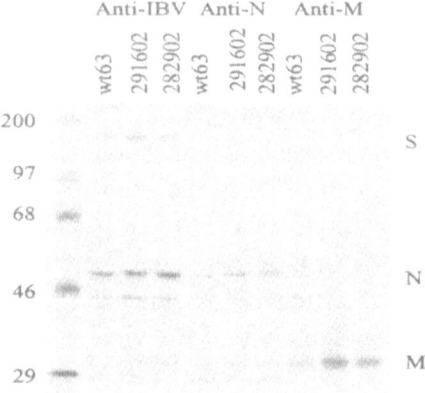

Figure 2. Structural protein profiles of plaque-purified wild type wt63, ts mutants ts291602 and ts282902. Labeled viral proteins were immuno-precipitated with antisera as indicated and separated on a 12.5% SDS polyacrylamide gel. Spike (S), nucleocapsid (N) and membrane (M) proteins are indicated. Numbers indicate molecular masses in kilodaltons.

3.4   Sequence analysis of the wild type virus and ts mutants

Nucleotide sequences of the wild type virus and ts mutants were determined using RT-PCR products of the three viruses. An insertion of 9 nucleotides, located near the 5'-end of the N gene of both mutants, was identified. This resulted in an in-frame Arg-Thr-Leu insertion in the N protein between the tenth and the eleventh residues of the wild type (Table 2). The insertion corresponds to the slow migration of the N protein of the ts mutants. In addition, three point mutations (Table 2), which caused two amino acid substitutions (a Gln^{294} to Leu^{294} and an Ile^{769} to Met^{769}) in the S protein of both mutants, were identified. To date, sequences of the 5'- and 3'-NCR, half of the 1a region, whole 1b and structural regions of the wild type and ts mutant have been determined and no other mutations, insertions and deletions of the deduced gene products were identified except those mentioned above.

Table 2. Sequence analysis of the plaque-purified wild type and ts mutants.

Nucleotide position	Wt63		ts291602		ts282902	
	Nt[#]	Aa*	Nt	Aa	Nt	Aa
21247/8	CA	Q	UU	L	UU	L
22674	A	I	G	M	G	M
25902/3	-	-	AGGACCUUA	RTL	AGGACCUUA	RTL

[#] nucleotide; * amino acid; - locus corresponding to the insertion in the mutants.

4.   CONCLSION

In this study, ts mutants emerged in the process of passaging the wild type IBV at progressively low temperatures. Under this selective pressure, ts mutants dominated soon in the viral population when the temperature shifted from 29^0C to 28^0C. To date, 80% of the genomes of the mutants have been sequenced and two mutations, Glu^{294} to Leu^{294} and Ile^{769} to Met^{769}, in the S protein and an Arg-Thr-Leu insertion in the N protein of the ts mutants have been identified. The insertion of the mutants was confirmed by PAGE analysis. It is noted that synthesis of the viral subgenomic RNAs of the ts mutants was hampered at the nonpermissive temperature, although no mutations were identified in the 1b polyprotein including the polymerase domains. Further studies are underway to map the defective gene(s) in structural and/or nonstructural regions for ts phenotypes and to understand the mechanisms involved.

REFERENCES

Boursnell, M.E., Brown, T.D., Foulds, I.J., Green, P.F., Tomley, F.M., & Binns, M.M. 1987, Completion of the sequence of the genome of the coronavirus avian infectious bronchitis virus. *J. Gen. Virol.* **68** (Pt 1): 57-77.

Lai, M..M.C., & Cavanagh, D., 1998, The Molecular Biology of Coronaviruses. In *Advances in Virus Research* **48**: 1-99.

Liu, D.X., Shen, S., Xu, H.Y., & Wang, S.F., 1998, Proteolytic mapping of the coronavirus infectious virus 1b polyprotein: evidence for the presence of four cleavage sites of the 3C-like proteinase and identification of two novel cleavage products. *Virology* **246**: 288-297.

Luytjes, W., Gerritsma, H., Bos, E., & Spaan, W., 1997, Characterization of two temperature-sensitive mutants of coronavirus mouse hepatitis virus strain A59 with maturation defects in the spike protein. *J. Virol.* **71**(2): 949-955.

Masters, P.S., Koetzner, C.A., Kerr, C.A., & Heo, Y., 1994, Optimization of targeted RNA recombination and mapping of a novel nucleocapsid gene mutation in the coronavirus mouse hepatitis virus. *J. Virol.* **68**:328-337.

Shen, S., Burke, B., & Desselberger, U., 1994, Rearrangement of the VP6 gene of a group A rotavirus in combination with a point mutation affecting trimer stability. *J.Virol.* **68**:1682-1688.

Shen, S., McKee, T.A., Wang, Z.D., Desselberger, U., & Liu, D.X., 1999, Sequence analysis and *in vitro* expression of genes 6 and 11 of an ovine group B rotavirus isolate, KB63: evidence for a non-defective, C-terminally truncated NSP1 and a phosphorylated NSP5. *J. Gen. Viral.* **80**:2077-2085.

Shen, S., & Liu, D.X., 2000, Determination of the complete nucleotide sequence of a vaccine strain of porcine reproductive and respiratory syndrome virus and identification of the NSP2 gene with an unique insertion. *Arch. Virol.* **145**:871-883.

Identification of a Noncanonical Transcription Initiation Site for Transcription of a Subgenomic mRNA of Mouse Hepatitis Virus

XUMING ZHANG

Department of Microbiology and Immunology, University of Arkansas for Medical Sciences, Little Rock, Arkansas 72205

1.   INTRODUCTION

Subgenomic mRNA transcription of mouse hepatitis virus (MHV) involves the interaction between the leader and the intergenic (IG) sequence of the template. All mRNAs contain multiple open reading frames (ORF) with the exception of the smallest mRNA. In general, only the 5'-most ORF of each mRNA is translated into a protein via the cap-dependent ribosomal scanning mechanism, while the down-stream ORFs are not translatable. The E (envelope) protein of MHV is a structural protein (Yu et al., 1994), essential for virion assembly. It is translated from the downstream ORF of the bicistronic mRNA5 *in vitro* (Thiel and Siddell, 1995), suggesting that translation of the E protein is cap-independent, possibly via an internal ribosomal entry site (IRES). Subsequently, the IRES has been mapped through a series of deletion mutants in an in vitro translation assay to be located between ≈100-nt upstream and 180-nt downstream of the ORF5b (Jendrach et al., 1999). The gene encoding the E in other coronaviruses includes ORF5b in bovine coronavirus (BCoV), ORF4 in transmissible gastroenteritis virus (TGEV) and human coronavirus (HCoV) 229E, and ORF3c in avian infectious bronchitis virus (IBV). Interestingly, while the ORF4 of TGEV and HCoV-229E, and ORF5b of BCoV are translated via the cap-dependent mechanism from a distinct mRNA species, mRNA4 and

The Nidoviruses (Coronaviruses and Arteriviruses).
Edited by Ehud Lavi *et al.*, Kluwer Academic/Plenum Publishers, 2001.

mRNA5-1, respectively, the ORF3c of IBV and ORF5b of MHV are translated via an IRES from a downstream ORF of mRNA3 and mRNA5, respectively (see review by Lai and Cavanagh, 1997). More intriguingly, within TGEV strains, the ORF3b of Muller strain is translated from mRNA3 via internal ribosomal entry whereas that of the Purdue strain is translated from mRNA3-1, an mRNA species differing from mRNA3, via cap-dependent mechanism (O'Connor and Brian, 2000). It is not known, however, why different strains of the same TGEV or different coronaviruses evolved such distinct mechanisms in regulating the expression of the same genes.

Here it is reported that a novel mRNA was identified in JHM2c-infected cells. The leader-body joining site for this mRNA is located ≈150-nt down stream of the authentic IG for mRNA5. When this sequence was placed in front of the chloramphenicol acetyl-transferase (CAT) gene in the defective-interfering (DI) RNA-CAT reporter, it directed the synthesis of a subgenomic CAT-containing mRNA and the expression of CAT activity, thus confirming that the non-canonical sequence serves as a promoter for subgenomic mRNA transcription.

2. MATERIALS AND METHODS

2.1 Cells and Virus

The murine astrocytoma cell line DBT (Hirano et al., 1974) was used for virus growth, virus infection and RNA transfection. The naturally occurring small plaque mutant JHM2c of MHV (Makino et al. 1984) were used throughout this study.

2.2 Reverse Transcription and Polymerase Chain Reaction (RT-PCR) and Cloning of Viral Subgenomic mRNAs

For detection of viral subgenomic mRNAs, DBT cells were infected with JHM2c at a multiplicity of infection (m.o.i.) of 5. Intracellular RNAs were isolated at 7 h postinfection (p.i.) (Zhang et al., 1994) and used for cDNA synthesis by RT with an antisense primer 3'IG5-300 (5'-GCG TAG GCC GTG AAG CTA-3'). This primer is complementary to a sequence approximately 300 nucleotides (nt) downstream of the IG5 (between genes 6 and 5). An additional sense primer (5'L9) specific to the leader (Zhang et al., 1994) was used for PCR amplification. The conditions for the RT-PCR were essentially the same as described previously (Zhang and Lai, 1994). Briefly, the RT reaction was carried out at 42 °C for 90 min, and the PCR was performed in a thermocycler (DNA Engine, M.J. Research Inc.) for 30 cycles. The condition for each cycle was: denaturation at 95 °C for 30 seconds, annealing at 62 °C for 1 minute, and extension at 72 °C for 1 minute. For detection of subgenomic mRNAs containing the CAT reporter gene, RT-PCR was carried out using a primer complementary to the 3'-end of the CAT ORF and 5'L9. PCR products were analyzed by agarose gel electrophoresis either directly or after cloning and restriction enzyme digestion as indicated. PCR products were directly cloned into the pTOPO2.1 TA cloning vector (In Vitrogen).

2.3   Analysis and Sequencing of cDNA Clones

All cDNA clones were analyzed by restriction enzyme digestion and agarose gel electrophoresis. Their sequences were determined with the automatic DNA sequencer (Model Prism 377, ABI) with either the T7 promoter primer or M13 reverse primer.

2.4   Plasmid Constructions

For generating DI CAT reporters containing TIS5-1 (13-nt core sequence) and its mutant, DNAs were synthesized from the pDECAT2-1 DNA templates by PCR with the primer pairs 5'-SpeIG5-1(13)CAT (5'-TT<u>A CTA GT</u>G CTT CCA ATT TAA ATG GAG AAA AAA AT-3') and 3'-CAT542 (Zhang et al., 1994). The 5'-primer contains an *Spe*I site at the 5'-end (underlined), the 13-nt core sequence (bold case) and the first 14-nt of the CAT ORF at the 3'-end. PCR fragments were digested with *Spe*I and *Bsp*EI, and directionally cloned into pDECAT2-1, resulting in pTIS5-1CAT (Fig. 2).

2.5   In Vitro Transcription, RNA Transfection, and CAT Assay

In vitro RNA transcription, RNA transfection and CAT assay were carried out as described previously (Zhang et al., 1994).

3.   RESULTS AND DISCUSSION

3.1   Identification of a Noncanonical Initiation Signal for Transcription of a Novel mRNA Species of MHV JHM2c.

Previous studies have identified the heterogeneity of subgenomic mRNA2-1 of JHM2c (Zhang and Lai, 1994). However, it was not clear whether such heterogeneity is unique to mRNA2-1 or it is a general phenomenon to all mRNA species of JHM2c. To address this question, I began to analyze the structure of all mRNA species in JHM2c-infected cells. Results from such investigation are reported in the accompanying chapter. During the course of this study, I unexpected found a novel species of subgenomic mRNA. When intracellular viral RNAs were isolated and cDNAs corresponding to the 5'-end 370-nt of mRNA5 were synthesized with RT-PCR, two species of the cDNA were identified: the major species (15 of 20 clones) has a size of 370-nt and the minor one (5 out of 20 clones) contains ≈220-nt. This result suggests that the consensus IG5 sequence is the major site for leader-body fusion of mRNA5 and that a sequence downstream may serve as a second site for leader-body fusion for a minor mRNA species.

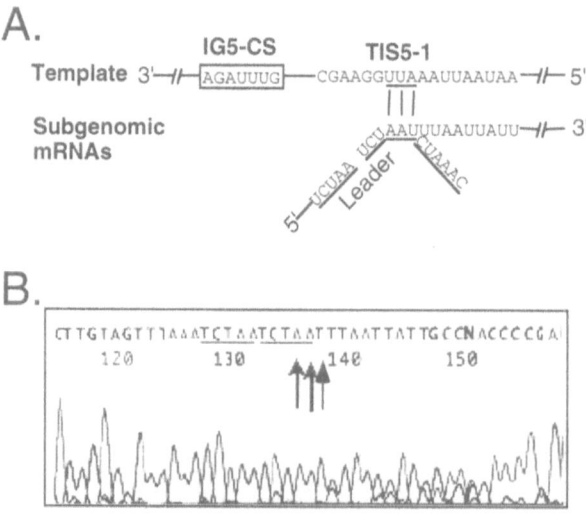

Figure 1. Identification of a noncanonical sequence for leader-body fusion of a novel subgenomic mRNA species. (A) mRNA leader-body fusion site. Template is in (-) sense; CS, consensus sequence; TIS, transcription initiation sequence. (B) Sequence profile showing the leader-body fusion site.

To confirm this finding, all 5 smaller clones representing the minor mRNA species were sequenced. As shown in Fig. 1, all 5 clones had a leader fused to a non-consensus sequence approximately 150-nt downstream of the authentic consensus IG5 sequence, with only three nucleotides complementary to the template. These data indicate that the noncanonical downstream sequence possibly serves as a site for leader-fusion for a smaller subgenomic mRNA species, which is termed mRNA5-1. It also suggests that RNA base-pairing may not be the major determinant for subgenomic mRNA initiation.

3.2 The Noncanonical Transcriptional Initiation Signal of JHM2c is a Functional Promoter for the Synthesis of a Subgenomic mRNA in a DI RNA-CAT Reporter System

To examine whether the non-canonical sequence is a functional promoter, the core sequence of 13-nt containing the leader-body fusion site for mRNA5-1 was placed in front of a CAT reporter gene in a DI RNA vector (Zhang et al., 1994). As shown in Fig. 2, the CAT activity expressed from TIS5-1CAT RNA-transfected cells was almost as high as that from IG7-CAT, which contains the promoter sequence for transcription of mRNA7 (Zhang et al., 1994). This result indicates that the 13-nt noncanonical sequence can drive the expression of the CAT gene in the DI RNA vector.

Because it is known that ORF5b is translated from mRNA5 via an IRES sequence, it is important to determine whether the CAT activity resulted from translation of the bicistronic DI RNA or of a separate subgenomic mRNA, even though this 13-nt does not contain the IRES sequence identified by Jendrach et al. (1999). To directly identify the CAT-containing subgenomic mRNAs, RNAs were isolated from MHV-infected and TIS5-1CAT RNA-transfected cells. RNAs isolated from IG7-CAT RNA-transfected cells and from cells without transfection were used as positive and negative controls, respectively. Subgenomic mRNAs were then amplified by RT-PCR with a CAT-specific antisense primer and a leader-specific sense primer. As shown in Fig. 2B, a specific subgenomic mRNA containing the CAT gene was identified in TIS5-1CAT RNA- and IG7-CAT RNA-transfected cells but not in the mock-transfected cells, indicating that a CAT-specific subgenomic mRNA was transcribed from the 13-nt sequence in TIS5-1CAT reporter. This result demonstrates that the noncanonical sequence serves as an initiation signal (promoter) for subgenomic mRNA transcription in a DI RNA-reporter system.

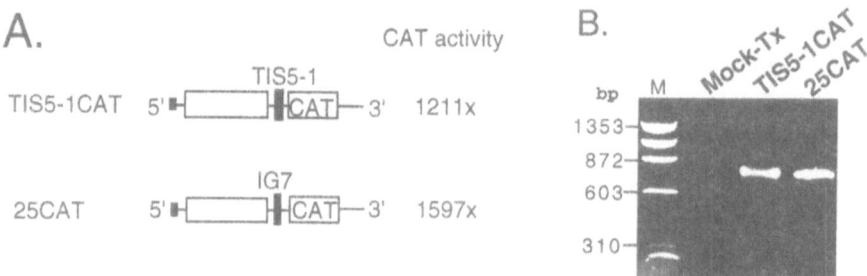

Figure 2. The noncanonical sequence serves as a promoter for transcription of the CAT-containing subgenomic mRNA in the DI CAT-reporter system. (A) TIS5-1, the newly identified transcription initiation sequence; IG7, intergenic sequence for mRNA7; CAT activity is the fold-increase above the background (1x). (B) RT-PCR results showing the specific amplification of subgemonic mRNAs as indicated.

The identification of a noncanonical sequence as a transcription initiation signal has two important implications in coronavirus gene expression. Firstly, coronavirus subgenomic mRNA transcription usually initiates at the IG consensus sequence. Base pairing between the IG and the leader has been shown to be very important for this discontinuous synthesis. The finding that noncanonical sequence can serve as the leader-body fusion site suggests that base-pairing may not be the determining factor in the selection of the initiation site. Instead, it suggests that RNA secondary structure or protein-RNA interaction is likely involved in the leader-body fusion. It would be interesting to further investigate these possibilities. However, because only a quarter clones represent mRNAs initiated from the downstream noncanonical sequence, the current data indicate that the consensus IG5 is a more preferred initiation site as compared to the downstream noncanonical sequence. Secondly, the present finding raises the possibility that ORF5b of MHV JHM2c can be translated from the 5'-most ORF of mRNA5-1. Indeed, we have recently observed that the E protein was translated from mRNA5-1 in an in vitro translation experiment (Zhang et al., unpublished results). Thus, it is likely that ORF5b of MHV can be translated from mRNA5 via internal entry of ribosome as well as from mRNA5-1 via cap-dependent mechanism. Why MHV requires two mechanisms to express the same protein remains unknown at present. But it may suggest that such a protein (E protein) is important for the viral life cycle. Indeed, the E protein is essential for virion assembly. In the case of TGEV, the ORF3b is translated from the second ORF of mRNA3 via internal entry of ribosome for the Muller strain, but it is expressed from the first ORF of mRNA3-1 via cap-dependent mechanism for the Purdue strain (O'Connor and Brian, 2000). These findings indicate that coronaviruses evolve diverse mechanisms for regulating their gene expression.

ACKNOWLEDGMENTS

This work was supported by Public Health Services grants AI 41515 and AI 47188 from the National Institutes of Health, U.S.A. (to X.M.Z.).

REFERENCES

Fisher, F., Stegen, C.F., Koetzner, C.A., and Masters, P.S., 1997, Analysis of a recombinant mouse hepatitis virus expressing a foreign gene reveals a novel aspect of coronavirus transcription. *J. Virol.* **71**: 5148-5160.

Hirano, N., Fujiwara, K., Hino, S., and Matsumoto, M., 1974, Replication and plaque formation of mouse hepatitis virus (MHV-2) in mouse cell line DBT culture. *Arch. Gesamte Virusforsch.* **44**: 298-302.

Jendrach, M., Thiel, V., and Siddell, S., 1999, Characterization of an internal ribosome entry site within mRNA5 of murine hepatitis virus. *Arch. Virol.* **144**:921-933.

Lai, M.M.C., and Cavanagh, D., 1997, The molecular biology of coronaviruses. *Adv. Virus Res.* **48**:1-100.

Lee, H.J., Shieh, C.K. Gorbalenya, A.E., Koonin, E.V., La Monica, N., Tuler, J., Bagdzhadzhyan, A., and Lai, M.M.C., 1991, The complete sequence (22 kilobases) of murine coronavirus gene 1 encoding the putative proteases and RNA polymerase. *Virology* **180**: 567-582.

Liao, C. L., and Lai, M.M.C., 1994, Requirement of the 5'-end genomic sequence as an upstream *cis*-acting element for coronavirus subgenomic mRNA transcription. *J. Virol.* **68**: 4727-4737.

Makino, S., Taguchi, F., Hirano, N., and Fujiwara, K., 1984, Analysis of genomic and intracellular viral RNAs of small plaque mutants of mouse hepatitis virus, JHM strain. *Virology* **39**:138-151.

O'Connor, J.B. and Brian, D.A., 2000, Downstream ribosomal entry for translation of coronavirus TGEV gene 3b. *Virology* **269**:172-182.

Pachuk, C., Bredenbeek, P.J., Zoltick, P.W., Spaan, W.J.M., and Weiss, S.R., 1989, Molecular cloning of the gene encoding the putative polymerase of mouse hepatitis coronavirus strain A59. *Virology* **171**:141-148.

Thiel, V., and Siddell, S., 1995, Translation of the MHV sM protein is mediated by the internal entry of ribosomes on mRNA5. In Corona- and Related Viruses (P.J. Talbot and G.A. Levy, eds.), Plenum Press, New York, pp.311-315.

Yu, X., Bi, W., Weiss, S.R., and Leibowitz, J.L., 1994, Mouse hepatitis virus gene 5b protein is a new virion envelope protein. *Virology* **202**:1018-1023.

Zhang, X. M., and Lai, M.M.C., 1994, Unusual heterogeneity of leader-mRNA fusion in a murine coronavirus: Implications for the mechanism of RNA transcription and recombination. *J. Virol.* **68**:6626-6633.

Zhang, X. M., Liao, C.L., and Lai, M.M.C., 1994, Coronavirus leader RNA regulates and initiates subgenomic mRNA transcription both *in trans* and *in cis*. *J. Virol.* **68:**4738-4746.

Infectious Bronchitis Virus Envelope Protein Targeting: Implications for Virus Assembly

EMILY CORSE AND CAROLYN E. MACHAMER
Department of Cell Biology and Anatomy, The Johns Hopkins University School of Medicine, Baltimore, MD

1.   INTRODUCTION

Coronaviruses acquire their membrane envelope by budding into the lumen of Golgi and pre-Golgi compartments. After budding, virions are thought to move in vesicles through the secretory pathway, and exit the cell when these vesicles fuse with the plasma membrane (Holmes *et al.*, 1981, Tooze *et al.*, 1987). The specific compartment into which coronaviruses bud is the *cis*-Golgi network (CGN), also known as the endoplasmic reticulum-Golgi intermediate compartment (Krijnse Locker *et al.*, 1994). Just as enveloped viruses that bud from the plasma membrane must direct the accumulation of their envelope proteins at the cell surface, coronaviruses must localize their envelope proteins to the membranes of the CGN.

The coronavirus avian infectious bronchitis virus (IBV) has three known membrane proteins. Studies of the intracellular localizations of the individual viral envelope proteins have revealed that the spike (S) protein is transported to the plasma membrane when expressed alone (Vennema *et al.*, 1990), while the matrix (M) protein is found in the CGN and *cis*-Golgi in the absence of virus infection (Machamer *et al.*, 1990). The envelope (E) protein, a recently discovered small transmembrane protein, is also localized to the Golgi complex when expressed alone (Corse and Machamer, 2000). Work with other coronaviruses suggests that the E protein has a critical role in virus assembly (Vennema *et al.*, 1996, Baudoux *et al.*, 1998, Fischer *et al.*, 1999).

The Nidoviruses (Coronaviruses and Arteriviruses).
Edited by Ehud Lavi *et al.*, Kluwer Academic/Plenum Publishers, 2001.

We are interested in understanding the mechanism by which coronavirus envelope proteins are collected in the CGN for virus assembly. In this study we examined the localizations of the IBV E and IBV M proteins when they are expressed alone and together by immunoelectron microscopy. Preliminary data suggests that the IBV M protein restricts the localization of the IBV E protein to the CGN. These results have interesting implications for coronavirus assembly.

2.   MATERIALS AND METHODS

The antibodies used in this study have been described previously (Corse and Machamer, 2000). Our peroxidase immunoelectron microscopy (IEM) protocol is based on that described in Brown and Farquhar, (1989), with the exception that all fixations were performed using microwave irradiation (Cluett and Machamer, 1996). BHK-21 cells were infected with recombinant vaccinia viruses vvIBVE and vvIBVM as previously described (Corse and Machamer, 2000). At 4 h postinfection the cells were fixed, stained, and processed for immunoelectron microscopy. Ultrathin (80 nm) sections were viewed on a Phillips CM 120 transmission electron microscope and photographed at 60 kV.

3.   RESULTS

We have shown that when the IBV E protein is expressed in cells by itself, it exhibits a juxtanuclear staining pattern that overlaps with a Golgi stack marker protein as well as resident proteins of the *cis*-Golgi complex and *trans*-Golgi network (Corse and Machamer, 2000). This suggests that the IBV E protein is distributed throughout the Golgi apparatus when expressed alone. When E is expressed with IBV M, the distribution of the two proteins completely overlap when examined by confocal microscopy (Corse and Machamer, 2000). Since IBV M is localized to the *cis*-Golgi when expressed alone, (Machamer *et al.*, 1990), the coexpression experiment suggested that E and/or M had an altered localization when the two proteins were coexpressed.

In order to examine the effects of the IBV E and M proteins on each others' localizations more closely, we performed immunoperoxidase electron microscopy (IEM). The IBV M protein was labeled in cells expressing M alone or M and IBV E together. In cells expressing M alone (Fig. 1, top panel), the label is present in a compartment adjacent to Golgi stacks, and in the nearest cisterna. This staining pattern most likely corresponds to the *cis*-Golgi network and *cis*-Golgi complex (Machamer *et al.*, 1990). The staining pattern of M protein in cells also expressing IBV E (Fig. 1, bottom panel) is similar to that in cells with only M protein, but it typically shows a broader distribution and may be restricted to the compartment adjacent to Golgi stacks.

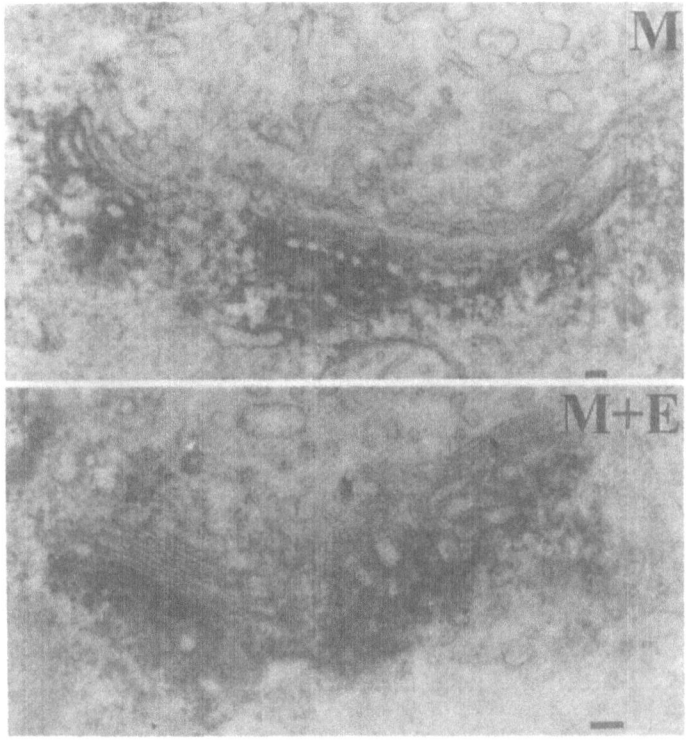

Figure 1. IBV E does not drastically alter the *cis*-Golgi localization of IBV M. BHK cells were infected with recombinant vaccinia viruses encoding M (top panel) or M and E (bottom panel) and processed for peroxidase IEM using anti-M antibody. Bar = 100 nm.

We also examined the effect of IBV M on the localization of IBV E by labeling the E protein in cells expressing E alone or E and M together. We observed that in cells expressing E alone, the peroxidase reaction product was distributed through Golgi stacks (Fig. 2, top panel), while in cells that were coexpressing the E and M proteins (Fig. 2, bottom panel), the product appeared only in a compartment adjacent to Golgi stacks. This suggests that the IBV M restricts the localization of IBV E, possibly to a pre-Golgi compartment similar to the budding compartment. Quantification of this result will require immunogold electron microscopy, which is currently in progress.

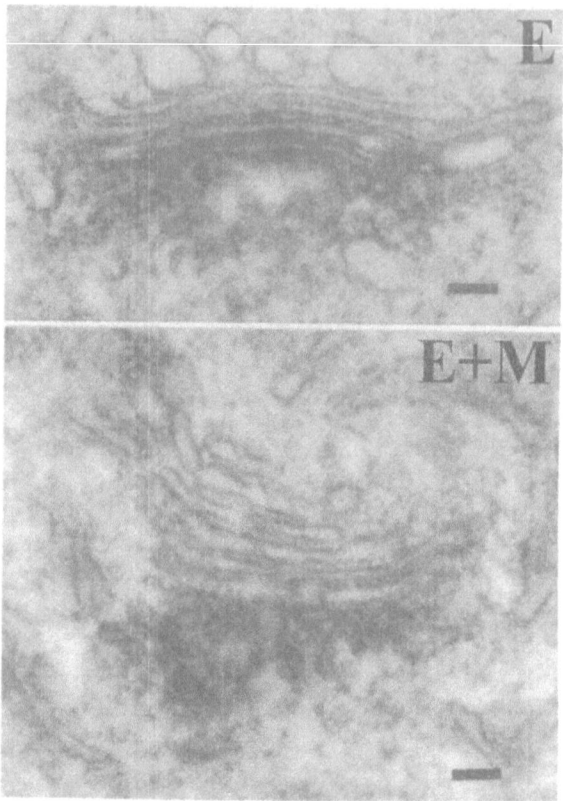

Figure 2. IBV M restricts the localization of IBV E. BHK cells were infected with recombinant vaccinia viruses encoding E (top panel) or E and M (bottom panel) and processed for peroxidase IEM using anti-E antibody. Bar = 100 nm.

4.   DISCUSSION

We were intrigued by the observation that the distribution of the IBV E protein partially overlapped that of markers of the CGN, stack and *trans*-Golgi network when expressed by itself, while it exactly colocalized with IBV M when the two proteins were coexpressed (Corse and Machamer, 2000). We determined the localizations of the IBV E and M proteins when they were expressed alone and together by immunoelectron microscopy in order to examine this observation more closely.

The localization of the IBV M protein was similar in the presence and absence of IBV E, suggesting that its *cis*-Golgi distribution is not drastically altered by E. However, the staining pattern appeared to be broader and excluded from the first Golgi cisterna in coexpressing cells. The IBV E protein was present throughout Golgi stacks when it was expressed alone. This is consistent with previous confocal microscopy results in which the distribution of E was shown to overlap with that of three marker proteins representative of different parts of the Golgi (Corse and Machamer, 2000). When expressed with M, however, E was no longer found in Golgi stacks but instead its staining pattern was much like that of M itself.

These results suggest that the IBV M protein may retain the IBV E protein near the coronavirus budding site. We are currently assessing if these proteins interact directly. The restricted localization of IBV E in the presence of IBV M is interesting in terms of recent results involving virus-like particle formation. When the M and E proteins of three other coronaviruses are coexpressed in cells, virus-like particles that are similar in size and density to real virions are released from the cells (Vennema *et al.*, 1996, Baudoux *et al.*, 1998, Godeke *et al.*, 2000). When the IBV M and E proteins are coexpressed in cells, however, only very small amounts of M and E are released as sedimentable particles (Corse and Machamer, 2000). Thus while IBV M and IBV E seem to be in a pre-Golgi compartment when they are coexpressed, this colocalization is not sufficient for significant release of virus-like particles.

What potential differences between IBV and other coronaviruses could account for the observed discrepancy in efficiency of virus-like particle formation? Interestingly, IBV mRNA3, the subgenomic RNA containing the E gene, is tricistronic and the two open reading frames upstream of the E gene, 3a and 3b, are unique to IBV. The 3a and 3b proteins have been shown to be expressed in IBV-infected cells (Liu *et al.*, 1991). The 3a protein is predicted to be extremely hydrophobic. We have cloned the 3a and 3b genes and preliminary results suggest that the 3a protein is found in the Golgi region when expressed by itself. We plan to test the possibility that the 3a and/or 3b proteins are involved in IBV assembly.

ACKNOWLEDGMENTS

We thank Mike Delannoy, Brad Harris, and Carol Cooke for their electron microscopy expertise. This work was supported by National Institues of Health grant GM42522.

REFERENCES

Baudoux, P., Carrat, C., Besnardeau, L., Charley, B., and Laude, H., 1998, Coronavirus pseudoparticles formed with recombinant M and E proteins induce alpha-interferon synthesis by leukocytes. *J. Virol.* **72**:8636-8643.

Brown, W.J., and Farquhar, M.G., 1989, Immunoperoxidase methods for the localization of antigens in cultured cells and tissue sections by electron microscopy. *Methods Cell Biol.* **31**:553-569.

Cluett, E.B., and Machamer, C.E., 1996, The envelope of vaccinia virus reveals an unusual phospholipid in Golgi complex membranes. *J.Cell Sci.* **109**:2121-2131.

Corse, E., and Machamer, C.E., 2000, Infectious bronchitis virus E protein is targeted to the Golgi complex and directs release of virus-like particles. *J. Virol.* **74**:4319-4326.

Fischer, F., Stegen, C.F., Masters, P.S., and Samsonoff, W.A., 1998, Analysis of constructed E gene mutants of mouse hepatitis virus confirms a pivotal role for E protein in coronavirus assembly. *J. Virol.* **72**:7885-7894.

Godeke, G.-J., de Haan, C.A.M., Rossen, J.W.A., Vennema, H., and Rottier, P.J.M., 2000, Assembly of spikes into coronaviruses particles is mediated by the carboxy-terminal domain of the spike protein. *J. Virol.* **74**:1566-1571.

Holmes, K.V., Doller, E.W., and Sturman, L.S., 1981, Tunicamycin resistant glycosylation of a coronavirus glycoprotein: demonstration of a novel type of viral glycoprotein. *Virol.* **115**:334-344.

Krijnse Locker, J., Ericsson, M., Rottier, P.J.M, and Griffiths, G., 1994, Characterization of the budding compartment of mouse hepatitis virus: evidence that transport from the RER to the Golgi complex requires only one vesicular transport step. *J. Cell Biol.* **124**:55-70.

Liu, D.X., Cavanagh, D., Green, P., and Inglis, S.C., 1991, A polycistronic mRNA specified by the coronavirus infectious bronchitis virus. *Virol.* **184**:531-544.

Machamer, C.E., Mentone, S.A., Rose, J.K., and Farquhar, M.G., 1990, The E1 glycoprotein of an avian coronavirus is targeted to the cis Golgi complex. *Proc. Natl. Acad. Sci. USA.* **87**:6944-6948.

Tooze, J., Tooze, S.A., and Fuller, S.D., 1987, Sorting of progeny coronavirus from condensed secretory proteins at the exit from the *trans*-Golgi network of AtT20 cells. *J. Cell Biol.* **105**:1215-1226.

Vennema, H., Heijnen, L., Zijderveld, A., Horzinek, M.C., and Spaan, W.J.M., 1990, Intracellular transport of recombinant coronavirus spike proteins: implications for viral assembly. *J. Virol.* **64**:339-346.

Vennema, H., Godeke, G.-J., Rossen, J.W.A., Voorhout, W.F., Horzinek, M.C., Opstelten, D.-J.E., and Rottier, P.J.M., 1996, Nucleocapsid-independent assembly of coronavirus-like particles by co-expression of viral envelope protein genes. *EMBO J.* **15**:2020-2028.

Characterization of Nucleocapsid-M Protein Interaction in Murine Coronavirus

[1,2]KRISHNA NARAYANAN AND [1,2]SHINJI MAKINO
[1]Department of Microbiology, The Institute for Cellular and Molecular Biology, The University of Texas at Austin, Austin, Texas 78712: [2]Department of Microbiology and Immunology, The University of Texas Medical Branch at Galveston, Galveston, Texas 77555-1019.

1. INTRODUCTION

Coronavirus is a large enveloped virus containing a 27-32 kb-long single-stranded, positive-sense RNA genome (Lee *et al 1991*). Coronaviruses generally have three envelope proteins S, M, and E. N protein and the viral genomic RNA form a helical nucleocapsid (Macnaughton *et al* 1978).

Coronavirus assembly is presumably dependent on protein-protein and protein-RNA interactions. Interactions between viral envelope protein(s) and the viral internal component are rather poorly characterised. Sturman *et al.* showed that MHV envelope glycoproteins (M and S) and the MHV nucleocapsid are separated by solubilization of the viral membrane with non-ionic detergent NP40 at 4°C followed by sucrose gradient centrifugation. Incubation of the NP40-disrupted MHV at 37°C resulted in the association of M protein and N protein-genomic RNA complex (Sturman *et al* 1980), indicating that this interaction is temperature-dependent.

The present study examined the interaction between the viral envelope protein(s) and viral internal component in virus particles. We showed the interaction between M protein and N protein-genomic RNA complex in mature virus particles. This interaction was ionic in nature. Characterisation of bromelain-treated MHV showed that the interaction between M protein and N protein-genomic RNA complex in bromelain-treated MHV was not disrupted under a high salt condition.

The Nidoviruses (Coronaviruses and Arteriviruses).
Edited by Ehud Lavi *et al.*, Kluwer Academic/Plenum Publishers, 2001.

2.   MATERIALS AND METHODS

2.1   Viruses and cells

The plaque-cloned A59 strain of MHV (Baric *et al* 1990) was used. Mouse DBT cells (Hirano *et al* 1974) were used for the growth of viruses.

2.2   Labelling of virion proteins and purification of viruses

For labelling of virion proteins, [^{35}S] methionine was added to virus-infected cells at 7.5 h p.i. and culture fluids were collected 12 h p.i. (Makino *et al* 1991). MHV particles were purified by sucrose gradient centrifugation and viral proteins, from purified virus particles, were analysed by SDS-PAGE as described previously (Kim *et al* 1997).

2.3   Disruption of virion with NP40

To disrupt virion under a low salt condition, purified viruses were pelleted and incubated in a low salt buffer (NTE buffer; 0.1 M NaCl, 0.01 M Tris-HCl [pH 7.5], 0.001 M EDTA) containing 0.25% NP40, for 30 min at 4°C. For high salt treatment, a high salt buffer (NTE buffer + 0.25 M KCl), containing 0.25% NP40, was used. The detergent-treated viruses were then layered over a 10-65% discontinuous sucrose gradient, made in NTE buffer, and sedimented at 38,000 rpm for 5 h at 4°C. Fractions were collected from the bottom of the gradient and the viral proteins in these fractions were pelleted by centrifugation at 38,000 rpm for 2.5 h at 4°C on a Beckman SW40 rotor.

2.4   Characterisation of virion RNA

Virus-specific RNAs were extracted from the pelleted sucrose gradient fractions and Northern blot analysis was performed with a ^{32}P-labeled, random-primed probe corresponding to the 5'-end of MHV genomic RNA as described previously (Fosmire *et al* 1992).

3.   RESULTS

We characterised the interaction between viral envelope protein(s) and the viral internal component. [^{35}S] methionine-labelled, purified MHV was treated with a low salt buffer, containing NP40. SDS-PAGE analysis of pelleted proteins, in each fraction, showed the cosedimentation of both M protein and N protein (Fig. 1A). The highest amount of both proteins was near the bottom of the gradient. S protein was present only near the top of the gradient. To examine the location of viral genomic RNA in the gradient, nonradiolabeled purified MHV was treated under the same condition. Northern blot analysis of virion RNA, extracted from the pelleted fractions, showed that the distribution of virion RNA was similar to that of N and M proteins (Fig. 1B). These data indicated that treatment of MHV in a low salt buffer, containing NP40, resulted in the aggregation of genomic RNA, N protein and M protein, which sedimented near the bottom of the gradient. S protein was separated from the aggregates of M protein, N protein and genomic RNA. Cosedimentation of M protein and N protein-genomic RNA complex was due to the interaction between M protein and N protein-genomic RNA complex, because anti-N protein antibody coprecipitated both M protein and MHV genomic RNA from the sucrose fractions (data not shown).

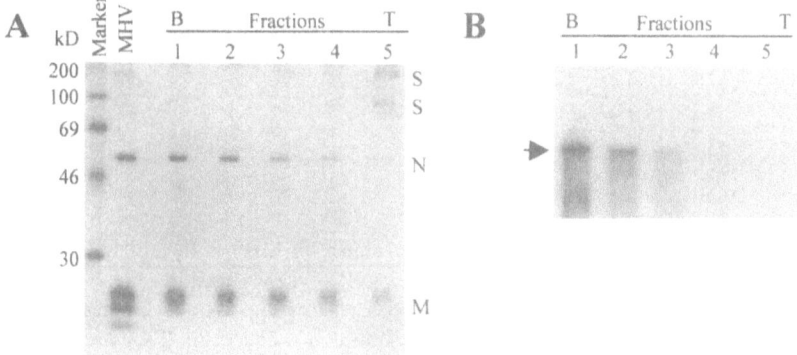

Figure 1. Characterisation of interaction after treatment of MHV with a low salt buffer, containing NP40. (A) ^{35}S-labeled, partially purified MHV was incubated in a low salt buffer, containing NP40, as described in Materials and Methods, and analysed by SDS-PAGE. (B) Nonradiolabeled MHV was treated under the same condition and virion RNA was analysed by Northern blot analysis. Arrow; MHV genomic RNA. B and T represent the bottom and the top of the gradient, respectively.

Next we attempted to separate envelope M protein from the aggregation of M protein, N protein and viral genomic RNA by incubating MHV in a high salt buffer, containing NP40. If the interaction between envelope M protein and the N protein-genomic RNA complex is ionic in nature, then this interaction may be susceptible to a high salt treatment. [35S] methionine-labelled purified MHV was incubated in a high salt buffer, containing NP40. SDS-PAGE analysis of each fraction showed that M protein did not cosediment with N protein; major peak of N protein was in fractions 2 and 3 while that of M protein was in fractions 5 and 6 (Fig. 2A). S protein sedimented to the top of the gradient. Northern blot analysis of each fraction showed that most of the viral genomic RNA cosedimented with N protein to fraction 2 (Fig. 2B). These data indicated that incubation of MHV under a high salt condition did not disrupt the interaction between viral genomic RNA and some, but not all, N protein, while the same incubation condition disrupted most of the interaction between M protein and the N protein-genomic RNA complex. Susceptibility, of the interaction between M protein and N protein-genomic RNA complex to a high concentration of salt, has not been reported previously in coronaviruses; this susceptibility suggests an ionic interaction between M protein and N protein-genomic RNA complex.

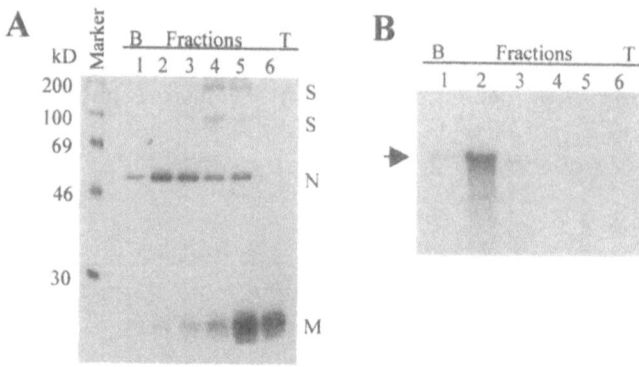

Figure 2. Characterisation of interaction after treatment of MHV with a high salt buffer, containing NP40. (A) [35S]-labeled partially purified MHV was incubated in a high salt buffer, containing NP40, and treated as described in the legend of Fig. 1A. (B) Nonradiolabeled MHV was treated under the same condition as described in A and virion RNA was analysed by Northern blot analysis. Arrow; MHV genomic RNA.

Next we characterised the interaction between M protein and N protein-genomic RNA complex in bromelain-treated MHV. Previous studies have shown that bromelain-treated MHV has an intact N protein and an 18 kD M protein (Makino *et al* 1983); most probably the 18 kD protein represented M protein lacking the N-terminal ectodomain (Makino *et al* 1983). [35S] methionine-labelled, partially purified virus was treated with bromelain and the bromelain-treated virus suspension was purified as described previously (Makino *et al* 1983). Purified bromelain-treated MHV was incubated in the low salt buffer, containing NP40, or in the high salt buffer, containing NP40. SDS-PAGE analysis of gradient fractions showed that the 18 kD M protein fragment and N protein cosedimented, after treatment with low salt buffer as well as high salt buffer (Fig. 3). These data indicated that the interaction between 18 kD M protein fragment and N protein-genomic RNA complex was not disrupted under high salt condition. Characterisation of viral genomic RNA showed that genomic RNA cosedimented with N protein under both buffer conditions (data not shown).

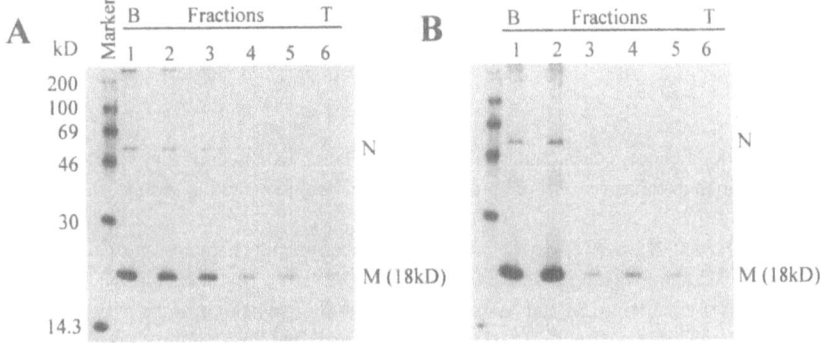

Figure 3. Characterisation of bromelain-treated MHV. ^{35}S-labeled, bromelain-treated MHV was incubated in a low salt buffer, containing NP40 (A) or in a high salt buffer, containing NP40 (B) as described in the legend of Fig.1A and samples were analysed by SDS-PAGE.

4. DISCUSSION

The present study demonstrated the interaction between M protein and N protein-genomic RNA complex in mature virus particles; this interaction appeared to be ionic in nature and was not disrupted under low salt condition. Our recent data revealed that interaction between M protein and N protein-genomic RNA complex exists in MHV-infected cells (Narayanan *et al* in press); the present study demonstrated that the interaction between M protein and N protein-genomic RNA complex, initiated in MHV-infected cells, was maintained in mature MHV particles. Our data in the present study were consistent with the data presented by Sturman *et al.* (Sturman *et al* 1980).

Interaction between M protein and N protein-genomic RNA complex in bromelain-treated MHV was not disrupted under a high salt condition, indicating that removal of the short ectodomain of M protein affected the property of its cytoplasmic tail, which probably interacts with N protein-genomic RNA complex. Removal of the envelope M protein ectodomain may extensively alter M protein conformation, and this altered conformation may stabilise the interaction between M protein and N protein-genomic RNA complex.

ACKNOWLEDGEMENTS

This work was supported by the Public Health Service Grant AI29984 from the National Institutes of Health.

REFERENCES

Baric, R. S., Fu K., Schaad, M. C., and Stohlman, S. A., 1990, Establishing a genetic recombination map for murine coronavirus strain A59 complementation groups. *Virology* **177**:646-656.

Fosmire, J. A., Hwang K., and Makino, S., 1992, Identification and characterisation of a coronavirus packaging signal. *J. Virol.* **66**:3522-3530.

Hirano, N., Fujiwara K., Hino, S., and Matsumoto, M., 1974, Replication and plaque formation of mouse hepatitis virus (MHV-2) in mouse cell line DBT culture. *Arch. Gesamte. Virusforch.* **44**:298-302.

Kim, K. H., Narayanan, K., and Makino, S., 1997, Assembled coronavirus from complementation of two defective interfering RNAs. *J. Virol.* **71**:3922-3931.

Lee, H.-J., Shieh, C.-K.,. Gorbalenya, A. E., Eugene, E. V., La Monica, N., Tuler, J., Bagdzhadzhyan, A., and Lai, M. M. C., 1991, The complete sequence (22 kilobases) of murine coronavirus gene 1 encoding the putative proteases and RNA polymerase. *Virology* **180**:567-582.

Macnaughton, M. R., Davies, H. A., and Nermut, M. V., 1978, Ribonucleoprotein-like structures from coronavirus particles. *J. Gen. Virol.* **39**:545-549.

Makino, S., Taguchi, F., Hayami, M., and Fujiwara, K., 1983, Characterisation of small plaque mutants of mouse hepatitis virus, JHM strain. *Microbiol. Immunol.* **27**:445-454.

Makino, S., Joo, M., and Makino, J. K., 1991, A system for study of coronavirus mRNA synthesis: a regulated, expressed subgenomic defective interfering RNA results from intergenic site insertion. *J. Virol.* **65**:6031-6041.

Narayanan, K., Maeda, A., Maeda J., and Makino, S., Characterisation of the coronavirus M protein and nucleocapsid interaction in infected cells. In press.

Sturman, L. S., Holmes, K. V., and Behnke, J., 1980, Isolation of coronavirus envelope glycoproteins and interaction with the viral nucleocapsid. *J. Virol.* **33**:449-462.

Production, Characterization, and Uses of Monoclonal Antibodies against Porcine Reproductive and Respiratory Syndrome Virus 3' Untranslated Region and Nucleoprotein RNA Binding Proteins

KUMAR SHANMUKHAPPA, FAHAD MAJHDI, AND SANJAY KAPIL

Department of Diagnostic Medicine -Pathobiology, College of Veterinary Medicine, Kansas State University, Manhattan, KS 66506, USA. Telephone (785) 532-4457; Fax: (785) 532-4481; E-mail: kapil@vet.ksu.edu

1. INTRODUCTION

Porcine reproductive and respiratory syndrome virus (PRRSV) belongs to the *Arteriviridae* family, order *Nidovirales*. It has a positive sense RNA genome of 15 kb in size and encodes eight genes.[1] The coding region of the genome is flanked by 5'& 3' untranslated regions (UTRs) of 156-220 nucleotides and 59-117 nucleotides, respectively.[2] The PRRSV undergoes discontinuous transcription, which is transcribed as a 3'- coterminal nested set of functionally monocistronic mRNAs.[3] These mRNAs have a common 5' leader sequence that is joined to the body of the mRNA by discontinuous transcription through junction sequence, UUAACC.[4,5] Both full-length RNA and subgenomic –ve sense RNA are produced in the first transcription. The transcription is initiated at the 3' UTR, and then it jumps to the intergenic sequence to join to the body of the RNA, where the intergenic sequences act as the transcription initiators. Here the 3'UTR plays an important role in regulation of transcription. The UTR's interacts with the intergenic sequences of RNA. This interaction has been noted to occur in both *cis* and *trans* fashion in the case of corona virus.[6] Studies have demonstrated that the 3' UTR of mouse hepatitis virus (MHV), is required for transcription of its genome.[7] This may occur by the interaction of the 3'

The Nidoviruses (Coronaviruses and Arteriviruses).
Edited by Ehud Lavi *et al.*. Kluwer Academic/Plenum Publishers. 2001.

UTR with upstream transcription-regulatory sequences mediated by RNA-protein or RNA-protein-protein interactions involving viral and/or cellular proteins.[8]

Host cell proteins are reported to bind 3'UTR and regulate the viral replication MHV.[9] Recently, Meat Animal Research Clone 145 (MARC-145) cell proteins that bind to 3'-UTR of PRRSV have been detected (Fahad and Kapil unpublished data) by UV cross-linking and Northwestern blotting. In this study, we have produced and characterized Mabs specific to several of these MARC-145 cell proteins, we also report the subcellular localization of the RNA binding proteins.

2.   MATERIALS AND METHODS

The MARC 145 cells were centrifuged at 2,000 x g for 15 min at 4°C, and were resuspended in 400 μl of extraction buffer (10 mM HEPES [pH 7.9]; 10 mM KCl; 0.1 mM EDTA; 1 mM DTT; 0.5 mM PMSF) and allowed to swell on ice for 15 min. Cells were lysed by adding 25 μl of 10% Nonidet P-40 followed by vortexing for 1 min. Cytoplasmic protein extract in the supernatant was collected by centrifugation at 12,000 x g for 1 min. The radiolabelled ([α-^{32}P]UTP) RNA transcript containing 3' UTR and the nucleoprotein gene was prepared from cDNA in pCR II ^{TM} Vector (Fahad and Kapil 1999 unpublished data) by using standard procedures. The cDNA was linearized by digestion with BamH1 and then transcribed *in vitro* using T$_7$ RNA polymerase (Riboscribe^{TM}, EpicentreTechnologies, Madison,WI) following manufacturer's instructions.

Cytoplasmic proteins of MARC cells were electrophoresed in a 10% discontinuous sodium dodecyl sulfate-polyacrylamide gel (SDS-PAGE) and blotted on a nitrocellulose membrane at 100 V for 1 hr. The membrane was incubated for 30 min with 6 M guanidine-HCl, which was replaced every 10 min with an equal volume of SBB binding buffer (0.05 M NaCl; 10 mM Tris [pH 7]; 1 mM EDTA; 0.02% polyvinyl pyrrolidone; 0.02% bovine serum albumin; 0.02% Ficoll) to allow for gradual renaturation. Prehybridization and hybridization were carried out in SBB buffer with 100 μg/ml of denatured salmon sperm DNA and 100 μg/ml yeast t-RNA for 30 min. During hybridization was done overnight at room temperature using 500,000 cpm/ml of probe. Posthybridization washes in SBB buffer were done 3 times for 30 min each and visualized by autoradiography.

After autoradiography, the desired protein band on the X-ray film was aligned to the Ponceau-S stained nitrocellulose membrane and cut from the membrane. The protein was eluted from the membrane with 500 μl of buffer (1% Triton X-100 in 50mM Tris-HCl [pH 9.5])/cm² of the membrane and injected into BALB/c mice along with RIBI adjuvant. (RIBI ImmunoChem Research, Inc. Hamilton, MT). The mice were euthanized and the spleen cells were collected and fused with Ag-8 cells following standard protocol.[10]

The cytoplasmic extract (10 ng/100 μl diluted in 0.05 M carbonate coating buffer [pH 9.6]) was coated on Immunolon-1 96 well plates (Dynatech Technologies, Chantilly, VA) overnight at 4°C. The plates were washed five times in 0.01 M PBS with 0.05% Tween-20 (Sigma Chemical, St. Louis, MO), and blocked with 0.5% glycine in 0.01 PBS (pH 7) at 37°C for 30 min. Fifty μl of Mab supernatant was added to each well, and incubated at 37°C for 30 min. After 5 washes, 50 μl of a 1:10,000 dilution (in 0.01 M PBS [pH 7]) of horse anti-mouse HRPO-conjugate (Vector Laboratories,Inc, Burlingame, CA) was added to each well and incubated at 37°C for 30 min. Plates were washed again five times, then 50 μl of 3,3',5,5'-tetramethylbenzidine peroxidase substrate (TMB; Kirkegaard & Perry Laboratories, Gaithersburg, MD) was added. After a 15 min incubation at 37°C, the absorbance was read at 650 nm in a 96-well plate spectrophotometer (Labtec, Salzburg, Austria). Western blot was performed using Mab supernatant on Marc cytoplasmic proteins.

To 50 μl of MARC-145 cytoplasmic extract, 50 μl of Mab supernatant was added, and the mixture was rocked overnight at 4°C. The immunocomplexes were precipitated on ice for 2 h with addition of 4 μl of formalin-fixed S. *aureus* cells. The immunoprecipitate was centrifuged at 4,000 x *g* for 10 min. The pellet was washed once in cold TSA (0.05 M Tris-HCl [pH 8.0]; 0.15M NaCl; 0.025 % NaN₃) containing 1% Triton X-100 and 1% SDS; once in cold TSA alone, and twice in 10 mM Tris-HCl pH 7.5 and 1 mM EDTA by centrifugations at 12,000 x *g* for 10 min. The pellet was suspended in 20 μl of 2 X SDS loading buffer and a Northwestern blot was carried out as described above.Indirect fluorescent antibody testing (FAT) was done to identify the subcellular localization of the RNA binding proteins. Monolayers of MARC cells were grown in wells of Teflon-coated slides, then they were fixed in ice cold acetone, air dried, and stored at -20°C. The slides were thawed at room temperature, 50 μl of Mab supernatant was added to each well, and slides were incubated at 37°C for 30 min. Then the slides were washed in 0.01 m PBS five times and incubated with 50 μl of anti-mouse FITC conjugate at 37°C for 30 min. Following five washes in 0.01 M PBS (pH 7.0), buffered glycerol mounting medium was added, coverslipped and observed by fluorescent microscopy.

3.   RESULTS

In North western blot, MARC 145 cytoplasmic proteins of 158, 148, 137, 124, 114, 101, 92, 81, 75, 67, 52, and 24 kD (data not shown) were found to bind with the 3' UTR PRRSV probe confirming earlier observations (Fahad and Kapil, unpublished data). The RNA binding protein of approximately 67 kD was injected into mice and the Mabs were produced. 11 Mabs were found positive by ELISA (Fig 1).

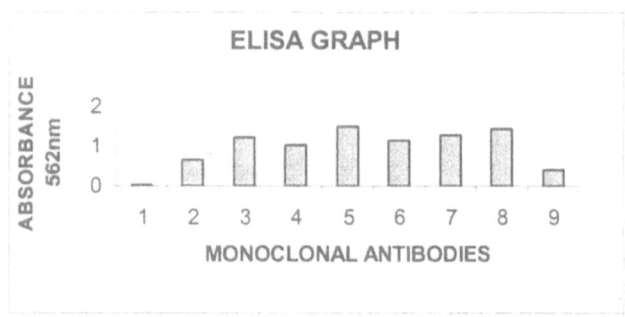

Figure 1. The graph shows the reactivity of the Mabs to the MARC cell lysates. Lane 1 negative control, LANES 2-9 show, the Mabs: 1E9, 2A5, 3E7, 3G3, 4F6, 7G10, 8A7, 9H4, 12E 9.

Immunoprecipitation was performed on these Mabs and found that the 6 Mabs precipitated MARC cell proteins all of which had the 67 kD protein while one Mab 7G10 immunoprecipitated a complex of proteins of 45 kD, 37 kD & 27 kD (Fig 2). In western blotting the Mab detected two bands of approximately 67 kD and 55 kD protein (data not Shown). FAT on MARC 145 cells was performed to know the subcellular localization of these RNA binding proteins. Mabs showed diffuse cytoplasmic staining and intense perinuclear staining to one side of the nucleus opposite to the nucleolus. One clone showed some faint nucleolar staining (data not shown).

4.   DISCUSSION

Numerous studies have reported the identification of regulatory sequences present in the 5' or 3' UTR of viral and cellular mRNAs that are involved in transcriptional control.[7,8] The regulatory regions are presumed to interact with RNA binding proteins that are associated with transcription, splicing, stability, transport and translation of RNA. The detection by Northwestern blot of 11 MARC 145 cell cytoplasmic proteins that interact with 3'UTR of PRRSV agrees with earlier observations (Fahad and Kapil, unpublished data). The detection of 67 kD and 55 kD proteins by all Mabs western blot in

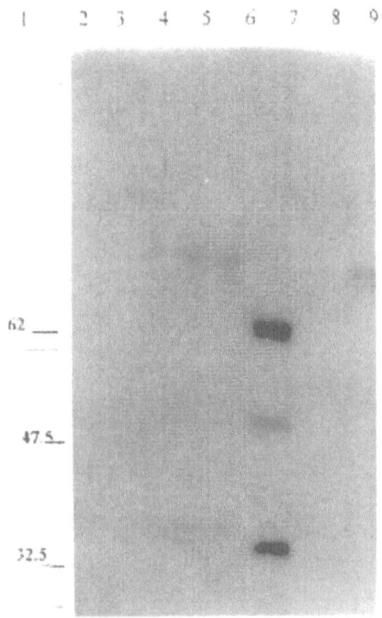

Figure 2. Immunoprecipitation assay. Lane 1 prestained kaleidoscope marker, 2 Negative control, 3: 1E9,4:2A5,5:3E7, 6:3G3,7:7G10,8:9H4,9:12E9.

western blot assays. This could indicate that might be subunits of the same protein. Immunoprecipitation, most of the clones showed the band of 67 kD while one clone, detected multiple proteins of 45, 37 and 27 kD. This could have been due to protein-protein interactions as Mabs essentially recognize the single protein. Because we carried out immunoprecipitation in the native condition, protein-protein interactions might have occurred causing the formation of complex precipitate. The multiple protein bands also suggest that RNA interacts with the 67 kD protein which, alone or in concert with other proteins, may either facilitate or hinder the formation of specialized complexes at particular sites on the RNA by protein-protein interactions which also can modify the RNA structure. These interactions can have a far reaching influence on transcription and translation of PRRSV RNA. The identification of these proteins using Mabs can facilitate our understanding of the process of discontinuous transcription seen in PRRSV.

Immunohistochemistry/ FAT of MARC 145 cells with all the MAbs revealed a similar pattern of staining that was predominantly cytoplasmic with perinuclear localization to one pole of the nucleus. This perinuclear staining may be related to the formation of double membrane vesicles (DMVs) 3-6 hrs after porcine alveolar macrophages have been infected with

PRRSV. These DMVs also have been detected in other arteriviruses [5] but their origin and function are unclear. They do not appear to be involved in virion assembly, and their roles in viral RNA synthesis and transcription are not fully known. In the case of equine arteritis virus, replicase subunits containing putative RNA polymerase and helicase functions are membrane associated and localized to the perinuclear region where the DMV are seen.[5] These DMVs, RNA binding proteins, viral RNA, and RNA polymerase may interact in the replication of the PRRS virus. Work is currently underway to characterize these RNA binding proteins using Mabs and to unfold the series of cellular and viral molecular interactions involved in PRRSV replication.

REFERENCES

1. Cavanagh D: Nidovirales: A new order comprising Coronaviridae and Arteriviridae. Arch. Virol. 1997;142:629-633.
2. Muelenberg JJM, Bos-de Ruetzer JNA, Wensvoort G, and Moormann RJM: Infectious transcripts from cloned genome-length cDNA of porcine reproductive and respiratory syndrome virus. J. Virol. 1998;72:380-387.
3. Kuo LL, Harty JT, Erickson L, Palmer GA, and Plagemann PGW: A nested set of eight RNAs is formed in macrophages infected with lactate dehydrogenase-elevating virus. J. Virol. 1991;65: 5118-5123.
4. Muelenberg JJM, de Meijer EJ, and Moormann RJM: Subgenomic RNAs of Lelystad virus contain a conserved leader-body junction sequence. J. Gen. Virol. 1993;74:1697-1701.
5. Snijder EJ, and Muelenberg JJM: The molecular biology of arteriviruses. J. Gen. Virol. 1998; 79:961-979.
6. Zhang X, Liao CL, and Lai MMC: Coronavirus leader RNA regulates and initiates subgenomic mRNA transcription both in trans and in cis.. J. Virol.1994;68: 4738-4746.
7. Hsue B, Masters PS: A bulged stem-loop structure in the 3' untranslated region of the genome of the coronavirus mouse hepatitis virus is essential for replication. J. Virol. 1997;71: 7567-7578.
8. Lai MMC, Liao CL, Lin YJ, and Zhang X: Coronavirus: how a large RNA viral genome is replicated and transcribed. Infect. Agents. Dis.1994;3: 98-105
9. Hsin-Pai Li, Xuming Zhang, Roger Duncan, Lucio Comai, and Michael M. C. Lai: Heterogeneous nuclear ribonucleoprotein A1 binds to the transcription-regulatory region of mouse hepatitis virus RNA. Proc. Natl. Acad. Sci. USA.1997;94:9544-9549.
10. Goding J: Purification, fragmentation and isotopic labelling of monoclonal antibodies. In: *Monoclonal antibodies. Principles and practice,* Goding J (Ed.). Academic Press, San Diego, CA, 1996, pp. 104-141.

The Membrane M Protein of the Transmissible Gastroenteritis Coronavirus Binds to the Internal Core through the Carboxy-Terminus

DAVID ESCORS, JAVIER ORTEGO, AND LUIS ENJUANES[1]
[1]*Department of Molecular and Cell Biology, Centro Nacional de Biotecnología, CSIC, Campus Universidad Autónoma, 28049 Madrid, Spain.*

1. INTRODUCTION

An internal core was recently described in two coronaviruses, the tramsmissible gastroenteritis coronavirus (TGEV) and the murine hepatitis virus (MHV) (Risco et al., 1996). The core has a spherical and possibly icosaedral shape. Purified cores included both the nucleoprotein and the membrane protein, as well as the viral genome. The presence of the M protein in purified cores was unexpected since this has been considered an integral membrane protein. This observation could be explained if the M protein molecules embedded in the viral envelope, with the intravirion carboxy-terminus, interact with the internal core by an intravirion domain. This hypothesis seems feasible since it was shown that the M protein interacts with the viral nucleocapsids in MHV virions (Sturman et al., 1980). However, further evidence for this interaction is required.

The Nidoviruses (Coronaviruses and Arteriviruses).
Edited by Ehud Lavi *et al.*, Kluwer Academic/Plenum Publishers, 2001.

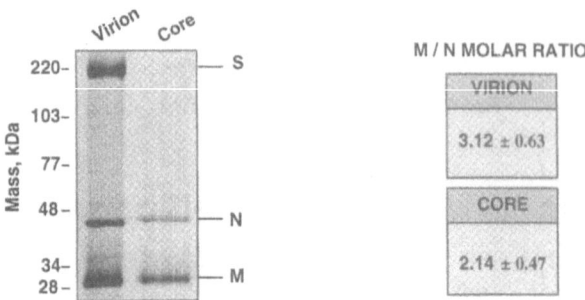

Figure 1. The protein composition of purified virions and cores was analyzed by SDS-PAGE and silver staining. The M to N molar ratio is shown and the difference was statistically significant (P=0.025).

2. MATERIALS AND METHODS

2.1 Purification and analysis of virions, cores and nucleocapsids

TGEV virions (PUR-MAD strain) were purified from infected ST cells (Jiménez et al., 1986). Cores were purified as described before (Risco et al., 1996) and nucleocapsids were obtained by disrupting the cores with 300 mM of KCl (Escors et al., 2000). The protein composition was analyzed by SDS-PAGE and silver staining. The gels were scanned and the protein bands quantified using the NIH-Image software. The M to N molar ratio was statistically compared in virions and cores.

2.2 *In vitro* interaction of the M protein with purified nucleocapsids

The M gene and several deletion mutants obtained by standard PCR techniques were expressed in a rabbit reticulocyte lysate in the presence of ^{35}S-labeled methionine/cysteine. A nucleoprotein-specific monoclonal antibody (mAb) was conjugated to protein G-Sepharose beads and was complexed to purified nucleocapsids. This complex was used to immunoprecipitate the wild-type and mutant proteins (Escors et al., 2000).

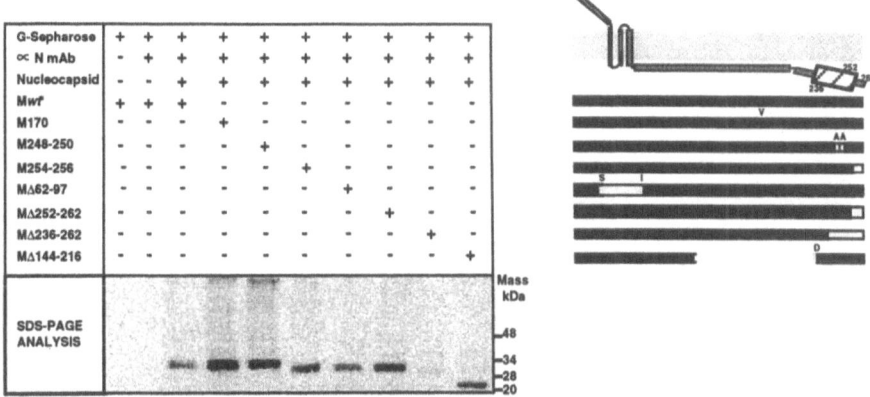

G-Sepharose	+	+	+	+	+	+	+	+	+	+
∝ N mAb	-	+	+	+	+	+	+	+	+	+
Nucleocapsid	-	-	+	+	+	+	+	+	+	+
Mwt	+	+	+	-	-	-	-	-	-	-
M170	-	-	-	+	-	-	-	-	-	-
M248-250	-	-	-	-	+	-	-	-	-	-
M254-256	-	-	-	-	-	+	-	-	-	-
MΔ62-97	-	-	-	-	-	-	+	-	-	-
MΔ252-262	-	-	-	-	-	-	-	+	-	-
MΔ236-262	-	-	-	-	-	-	-	-	+	-
MΔ144-216	-	-	-	-	-	-	-	-	-	+

SDS-PAGE ANALYSIS

Mass kDa
48
34
28
20

Figure 2. The binding of labeled M protein and several protein mutants was analyzed in the binding assay. Point mutations are shown as bars (M170 and M248-250) and deletions in white boxes (MΔ62-97, MΔ252-262, M236-262 and MΔ144-216). A deletion from residue 236 to the carboxy-terminus (MΔ236-262) abolished the binding to nucleocapsids. The identified interaction domain (residues 236 to 252) is shown as a dashed box in the carboxy-terminus.

3.   RESULTS

Cores were purified from TGEV virions by disrupting the viral envelope with NP-40 and sucrose gradient centrifugation. Two thirds of the M protein molecules co-purified with TGEV cores (Figure 1). The binding of the M protein to cores was resistant to a wide range of salt concentrations (NaCl and KCl <200 mM), reducing agents (2-mercaptoethanol), and detergent (Triton X-100). These data suggested that the binding was specific (Escors et al., 2000). In addition, it was observed a correlation between the loss of the M protein and the release of the nucleocapsid (data not shown). The specificity of the M protein to nucleocapsid binding was studied by analizing the binding of M protein mutants to nucleocapsid (Figure 2). The results were compatible with an interaction of an M protein domain of 16 aminoacids with nucleocapsid (residues 236 to 252). The specificity of the binding was confirmed by two independent approaches (Escors et al., 2000). The first one was based on the efficient inhibition of the M protein binding by a synthetic peptide identical to the interaction domain previously identified. The second one was based on the specific inhibition of the binding by M-specific monoclonal antibodies directed to the carboxy-terminus (data not shown).

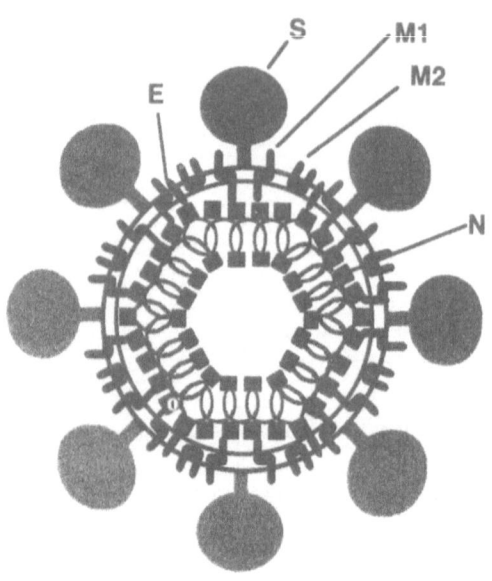

Figure 3. Structural model proposed for the TGEV virion. Virions contain an internal core made of the viral genome, nucleoprotein (N) and the carboxy-terminus of the membrane protein (M). The internal core is surrounded by a lipid envelope including the spike protein (S), the envelope protein (E) and the membrane protein. The M molecules in a Nexo-Cendo (M1) topology directly interact with the N protein through the carboxy-terminus. The M protein in a Nexo-Cexo topology (M2) remains embedded in the viral envelope and is lost during core purification.

4.   CONCLUSIONS

The experimental evidence obtained so far in TGEV virions is compatible with the structural model proposed in figure 3 (Escors et al., 2000). According to this model the membrane M protein adopts two topologies within the virus envelope, a Nexo-Cendo and a Nexo-Cexo topology (Risco et al., 1995). The M protein in a Nexo-Cendo topology interacts with the internal core by the carboxy-terminus. One third of the M protein molecules were specifically lost during the purification process and there is biochemical evidence showing that these molecules correspond to those in a Nexo-Cexo topology (unpublished results). Therefore, the M protein specifically interacts from the viral envelope with the internal core by 16 aminoacids located in the carboxy-terminus, contributing to core stability (Escors et al., 2000). A coronavirus cDNA clone has been recently engineered as an infectious bacterial artificial chromosome (Almazán et al., 2000). Mutations in the M protein interaction domain will be introduced in the viral genome and their effects on TGEV morphogenesis and structure will be tested.

ACKNOWLEDGMENTS

This work has been supported by grants from the Comisión Interministerial de Ciencia y Tecnología (CICYT), La Consejería de Educación y Cultura de la Comunidad de Madrid, and the European Communities (Key Action 2: Infectious Diseases). DEM received a fellowship from the Spanish Department of Science and Technology.

REFERENCES

Almazán, F., González, J.M., Penzes, Z., Izeta, A., Calvo, E., Plana, J., and L. Enjuanes., 2000. Engenearing the largest RNA virus genome as an infectious bacterial artificial chromosome. PNAS. **97**, 5516-5521.

Escors, D., Ortego, J., and L. Enjuanes., 2000. The membrane M protein carboxy-terminus interacts with transmissible gastroenteritis coronavirus core and is essential for core stability. Submitted for publication.

Jiménez, G., I. Correa, M.P. Melgosa, M.J. Bullido, and L. Enjuanes., 1986. Critical epitopes in transmissible gastroenteritis virus neutralization. J. Virol. **60**, 131-139.

Risco, C., I. M. Antón, L. Enjuanes, and J.L. Carrascosa., 1996. The transmissible gastroenteritis coronavirus contains a spherical core shell consisting of M and N proteins. J. Virol. **70**, 4773-4777.

Risco, C., I. M. Antón, C. Suñé, A.M. Pedregosa, J.M. Martín-Alonso, F. Parra, J. L., Carrascosa, and L. Enjuanes., 1995. Membrane protein molecules of the transmissible gastroenteritis coronavirus also expose the carboxy-terminal region on the external surface fo the virion. J. Virol. **69**, 5269-5277.

Sturman, L.S., K.V. Holmes, and J. Behnke., 1980. Isolation of coronavirus envelope glycoproteins and interaction with the viral nucleocapsid. J. Virol. **33**, 449-462.

Physical Interaction Between the Membrane (M) and Envelope (E) Proteins of the Coronavirus Avian Infectious Bronchitis Virus (IBV)

K. P. LIM, H. Y. XU, AND D. X. LIU
Institute of Molecular Agrobiology, The National University of Singapore, 1 Research Link, Singapore 117604

1. INTRODUCTION

Coronavirus avian infectious bronchitis virus (IBV) encodes at least four structural proteins, i.e. the nucleocapsid (N), membrane (M), spike (S) and envelope (E) proteins. The IBV E protein was first identified in 1990 (Liu et. al., 1990) and demonstrated to be associated with the virion envelope (Liu and Inglis, 1991). Several studies on the two other members of the coronavirus family, namely Mouse Hepatitis Virus (MHV) and Transmissible Gastroenteritis Virus (TGEV), have indicated that E is likely to play a crucial role in viral assembly (Fischer et. al., 1998; Baudoux et. al., 1998). Together with M, E was demonstrated to be the minimal machinery required for the formation of Virus-Like Particles (VLPs) (Vennema et al, 1996). Recent reports took a step further in demonstrating that expression of E alone could lead to the release of VLPs (Maeda et. al., 1999; Raamsman et. al., 2000; Corse and Machamer, 2000).

The Nidoviruses (Coronaviruses and Arteriviruses).
Edited by Ehud Lavi *et al.*, Kluwer Academic/Plenum Publishers, 2001.

2.   MATERIAL AND METHODS

2.1   Transient expression of IBV sequences in Cos-7 cells

IBV sequences placed under the control of a T7 promoter were transiently expressed in mammalian cells using the recombinant vaccinia virus (vTF7-3) system as described before (Liu et al., 1991). In this study, the transfection reagent used was Lipofectin (Life Technologies).

2.2   Radioimmunoprecipitation

Media of transfected cells were collected and mixed with 5 X radioimmunoprecipitation assay (RIPA) buffer and precleared by centrifugation at 4,000 g for 30 min at 4°C in a microfuge. Cells were lysed with 1 X RIPA buffer and precleared by centrifugation at 12,000 rpm. Immunoprecipitation with anti-E and anti-M rabbit polyclonal antisera (Liu and Inglis, 1991; Ng and Liu, 2000) and anti-T7 (Novagen) was carried out as previously described (Liu et. al., 1991).

2.3   SDS-polyacrylamide gel electrophoresis (SDS-PAGE)

Electrophoresis of viral polypeptides was performed on SDS-17.5 % polyacrylamide gels (Laemmli, 1970). The [^{35}S]-labeled polypeptides were detected by autoradiography.

2.4   Immunofluorescence and confocal microscopy

IBV sequences were transiently expressed in Cos-7 cells grown on 4-well chamber slides (IWAKI). At 5 h post-transfection, cells were rinsed with phosphate-buffered saline (PBS) and subjected to fixation using 4% paraformaldehyde for 15 mins and then permeabilized with 0.2% Triton-X. Fluorescence staining was performed by incubating cells with either antibody or a mixture of both primary antibodies [rabbit anti-M (1:30) or mouse anti-T7 (1:200)] for 1 h at room temperature, followed by fluorescein isothiocyanate (FITC)- or tetramethyl rhodamine isocyanate (TRITC)-conjugated secondary antibodies for 1 h at 4°C. Goat anti-rabbit was used at 1:400 (Sigma) and goat anti-mouse IgG at 1:20 (DAKO). Images were viewed and collected with a confocal laser-scanning microscope (Zeiss).

2.5   Polymerase Chain Reaction (PCR)

Appropriate primers and template DNAs were used in amplification reactions with Pfu DNA polymerase (Stratagene) under the standard buffer conditions with 2 mM MgCl$_2$.

2.6   Construction of plasmids

Plasmid pIBVM-1 which covers the IBV sequence between nucleotides 24498 and 25159 was constructed by cloning a PvuII/SacI-digested PCR fragment, generated using primers LDX59 (5'-CAGCAA<u>CAGCTG</u>AAG-ATGCCCAACG-3') and LDX60 (5'-CTACACAC<u>GAGCTC</u>TTATGTG-TAAAGA-3') into PvuII/SacI digested pKTO vector. This plasmid was digested with BglII and EcoRI to give a 661 bp fragment which was subsequently cloned into a BglII/EcoRI-digested pGFPC1 vector (CLONTECH), to give pGFPM. A 735 bp fragment, obtained by PCR using LDX55 (5'-GATTGTTCAGG<u>CCATGG</u>TGAATTTATTGAA-3') and XIANG8 (5'-GCACCATTGGCACACTC-3'), was digested with NcoI and BamHI and ligated into NcoI/BamHI-digested pKTO, resulting in pIBVE which contained the IBV sequence between nucleotides 24205 and 24795. The same PCR fragment was digested with BglII and SmaI and cloned into pT7Φ10 vector, to give pT7E fusion construct. Plasmid pT7Φ10, which contains the T7 tag was constructed by cloning an NcoI/HindIII fragment from pET3d (Novagen) into NcoI/HindIII digested pKTO vector. Deletions made in the E protein was indicated in Figure 2a.

3.   RESULTS

3.1   The putative second transmembrane domain of E was responsible for the physical interaction with M

While coronavirus M and E proteins were shown to be the minimal essential components for VLP formation, we demonstrated that antiserum raised against M (anti-M) could coimmunoprecipitate a T7-tagged E (T7E) (Fig.1a, lane 5). Specific anti-T7 antibody could also coimmunoprecipitate M when cells were co-transfected with plasmids expressing T7E and M.

E consists of 109 amino acids. To determine the domain(s) required for the physical interaction with M, a series of deletion mutants of E, as shown in Fig. 1b, was constructed. T7EΔNT contains a deletion at the N-terminus between amino acid residues 1 and 14; T7EΔTM1 contains a deletion between nucleotides 18 and 33; T7EΔTM2a has a deletion between residues 34 and 51, whereas T7EΔTM2b contains a deletion between nucleotides 50 and 63. T7EΔmid contains deletion between residues 67 and 84, and T7EΔCT is a mutant E lacking the C-terminus from residues 83 onwards. As predicted by a computer programme, T7EΔTM1 deletes the first transmembrane domain of E, while T7EΔTM2a and T7EΔTM2b delete the first and second half of the second putative transmembrane domain, respectively, and T7ΔCT deletes a putative ER localization signal (data not shown).

Using the coimmunoprecipitation approach, we were able to determine that the second putative transmembrane domain of E was likely to be responsible for complex formation in vivo with M. Most mutants could be coimmunoprecipitated with M, except for T7EΔTM2a (Fig. 1c, lanes 5 and 12), where only trace amounts of T7EΔTM2a and M could be detected using anti-M and anti-T7, respectively.

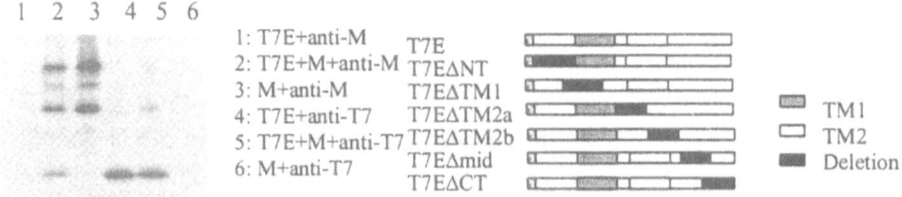

Figure 1a. CoIP of M and T7E   *Figure 1b.* Diagram of various deletion mutants

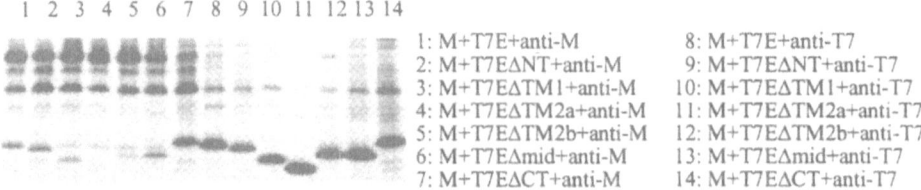

Figure 1c. Coimmunoprecipitation of M and mutant E proteins

3.2   Effects of deletion within E on the release of VLPs

Figures 2a and 2b show that by tagging of the T7 tag to E did not affect the VLP release. Figure 2c shows the effects of deletion within E on VLP release. At least a 50 % drop in VLP detection was observed when the N-terminal 14 amino acids (lane 12), the first putative transmembrane domain (lane 13) and the second putative transmembrane domain (lane 14) were deleted. Deletion of the C-terminal 26 amino acids (lane 17) had the most detrimental effect on VLP release.

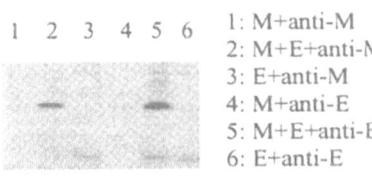

1: M+anti-M
2: M+E+anti-M
3: E+anti-M
4: M+anti-E
5: M+E+anti-E
6: E+anti-E

Figure 2a. Release of E from cells expressing E

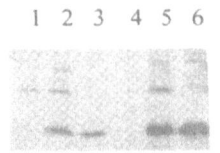

1: M+anti-M
2: M+T7E+anti-M
3: T7E+anti-M
4: M+anti-T7
5: M+T7E+anti-T7
6: T7E+anti-T7

Figure 2b. Release of T7E from cells expressing T7E

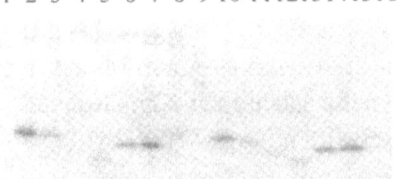

1. total M   9: M+anti-M
2. total M+T7E   10: T7E+anti-T7
3. total T7EΔNT   11: T7EΔNT+anti-T7
4. total T7EΔTM1   12: T7EΔTM1+anti-T7
5. total T7EΔTM2a   13: T7EΔTM2a+anti-T7
6. total T7EΔTM2b   14: T7EΔTM2b+anti-T7
7. total T7EΔmid   15: T7EΔmid+anti-T7
8. total T7EΔCT   16: T7EΔCT+anti-T7

Figure 2c. Release of E and its deletion mutants

3.3 M could be incorporated into VLPs by non-specific interactions with E

Although E was demonstrated to physically interact with M via the second putative transmembrane domain, we do not rule out the possibility that both proteins could interact by weak hydrophobic interactions. To detect trace amounts of M incorporated into VLPs, we performed immunoprecipitation of precleared media using a mixture of anti-M and anti-T7 antisera. Bands corresponding to M were coimmunoprecipitated with T7E (Fig.3 lane 9), T7EΔNT (lane 10), T7EΔTM2b (lane 13) and T7EΔmid (lane 14), indicating that M was released into the media. Taken into account that less VLPs containing T7EΔTM1 (Fig.2c, lane 13) and T7ΔCT (lane 17) were detected, the amounts of M incorporated by these two mutants (Fig. 3, lane 11 and 15) were comparable to T7E (lane 9). However, trace amounts of M protein could also be coimmunoprecipitated with T7EΔTM2a (which lacks the second putative transmembrane domain required for interaction).

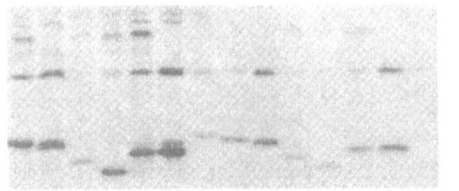

1: total M+T7E   8: M+T7E+anti-M anti-T7
2: total M+T7EΔNT   9: M+T7EΔNT+anti-M+anti-T7
3: total M+T7EΔTM1   10: M+T7EΔTM1+anti-M+anti-T7
4: total M+T7EΔTM2a   11: M+T7EΔTM2a+anti-M+anti-T7
5: total M+T7EΔTM2b   12: M+T7EΔTM2b+anti-M+anti-T7
6: total M+T7EΔmid   13: M+T7EΔmid+anti-M+anti-T7
7: total M+T7EΔCT   14: M+T7EΔCT+anti-M+anti-T7

Figure 3. Release of VLPs with M incorporated

3.4   E protein is able to retain M in the compartment it resides in

Previous reports showed that M was localized to the Golgi compartment (Machamer et.al., 1990). Similar observations were shown in Figure 4A. The staining patterns of T7E, T7EΔTM1, T7EΔ2a and T7EΔCT were shown in B to E. Colocalization of T7E (Fig. 5A) and T7EΔTM2a (5C) with the endoplasmic reticulum (ER), were indicated by the merged images (F, C and I). The ER was stained using R6 (Rhodamine B hexylester chloride, Molecular Probes). However, mutants T7EΔTM (D) and T7EΔCT (J) were not localized to the ER (F and L).

The physical interaction between M and E was further indicated by the colocalization of M and the deletion mutants of E. Coexpression of M and E in Cos-7 cells led to the detection of M in the same subcellular compartment as T7E (Fig.6A to C). When expressed together with mutant T7EΔTM1 (G), M displayed a cytoplasmic staining (E and F) instead. M was also observed to colocalize with T7EΔTM2a and T7EΔCT (6G to I; J to L).

To further confirm that this interaction, cells were cotransfected with GFPM and T7E or T7EΔTM1. When expressed alone, GFPM showed a profile similar to the wildtype M (7A). Cotransfection with T7E resulted in a perinuclear staining (7B), while transfection with T7EΔTM1 resulted in a diffuse staining (7C). These patterns are clearly different from the cells expressing GFPM alone.

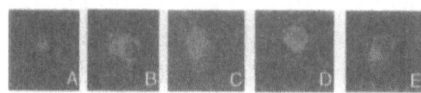

Figure 4. Subcellular localization of M(A) and T7E(B), T7EΔTM1(C), T7EΔTM2a(D) and T7EΔCT(E) in transfected cells.

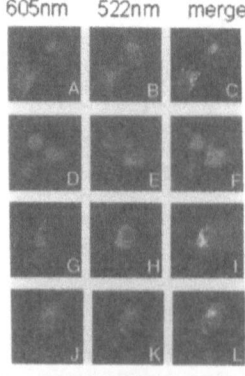

Figure 5. Colocalization of T7E(A), T7EΔTM1(D), T7EΔTM2a(G) and T7EΔCT(J) with the ER, (B, E, H and J), which is stained by R6. Images C, F, I and L refer to merged images.

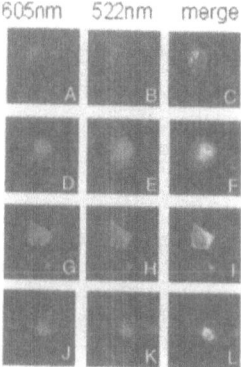

Figure 6. Dual labeling of T7E(A), T7EΔTM1(D), T7EΔTM2a(G) and T7EΔCT(J) with M protein, (B, E, H and J). Images C, F, I and L refer to colocalization between M and E.

Figure 7. GFP staining of M when expressed alone (A), and coexpressed with T7E(B) or T7EΔTM1(C)

4. DISCUSSION AND CONCLUSION

In this report, we presented two lines of evidence that E could physically interact with M. First, in the presence of detergent and high salt content, both M and E could be coimmunoprecipitated by either specific antiserum from cells overexpressing both proteins. This method also allowed us to determine that the predicted second transmembrane domain of E (T7EΔTM2a) was responsible for the interaction with M. However, we observed that when coexpressed with T7EΔTM2a, M could still be incorporated into VLPs, possibly via weak non-specific interactions. Similar conclusions could be drawn from subcellular localization studies, which provided the second line of evidence. When expressed alone, M was seen to be localized at the Golgi (Klumperman et. al., 1994), while at the same time, E (T7E) was observed to be located in the ER, as indicated by the staining of R6. However in cells expressing both M and E, an ER-like staining pattern of M was detected instead, suggesting that E could retain M in the ER upon physical interaction, near the site of virus budding (Klumperman et. al., 1994; Chen and Itakura, 1996). This was further reinforced by the observation that M displays a diffuse cytoplasmic staining similar to that of an E mutant with the first predicted transmembrane deleted (T7EΔTM1) upon coexpression. It was likely that this mutant had a defect in membrane insertion, and hence could not induce proper viral budding. An interesting mutant is T7EΔCT, which did not show colocalization pattern with the R6-stained ER compartment. The mutant protein is localized to the Golgi complex instead. Computer prediction suggested the presence of a potential ER retention signal (KDEL) at the C-terminal region of E. Deletion of this signal in T7EΔCT changes the subcellular localization of E.

REFERENCES

Baudoux, P., Carrat, C., Besnardeau, L., Charley, B., and Laude, H. 1998. Coronavirus pseudoparticles formed with recombinant M and E proteins induce alpha interferon synthesis by leukocytes. J. Virol. 72: 8636-8643.

Chen, B. Y., and Itakura, C. 1996. Cytopathology of chick renal epithelial cells experimentally infected with avian infectious bronchitis virus. Avian Pathology.25: 675-690.

Corse, E., and Machamer, C. E. 2000. Infectious bronchitis virus E protein is targeted to the golgi complex and directs release of virus-like particles. J. Virol. 74: 4319-4326.

Fischer, F., Stegen, C. F., Masters, P. S., and Samsonoff, W. A. 1998. Analysis of constructed E gene mutants of mouse hepatitis virus confirms a pivotal role for E protein in coronavirus assembly. J. Virol. 72: 7885-7894.

Furest, T. R., Niles E. G., Studier, F. W., and Moss, B. 1986. Eukaryotic transient expression system based on recombinant vaccinia virus that synthesis bacteriophage T7 RNA polymerase. Proc. Natl. Acad Sci. USA 83: 8122-8127.

Klumperman, J., Locker, J. K., Meijer, A., Horzinek, M., C., Geuze, H. J., and Rottier., P. J. M. 1994. Coronavirus accumulates beyond the site of virion budding. J. Virol. 68: 6523-6534.

Laemmli, U. K. 1970. Cleavage of structural proteins during the assembly of the head of bacteriophage T4. Nature 227: 680-685.

Liu, D. X., Cavanagh, P. G., and Inglis, S. C. 1991. A polycistronic mRNA specified by the coronavirus infectious bronchitis virus. Virology 184: 531-544.

Liu, D. X., and Inglis, S. C. 1991. Association of the infectious bronchitis virus 3c protein with the virion envelope. Virology 185: 911-917.

Machamer, C. E., Mentone S. A., Rose, J. K., and Farquhar, M. G. 1990. The E1 glycoprotein of an avian coronavirus is targeted to the cis golgi complex. Proc. Natl. Acad Sci. USA 87(18): 6944-6948.

Maeda, J., Maeda, A., and Makino, S. 1999. Release of coronavirus E protein in membrane vesicles from virus-infected cells and E protein-expressing cells. Virology 263: 265-272.

Ng, L. F. P., and Liu, D. X. 2000. Further characterization of the coronavirus infectious bronchitis virus 3C-like proteinase and determination of a new cleavage site. Virology 272: In press.

Raamsman, M. J. B, Locker, J. K., de Hooge, A. , de Vries, A. A. F., Griffiths, G., Vennema, H., and Rottier., P. J. M. 2000. Characterization of coronavirus mouse hepatitis virus strain A59 small membrane protein E. J. Virol. 74: 2333-2342.

Vennema, H., Godeke G-J, Rossen, J. W. A., Voorhout, W. F., Horzinek, M., C., Opstelten, D-J, E., and Rottier., P. J. M. 2000. Nucleocapsid-independent assembly of coronavirus-like particles by co-expression of viral envelope protein genes. EMBO J. 15: 2020-2028.

Mitochondrial Aconitase Binds to the 3'-UTR of Mouse Hepatitis Virus RNA

SANTOSH K. NANDA AND JULIAN L. LEIBOWITZ

Department of Pathology and Laboratory Medicine, Texas A&M University System Health Science Center, College Station, Texas 77843-1114

1. INTRODUCTION

Cis-acting signals located in the 3'-UTR of the viral genome were first identified by deletion analyses of MHV defective interfering (DI)-RNAs (Kim, Jeong, and Makino, 1993; Lin and Lai, 1993). The *cis*-acting signals for the synthesis of minus-strand RNA are contained within the last 55 nucleotides (nt) plus the poly (A) tail (Lin, Liao, and Lai, 1994). Our lab has previously shown that host cell proteins specifically bind to two distinct sites within the MHV 3'UTR (Yu and Leibowitz, 1995b). In the current work, we show that the RNA-protein (RNP) complex formed within the last 42 nt of the genomic RNA contains four proteins of molecular mass 90, 70, 58 and 40 kDa and identify the 90 kDa protein as mitochondrial aconitase.

2. MATERIALS AND METHODS

2.1 RNase protection/gel mobility shift and UV cross-linking assays

Cytoplasmic extracts were prepared from Dounce homogenized 17Cl-1 cells by a modification of a previously described method (Yu and Leibowitz,

The Nidoviruses (Coronaviruses and Arteriviruses).
Edited by Ehud Lavi *et al.*, Kluwer Academic/Plenum Publishers, 2001.

1995b). RNase protection/gel mobility shift and UV cross-linking assays were performed as described (Yu and Leibowitz, 1995b). Unlysed cells and nuclei were removed by centrifugation at 750g for 10 min. The supernatant was centrifuged at 10,000g for 30 min at 4°C and this supernatant was stored as the post-mitochondrial fraction.

2.2   Purification of RNA binding proteins

Proteins which interacted with the last 42 nt of the MHV genome [3'(+)42] were purified from cytoplasmic lysates by sequential batch fractionation over High Q and High S ion exchange matrices (Bio-Rad) and a heparin agarose matrix (Sigma). At each step samples were assayed for RNA binding activity and analyzed by SDS-PAGE.

A biotinylated synthetic RNA corresponding to nt 42–5 at the 3' end of the MHV genome [position 1 is the first nt upstream of the 3' poly(A) tail] was purchased from Dharmacon Research. The RNA solution was adjusted to 100 mM KCl, 5 mM $MgCl_2$ and 1 mM DTT and bound to 1 mg of BioMag Streptavidin beads (PerSeptive Biosystems). The eluate from the heparin agarose matrix was added to the beads for 2 h at 4°C. After four washes proteins were eluted with 2 M KCl or by boiling in SDS-PAGE loading buffer.

2.3   Peptide sequencing

The partially purified 90 kDa protein was resolved from other proteins by SDS-PAGE, located by staining, and was cut out from the gel. The gel slice was digested with trypsin and the resulting tryptic peptides purified by HPLC and analyzed by MALDI mass spectrometry. Two peptides were subjected to sequential Edman degradation.

3.   RESULTS

3.1   Partial purification of the MHV-JHM 3' (+) RNA binding proteins

Cytosolic extracts from murine 17Cl-1 cells were assayed with a ^{32}P-labeled transcript corresponding to nt 16-84 upstream from the 3' end of the MHV-JHM genome. Three RNA-protein complexes formed, with the slowest migrating complex, complex 1, being the most abundant (Figure 1A). The specificity of the RNA-protein binding was confirmed by competition experiments (not shown). Molecular masses of cytoplasmic proteins that bind to the MHV-JHM 3'(+)42 RNA were estimated by UV-induced cross-linking assays. Four proteins of 90, 70, 58 and 40 kDa were consistently detected (Fig. 1B). The 90 kDa protein was the most prominent. *In situ* cross-linking (not shown) revealed that complex 1 contained the same four protein species.

RNA-binding proteins were purified from cytoplasmic lysate as described in Materials and Methods and assayed by gel mobility shift/RNase protection (Fig. 1A), UV cross-linking, and SDS-PAGE (Fig. 1B). Virtually all of the RNA binding activity was in the High Q matrix flow through fraction. This fraction was bound to High S matrix and RNA binding activity was then eluted with 150 mM KCl. This fraction was subjected to non-specific affinity chromatography with heparin-agarose. This purification scheme enriched for RNA-binding proteins. A 90 kDa protein present in material purified by this method was well resolved from other proteins and strongly labeled by UV cross-linking   (Fig. 1B). Two-dimensional gel electrophoresis of this material indicated that this 90 kDa band was a single protein (not shown).

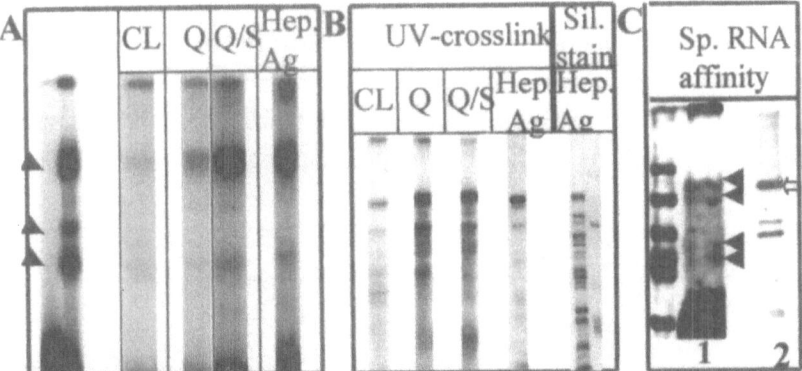

Figure 1. Panel A. A standard RNase protection gel mobility shift assay in on the left. The enrichment of binding activity during purification from crude lysate (CL) and from matrices as described in the text. Panel B. UV cross-linking patterns obtained during purification. Panel C. Affinity purified material visualised by silver staining (lane 1) and by Western Blot with anti-m-aconitase antibody (lane 2). Molecular wt. markers are in the left lane.

3.2   Identification of the 90 kDa MHV-JHM 3' (+) RNA binding protein

A large scale preparation of heparin agarose affinity matrix purified material was reduced, carboxymethylated and subjected to SDS-PAGE. The 90 kDa band was excised from the gel, digested with trypsin, fractionated by HPLC and analyzed by MALDI-mass spectrometry. The three best fits with the MS data were bovine, swine, and human mitochondrial aconitases (m-aconitase) with 63%, 54%, and 54% masses matched respectively. Two peptides were sequenced and yielded the sequences IVYGHLDDPANQEIER and LTIQGLK. A BLAST search of the Swiss-Prot database revealed 100% sequence identity with residues 69-83 and 724-730 of human m-aconitase.

Proteins binding to the MHV 3' (+)42 protein binding element were further purified by affinity chromatography. Ninety, 70, 58, and 40 kDa proteins were eluted from the specific RNA affinity matrix (Fig. 2A). The 90 kDa protein co-electrophoresed with purified bovine m-aconitase and was immunoreactive with an anti-m-aconitase antibody (Fig. 1C). The 90 kDa protein contained in RNA-protein complex 1 isolated by native gel electrophoresis was also recognized by this antibody (not shown). These results strongly support the identification of the 90 kDa protein in RNA-protein complex 1 as m-aconitase.

RNase protection/gel mobility shift reactions were carried out with purified bovine mitochondrial holo- and apo-aconitase but failed to detect binding to the MHV 3'(+)42 RNA. When a UV cross-linking step was added prior to electrophoresis, only the mitochondrial apo-aconitase formed a complex. This complex co-electrophoresed with complex 2 formed with cytoplasmic lysate (Fig. 2A). Specificity of binding of apo-aconitase to MHV 3'(+)42 RNA was verified by competition (Fig. 2A).

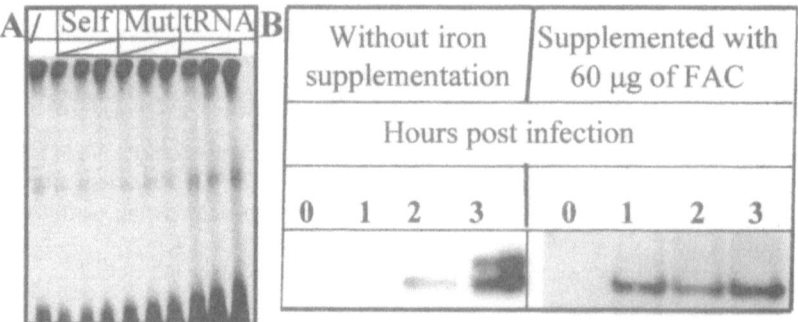

Figure 2. Panel A. Specific binding of m-apo-aconitase in the presence of competitor RNAs. B. Western Blot analysis of N protein levels in control and FAC treated cells.

3.3   Iron supplementation increases the level m-aconitase and N protein

The 5' UTR of m-aconitase contains an IRE which links cellular iron status and m-aconitase expression (Kim *et al.*, 1996). To seek functional effects of the interaction of m-aconitase with the MHV 3'(+)42 protein binding element, we examined the effect of ferric ammonium citrate (FAC) supplementation on virus replication. The ability of MHV to form plaques was unchanged by FAC treatment of cells. Immunoblots showed that iron supplementation increased N protein accumulation at 1-3 h p.i. compared to untreated cultures (Fig. 2B).

4.   DISCUSSION

Two protein binding elements have been mapped within the MHV 3 (+) UTR and are implicated in viral replication (Liu, Yu, and Leibowitz, 1997; Yu and Leibowitz, 1995a; Yu and Leibowitz, 1995b). In this work we have demonstrated that there are at least four protein components in the RNP complex formed at the 3' most *cis*-element in the MHV-JHM 3' UTR; m-aconitase is one component of the complex.

Mammalian cells have two aconitases encoded by separate nuclear genes: the mitochondrial enzyme and cytoplasmic aconitase, better known as iron regulatory protein (IRP). The IRP has 30% amino acid identity with m-aconitase (Kennedy *et al.*, 1992) and ~56% overall sequence similarity. Activity of the enzymes depend on the presence of an iron sulfur [4Fe-4S] cluster in the catalytic center. IRP is a conditional cytoplasmic mRNA-binding protein which interacts with iron-responsive elements (IREs) located in the 5' UTR of ferritin mRNA and the 3' UTR of transferrin receptor (TfR) mRNA and coordinates post-transcriptional regulation of cellular iron metabolism (Klausner, Rouault, and Harford, 1993). Disassembly of the IRP iron-sulfur cluster upon iron starvation yields the RNA-binding form of the IRP. We have demonstrated that only the m-aconitase apoprotein binds to the MHV 3'(+)42 RNA, and that this binding is specific. M-aconitase, like its IRP homolog, is a bifunctional protein whose dynamic [4Fe-4S] cluster determines if it functions as an enzyme or as a RNA binding protein.

It is surprising that m-aconitase, a mitochondrial matrix protein, binds to a viral RNA that replicates in the cytoplasm. RNA binding activity did not partition with the majority of m-aconitase detected in mitochondrial lysates (not shown). We believe that m-aconitase interacts with MHV RNA prior to importation into mitochondria. Alternatively, many mitochondrial proteins have been demonstrated to be present outside of mitochondria under certain conditions (Soltys and Gupta, 1999).

Iron supplementation increased expression of N protein as early as one h p.i. with a corresponding change in MHV-specific mRNA levels (not shown). The effect on viral protein synthesis parallels a modest increase in m-aconitase level in iron treated cells. We believe that m-aconitase binding to the 3' UTR increases the expression of viral proteins, similar to the role of IRP in regulating TfR (Klausner, Rouault, and Harford, 1993).

ACKNOWLEDGMENT

This work was supported by National Multiple Sclerosis Society grant RG2203-B-6 and a gift from the Stearman family.

REFERENCES

Kennedy, M.C., Mende-Mueller, L., Blondin, G.A., and Beinert, H. (1992). Purification and characterization of cytosolic aconitase from beef liver and its relationship to the iron-responsive element binding protein. *Proc Natl Acad Sci U S A* **89**(24), 11730-4.

Kim, H.Y., LaVaute, T., Iwai, K., Klausner, R.D., and Rouault, T.A. (1996). Identification of a conserved and functional iron-responsive element in the 5'-untranslated region of mammalian mitochondrial aconitase. *J. Biol. Chem.* **271**, 24226-24230.

Kim, Y.-N., Jeong, Y.S., and Makino, S. (1993). Analysis of cis-acting sequences essential for coronavirus defective interfering RNA replication. *Virology* **197**, 53-63.

Klausner, R.D., Rouault, T.A., and Harford, J.B. (1993). Regulating the fate of mRNA: the control of cellular iron metabolism. *Cell* **72**(1), 19-28.

Leibowitz, J.L., Wilhelmsen, K.C., and Bond, C.W. (1981). The virus-specific intracellular RNA species of two murine coronaviruses: MHV-A59 and MHV-JHM. *Virology* **114**, 39-51.

Lin, Y., Liao, C., and Lai, M.M.C. (1994). Identification of the cis-acting signal for minus-strand RNA synthesis of a murine coronavirus: implications for the role of minus-strand RNA in RNA replication and transcription. *J. Virol.* **68**(12), 8131-8140.

Lin, Y.-J., and Lai, M.M.C. (1993). Deletion mapping of a mouse hepatitis virus defective interfering RNA reveals the requirement of an internal and discontinous sequence for replication. *J. Virol.* **67**, 6110-6118.

Liu, Q., Yu, W., and Leibowitz, J.L. (1997). A specific host cellular protein binding element near the 3' end of mouse hepatitis virus genomic RNA. *Virology* **232**, 74-85.

Soltys, B.J., and Gupta, R.S. (1999). Mitochondrial-matrix proteins at unexpected locations: are they exported? *Trends Biochem. Sci.* **24**, 174-177.

Yu, W., and Leibowitz, J.L. (1995a). A conserved motif at the 3' end of mouse hepatitis virus genomic RNA required for host protein binding and viral RNA replication. *Virology* **214**, 128-138.

Yu, W., and Leibowitz, J.L. (1995b). Specific binding of host cellular proteins to multiple sites within the 3' end of mouse hepatitis virus genomic RNA. *J. Virol.* **69**, 5033-5038.

The Cell Biology of Coronavirus Infection

ERIK PRENTICE AND MARK R. DENISON
Departments of Microbiology & Immunology and Pediatrics and The Elizabeth B. Lamb Center for Pediatric Research, Vanderbilt University Medical Center, Nashville, TN 37232

1. INTRODUCTION

Several aspects of coronavirus interaction with host cells have been identified, the best characterized of which involve virus entry and virion assembly (for review, see Holmes and Lai, 1996). These accomplishments are critical in the understanding of virus-cell interactions; however, since such interactions occur in four dimensions (over time in the three-dimensional space of the cell) they are important landmarks on a path that for the most part remains unilluminated. To understand the dynamic relationship of the virus to the cell it is important to use approaches that can observe interactions in the context of the entire cell and throughout the course of infection.

More recently, several groups have begun to investigate the formation and fate of coronavirus replication complexes in virus-infected cells (Bi *et al.*, 1998; Denison *et al.*, 1999; Shi *et al.*, 1999; van der Meer *et al.*, 1999). Our studies on the cell biology of coronavirus infection have been based on the observation that viral RNA synthesis localizes to punctate cytoplasmic complexes that are distinct from known sites of virion assembly in the ER-Golgi-intermediate complex (ERGIC) (Bost *et al.*, 2000; Sims, Ostermann, and Denison, 2000). These results have raised questions concerning the mechanisms by which the newly synthesized viral RNA is delivered to sites of virion assembly. Answers to these questions will allow us to define the viral and cellular proteins involved in replication and transport of RNA. We have developed methods for imaging multiple MHV proteins, cellular proteins, and cytoskeletal elements MHV cells, both in fixed and live cells.

The Nidoviruses (Coronaviruses and Arteriviruses).
Edited by Ehud Lavi *et al.*, Kluwer Academic/Plenum Publishers, 2001.

We have used confocal microscopy to generate three-dimensional reconstructions of infected cells and have investigated virus-induced changes in cells over time using video microscopy. We have demonstrated that the replication complexes vary during the course of infection, and that it is critical to observe the entire context of the cell over time to draw accurate conclusions concerning the formation, function and fate of viral replication complexes.

2.   METHODS AND MATERIALS

2.1   Virus Infection and Immunofluorescence

DBT cells were grown on glass coverslips as previously described (Bost *et al.*, 2000). In general, cells were seeded at lower densities and grown for 36-48 h prior to infection. Cells were infected at moi's of 10-20 pfu/cell. Cells were fixed and permeabilized by aspiration of media and direct addition of 100% methanol (-20°C) to the well. Cells were fixed for at least 1 hr prior to processing for immunofluorescence. Cells were processed for immunofluorescence as previously described (Denison *et al.*, 1999).

2.2   Confocal Microscopy

All fixed cell images were acquired on Zeiss LSM 410 laser scanning confocal microscope (Denison *et al.*, 1999). Unless otherwise stated, images were acquired using lasers at 488nm(green) and 568nm (red). Images were obtained with brightness and contrast settings to optimize the resolution and maximize the signal-to-noise ratio. Images were acquired using a 40x, NA 1.3 oil objective at a magnification of 3.1 to obtain resolution in x,y and z dimensions of ~0.2 μm/pixel. The confocal pinhole was set to optimize the confocal "slice" size to create a non-overlapping volume set.

2.3   Live Cell Imaging

DBT cells were grown on glass 40mm glass coverslips and infected as above. Prior to imaging the coverslip was transferred to a Bioptechs FCS live-cell imaging chamber. Cells were simultaneously imaged using Nomarski DIC and fluorescence. Cells were transfected with constructs expressing viral N fused to enhanced green fluorescent protein (GFP). For experiments in this report, images were obtained at 30 second intervals for up to 8 h, beginning 3-4 h p.i. For maximum brightness over time, fluorescent signal was acquired using "nonconfocal" settings.

2.4 Post-Acquisition Image Processing

Images were acquired in the LSM software as 512 x 512 pixel TIFF images. Image modifications were made using Adobe Photoshop 5.0, NIH image 1.62, Adobe Premiere 4.0, and Apple Quicktime 4.0 Pro. Original data was archived as LSM TIFFs prior to any image modification. To reconstruct 3D images of fixed cells, "stacks" of individual confocal images were constructed, and rotations of the x,y,z volume were made.

3. RESULTS AND DISCUSSION

3.1 3D Reconstructions of MHV-Infected Cells Show the Relationships of Gene 1 and Structural Proteins

When MHV infected cells were dual-probed for the gene 1 protein p65 and the structural M protein at 6.5 h p.i. the proteins appeared to be partially co-localized when viewed by nonconfocal indirect immunofluorescence.

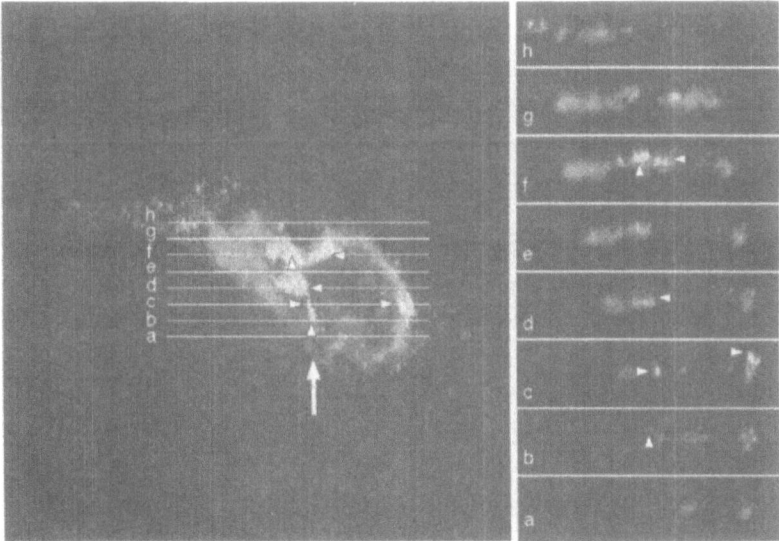

Figure 1. Three dimensional reconstruction of MHV infected DBT cell. The left image shows the x/y image with a flattened "z projection" of 70 "slices" (14μm in z depth). Labeling for p65 as medium grey and M is white. Horizontal lines (a-h) represent location of reslicing of the x/y/z volume set perpendicular or "into" the plane of the image. The respective x/z slices are seen to the right as viewed from the direction of the white arrow in the image to the left. The location of foci of M on both sets of images are shown by the white arrowheads. The "bottom" of each image to the left represents the plane of the coverslip.

However, when the same cell was imaged using confocal settings, the apparent coincidence of signals was eliminated and p65 and M had entirely distinct patterns of distribution with little or no overlap. To determine the three-dimensional relationship of M and p65, 70 images were obtained at 0.2μm intervals in Z, the images were "stacked" and the stack or volume data was used to analyze the relationship of p65 and M (Fig. 1). The volume was analyzed in rotations and stereo images, or post image "reslicing" of the cell in different planes. When the dual-labeled, infected cell was analyzed in this way, it was immediately apparent that the proteins were localized to extensive structures that were organized in distinct parts of the cell. P65 localized to punctate structures that were generally in the peripheral cytoplasm and not tightly associated with the nucleus. In contrast, M was localized to structures closely juxtaposed to the nucleus and predominantly eccentrically localized to one side of the nucleus, consistent with the distribution of the (Fig. 1). In addition, the reconstructions demonstrated that the perinuclear distribution of M was "above" the cytoplasmic foci of p65 when the coverslip was viewed as the "bottom" of the cell. These reconstructions also revealed the relative height in z of the nucleus compared to the more distributed cytoplasm.

3.2   Live Cell Imaging of MHV-Infected Cells

The distribution of gene 1 and structural proteins showed consistent patterns across entire monolayers, but it was also obvious that there was evolution of the pattern depending on the time p.i. and on whether the cells were involved in syncytia formation. We used a live cell imaging closed chamber system to analyze changes occurring in infected cells over time (Fig. 2). To assess the directionality of fusion and distribution of intracellular contents, we overexpressed N as a fusion with enhanced GFP in the context of infection. What we observed was active movement and division of cells in the monolayer for the first four hours followed by loss of sharpness of cell membrane margins of closely approximated cells. This was succeeded by rapid fusion of cell membranes and coalescence of nuclei and all easily visible intracellular membrane-bound organelles to the center of syncytia. In addition, we observed that "recruitment" of cells into syncytia was often initiated by contact of extended processes, followed by fusion at the point of contact and rapid movement of the uninvolved cell toward the multinucleated cell. Our results suggested that the process of syncytia formation involved the linking of cytoskeletal elements and was active rather than passive. Although syncytia have not been thought to be critical in the normal lifecycle and pathogenesis of MHV infection, our images raise the possibility that cell contact through extended filopodia might play a role in cell-to-cell transmission of virus. GFP:N was not

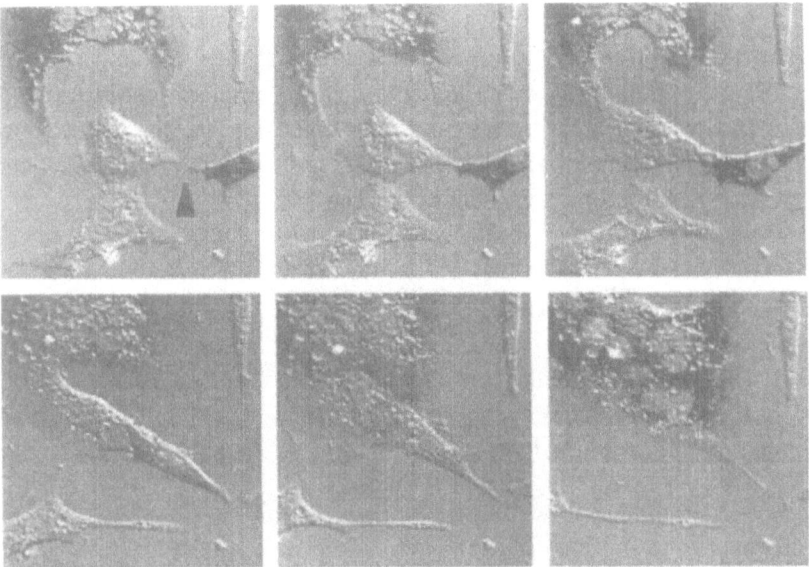

Figure 2. Confocal Video Microscopy of MHV-infected DBT cells expressing an N-GFP fusion protein. Cells were transfected with constructs expressing N protein fused to enhanced green fluorescent protein (GFP:N) for 48 h, infected, and imaged from 4-12 h p.i. Simultaneous DIC and fluorescent images obtained every 30 seconds. Equally spaced images are shwon. The black signal in cells and syncytia indicates expression of of GFP-N. The black arrowhead shows the sites of filopodia contact, directional recruitement, and incorporation of cell contents.

detected in the nuclei of tranfected cells, but was detected in the cytoplasm in both a diffuse cytoplasmic distribution and in punctate foci of various size that were never seen with GFP expression alone. The punctate character of GFP:N foci was retained even during the alterations of cells during syncytia formation. Finally, rapid directional redistribution of the contents of one cell into a syncytium was frequently observed, supporting the concept of active involvement of cytoskeletal elements such as actin and/or microtubules in MHV syncytia formation.

4. SUMMARY

The ability to obtain entire volume data on infected cells will allow us to define much more accurately the interactions of viral proteins with host cell structures such as ER, Golgi, and cytoskeletal elements. In addition, the demonstrated ability to express viral proteins fused to fluorescent markers in in live cells will allow us to follow specific proteins or complexes during the course of infection and to determine if exogenously expressed proteins are able to target to sites of active viral replication. This in turn will allow new approaches to the study of viral and cellular protein-protein interactions, as methods to study the biology and pathogenesis of MHV infection at a cellular level. Finally, the approaches described here will allow us to define protein complementation of defective viruses at a cellular level, rather than being dependent on population measurements of RNA, protein, or progeny virus.   By combining these approaches with available biochemical and molecular biological approaches and the emerging reverse genetic and recombinant genetic approaches, rapid progess in understanding the details of coronavirus-cell interactions should be possible.

ACKNOWLEDGMENTS

This work was supported in part by NIH grants' AI-26603 and AI-01479 (M. Denison).   We thank Jonathan Sheehan and David Piston in the Molecular Imaging Shared Resource of the Vanderbilt-Ingram Cancer Center (IP30CA68485)

REFERENCES

Bi, W., Pinon, J. D., Hughes, S., Bonilla, P. J., Holmes, K. V., Weiss, S. R., and Leibowitz, J. L. (1998). Localization of mouse hepatitis virus open reading frame 1a derived proteins. *J. Neurovirology* **4**, 594-605.

Bost, A. G., Carnahan, R. H., Lu, X.-T., and Denison, M. R. (2000). Four proteins processed from the replicase gene polyprotein of mouse hepatitis virus colocalize in the cell periphery and adjacent to sites of virion assembly. *J Virol* **74**, 3379-3387.

Denison, M. R., Spaan, J. M., van der Meer, Y., Gibson, C. A., Sims, A. C., Prentice, E., and Lu, X. T. (1999). The putative helicase of the coronavirus mouse hepatitis virus is processed from the replicase gene polyprotein and localizes in complexes that are active in viral RNA synthesis. *J. Virol.* **73**, 6862-6871.

Holmes, K. V., and Lai, M. M. C. (1996). Coronaviridae: The viruses and their replication. *In* "Virology" (B. N. Fields, D. M. Knipe, and P. M. Howley, Eds.), Vol. 1, pp. 1075-1093. 2 vols. Lippincott-Raven Publishers, Philadelphia.

Shi, S. T., Schiller, J. J., Kanjanahaluethai, A., Baker, S., Oh, J., and Lai, M. M. C. (1999). Colocalization and membrane association of murine hepatitis virus gene 1 products and de novo-synthesized viral RNA in infected cells. *J. Virol.* **73**, 5957-5969.

Sims, A. C., Ostermann, J., and Denison, M. R. (2000). Mouse hepatitis virus replicase proteins associate with two distinct populations of intracellular membranes. *J Virol* **74**, 5647-5654.

van der Meer, Y., Snijder, E. J., Dobbe, J. C., Schleich, S., Denison, M. R., Spaan, W. J. M., and Krinjnse Locker, J. (1999). The localization of mouse hepatitis virus nonstructural proteins and RNA synthesis indicates a role for late endosomes in viral replication. *J. Virol.* **73**, 7641-57.

Induction of Apoptosis in Murine Coronavirus-Infected 17Cl-1 Cells

[1]CHUN-JEN CHEN, [2]SUNGWHAN AN, AND [1,2]SHINJI MAKINO
[1]Department of Microbiology and Immunology, The University of Texas Medical Branch at Galveston, Galveston, Texas 77555-1019: [2]Department of Microbiology, and Institute for Cellular and Molecular Biology, The University of Texas at Austin, Austin, Texas 78712

1.   INTRODUCTION

Apoptosis is an important process in the development and homeostasis of multicellular organisms (Jacobson *et al* 1997). In most cases, apoptosis is executed by activating a proteolytic system involving a family of caspases. Caspases participate in a cascade that is triggered in response to proapoptotic signals and culminates in cleavage of a set of proteins, resulting in cell death (Cohen 1997). Apoptosis also represents a highly efficient defence mechanism against virus infection; apoptosis aids in removal of viral proteins and nucleic acids by the infected host.

Eleouet *et al* (1998) demonstrated that infection of coronavirus transmissible gastroenteritis virus (TGEV) induces caspase-dependent apoptosis in several cell lines. Belyavskyi *et al* (1998) showed that mouse hepatitis virus (MHV) strain 3 (MHV-3) infection of cultured macrophages induces apoptosis, while it is not clear whether MHV-3-induced apoptosis is caspase-dependent. In the present study we demonstrated that MHV infection in an established cultured cell line also induced caspase-dependent apoptosis.

The Nidoviruses (Coronaviruses and Arteriviruses).
Edited by Ehud Lavi *et al.*, Kluwer Academic/Plenum Publishers, 2001.

2.   MATERIALS AND METHODS

2.1   Viruses and cells

Plaque-cloned A59 strain of MHV was used. Mouse 17Cl-1 cells were cultured in Dulbecco's modified eagle's medium (DMEM) (with sodium pyruvate, JRH Biosciences) containing 10% fetal calf serum (FCS). After MHV infection, FCS concentration was reduced to 2%.

2.2   DNA fragmentation assay

Low-molecular weight DNA was extracted as described by Hinshaw *et al* (1994).

2.3   Hoechst dye staining

At various times postinfection (p.i.) of MHV, cells were collected, washed with phosphate buffered saline (PBS), and resuspended in 100 μl PBS. Cells were fixed by first slowly adding 200 μl of fixing solution (methanol: acetic acid = 3: 1) and then adding additional fixing solution up to 2 ml. After incubation at room temperature for at least 1 h, the cells were collected by centrifugation and resuspended in 400 μl of fixing solution. Cell suspensions were deposited on microscopic slides and air-dried. Cells were stained with a drop of staining solution (50% glycerol, 50% 0.1 M Tris-HCl [pH 7.4], 1 μg/ml Hoechst 33342), coversliped, and observed under a fluorescent microscope.

2.4   Effect of caspase inhibitors on apoptosis induction

17Cl-1 cells were incubated for 2 h with 80 μM Z-Asp-Glu-Val-Asp-fmk (Z-DEVD-fmk) (Enzyme Systems Products, Livermore, CA) prior to MHV infection. After MHV infection, cells were incubated in the presence of 80 μM of Z-DEVD-fmk. At 24 h p.i., the Z-DEVD-fmk was replenished. At 48 h p.i., apoptosis was measured by DNA fragmentation assay.

3.   RESULTS

To investigate whether MHV infection in established cell lines induced apoptosis, we examined MHV-infected 17Cl-1 cells for the presence of internucleosomal DNA cleavage, an event commonly observed in apoptotic cells. At various times p.i., cells attached to the plates were scraped by using a rubber policeman and combined with those floating in the medium. Low-molecular-weight apoptotic DNA fragments were extracted and then separated by agarose gel electrophoresis (Fig. 1A). A sign of internucleosomal DNA cleavage, which appeared as a low level of smearing of DNA in agarose gel electrophoresis, was first detected at 36 h p.i. At 62 h p.i., a DNA ladder was evident in MHV-infected cells but not in the mock-infected cells. DNA ladder was also detected in MHV-JHM-infected cells (data not shown). We used the terminal deoxynucleotidyl transferase-mediated dUTP nick-end-labeling (TUNEL) assay (ApoAlert DNA Fragmentation Assay Kit, Clontech) to further confirm DNA fragmentation in MHV-infected 17Cl-1 cells. At 36 h p.i., approximately one-quarter of the infected cells were TUNEL positive. Very few TUNEL positive cells were observed in the mock-infected cultures (data not shown). To further confirm that MHV-infected 17Cl-1 cells were undergoing apoptosis, MHV-infected 17Cl-1 cells were stained with the fluorescent dye Hoechst 33342. In the mock-infected culture, the majority of cells contained intact nuclei at 68 h p.i., while at the same time MHV-infected 17Cl-1 cultures showed chromatin condensation (Fig. 2). These data demonstrated that apoptosis was induced in MHV-infected 17Cl-1 cells, and the strong apoptotic signs were evident late in infection.

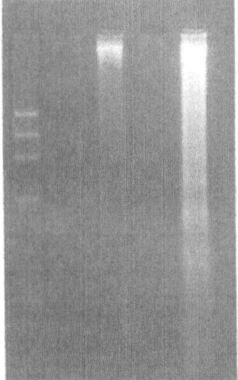 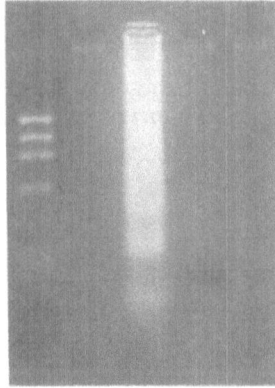

Figure 1. DNA fragmentation analysis. (A) 17Cl-1 cells were mock-infected or infected with MHV-A59. Low-molecular-weight DNA was extracted from cells at indicated time after infection. (B) Culture fluid containing MHV-A59 was UV-irradiated for indicated time at 0°C, and then inoculated into 17Cl-1 cells. Low-molecular-weight DNA was isolated at 46 h p.i. Samples were electrophoresed in 2% agarose gels and visualized with ethidium bromide.

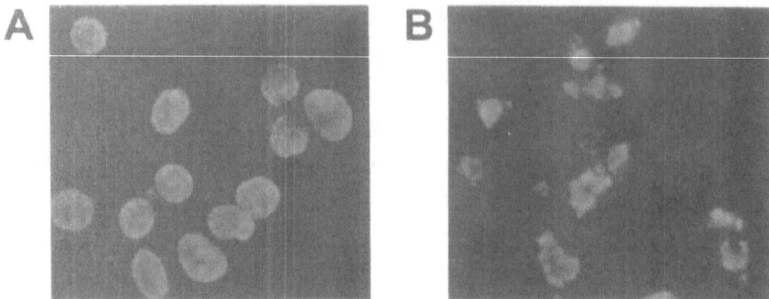

Figure 2. Chromatin condensation in 17Cl-1 cells infected with MHV. 17Cl-1 cells were mock-infected (A) or infected with MHV-A59 (B). At 68 h p.i., mock-infected cells and floating cells in MHV-infected culture media were collected and stained with Hoechst 33342.

To test whether binding of MHV to MHV receptors or some unidentified substances, other than MHV, present in the inoculum induced apoptosis, MHV was inactivated by irradiation of inoculum with UV-light and then added to 17Cl-1 cells. The infectivity of UV-irradiated samples was less than 1 PFU. After incubation for 1 h at 37°C, the inoculum was removed and the cells were incubated for up to 46 h. Neither CPE (data not shown) nor internucleosomal DNA cleavage (Fig. 1B) was observed in the cells that underwent this treatment, demonstrating that binding of MHV to MHV receptors alone or unidentified substances, which may be present in the inoculum, did not induce apoptosis. Replication of MHV was necessary for inducing apoptosis.

In the very late stage of this study we noticed that changing cell culture condition significantly affected the induction of apoptosis. When MHV-infected 17Cl-1 cells were incubated in a medium containing DMEM from a different vendor (Gibco/BRL), DNA ladder was clearly detected as early as 24 h p.i. (data not shown). There were several differences in the contents of two batches of DMEMs used in this study; one major difference was that DMEM from Gibco/BRL lacked sodium pyruvate, while DMEM used in most of our other experiments contained sodium pyruvate.

We examined whether apoptosis was induced in another MHV-susceptible cell line, DBT. Floating fused MHV-A59-infected DBT cells and the small number of cells that were still attached to the plates were collected at 12 h, 36 h, and 48 h p.i. Gel electrophoresis of low-molecular-weight DNA extract from these samples and TUNEL assay at 12 h p.i. showed no sign of DNA fragmentation (data not shown), demonstrating that MHV infection in DBT cells did not induce apoptosis.

To determine whether MHV-induced apoptosis was caspase-dependent, we studied the effect of Z-DEVD-fmk, an irreversible and cell-permeable inhibitor of caspase-3, on the induction of apoptosis in MHV-A59-infected 17Cl-1 cells. DNA fragmentation assay showed that addition of Z-DEVD-fmk, at a final concentration of 80 μM, strongly inhibited internucleosomal DNA cleavage (Fig. 3A), demonstrating that MHV-induced apoptosis in 17Cl-1 cells was caspase-dependent. One step MHV growth curves in 17Cl-1 cells in the presence and absence of Z-DEVD-fmk showed no significant difference, indicting that suppression of caspase-3 activity did not affect MHV growth (Fig. 3B).

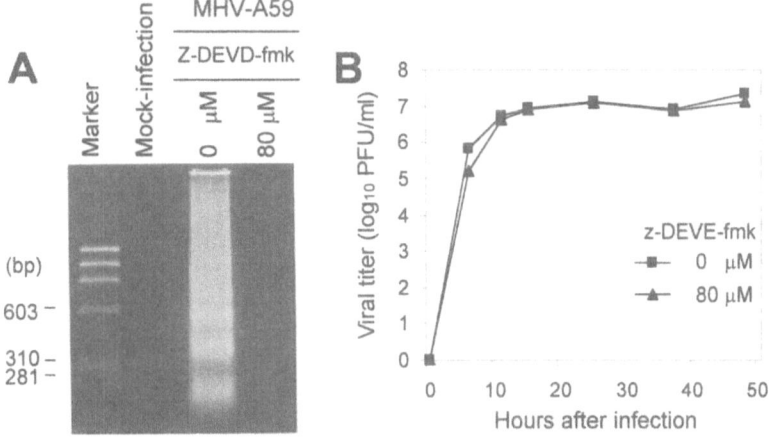

Figure 3. Effect of Z-DEVD-fmk on MHV-induced apoptosis. (A) 17Cl-1 cells were mock-infected or incubated with medium lacking Z-DEVD-fmk or that containing Z-DEVD-fmk at a concentration of 80 μM. At 48 h p.i., low-molecular-weight DNA was extracted and separated in a 2% agarose gel. (B) Effect of Z-DEVD-fmk on one-step growth curve of MHV-A59 in 17Cl-1 cells.

4. DISCUSSION

Induction of apoptosis in MHV-infected 17Cl-1 cells was demonstrated by agarose gel electrophoresis of small DNA fragments, TUNEL assay, and Hoechst staining. Our present data and previous findings of apoptosis in TGEV-infected cultured cells (Eleouet *et al* 1998) and in MHV-3-infected cultured macrophages (Belyavskyi *et al* 1998) established that apoptosis is one mechanism of cell death in coronavirus-infected cells. MHV-induced apoptosis was suppressed by caspase-3 inhibitor Z-DEVD-fmk, demonstrating that MHV-induced apoptosis was caspase-dependent. TGEV-mediated apoptosis is also caspase-dependent (Eleouet *et al* 1998). These data indicate that coronavirus infection induces caspase-dependent apoptosis (Cohen 1997).

MHV-infected DBT cells did not show any significant sign of apoptosis throughout infection, indicating that apoptosis was not a universal event among MHV-infected established cell lines. DBT cells are derived from an astrocytoma isolated from CDF1 mice (Hirano *et al* 1974), while 17Cl-1 cells are a line of spontaneously transformed mouse BALB/c 3T3 cells (Sturman and Takemoto 1972); DBT cells and 17Cl-1 cells are derived from different organs of different strains of mice. The difference in the apoptotic response to MHV infection between DBT cells and 17Cl-1 cells may be determined by the difference in the origin of organs and/or difference in the strain of mice.

ACKNOWLEDGMENTS

This work was supported by Public Health Service Grant AI29984 from National Institutes of Health.

REFERENCES

Belyavskyi, M., Belyavskaya, E., Levy, G. A., and Leibowitz, J. L., 1998, Coronavirus MHV-3-induced apoptosis in macrophages. *Virology* **250**: 41-49.

Cohen, G. M., 1997, Caspases: the executioners of apoptosis. *Biochem. J.* **326**: 1-16.

Eleouet, J.-F., Chilmonczyk, S., Besnardeau, L., and Laude, H., 1998, Transmissible gastroenteritis coronavirus induces programmed cell death in infected cells through a caspase-dependent pathway. *J. Virol.* **72**:4918-4924.

Hinshaw, V. S., Olsen, C. W., Dybdahl-Sissoko, N., and Evans, D., 1994, Apoptosis: a mechanism of cell killing by influenza A and B viruses. *J. Virol.* **68**: 3667-3673.

Hirano, N., Fujiwara, K., Hino, S., and Matsumoto, M., 1974, Replication and plaque formation of mouse hepatitis virus (MHV-2) in mouse cell line DBT culture. *Arch. Fesamte Virusforch.* **44**: 298-302.

Jacobson, M. D., Weil, M., and Raff, M. C., 1997, Programmed cell death in animal development. *Cell* **88**: 347-354.

Sturman, L. S., Holmes, K. V., and Behnke, J., 1980, Isolation of coronavirus envelope glycoproteins and interaction with the viral nucleocapsid. *J. Virol.* **33**: 449-462.

Specific Cleavage of 28S Ribosomal RNA in Murine Coronavirus-Infected Cells

[1,2] SANGEETA BANERJEE, [1] SUNGWHAN AN, AND [1,2] SHINJI MAKINO
[1]*Department of Microbiology and Institute of Cellular and Molecular Biology, The University of Texas at Austin, Austin, Texas 78712:* [2]*Department of Microbiology and Immunology, The University of Texas Medical Branch, Galveston, Texas 77555*

1. INTRODUCTION

Mouse hepatitis virus (MHV), a prototypic coronavirus, causes gastrointestinal and upper respiratory tract illnesses in animals and man. The 32 kb-long MHV genome is a positive-sense, single-stranded RNA (Lai *et al* 1981, Lee *et al* 1991) that encodes 11 open reading frames, expressed through the production of a genomic-size and six to eight species of subgenomic mRNAs (Lai *et al* 1981). Genomic-size mRNA encodes the 5'-most 22 kb-long gene 1, which encodes the RNA polymerase function (Lee *et al* 1991). MHV contains three envelope proteins, S, M and E proteins. S protein binds to the coronavirus receptor (Dveskler *et al* 1991). M and E proteins are important for the formation of the MHV envelope (Kim *et al* 1997).

Extensive morphological and physiological changes occur in coronavirus-infected cells. Some of these changes contribute to damage of cells and tissues. The specific basis, for these deleterious effects on host cells, is not well understood. MHV-infection causes host protein translation inhibition (Hilton *et al* 1986, Tahara *et al* 1994). Inhibition of host protein synthesis accompanies an increase in MHV protein synthesis (Siddell *et al* 1981). Some host mRNAs are degraded in MHV-infected cells, while others are transcriptionally upregulated (Kyuwa *et al* 1994). The mechanism of selective MHV-specific protein synthesis in infected cells, concomitant with

The Nidoviruses (Coronaviruses and Arteriviruses).
Edited by Ehud Lavi *et al.*, Kluwer Academic/Plenum Publishers, 2001.

host protein inhibition, is poorly characterised, although it has been suggested that binding of N protein to MHV leader sequences may act as a strong translation initiation signal (Tahara *et al* 1994, Tahara *et al* 1998).

In the present study, we described a hitherto unknown phenomenon in MHV-infected cells: the specific cleavage of 28S ribosomal RNA (rRNA). This phenomenon was not restricted to any particular MHV strain or host cell type. Our data also suggested that the observed cleavage was probably not regulated by the apoptotic pathway. Hence we demonstrated that MHV-induced 28 rRNA cleavage was different from other known ribosomal RNA cleavage events.

2.   MATERIALS AND METHODS

2.1   Viruses and cells

The plaque-cloned A59 strain of MHV (Lai *et al* 1981), MHV-JHM and MHV-2 were used. Mouse DBT cells were used for MHV growth.

2.2   Northern (RNA) blotting

Northern blot analysis was performed using [^{32}P]-γATP-labeled oligonucleotide probes as previously described (Makino *et al* 1991). Probe 1 binding to nucleotides 1532-1551 from the 5'-end of mouse 28S rRNA, was used to detect 28S rRNA and its cleavage products. Probe 2 and probe 3, binding to nucleotides 921-940 and nucleotides 1846-1870 from the 5'-end of mouse 18S rRNA, respectively, were used to detect 18S rRNA. All hybridisation were performed at 60°C.

3.   RESULTS

Agarose gel electrophoresis of intracellular RNA from MHV-A59-infected DBT cells showed a major band that migrated between the 28S and 18S rRNAs. No MHV-specific probe hybridised with this RNA band and its size differed from all MHV subgenomic mRNAs (data not shown). Northern blot analysis of intracellular RNA from MHV-infected cells, using a 28S rRNA-specific probe 1 (Fig. 1A) revealed a major 28S rRNA cleavage product, 28S-CL1 and four minor 28S rRNA bands, CL2-CL5, in addition to the intact 28S rRNA. 28S-CL1 appeared as early as 4 h p.i. (data not shown) and it remained a major cleavage product until 12 h p.i. 28S-CL-3 and 28S-CL4 accumulated late in infection. The amount of intact 28S rRNA decreased significantly with increasing time of infection. In contrast, 18S rRNA did not undergo cleavage in MHV-infected DBT cells (Fig. 1B). The amount of 18S rRNA in MHV-A59-infected cells remained similar to that in uninfected cells, until 16 h p.i., when it decreased slightly.

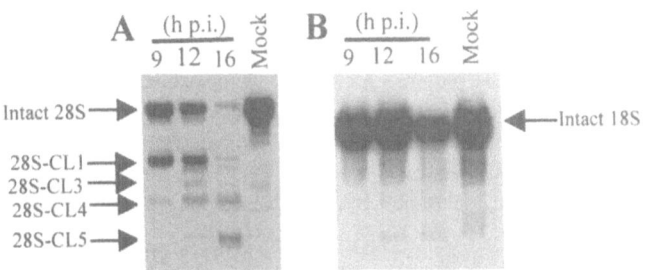

Figure 1. Specific degradation of 28S rRNA in MHV-infected DBT cells. Cytoplasmic RNA was extracted from MHV-infected cells at indicated times and Northern blot analysis was performed using oligonucleotide probes against mouse 28S rRNA (A) and 18S rRNA (B).

Infecting DBT cells with UV-inactivated MHV inoculum did not cause any 28S rRNA cleavage (Fig. 2A), demonstrating that binding of MHV to MHV receptors alone or other unidentified substances, did not induce 28S rRNA cleavage. Induction of 28S rRNA cleavage required MHV replication. Comparing the kinetics of 28S rRNA cleavage with other cytopathic effects in MHV-infected cells, we found 28S rRNA cleavage started earlier than the onset of MHV-A59-induced cell fusion. Infection with a nonfusogenic MHV strain, MHV-2, also caused 28S rRNA cleavage, in DBT cells (data not shown). These data demonstrated that MHV-induced 28S rRNA cleavage was not directly related to cell fusion. A substantial increase in the amount of 28S-CL1, in MHV-infected cells, preceded the peak of MHV RNA synthesis and the cleavage continued at a time when MHV RNA synthesis declined (Sawicki and Sawicki 1986).

We examined the relationship between 28S rRNA cleavage and host protein synthesis in MHV-infected cells. Consistent with previous studies (Siddell *et al* 1981, Tahara *et al* 1994), host protein synthesis inhibition in MHV-infected cells was evident by 6 h p.i. (data not shown); 28S-CL1 appeared as a major cleavage product starting 5 h p.i. Hence, the 28S-CL1 cleavage product needed to accumulate before host protein synthesis inhibition could occur. Thereafter, both 28S rRNA cleavage and host protein synthesis inhibition continued as infection proceeded.

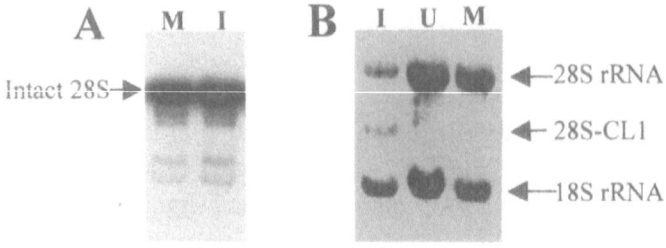

Figure 2. MHV replication is required for mature cytoplasmic 28S rRNA cleavage. (A) Cytoplasmic RNA from mock-infected (M) or UV-irradiated MHV-A59-infected (I) DBT cells were extracted at 8 h p.i. and analysed by Northern blot. (B) Ribosomal precursors were labelled with ^{3}H-uridine and chased for 13 hr. Then, cells were mock-infected (M) or infected with MHV-A59 (I). 8 h p.i. cytoplasmic RNA was extracted. U= Uninfected.

To determine whether mature cytoplasmic rRNA is cleaved in MHV-infected cells, precursor rRNA molecules, in DBT cells, were labelled with ^{3}H-uridine and allowed to be processed and transported to the cytoplasm. Cells were then infected with MHV-A59 and labelled, mature cytoplasmic RNA were analysed by electrophoresis (Fig. 2B). Mock-infected cells showed intact 28S rRNAs while MHV-infected cells showed a reduced amount of intact 28S rRNA and the presence of 28S-CL1. We concluded that MHV infection induced cleavage of mature 28S rRNA, that was processed and transported to the cytoplasm, prior to MHV infection.

28S rRNA cleavage was not confined to any particular cell type or MHV strain. MHV-A59-infection to DBT and 17Cl-1 cells produced all 28S rRNA cleavage products; the kinetics of appearance of the cleavage products and reduction of intact 28S rRNA in MHV-A59-infected DBT and 17CL-1 cells were similar (data not shown). MHV-JHM-infection to DBT and 17Cl-1 cells also induced similar 28S rRNA cleavage products but progress of MHV-JHM-induced 28S rRNA cleavage was slow (data not shown); A lower level of MHV-JHM replication efficiency in infected cells may be related to less efficient 28S rRNA cleavage.

There have been some reports of rRNA degradation occurring in apoptotic cells. MHV-infected 17Cl-1 cells undergo apoptosis, but not MHV-infected DBT cells (An *et al* 1999). Nevertheless, 28S rRNA cleavage occurred in both DBT and 17Cl-1 cells (Fig. 1, and data not shown). If apoptosis regulates MHV-induced 28S rRNA cleavage then blocking apoptosis should block 28S rRNA cleavage. Accordingly we treated MHV-infected 17Cl-1 cells with an irreversible caspase-3 inhibitor, Z-DEVD-fmk, that effectively blocks apoptosis. We found that blocking apoptosis in MHV-infected 17Cl-1 cells did not block or delay 28S rRNA cleavage (Fig. 3). Thus, MHV-induced 28S rRNA cleavage was upstream of caspase-activation, or apoptosis induction and 28S rRNA cleavage occurred via two independent pathways.

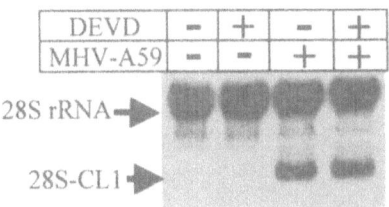

Figure 3. Blocking apoptosis does not block MHV-induced 28S rRNA cleavage. 17Cl-1 cells were mock-treated or treated with 80 μM Z-DEVD-fmk and mock-infected or infected with MHV-A59. 8 h p.i. cytoplasmic RNA was extracted and analyzed by Northern blot.

4. DISCUSSION

We report a novel MHV-induced 28S rRNA cleavage in the present study. The MHV-induced rRNA cleavage occurred only in 28S rRNA, and not in 18S rRNA. Specific cleavage of 28S rRNA required ongoing MHV replication since UV-inactivated MHV failed to induce 28S rRNA cleavage. Fully processed, mature cytoplasmic 28S rRNA underwent cleavage. Currently, we do not know which step of the viral life cycle or specific viral factor(s) causes this 28S rRNA cleavage. MHV infection to all susceptible cell lines, using several different MHV strains, induced 28S rRNA cleavage. MHV-induced 28S rRNA cleavage was independent of virus-induced CPE. and it differed from apoptosis-related rRNA degradation, in that blocking apoptosis had no effect on the cleavage of 28S rRNA.

MHV-induced 28S rRNA cleavage could occur via the activation of a cellular or viral RNase. That same RNase may degrade host mRNAs, which maybe partly responsible for the host protein shut-off in MHV-infected cells (Hilton *et al* 1986). Alternatively, binding of MHV factors or host factors may alter the structure of 60S ribosome, allowing a cellular or viral RNase to access specific regions of 28S rRNA and cleave it.

28S rRNA being an integral component of the 60S ribosomal subunit, MHV-induced 28S rRNA cleavage may affect protein synthesis by rendering polysomes, containing cleaved 28S rRNA, translationally inactive. Polysomes with intact 28S rRNA maybe used for MHV-specific protein synthesis, resulting in host mRNA translation inhibition. Alternatively, ribosomes, containing 28S cleavage products, maybe structurally altered to better translate increasing amounts of MHV-specific mRNAs. Thus, MHV-induced 28S rRNA cleavage may alter the cellular environment and translation machinery to slow down host protein synthesis and allow efficient expression of viral proteins.

ACKNOWLEDGMENTS

This work was supported by Public Health Service grant AI29984 from the National Institutes for Health.

REFERENCE

An, S., Chen, C.-J., Yu, X., Leibowitz, J. L., and Makino, S., 1999, Induction of apoptosis in murine coronavirus-infected cultured cells and demonstration of E protein as an apoptosis inducer. J. Virol. 73:7853-7859.

Dveksler, G. S., Pensiero, M. N., Cardellichio, C. B., Williams, R. K., Jiang, G.-S., Holmes, K. V., and Dieffenbach, C. W., 1991, Cloning of the mouse hepatitis virus (MHV) receptor: expression in human and hamster cell lines confers susceptibility to MHV. *J. Virol.* **65**:6881-6891.

Hilton, A., Mizzen, L., Macintyre, G., Cheley, S., and Anderson, R., 1986, Translational control in murine hepatitis virus infection. *J. Gen. Virol.* **67**:923-932.

Kim, K.-H., Narayanan, K., and Makino, S., 1997, Assembled coronavirus from complementation of two defective interfering RNAs. *J. Virol.* **71**:3922-3931.

Kyuwa, S., Cohen, M., Nelson, G.W., Tahara, S. M., and Stohlman, S. A., 1994, Modulation of cellular macromolecular synthesis by coronavirus:Implications for pathogenesis. *J. Virol.* **68**:6815-6819.

Lai, M. M. C., Brayton, P. R., Armen, R. C., Patton, C. D., Pugh, C., and Stohlman, S. A., 1981, Mouse hepatitis virus A59: mRNA structure and genetic localisation of the sequence divergence from hepatotropic strain MHV-3. *J. Virol.* **39**:823-834.

Lee, H.-J., Shieh, C.-K., Gorbalenya, A. E., Koonin, E. V., La Monica, N., Tuler, J., Bagdzhadzhyan, A., and Lai, M. M. C., 1991, The complete sequence (22 kilobases) of murine coronavirus gene 1 encoding the putative proteases and RNA polymerase. *Virology* **180**:567-582.

Makino, S., Joo, M., and Makino, J. K., 1991, A system for study of coronavirus mRNA synthesis: a regulated, expressed subgenomic defective interfering RNA results from intergenic site insertion. *J. Virol.* **65**:6031-6041.

Sawicki, S. G., and Sawicki, D. L., 1986, Coronavirus minus-strand RNA synthesis and effect of cyclohexamide on coronavirus RNA synthesis. *J.Virol.* **57**:328-334.

Siddell, S., Wege, H., Barthel, A., and ter Meulen, V., 1981, Coronavirus JHM: Intracellular protein synthesis. *J. Gen. Virol.* **53**:145-155.

Tahara, S. M., Dietlin, T. A., Bergmann, C. C., Nelson, G. W., Kyuwa, S., Anthony, R. P., and Stohlman, S. A., 1994, Coronavirus translation regulation: leader affects mRNA efficiency. *Virology* **202**:621-630.

Tahara, S. M., Dietlin, T. A., Nelson, G. W., Stohlman, S. A., and Manno, D. J., 1998, Mouse hepatitis virus nucleocapsid protein as a translational effector of viral mRNAs. *Adv. Exp. Med. Biol.* **440**:313-318.

Homotypic Interactions of the Nucleocapsid Protein of Porcine Reproductive and Respiratory Syndrome Virus (PRRSV)

DONGWAN YOO* AND SARAH WOOTTON
*Department of Pathobiology, Ontario Veterinary College, University of Guelph, Guelph, Ontario N1G 2W1, CANADA. *Email:dyoo@uoguelph.ca*

1.   INTRODUCTION

Porcine reproductive and respiratory syndrome virus (PRRSV) is able to code for six major structural proteins, GP2 through GP6 and a nucleocapsid protein. Open reading frame 2 coding for the GP2 protein contains a small internal open reading frame in the +3 frame, and this internal open reading frame has been shown to be functional in that it directs the synthesis of a 73 amino acid polypeptide. Thus to date, seven structural proteins have been reported for PRRSV. The nucleocapsid (N) protein is a small, basic protein of 123 amino acids. The N protein appears to be highly immunogenic inducing a strong antibody response upon PRRSV infection. Ultrastructural examination of cells infected with PRRSV shows that two distinct forms of particles exist. One form is the fully matured virion particle surrounded by a lipid envelope and the other form is an unenveloped virus-like particle. This suggests that virus-like particles may be a multimerized form of the N protein. Using a series of deletions and truncations of the N protein expressed in HeLa cells and conformation-dependent monoclonal antibodies (Mabs), it has been shown that the 11 most C-terminal amino acids play a major role for maintenance of the conformational structure of the N protein (Wootton et al 1998). It has been predicted that the stretch of amino acids at the C-terminus is able to form a strong β-sheet followed by a coiled structure

The Nidoviruses (Coronaviruses and Arteriviruses).
Edited by Ehud Lavi *et al.*, Kluwer Academic/Plenum Publishers, 2001.

(Figure 1), suggesting that the C-terminal 11 amino acids may mediate the N protein-N protein interactions during multimerization. In the present study, we examined the role of individual amino acids for the overall protein conformation and studied intermolecular interactions of the N protein.

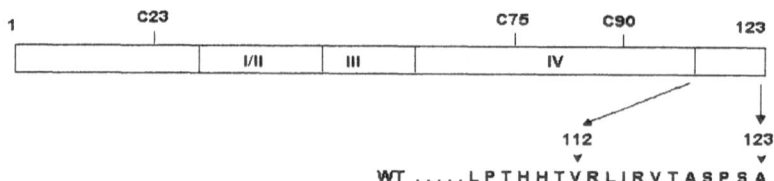

Figure 1. Amino acid sequence at the C-terminus of the nucleocapsid protein of PRRSV isolate PA8. Four major antigenic domains are indicated as I, II, III, and IV. Cysteine residues are present at positions 23, 75, and 90.

2.   MATERIALS AND METHODS

2.1   Cells, viruses, and plasmids

HeLa cells were maintained in Dulbecco's modified Eagle's medium with 10% fetal bovine serum. Marc-145 cells were maintained in Eagle's minimal essential medium with 4% serum. The PRRSV PA8 strain was grown in Marc-145 cells and vTF7-3 vaccinia virus was grown in HeLa cells. The N gene was cloned into pCITE vector (Novagene) and used for transient expression studies. For bacterial expression of the N protein, pGEX-3X (Pharmacia) was used for GST-N fusion construction.

2.2   Protein expression

HeLa cells were infected with recombinant vaccinia virus expressing T7 RNA polymerase (vTF7-3) and transfected for 8 hrs with appropriate plasmid DNA using lipofectamine (Gibco BRL). The cells were radiolabelled at 10 hrs post-infection-transfection for 16 hrs with [^{35}S] methionine at a concentration of 50 µCi/ml. Cell lysates were prepared and subjected to immunoprecipitation followed by image capture using PhosphoImager (Molecular Dynamics). For prokaryotic expression, E. coli cultures were induced with IPTG, and the expressed protein was prepared as cell lysates by using non-ionic detergents.

2.3 Chloramphenicol acetyltransferase (CAT) assays

The N gene and its mutant derivatives were cloned into both the pM and pVP16 plasmids (Clontech). Cells grown on 35mm dishes were cotransfected with pM-N, pVP16-N, and pG5CAT plasmids for 36 hrs. Cell lysates were prepared in the absence of SDS, and CAT assays were performed using the nonradioactive FLASH CAT assay kit (Stratagene).

3. RESULTS AND DISCUSSION

3.1 Role of beta-sheet Structure at C-terminus for Protein Conformation

To dissect the role of the β-sheet present at the C-terminus of the N protein, individual amino acids within the C-terminal 11 most residues were substituted with amino acids considered to be β-sheet breakers. A total of eight mutants were constructed: V112P (substitution of valine at position 112 with proline), R113D, R113P, L114P, I115P, R116P, T118S, and P121A. The mutant proteins were expressed in HeLa cells using T7 vaccinia and were immunoprecipiated using Mabs 5H2, SDOW17, SR30, and EP147. Mutations at 114, 115, and 116 (L114P, I115P, R116P) abolished immunoreactivities with all three (SDOW17, SR30, EP147) of the C-terminal dependent conformational Mabs (Yoo and Wootton 2000). Mab 5H2, which was considered to be C-terminal independent, reacted with all of the mutants, including the wild type N protein. In T118S and P121S, mutations were introduced downstream of the β-sheet, and these mutations did not affect the immunoreactivities. These results demonstrate that the β-sheet structure present at the C-terminus of the N protein plays an important role in establishing the overall conformation of the protein.

3.2 Homodimer Formation of N Protein

Three cysteines have been identified at positions 23, 75, and 90 in the N protein (Figure 1), and these cysteines appear to be conserved in all PRRSV of the American genotype. In contrast, the European type PRRSV contains only two cysteines at positions 23 and 75, and the cysteine at 90 is not found in any isolates of the European type. We examined whether the N protein formed a cysteine-mediated homodimer. Under non-reducing conditions, a protein of approximately 30 kDa was identified, and this protein is considered to be a dimerized form of the N protein (Figure 2A). However, when cell lysates were prepared in the presence of alkylating agent, the

30kDa protein was not detected under non-reducing conditions (Figure 2B). This finding suggests that the N protein dimeric formation may be a non-specific interaction due to the oxidation process during the preparation of cell lysates. Similar results were observed by other researchers (Mardassi et al., 1995; Meulenberg et al., 1996). Thus, we examined the structural nature of the N protein in virions. Virus-infected cells were pulse-labelled and chased for 24 hrs. Free virus particles were harvested and immunoprecipiated under non-reducing conditions in the presence of an alkylating agent. The N protein was found to be dimerized in this experiment (data not shown), suggesting that the dimeric form of the N protein is in fact incorporated into virions and thus may have a specific function in the life cycle of PRRSV.

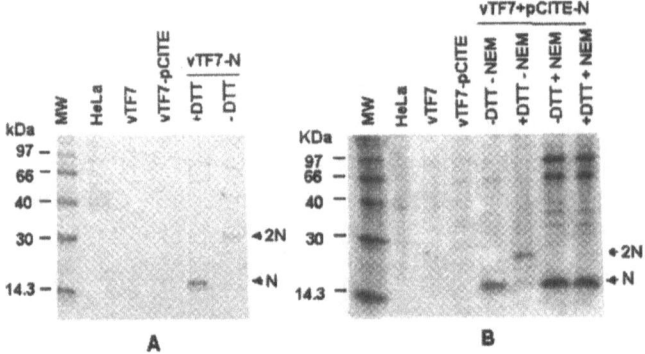

Figure 2. Homodimer formation of the N protein of PRRSV in the absence (A) or presence (B) of alkylating agent NEM. Arrow heads indicate monomeric (N) or dimeric (2N) forms of the N protein identified under reducing (+) and non-reducing (-) conditions.

3.3   Homotypic Interactions of N Proteins

To study further the N-N interactions, a co-immunoprecipitation study was conducted. The N gene, fused downstream of the glutathione S-transferase (GST) gene, was cloned into the eukaryotic expression vector pCITE thereby producing a GST-N fusion construct. HeLa cells were transfected with both N and GST-N genes and radiolabelled with [35S] methionine. Cell lysates were immunoprecipitated using anti-GST antibody (Sigma). When using an anti-GST antibody, both N and GST-N fusion proteins were specifically precipitated (data not shown), demonstrating that the N proteins in question interact.

To further confirm N-N interactions, a GST bead-binding assay was performed. The GST-N fusion construct was expressed in E. coli, and lysate was prepared. The lysate was then incubated with Sepharose beads coupled with glutathoine (Pharmacia), and the reaction was further incubated with the [^{35}S] labelled N protein expressed in HeLa cells. Unbound proteins were washed off from the beads, and proteins specifically bound to the beads were analysed by SDS-PAGE. The radiolablled N protein was pulled down by the beads coupled with GST-N protein (data not shown). Thus, the bead-binding assay confirms that the N protein interacts with itself.

The N-N interactions were also examined in cells in vivo. For in vivo studies, the mammalian two-hybrid system was employed. The N gene was cloned behind the yeast GAL4 DNA binding domain and the herpes simplex virus VP16 activation domain. The two constructs were co-transfected with a CAT expression plasmid in which the CAT gene was placed under the adenovirus E1b promoter. Lysates were prepared from cells transfected with all three plasmids, and CAT activity was measured by thin layer chromatography (Figure 3). As shown in lane 4, a significant level of CAT activity was measured in cells co-transfected with pM-N and pVP16-N, both of which were fused with the PRRSV N gene, indicating that the CAT gene expression was mediated by N protein-N protein interactions within the transfected cell.

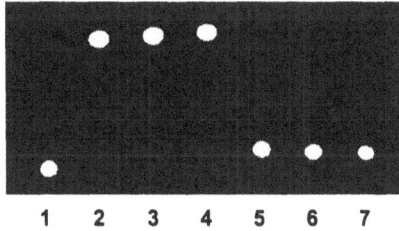

Figure 3. Chloramphenicol acetyltransferase activities mediated by the interaction of the PRRSV N protein. Lanes: 1, untransfected cells lysate; 2, co-transfection of pM3 and pVP16 as positive control; 3, pM-p53 and pVP16-SV40-TAg as positive control; 4, pM-PRRSV-N and pVP16-PRRSV-N cotransfection; 5, pM alone as negative control; 6, pVP16 alone as negative control; pG5CAT alone as negative control.

In the present study, we have shown the homotypic interaction of the PRRSV N protein by GST bead-binding assay and co-immunoprecipitation assay. The interaction was also demonstrated in vivo using the mammalian two- hybrid assay. It is possible that N-N interactions may, at least initially, be mediated by the putative β-sheet located within the 11 most C-terminal amino acids. Alternatively, the C-terminal 11 amino acids may be indirectly involved in N-N interactions by determining the overall protein conformation, and thus the overall structure of the protein becomes important for N-N interactions. In either case, it seems clear that such a short stretch of C-terminal amino acids plays an important role in the structure-function of the N protein. The interaction that we have demonstrated in this study is likely an initial step in the multimerization of the capsid protein, leading to the nucleocapsid assembly and the virus maturation.

ACKNOWLEDGMENTS

This work was supported by Vetrepharm Animal Health, Ontario Pork, and Ontario Ministry of Agriculture Food and Rural Affairs.

REFERENCES

Mardassi, H., Massie, B., and Dea, S. 1996. *Virology* **221**: 98-112.
Meulenberg, J. J. M., Peterson-Den Bester, A., De Kluyver, E. P., Moorman, R. J. M., Schaaper, W. M. M., Wensvoort, G. 1995. *Virology* **206**: 155-163.
Wootton, S., Nelson, E., and Yoo, D. 1998. *Clin. Diag. Lab. Immunol.* **5**: 773-779.
Yoo D., and Wootton, S. 2000. *Vet. Res.* **31**: 24-25.

Interactions of Cellular Proteins with the Positive Strand of 3'-Untranslated Region RNA and the Nucleoprotein Gene of Porcine Reproductive and Respiratory Syndrome Virus

FAHAD MAJHAD I AND SANJAY KAPIL

Department of Diagnostic Medicine-Pathobiology, College of Veterinary Medicine, Kansas State University, Manhattan, KS 66506

## 1.	INTRODUCTION

Porcine reproductive and respiratory syndrome virus (PRRSV) is a member of the Arteriviridae family, which includes enveloped and positive-sense single-stranded RNA viruses (Meulenberg *et al.*, 1993). The size of the viral genome is approximately 15 kb, and eight functional genes have been identified, five of which encode viral structural proteins (Meulenberg *et al.*, 1997). The organization, structure, and replication strategies of the viral genome are similar to those of coronaviruses (Snijder and Spaan, 1995). Interaction of the N protein with the 3'-end of the viral genome suggests that the N protein may regulate viral RNA synthesis. Several cellular proteins have been reported to bind to the 3'-untranslated region (UTR) of RNA viruses (Nakhasi *et al.*, 1991; Pardigon and Strauss, 1992).

The first step in arteriviral RNA replication is the synthesis of negative-strand RNA from the viral genome. Negative-strand RNA then serves as template for the synthesis of positive-strand RNA and transcription of subgenomic mRNAs (de Vries *et al.*, 1997; Snijder and Meulenberg, 1998).

The Nidoviruses (Coronaviruses and Arteriviruses).
Edited by Ehud Lavi *et al.*, Kluwer Academic/Plenum Publishers, 2001.

Previous studies with positive-stranded RNA viruses have suggested that viral RNA synthesis is mediated by multiple RNA regions. The 3'-end of viral RNA has the sequence necessary for initiation of the (-) strand RNA synthesis (Lin and Lai, 1993; Lin *et al.*, 1994). Cellular proteins have been shown to regulate replication of the viral genome of several RNA viruses by binding to the 3'(+) UTR of the viral RNA (Lahser *et al.*, 1993; O'Neill and Palese, 1994; Sriskanda *et al.*, 1996). These findings present strong evidence that interactions between these cellular proteins and specific viral RNA regions are necessary for viral RNA synthesis.

In this study, by performing Northwestern blot assay with cytoplasmic extracts of MARC-145 cells, we detected multiple cellular proteins that bind to the 3'UTR/N of PRRSV RNA genome.

2. MATERIALS AND METHODS

Extracts from MARC-145, Baby Hamster Kidney (BHK-21), and porcine alveolar macrophage (PAM) cell monolayers were prepared as previously described (Zhang and Lai, 1995). The viral RNA of PRRSV was reverse transcribed with reverse transcriptase (RT), then the cDNA was amplified by the polymerase chain reaction (PCR) with genomic sense primer for 3' UTR/N and anti-genomic sense primer. The PCR product was directly ligated into the pCR™II plasmid cloning vector (Invitrogen, Carlsbad, CA). *In vitro* synthesis of RNA transcripts and unlabeled RNA competitors were carried out according to the manufacturer's procedure (Promega, Madison, WI). The ^{32}P-labeled RNAs and unlabeled RNA competitors were extracted with phenol-chloroform, precipitated with ethanol, then dissolved in TE buffer (10 mM Tris-HCl [pH 8.0]; 1mM EDTA). Protein extracts (10 µg) from MARC-145 cells were separated by electrophoresis in 10% SDS-PAGE. Electrophoresis was carried out at 200 V for 45 min, then the protein bands were transferred electrophoretically onto a polyvinylidene difluoride membrane (Gelman Sciences, Ann Arbor, MI). The membrane blots were washed for 30 min at room temperature with SBB probe buffer (0.1 M of NaCl; 10 mM Tris [pH 7]; 1 mM EDTA; 0.02% BSA; 0.02% Ficoll; 0.02% polyvinyl pyrrolidone). The blot was incubated in a solution containing SBB buffer with 10 µg/ml of t-RNA and 100 µg/ml denatured salmon sperm DNA (SS DNA) to block nonspecific binding. The ^{32}P-labeled RNA probe (8×10^4cpm/ng) was added to the SBB buffer and incubated for 1 h at room temperature. For the competition Northwestern assay, a 1-fold, 10-fold and 100-fold excess of unlabeled specific competitor RNAs and non-specific tRNA and ssDNA were added to the reaction mixture prior to addition of ^{32}P-labeled RNA probe.

3. RESULTS

3.1 Detection of cellular proteins that bind to the 3'UTR/N of the PRRSV positive-strand RNA

RNA overlay protein blot analysis (ROPBA) was used to determine if protein from MARC-145, PAM, or BHK-21 extracts bound to the 3'-end of the PRRSV positive-strand RNA. The (+)-sense RNA probe used in this experiment was 319 nt long and had the 3'UTR/N of the PRRSV positive-strand RNA. Bands of approximately 53, 50, 46, 37, 33, 28, 26, 23 21 and 18 kDa were detected (Fig. 1).

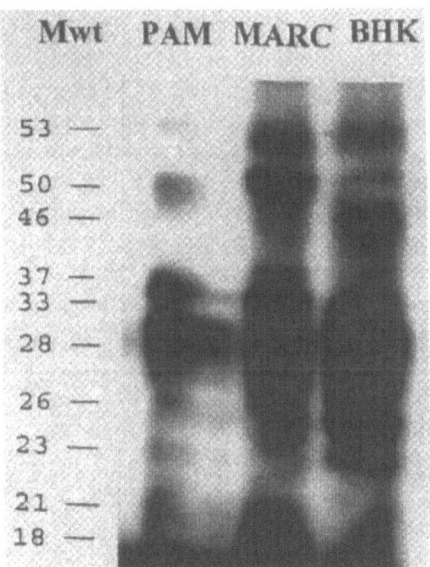

Figure 1. Detection and comparison of the interactions between PRRSV 3'UTR/N RNA and cytoplasmic protein extracts of PAM, MARC-145 and BHK-21 cells by Northwestern blot assay. Cytoplasmic proteins (10 μg/sample) were probed with ^{32}P-labeled RNA (8×10^4cpm/ng) and incubated with the blot.

To determine whether the protein binding was specific in the Northwestern assay, increasing amounts (1-, 10-, and 100-fold) of unlabeled competitor RNA, non-specific tRNA and ssDNA were preincubated with the cytoplasmic proteins (Figs. 2 and 3; Lanes 1-6). The results indicated an inhibition of protein binding of unlabeled specific homologus competitor RNA, suggesting that the binding is specific (Fig. 2 and 3; Lanes 1-3) but is not affected by heterologous nonspecific competitor RNA (Figs. 2 and 3; Lanes 4-6).

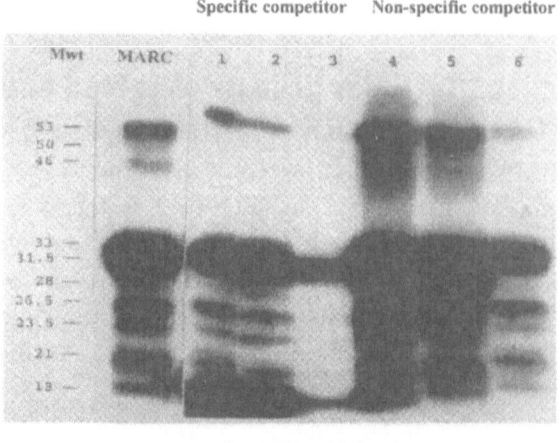

Figure 2. Northwestern blot and competition assay of cytoplasmic extract of MARC-145 cells with PRRSV 3'UTR/N RNA. Cytoplasmic proteins (10 µg/sample) were probed with ^{32}P-labeled RNA (1×10^6 cpm/200 ng) and incubated with the blot after addition of unlabeled competitors. Lane MARC: cytoplasmic extract no competitor was added. Lanes 1-3: increased amounts of specific unlabeled competitor RNA [200 ng, 2 µg and 20 µg, respectively]. Lanes 4-6: increased amounts of non-specific competitor ssDNA and tRNA [400 ng, 4 µg and 40 µg, respectively]. The approximate molecular mass (kDa) of each protein is indicated on the left.

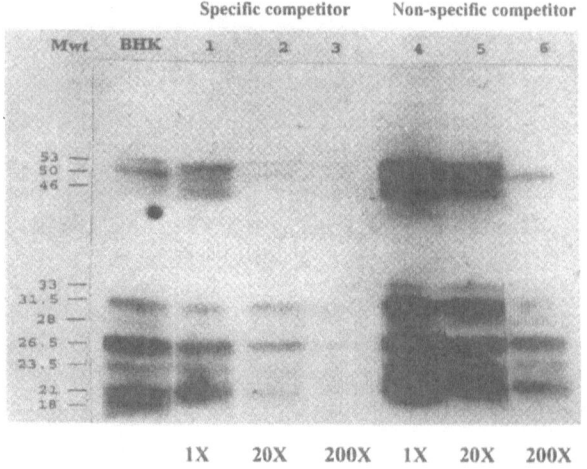

Figure 3. Northwestern blot and competition assay of cytoplasmic extract of BHK-21 cells with PRRSV 3'UTR/N RNA. Cytoplasmic proteins (10 µg/sample) were probed with ^{32}P-labeled RNA (1×10^6 cpm/200 ng) and incubated with the blot after addition of unlabeled competitors. Lane BHK-21: cytoplasmic extract with no competitor added. Lanes 1-3: increasing amounts of specific unlabeled competitor RNA [200 ng, 2 µg and 20 µg, respectively]. Lanes 4-6: increasing amounts of nonspecific competitor ssDNA and tRNA [400 ng, 4 µg and 40 µg, respectively]. The approximate molecular mass (kDa) of each protein is indicated on the left.

4.   DISCUSSION

Very little is known about the nature and the mechanism of PRRSV genome synthesis, arteriviruses may replicate their own RNA through a complex process in which both viral and cellular proteins participate. Involvement of cellular proteins in RNA replication has been demonstrated previously in several RNA viruses including Qβ phage, Sindbis virus, brome mosaic virus, influenza virus, cucumber mosaic virus, poliovirus, and potato virus X (Hwang and Brinton, 1998). Cellular proteins have been shown to interact specifically to the 3'(+) UTR of viral RNA and regulate viral genome replication [5, 11, 15]. The poly-A tail in arteriviruses is not well studied and its role in PRRSV replication has not yet been delineated. However, the 3'UTR of mouse hepatitis virus (MHV) RNA has been reported to bind a 57 kDa cellular protein which has been identified as the polypyrimidine-tract-binding protein (PTB) (Huang and Lai, 1999).

In this study, using the Northwestern assay, we investigated the interactions of cytoplasmic proteins from PRRSV susceptible PAM cells and MARC-145 cells, and of unsusceptible BHK-21 cell with the 3' UTR/N RNA of PRRSV. We found similar binding patterns among permissive and non-permissive cell lines, suggesting that the differences between these cells might be at the level of cell receptor (Kreutz, 1998). Ten proteins of similar molecular masses were detected in both permissive MARC-145 cells and non-permissive BHK-21 cells (Figs. 2 and 3). Our results are in agreement with a previous finding (Kreutz, 1998) where it was shown that infectious PRRSV was recovered from PRRSV RNA transfected BHK-21 cells thus, the cellular protein may play a role in viral replication. Northwestern technique was used to study the interaction of cellular proteins with the positive strand of 3'-UTR and 3'-UTR/N gene of PRRSV. All proteins bound to the 3'UTR of the PRRSV but not to the nucleoprotein sequences. The bands are consistent when both 3'UTR and 3'UTR/N probes were used (data not shown). We were able to identify similar bands (53, 50, 46, 37, 33, 31, 28, 26, 23, 21 and 18 kDa) with both probes suggesting the importance of the 3'UTR for interaction with cellular proteins (Figs. 1-3). Also, we studied the effect that salts (NaCl, KCl, $ZnCl_2$, and $MnCl_2$ at 50mM, 100mM, 150mM and 200mM) have on RNA binding of MARC and BHK cytoplasmic protein extracts using Northwestern assay. We found that increasing concentrations of NaCl reduced the proteins' RNA binding affinities whereas increasing concentrations of KCl had no effect on binding (data not shown). At this point, it is unclear as to how tightly these proteins must bind to the RNA, which could affect the interaction (binding) stability. We strongly believe that at least one of these proteins may serve as a target for virus replication playing an important role in viral pathogenicity. PETA-

3/ CD151 a transmembrane protein with 23 kDa, that interacts with MARC-145 cellular protein, was cloned and identified (Shanmukhappa, K. and S. Kapil, unpublished data). It specifically bound to 3'UTR/N in Northwestern assay. This protein was detected in our investigation, suggesting the importance that may play in viral replication.

REFERENCES

de Vries, A. A. F., M. C. Horzinek, P. J. M. Rottier, and R. J. de Groot. 1997. The genome organization of the Nidovirales: similartities and differences between arteri-, toro-, and coronaviruses. *Semin. Virol.* **8**: 33-47.

Hwang, Y. K., and M. A. Brinton. 1998. A 68-Nucleotide sequence within the 3' noncoding region of simian hemorrhagic fever virus negative-strand RNA binds to four MA104 cell proteins. *J. Virol.* **72**: 4341-4351.

Huang, P. and M. M. C. Lai. 1999. The complementary strand of the mouse hepatitis virus 3'untranslated region binds polypyrimidine-tract-binding protein. ASV 18th annual meeting W21-9 page 96 (University of Massachusetts, Amherst, Massachusetts, July 10-14, 1999).

Kreutz, L. C. 1998. Cellular membrane factors are the major determinants of porcine reproductive and respiratory syndrome virus tropism. *Virus. Res.* **53**:121-128.

Lahser, F. C., L. E. Marsh, and T. C. Hall. 1993. Contributions of the brome mosaic virus RNA-3 3'-nontranslated region to replication and translation. *J. Virol.* **67**:3295-3303.

Lin, Y. J., and M. M. C. Lai. 1993. Deletion mapping of a mouse hepatitis virus defective interfering RNA reveals the requirement of an internal and discontiguous sequence for replication. *J. Virol.* **67**:6110-6118.

Lin, Y. J., C. L. Liao, and M. M. C. Lai. 1994. Identification of the *cis*-acting signal for minus-strand RNA synthesis of a murine coronavirus: implication for the role of minus-strand RNA in RNA replication and transcription. *J. Virol.* **68**:8131-8140.

Meulenberg, J. J. M., E. J. de Mejjer, and R. J. M. Moormann. 1993. Subgenomic RNAs of Lelystad virus contain a conserved leader-body junction sequence. *J. Gen. Virol.* **74**:1697-1701.

Meulenberg, J. J. M., A. P. van Nieuwstadt, A. van Essen-Zanbergen, and J. P. Langeveld. 1997. Posttranslational processing and identification of a neutralization domain of the GP4 protein encoded by ORF4 of Lelystad virus. *J. Virol.* **71**:6061-6067.

Nakhasi, H., X. Q. Cao, T. A. Rouault, and T. Y. Liu. 1991. Specific binding of host cell proteins to the 3'-terminal stem-loop structure of rubella virus negative-strand RNA. *J.Virol.* **65**:5961-5967.

O'Neill, R., and P. Palese. 1994. Cis-acting signals and trans-acting factors involved in influenza virus RNA synthesis. *Infect. Agents. Dis.* **3**:77-84.

Pardigon, N., and J. H. Strauss. 1992. Cellular proteins bind to the 3' end of Sindbis virus minus strand RNA. *J. Virol.* **66**:1007-1015.

Snijder, E. J., and W. J. M. Spaan. 1995. The coronaviruslike superfamily. In *The coronaviridae*, p. 239-255. S. G. Siddell (ed.). Plenum Press. New York, NY.

Snijder, E. J., and J. J. M. Meulenberg. 1998. The molecular biology of arteriviruses. *J. Gen. Virol.* **79**:961-979.

Sriskanda, V. S., G. Pruss, X. Ge, and V.B. Vance. 1996. An eight-nucleotide sequence in
  the potato virus X 3' untranslated region is required for both host protein   binding and
  viral multiplication. *J. Virol.* **70**:5266-5271.

Zhang, X., and M. C. Lai. 1995. Interactions between the cytoplasmic proteins and the
  intergenic(promoter) sequence of mouse hepatitis virus RNA: correlation with the
  amounts of subgenomic mRNA transcribed. *J. Virol.* **69**:1637-1644.

Cloning and Identification of MARC-145 Cell Proteins Binding to 3' UTR and Partial Nucleoprotein Gene of Porcine Reproductive and Respiratory Syndrome Virus

KUMAR SHANMUKHAPPA AND SANJAY KAPIL[1]

[1]Department of Diagnostic Medicine -Pathobiology, College of Veterinary Medicine, Kansas State University, Manhattan, KS 66506, USA.Telephone (785)532-4457; Fax:(785)532-4481; E-mail: kapil@vet.ksu.edu

1. INTRODUCTION

The PRRSV has a (+) sense RNA genome 15 kb in size, with 5' and 3' untranslated regions (UTR) of 156-220 and 59-117 nucleotides flanking the viral genome (Snijder and Meulenberg, 1998). Virus grows in MARC, CL2621 and MA-104 cell lines. BHK-21 cells when transfected with PRRSV RNA, produced new infectious virus (Meulenberg *et al.*, 1998). PRRSV enters cells by standard endocytosis route and low pH in these vesicles is required for the viral entry (Kreutz and Ackermann, 1996).

Replication of PRRSV proceeds by discontinuous transcription forming a 3'-coterminal nested set of monocistronic mRNAs (Snijder and Meulenberg, 1998). The leader sequences and body of RNA are reported to have *cis* and *trans* interaction in mouse hepatitis virus (MHV) (Zhang *et al.*, 1994). The 3'UTR of MHV is shown to be essential for the transcription of the genome (Lin *et al.*, 1996). The 3'UTR may interact with the upstream regulatory sequences during replication either by sequence complementarity or through the regulatory proteins. But there is very limited sequence complementarity between the 3'UTR and upstream regulatory sequences. So the main

The Nidoviruses (Coronaviruses and Arteriviruses).
Edited by Ehud Lavi *et al.*, Kluwer Academic/Plenum Publishers, 2001.

641

mechanism may be mediated through RNA-protein interactions. The proteins involved are may be viral or cellular proteins.

Previous studies in our laboratory have identified 11 MARC cell proteins that bind to the 3'UTR of PRRSV (Fahad and Kapil, unpublished data). In the present study, we used RNA ligand screening of a MARC cell expression library to identify those cellular proteins that bind to 3'UTR of PRRSV. Here we also report for the first time the RNA binding property of a tetraspanin molecule, CD 151. This molecule also renders BHK-21 cells susceptible to PRRSV infection.

2.   MATERIALS AND METHODS

The 3'UTR was cloned into pCR II vector by RT-PCR amplification of the PRRSV RNA by TA cloning [Forward primer, 5'-CCCCATTTTCCT-CTAGCGACTG-3' and Reverse primer, 5'-CGGCCGCATGGTTCTCGC-CAAT-3']. The Radiolabelled ([α-^{32}P]UTP) 3'UTR RNA was prepared by *in vitro* transcription using T7 RNA transcription kit, Riboscribe^{TM} (Epicentre Technologies, Madison,WI). The MARC cell line cDNA library was prepared in the λ ZAP Express vector using a ZAP Express cDNA synthesis kit (Stratagene, La Jolla, CA). The cDNA library was screened by northwestern blotting using 3'UTR PRRSV (^{32}P-UTP) RNA probe as described previously (Sagesser *et al.*, 1997). The insert in the pBK-CMV vector was sequenced with T7 and T3 primers using sequitherm EXCEL II DNA sequencing kit (Epicentre Technologies, Madison WI).
The BHK-21 and MARC cell lines were transfected with plasmid DNA using Lipofectamine reagent (Life Technologies, Inc., Gaithersburg, MD). For stable transfection G418 (1mg/ml for BHK-21 cells and 0.7 mg/ml for MARC cells) was used, half the selection dose was used for CD 151 expression. Immunoprecipitation was carried out with CD151 (clone 14A2.H1, Pharmingen International, San Diego, CA.) and β-galactosidase monoclonal antibodies (Mab). Northwestern blot on immunoprecipitate was carried out as described previously (Sagesser *et al.*, 1997).

Stable transfected MARC cells were infected with PRRSV at MOI of 0.1 at 37°C for 1 hr. The amount of virus in the supernatant was titrated by plaque assay using parent MARC cells. To test the susceptibility of stable transfected BHK-21 cells to PRRSV infection, cells were infected at MOI of 0.01. Infectivity was checked by immunohistochemistry 24-hrs post infection by using PRRSV nucleoprotein SR 30 Mab (Rural technologies, Brookings, SD.). The anti-mouse IgG biotinylated antibody (Vector Labs, Burlingame, CA) was used as secondary antibody. The DAB substrate was used for detection and examined by light microscopy.

3. RESULTS

The single reacting clone was obtained by repeated plaque purification and rescreening five times. On sequencing and BLAST search, the insert matched with the CD 151, a tetraspanin molecule. Multiple sequence analysis showed very high homology (93%) with platelet endothelial tetraspanin antigen-3 (PETA-3) which is a human protein. The CD 151 was obtained after three independent screenings of the library. This proves that CD 151 binds to 3' UTR of PRRSV with high affinity and avidity. The RNA binding activity of CD 151 checked by immunoprecipitation followed by northwestern blotting and is shown in (Fig 1).

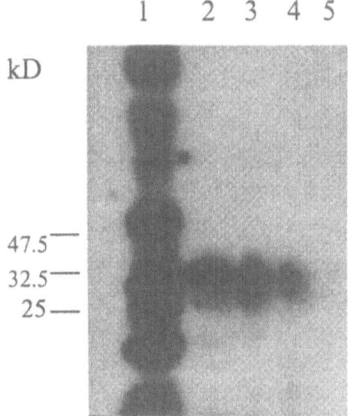

Figure 1. North-westernblot of the expressed protein. Lane1 MARC lysate. Lanes 2-5 northwestern blot after immunoprecipitation with CD 151 Mab 2;MARC transfected, 3; BHK-21 transfected, 4; MARC untransfected, 5; BHK-21 transfected.

Northwestern blot of the immunoprecipitated protein was also probed with to 3' UTR alone without portion of nucleoprotein (data not shown). This proves that CD 151 also binds 3' UTR.

3.1 Presence of CD 151 in different cell lines

To determine the possible relation between the CD 151 and susceptibility of cell lines to PRRSV infection RT-PCR was done using CD 151 specific primers. The CD 151 is present in MA 104, MARC, vero and COS-7 cell lines while it is absent in BHK-21 and HRT cell lines (Fig 2). The CD 151 is present in the susceptible cell lines like MARC, MA 104 and CL2621 while it is absent in BHK-21 cell line. This tells us that CD 151 is one of the factors in determining the susceptibility to PRRSV infection with involvement of other factors.

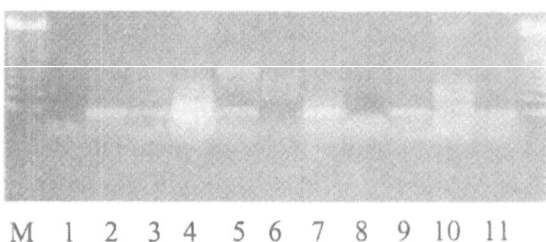

Figure 2. RT-PCR result showing the amplification of CD 151 of 105 bp. M; 123 bp ladder, 1; –ve control, 2; MARC, 3; MA-104, 4; CL 2621, 5; VERO, 6; BHK-21, 7;ST, 8; MDBK, 9; ST-K, 10; COS, 11; BHK-21 (stable transfected).

3.2   Effect of CD 151 level on PRRSV infection level

Virus amplification assay was done to check the effect of levels of CD 151 on the PRRSV replication. After infecting MARC cells (both parent and overexpressing CD 151) with plaque purified PRRSV. The amplification of PRRSV was measured by plaque assay using parent MARC cells. There was atleast 100-fold increase in the amount of virus in MARC cells overexpressing CD 151 as compared to parent MARC cells. This indicates that CD 151 might have increased the levels of PRRSV infection either by promoting viral entry or replication.

To further test the influence of CD 151 on PRRSV, the stable transfected BHK-21 cells were used in the study. The BHK-21 cells lack CD 151 (Fig 3) and don't permit the entry of the PRRSV but upon transfection of viral RNA, they do produce the infectious progeny. The stable transfected BHK-21 cells expressing CD 151 become susceptible to PRRSV infection as tested by immunohistochemistry. The stable transfected cells were positive for both CD 151 and PRRSV, while the transfected cells were negative for both (Data not shown). This proves that CD 151 might have facilitated the PRRSV infection in BHK-21 cells.

4.   DISCUSSION

Here we report the identification of a novel host cell protein, CD 151 binding to 3'UTR of PRRSV. The CD 151 is a transmembrane glycoprotein, belonging to the tetraspanin or transmembrane 4 superfamily of cellular proteins. These have four highly conserved hydrophobic domains spanning the lipid bilayer, two extracellular domains between them. N and C terminals are found in the cytoplasm (Fitter *et al.*, 1995). The CD 151 shares 98% homology with PETA-3. This protein is a 27 kD glycoprotein present in endothelial cell membranes (Fitter *et al.*, 1995), and also in epithelium, lung, muscle, kidney in glomeruli and mainly in platelets and megakaryocytes (Sincock *et al.*, 1997). The same protein (SFA-1) has been identified to be upregulated by the human T cell leukemia virus type 1 in SF-HT cells (Hasegawa *et al.*, 1996). The CD 151 has been shown to play a role in endothelial cell motility and bring about platelet aggregation and mediator release (Ashman *et al.*, 1991). These properties of the CD 151 may account for the vascular lesions seen in the PRRSV infection. The 66% of the protein is in intracellular compartments localized to perinuclear vesicles (Sincock *et al.*, 1999). Arteriviruses form double membrane vesicles (DMV) at 3-6 hrs post-infection (Snijder and Meulenberg, 1998). These DMVs are predicted to provide a suitable micro-environment for viral RNA synthesis or may facilitate the recruitment of host cell proteins for viral replication and transcription. In equine arteritis virus several replicase subunits are anchored to the intracellular membranes (Pedersen *et al.*, 1999). So we propose that the CD 151 with its well established role in protein-protein interactions, might be involved in the PRRSV replication through the interaction with 3'UTR and might recruit other cellular and viral proteins to DMV.

PRRSV is reported to replicate in MARC, MA 104, CL 2621, and primary cell cultures of alveolar macrophages. Not all strains grow in all these cell types. This could be due to differences in intrinsic factors. The BHK-21 cells are not susceptible to PRRSV infection while the transfection of either the viral RNA or the infectious cDNA clone produces the infectious viral progeny (Meulenberg *et al.*, 1998). In our study, we have shown that BHK-21 cells lack CD 151 by RT-PCR and Immunohistochemistry. But the stable transfected BHK-21 cells expressing the CD 151 become permissive to PRRSV. We speculate that CD 151 might have some kind of RNA transporter role. It is known that the PRRSV entry is mediated by endocytosis, and that the decrease of pH in internal components of the cells is essential for the viral entry (Kreutz and Ackermann, 1996). Here we propose that the CD 151 through the interaction with the RNA of 3'UTR PRRSV might act as the viral RNA transporter molecule into the cell.

REFERENCES

Ashman, L. K., G. W. Aylett, P. A. Mehrabani, L. J. Bendall, S. Niutta, A. C. Cambareri, S. R. Cole, and M. C. Berndt. 1991. The murine monoclonal antibody, 14A2.H1, identifies a novel platelet surface antigen. Br J Haematol. **79**:263-70.

Fitter, S., T. J. Tetaz, M. C. Berndt, and L. K. Ashman. 1995. Molecular cloning of cDNA encoding a novel platelet-endothelial cell tetra-span antigen, PETA-3. Blood. **86**:1348-55.

Hasegawa, H., Y. Utsunomiya, K. Kishimoto, K. Yanagisawa, and S. Fujita. 1996. SFA-1, a novel cellular gene induced by human T-cell leukemia virus type 1, is a member of the transmembrane 4 superfamily [published erratum appears in J Virol 1997 Feb;71(2):1737]. J Virol. **70**:3258-63.

Kreutz, L. C., and M. R. Ackermann. 1996. Porcine reproductive and respiratory syndrome virus enters cells through a low pH-dependent endocytic pathway. Virus Res. **42**:137-47.

Lin, Y. J., X. Zhang, R. C. Wu, and M. M. Lai. 1996. The 3' untranslated region of coronavirus RNA is required for subgenomic mRNA transcription from a defective interfering RNA. J Virol. **70**:7236-40.

Meulenberg, J. J., J. N. Bos-de Ruijter, G. Wensvoort, and R. J. Moormann. 1998. An infectious cDNA clone of porcine reproductive and respiratory syndrome virus. Adv Exp Med Biol. **440**:199-206.

Pedersen, K. W., Y. van der Meer, N. Roos, and E. J. Snijder. 1999. Open reading frame 1a-encoded subunits of the arterivirus replicase induce endoplasmic reticulum-derived double-membrane vesicles which carry the viral replication complex. J Virol. **73**:2016-26.

Sagesser, R., E. Martinez, M. Tsagris, and M. Tabler. 1997. Detection and isolation of RNA-binding proteins by RNA-ligand screening of a cDNA expression library. Nucleic Acids Res. **25**:3816-22.

Sincock, P. M., S. Fitter, R. G. Parton, M. C. Berndt, J. R. Gamble, and L. K. Ashman. 1999. PETA-3/CD151, a member of the transmembrane 4 superfamily, is localised to the plasma membrane and endocytic system of endothelial cells, associates with multiple integrins and modulates cell function. J Cell Sci. **112**:833-44.

Sincock, P. M., G. Mayrhofer, and L. K. Ashman. 1997. Localization of the transmembrane 4 superfamily (TM4SF) member PETA-3 (CD151) in normal human tissues: comparison with CD9, CD63, and alpha5beta1 integrin. J Histochem Cytochem. **45**:515-25.

Snijder, E. J., and J. J. Meulenberg. 1998. The molecular biology of arteriviruses. J Gen Virol. **79**:961-79.

Zhang, X., C. L. Liao, and M. M. Lai. 1994. Coronavirus leader RNA regulates   initiates subgenomic mRNA transcription both in trans and in cis. J Virol. **68**:4738-46.

Identification of Cell Proteins that Bind to the SHFV 3' (+)NCR

TARONNA R. MAINES AND MARGO A. BRINTON

Department of Biology, Georgia State University, 402 Kell Hall, 24 Peachtree Center Avenue, Atlanta, Georgia 30303

1.   INTRODUCTION

African primates are the natural hosts for simian hemorrhagic fever virus (SHFV), a member of the family *Arteriviridae*. SHFV causes asymptomatic, persistent or acute self-limiting infections in patas monkeys, African green monkeys, and baboons. In contrast, SHFV infection in macaque monkeys causes a hemorrhagic fever that results in death within two weeks. SHFV spreads quickly through macaque colonies either by aerosol or direct contact with body fluids. In primate holding facilities infection of macaques with SHFV has been initiated by inadvertent transfer of virus from African primates by humans or by housing African primates and macaque species in close contact. SHFV was first isolated in 1964 after it had caused outbreaks in macaque colonies in the United States, Russia, and Europe (London, 1977). Subsequent outbreaks in two U.S. primate facilities occurred in 1989 and 1996 (Hayes et al., 1992; Jahrling, 1990; Rollin et al., 1999).

The genome of SHFV is a single-stranded, positive-sense RNA of about 15.7 kb with a 5' cap and a 3' poly(A) tract (Sagripanti, 1984). The molecular mechanisms used for SHFV RNA transcription and replication initiation are not well understood. It has been shown in several other positive strand virus systems that cell proteins are required components of viral replication complexes (Landers et al., 1974; Hayes and Buck, 1990). Here we report the identity of host cell proteins that interact with the SHFV

The Nidoviruses (Coronaviruses and Arteriviruses).
Edited by Ehud Lavi *et al.*, Kluwer Academic/Plenum Publishers, 2001.

3' (+)NCR RNA, the region of the genome from which minus-strand RNA synthesis initiates.

2.   MATERIALS AND METHODS

2.1   Cells and virus

MA104 cells were cultured in minimal essential medium with 10% fetal bovine serum at 37°C in a 3% CO_2 atmosphere.   Confluent MA104 monolayers were infected with SHFV, strain LVR 42-0/M6941 (American Type Culture Collection), at an MOI of 0.2.   Stock pools of SHFV were prepared in MA104 cells.

2.2   *In vitro* synthesis of RNA transcripts

An SHFV 3' (+) NCR cDNA was generated from purified SHFV RNA by RT-PCR and cloned into pCR2.1 (Invitrogen).   This plasmid (p3NCR) was used to generate PCR products which were then used as templates for the synthesis of the RNA probes.   The PCR sense primer included a T7 promotor to facilitate *in vitro* transcription of positive-sense RNA with T7 RNA polymerase.

2.3   Gel mobility shift assay

^{32}P-labeled SHFV 3' (+)NCR RNA was incubated with MA104 S100 cytoplasmic extracts for 20 minutes at ambient temperature prior to analysis of RNA-protein complexes on 10% nondenaturing polyacrylamide gels. For competition gel mobility shift assays, either a specific (unlabeled SHFV3' (+)NCR RNA) or a nonspecific (tRNA, poly IC, or WNV 3' (+)SL RNA) competitor RNA was included in the binding reactions.

2.4   UV-induced crosslinking assay

^{32}P-labeled SHFV 3' (+)NCR RNA was incubated with MA104 S100 cytoplasmic extracts for 20 minutes at ambient temperature. The reactions were exposed to ultraviolet light for 30 minutes on ice and then RNase A was added.   Crosslinked RNA-protein complexes were precipitated with acetone/methanol (1:1), analyzed by SDS-PAGE, and detected by autoradiography.

2.5 Immunoprecipitation of crosslinked proteins

UV-induced crosslinking reactions were incubated for 1 hour at 4°C with protein A-Sepharose (Amersham Pharmacia Biotech) and then conjugated to anti-PTB antibody or anti-hnRNPA1 (as a control) antibody. Immune complexes were pelleted, washed, heated to 100°C for 5 minutes, analyzed by SDS-PAGE, and detected by autoradiography.

2.6 RNA affinity chromatography

MA104 S100 cytoplasmic extracts were incubated with various nonspecfic competitor RNAs and passed over a column containing SHFV 3′ (+)NCR RNA linked to an agarose matrix (Amersham Pharmacia Biotech). Proteins were eluted from the column with increasing concentrations of NaCl. The presence of viral RNA binding proteins in the eluate was determined with a UV-induced crosslinking assay.

3. RESULTS

3.1 Detection of MA104 cell proteins that specifically interact with the SHFV 3′ (+)NCR

A single RNA-protein complex was detected using SHFV-infected and uninfected MA104 S100 cytoplasmic extracts and ^{32}P-labeled-SHFV 3′ (+)NCR RNA in a gel mobility shift assay. This RNA-protein interaction was shown to be specific using a competition gel mobility shift assay. Complete competition was observed when a 30-fold molar excess of unlabeled SHFV3′ (+)NCR RNA was used as competitor. In contrast, no competition was observed when either a 250-fold molar excess of tRNA or poly IC was used as competitor or when a 150-fold molar excess of WNV3′ (+) stemloop RNA was used as competitor (data not shown).

3.2 Determination of the molecular masses of the cellular proteins that interact with the SHFV3′ (+)NCR

The molecular masses of the proteins present in the single RNA-protein complex were determined by UV-induced crosslinking assays. One strong (56 kDa) and one weak (42 kDa) protein band were detected using both

SHFV-infected and uninfected MA104 cell extracts suggesting that these proteins are most likely cellular proteins (data not shown).

3.3   Identification of p56 and p42

Since polypyrimidine tract-binding protein (PTB) has a molecular mass similar to that of p56 and PTB had previously been shown to specifically interact with the terminal regions of several other viral RNAs (as reviewed by Lai, 1998; Huang and Lai, 1999; Li et al., 1999) as well as with the SHFV 3′ (-)NCR RNA (Hwang and Brinton, 1998), immunoprecipitation of UV crosslinked proteins in S100 extracts was performed using an anti-PTB antibody obtained from Dr. James Patton. p56 crosslinked to ^{32}P-labelled SHFV 3′ (+)NCR RNA was immunoprecipitated by the anti-PTB antibody.

To further confirm the identity of p56 as PTB, a recombinant His-PTB cDNA clone was obtained from Dr. James Patton. His-tagged PTB was expressed in *E. coli* and purified using a His-Bind column (Novagen). His-PTB protein was shown to interact in a specific manner with the SHFV 3′ (+)NCR RNA using a competition gel-mobility shift assay (Figure 1A).

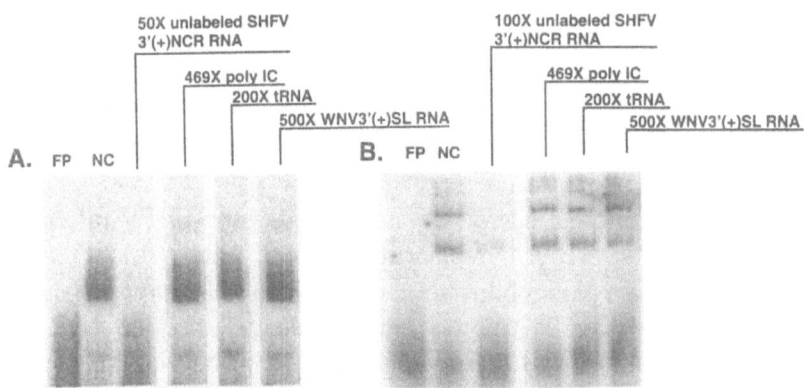

Figure 1. Competition gel mobility shift assay. ^{32}P-labeled SHFV 3′ (+)NCR RNA and (A.) purified His-PTB (55 ng) or (B.) purified aldolase (15 ng) were incubated with or without competitor RNA. Unlabeled SHFV 3′ (+)NCR RNA was used as the specific competitor and either poly IC, tRNA, or West Nile virus 3′ (+)stemloop RNA was used as the non-specific competitor. The RNA-protein complexes were resolved by 10% nondenaturing PAGE.

RNA affinity chromatography was used to partially purify p56 and p42. Both p56 and p42 were detected in the column eluate by a UV-induced crosslinking assay. p56 and p42 in the column eluate were distinguished from the background protein bands on a Coomassie blue-stained gel and were excised from the gel. The p42 protein band was sent to the Beckman Research Institute, City of Hope for sequencing. In-gel tryptic digestion was performed and eight peptides were selected for sequencing. The sequences of all eight peptides showed 100% identity with fructose-bisphosphate aldolase A, which has a molecular mass of 39 kDa. Rabbit fructose-bisphosphate aldolase A was obtained from Calbiochem and in a competition gel-mobility shift assay, the purified protein interacted in a specific manner with the SHFV 3' (+)NCR RNA (Figure 1B). p56 from the column eluate will also be sent for peptide sequencing.

4. DISCUSSION

We have shown that two MA104 cellular proteins (56 kDa and 42 kDa) interact specifically with the SHFV 3' (+)NCR RNA. Immunoprecipitation of viral RNA crosslinked to p56 with anti-PTB antibody suggested that p56 was PTB. The sequences of peptides obtained from p42 revealed that it was fructose-bisphosphate aldolase A. Competition gel mobility shift assays with either purified PTB or aldolase confirmed that the interactions between the SHFV 3' (+)NCR RNA and these proteins are specific.

In cells, PTB functions in the regulation of alternative pre-mRNA splicing and cap-independent translation (Kramer, 1996). In addition, it has previously been shown to specifically interact with the terminal regions of several other viral RNAs including MHV, HIV, HTLV-2, HAV, poliovirus, and hepatitis C virus and may play a role in their replication/transcription or translation (as reviewed by Lai, 1998; Huang and Lai, 1999; Li et al., 1999).

Fructose-bisphosphate aldolase A, an enzyme that participates in the glycolytic pathway, has not previously been reported to interact with a viral or a cellular RNA. However, another enzyme in the gycolytic pathway, glyceraldehyde 3-phosphate dehydrogenase (GAPDH), was reported to bind to the 3' (+)UTR and 3' (-)UTR of the human parainfluenza virus-3 (HPIV3) and to the HAV 5' (+) RNA. Interestingly, PTB was also found to bind to the HAV 5' (+)UTR (De et al., 1996; Schultz et al., 1996).

The identities of the six cell proteins shown to bind to the MHV 3' (+)UTR have not yet been reported. A repeated 11 nt sequence within the MHV 3' (+)UTR was shown to be needed for host cell protein binding and RNA replication (Yu and Leibowitz, 1995a; Yu and Leibowitz, 1995b). This repeated sequence was not found in the SHFV 3' (+)NCR. However, a 4 nt repeated sequence (5' A-U/C-U-A3 ') occurs at four locations within the SHFV 3' (+)NCR and each of these repeats is predicted to be single-stranded. Preliminary mapping data suggest that these repeat sequences may be important for PTB binding.

The significance of the observed interactions between the SHFV 3' (+)NCR RNA and PTB and aldolase is currently not known. Experiments are underway to further map the binding sites for PTB and aldolase within the SHFV 3' NCR RNA as well as to map *cis*-acting RNA sequences and/or structures required for (-) RNA synthesis.

ACKNOWLEDGEMENTS

Funding for this research was provided by the Georgia State University Research Foundation.

REFERENCES

De, B.P., Lesson, A., and Banerjee, A.K. 1996. Specific interaction in vitro and in vivo of glyceraldehyde-3-phosphate dehydrogenase and La protein with cis-acting RNAs of human parainfluenza virus type 3. J. Biol.Chem. 271:24728-24735.

Hayes, C.G., Burans, J.P., Ksiazak, T.G., Del Rosario, R.A., Miranda, M.E.G., Manaloto, C.R., Barrientos, A.B., Robles, C.G., Dayrit, M.M., and Peters C.J. 1992. Outbreak of fatal illness among captive macaques in the Philippines caused by an Ebola-related filovirus. Am. J. Trop. Med. Hyg. 46:664-671.

Hayes, R.J. and Buck, K.W. 1990. Complete replication of a eukaryotic virus RNA in vitro by a purified RNA-dependent RNA polymerase. Cell 63:363-368.

Huang, P. and Lai, M.M.C. 1999. Polypyrimidine tract-binding protein binds to the complementary strand of the mouse hepatitis virus 3' untranslated region, thereby altering RNA conformation. J. Virol. 73:9110-9116.

Hwang, Y.K. and Brinton, M.A. 1998. A 68-nucleotide sequence within the 3' noncoding region of simian hemorrhagic fever virus negative-strand RNA binds to four MA104 cell proteins. J. Virol. 72:4341-4351.

Jahrling, P.B., Geisbert, T.W., Dalgard, D.W., Johnson, E.D., Ksiazek, T.G., Hall, W.C., and Peters C.J. 1990. Preliminary report: Isolation of Ebola virus from monkeys imported to U.S.A. Lancet 335:502-505.

Kramer, A. 1996. The structure and function of proteins involved in mammalian pre-mRNA splicing. Annu.Rev. Biochem. 65:367-409.

Lai, M.M.C. 1998. Cellular factors in the transcription and replication of viral RNA genomes: a parallel to DNA-dependent RNA transcription. Virology 244:1-12.

Landers, T.A., Blumenthal, T., and Weber, K. 1974. Function and structure in ribonucleic acid phage Qβ ribonucleic acid replicase. J Biol Chem 249:5801-5808.

Li, H.P., Huang, P, Park, S, and Lai, M.M.C. 1999. Polypyrimidine tract-binding protein binds to the leader RNA of mouse hepatitis virus and serves as a regulator of viral transcription. J. Virol. 73:772-777.

London, W.T. 1977. Epizootiology, transmission and approach to prevention of fatal simian hemorrhagic fever in rhesus monkeys. Nature 268:344-345.

Rollin, P.E., Williams, R.J., Bressler, D.S., Pearson, S., Cottingham, M., Pucak, G., Sanchez, A., Trappier, S.G., Peters, R.L., Greer, P.W., Zaki, S., Demarcus, T., Hendricks, K., Kelley, M., Simpson, D., Geisbert, T.W., Jahrling, P.B., Peters, C.J., and Ksiazek, T.G. 1999. Ebola (subtype Reston) virus among quarantine nonhuman primates recently imported from the Philippines to the United States. J. Infect. Dis. 179:S108-S114.

Sagripanti, J.L. 1984. The genome of simian hemorrhagic fever virus. Arch. Virol. 82:61-72.

Yu, W. and Leibowitz, J.L. 1995a. Specific binding of host cellular proteins to multiple sites within the 3' end of mouse hepatitis virus genomic RNA. J Virol 69:2016-2023.

Yu, W. and Leibowitz, J.L. 1995b. A conserved motif at the 3' end of mouse hepatitis virus genomic RNA required for host protein binding and viral RNA replication. Virology 214:128-138.

MHV-A59 Gene 1 Proteins are Associated with Two Distinct Membrane Populations

MARK R. DENISON AND AMY C. SIMS

Departments of Pediatrics and Microbiology and The Elizabeth B. Lamb Center for Pediatric Research, Vanderbilt University Medical Center, Nashville, TN 37232

1.   INTRODUCTION

All of the stages of coronavirus replication that have been investigated have been shown to occur on or within intracellular membranes. We and others have shown that mouse hepatitis virus (MHV) RNA synthesis occurs in association with intracellular membranes (Bi *et al.*, 1998; Denison *et al.*, 1999; Dennis and Brian, 1982; Sethna and Brian, 1997; Shi *et al.*, 1999; van der Meer *et al.*, 1999). Several proteins processed from the gene 1 (replicase gene) polyprotein have been shown by immunofluorescence and electron microscopic approaches to be associated with intracellular membranes. Specifically, membranes containing markers for late endosomes have been shown to be sites of localization of newly synthesized viral RNA as well as at least one of the mature gene 1 proteins (van der Meer *et al.*, 1999). However, different patterns of gene 1 protein localization and interaction have been reported in the settings of distinct cell types, experimental approaches and virus strains. Most recently, it has been shown by confocal microscopic analysis of MHV-A59 infected cells that multiple replicase gene proteins as well as structural proteins may not completely colocalize but rather are organized in closely associated or "interdigitated" membranes, suggesting that proteins involved in MHV RNA synthesis may be localized to more than one membrane population (Bost *et al.*, 2000).

In this report we have used biochemical fractionation of MHV-A59 infected cells to determine if gene 1 proteins, MHV structural proteins M

The Nidoviruses (Coronaviruses and Arteriviruses).
Edited by Ehud Lavi *et al.*, Kluwer Academic/Plenum Publishers, 2001.

and N, and viral RNA localize to one or more membrane populations. In addition, we have defined nature of the membranes to which the viral proteins localize using both enzymatic and marker protein analyses. We have shown that gene 1 proteins localize to at least two biochemically distinct membrane populations, only one of which is the location of newly synthesized viral RNA.

2.   METHODS AND MATERIALS

2.1   Virus Infection and Radiolabeling

DBT cells were infected with MHV-A59 as previously described (Sims *et al.*, 2000). Cells were propagated on 150 cm^2 flasks and were infected for 6 hr, with actinomycin D added at 2.5h post infection. For radiolabeling and lysis, cells were removed from the flasks by trypsinization and 1 x 10^8 cells were suspended in 2ml of DMEM lacking methionine and cysteine but containing 2% FCS and isotope for labeling protein [^{35}S met/cys] or RNA [^{3}H uridine]. All labeling was performed for 1-2 hrs between 6 and 8 h p.i.

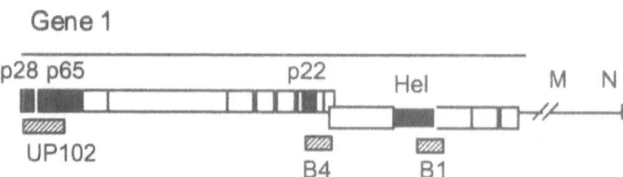

Figure 1. Organization of the MHV genome. The location of the domains within gene 1 encoding p28, p65, p22 and Hel are shown as black boxes within the gene 1 coding region. The hatched boxes show the polypeptides used to generate the rabbit polyclonal sera used in this study directed against the proteins shown in black. Antibodies against M and N are described in the text.

2.2   Cell Fractionation, Antibodies and Immunoprecipitation

Following infection and radiolabeling, cells were lysed using a ball bearing homogenizer in a sucrose-Tris buffer in the absence of detergent (Sims *et al.*, 2000). Differential centrifugation of the lysed cells was performed to obtain pellets and cytosol at 1000 x g (P1 or nuclear pellet), 2,300 x g (P2.3), 100,000 rpm (P100), and the residual cytosol (S100). The membranes in the P100 pellet were fractionated on a 10 to 30% Iodixanol (Optiprep-Nycomed) gradient. Both crude differential fractionation pellets, S100, and gradient fractions were analyzed for the presence of MHV proteins by immunoprecipitation in TTK buffer containing 1% Triton X100. Antibodies used in these experiments were rabbit polyclonal sera directed against p28 and p65 (UP102) (Denison *et al.*, 1995), p22 (B4) (Lu *et al.*, 1998), Hel (B1) (Denison *et al.*, 1999), and mouse monoclonal antibodies directed against M (J.1.3) and N (J.3.3) obtained from John Fleming (Fig. 1).

2.3 Assays for Cellular Proteins

Iodixanol gradient fractions were assessed for endosomes/lysosomes (aLAMP-1 antibodies) Golgi membranes (galactosyltransferase activity), endoplasmic reticulum (NADPH-cytochrome C reductase activity), as previously described (Sims *et al.*, 2000).

3. RESULTS

3.1 Association of MHV Replicase Proteins with Membranes

Antibodies directed gene 1 proteins and the structural proteins M and N were used to immunoprecipitate proteins both from crude differential fractionation pellets and gradient fractions. When the crude pellets and cytosol were analyzed, the gene 1 proteins p28, p65, p22, and Hel, as well as the structural M protein were detected almost exclusively to the high speed P100 pellet. N was most abundant in P100, but was also readily detected in the nuclear pellet, low speed P2.3 pellet and the S100 cytosol. The P100 pellet is post mitochondrial, and has been shown to contain small membranes of endosomes, lysosomes, ER, and Golgi. Thus these results were consistent with previous studies showing localization of M to Golgi and of gene 1 proteins to endosomes, and possible to ER. Since all of the replicase proteins studied localized to P100, this pellet was selected for further fractionation on the iodixanol gradient, as well as for studies of RNA localization

3.2 MHV Proteins Segregate to Distinct Membranes

The P100 pellets were fractionated on the iodixanol gradient and separated into 10 fractions before immunoprecipitation. A summary of the results is shown in Fig. 2. A clear pattern emerged with the proteins localizing toward the two extremes of the gradient. P28 and helicase were detected only in fractions 7 through 10 (most dense), with peak of detection in fractions 8 and 9. N was also most abundant in fractions 8 and 9, although it had a broader distribution on the gradient, being readily detectable in fractions 5 through 9. The pattern with p22 and p65 was remarkably distinct, with peaks in fractions 2 and 3 and detectability in fractions 1 through 5 (least dense).

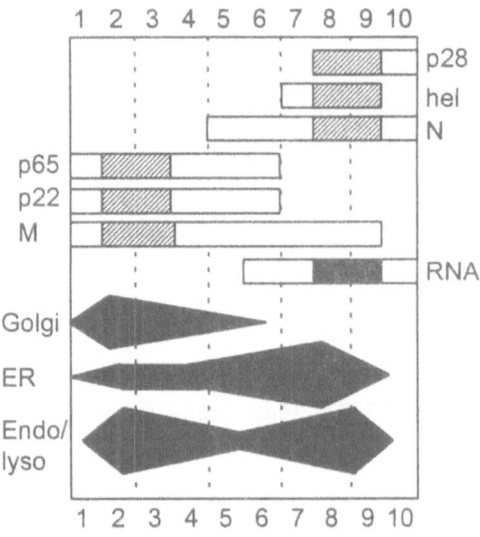

Figure 2. Fractionation of gene 1 proteins, structural proteins, viral RNA and cellular marker proteins on an iodixanol gradient. Fraction numbers are above and below the figure. White rectangles show the extent of detectability on the gradient. Hatched boxes show areas of maximal detection. Black box shows area of maximal RNA detection. Black polygons show relative detection of markers for Golgi, ER and endo/lysosomes

3.3   Viral RNA Localizes to Membranes with p28, Hel, and N

MHV specific, actinomycin D resistant viral RNA synthesized between 6 and 7 h p.i. was predominantly detected in the P100 crude fraction, as determined by total TCA precipitable [3H]uridine incorporation. When the P100 pellet was fractionated, viral RNA was detected only in fractions 6 through 10, with a very distinct peak in fractions 8 and 9.

3.4 Identification of Membrane Proteins in Gradient Fractions

The enzymatic marker for Golgi (Gal-T) was concentrated in fraction 2 and 3, although as expected, some activity above baseline was detected in almost all fractions. This was consistent with localization of M to the same fractions. The ER marker (NADPH cyt C Red) showed a detectable level across the gradient but a distinct peak in fractions 6 through 8. Finally, the protein marker for endosomes/lysosomes (LAMP 1) showed a bimodal pattern, with clear peaks in fractions 1 through 3 and 8 through 10, respectively. Thus the marker for endo/lysosomes was present in both membrane populations containing gene 1 proteins, and the ER marker overlapped only with those containing p28, N, hel and viral RNA.

4. DISCUSSION

The results of this study demonstrate that gene 1 and structural proteins may segregate to distinct membrane populations during infection. If the proteins analyzed in this study are representative of the remainder of gene 1 proteins, then two different membrane-associated complexes will be the sites of gene 1 protein localization and function. This outcome was suggested in our previous confocal microscopic study that demonstrated that several gene 1 proteins did not completely colocalize but rather colocalized to different extents and demonstrated areas of close approximation and interdigitation at the resolution of light (Bost *et al.*, 2000). The present study confirms and extends those results by showing that several membrane-associated gene 1 proteins localize to distinct membranes that are readily separable using the least stringent approaches to cell lysis and fractionation. In fact, given that even cell marker proteins show some distribution across the entire gradient under these conditions, it was remarkable that the gene 1 protein/membrane populations showed no overlap on the gradient. This result suggests that the interaction of the membrane/protein populations may serve different functions, or may interact to mediate one function, specifically viral RNA synthesis, in a manner that is easily altered and may change over time. At the least, it appears that the interaction of these populations observed by both light and electron microscopy is not mediated by covalent or strong hydrophobic forces that are difficult to disrupt.

It was also interesting that all viral RNA synthesized during this same period of time (6 through 8h p.i.) was detected only in the fractions containing p28, helicase, and the peak of N. The presence of hel in these fractions suggests that this membrane population might be the site for RNA synthesis. However, the localization of the RdRp (polymerase) has not been determined using these approaches. Thus it is possible that replication may occur in the interface of the closely approximated membrane/protein complexes, and that newly synthesized RNA may then target to the membranes along with hel and N.

Finally, the cell marker experiments along with previous data support the conclusion that the membranes containing p22 and p65 are derived from late endosomes. The membranes containing p28, hel, N and RNA either contain a mixed population of ER and lysosomal membranes or possibly ER alone. The latter possibility is suggested by the fact that proteins such as hel do not colocalize with lysosomes as determined by markers for acidic compartments (data not shown). Together these data suggest a model in which the interface of different membrane/protein complexes forms a "macro" complex where protein expression, protein processing, and RNA synthesis occurs.

ACKNOWLEDGMENTS

This work was supported by NIH grants AI-26603 and AI-01479 (M. Denison).

REFERENCES

Bi, W., Pinon, J. D., Hughes, S., Bonilla, P. J., Holmes, K. V., Weiss, S. R., and Leibowitz, J. L. (1998). Localization of mouse hepatitis virus open reading frame 1a derived proteins. *J. Neurovirology* **4**(6), 594-605.

Bost, A. G., Carnahan, R. H., Lu, X.-T., and Denison, M. R. (2000). Four proteinss processed from the replicase gene polyprotein of mouse hepatitis virus colocalize in the cell periphery and adjacent to sites of virion assembly. *J Virol* **74**(7), 3379-3387.

Denison, M. R., Hughes, S. A., and Weiss, S. R. (1995). Identification and characterization of a 65-kDa protein processed from the gene 1 polyprotein of the murine coronavirus MHV-A59. *Virology* **207**(1), 316-20.

Denison, M. R., Spaan, J. M., van der Meer, Y., Gibson, C. A., Sims, A. C., Prentice, E., and Lu, X. T. (1999). The putative helicase of the coronavirus mouse hepatitis virus is processed from the replicase gene polyprotein and localizes in complexes that are active in viral RNA synthesis. *J. Virol.* **73**, 6862-6871.

Dennis, D. E., and Brian, D. A. (1982). RNA-dependent RNA polymerase activity in coronavirus- infected cells. *Journal of Virology* **42**(1), 153-64.

Lu, X. T, Sims, A., and Denison, M. R. (1998). Mouse hepatitis virus 3CLpro cleaves a 22 kDa protein from the ORF 1a polyprotein in virus-infected cells and in vitro. *J Virol* **72**(3), 2265-2271.

Sethna, P. B., and Brian, D. A. (1997). Coronavirus genomic and subgenomic minus-strand rnas copartition in membrane-protected replication complexes. *Journal of Virology* **71**(10), 7744-7749.

Shi, S. T., Schiller, J. J., Kanjanahaluethai, A., Baker, S., Oh, J., and Lai, M. M. C. (1999).

Colocalization and membrane association of murine hepatitis virus gene 1 products and de novo-synthesized viral RNA in infected cells. *J. Virol.* **73,** 5957-5969.

Sims, A. C., Ostermann, J., and Denison, M. R. (2000). Mouse hepatitis virus replicase proteins associate with two distinct populations of intracellular membranes. *J Virol* **74**(12), 5647-5654.

van der Meer, Y., Snijder, E. J., Dobbe, J. C., Schleich, S., Denison, M. R., Spaan, W. J. M., and Krinjnse Locker, J. (1999). The localization of mouse hepatitis virus nonstructural proteins and RNA synthesis indicates a role for late endosomes in viral replication. *J. Virol.* **73**(9), 7641-57.

Differential Expression of Tumor Necrosis Factor in Primary Glial Cell Cultures Infected with Demyelinating and Non-Demyelinating MHVs

LI FU, JAYASRI DAS SARMA, AND EHUD LAVI

Department of Pathology and Laboratory Medicine, University of Pennsylvania, School of Medicine, Philadelphia, PA, USA

1.   INTRODUCTION

TNF-α is a monocyte/macrophage derived cytokine with multiple biological activities that are antitumorigenic, cytotoxic, and pro-inflammatory in nature. TNF-α mediates cellular response through two distinct cell surface receptors, TNF-R1(p55) and TNF-R2(p75) (Barbara *et al* 1996). TNF-α appears to have a diverse range of functions in the CNS due to its effect on astrocytes and oligodendrocytes. These include: induction of MHC class I Ag on astrocytes (Lavi *et al* 1988; Mauerhoff et al 1988), induction of ICAM-1, up-regulation of MHC class II Ag induced by IFN-γ and/or virus on astrocytes, stimulation of IL-6 secretion by astrocytes, myelin damage (Selmaj and Raine 1988) and lysis of oligodendrocytes (Robbins *et al* 1987).

The multiple effects of this cytokine on various cell populations in the CNS, including autocrine stimulation of astrocytes, suggest that TNF-α may have a central role in augmenting intracerebral immune responses and inflammatory demyelination (Chung and Benveniste 1990; Chung *et al* 1991). The hypothesis of our study is that TNF-α may play a role in MHV-A59 induced demyelination. To test this hypothesis we infected astrocyte cultures and mice with viruses of different biologic phenotypes (demyelination-positive MHV-A59, and demyelination-negative MHV-2)

The Nidoviruses (Coronaviruses and Arteriviruses).
Edited by Ehud Lavi *et al.*, Kluwer Academic/Plenum Publishers, 2001.

663

and examined the levels of TNF-α mRNA and protein by quantitative RT-PCR and ELISA, respectively.

2.   MATERIALS AND METHODS

2.1   Primary astrocytes cultures

Primary astrocytes cultures were prepared from neonatal C57BL/6 mouse brains following removal of the meninges (Mauerhoff *et al* 1988; Shahar *et al* 1989) and maintained in DMEM, 6 g/liter glucose, and 20% FBS. After the second passage the FBS concentration was reduced to 10%. The astrocytes were stained with monoclonal antibody to GFAP (1:4) for 30 min at room temperature, followed by 30 min incubation with Goat anti-mouse Ig-FITC (1:20). Astrocytes cultures were routinely > 95% positive for GFAP. The experiments were performed between 3-5 passages.

2.2   Mice, and viruses

Four- week-old C57BL/6 mice (Jackson Laboratories Bar Harbor, ME) were used in these experiments. The following viruses were used: MHV-A59 (Budzilowicz *et al* 1985; Lavi *et al* 1984), MHV-2 (Keck *et al* 1988).

2.3   TNF-a production by astrocytes

Primary mouse astrocytes were resuspended in DMEM containing 10% FBS, and plated at 1 X 10^6 cell/well into 6 well plates (Costa, cambridge, MA). The plates were incubated overnight to allow recovery of the cells from trypsinization and to assure adherence of the astrocytes. When the astrocytes reached confluence, the original media was aspirated off, and 2 ml 1 X DMEM containing 2% FBS was added to the wells. Astrocytes were infected with MHV-A59 (m.o.i.=1) or MHV-2 (m.o.i.=1) for one hour followed by washing with PBS. Alternatively, cultures were treated with LPS (1μg/ml), or a combination of the above (1 hour infection, followed by LPS treatment). Cultures were maintained for 12-18 hours following treatment, then supernatants were collected, centrifuged to remove contaminating cells, and stored at -70°C until use.

2.4 Infection of mice

Viruses were diluted in PBS containing 0.75% BSA. Mice were anesthetized with methoxyflurane (Methofane, Pittman-Moore, Mundelein, IL), then 25μl of diluted viruses were injected into the left cerebral hemisphere at the following doses: MHV-A59 3000pfu/ml, MHV-2 1000pfu/ml.

2.5 Quantitative PCR analysis

Livers, brains, and spinal cords were obtained from mice at 5 days (acute phase) or 30 days (chronic phase) post-infection. Total RNA was extracted from 1×10^7 Cells or 1g tissue using QIAGEN RNA easy Kit, then tested by qRT-PCR using a mimic method based on competitive PCR with non-homologous internal standards called PCR MIMICs (Clontech, Palo Alto, CA). Each PCR MIMIC consists of a heterologous DNA fragment with primer templates that are recognized by a pair of gene-specific primers. Thus, these templates " mimic " the target and are amplified during PCR.

Two composite primers are used. Each composite primer has the target gene primer sequence attached to a short, 20-nucleotide stretch of sequence designed to hybridize to opposite strands of a MIMIC DNA fragment. The desired primer sequences are thus incorporated during PCR reaction. A dilution of the first PCR reaction is then amplified again using only the gene specific primers. This ensures that all PCR MIMIC molecules have the gene specific primer sequences. Following the second PCR amplification the PCR MIMIC is purified by passage through CHROMA SPIN+TE-100 Columns. The yield of PCR MIMIC is calculated and diluted to 100 attomole/μl. Serial dilutions of PCR MIMICs are added to PCR amplification reactions containing constant amounts of experimental cDNA sample. Thus, PCR MIMIC and target templates, compete for the same primers in the same reactions. By knowing the amount of PCR MIMIC added to the reaction, we can determine the amount of target template, thus the initial mRNA levels.

2.6 TNF-α protein assay

Samples were assayed by ELISA according to the modified protocol from Pharmingen. The 96 well plates were read on a Titertek Multiscan using a wavelength of 405nm.

2.7   Measurement of viral titers

Samples of media from infected cultures were collected at 48 hours post infection and stored at -70^0C until tested.   Viral titers were measured by a 6-well-plate plaque assay of duplicate 10-fold dilutions of the samples.

3.   RESULTS

3.1   Detection of TNF-α mRNA in MHV-A59 and MHV-2 infected astrocytes by competitive RT-PCR

RNA was isolated from uninfected MHV-A59-infected or MHV-2-infected astrocytes. We synthesized the corresponding cDNA by reverse transcription, and then used PCR to amplify a specific sequence of the TNF-α cDNA and TNF-α MIMIC cDNA. The amplified TNF-α cDNA was detected in astrocytes infected with MHV-A59 and MHV-2 but not in uninfected astrocyte cultures. Competitive PCR analysis showed that the level of TNF-α mRNA in MHV-A59 infected astrocytes was 80-fold higher than in astrocytes infected with MHV-2. To rule out that the differences were due to differences in viral replication in the cultures viral titers were measured and found to be similar (approximately 10^5PFU/ml in both infections).

3.2   Detection of TNF-α mRNA in tissues of mice infected with MHV-A59 and MHV-2

Mice were injected intracerebrally with MHV-A59 or MHV-2 and sacrificed at 5 or 30 days post infection. RNA was isolated from the livers, brains and spinal cords. Competitive RT-PCR was used to compare the level of TNF-α mRNA in tissues infected with the different viruses.

As shown in table 1, the amplified TNF-α mRNA was detected in livers, brains and spinal cords in both MHV-A59 and MHV-2 infected mice at 5 days post infection, but not in uninfected mice. TNF-α mRNA in the MHV-A59 infected brain and spinal cord was 2.5 and 10 fold higher than the brain and spinal cord of MHV-2 infected mice respectively. Furthermore, at 30 days post infection, TNF-α mRNA was only detected in the MHV-A59 infected brains and spinal cords but not in MHV-2 infected mice.

3.3 The effect of MHV-A59 and MHV-2 infection on TNF-α production in mouse astrocytes cultures

Astrocytes were incubated with either MHV-A59 or MHV-2 (m.o.i.=1) for 1 hour, then viruses were washed out and cultures were maintained for the interval of 18 hours. In this experiment, the LPS (1μg/ml) was used as a positive control and also combined with MHV-A59 or MHV-2 separately. After 18 hours, the supernatants were harvested and assayed for the concentration of soluble TNF-α. Unstimulated astrocytes did not produce detectable levels of TNF-α. MHV-2 alone did not induce astrocytes to secrete TNF-α, but when combined with LPS the induction of TNF-α was significantly increased. Astrocytes could also be stimulated by a combination of LPS and MHV-A59 to secrete TNF-α. The TNF-α secretion stimulated by MHV-A59 was significantly higher than LPS alone or MHV-2 combined with LPS.

Table 1. TNF-α mRNA Detected by Competitive PCR in Mice and astrocyte cultures.

		Control	A59	MHV-2
Astrocytes		–	80X	1X
5D	Liver	–	10X	10X
	Brain	–	5X	2X
	SC	–	2X	0.2X
30D	Liver	–	1X	0.1X
	Brain	–	+	–
	SC	–	+	–

Astrocyte cultures were prepared from newborn C57BL/6 mouse brains, infected at m.o.i=1 of each virus and tested for TNF-α mRNA at 48 hours post infection. Tissues were sampled at 5 days post infection (peak of acute phase) and at 30 days post infection (during the chronic inflammatory stage). SC: Spinal cord. The units are all arbitrary relative units.

4. DISCUSSION

4.1 Demyelination positive virus MHV-A59 induced TNF-α mRNA at a significantly higher level than demyelination negative virus MHV-2 in both astrocytes culture and in mouse tissues. The difference was most significant in the CNS.

4.2 ELISA assay results suggest that the secretion of the TNF-α protein into the medium of infected cultures is also more significantly up-regulated in MHV-A59 infected astrocytes than in MHV-2 infected astrocytes. The level of viral replication in both cases is similar, suggesting that there is an intrinsic difference between these two viruses in their interaction with astrocytes.

4.3   These findings suggest that TNF-α may play a role in neurotropism and demyelination, which may explain the differential biologic phenotypes of these two closely related strains of viruses.

ACKNOWLEDGMENTS

Supported by grants from the National Multiple Sclerosis Society (RG 2615-B-2) and the NIH (NS30606). We thank Donna Bauhof for critical review of the manuscript and Elsa Aglow for histology expertise.

REFERENCES

Barbara, J. A. J., X. V. Ostande, and A. F. Lopez. 1996. Tumor necrosis factor-alpha: the good, the bad and potentially very effective. Immunology and Cell Biology. **74**:434-443.

Budzilowicz, C. J., S. P. Wilczynski, and S. R. Weiss. 1985. Three intergenic regions of mouse hepatitis virus strain A59 genome RNA contain a common nucleotide sequence that is homologous to the 3' end of the viral mRNA leader sequence. J. Virol. **53**:834-840.

Chung, I. L. Y., and E. N. Benveniste. 1990. Tumor necrosis factor - a production by astrocytes. J. Immunol. **144**:2999-3007.

Chung, I. L. Y., J. G. Norris, and E. N. Benveniste. 1991. Differential tumor necrosis factor a expression by astrocytes from experimental allergic encephalomyelitis-susceptible and resistant rat strains. J. Exp. Med. **173**:801-811.

Keck, J. G., L. H. Soe, S. Makino, S. A. Stohlman, and M. M. C. Lai. 1988. RNA recombination of murine coronavirus: recombination between fusion-positive mouse hepatitis virus A59 and fusion-negative mouse hepatitis virus 2. J. Virol. **62**:1989-1998.

Lavi, E., D. H. Gilden, Z. Wroblewska, L. B. Rorke, and S. R. Weiss. 1984. Experimental demyelination produced by the A59 strain of mouse hepatitis virus. Neurology. **34**:597-603.

Lavi, E., A. Susumura, D. M. Murasko, E. M. Murray, D. H. Silberberg, and S. R. Weiss. 1988. Tumor necrosis factor induces expression of MHC class I antigens on mouse astrocytes. J. Neuroimmunol. **18**:245.

Mauerhoff, T., R. Pujol-Borrell, R. Mirakan, and G. F. Bottazzo. 1988. Differential expression and regulation of major histocompatibility complex (MHC) products in neural and glial cells of the human fetel brain. J Neuroimmunol. **18**:271.

Robbins, D. S., Y. Shirazi, B. E. Drysdale, A. Leiberman, H. S. Shin, and M. L. Shin. 1987. Production of cytotoxic factor for oligodendrocytes by stimulated astrocytes. J. Immunol. **139**:2593.

Selmaj, K. W., and C. S. Raine. 1988. Tumor necrosis factor mediates myelin and oligodendrocytes damage in vitro. Ann. Neurol. **23**:339.

Shahar, A., J. D. Vellis, A. Vermadakis, and B. Haber (ed.). 1989. A dissection and tissue culture mannual of the nervous system. Alan R. Liss, Inc.,, New York.

Stohlman, S. A., and L. P. Weiner. 1981. Chronic central nervous system demyelination in mice after JHM virus infection. Neurology. **31**:38-44.

Infectious Bronchitis Virus Nucleocapsid Protein Interactions with the 3' Untranslated Region of Genomic RNA Depend on Uridylate Bases

[1]ELLEN W. COLLISSON, [1]MINGLONG ZHOU, [2]PAUL GERSHON, AND [1]JYOTHI JAYARAM

[1]*Department of Veterinary Pathobiology, Texas A&M University, College Station, TX;* [2]*Institute of Biosciences and Technology, Texas A&M University, Houston, TX*

1. INTRODUCTION

Infectious bronchitis virus (IBV) is a highly contagious virus of worldwide economic concern in poultry. Infection causes respiratory disease, as well as kidney lesions, reproductive problems and gastrointestinal disfunction. The virion consists of four structural proteins: the membrane, spike, nucleocapsid (N), and the small envelope proteins. The IBV N protein, a basic, phosphorylated protein of 409 residues, is highly conserved especially within the central or middle region among IBV strains, (Williams *et al*, 1992).

In addition to being closely associated with the RNA genome, the coronavirus nucleocapsid protein has been suggested to have multiple functions (Robbins *et al.,* 1986; Compton *et al.,* 1987; Stohlman *et al.,* 1988; Baric *et al.,* 1988). The MHV nucleocapsid protein specifically binds to small leader-containing RNA. Within the cytosol of MHV infected cells, it can interact with membrane-bound small leader RNA in transcription complexes (Stohlman *et al.,* 1988). Anti-nucleocapsid monoclonal antibodies reportedly precipitate both full-length and subgenomic mRNA, as well as replicative intermediate RNA (Baric *et al.,* 1988). The amount of N protein found associated with the genome and the putative functions of this

The Nidoviruses (Coronaviruses and Arteriviruses).
Edited by Ehud Lavi *et al.*, Kluwer Academic/Plenum Publishers. 2001.

protein suggest that it should readily associate with additional regions of coronavirus RNA.

Overall, functions of the coronavirus N depend on its interaction with viral RNA. IBV N has been shown to bind to sequences representing the 3' untranslated region (UTR) of genomic or subgenomic RNA (Zhou *et al*, 1996). Specificity was suggested by preferential interactions with selected regions of the 3' UTR and the absence of interactions with ribosome and yeast tRNA. A region lying between 215 and 78 nt from the 3' end of genomic and subgenomic RNA did not shift when exposed to N, unlike the other similarly sized oligonucleotides spanning the 3' UTR. This region overlapped with CD, corresponding to the region lying 155 nt within the UTR at the ultimate 3' end. It was further shown that the amino and carboxyl regionw, but not the middle region, were able to bind to the 3' UTR (Zhou and Collisson, 2000). The binding site of the amino domain required more than the first 91 residues. In this report, sequences required for recognition by the IBV N and the UTR were investigated.

2.   MATERIALS AND METHODS

2.1   Oligonucleotides

cDNA templates for RNA oligonucleotides corresponding to sequence within the 3' UTR of genomic IBV RNA were generated by PCR and incorporated the T7 RNA polymerase promoter. The oligonucleotides were synthesized by in vitro transcription with T7 RNA polymerase (Promega Corporation, Madison, WI). These RNA probes were labeled with ^{32}P by incorporating the ^{32}P-CTP during RNA transcription. A 40mer poly-U was synthesized and labeled with ^{32}P using T4 polynucleotide kinase (Promega Corporation, Madison, WI).

2.2   Polypeptides

Preparation of the N protein and polypeptides has been described in Zhou & Collisson (2000) and Zhou *et al* (1996). Briefly, the encoding nucleotide sequences were amplified by PCR from template plasmid and cloned into *Bam*HI and *Hin*dIII sites in the pQE8 vector (Qiagen Inc., Chatsworth, CA). Expression of protein from pQE8 resulted in fusion polypeptides with amino terminus six histidine tags. The polypeptides were purified first with a N^{+2}-NTA resin and then with Sephadex G200 or Bio-gel P60 (Ausubel *et al*, 1987).

2.3 Gel shift assays

Protein-RNA interactions were analyzed by a modified gel-shift assay (Zhou *et al*, 1996). RNA and varying concentrations of nucleocapsid protein were co-incubated for 20 min at room temperature in 10 μl of gel shift buffer (25mM Hepes, 25 mM EDTA, 150 mM NaCl, 5 mM DTT, 10% glycerol and 20 units Rnase Inhibitor, Boehringer Mannheim Corp., Indianapolis, IN). After adding 1 μl 10x sample buffer, the reaction mixtures were loaded onto a 1% agarose gel and electrophoresed at 60V in 1x Tris-borate-EDTA (Sambrook *et al.*, 1989). Gels were then dried and autoradiographed.

3. RESULTS

cRNA was synthesized that corresponded to regions of the 3' UTR of IBV RNA. The sequences that were selected incorporated putative stem-loop structures (Fig 1). These oligonucleotides were used for gel shift assays to evaluate binding whole N and the amino, middle and carboxyl domains of N (Zhou and Collisson, 2000). In previous studies, 5 nM of IBV N protein and the amino and carboxyl regions had been found to shift RNA encoding the N gene to the 3' end of the genome and CD RNA, the ultimate 3' terminal 155 nucleotides (Zhou and Collisson, 2000). Interactions between the N polypeptides and RNA sequences within the 3' UTR RNA's were examined using 5 nM of IBV N protein (Fig 2) and N polypeptides (Table 1). The intact N protein, the carboxyl terminal polypeptide (C140) or the amino region (A171), shifted RNA1, RNA2, RNA3, RNA4, RNA5 and RNA7 probes very effectively. The shift of RNA7 with varying concentrations of the amino, middle and carboxyl regions of N are shown in Figure 3a. RNA 6 did not shift when exposed with N (Fig 3b) or the N polypeptides. This oligonucleotide corresponded to a region found within the larger 155 nt EF that also did not interact with N on gel shift assays.

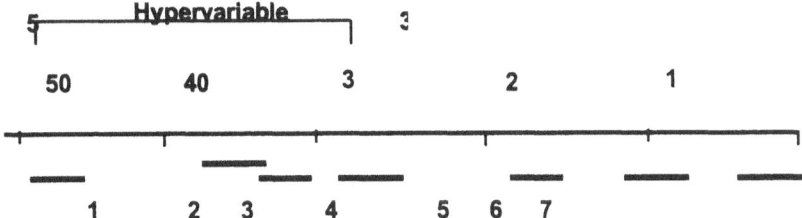

Figure 1. Schematic of synthesized 3' UTR sequences used for studies identifying interactions with the IBV N protein and polypeptides.

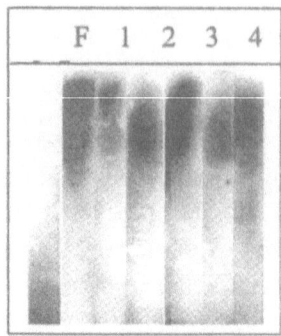

Figure 2. Gel shift of 1 ng of ^{32}P labeled oligonucleotide probe representing sequences from the 3' UTR of the IBV genome. The numbers above each lane indicate the oligonucleotide, as shown in Figure 1, used to interact with the 5nM of N protein. Free RNA1 in the absence of protein is shown in lane F.

Varying concentrations of polypeptide used in gel shift assays also demonstrated interactions with N and N polypeptides. The gel shifts of labeled RNA 7 when exposed to N, the amino region or the carboxyl region are shown in Fig 2. The polypeptide corresponding to the middle region of N did not shift any of these RNA probes (Table 1; Fig 2). In contrast, no interaction was observed between N or N polypeptides at any concentration of RNA 6 examined (Fig 3a). The single obvious difference between RNA 6 and the oligonucleotides that did bind to the N polypeptides was the low number of uridylates (U) in RNA 6. RNA 6 had only 7%, whereas from 24% to 58% of the other oligonucleotides consisted of U.

Table 1. Interactions of N polypeptides with oligonucleotides in Fig 1.

RNA	Amino	Middle	Carboxyl	Whole N	%U*
1	+	-	+	+	41 (41)
2	+	-	+	+	52 (41)
3	+	-	+	+	47 (40)
4	+	-	+	+	24 (41)
5	+	-	+	+	58 (41)
6	-	-	-	-	7 (43)
7	+	-	+	+	36 (53)

* () indicate the number of bases within each oligonucleotide.

Gel shifts of additional oligonucleotides with N and the N domain fragments were examined. No detectable interactions were observed by gel shift assays between N polypeptides and an oligonucleotide synthesized with A, C, and G bases and no U's (Fig 3). However, another probe was synthesized with two U triplets placed within the ACG repeating 40-mer nucleotide sequence. Binding was observed with the ACG incorporated with the U triplets.

As the U bases seemed to be important in the interactions with the N protein, a 40mer poly-U sequence was used to examine potential interactions in the absence of A, G, C and the absence of any secondary structure. In the presence of N, the labeled poly-U sequence shifted within the agarose gel.

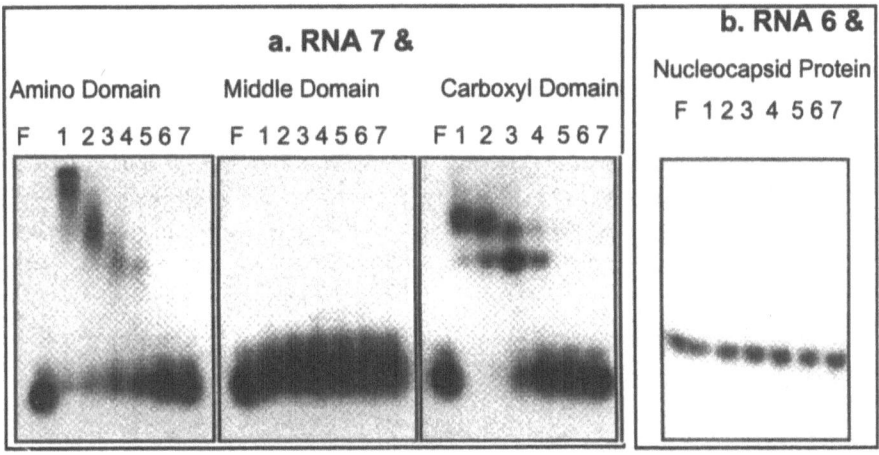

Figure 3. Gel shift of [2]P-labeled RNA7 with 3μg of the three regions of the N protein and RNA6 with 3μg of whole N protein. RNA in lanes 1 to 7 were exposed to serial 2-fold dilutions of polypeptide with 1 ng of free probe in lane 1. F indicates labeled RNA that was not exposed to protein.

4.   DISCUSSION

RNA structures recognized by various viral or cellular proteins are tremendously varied, ranging from single-strands, hairpins to seemingly complex irregular helices and cation dependent tertiary structures (Draper, 1995). A bulged stem-loop structure in the 3' non-translated region on the MHV genome was shown to be critical for negative RNA synthesis and cellular protein interaction, and a stem-loop structure in polymerase gene was shown to be critical for packaging (Hsue and Masters, 1997; Yu and Leibowitz, 1995; Fosmire *et al.,* 1992). In addition, stem-loops on the TAR of HIV-1 RNA, U1 snRNA, poliovirus 5' noncoding region and 16S ribosomal RNA have all been implicated as critical for recognition of proteins (Weeks and Crothers, 1991; Hall and Stump, 1992). It is anticipated that the interaction between N and the 3' UTR also involves secondary structure, such as a stem-loop. Although the buffer conditions in these studies were not the most favorable for maintaining secondary structure, these studies do provide evidence that the presence of uridylate bases contribute to the successful binding of N to sequences within the 3'

UTR. Uridylate-rich sequences have also been identified as protein binding sites for poly (A) polymerase from vaccinia virus (Gershon and Moss, 1993, Deng and Gershon, 1997). The eight-nucleotide sequence (UAUUUUCU) in the potato X 3' untranslated region (UTR) is required for both host protein binding and viral replication (Sriskanda *et al*, 1996). Proteins in cellular extracts bind to AU-rich repeats in the 3'-UTR of human TNF-alpha mRNA and the binding activity correlates the expression level of the TNF-alpha in human epithelial cancer cells (Wang *et al*, 1998).

An overall influence of secondary or tertiary structure may also impact binding. Interactions may be demonstrated with small simple units of RNA, but in fact, be optimized with larger regions that, for example, confer more tertiary stability. Often, function depends on protein-induced alterations in RNA that create or expose sites necessary for binding to additional proteins (Draper, 1995).

Although RNA structures can be identified by computer analysis, predicting structures or sequences that are responsible for protein interactions is not always successful. The stability of the RNA structures may be directly related to the biological functions. The 3' UTR of mRNA has been shown to differentially regulate gene expression (Conne *et al*, 2000). The AUUUA pentanucleotide sequence is associated interactions between mRNA and RNA-binding proteins. Defective interactions lead to abnormal stabilization (Chen and Shyu, 1995; Peng *et al*, 1998). It is possible that N interacts with the U triplets so as to inhibit interactions with these proteins and stabilize viral RNA.

ACKNOWLEDGEMENTS

This research was supported by the U. S. Poultry & Egg Association, grant #142; USDA Animal Health (Section 1433), # TEXO-6824, Texas A&M University Interdisciplinary Program, 1998, and Texas Technology Research program #999902-212.

REFERENCES

Ausubel, F. M., Brent, R., Kingston, R., Moore, D. D., Seidman, J. G., Smith, J. A. and Struhl, K., 1994, *Current protocols in molecular biology*. 18.5.3-4. John Wiley & Son, inc.

Baric, R. S., Stohlman, S. A., Razavi, M. K., and Lai, M. M. C., 1985, Characterization of leader-related small RNAs in Coronavirus-infected cells. Further evidence for leader-primed mechanism of transcription. *Virus Res*. **3**: 19-33.

Masters, P. S., 1992, Localization of an RNA-binding domain in the nucleocapsid protein of the coronavirus mouse hepatitis virus. *Arch. Virol.* **125**: 141-160.

Chen, C.Y. & Shyu, A.B., 1995, AU-rich elements: characterization and importance in mRNA degradation. Trends Biochem. Sci. 20:465-470.

Compton, S. R., Rogers, D. B., Holmes, K. V., Fertsch, D., Remenick, J. and McGowan, J. J., 1987, In vitro replication of mouse hepatitis virus strain A59. *J. Virol.* **61**:1814-1820.

Conne, B., Stutz, A. & Vassalli, J.-D., 2000, The 3' untranslated region of messenger RNA : A molecular 'hotspot' for pathology? Nature Medicine 6:637-641.

Peng, S.S., Chen, C.Y., Xu, N. & Shyu, A.B., 1998, RNA stabilization by the AU-rich element binding protein, HuR, an EAV protein. EMBO J. 17:3461-3470.

Robbins, S. G., Frana, F. M., McGowan J. J., Boyle, J. F., Holmes, K. V., 1986, RNA-binding proteins of coronavirus MHV: detection of monomeric and multimeric N protein with an RNA overlay protein blot assay. *Virology.* **150**: 402-410.

Stohlman, S. A., Baric, R. S., Nelson, G. N., Soe, L. H., and Welter, L. M., 1988, Specific interaction between Coronavirus leader RNA and Nucleocapsid Protein. *J. Virology.* **62**: 4288-4295.

Williams, A. K., Wang, L., Sneed, L. W. and Collisson, E. W., 1992, Comparative analysis of the nucleocapsid genes of several strains of infectious bronchitis virus and other coronaviruses. *Virus Res.* **25**: 213-222

Williams, A. K., Wang, L., Sneed, L. W. and Collisson, E. W., 1993, Analysis of a hypervariable region in the 3' non-coding end of the infectious bronchitis virus genome. *Virus Res.* **28**: 19-27

Zhou, M. & Collisson, E.W., 2000, The amino and carboxyl domains of the infectious bronchitis virus nucleocapsid protein intereract with 3' genomic RNA. Virus Res, 67:31-39.

Zhou, M., Williams, A. K., Chung, S., Li, W. and Collisson, E. W., 1996, The infectious bronchitis virus nucleocapsid protein binds RNA sequences in the 3' terminus of the genome. *Virology.* **217**: 191-199.

Induction of Apoptosis in MRC-5, Diploid Human Fetal Lung Cells after Infection with Human Coronavirus OC43

ARLENE R. COLLINS
Department of Microbiology, State University of New York at Buffalo, Buffalo NY USA 14214

1.   INTRODUCTION

Human coronaviruses (HCoV) cause upper respiratory tract infections manifested by symptoms of the common cold including rhinitis, tussis, fever, headache and myalgia (Hruskova *et al* 1990). HCoV respiratory illness is primarily virus-mediated; immunologic events have not been postulated to play a role (Makela *et al* 1998). Evidence showing the presence of viral genome in spinal fluid and brain suggests that HCoV may play an etiologic role in multiple sclerosis (Stewart *et al* 1992, Cristallo *et al* 1997). However, the pathogenesis of HCoV infections is poorly understood. In human embryonic tracheal organ cultures, HCoV causes a slow patchy destruction of the ciliated epithelial cells and in respiratory epithelial tissue cultures, the cytopathic effect is subtle, evident only by vacuolization and spindling.

Since apoptosis is a mechanism of cell death for other coronaviruses such as transmissible gastroenteritis virus (TGEV) (Eleouet *et al* 1998) and mouse hepatitis virus (MHV) (An *et al* 1999), we examined infected MRC-5 cells with HCoV and examined the cells for changes associated with programmed cell death. DNA fragmentation and formation of apoptotic bodies was observed in cells infected with HCoV-OC43.

The Nidoviruses (Coronaviruses and Arteriviruses).
Edited by Ehud Lavi *et al.*, Kluwer Academic/Plenum Publishers, 2001.

677

2.   MATERIALS AND METHODS

2.1   Viruses and Cells

HCoV-229E and HCoV-OC43 strains that had been repeatedly passaged in M-7 and MRC-5, human lung fetal cells were used in this study.

2.2   Assessment of DNA Degradation

HCoV-OC43-infected and uninfected MRC-5 cells ($2x10^6$ cells) were harvested after six days of exposure to virus. DNA was isolated with DNAzol (Molecular Research Center, Cincinnati, OH), incubated with RNAse A (1µg/ml) at 50°C for 1 h, and then loaded onto a 1.2% agarose gel containing ethidium bromide. After electrophoresis, the gel was photographed and examined for the presence of DNA laddering (Jan and Griffin 1999).

2.3   Visualization of Apoptotic Nuclei by Ethidium Bromide Staining

To detect DNA condensation and fragmentation, the cells were fixed in cold acetone, stained for 10 minutes with 100µg/ml ethidium bromide in PBS and washed twice in PBS. Microscopic visualization of apoptotic nuclei was performed by light microscopy using epiilumination and a filter for fluorescein.

2.4   Immunofluorescence

After fixing the cells in cold acetone, viral antigen was detected by indirect immunofluorescent staining using monoclonal antibody 4B6.1 to HCoV-OC43 nucleocapsid as the primary antibody followed by fluorescein-conjugated goat anti-mouse immunoglobulin (Vector, Burlingame CA).

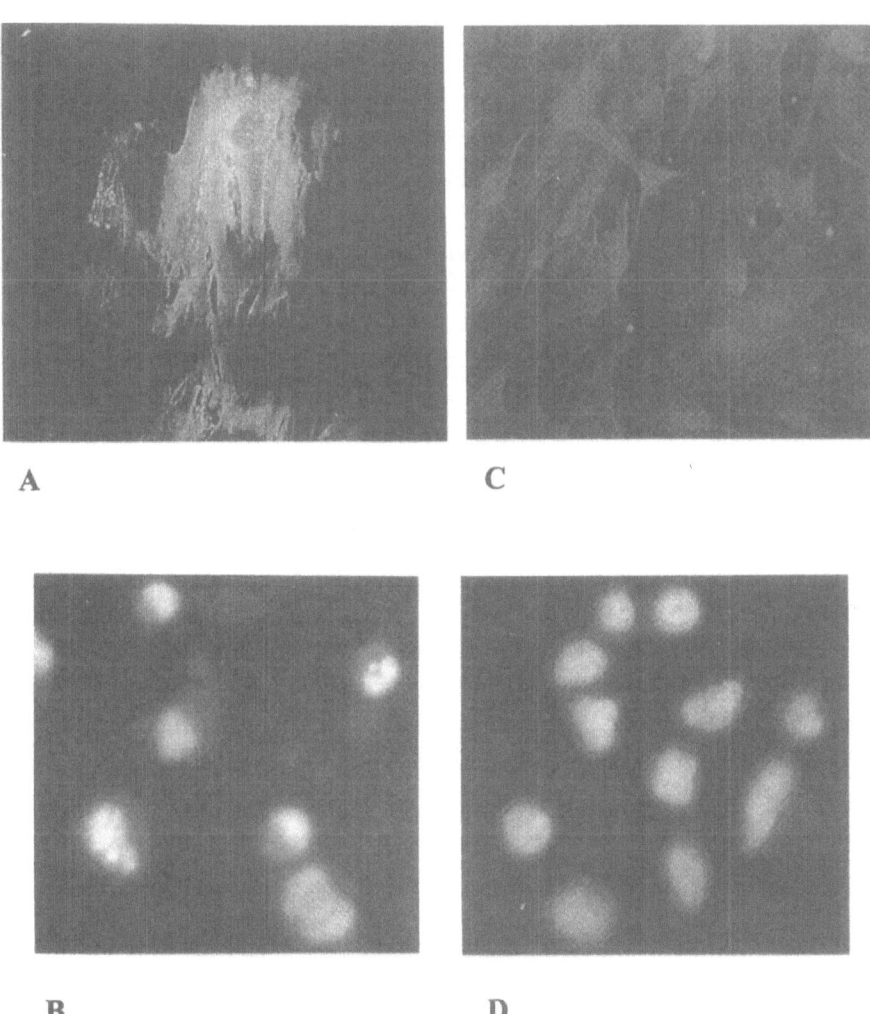

Figure 1. Viral antigen expression and induction of apoptosis in MRC-5 cells infected with HCoV-OC43 at a moi of 1and incubated at 33°C. Panels A and C: Immunofluorescent staining for viral nucleocapsid antigen with mAb 4B6.1. Panels B and D: Ethidium bromide staining for DNA fragmentation. Panels A and B are infected cells. Panels C and D are mock-infected cells.

3.   RESULTS

In order to assess the cytologic changes in MRC-5 cells after infection by HCoV-OC43, we exposed semi-confluent monolayer of cells to virus at a moi of 1 and incubated the cells at 33°C. Three days after infection, swollen, rounded, granular cells were observed. These cells showed immunuofluorescent staining for viral nucleocapsid (Fig1 A). By six days post-infection, many cells became detatched and those cells remaining were further enlarged and granular. The adherent cells were fixed in acetone and the nuclear DNA was stained with ethidium bromide. We observed DNA condensation and formation of apoptotic bodies in the nucleus (Fig 1B). Further confirmation of DNA fragmentation was obtained by extracting the DNA from a pool of detached and adherent cells infected for six days and assessing for DNA ladder formation after gel electrophoresis. Fragmentation was present in DNA from HCoV-OC43-infected cells, but not in mock-infected cells (Fig.2). Interestingly, DNA fragmentation was observed at six days post-infection indicating that the onset of apoptosis was two to three days earlier, coinciding with the time of maximum viral yield. The yield of infectious HCoV-OC43 was greater in the cell-associated than in the cell-free virus fraction suggesting that accumulation of virus or viral protein(s) caused induction of apoptosis. In HCoV-229E infected MRC-5 cells maximum virus yield was fivefold less and was higher in the cell-free than in the cell-associated fraction. The viral titer reached its peak earlier in HCoV-229E than in HCoV-OC43 infection and the nucleus did not appear to undergo apoptotic condensation.

Table 1. Characteristics of HCoV-OC43-infected MRC-5cells

Property	Characteristic
Cytopathic effect	Granular degeneration, enlargement
Viral titer	2.8×10^6 pfu/ml
Maturation and release	More cell associated
DNA fragmentation	yes
Cell death	apoptosis

4.   DISCUSSION

We found evidence of apoptotic death in MRC-5 cells six days after infection with HCoV-OC43, in cells that contained large amounts of intracellular virus. The events that triggered apoptosis have yet to be determined. In MHV-induced apoptosis, the coronavirus E protein alone was responsible (An *et al* 1999). The E protein of MHV, found on gene5, is expressed from an internal ribosome entry structure (IRES) which could

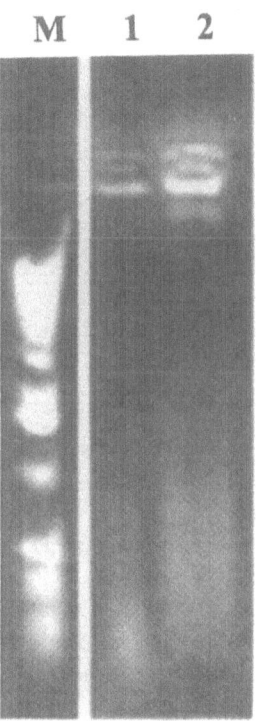

Figure 2. Gel electrophroresis of DNA for assessment of fragmentation in MRC-5 cells. Lane M: 1kb markers, lane1: mock, lane 2: infected with HCoV-OC43.

account for its accumulation leading to induction of apoptosis (Jendrach *et al* 1999). Induction of apoptosis by HCoV-OC43 may play a role in evasion of immune recognition and suppression of host-mediated responses, events that favor establishment of persistent infection. Other mechanisms favoring persistence include protection of the infected cell from complement-mediated lysis by regulators of complement activation and antigenic variation of the virus (Varsano *et al* 1995, Collins *et al* 1986, Arbour *et al* 1999). In comparison, HCoV-229E infection in MRC-5 cells proceeds more rapidly without programmed cell death. HCoV-229E infected cells may be more sensitive to anti-apoptotic BCL-2 proteins (Bonati *et al* 1996). Interestingly in a study of primary infections showing a seroconversion to the virus, HCoV-229E acute respiratory disease included conjunctivitis and lymphadenitis, symptoms that were not associated with HCoV-OC43 illness (Hruskova *et al* 1990). Further examination of HCoV pathogenesis is important in the management of childhood asthma and older people with chronic illnesses.

ACKNOWLEDGEMENTS

The author wishes to thank Drs. Shinpei Ohki and .Michio Ito for many helpful discussions.

REFERENCES

An, S., Chen, C-J., Xin, Y., Leibowitz, J.L. and S. Makino. 1999. Induction of apoptosis in murine coronavirus-infected cultured cells and demonstration of E protein as an apoptosis inducer. *J. Virol.* 73: 7853-7859.

Arbour, N., Côté, G., Lachance, C., Tardieu, M., Cashman, N.R., and P. J. Talbot. 1999. Acute and Persistent Infection of Human Neural Cell Lines by Human Coronavirus OC43. *J. Virol.* 73: 3338-3350.

Bonati, A., Albertini, R., Garau, D., Pinelli, S., Lunghi, P., Almici, C., Carlo-Stella, C., Rizzoli, V., and P. Dall'aglio. 1996. BCL2 oncogene protein expression in human hematopoietic precursors during fetal life. *Exp.Hematol.* 24:459-465.

Collins, A.R. and O. Sorensen. 1986 Regulation of viral persistence in human glioblastoma and rhabdomyosarcoma cells infected with coronavirus OC43. *Microbial Path.* 1: 573-583.

Cristallo, A., Gambaro, F., Biamonte G., Ferrante, P., Battaglia, M. and P.M. Cereda. 1997. Human coronavirus polyadenylated RNA sequences in cerebrospinal fluid from multiple sclerosis patients. *New Microb.* 20: 105-114.

Eleouet,J., Chilmonczyk,S., Besnardeau,L., and H. Laude. 1998. Transmissible gastroenteritis coronavirus induces programmed cell death in infected cells through a caspase-dependent pathway. *J.Virol.* 66: 4918-4924.

Fischer, F., Stegen, C.R., Masters, P.S. and W.A. Samsonoff. 1998. Analysis of constructed E gene mutants of mouse hepatitis virus confirms a pivotal role for E protein in coronavirus assembly. *J.Virol.* 72: 7885-7894.

Hruskova, J.,Heinz, F., Svandova, E. and S. Pennigerova. 1990. Antibodies to human coroanviruses 229e and OC43 in the population of C.R. *Arch. Virol.* 34: 346-352.

Jan, J. and D.E. Griffin. 1999. Induction of apoptosis by Sindbis virus occurs at cell entry and does not require virus replication. *J.Virol.* 73: 10296-10302.

Jendrach, M., Theil. V., and S. Siddell. 1999. Characterization of an internal ribosome entry site within mRNA5 of murine hepatitis virus. *Arch.Virol.* 144: 921-933.

Makela, M.J., Puhakka, T., Ruuskanen, O., Leinonen, L.M., Saikku, P., Kimpimaki, M., Blomqvist, S., Hyypia, T. and P. Arstila. 1998. Viruses and bacteria in the etiology of the common cold. *J. Clin. Microbiol.* 36: 539-542.

Stewart, J.N., Mounir,S. and PJ. Talbot. 1992. Human coronavirus gene expression in the brains of multiple sclerosis patients. *Virol.* 191:502-505.

Varsano, S., Frolkis, I., and D. Ophir. 1995. Expression and distribution of cell-membrane complement regulatory glycoproteins along the human respiratory tract. *Am. J. Respir. Crit. Care Med.* 152: 1087-93.

Effects of Heparin on the Entry of Porcine Reproductive and Respiratory Syndrome Virus into Alveolar Macrophages

N. VANDERHEIJDEN, P. DELPUTTE, H. NAUWYNCK, AND M. PENSAERT

Laboratory of Virology, Faculty of Veterinary Medicine, Ghent University, Salisburylaan 133, B-9820 Merelbeke, Belgium

1. INTRODUCTION

Porcine reproductive and respiratory syndrome (PRRS) is an infectious disease of swine that causes abortion and respiratory disorders, leading to important economic losses (Rossow, 1998). The PRRS virus (PRRSV) was first isolated in 1991 in the Netherlands (Wensvoort *et al.*, 1991) and is a member of the *Arteriviridae* family, that was recently incorporated in the order *Nidovirales* (Cavanagh, 1997). Serological and genomic studies have revealed major strain differences among isolates, most markedly between those from the USA and Europe (Wensvoort *et al.*, 1992; Bautista *et al.*, 1993; Murtaugh, 1995).

PRRSV demonstrates a high tropism for cells of the monocyte/macrophage lineage, both *in vitro* and *in vivo* (Rossow *et al.*, 1996; Duan *et al.*, 1997). Nevertheless, besides primary cultures of porcine alveolar macrophages (PAM), MARC-145 cells, a subclone of monkey kidney MA-104 cells, also support PRRSV replication (Wensvoort *et al.*, 1991; Kim *et al.*, 1993).

A putative receptor for PRRSV has been identified as a 210 kDa protein on alveolar macrophages using monoclonal antibodies (Duan *et al.*, 1998). However, these antibodies were not able to prevent completely the binding

The Nidoviruses (Coronaviruses and Arteriviruses).
Edited by Ehud Lavi *et al.*, Kluwer Academic/Plenum Publishers, 2001.

of the virus to the PAM. This observation is indicative for the existence of a co-receptor. The 210 kDa protein was not detected on MARC-145 cells. Jusa *et al.* (1997) demonstrated that addition of heparin to two Japanese strains (EDRD-1 and 5-53) of PRRS virus induced an important reduction in the number of plaques formed upon inoculation of MARC-145 cells. In addition, pre-treatment of MARC-145 cells with heparinase I reduced the infectivity of these strains. These data indicate that a heparin-like molecule present on the surface of MARC-145 cells is involved in the viral entry process.

In this study, we analyzed the effects of heparin and heparinase I on the entry of PRRSV into its main *in vivo* target cell, the porcine alveolar macrophage.

2.　　MATERIAL AND METHODS

2.1　　Viruses and cells

PAM were obtained from 4- to 6-week-old conventional Belgian Landrace pigs from a PRRSV-negative herd according to the method previously described by Wensvoort *et al.* (1991). The European prototype Lelystad strain (LV) was kindly provided by G. Wensvoort, the 94V360 was isolated from the lungs of a pig with respiratory problems in Belgium, and the American isolates US-5, US-11, and US-15 were isolated from outbreaks of reproductive disorders in Iowa. All the PRRSV isolates used in this study were adapted to MARC-145 cells (4 passages for 94V360 and LV, 3 passages for US-5, 1 passage for US-11 and US-15). The Pseudorabies virus Kaplan strain (PRV) and the PRV gC-deleted (PRVgC⁻) mutant were kindly provided by T. Mettenleiter (Insel Riems, Germany).

2.2　　Incubation of viruses or cells with heparin

Viruses were incubated with 0, 50, or 500 Units/ml heparin (Sigma) for 1h at 37°C before inoculation on PAM (MOI 1 or 0.1). One hour after virus adsorption, cells were washed to remove heparin, fresh medium was added to the cells which were incubated for 9h at 37°C. In another experiment, PAM were incubated for 1h at 37°C with 0, 50, or 500 Units/ml heparin, washed and inoculated for 1h with viruses. Fresh medium was added and cells were incubated for 9h at 37°C. Cells were fixed and infected cells, stained by immunocytochemistry, were counted.

2.3   Treatment of cells with heparinase I

Cells were incubated for 1h at 37°C with 2 different concentrations of heparinase I (Sigma), and were washed before being inoculated with the viruses and treated as above.

3.   RESULTS

3.1   Effect of heparin on the entry of PRRSV into PAM

The effect of the addition of heparin to PRRSV during adsorption on PAM was studied, and wild-type as well as gC deleted PRV strains were used respectively as positive (heparin sensitive; Mettenleiter *et al.*, 1990) and negative (heparin resistant; Mettenleiter *et al.*, 1990) controls. PAM are highly permissive to PRV infection (Iglesias *et al.*, 1989). After treatment with 50U/ml or 500U/ml heparin, the number of PRRSV infected cells was decreased respectively by 21±3% (mean of triplicates ± standard deviation) and 24±4% with LV, 15±7% and 37±12% with 94V360, 31±6% and 61±7% with US-5, 30±10% and 51±3% with US-11, 28±9% and 41±9% with US-15.

These results were all significantly different from the values obtained without heparin (*P*<0,05) except for strains 94V360, US-11 and US-15 at 50U/ml heparin. Infection with PRV was already reduced by 79±6% at 50U/ml of heparin (*P*<0,01) whereas no significant reduction of infection was observed with PRV gC⁻ (Figure 1). These results indicate that infection of PAM with PRRSV can be reduced with heparin. Statistical analysis of additional data indicates that a significant difference (*P*<0,05) is obtained when the mean percentages of reduction of infection are compared pairwise between the American US-5 strain and the two European strains used in this study. These results show that the sensitivity to heparin treatment is variable from strain to strain and is less important than with PRV.

When PAM were pre-incubated with heparin, washed and subsequently inoculated with virus, no significant effect of heparin was observed with any of the PRRSV strains tested nor with PRV gC⁻ (Figure 2). Thus, in order to be effective, heparin must be present in the culture medium during the viral adsorption period.

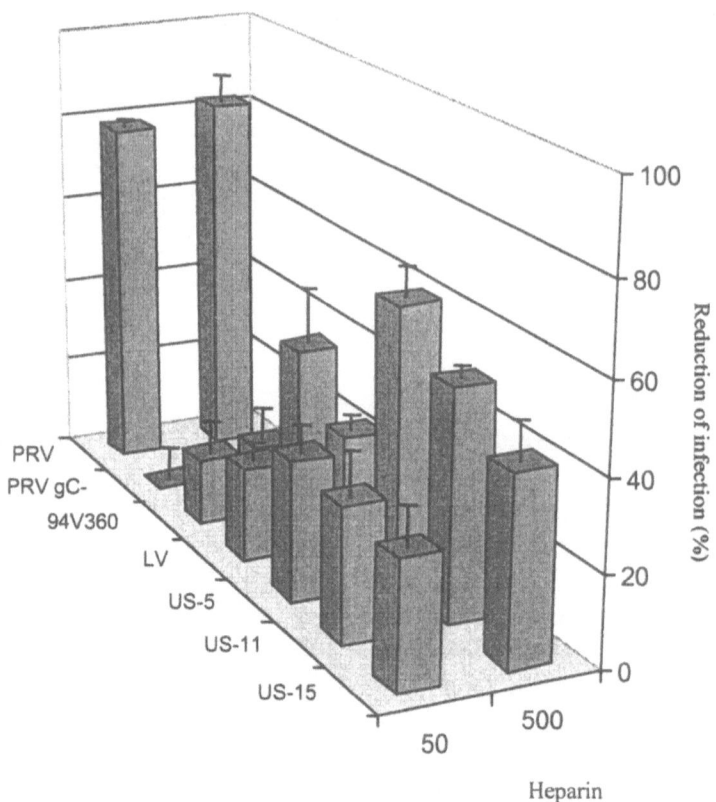

Figure 1. Effect of addition of heparin during virus adsorption on PAM. Infected cells were detected by immunocytochemistry and counted 10h post infection. The reduction of infection is expressed as a percentage of the average number of infected cells obtained in the absence of heparin.

3.2   Effect of heparinase treatment of PAM on PRRSV entry

PAM were treated with different concentrations of heparinase I prior to infection with PRRSV. Treatment with 1U/ml of heparinase resulted in a significant reduction in the number of infected cells only with US-5. Treatment with 3U/ml of heparinase resulted in a significant reduction with the three PRRSV strains tested (Figure 3). The reduction observed with LV (19±2%, $P<0,05$) and 94V360 (37±4%, $P<0,05$) was less pronounced than with US-5 (61±7%, $P<0,01$). No significant effect was detected with PRV gC⁻.

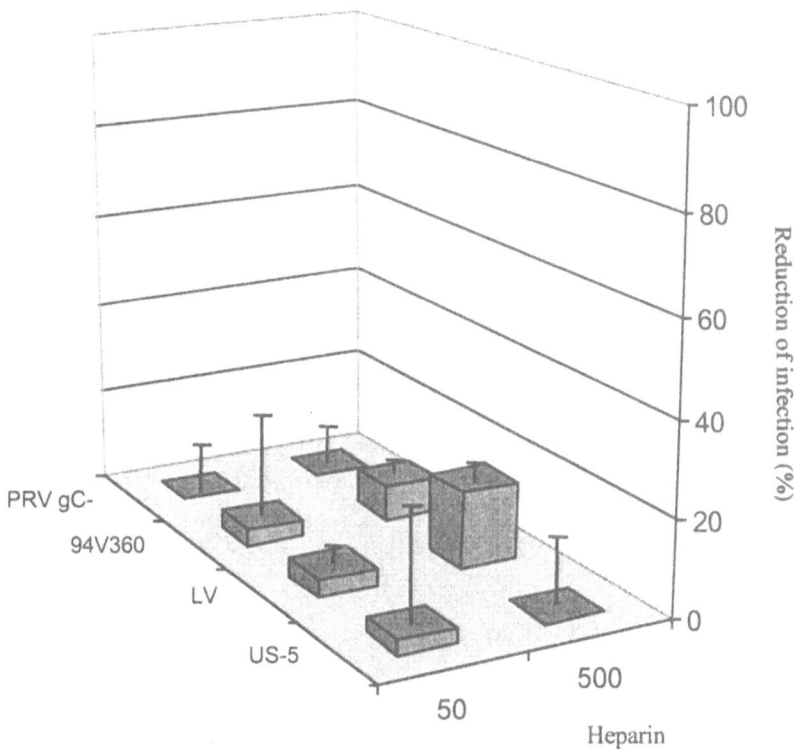

Figure 2. Effect of pre-incubation of PAM with heparin before virus adsorption. Infected cells were detected by immunocytochemistry and counted 10h post infection. The reduction of infection is expressed as a percentage of the average number of infected cells obtained in the absence of heparin.

4. DISCUSSION

Our results indicate that heparin-like molecules are also playing a role in the entry of PRRSV into alveolar macrophages. However, the effect was less pronounced that observed by Jusa *et al.* (1997) with MARC-145 cells, where addition of 50U/ml of heparin induced a 93% reduction in the number of PRRSV plaques. These differences could be due to (1) the viral strain, (2) the cell type, and/or (3) the experimental conditions used:

(1) Our results show that the effect of heparin on PRRSV binding in PAM is strain-dependent, with the American strain US-5 being significantly more sensitive to heparin than the two European strains tested. Also, the strains (EDRD-1 and 5-53) shown to be sensitive to heparin (Jusa *et al.*

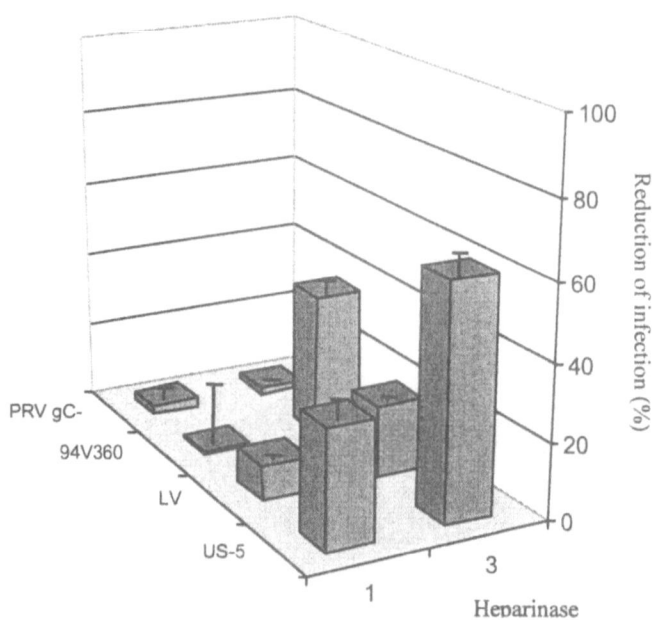

Figure 3. Effect of heparinase treatment of PAM before virus adsorption. Infected cells were detected by immunocytochemistry and counted 10h post infection. The reduction of infection is expressed as a percentage of the average number of infected cells obtained in the absence of heparinase.

1997) were Japanese isolates, which are more serologically and genetically related to American type than to European type PRRSV (Murakami *et al.*, 1994; Saito *et al.*, 1996). Altogether, these results suggest that American isolates are more sensitive than European isolates to heparin inhibition, which may reflect amino acid differences in the viral protein domain(s) involved in the attachment/entry process.

(2) Whereas no significant reduction of infection was observed at 50U/ml heparin with the Belgian strain 94V360 on PAM (Figure 1), preliminary results indicate a significant reduction of infection (60±18% with $P < 0,01$) on MARC-145 cells using the same isolate and heparin concentration. This additional result indicates that, for a given PRRSV strain, infection of MARC-145 cells is more sensitive to heparin inhibition than infection of PAM and suggests that a heparin-like molecule on MARC-145 cells plays a more preponderant role in the entry process than an equivalent molecule on PAM.

(3) Although our experimental conditions differed from Jusa *et al.*, 1997 (infected cells were counted after a single round of virus replication in our

study), we obtained the same results with the control PRV strain, suggesting that our experimental conditions may not account for the observed differences in heparin sensitivity.

ACKNOWLEDGMENTS

This work was supported by the Belgian Ministry of Agriculture. We thank Chantal Vanmaercke for excellent technical assistance.

REFERENCES

Bautista, E.M., Goyal, S.M., and Collins, J.E., 1993, *J. Vet. Diagn. Invest.* **5**, 612-614.

Cavanagh, D., 1997, *Arch. Virol.* **142**, 629-633.

Duan, X., Nauwynck, H.J., and Pensaert, M.B., 1997, *Vet. Microb.* **56**, 9-19.

Duan, X., Nauwynck, H., Favoreel, H., and Pensaert, M. 1998, *J.Virol.* **72**, 4520-4523.

Iglesias, G., Pijoan, C., Molitor, T., 1989, *J. Leukoc. Biol.* **45**, 410-415.

Jusa, E., Inaba, Y., Kouno, M., and Hirose, O., 1997, *Am. J. Vet. Res.* **58**, 488-491.

Kim, H.S., Kwang, J., Yoon, I.J., Joo, H.S., and Frey, M.L., 1993, *Arch. Virol.* **133**, 477-483.

Mettenleiter, T., Zsak, L., Zuckermann, F., Sugg, N., Kern, H., and Ben-Porat, T., 1990, *J. Virol.* **64**, 278-286.

Murakami, Y., Kato, A., Tsuda, T., Morozumi, T., Miura, Y., and Sugimura, T., 1994, *J. Vet. Med. Sci.* **56**, 891-894.

Murtaugh, M.P., Elam, M., Kakach, L., 1995, *Arch. Virol.* **140**, 1451-1460.

Rossow, K.D., Benfield, D.A., Goyal, S.M., Nelson, E.A., Christopher-Hennings, J., Collins, J.E., 1996, *Vet Pathol* **33**, 551-5566

Rossow, K.D., 1998, *Vet. Pathol.* **35**:1-20.

Saito, A., Kanno, T., Murakami, Y., Muramatsu, M., and Yamaguchi, S., 1996, *J. Vet. Med. Sci.* **58**, 377-380.

Wensvoort, G. Terpstra, C., Pol, J., ter Laak, E., Bloemraad, M., de Kluyver, E., Kragten, C., van Buiten, L., den Besten, A., Wagenaar, F., Broekhuijsen, J., Moonen, P., Zetstra, T., de Boer, E., Tibben, H., de Jong, M., van't Veld, P., Groenland, G., van Gennep, J., Voets, M., Verheijden, J., and Braamskamp, J., 1991, *Vet. Q.* **13**, 121-130.

Wensvoort, G., De Kluijver, E.P., Luijtze, E.A., and den Besten, A., 1992, *Proceedings of 12th Congress of The International Pig Veterinary Society*, The Hague, Netherlands, August 17-20 1992, 113.

Apoptosis in the Lungs of Pigs During an Infection with a European Strain of Porcine Reproductive and Respiratory Syndrome Virus

G.G. LABARQUE, H.J. NAUWYNCK, K. VAN REETH, AND M.B. PENSAERT

Laboratory of Virology, Faculty of Veterinary Medicine, Ghent University, Salisburylaan 133, B-9820 Merelbeke, Belgium

1. INTRODUCTION

It has been shown, *in vitro*, that porcine reproductive and respiratory syndrome virus (PRRSV) induces apoptosis during its replication in both MA-104 cells and porcine alveolar macrophages (Suarez *et al.*, 1996). Since transfection of mammalian cells with ORF5 of PRRSV leads to apoptosis it was suggested that the 25-kDa glycosylated membrane protein GP5 of PRRSV is responsible for this phenomenon. Apoptosis has also been demonstrated *in vivo* with North-American strains of PRRSV (Sur *et al.*, 1997, 1998; Sirinarumitr *et al.*, 1998). Apoptotic cells were localized in germinal epithelial cells of testicles (Sur *et al.*, 1997), in lungs and in lymphoid tissues of pigs (Sirinarumitr *et al.*, 1998; Sur *et al.*, 1998). The apoptotic cells were morphologically recognized as alveolar and pulmonary intravascular macrophages and mononuclear cells in the alveolar septa in lungs and as macrophages and mononuclear cells in lymph nodes. Apoptotic cells were more abundant than PRRSV-infected cells in all tissues and double-labeling experiments indicated that the majority of apoptotic cells were uninfected bystander cells.

In this study, apoptosis in lungs and bronchoalveolar lavage (BAL) cells was investigated with a European strain of PRRSV (Lelystad virus) and it

The Nidoviruses (Coronaviruses and Arteriviruses).
Edited by Ehud Lavi *et al.*, Kluwer Academic/Plenum Publishers, 2001.

was questioned whether apoptotic cells are virus-infected or not. The apoptotic cells were further phenotypically characterized using the monoclonal antibody (MAb) 41D3 (Duan *et al.*, 1998). This MAb reacts specifically with the 210-kDa putative PRRSV receptor, which is present on well-differentiated lung macrophages but not on peripheral blood mononuclear cells. Also, an attempt was made to clarify a possible role of apoptosis in the reduction of the population of differentiated macrophages during the first two weeks of a PRRSV infection (Labarque *et al.*, 2000).

2.   MATERIALS AND METHODS

2.1   Virus strain

PRRSV (Lelystad) (Wensvoort *et al.*, 1991) was used in the present study. Virus used for inoculation was at the fifth passage in pulmonary alveolar macrophages (PAMs) from four- to six-week-old gnotobiotic pigs.

2.2   Pigs and inoculation

A total of twenty-four caesarean-derived colostrum-deprived (CDCD) pigs were used. They were housed in isolation facilities. Twenty-two pigs were intranasally inoculated at the age of 4 to 5 weeks with $10^{6.0}$ TCID$_{50}$ Lelystad virus in 3ml phosphate buffered saline (PBS) (1.5 ml in each nostril). The remaining two pigs were left uninoculated and served as negative controls. One to three of the PRRSV-inoculated pigs were euthanatized at 1, 3, 5, 7, 9, 14, 20, 25, 30, 35, 40 and 52 days post inoculation (PI) by intraperitoneal injection with an overdose of barbiturates (Natriumpentobarbital® 20%, IC KELA).

The control pigs were euthanatized at 4 and 5 weeks of age. The right lung was used for bronchoalveolar lavage and samples from the left lung lobes were collected for virological examinations and detection of apoptosis.

2.3   Collection of samples

The right lung was lavaged with 60 to 120 ml of Dulbecco's PBS without Ca^{2+} and Mg^{2+} via an 18-gauge blunt needle inserted through the trachea. The left main bronchus was cross-clamped to prevent lung lavage fluid from entering the left lung. About 75 to 90% of the initial volume of the lavage fluid was recovered. The BAL fluid was centrifuged (400xg, 10 minutes, 4°C) to separate the cells and the cell-free lavage fluid. Cell pellets were resuspended in PBS and the total number of BAL cells was determined. Cytocentrifuge preparations of BAL cells were made by centrifuging at 140xg for 5 minutes. One preparation was fixed in 4% paraformaldehyde for 10 minutes at room temperature for detection of apoptosis and another was fixed in acetone for 20 minutes at $-20°C$ to determine the percentage of infected cells using a streptavidin-biotin immunofluorescence technique, as described by Labarque *et al.* (2000).

Tissue samples from the left lung lobes were embedded in methylcellulose medium and frozen at $-70°C$. Cryostat sections (5 to 8 μm) were made and fixed in 4% paraformaldehyde for 10 minutes at room temperature for detection of apoptosis and in acetone for 20 minutes at $-20°C$ for quantitation of PRRSV-infected cells.

2.4   Detection of apoptosis

In order to detect apoptosis, cytocentrifuge preparations of BAL cells and cryostat sections of lung tissue were processed for enzyme terminal deoxynucleotidyl transferase (TdT)-mediated dUTP nick end labeling (TUNEL) using an In Situ Cell Death Detection Kit, Fluorescein (Boehringer Mannheim) according to the manufacturer's instructions. Briefly, cytocentrifuge preparations and cryostat sections were treated with Triton X-100 (0.1%) at 4°C for two minutes. Then, the cytocentrifuge preparations and cryostat sections were subjected to an enzymatic incorporation of digoxygenin-labeled nucleotide with TdT. Finally, the preparations were washed with PBS, mounted in a glycerin-PBS solution (0.9:0.1, v/v) with 2.5% 1,4-diazabicyclo(2.2.2) octane (DABCO) (Janssen Chimica) and TUNEL-positive cells were detected and counted by fluorescence microscopy (Leica DM RBE, Wild Leitz).

2.5   Double-labeling experiments

A first double-labeling experiment was conducted to determine whether the apoptotic cells were PRRSV-infected or not. Briefly, cytocentrifuge preparations and cryostat sections were treated with Triton X-100 (0.1%) at 4°C for two minutes. The cytocentrifuge preparations and cryostat sections were first incubated with a pool of MAbs against the PRRSV nucleocapsid protein (dilution 1/100 of WBE1 and WBE4-6) (Drew *et al.*, 1995). Subsequently, the preparations were subjected to an enzymatic incorporation of digoxygenin-labeled nucleotide with TdT and incubated with 1/100 goat anti-mouse TexasRed (Amersham). Finally, the preparations were washed with PBS, mounted in DABCO and TUNEL-positive and/or PRRSV-infected cells were detected and counted by fluorescence microscopy (Leica DM RBE, Wild Leitz).

A second double-labeling experiment was conducted to determine whether the apoptotic cells were 41D3-positive cells. Briefly, cytocentrifuge preparations and cryostat sections were treated with Triton X-100 (0.1%) at 4°C for two minutes. The cytocentrifuge preparations and cryostat sections were first incubated with MAb 41D3 (dilution 1/100) (Duan *et al.*, 1998). Subsequently, the preparations were subjected to an enzymatic incorporation of digoxygenin-labeled nucleotide with TdT and incubated with 1/100 goat anti-mouse TexasRed (Amersham). Finally, the preparations were washed with PBS, mounted in DABCO and TUNEL-positive and/or 41D3-positive cells were detected and counted by fluorescence microscopy (Leica DM RBE, Wild Leitz).

3.   RESULTS

The evolutions of the number of viral antigen- and TUNEL-positive cells in lung tissue and BAL cells throughout a PRRSV infection are presented in Figures 1 and 2.

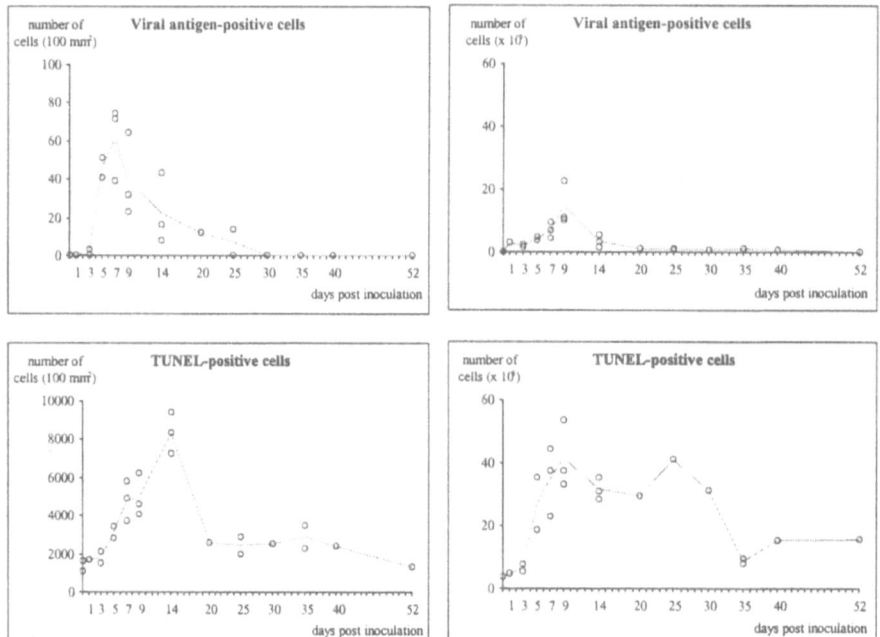

Figure 1. Evolution of the number of viral antigen-positive cells and the number of TUNEL-positive cells in lung tissue (100 mm²) throughout a PRRSV infection.

Figure 2. Evolution of the number of viral antigen-positive BAL cells and the number of TUNEL-positive BAL cells (x10⁶) throughout a PRRSV infection.

Each point represents the individual value of one euthanatized pig.
The continuous line represents the mean of the individual values at each time point.

Viral antigen-positive cells in lung tissue were observed from 3 until 25 days PI with a maximal number of 61 cells /100 mm^2 lung tissue at 7 days PI. PRRSV-infected BAL cells were first observed at 1 day PI (3 x10^6 cells), increased to a maximum of 15 x10^6 cells at 9 days PI, decreased to 3 x10^6 cells at 14 days PI and remained at levels of 0.5-1 x10^6 cells until 40 days PI. Viral antigen-positive cells were not observed in lung tissue and BAL cells of uninoculated control pigs.

TUNEL-positive cells were detected in lung tissue and BAL cells of both uninoculated and PRRSV-inoculated pigs. In uninoculated pigs, the mean number was 1350 cells /100 mm^2 lung tissue and 3.6 x10^6 in the BAL cell population. In PRRSV-infected pigs, the mean number in lung tissue was normal until 3 days PI, increased thereafter to a 5-fold at 14 days PI and sharply decreased to numbers similar to those of the uninoculated controls starting at 20 days PI. In BAL cells, the mean number was similar to that of the uninoculated controls during the first 3 days PI. Mean numbers increased from 6.6 x10^6 at 3 days PI to 41.4 x10^6 at 9 days PI and then remained at high levels until 30 days PI with numbers ranging between 29.5 and 41.2 x10^6. From 30 days PI, mean numbers dropped to numbers similar to those of the uninoculated control pigs. One pig, euthanatized at 25 days PI, had an extreme high number of TUNEL-positive BAL cells (85 x10^6). This value was not included in the calculation of the mean value.

Double-labeling experiments revealed that the majority of TUNEL-positive cells were uninfected cells. The mean percentages of PRRSV-infected cells, which were TUNEL-positive, ranged from 20 to 34% in lung tissue and from 21 to 25% in the BAL cell population between 5 and 14 days PI.

Double-labeling experiments with MAb 41D3 revealed that 74 to 85% of the apoptotic cells were expressing the putative PRRSV receptor on their cell membrane.

4.   DISCUSSION

In the present study, it was shown that apoptosis occurs in lungs during a PRRSV infection in both infected and uninfected bystander cells.

A general feature of apoptotic cell death is that it does not induce severe inflammation and massive neutrophil infiltration into the lungs. This may explain why only a very mild lung inflammation is present during an infection with PRRSV (Lelystad virus) (Pol *et al.*, 1991) and why only low percentages of neutrophils are present in BAL fluids (Van Reeth *et al.*, 1999 ; Labarque *et al.*, 2000). Further, apoptosis of virus-infected cells may be an efficient mechanism by which the virus escapes from humoral immunity because progeny virus, which is present in membrane-bound apoptotic bodies, can be taken up by neighbouring cells while protected from antibodies, favouring persistence of the virus in its host. This phenomenon has already been demonstrated for chicken anaemia virus (Jeurissen *et al.*, 1992). PRRSV may use this type of immune-evasion to persist in the respiratory tract until 40 (Labarque *et al.*, 2000) to 49 days after inoculation (Mengeling *et al.*, 1995) despite the presence of antibodies in sera and BAL fluids from 9 days PI (Labarque *et al.*, 2000).

PRRSV causes a reduction of the population of 41D3-positive well-differentiated lung macrophages during the first two weeks of infection (Labarque *et al.*, 2000). The reduction of this cell population can be largely attributed to cell lysis due to virus replication and apoptosis. The highest number of PRRSV-infected cells in lungs and BAL fluids was indeed detected between 3 and 14 days PI and a marked increase of apoptosis was demonstrated during the same time period in both lung tissue and BAL cells.

A relation was found between the peak of virus replication and the onset of apoptosis. The basis for the relation could be the production of a molecule that causes apoptosis. This can be a viral antigen, such as GP5 which has already been associated with the induction of apoptosis (Suarez *et al.*, 1996) or a product induced by the virus infection, such as Interleukine-1 (IL-1). IL-1 has already been demonstrated in BAL fluids of PRRSV-inoculated pigs starting from 3 days PI (Van Reeth *et al.*, 1999). Whether release of IL-1 is responsible for or coincides with the induction of apoptosis in the lungs of PRRSV-infected pigs needs further examination.

In lung tissue, the number of apoptotic cells dropped to a normal level starting from 14 days PI, while the number of apoptotic BAL cells remained at a rather high level until 30 days PI. The earlier time point at which the number of apoptotic cells dropped to normal values in lung tissue may be partly explained by the more rapid clearance of virus.

ACKNOWLEDGMENTS

The authors would like to thank Fernand De Backer, Tini De Lausnay, Chris Bracke and Lieve Sys for their excellent technical assistance. The authors would also like to thank Dr. T. Drew for his kind gift of the monoclonal antibodies WBE1 and WBE4-6 and Dr. G. Wensvoort for the supply of the Lelystad isolate of PRRSV.

REFERENCES

Drew, T.W., Meulenberg, J.J.M., Sands, J.J., and Paton, D.J., 1995, *J. Gen. Virol.*, **76**: 1361-1369.

Duan, X., Nauwynck, H.J., Favoreel, H.W., and Pensaert, M.B., 1998, *J. Virol.*, **72**: 4520-4523.

Jeurissen, S.H., Wagenaar, F., Pol, J.M., van der Eb, A.J., and Noteborn, M.H., 1992, *J. Virol.*, **66**: 7383-7388.

Labarque, G.G., Nauwynck, H.J., Van Reeth, K., and Pensaert, M.B., 2000, *J. Gen. Virol.*, **81**: 1327-1334.

Mengeling, W.L., Lager, K.M., and Vorwald, A.C., 1995, *J. Vet. Diagn. Invest.*, **7**: 3-16.

Pol, J.M.A., van Dijk, J.E., Wensvoort, G., and Terpstra, C., 1991, *Vet. Q.*, **13(3)**: 137-143.

Sirinarumitr, T., Zhang, Y., Kluge, J.P., Halbur, P.G., and Paul, P.S., 1998, *J. Gen. Virol.*, **79**: 2989-2995.

Suarez, P., Diaz-Guerra, M., Prieto, C., Esteban, M., Castro, J.M., Nieto, A., and Ortin, J., 1996, *J. Virol.*, **70(5)**: 2876-2882.

Sur, J.-H., Doster, A.R., Christian, J.S., Galeota, J.A., Wills, R.W., Zimmerman, J.J., and Osorio, F.A., 1997, *J. Virol.*, **71(12)**: 9170-9179.

Sur, J.-H., Doster, A.R., and Osorio, F.A., 1998, *Vet. Pathol.*, **35**: 506-514.

Van Reeth, K., Labarque, G., Nauwynck, H., and Pensaert, M., 1999, *Res. Vet. Sci.*, **67**: 47-52.

Wensvoort, G., Terpstra, C., Pol, J.M.A., ter Laak, E.A., Bloemraad, M., de Kluyver, E.P., Kragten, C., van Buiten, L., den Besten, A., Wagenaar, F., Broekhuijsen, J.M., Moonen, P.L.J.M., Zetstra, T., de Boer, E.A., Tibben, H.J., de Jong, M.F., van 't Veld, P., Groenland, G.J.R., van Gennep, J.A., Voets, M.T., Verheijden, J.H.M., and Braamskamp, J., 1991, *Vet. Q.*, **13**: 121-130.

Participants of the VIIIth International Symposium of Nidoviruses.

699

VIIIth International Symposium of Nidoviruses
List of Participants

Fernando Almazan
Department of Molecular and Cell Biology
Centro Nacional de Biotecnología. CSIC.
Campus Universidad Autónoma   Ph: 34-1-585 45 26
Cantoblanco, 28049 Madrid   Fax: 34-1-585 45 06
Spain

Sara Alonso
Department of Molecular and Cell Biology
Centro Nacional de Biotecnología. CSIC.
Campus Universidad Autónoma   Ph: 34-1-585 45 26
Cantoblanco, 28049 Madrid   Fax: 34-1-585 45 06
Spain   e-mail: salonso@cnb.uam.es

Jean-Christophe Audonnet
MERIAL
Lyon Gerland Laboratory
254, rue Marcel Merieux   Ph: 33 (0)4 72 72 33 79
69007 LYON Cedex   Fax: 33 (0)4 72 72 34 97
France   e-mail: jean-christophe.audonnet@merial.com

Susan C. Baker
Department of Microbiology and Immunology
Loyola University Chicago, School of Medicine
2160 South First Avenue
Bldg 105, RM 3846   Ph: 708-216 6910
Maywood - Illinois 60153   Fax: 708-216 9574
USA   e-mail: sbaker1@luc.edu

Udeni B.R. Balasuriya
University of California, Davis
Dept of Pathology, Microbiology & Immunology,
School of Veterinary Medicine
One Shields Av.   Ph: 530-752-1163
Davis, CA 95616   Fax: 530-752-3349
USA   e-mail: ubbalasuriya@ucdavis.edu

Sangeeta Banerjee
Department of Microbiology and Immunology
The University of Texas Medical Branch at Galveston
301 University Blvd.
Route 1019   Ph: 409-772-8172
Galveston, TX 77555   Fax: 409-772-5065
USA   e-mail: sabanerj@utmb.edu

Annette Barfoed
Virology Res. Ctr
Institut Armand Frappier
531, boul des Prairies   Ph: 514-687-5010
Laval, PQ   Fax: 514-686-5303
Canada   e-mail: annettegrubbe@hotmail.com

Ralph S. Baric
Epidemiology
University of North Carolina   Ph: 919-966-3895
Chapel Hill, NC 27599   Fax: 919-966-2089
USA   e-mail: rbaric@sophia.sph.unc.edu

Elida M. Bautista
Elanco Animal Health
2001 W. Main Street   Ph: 317-277-0925
Drop Code GL14   Fax: 317-277-4522.
Greenfield, IN 46140   e-mail: bautista_elida_m@Lilly.com
USA

Conni Bergmann
Department of Neurology
University of Southern California
School of Medicine
1330 San Pablo Street MCH 142   Ph: 323-442-1062
Los Angeles, CA 90033   Fax: 323-225-2369
USA   e-mail: cbergman@hsc.usc.edu

Melissa and John Bien
South Dakota State University
Bio/Micro Dept NPB252   Ph: 605-688-5109
Brookings SD 57007   Fax: 605-688-5624
USA

Dianna Blau
University of Colorado Health Sciences Center
Department of Microbiology
4200 East 9th Avenue, Campus Box B-175   Ph: 303-315-7329
Denver, Colorado 80262   Fax: 303-315-6785
USA   e-mail: dianna.blau @uchsc.edu

Aurelio M. Bonavia
Microbiology
U. Of Colorado HSC   Ph: 303-315-7318
4200 E. 9Th Ave., B175   Fax: 303-315-6785
Denver, CO 80262   e-mail: aurelio.bonavia@uchsc.edu
USA

Pedro J. Bonilla
Department of Molecular Virology and Microbiology
Mail Stop BCM-385
Baylor College of Medicine
One Baylor Plaza   Ph: 713-798-3608
Houston, TX 77030-3411   Fax: 713-798-5075
USA   e-mail: pbonilla@melnick.mvir.bcm.tmc.edu

Berend Ian Boach
Institute of Virology
Utrecht University   Ph: 31 30-2534195
Yaleaan 1, 3584 CL Utrecht   Fax: 31 30-2536723
The Netherlands   e-mail: b.j.boach@vet.uu.nl

Annie Boucher
Centre de recherché en santé humaine
INRS- Institut Armand-Frappier
531 Boulevard des Prairies
Ville de Laval   Ph: 514-687 5010. Ext 4406
Québec, H7N 4Z3   Fax: 514-686 5531
Canada   e-mail: annie.boucher@inrs-iaf.uquebec.ca

David A. Brian
Microbiology
Univ. Of Tennessee   Ph: 423-974-4030
Knoxville, TN 37996-0845   Fax: 423-974-4007
USA   e-mail: brian@utk.edu

Margo A. Brinton
Department of Biology
Georgia State University
P.O. Box 4010   Ph: 404- 651 3113
Atlanta GA 30302-4010   Fax: 404- 651 2509
USA   e-mail: biomab@panthet.gsu.edu

Paul Britton
Division of Molecular Biology
Institute for Animal Health
Compton, Newbury   Ph: 44-1635 578411
Berkshire RG20 7NN   Fax: 44-1635 577263
United Kingdom   e-mail: paul.britton@bbsrc.ac.uk

Michael J. Buchmeier
Neuropharm/Virology
The Scripps Research Inst   Ph: 619-784-7056
10550 N. Torrey Pines Rd.   Fax: 619-784-7369
La Jolla, CA 92037   e-mail: buchm@scripps.edu
USA

Jay G. Calvert
Pfizer Central Research
Eastern Point Road   Ph: 860-441 7978
Groton, CT 06340   Fax: 860-441 8739
USA   e-mail: jay_calvert@groton.pfizer.com

Rosa Casais
Institute for Animal Health
LAB G2B, I.A.H., Compton,   Ph: 44 1635 577 274
Newbury RG20 7NN   Fax: 44 1635 577 263
United Kingdom   e-mail: rosa.casais@bbsrc.oc.uk

Dave Cavanagh
Institute for Animal Health
Compton Laboratory
Compton, Nr. Newbury   Ph: 44-1-635-577273
Berkshire RG20 7NN   Fax: 44-1-635- 577263
United Kingdom   e-mail: dave.cavanagh@bbsrc.ac.uk

Kevin W. Chang
Dept. of Microbiology and Immunology
LSU Health Sciences Center
1501 Kings Highway   Ph: (318) 675-6685
Shreveport, LA 71130   Fax: (318) 675-5764
USA   e-mail: kchang@lsumc.edu

Chun-Jen Chen
Department of Microbiology and Immunology
The University of Texas Medical Branch at Galveston
301 University Blvd.   Ph: 409-772-8172
Route 1019   Fax: 409-772-50650
Galveston, TX 77555
USA

Arlene R. J. Collins
Dept. Of Microbiology
School of Medicine and Biomedical Sciences
138 Farber Hall
3435 Main St.   Ph: 716-829 2161
SUNY at Buffalo   Fax: 716-829 2158
Buffalo NY 14214   e-mail: acollins@ubvms.edu
USA

Ellen W. Collisson
Department of Veterinary Pathobiology
Texas A&M University
College Station   Ph: 979-845 4122
Texas 77843-4467   Fax: 979-845 9972
USA   e-mail: ecollisson@evm.tamu.edu

Emily Corse
Dept of Cell Biology and Anatomy
Johns Hopkins University
725 Wolfe St.   Ph: 410-955-1809
Baltimore, MD 21205-2105 Fax: 410-955-4129
USA   e-mail: ecorse@jhmi.edu

Jeff Cowley
CSIRO Tropical Agriculture
Private Bag No. 3   Ph: 61-7-3214-2855
Indooroopilly, QLD 4068   Fax: 61-7-3214-2881
Australia   e-mail: Jeff.cowley@tag.csiro.au

Kristopher Curtis
University of North Carolina- Chapel Hill
343 Cobblestone Ct.   Ph: 919-924-1428
Chapel Hill NC 27514
USA   e-mail: kmcurtis@med.unc.edu

Girish C. Daginakatte
Department of Diagnostic Medicine/Pathology
College of Veterinary Medicine
1800 Denison Avenue   Ph: 785-532-4471
Manhattan, KS 66506 –5606   Fax: 785-532-4829
USA   e-mail: girish@ksu.edu

Jayasri Das Sarma
University of Pennsylvania School of Medicine
Institute for Environmental Medicine
1 John Morgan Bldg/6068
3620 Hamilton Walk   Ph: (215)-898-9093
Philadelphia, PA 19104   Fax: (215)-898-0868
USA   e-mail: sarmad@mail.med.upenn.edu

Linda de Groot
Department of Microbiology
University of Pennsylvania
School of Medicine
202F Johnson Pavilion
36th Street and Hamilton Walk   Ph: 215-898 4672
Philadelphia, Pa 19104-6076   Fax: 215-573-4858
USA

Raoul de Groot
Institute of Virology
Faculty of Veterinary Medicine
Utrecht University
Yalelaan 1   Ph: 31-30-2532462
3584 CL, Utrecht   Fax: 31-30-2536723
The Netherlands

Xander de Haan
Institute of Virology
Faculty of Veterinary Medicine
Utrecht University
P.O. Box 80.165   Ph: 31-30-2534195
3508TD Utrecht   Fax: 31-30-2536723
The Netherlands   e-mail: x.haan@vet.uu.nl

Serge Dea
Virology Res. Ctr
Institut Armand Frappier
531, boul des Prairies   Ph: 514-687-5010
Laval, PQ   Fax: 514-686-5303
Canada   e-mail: serge_dea@iaf.uquebec.ca

Ruitang Deng
Pfizer Central Research
Eastern Point Road   Ph: 860-441-7814
Groton CT 06340   Fax: 860-441-8739
USA   e-mail: ruitang_deng@groton.pfizer.com

Mark R. Denison
Pediatric Infectious Diseases
Vanderbilt University Medical Center
D 7235 MCN, 1162 21st Avenue South   Ph: 615- 3439881
Nashville, TN 37232-2581   Fax: 615- 343 9723
USA   e-mail: denison@ctrvax.vanderbilt.edu

Brian Dove
Institute for Animal Health, Compton Laboratory
Compton, Nr. Newbury   Ph: 44-1-635-577273
Berkshire RG20 7NN   Fax: 44-1-635- 577263
United Kingdom   e-mail: brian.dove@bbsrc.ac.uk

Knut Elbers
Boehringer Ingelheim Vetmedica GmbH
Corporate Research and Development   Ph: 49 61 32 77 25 09
D55216 Ingelheim am Rhein   Fax: 49 61 32 77 38 96
Germany   e-mail: elbers@ing.boehringer-ingelheim.com

Luis Enjuanes
Centro Nacional de Biotecnologia
Department of Molecular and Cell Biology   Ph: 34-91- 585 45 55
Campus Univ. Autonoma, Cantoblanco   Fax: 34-91- 585 45 55/06
28049 Madrid,   e-mail: L.Enjuanes@cnb.uam.es
Spain

David Escors
Department of Molecular and Cell Biology
Centro Nacional de Biotecnología. CSIC.
Campus Universidad Autónoma   Ph: 34-91-585 45 26
Cantoblanco, 28049 Madrid   Fax: 34-91-585 45 06
Spain

Sharon Evans
Institute for Animal Health
Compton, Nr. Newbury   Ph: 44-1-635-578411
Berkshire RG20 7NN   Fax: 44-1-635- 577263
United Kingdom   e-mail: sharon.evans@bbsrc.ac.uk

Kay S. Faaberg
Department of Veterinary Pathobiology
University of Minnesota
1971 Commnwealth Ave, VSB Rm.237   Ph: 612-624 9746
St. Paul, MN 55108   Fax: 612-625 5203
USA   e-mail: kay@lenti.med.umn.deu

Ying Fang
235 Northern Plain Biostress Laboratories
Department of Biology/Microbiology
South Dakota State University   Ph: 605-688-6141
Brookings, SD 57007   Fax: 605-688-6677
USA   e-mail: T5BZ@sdsumus.sdstate.edu

Rachel Farwell
South Dakota State University
Bio/ Micro Dept. NPB252   Ph: 605-688-5109
Brookings SD 57007   Fax: 605-688-5624
USA

Li Fu
Division of Neuropathology
University of Pennsylvania
School of Medicine
612 Stellar-Chance Labs.
422 Curie Blvd.
Philadelphia   Ph: 215-898-4719
Pennsylvania 19104-6100   Fax: 215-898-9969
USA   e-mail: lifu@mail.med.upenn.edu

Carl A. Gagnon
Virology Res. Ctr
Institut Armand Frappier
531, boul des Prairies, Ed #27   Ph: 514-687-5010
Laval, PQ   Fax: 514-686-5627
Canada   e-mail: carl_gagnon@iaf.uquebec.ca

Thomas M. Gallagher
Loyola University Medical Center
Department of Microbiology and Immunology
2160 South First Avenue   Ph: 708-216-4850
Maywood - Illinois 60153   Fax: 708-216-9574
USA   e-mail: tgallag@luc.edu

Jim Gombold
Dept. of Microbiology and Immunology
LSU Health Sciences Center   Ph: (318) 675-6684
Shreveport, LA 71130   Fax: (318) 675-5764
USA   e-mail: jgombo@lsumc.edu

José M. González Martínez
Department of Molecular and Cell Biology
Centro Nacional de Biotecnología. CSIC.
Campus Universidad Autónoma   Ph: 34-1-585 45 26
Cantoblanco, 28049 Madrid   Fax: 34-1-585 45 06
Spain   e-mail: jmgonzalez@cnb.uam.es

Alexander E. Gorbalenya
Advanced Biomedical Computing Center   Ph: 301-846-1991 (office)
430 Miller Dr. Rm. 235   Ph: 301-846-5763 (secretary)
SAIC/NCI-FCRDC, P.O. Box B   Fax: 301-846-5762
Frederick, MD 21702-1201   e-mail: gorbalen@ncifcrf.gov
USA

Rainer Gosert
Institute for Medical Microbiology
Department of Virology   Ph: (061) 267-3289
University of Basel   Fax: (061) 267-3298
Petersplatz 10, 4003
Basel, Switzerland   e-mail: Rainer.Gosert@unibas.ch

Hélène Groot Bramel-Verheije
ID-Lelystad, Department of Mammalian Virology
P.O.Box 65   Ph: 31 320 238805
8200 AB Lelystad   Fax: 31 320 328668
The Netherlands   e-mail: m.h.grootbramel-verheije@id.wag-ur.nl

Bert-Jan Haijema
Institute of Virology
Utrecht University   Ph: 31 30 2534195
Yaleaan 1, 3584 CL Utrecht   Fax: 31 30 2536723
The Netherlands   e-mail: b.haijema@vet.uu.nl

Jeffrey Hayes
Animal Disease Diagnostic Laboratory
Ohio Department of Agriculture
8995 East Main Street   Ph: 614-728-63000
Reynoldsburg, Ohio 43068   Fax: 614-728-6310
USA   e-mail: hayes@odnt.agri.state.oh.us

Jodi F. Hedges
University of California, Davis,
1126 Haring Hall,
Department of Pathology, Microbiology and Immunology,
School of Veterinary Medicine,   Ph: 530-752-1163
Davis, CA 95616   Fax: 530-752-3349
USA   e-mail: jfhedges@ucdavis.edu

Susan T. Hingley
Department of Microbiology and Immunology
Philadelphia College of Osteopathic Medicine
4170 City Avenue   Ph: 215-871-6854
Philadelphia, PA 19131   Fax: 215-871-6869
USA   e-mail: susanh@pcom.edu

Norio Hirano
Department of Veterinary Microbiology
Iwate University   Ph: 81-19-621 6222
Morioka 020   Fax: 81-19-621 6231
Japan   e-mail: nhirano@msv.cc.iwate-u.ac.jp

Brenda G. Hogue
Baylor College of Medicine
One Baylor Plaza
Department of Microbiology and Immunology   Ph: 713-798-6412
Houston - Texas 77030   Fax: 713-798-7375
USA   e-mail: bhogue@bcm.tmc.edu

Kathryn V. Holmes
University of Colorado Health Sciences Center
Department of Microbiology
4200 East 9th Avenue, Campus Box B-175   Ph: 303-315-7329
Denver, Colorado 80262   Fax: 303-315-6785
USA   e-mail: kathryn.holmes@uchsc.edu

Dennis Horter
Iowa State University
Department of Veterinary Microbiology
and Preventative Medicine
College of Veterinary Medicine   Ph: 515-294-9344
Ames Iowa
USA   e-mail: dhorter@iastate.edu

Hélène Jacomy
Centre de recherché en santé humaine
INRS- Institut Armand-Frappier
531 Boulevard des Prairies
Ville de Laval   Ph: 514-687 5010. Ext 4406
Québec, H7N 4Z3   Fax: 514-686 5531
Canada   e-mail: helene.jacomy @inrs-iaf.uquebec.ca

Reed Johnson
TAMU Health Science Center
208 Reynolds Medical Bldg   Ph: 409-826-1376
College Station, TX 77843-1114   Fax: 409-826-1299
USA   e-mail: rfj8636@medicine.tamu.edu

Amornrat Kanjanahaluethai
Department of Microbiology and Immunology
Loyola University Chicago
School of Medicine
2160 South First Avenue, Bldg 105, RM 3846 Ph: 708-216 6910
Maywood - Illinois 60153   Fax: 708-216 9574
USA   e-mail: Akanjan1@wpo.it.luc.edu

Sanjay Kapil
Vet. Diag. Med./Pathobiology
Kansas State University
College of Vet. Med   Ph: 913-532-5650
Manhattan, KS 66506-5601   Fax: 913-532-4481
USA   e-mail: kapil@vet.ksu.edu

Taeg Kim
South Dakota State University
Bio/ Micro Dept. NPB252   Ph: 605-688-5109
Brookings SD 57007   Fax: 605-688-5624
USA

Gus Kousoulas
Department of Veterinary Microbiology and Parasitology
School of Veterinary Medicine
Louisiana State University   Ph: 225-346-3345
Baton Rouge, LA 70803   Fax: 225-346-5715
USA   e-mail: VTGUESK@lsu.edu

Andreas F. Kolb
Cell Physiology Group
Hannah Research Institute   Ph: +44-1292-674020
Ayr, KA6 5HL   Fax: +44-1292-674003
United Kingdom   e-mail: kolba@hri.sari.ac.uk

Michiel V. Kroese
Institute of Animal Science and Health
ID-Lelystad
P.O. Box 65,   Ph: 31 320 238897
8200 AB Lelystad,   Fax: 31 320 23 8668
The Netherlands   e-mail: m.v.kroese@id.wag-ur.nl

Shigeru Kyuwa
Department of Animal Pathology
Institute of Medical Science
University of Tokyo
4-6-1 Shirokanedai   Ph: 81-3-5449-5753
Minato-ku, Tokyo 108   Fax: 81-3-5449-5455
Japan   e-mail: kyuwa@ims.u-tokyo.ac.jp

Thomas E. Lane
Department of Molecular Biology and Biochemistry
UC Irvine
3205 Bio Sci II   Ph: 949-824-5878
Irvine, CA 92697-3900   Fax: 949-824-8551
USA   e-mail: tlane@uci.edu

Ehud Lavi
Division of Neuropathology
University of Pennsylvania
School of Medicine
613 Stellar-Chance Labs.
422 Curie Blvd.
Philadelphia   Ph: 215-898-8198
Pennsylvania 19104-6100   Fax: 215-898-9969
USA   e-mail: lavi@mail.med.upenn.edu

Julian L. Leibowitz
Dept. of Pathology and Lab. Medical
Texas A&M Health Science Center
208 Joe H. Reynolds Medical Building   Ph: 409-845-7288
College Station. TX 77843-1114   Fax: 409-862-1299
USA   e-mail: jleibowitz@tamu.edu

Koh Pang Lim
Institute of Molecular Agrobiology
59 A The Flemings   Ph: 65-7719834
I Science Park Drive   Fax: 65-7742857
Singapore 118240   e-mail: imalimkp@leonis.nus.sg

Ding Xiang Liu
Institute of Molecular Agrobiology
National University of Singapore
59 A The Flemings   Ph: 65-7719834
I Science Park Drive   Fax: 65-7742857
Singapore 118240   e-mail: imaliudx@leonis.nus.sg

Carolyn Machamer
Dept of Cell Biology and Anatomy
Johns Hopkins University
725 Wolfe St.   Ph: 410-955-1809
Baltimore, MD 21205-2105   Fax: 410-955-4129
USA   e-mail: carolyn_machamer@qmail.bs.jhu.edu

N. James MacLachlan
1126 Haring Hall,
Dept of Pathology, Microbiology and Immunology,
School of Veterinary Medicine
University of California, Davis
One Shields Av.   Ph: 530-752-1385
Davis, CA 95616   Fax: 530-752-3349
USA   e-mail: njmaclachlan@ucdavis.edu

Taronna Maines
Department of Biology
Georgia State University
P.O. Box 4010   Ph: 404- 651 3113
Atlanta GA 30302-4010   Fax: 404- 651 2509
USA   e-mail: biomab@panthet.gsu.edu

Fahad Majhdi
Department of Diagnostic Medicine/Pathology
College of Veterinary Medicine
1800 Denison Avenue   Ph: 785-532-4463
Manhattan, KS 66506 –5606   Fax: 785-532-4824
USA   e-mail: Almajhdi@ksu.edu

Shinji Makino
Department of Microbiology
The University of Texas at Austin
24th at Speedway, ESB 304   Ph: 512-471-6876
Austin - Texas 78712   Fax: 512-471-7088
USA   e-mail: makino@mail.utexas.edu

Norman W. Marten
Department of Neurology
University of Southern California
School of Medicine
1330 San Pablo Street MCH 142   Ph: 323-442-3367
Los Angeles, CA 90033   Fax: 323-225-2369
USA   e-mail: marten@hsc.usc.edu

Paul S. Masters
Wadsworth Center for Lab. & Research
NY State Department of Health
P.O. Box 22002
New Scotland Ave   Ph: 518- 474 1283
Albany NY 12201-2002   Fax: 518- 473 1326
USA   e-mail: masters@wadsworth.org

Shutoku Matsuyama
Institute of Neuroscience
NCNP 4-1-1 Ogawahigashi, Kodaira   Ph: 81 42-341-2711 (ext 5275)
Tokyo 187-8502   Ph: 81 42 346-1754
Japan   e-mail: matsuyama@ncnp.go.jp

Amy Matthews
Department of Microbiology
University of Pennsylvania
School of Medicine
323 Johnson Pavilion   Ph: 215-898-3461
Philadelphia, PA 19104-6076   Fax: 215-573-4666
USA   e-mail: amatth13@dolphin.upenn.edu

Michael Murtaugh
University of Minnesota
Department of Pathobiology, VSB
Science Building
RM.239B, 1971 Commonwealth Avenue   Ph: 612-625 6735
St. Paul, MN 55108   Fax: 612-625 5203
USA   e-mail: murtaugh@biosci.cbs.umn.edu

Santosh Nanda
TAMU Health Science Center   Ph: 409-826-1376
208 Reynolds Medical Bldg   Fax: 409-826-1299
College Station, TX 77843-1114   e-mail: Nanda@medicine.tamu.edu
USA

Syad A. Naqi
Dept. of Microbiology and Immunology
College of Veterinary Medicine
Cornell University   Ph: 607-253-4045
Ithaca, NY 14853   Fax: 1-607 253 3369
USA   e-mail: san7@cornell.edu

Neal Nathanson
Office of AIDS Research
National Institutes of Health
Building 31, Room 4B54   Ph: 310-402-3357
Bethesda, Maryland 20892   Fax: 301-402-3360
USA   e-mail: billingr@od.nih.gov

Sonia Navas-Martin
Department of Microbiology
University of Pennsylvania
School of Medicine
202F Johnson Pavilion
36th Street and Hamilton Walk   Ph: 215-898 4672
Philadelphia, PA 19104-6076   Fax: 215-573-4858
USA

Krishna Narayanan
Department of Microbiology and Immunology
The University of Texas Medical Branch at Galveston
301 University Blvd.   Ph: 409-772-8172
Route 1019, Galveston, TX 77555   Fax: 409-772-50650
USA

Hans Nauwynck
Laboratory of Virology
Faculty of Veterinary Medicine
University of Gent
Salisburylaan 133   Ph: 32 9 264 73 73
9820 Merelbeke   Fax: 32 9 264 74 95
Belgium   e-mail: Hans.Nauwynck@rug.ac.be

Eric A. Nelson
Veterinary Science
South Dakota State U.
PO Box 2175   Ph: 605-688-5171
Brookings, SD 57007-1396   Fax: 605-688-6003
USA   e-mail: nelsone@mg.sdstate.edu

Benjamin Neuman
Division of Molecular Biology
Institute for Animal Health
Compton, Newbury   Ph: 44-1635 578411
Berkshire RG20 7NN   Fax: 44-1635 577263
United Kingdom

Lisa F. P. Ng
Institute of Molecular Agrobiology
59A The Fleming. N° 1 Science Park Drive   Ph: 65-77 19 834
Singapore Science Park   Fax: 65-77 42 857
Singapore. 118240   e-mail: imangl@leonis.nus.sg

Evelena Ontiveros
Department of Pediatrics
Medical Laboratories 207
University of Iowa   Ph: 319- 335 8549
Iowa City, IA 52242   Fax: 319- 356 4855
USA   e-mail: evelena-ontiveros@uiowa.edu

Beatriz Parra
Neurology Department
University of Southern California
Keck School of Medicine
1333 San Pablo St. MCH 142   Ph: 323-225-2369
Los Angeles, CA 90033   Fax: 323-442-2369
USA   e-mail: bparra@hsc.usc.edu

Alexander Pasternak
Department of Virology
Institute of Medical Microbiology
Leiden University
AZL building 1, Room P4-26   Ph: 31- 71 - 526 1657
P.O. Box 9600, 2300 RC Leiden   Fax: 31- 71 - 526 6761
The Netherlands

Doug Pearce
Pfizer Inc.   Ph: 860-441-0879
Groton, CT 06340   Fax: 860-715-7979
USA   e-mail: Douglas_pearce@groton.pfizer.com

Stanley Perlman
Department of Pediatrics
Medical Laboratories 207
University of Iowa   Ph: 319- 335 8549
Iowa City, IA 52242   Fax: 319- 356 4855
USA   e-mail: stanley-perlman@uiowa.edu

Joanna Phillips
Department of Microbiology
University of Pennsylvania School of Medicine
202F Johnson Pavilion
36th Street and Hamilton Walk   Ph: 215-898 4672
Philadelphia, Pa 19104-6076   Fax: 215-573-4858
USA   e-mail: phillipj@mail.med.upenn.edu

Peter G. W. Plagemann
University of Minnesota
1955 Cleveland Av.   Ph: 612-624-3187
St. Paul MN 55113   Fax: 612-626-0623
USA   e-mail: veglahn@mail.ahc.umn.edu

Susan Ropp
South Dakota State University
2309 42nd St.   Ph: 605-688-5171
Brookings SD 57006   Fax: 605-688-6003
USA   e-mail: Susan_Ropp@sdstate.edu

Peter J. M. Rottier
Institute of Virology, Veterinary Faculty
Utrecht University   Ph: 31-30-2532462
Yalelaan 1, 3584 CL, Utrecht   Fax: 31-30-2536723
The Netherlands   e-mail: P.Rottier@vetmic.dgk.ruu.nl

Raymond Rowland
South Dakota State University
Biology, Microbiology Department NPB 252
P.O. Box 2104 D   Ph: 605-688 5982
Brookings - South Dakota 57007   Fax: 605-688 5624
USA   e-mail: rowlandr@mg.sdstate.edu

Dorothea L. Sawicki
Medical College of Ohio
Department of Microbiology
3000 Arlington Avenue   Ph: 419-381-4337
Toledo - Ohio 43614   Fax: 419-381-3002
USA   e-mail: sawickid@opus.mco.edu

Stanley G. Sawicki
Department of Microbiology
Medical College of Ohio
3000 Arlington Avenue   Ph: 419-381-3928
Toledo - Ohio 43614   Fax: 419-381-3002
USA   e-mail: sawickis@opus.mco.edu

Talya Schwartz
Division of Neuropathology
University of Pennsylvania School of Medicine
612 Stellar-Chance Labs.
422 Curie Blvd.
Philadelphia   Ph: 215-898-4719
Pennsylvania 19104-6100   Fax: 215-898-9969
USA   e-mail: s2t2@hotmail.com

Christel Schwegmann
Institut fuer Virologie
Tieraerztliche Hochschule Hannover
Buenteweg 17   Ph: 0049-511-953-8857
30559 Hannover   Fax: 0049-511-953-8898
Germany   e-mail: Christel.Schwegmann@gmx.de

Sang Heui Seo
College of Vet Med
Univ. of Minnesota
1365 Gortner Ave.   Ph: 612-625-1982
225 Vet. Teaching Hosp.   Fax: 612-625-6241
St. Paul, MN 55108   e-mail: seoxx007@tc.umn.edu
USA

Su Hun Seo
Department of Microbiology
University of Pennsylvania School of Medicine
202F Johnson Pavilion
36th Street and Hamilton Walk   Ph: 215-898-4672
Philadelphia, Pa 19104-6076   Fax: 215-573-4858
USA   e-mail: seosh@mail.sas.upenn.edu

Anja Seybert
ICRF
Clare Hall Laboratories
Molecular Enzymology Laboratory / D 30
Blanche Lane   Ph: 44 207-269-3976
South Mimms, EN6 3LD   Fax: 44 207-269-3803
UK   email: A.seybert@icrf.icnet.uk

Kumar S. Shanmukhappa
Department of Diagnostic Medicine/Pathology
College of Veterinary Medicine
1800 Denison Avenue   Ph: 785-532-4471
Manhattan, KS 66506 –5606   Fax: 785-532-4829
USA   e-mail: Kumar@ksu.edu

Xiaolan Shen
Wadsworth Center for Laboratories and Research
NYSDOH, David Axelrod Institute
New Scotland Avenue
P.O. Box 22002   Ph: 518-473-5820
Albany, N.Y. 12201-2002   Fax: 518-474-3181
USA   e-mail: shenx@wadsworth.org

Stephanie Shi
Molecular Microbiology and Immunology
University of Southern California
School of Medicine, 503-HMR
2011 Zonal Avenue   Ph: 323-442-1748
Los Angeles, CA 9003   Fax: 323-342-9555
USA   e-mail: tshi@hsc.usc.edu

Shuo Shen
Institute of Molecular Agrobiology
1 Research Link   Ph: 65-8727469
National University of Singapore   Fax: 65-8727007
Singapore 117604   e-mail: shenshuo@ima.org.sg

Stuart G. Siddell
Institute of Virology
Univ. Of Wuerzburg
Versbacher Str. 7   Ph: 49-931-2013896
D97078 Wurzburg   Fax: 49-931-2013934
Germany   e-mail: siddell@vim.uni-wuerzburg.de

Olga Slobodskaya
NCI / NIH
535 Sultan Street NCI-FCRDC
P.O. Box B   Ph: 301-846-1857
Frederick MD 21702-1201   Fax: 301-846-7146
USA   e-mail: oslobodskaya@mail.ncifcrf.gov

Eric J. Snijder
Department of Virology
Institute of Medical Microbiology
Leiden University
AZL building 1, Room P4-26   Ph: 31- 71 - 526 1657
P.O. Box 9600, 300 RC Leiden   Fax: 31- 71 - 526 6761
The Netherlands   e-mail: snijder@virology.azl.nl

Isabel Sola
Department of Molecular and Cell Biology
Centro Nacional de Biotecnología. CSIC.
Campus Universidad Autónoma   Ph: 34-91-585 45 26
Cantoblanco, 28049 Madrid   Fax: 34-91-585 45 06
Spain   e-mail: isola@cnb.uam.es

Willy J. M. Spaan
Department of Virology
University of Leiden
P.O. Box 9600   Ph: 31- 71- 5261652
2300 AH Leiden   Fax: 31-71- 5266761
The Netherlands   e-mail: spaan@virology.azl.nl

Jeannie Spagnolo
Baylor College of Medicine
One Baylor Plaza
Department of Microbiology & Immunology   Ph: 713-798-6429
Houston - Texas 77030   Fax: 713-798-7375
USA   e-mail: js691796@melnick.mvir.bcm.tmc.edu

Ann Sperlich
235 Northern Plain Biostress Laboratories
Department of Biology/Microbiology
South Dakota State University   Ph: 605-688-6141
Brookings, SD 57007   Fax: 605-688-6677
USA

Konrad Stadler
Institute of Virology
Department of Infectious Diseases and Immunology
Veterinary Faculty
University of Utrecht
Yalelaan 1   Ph: 31 30 253 2463
3584 CL Utrecht   Fax: 31 30 253 6723
The Netherlands   e-mail: K.Stadler@vet.uu.nl

Stephen A. Stohlman
Neurology Department
University of Southern California
Keck School of Medicine
1333 San Pablo St. MCH 142   Ph: 323-225-2369
Los Angeles, CA 90033   Fax: 323-442-2369
USA   e-mail: stohlman@hsc.usc.edu

Fumihiro Taguchi
National Institute of Neuroscience,
NCNP
4-1-1 Ogawahigashi
Kodaira   Ph: 81-423 41 2711
Tokyo 187   Fax: 81-423 46 1754
Japan   e-mail: taguchi@ncnaxp.ncnp.go.jp

Pierre J. Talbot
Institut Armand-Frappier
531 Boulevard des Prairies
Ville de Laval   Ph: 514-687 5010. Ext 4406
Québec, H7N 4Z3   Fax: 514-686 5531
Canada   e-mail: pierre.talbot@iaf.uquebec.ca

Volker Thiel
Institut fur Virologie
Universitat Wurzburg
Versbacher Str. 7   Ph: 931-201 3966
97078 Wurzburg   Fax: 931-201 3934
Germany   e-mail: v.thiel@rzbox.uni-wuerzburg.de

Dominic Therrien
Virology Res. Ctr
Institut Armand Frappier
531, boul des Prairies   Ph: 450-687-5010 ext. 4380
Laval, PQ   Fax: 450-686-5627
Canada   e-mail:Dominic.Therrien@INRS-IAF.uquebec.ca

Larissa Thackray
University of Colorado Health Sciences Center
Department of Microbiology
4200 East 9th Avenue,   Ph: 303-315-7220
Denver, Colorado 80262   Fax: 303-313-6785
USA   e-mail: Larissa.thackray@uchsc.edu

Marieke Tijms
Department of Virology
Institute of Medical Microbiology
Leiden University
AZL building 1, room P4-26 P.O. Box 9600 Ph: 31- 71 - 526 1657
2300 RC Leiden   Fax: 31- 71 - 526 6761
The Netherlands

Melanie Tremblay
Human Health Research Center
INRS- Institute Armand-Frappier
531 boulevard des Prairies   Ph: 450-687-5010 (4282)
Laval, Quebec, H7V 1B7   Fax: 450-686-5501
Canada   e-mail: melanie.tremblay@inrs-iaf.uquebec.ca

Jean Tsai
Department of Microbiology
University of Pennsylvania School of Medicine
202F Johnson Pavilion
36th Street and Hamilton Walk   Ph: 215-898 4672
Philadelphia, PA 19104-6076   Fax: 215-573-4858
USA   e-mail: jct@mail.med.upenn.edu

Brian C. Turner
University of Colorado Health Sciences Center
Department of Microbiology
4200 East 9th Avenue,   Ph: 303-861-5867
Denver, Colorado 80262   Fax: 303-313-6785
USA   e-mail: brain.turner@uchsc.edu

Nathalie Vanderheijden
University of Ghent, Fac. Vet. Med.
Salisburylaan 133,   Ph: 32 9 264 73 75
B-9820 Merelbeke   Fax: +32 9 264 74 95
Belgium   e-mail: nvdheijd@bigben.vub.ac.be

Peter van Woensel
Intervet International bv,
Wim de Korverstraat 35,   Ph: 31 485 587790
5831 AN Boxmeer   Fax: 31 485 587339
The Netherlands   e-mail: Peter.vanWoensel@intervet.akzonobel.nl

Tao Wang
Dept. of Microbiology
Medical College of Ohio
P.O.Box 10008   Ph: 419-383-4277
Toledo Ohio 43614
USA   e-mail: twang@mco.edu

Brandon Warrick
235 Northern Plain Biostress Laboratories
Department of Biology/Microbiology
South Dakota State University   Ph: 605-688-6141
Brookings, SD 57007   Fax: 605-688-6677
USA

Susan Weiss
Department of Microbiology
University of Pennsylvania School of Medicine
203A Johnson Pavilion
36th Street and Hamilton Walk   Ph: 215-898 8013
Philadelphia, PA 19104-6076   Fax: 215-573-4858
USA   e-mail: weisssr@mail.med.upenn.edu

Siao-Kun (Jenny) Welch
Pfizer Central Research, MS 8288-15
Groton, CT 06340
USA   e-mail: jenny_welch@groton.pfizer.com

David Wentworth
University of Colorado Health Sciences Center
Department of Microbiology
4200 East 9th Avenue, Campus Box B-175   Ph: 303-315-7329
Denver, Colorado 80262   Fax: 303-315-6785
USA

Gregory Wu
2037 Medical Labs
University of Iowa   Ph: (319)-335-7576
Iowa City, Iowa 52242   Fax: 319- 356 4855
USA   e-mail: gregory-wu@uiowa.edu

Wai-Hong Wu
South Dakota State University
North Campus Drive
Vet Science Building SDSU   Ph: 605-688-5171
Brookings, SD 57007   Fax: 605-688-6003
USA   e-mail: waihongwa@hotmail.com

Guang Yang
Molecular Microbiology and Immunology
University of Southern California
2011 Zonal Av., HMR 504   Ph: 323-442-3504
Los Angeles, CA 90089   Fax: 323-342-9555
USA   e-mail: yang@hsc.usc.edu

Dongwan Yoo
Pathobiology, University of Guelph   Ph: 519-824-4120, X4729
Guelph, ON   Fax: 519-767-0809
Canada   e-mail: dyoo@ovcnet.uoguelph.ca

Kyoung-Jin Yoon
Department of Veterinary Diagnostic & Production Animal Medicine
College of Veterinary Medicine
Iowa State University
Ames, IA 50011
USA   e-mail: kyoon@iastate.edu

Shishan Yuan
Department of Veterinary Pathobiology
University of Minnesota
1971 Commonwealth Ave, VSB Rm.237   Ph: 612-624 9746
St. Paul, MN 55108   Fax: 612-625 5203
USA   e-mail: yuan0009@maroon.tc.umn.edu

Xuming Zhang
Department of Microbiology and Immunology
University of Arkansas for Medical Sciences
4301 Markham Slot 511   Ph: 501-686 5145
Little Rock, AR 72205   Fax: 501-686-5362
USA   e-mail: Zhangxuming@exchange.uams.edu

John Ziebuhr
Institut für Virologie
Universität Wurzburg
Versbacher Strasse 7   Ph: 49-931-201 3966
97078 Würzburg   Fax: 49-931-201 3934
Germany   e-mail: zieburhr@vim.uni-wuerzburg.de

Jiahao Zhou
University of Southern California
Department of Neurology and Pathology
1333 San Pablo St., MCH 142   Ph: 323-442-1060
Los Angeles, CA 90033   Fax: 323-225-2369
USA   e-mail: jiehao@hsc.usc.edu

INDEX